Mathematical Inequalities

North-Holland Mathematical Library

VOLUME 67

ELSEVIER

Amsterdam – Boston – Heidelberg – London – New York – Oxford – Paris – San Diego – San Francisco – Singapore – Sydney – Tokyo

Mathematical Inequalities

B.G. Pachpatte

57 Shri Niketan Colony
Near Abhinay Talkies
Aurangabad 431 001
Maharashtra
India

ELSEVIER

2005

Amsterdam – Boston – Heidelberg – London – New York – Oxford –
Paris – San Diego – San Francisco – Singapore – Sydney – Tokyo

ELSEVIER B.V.
Radarweg 29
P.O. Box 211
1000 AE Amsterdam
The Netherlands

ELSEVIER Inc.
525 B Street, Suite 1900
San Diego, CA 92101-4495
USA

ELSEVIER Ltd
The Boulevard, Langford Lane
Kidlington, Oxford OX5 1GB
UK

ELSEVIER Ltd
84 Theobalds Road
London WC1X 8RR
UK

First edition 2005

Library of Congress Cataloging in Publication Data
A catalog record is available from the Library of Congress.

British Library Cataloguing in Publication Data
A catalogue record is available from the British Library.

ISBN: 0-444-51795-2
Series ISSN: 0924-6509

Transferred to Digital Printing 2007

To the Memory of My Parents

Preface

Inequalities play an important role in almost all branches of mathematics as well as in other areas of science. The basic work "*Inequalities*" by Hardy, Littlewood and Pólya appeared in 1934 and the books "*Inequalities*" by Beckenbach and Bellman published in 1961 and "*Analytic Inequalities*" by Mitrinović published in 1970 made considerable contributions to this field and supplied motivations, ideas, techniques and applications. Since 1934 an enormous amount of effort has been devoted to the discovery of new types of inequalities and to the application of inequalities in many parts of analysis. The usefulness of mathematical inequalities is felt from the very beginning and is now widely acknowledged as one of the major driving forces behind the development of modern real analysis.

The theory of inequalities is in a *process* of continuous development state and inequalities have become very effective and powerful tools for studying a wide range of problems in various branches of mathematics. This theory in recent years has attracted the attention of a large number of researchers, stimulated new research directions and influenced various aspects of mathematical analysis and applications. Among the many types of inequalities, those associated with the names of Jensen, Hadamard, Hilbert, Hardy, Opial, Poincaré, Sobolev, Levin and Lyapunov have deep roots and made a great impact on various branches of mathematics. The last few decades have witnessed important advances related to these inequalities that remain active areas of research and have grown into substantial fields of research with many important applications. The development of the theory related to these inequalities resulted in a renewal of interest in the field and has attracted interest from many researchers. A host of new results have appeared in the literature.

The present monograph provides a systematic study of some of the most famous and fundamental inequalities originated by the above mentioned mathematicians and brings together the latest, interesting developments in this important research area under a unified framework. Most of the results contained here are only recently discovered and are still scattered over a large number of nonspecialist periodicals. The choice of material covers some of the most important results in the field which have had a great impact on many branches of mathematics. This work will be of interest to mathematical analysts, pure and applied mathematicians, physicists, engineers, computer scientists and other areas of science. For researchers working in these areas, it will be a valuable source of reference and inspiration. It could also be used as a text for an advanced graduate course.

The author acknowledges with great pleasure his gratitude for the fine cooperation and assistance provided by the staff of the book production department of Elsevier Science. I also express deep appreciation to my family members for their encouragement, understanding and patience during the writing of this book.

B.G. Pachpatte

Contents

Introduction

The usefulness of mathematical inequalities in the development of various branches of mathematics as well as in other areas of science is well established in the past several years. The major achievements of mathematical analysis from Newton and Euler to modern applications of mathematics in physical sciences, engineering, and other areas have exerted a profound influence on mathematical inequalities. The development of mathematical analysis is crucially dependent on the unimpeded flow of information between theoretical mathematicians looking for applications and mathematicians working in applications who need theory, mathematical models and methods. Twentieth century mathematics has recognized the power of mathematical inequalities which has given rise to a large number of new results and problems and has led to new areas of mathematics. In the wake of these developments has come not only a new mathematics but a fresh outlook, and along with this, simple new proofs of difficult results.

The classic work "*Inequalities*" by Hardy, Littlewood and Pólya appeared in 1934 and earned its place as a basic reference for mathematicians. This book is the first devoted solely to the subject of inequalities and is a useful guide to this exciting field. The reader can find therein a large variety of classical and new inequalities, problems, results, methods of proof and applications. The work is one of the classics of the century and has had much influence on research in several branches of analysis. It has been an essential source book for those interested in mathematical problems in analysis. The work has been supplemented with "*Inequalities*" by Beckenbach and Bellman written in 1965 and "*Analytic Inequalities*" by Mitrinović published in 1970, which made considerable contributions to this field. These books provide handy references for the reader wishing to explore the topic in depth and show that the theory of inequalities has been established as a viable field of research.

The last century bears witness to a tremendous flow of outstanding results in the field of inequalities, which are partly inspired by the aforementioned monographs, and probably even more so by the challenge of research in various branches of mathematics. The subject has received tremendous impetus from outside of mathematics from such diverse fields as mathematical economics, game theory, mathematical programming, control theory, variational methods, operation research, probability and statistics. The theory of inequalities has been recognized as one of the central areas of mathematical analysis throughout the last century and it is a fast growing discipline, with ever-increasing applications in many scientific fields. This growth resulted in the appearance of the theory of inequalities as an independent domain of mathematical analysis.

The Hölder inequality, the Minkowski inequality, and the arithmetic mean and geometric mean inequality have played dominant roles in the theory of inequalities. These and many other fundamental inequalities are now in common use and, therefore, it is not surprising that numerous studies related to these areas have been made in order to achieve a diversity of desired goals. Over the past decades, the theory of inequalities has developed rapidly and unexpected results were found, along with simpler new proofs for existing results, and, consequently, new vistas for research opened up. In recent years the subject has evoked considerable interest from many mathematicians, and a large number of new results has been investigated in the literature. It is recognized that in general some specific inequalities provide a useful and important device in the development of different branches of mathematics. We shall begin our consideration of results with some important inequalities which find applications in many parts of analysis.

The history of convex functions is very long. The beginning can be traced back to the end of the nineteenth century. Its roots can be found in the fundamental contributions of O. Hölder (1889), O. Stolz (1893) and J. Hadamard (1893). At the beginning of the last century J.L.W.V. Jensen (1905, 1906) first realized the importance and undertook a systematic study of convex functions. In the years thereafter this research resulted in the appearance of the theory of convex functions as an independent domain of mathematical analysis.

In 1889, Hölder [151] proved that if $f''(x) \geqslant 0$, then f satisfied what later came to be known as Jensen's inequality. In 1893, Stolz [412] (see [390,391]) proved that if f is continuous on $[a, b]$ and satisfies

$$f\left(\frac{x+y}{2}\right) \leqslant \frac{1}{2}\left[f(x)+f(y)\right], \tag{1}$$

then f has left and right derivatives at each point of (a, b). In 1893, Hadamard [134] obtained a basic integral inequality for convex functions that have an increasing derivative on $[a, b]$. In his pioneering work, Jensen [164,165] used (1) to define convex functions and discovered the great importance and perspective of these functions. Since then such functions have been studied more extensively, and a good exposition of the results has been given in the book "*Convex Functions*" by A.W. Roberts and D.E. Varberg [397].

Among many important results discovered in his basic work [164,165] Jensen proved one of the fundamental inequalities of analysis which reads as follows.

Let f be a convex function in the Jensen sense on $[a, b]$. For any points $x_1, \dots, x_n$ in $[a, b]$ and any rational nonnegative numbers $r_1, \dots, r_n$ such that $r_1 + \dots + r_n = 1$, we have

$$f\left(\sum_{i=1}^{n} r_i x_i\right) \leqslant \sum_{i=1}^{n} r_i f(x_i). \tag{2}$$

Inequality (2) is now known in the literature as Jensen's inequality. It is one of the most important inequalities for convex functions and has been extended and refined in several different directions using different principles or devices. The fundamental work of Jensen was the starting point for the foundation work in convex functions and can be cited as anticipation what was to come. The general theory of convex functions is the origin of powerful tools for the study of problems in analysis. Inequalities involving convex functions are the most efficient tools in the development of several branches of mathematics and has been given considerable attention in the literature.

One of the most celebrated results about convex functions is the following fundamental inequality.

Let $f : [a, b] \to \mathbb{R}$ be a convex function, where $\mathbb{R}$ denotes the set of real numbers. Then the following inequality holds

$$f\left(\frac{a+b}{2}\right) \leqslant \frac{1}{b-a} \int_a^b f(x)\,\mathrm{d}x \leqslant \frac{f(a) + f(b)}{2}. \tag{3}$$

Inequality (3) is now known in the literature as Hadamard's inequality. The left-hand side of (3), proved in 1893 by Hadamard [134] before convex functions had been formally introduced, for functions f with f' increasing on $[a, b]$, is sometimes called the Hadamard inequality and the right-hand side is known as the "Jensen inequality" or vice versa. There are also papers which attribute inequality (3) completely to Hadamard.

In view of the repeated mentioning of the inequality given in (3), it will be referred to it as to the "Hadamard inequality". In 1985, Mitrinović and Lacković [212] pointed out that the inequalities in (3) are due to C. Hermite who obtained them in 1883, ten years before Hadamard. Inequalities of the form (3) not only are of interest in their own right but also have important applications in various branches of mathematics. The last few decades have witnessed important advances related to inequalities (2) and (3) and numerous variants, generalizations and extensions of these inequalities have appeared in the literature.

One of the many fundamental and remarkable mathematical discoveries of D. Hilbert is the following inequality (see [141, p. 226]).

If $p > 1$, $p' = p/(p-1)$ and $\sum a_n^p \leqslant A$, $\sum b_n^p \leqslant B$, the summations running from 1 to ∞, then

$$\sum\sum \frac{a_m b_n}{m+n} \leqslant \frac{\pi}{\sin(\pi/p)} A^{1/p} B^{1/p}, \tag{4}$$

unless the sequence $\{a_m\}$ or $\{b_n\}$ is null.

The above result is known in the literature as Hilbert's inequality or Hilbert's double series theorem. The integral analogue of Hilbert's inequality can be stated as follows (see [141, p. 226]).

If $p > 1$, $p' = p/(p-1)$ and $\int_0^\infty f^p(x)\,\mathrm{d}x \leqslant F$, $\int_0^\infty g^{p'}(y)\,\mathrm{d}y \leqslant G$, then

$$\int_0^\infty \int_0^\infty \frac{f(x)g(y)}{x+y}\,\mathrm{d}x\,\mathrm{d}y \leqslant \frac{\pi}{\sin(\pi/p)} F^{1/p} G^{1/p'}, \tag{5}$$

unless $f \equiv 0$ or $g \equiv 0$.

The inequalities in (4) and (5) marked the beginning of a new era in the development of the theory of inequalities, which, within a few decades, was very successful and produced numerous variants, generalizations and applications. This work was inspired by the great mathematician D. Hilbert (see [141, p. 226]) whose fundamental contributions to many areas of mathematics are well known.

In the course of attempts to simplify the proofs of inequalities (4) and (5) Hardy [136] (see also [141, pp. 239–240]) discovered the following famous inequality.

If $p > 1$, $a_n \geqslant 0$, $A_n = a_1 + \cdots + a_n$, then

$$\sum_{n=1}^{\infty} \left(\frac{A_n}{n}\right)^p < \left(\frac{p}{p-1}\right)^p \sum_{n=1}^{\infty} a_n^p, \tag{6}$$

unless all the a_n's are zeros. The constant $(p/(p-1))^p$ is the best possible.

The most celebrated result corresponding to the series inequality (6) for integrals due to Hardy [136] is embodied in the following inequality.

If $p > 1$, $f(x) \geqslant 0$ and $F(x) = \int_0^x f(t)\,\mathrm{d}t$, then

$$\int_0^\infty \left(\frac{F}{x}\right)^p \mathrm{d}x < \left(\frac{p}{p-1}\right)^p \int_0^\infty f^p\,\mathrm{d}x, \tag{7}$$

unless $f \equiv 0$. The constant is the best possible.

Inequality (6) or its integral analogue given in (7) is now known in the literature as Hardy's inequality. Inequalities (6) and (7) are the most inspiring and fundamental inequalities in mathematical analysis. A detailed account on earlier developments related to inequalities (4)–(7) can be found in [141, Chapter IX]. Hardy's inequalities given in (6) and (7) were the major influences in the further development of the theory and applications of such inequalities. Since the appearance of inequalities (6) and (7), a large number of papers has appeared in the literature which deals with alternative proofs, various generalizations, extensions, and applications of these inequalities.

In the past several years there has been considerable interest in the study of integral inequalities involving functions and their derivatives. In 1960, Z. Opial [231] published a remarkable paper which contains the following integral inequality.

Let $y(x)$ be of class C^1 on $0 \leqslant x \leqslant h$ and satisfy $y(0) = y(h) = 0$ and $y(x) > 0$ in $(0, h)$. Then the following inequality holds

$$\int_0^h \left|y(x)y'(x)\right| \mathrm{d}x \leqslant \frac{h}{4} \int_0^h \left|y'(x)\right|^2 \mathrm{d}x. \tag{8}$$

The constant $\frac{h}{4}$ is the best possible.

In the same year, C. Olech [230] published a note which deals with a simple proof of Opial's inequality. Moreover, Olech showed that (8) is valid for any function $y(x)$ which is absolutely continuous on $[0, h]$ and satisfies the boundary conditions $y(0) = y(h) = 0$. From Olech's proof, it is clear that in order to prove (8), it is sufficient to prove the following inequality.

Let $y(t)$ be absolutely continuous on $[0, h]$ and $y(0) = 0$. Then the following inequality holds

$$\int_0^h \left|y(x)y'(x)\right| \mathrm{d}x \leqslant \frac{h}{2} \int_0^h \left|y'(x)\right|^2 \mathrm{d}x. \tag{9}$$

The constant $\frac{h}{2}$ is the best possible.

Inequality (8) is known in the literature as Opial's inequality and it is one of the most important and fundamental integral inequalities in the analysis of qualitative properties of solutions of ordinary differential equations. Since the discovery of Opial's inequality in 1960 an enormous amount of work has been done, and many papers which deal with new proofs, various generalizations, extensions and discrete analogues have appeared in the literature; see [4] and the references cited therein.

Motivated by a paper of H.A. Schwarz [404] published in 1885, in the year 1894, H. Poincaré established [389] (see also [211, p. 142]) the following fundamental inequality

$$\iint_T f^2(x,y)\,\mathrm{d}x\,\mathrm{d}y \leqslant \frac{7\sigma^2}{24}\iint_T\left[\left(\frac{\partial f}{\partial x}\right)^2+\left(\frac{\partial f}{\partial y}\right)^2\right]\mathrm{d}x\,\mathrm{d}y, \tag{10}$$

where T is a convex region and f is a function such that $\iint_T f(x,y)\,\mathrm{d}x\,\mathrm{d}y = 0$ and σ is the chord of that region.

In the same paper Poincaré gave an inequality analogues to (10) for a three-dimensional region. In view of the importance of the inequalities of the form (10) many authors have investigated different versions of the above inequality from different view points. The most useful inequality analogous to (10) which is now known in the literature as Poincaré inequality can be stated as follows.

If E is a bounded region in two or three dimensions and u is a sufficiently smooth function which vanishes on the boundary ∂E of E, then

$$\lambda\int_E u^2\,\mathrm{d}A \leqslant \int_E |\nabla u|^2\,\mathrm{d}A, \tag{11}$$

where λ denotes the smallest eigenvalue of the problem

$$\nabla^2 v + \lambda v = 0 \quad \text{in } E, \qquad v = 0 \quad \text{on } \partial E, \tag{12}$$

where $\nabla = (\frac{\partial}{\partial x_1}, \dots, \frac{\partial}{\partial x_n})$.

It is recognized that Poincaré-type inequalities provide, in general, a useful and important device in the study of qualitative as well as quantitative properties of solutions of partial differential equations. Because of their usefulness and importance, Poincaré-type inequalities have attracted much attention and generalizations to various aspects have been established in the literature. The discrete analogues of Poincaré-type inequalities have gained increasing significance in the last decades as is apparent from the large number of applications in the study of finite difference equations. Especially, in view of wider applications, the inequalities of the forms (10) and (11) have been generalized and sharpened from the very day of their discovery.

One of the most celebrated results discovered by S.L. Sobolev [410] is the following integral inequality (see [157, p. 101])

$$\int_{-\infty}^{\infty}\int_{-\infty}^{\infty} u^4\,\mathrm{d}x\,\mathrm{d}y \leqslant \frac{\alpha}{2}\left(\int_{-\infty}^{\infty}\int_{-\infty}^{\infty} u^2\,\mathrm{d}x\,\mathrm{d}y\right)\left(\int_{-\infty}^{\infty}\int_{-\infty}^{\infty} |\operatorname{grad} u|^2\,\mathrm{d}x\,\mathrm{d}y\right), \tag{13}$$

where $u(x,y)$ is any smooth function of compact support in two-dimensional Euclidean space E_2, $|\operatorname{grad} u|^2 = |\frac{\partial u}{\partial x}|^2 + |\frac{\partial u}{\partial y}|^2$ and α is a dimensionless constant.

Inequality (13) is known as Sobolev's inequality, although the same name is used also for the above inequality in n-dimensional Euclidean space. Inequalities of the forms (10), (11) and (13) or their variants have been applied with considerable success to the study of problems in the theory of partial differential equations and have established the foundations of the finite element analysis. There is vast literature which deals with various generalizations, extensions, and variants of these inequalities and their applications; see [3,120,121] and references therein.

It is well known that one of the important and effective techniques in the theory of differential equations is the comparison method (see [416]). Inequalities involving comparison of solutions of second-order differential equations provide a major tool in the study of second-order differential equations. In particular, the basic comparison results due to C. Sturm [414] (see also [145, pp. 334–336]) and that of A.J. Levin [187] have played an important role in the study of several qualitative properties of the solutions of certain second-order differential equations. These comparison results can be found in several classical books, see [145,416].

A useful tool for the study of the qualitative nature of solutions of ordinary linear differential equations of the second order is the fact that if $y(t)$ is a real-valued, absolutely continuous function on $[a,b]$ with $y'(t)$ of integrable square and $y(a)=0=y(b)$, then for s in (a,b) we have

$$\int_a^b \left[y'(t)\right]^2 \mathrm{d}t \geqslant \frac{4}{b-a} y^2(s). \tag{14}$$

Moreover, if $y(t) \not\equiv 0$ on $[a,b]$ the equality holds only if $s=(a+b)/2$ and $y(t) \equiv y(s)\{1-|(2t-a-b)/(b-a)|\}$. In particular, with the aid of this inequality one may show that if $p(t)$ is a real-valued continuous function such that the differential equation

$$y''(t)+p(t)y(t)=0 \tag{15}$$

has a nonidentically vanishing real-valued solution possessing two distinct zeros on $[a,b]$, then

$$\int_a^b p^+(t)\,\mathrm{d}t > \frac{4}{b-a}, \tag{16}$$

where $p^+(t)=\max\{p(t),0\}$, see [393–395].

Inequality (16) is due originally to Lyapunov [201] and it is known that the constant equal to 4 in (16) cannot, in general, be replaced by a larger one. One of the nice purposes of (16) is that a researcher may obtain a lower bound for the distance between two consecutive zeros of a solution of (15) by means of an integral measurement of p. The importance of this famous result of Lyapunov for the study of differential equations has been recognized since its discovery and has received extensive attention over the years, and a number of new Lyapunov-type inequalities which are quite useful in the study of various classes of second-order differential equations investigated in the literature.

The aforementioned inequalities play a fundamental role in different branches of mathematics, and in recent years has attracted the attention of a large number of researchers who are interested both in theory and in applications. The abundance of applications is stimulating a rapid development of the theory of these inequalities and, at present, this theory is one of the most rapidly developing areas of mathematical analysis. Over the years, generalizations, extensions, refinements, improvements, discretizations and new applications of these inequalities are constantly being found by researchers in various branches of mathematics. Although much progress in this field has been made in recent years, these results have not been readily accessible to a wider audience until now. These new developments has motivated the author to write a monograph devoted to the recent developments related to these most important inequalities in mathematics.

A major problem for anyone attempting an exposition related to the above inequalities is the vast extent of the literature. It would be neither easy nor particularly desirable to include everything that is known about these inequalities between the covers of one book, so in this monograph an attempt has been made to present a detailed account of the most inspiring and fundamental results related to the above inequalities which are mostly discovered over the most recent years. A list of applications related to these inequalities is nearly endless, and we are convinced that many new and beautiful applications are still waiting to be revealed. A detailed and comprehensive account of typical applications, together with a full bibliography, may be found in the various references given at the end.

This monograph consists of five chapters and an extensive list of references. Chapter 1 deals with important inequalities involving convex functions which find important applications in various branches of mathematics. It contains a detailed study of a wide variety of inequalities related to the well-known Jensen and Hadamard inequalities, that have recently entered the literature. Chapter 2 is devoted to a great variety of new and fundamental inequalities related to the well-known Hardy and Hilbert inequalities recently investigated in the literature and which will open up new vistas for further research in this field. Chapter 3 considers many new inequalities of the Opial type recently investigated in the literature and which involve functions of one or many independent variables and which has proven to be important in the theory of ordinary and partial differential equations. Chapter 4 presents a number of new inequalities related to the well-known inequalities of Poincaré and Sobolev which finds important applications in the study of partial differential equations and finite element analysis. Chapter 5 is concerned with basic inequalities developed in the literature related to the most important inequalities of Levin and Lyapunov which are useful in the study of differential equations. It deals with a number of new generalizations, extensions, and variants of the original Levin and Lyapunov inequalities. Each chapter ends with miscellaneous inequalities for further study and notes on bibliographies.

Chapter 1

Inequalities Involving Convex Functions

1.1 Introduction

The fundamental work of Jensen [164,165] in the years 1905, 1906 is the starting point of the systematic study of convex functions. Even before Jensen, the literature shows results which refer to convex functions. In fact the roots of such functions can be found in the work of Hölder [151] in 1889 and Hadamard [134] in 1893, although these roots were not explicitly specified in their works. As noted by Popoviciu [390, p. 48], Stolz [412] is the first to introduce convex functions in the year 1893. Starting from the pioneer papers of Jensen [164,165] there is remarkable interest in the theory of convex functions and these ideas are at the core of many problems in different branches of mathematics. Over the years several new inequalities involving convex functions which have important applications in various branches of mathematics have been developed. This chapter presents some basic inequalities involving convex functions which find significant applications in mathematical analysis, applied mathematics, probability theory, and various other branches of mathematics.

1.2 Jensen's and Related Inequalities

Let I denote a suitable interval of the real line $\mathbb{R}$. A function $f : I \to \mathbb{R}$ is called convex in the Jensen sense or J-convex or midconvex if

$$f\left(\frac{x+y}{2}\right) \leqslant \frac{f(x)+f(y)}{2} \tag{1.2.1}$$

for all $x, y \in I$. Jensen [164,165] is first to define a convex function by using inequality (1.2.1) and to draw attention to their importance. A function $f : I \to \mathbb{R}$

is called convex if

$$f(\lambda x + (1-\lambda)y) \leqslant \lambda f(x) + (1-\lambda) f(y) \tag{1.2.2}$$

for all $x, y \in I$ and $\lambda \in (0, 1)$. It is called strictly convex provided that inequality (1.2.2) is strict for $x \neq y$. If $-f : I \to \mathbb{R}$ is convex, then we say that $f : I \to \mathbb{R}$ is concave. Taking $\lambda = 1/2$ it shows that all functions satisfying (1.2.2) also satisfy (1.2.1), but the definitions are not equivalent since there are functions discontinuous on an open interval that satisfy (1.2.1), while all functions that satisfy (1.2.2) are continuous on open intervals. The definition of convex function has very natural generalization to real-valued functions defined on an arbitrary normed linear space L. We merely require that the domain U of f be convex. This response assures that for $x_1, x_2 \in U$, $\alpha \in (0, 1)$, f will always be defined at $\alpha x_1 + (1-\alpha)x_2$. We then define f to be convex on $U \subseteq L$ if

$$f(\alpha x_1 + (1-\alpha)x_2) \leqslant \alpha f(x_1) + (1-\alpha) f(x_2).$$

A detailed account on the various properties of convex functions can be found in [211,384,397].

In this section we shall deal with Jensen's and related inequalities involving convex functions investigated by various authors over the years, which find significant applications in various branches of mathematics.

We begin with Jensen's inequality, which is one of the basic and most important inequalities in mathematics.

THEOREM 1.2.1. *Suppose that f is J-convex and $I = [a, b]$. For any points $x_1, \ldots, x_n \in I$ and any rational nonnegative numbers, $r_1, \ldots, r_n$ such that $r_1 + \cdots + r_n = 1$, we have*

$$f\left(\sum_{i=1}^{n} r_i x_i\right) \leqslant \sum_{i=1}^{n} r_i f(x_i). \tag{1.2.3}$$

PROOF.

Case 1. For $n = 2$ and $r_1 = r_2 = 1/2$, we have (1.2.1). For $r_i = 1/n$, $i = 1, \ldots, n$, inequality (1.2.3) becomes

$$f\left(\frac{1}{n}\sum_{i=1}^{n} x_i\right) \leqslant \frac{1}{n}\sum_{i=1}^{n} f(x_i). \tag{1.2.4}$$

First we prove (1.2.4) by using an idea of a proof from Pečarić [370]. Suppose that (1.2.4) is valid for all k, $2 \leqslant k \leqslant n$. Denoting $x = \frac{1}{n+1}\sum_{i=1}^{n+1} x_i$, we have

$$\begin{aligned} f(x) &= f\left(\frac{1}{n+1}\sum_{i=1}^{n+1} x_i\right) \\ &= f\left(\frac{1}{2}\left(\frac{1}{n}\sum_{i=1}^{n} x_i + \frac{n-1}{n}x + \frac{1}{n}x_{n+1}\right)\right) \\ &\leqslant \frac{1}{2}\left(f\left(\frac{1}{n}\sum_{i=1}^{n} x_i\right) + f\left(\frac{n-1}{n}x + \frac{1}{n}x_{n+1}\right)\right) \\ &\leqslant \frac{1}{2}\left(\frac{1}{n}\sum_{i=1}^{n} f(x_i) + \frac{1}{n}\big((n-1)f(x) + f(x_{n+1})\big)\right), \end{aligned}$$

which is

$$f(x) \leqslant \frac{1}{n+1}\sum_{i=1}^{n+1} f(x_i),$$

and the proof of (1.2.2) is complete by induction.

Case 2. Since $r_1, \dots, r_n$ are nonnegative rational numbers there is a natural number m and nonnegative integers $p_1, \dots, p_n$ such that $m = p_1 + \dots + p_n$ and $r_i = \frac{p_i}{m}$, $i = 1, \dots, n$. Now, by Case 1, we have

$$\begin{aligned} &f\left(\frac{(x_1 + \dots + x_1) + \dots + (x_n + \dots + x_n)}{m}\right) \\ &\quad \leqslant \frac{(f(x_1) + \dots + f(x_1)) + \dots + (f(x_n) + \dots + f(x_n))}{m}, \end{aligned} \tag{1.2.5}$$

where in the first bracket there are p_1 terms, and so on, in the nth bracket p_n terms. Thus (1.2.5) reads as

$$f\left(\frac{1}{m}\sum_{i=1}^{n} p_i x_i\right) \leqslant \sum_{i=1}^{n} \frac{p_i}{m} f(x_i), \tag{1.2.6}$$

and by taking $p_i/m = r_i$ in (1.2.6) we get (1.2.3). The proof is complete. □

REMARK 1.2.1. Following Jensen, there came a series of papers giving conditions under which (1.2.3) is valid. If we remove some of the restrictions on the r_i, thereby increasing the kinds of combinations of points $x_1, \ldots, x_n$ under consideration, it is to be expected that the class of functions still satisfying Jensen's inequality will be smaller.

As an immediate consequence of Theorem 1.2.1 we have the following useful version of Jensen's inequality.

COROLLARY 1.2.1. *Let $f : U \subseteq L \to \mathbb{R}$ is a convex mapping on convex set U of real linear space L, x_i are in C, $i = 1, \ldots, n$, and $p_i \geqslant 0$ with $P_n = \sum_{i=1}^{n} p_i > 0$, then*

$$f\left(\frac{1}{P_n}\sum_{i=1}^{n} p_i x_i\right) \leqslant \frac{1}{P_n}\sum_{i=1}^{n} p_i f(x_i). \tag{1.2.7}$$

In [367] Pečarić has given a simple proof of the following form of the Jensen–Steffensen inequality.

THEOREM 1.2.2. *Let x and p be two n-tuples of real numbers such that x is nonincreasing, $x_i \in [a, b]$, $1 \leqslant i \leqslant n$, and $0 \leqslant P_k \leqslant P_n$, $k = 1, \ldots, n-1$, $P_n > 0$, where $P_k = \sum_{i=1}^{k} p_i$, $k = 1, \ldots, n$. Then for every real-valued convex function f defined on $[a, b]$,*

$$f\left(\frac{1}{P_n}\sum_{i=1}^{n} p_i x_i\right) \leqslant \frac{1}{P_n}\sum_{i=1}^{n} p_i f(x_i). \tag{1.2.8}$$

PROOF. Note that if each p_i is positive, inequality (1.2.8) follows easily from the definition of a convex function. A convex function f is characterized by having a supporting line at each point, that is,

$$f(z) - f(c) \geqslant M(z - c) \tag{1.2.9}$$

for all z and c, where M depends on c. (In fact $M = f'(c)$ where $f'(c)$ exists, and M is any number between $f'_-(c)$ and $f'_+(c)$ at the countable set where these are different.)

Using (1.2.9) we can easily obtain the following known inequalities:

$$\begin{aligned} &f(z) - f(y) \geqslant M(z - y), \quad z \geqslant y \geqslant c, \quad \text{and} \\ &f(z) - f(y) \leqslant M(z - y), \quad y \leqslant z \leqslant c, \end{aligned} \tag{1.2.10}$$

where M is defined as above.

Let x and p satisfy the conditions of the Jensen–Steffensen inequality. Let $\bar{x} = \frac{1}{P_n}\sum_{i=1}^{n} p_i x_i$ and let $\overline{P}_k = P_n - P_{k-1}$. Then

$$P_n(x_1 - \bar{x}_n) = \sum_{i=2}^{n} p_i(x_1 - x_i) = \sum_{j=2}^{n}(x_{j-1} - x_j)\overline{P}_j \geqslant 0,$$

so $\bar{x} \leqslant x_1$. Similarly,

$$P_n(\bar{x} - x_n) = \sum_{i=1}^{n-1} p_i(x_i - x_n) = \sum_{j=1}^{n-1}(x_j - x_{j+1})P_j \geqslant 0,$$

so $x_n \leqslant \bar{x} \leqslant x_1$. Let m be such that $\bar{x} \in [x_{m+1}, x_m]$. Hereafter, M is to be given its value at $c = \bar{x}$.

We can easily show that the following identity is valid

$$\begin{aligned}
&f\left(\frac{1}{P_n}\sum_{i=1}^{n} p_i x_i\right) - \frac{1}{P_n}\sum p_i f(x_i)\\
&= \sum_{i=1}^{m-1}\big(M(x_i - x_{i+1}) - f(x_i) + f(x_{i+1})\big)\frac{P_i}{P_n}\\
&\quad + \big(M(x_m - \bar{x}) - f(x_m) + f(\bar{x})\big)\frac{P_m}{P_n}\\
&\quad + \big(f(\bar{x}) - f(x_{m+1}) - M(\bar{x} - x_{m+1})\big)\frac{\overline{P}_{m+1}}{P_n}\\
&\quad + \sum_{i=m+1}^{n-1}\big(f(x_i) - f(x_{i+1}) - M(x_i - x_{i+1})\big)\frac{\overline{P}_{i+1}}{P_n}.
\end{aligned} \tag{1.2.11}$$

Now, using (1.2.10) and (1.2.11) we get (1.2.8). The proof is complete. □

Let $f : I \to \mathbb{R}$ be a real-valued function and $x = (x_1, \ldots, x_n) \in I^n$, the expression

$$f_{k,n} = f_{k,n}(x) = \frac{1}{\binom{n}{k}} \sum_{1 \leqslant i_1 < \cdots < i_k \leqslant n} f\left(\frac{1}{k}(x_{i_1} + \cdots + x_{i_k})\right)$$

is used by Gabler [123] to define "sequentially convex functions". These functions are a special case of convex functions. Gabler also gives the inequality

$$f_{k,n}(x) \geqslant f_{k+1,n}(x), \quad k = 1, \ldots, n-1, \tag{1.2.12}$$

for sequentially convex functions.

In the following theorem we present the result given by Pečarić in [375] which gives a sequence of interpolating inequalities for the well-known Jensen inequality for convex functions.

THEOREM 1.2.3. *Let $f : I \to \mathbb{R}$ be a convex function and $x = (x_1, \ldots, x_n) \in I^n$. Let*

$$\begin{aligned} f_{k,n} &= f_{k,n}(x, p) \\ &= \frac{1}{\binom{n-1}{k-1} P_n} \sum_{1 \leqslant i_1 < \cdots < i_k \leqslant n} (p_{i_1} + \cdots + p_{i_k}) f\left(\frac{p_{i_1} x_{i_1} + \cdots + p_{i_k} x_{i_k}}{p_{i_1} + \cdots + p_{i_k}} \right), \end{aligned}$$

where p_i are positive numbers and $P_n = \sum_{i=1}^{n} p_i$. Then

$$f_{k,n}(x, p) \geqslant f_{k+1,n}(x, p), \quad k = 1, \ldots, n-1, \tag{1.2.13}$$

is valid.

PROOF. Indeed, we have

$$\begin{aligned} &(p_{i_1} + \cdots + p_{i_{k+1}}) f\left(\frac{p_{i_1} x_{i_1} + \cdots + p_{i_{k+1}} x_{i_{k+1}}}{p_{i_1} + \cdots + p_{i_{k+1}}} \right) \\ &\quad = (p_{i_1} + \cdots + p_{i_{k+1}}) \\ &\qquad \times f\left(\frac{\sum_{j=1}^{k+1} (p_{i_1} + \cdots + p_{i_{k+1}} - p_{i_j}) \frac{p_{i_1} x_{i_1} + \cdots + p_{i_{k+1}} x_{i_{k+1}} - p_{i_j} x_{i_j}}{p_{i_1} + \cdots + p_{i_{k+1}} - p_{i_j}}}{\sum_{j=1}^{k+1} (p_{i_1} + \cdots + p_{i_{k+1}} - p_{i_j})} \right) \\ &\quad \leqslant (p_{i_1} + \cdots + p_{i_{k+1}}) \\ &\qquad \times \frac{\sum_{j=1}^{k+1} (p_{i_1} + \cdots + p_{i_{k+1}} - p_{i_j}) f\left(\frac{p_{i_1} x_{i_1} + \cdots + p_{i_{k+1}} x_{i_{k+1}} - p_{i_j} x_{i_j}}{p_{i_1} + \cdots + p_{i_{k+1}} - p_{i_j}} \right)}{\sum_{j=1}^{k+1} (p_{i_1} + \cdots + p_{i_{k+1}} - p_{i_j})} \\ &\quad = \frac{1}{k} \sum_{j=1}^{k+1} (p_{i_1} + \cdots + p_{i_{k+1}} - p_{i_j}) f\left(\frac{p_{i_1} x_{i_1} + \cdots + p_{i_{k+1}} x_{i_{k+1}} - p_{i_j} x_{i_j}}{p_{i_1} + \cdots + p_{i_{k+1}} - p_{i_j}} \right). \end{aligned}$$

Therefore, we have

$$\begin{aligned} f_{k+1,n} &= \frac{1}{\binom{n-1}{k}P_n} \sum_{1\leqslant i_1<\cdots<i_{k+1}\leqslant n} (p_{i_1}+\cdots+p_{i_{k+1}}) \\ &\qquad\qquad \times f\left(\frac{p_{i_1}x_{i_1}+\cdots+p_{i_{k+1}}x_{i_{k+1}}}{p_{i_1}+\cdots+p_{i_{k+1}}}\right) \\ &\leqslant \frac{1}{k\binom{n-1}{k}P_n} \sum_{1\leqslant i_1<\cdots<i_{k+1}\leqslant n} \sum_{j=1}^{k+1}(p_{i_1}+\cdots+p_{i_{k+1}}-p_{i_j}) \\ &\qquad\qquad \times f\left(\frac{p_{i_1}x_{i_1}+\cdots+p_{i_{k+1}}x_{i_{k+1}}-p_{i_j}x_{i_j}}{p_{i_1}+\cdots+p_{i_{k+1}}-p_{i_j}}\right) \\ &= \frac{1}{\binom{n-1}{k-1}P_n} \sum_{1\leqslant i_1<\cdots<i_k\leqslant n} (p_{i_1}+\cdots+p_{i_k}) f\left(\frac{p_{i_1}x_{i_1}+\cdots+p_{i_k}x_{i_k}}{p_{i_1}+\cdots+p_{i_k}}\right) \\ &= f_{k,n}. \end{aligned}$$

The proof is complete. □

REMARK 1.2.2. We note that inequality (1.2.12) is an interpolating inequality for Jensen's inequality for convex functions. Indeed, we have

$$\begin{aligned} f\left(\frac{1}{P_n}\sum_{i=1}^{n} p_i x_i\right) &= f_{n,n} \leqslant \cdots \leqslant f_{k+1,n} \leqslant f_{k,n} \leqslant \cdots \leqslant f_{1,n} \\ &= \frac{1}{P_n}\sum_{i=1}^{n} p_i f(x_i). \end{aligned} \tag{1.2.14}$$

The above results are also valid for convex functions defined on arbitrary real linear space, and if p_i, $i = 1, \ldots, n$, are rational numbers, they are also valid for midconvex functions defined on an arbitrary real linear space.

Let $f: C \subset X \to \mathbb{R}$ be a convex function on convex set C of real linear space X, $x_i \in C$ and $p_i \geqslant 0$, $i = 1, \ldots, n$, with $P_n = \sum_{i=1}^{n} p_i > 0$. Let T be a nonempty set and let m be a natural number with $m \geqslant 2$. Suppose that $\alpha_1, \ldots, \alpha_m : T \to \mathbb{R}$ are m functions with the property that $\alpha_i(t) \geqslant 0$ and $\alpha_1(t) + \cdots + \alpha_m(t) = 1$ for all t in T and $i = 1, \ldots, m$.

Consider the sequence of functions defined by (see [90])

$$F_1^{[m]}(t) = \frac{1}{P_n}\sum_{i_1=1}^{n} p_{i_1} f\left[\alpha_1(t)x_{i_1} + \left(\alpha_2(t)+\cdots+\alpha_m(t)\right)\frac{1}{P_n}\sum_{i=1}^{n} p_i x_i\right],$$

$$F_2^{[m]}(t) = \frac{1}{P_n^2}\sum_{i_1,i_2=1}^{n} p_{i_1}p_{i_2} f\left[\alpha_1(t)x_{i_1} + \alpha_2(t)x_{i_2} + \left(\alpha_3(t)+\cdots+\alpha_m(t)\right)\frac{1}{P_n}\sum_{i=1}^{n} p_i x_i\right],$$

$$\vdots$$

$$F_{m-1}^{[m]}(t) = \frac{1}{P_n^{m-1}}\sum_{i_1,\dots,i_{m-1}=1}^{n} p_{i_1}\cdots p_{i_{m-1}} f\left[\alpha_1(t)x_{i_1} + \cdots + \alpha_{m-1}(t)x_{i_{m-1}} + \alpha_m(t)\frac{1}{P_n}\sum_{i=1}^{n} p_i x_i\right]$$

and

$$F^{[m]}(t) = \frac{1}{P_n^m}\sum_{i_1,\dots,i_m=1}^{n} p_{i_1}\cdots p_{i_m} f\left(\alpha_1(t)x_{i_1} + \cdots + \alpha_m(t)x_{i_m}\right),$$

where t is in T, $n \geqslant 1$.

It is clear that the above mappings are well defined for all t in T.

The following theorem is proved in [90].

THEOREM 1.2.4. *Let f, p_i, x_i, $i = 1,\dots,n$, and m be as above. Then*

(i) *we have the inequalities*

$$f\left(\frac{1}{P_n}\sum_{i=1}^{n} p_i x_i\right) \leqslant F_1^{[m]}(t) \leqslant \cdots \leqslant F_{m-1}^{[m]}(t) \leqslant F^{[m]}(t) \leqslant \frac{1}{P_n}\sum_{i=1}^{n} p_i f(x_i) \tag{1.2.15}$$

for all t in T;

(ii) *if there exists a $t_0 \in T$ so that $\alpha_1(t_0) = \cdots = \alpha_p(t_0) = 0$, $1 \leqslant p \leqslant m-1$, then*

$$\inf_{t\in T} F_j^{[m]}(t) = F_j^{[m]}(t_0) = f\left(\frac{1}{P_n}\sum_{i=1}^{n} p_i x_i\right) \tag{1.2.16}$$

for all $1 \leqslant j \leqslant p$;

(iii) *if there exists a* $t_1 \in T$ *so that* $\alpha_p(t_1) = 1$, *then*

$$\sup_{t\in T} F_j^{[m]}(t) = \sup_{t\in T} F^{[m]}(t) = F_j^{[m]}(t_1) = \frac{1}{P_n}\sum_{i=1}^{n} p_i f(x_i) \tag{1.2.17}$$

and

$$\inf_{t\in T} F_q^{[m]}(t) = F_q^{[m]}(t_1) = f\left(\frac{1}{P_n}\sum_{i=1}^{n} p_i x_i\right), \tag{1.2.18}$$

for all $p \leqslant j \leqslant m-1$ *and* $1 \leqslant q \leqslant p-1$;

(iv) *if* T *is a convex subset of a linear space* Y *and* α_i, $i = 1, \ldots, n$, *satisfies the condition*

$$\alpha_i(\gamma t_1 + \beta t_2) = \gamma\alpha_i(t_1) + \beta\alpha_i(t_2) \tag{AF}$$

for all $t_1, t_2 \in T$ *and* $\gamma, \beta \geqslant 0$ *with* $\gamma + \beta = 1$, *then* $F_j^{[m]}$, $1 \leqslant j \leqslant m-1$, *and* $F^{[m]}$ *are convex mappings in* T.

PROOF. (i) By Jensen's inequality, we have

$$\begin{aligned}
F_1^{[m]}(t) &\geqslant f\left[\left(\frac{1}{P_n}\sum_{i_1=1}^{n} p_{i_1}x_{i_1}\right)\alpha_1(t) + \big(\alpha_2(t)+\cdots+\alpha_m(t)\big)\frac{1}{P_n}\sum_{i=1}^{n} p_i x_i\right] \\
&= f\left(\frac{1}{P_n}\sum_{i=1}^{n} p_i x_i\right)
\end{aligned}$$

for all t in T, which shows the first inequality in (1.2.15).

Now, suppose that $1 \leqslant j \leqslant m-2$ and $t \in T$. Then, by Jensen's inequality, we have

$$\begin{aligned}
&F_{j+1}^{[m]}(t) \\
&\quad= \frac{1}{P_n^{j+1}} \sum_{i_1,\ldots,i_{j+1}=1} p_{i_1}\cdots p_{i_{j+1}} f\Bigg[\alpha_1(t)x_{i_1} + \cdots + \alpha_{j+1}(t)x_{j+1} \\
&\qquad\qquad + \big(\alpha_{j+2}(t)+\cdots+\alpha_m(t)\big)\frac{1}{P_n}\sum_{i=1}^{n} p_i x_i\Bigg]
\end{aligned}$$

$$
\begin{aligned}
&\geqslant \frac{1}{P_n^j} \sum_{i_1,\dots,i_j=1}^n p_{i_1}\cdots p_{i_j} f\Bigg[\alpha_1(t)x_{i_1}+\cdots+\alpha_j(t)x_{i_j} \\
&\qquad\qquad + \left(\frac{1}{P_n}\sum_{i_{j+1}=1}^n p_{i_{j+1}}x_{i_{j+1}}\right)\alpha_{j+1}(t) \\
&\qquad\qquad + \big(\alpha_{j+2}(t)+\cdots+\alpha_m(t)\big)\frac{1}{P_n}\sum_{i=1}^n p_i x_i\Bigg] \\
&= F_j^{[m]}(t)
\end{aligned}
$$

which shows that the sequence $\{F_j^{[m]}(t)\}_{j=1}^{m-1}$ is monotonous nondecreasing for all t in T.

On the other hand, by Jensen's inequality, we also have

$$
\begin{aligned}
F^{[m]}(t) &\geqslant \frac{1}{P_n^{m-1}} \sum_{i_1,\dots,i_{m-1}=1}^n p_{i_1}\cdots p_{i_{m-1}} f\Bigg[\alpha_1(t)x_{i_1}+\cdots+\alpha_{m-1}(t)x_{i_{m-1}} \\
&\qquad\qquad + \left(\frac{1}{P_n}\sum_{i_m=1}^n p_{i_m}x_{i_m}\right)\alpha_{i_m}(t)\Bigg] \\
&= F_{m-1}^{[m]}(t)
\end{aligned}
$$

for all t in T.

Finally, by the convexity of f on C, one has

$$
f\big(\alpha_1(t)x_{i_1}+\cdots+\alpha_m(t)x_{i_m}\big) \leqslant \alpha_1(t)f(x_{i_1})+\cdots+\alpha_m(t)f(x_{i_m})
$$

for all t in T and $x_i \in C$, $i=1,\dots,m$. Multiplying by $p_{i_1},\dots,p_{i_m}$ and summing in $i_1,\dots,i_m$ from 1 to n, we get

$$
\begin{aligned}
&\sum_{i_1,\dots,i_m=1}^n p_{i_1}\cdots p_{i_m} f\big(\alpha_1(t)x_{i_1}+\cdots+\alpha_m(t)x_{i_m}\big) \\
&\quad\leqslant \sum_{i_1,\dots,i_m=1}^n p_{i_1}\cdots p_{i_m}\big[\alpha_1(t)f(x_{i_1})+\cdots+\alpha_m(t)f(x_{i_m})\big] \\
&\quad= \alpha_1(t)P_n^{m-1}\sum_{i_1=1}^n p_{i_1}f(x_{i_1})+\cdots+\alpha_m(t)P_n^{m-1}\sum_{i_m=1}^n p_{i_m}f(x_{i_m}) \\
&\quad= P_n^{m-1}\sum_{i=1}^n p_i f(x_i)
\end{aligned}
$$

for all $t \in T$, which is equivalent with the last part of inequality (1.2.15). The proof of statement (i) is finished.

(ii) If $\alpha_1(t_0) = \cdots = \alpha_p(t_0) = 0$, $1 \leqslant p \leqslant m-1$, then

$$\begin{aligned} F_p^{[m]}(t_0) &= \frac{1}{P_n^p} \sum_{i_1,\ldots,i_p=1}^{n} p_{i_1} \cdots p_{i_p} f\left((\alpha_{p+1}(t_0) + \cdots + \alpha_m(t_0)) \frac{1}{P_n} \sum_{i=1}^{n} p_i x_i \right) \\ &= \frac{1}{P_n^p} \sum_{i_1,\ldots,i_p=1}^{n} p_{i_1} \cdots p_{i_p} f\left(\frac{1}{P_n} \sum_{i=1}^{n} p_i x_i \right) \\ &= f\left(\frac{1}{P_n} \sum_{i=1}^{n} p_i x_i \right). \end{aligned}$$

Since

$$f\left(\frac{1}{P_n} \sum_{i=1}^{n} p_i x_i \right) \leqslant F_1^{[m]}(t_0) \leqslant \cdots \leqslant F_p^{[m]}(t_0),$$

statement (ii) is proved.

(iii) If $\alpha_p(t_1) = 1$, then $\alpha_s(t_1) = 0$ for all $s \neq p$, $1 \leqslant s \leqslant m$. Thus, for $p \leqslant j \leqslant m-1$, one has

$$\begin{aligned} F_j^{[m]}(t_1) &= \frac{1}{P_n^j} \sum_{i_1,\ldots,i_j=1}^{n} p_{i_1} \cdots p_{i_j} f(x_{i_j}) \\ &= \frac{1}{P_n} \sum_{i=1}^{n} p_i f(x_i) \\ &= F^{[m]}(t_1). \end{aligned}$$

If $1 \leqslant q \leqslant p-1$, we have

$$\begin{aligned} F_q^{[m]}(t_1) &= \frac{1}{P_n^q} \sum_{i_1,\ldots,i_q=1}^{n} p_{i_1} \ldots p_{i_q} f\left(\frac{1}{P_n} \sum_{i=1}^{n} p_i x_i \right) \\ &= f\left(\frac{1}{P_n} \sum_{i=1}^{n} p_i x_i \right), \end{aligned}$$

which shows the statements (1.2.17) and (1.2.18).

(iv) Let $\gamma, \beta \geqslant 0$ with $\gamma + \beta = 1$ and $t_1, t_2 \in T$. Then, by the convexity of f, we have

$$\begin{aligned}
&F_j^{[m]}(\gamma t_1 + \beta t_2) \\
&= \frac{1}{P_n^j} \sum_{i_1,\dots,i_j=1}^{n} p_{i_1} \cdots p_{i_j} f\Bigg[\alpha_1(\gamma t_1 + \beta t_2)x_{i_1} + \cdots + \alpha_j(\gamma t_1 + \beta t_2)x_{i_j} \\
&\qquad + \big(\alpha_{j+1}(\gamma t_1 + \beta t_2) + \cdots \\
&\qquad + \alpha_m(\gamma t_1 + \beta t_2)\big)\frac{1}{P_n}\sum_{i=1}^{n} p_i x_i\Bigg] \\
&= \frac{1}{P_n^j} \sum_{i_1,\dots,i_j=1}^{n} p_{i_1} \cdots p_{i_j} f\Bigg(\gamma\Bigg[\alpha_1(t_1)x_{i_1} + \cdots + \alpha_j(t_1)x_{i_j} \\
&\qquad + \big(\alpha_{j+1}(t_1) + \cdots + \alpha_m(t_1)\big)\frac{1}{P_n}\sum_{i=1}^{n} p_i x_i\Bigg] \\
&\qquad + \beta\Bigg[\alpha_1(t_2)x_{i_1} + \cdots + \alpha_j(t_2)x_{i_j} \\
&\qquad + \big(\alpha_{j+1}(t_2) + \cdots + \alpha_m(t_2)\big)\frac{1}{P_n}\sum_{i=1}^{n} p_i x_i\Bigg]\Bigg) \\
&\leqslant \gamma F_j^{[m]}(t_1) + \beta F_j^{[m]}(t_2)
\end{aligned}$$

for all $1 \leqslant j \leqslant m-1$, which shows that $F_j^{[m]}$ is convex on T.

The fact that $F^{[m]}$ is convex on T goes likewise, we omit the details. The proof of the theorem is finished. □

The classical inequality between the weighted arithmetic and geometric means states:

If $x_1, \dots, x_n$ and $p_1, \dots, p_n$ are positive real numbers, then

$$\left(\prod_{i=1}^{n} x_i^{p_i}\right)^{1/P_n} \leqslant \frac{1}{P_n}\sum_{i=1}^{n} p_i x_i, \tag{1.2.19}$$

where $P_n = \sum_{i=1}^{n} p_i$. Equality holds in (1.2.19) if and only if $x_1 = \cdots = x_n$.

The following corollary given in [90] improves inequality (1.2.19).

COROLLARY 1.2.2. *Let $f : C \subset X \to (0, \infty)$ be a convex function on a convex subset C of a linear space X which is also logarithmically concave on C. Then for all x_i, p_i, α_j and m as above, we have the following refinement of the arithmetic mean–geometric mean inequality:*

$$
\begin{aligned}
\frac{1}{P_n}\sum_{i=1}^{n} p_i f(x_i) &\geqslant F^{[m]}(t) \geqslant F_{m-1}^{[m]}(t) \geqslant \cdots \geqslant F_2^{[m]}(t) \\
&\geqslant F_1^{[m]}(t) f\left(\frac{1}{P_n}\sum_{i=1}^{n} p_i x_i\right) \\
&\geqslant G_1^{[m]}(t) \geqslant G_2^{[m]}(t) \geqslant \cdots \geqslant G_{m-1}^{[m]}(t) \geqslant G^{[m]}(t) \\
&\geqslant \left(\prod_{i=1}^{n} [f(x_i)]^{p_i}\right)^{1/P_n}
\end{aligned}
\tag{1.2.20}
$$

for all t in T, where

$$
G_1^{[m]}(t) = \left(\prod_{i_1=1}^{n} f^{p_{i_1}}\left[\alpha_1(t)x_{i_1} + \left(\alpha_2(t) + \cdots + \alpha_m(t)\right)\frac{1}{P_n}\sum_{i=1}^{n} p_i x_i\right]\right)^{1/P_n},
$$

$$
\begin{aligned}
G_2^{[m]}(t) = \Bigg(\prod_{i_1,i_2=1}^{n} f^{p_{i_1}p_{i_2}}\Bigg[&\alpha_1(t)x_{i_1} + \alpha_2(t)x_{i_2} \\
&+ \left(\alpha_3(t) + \cdots + \alpha_m(t)\right)\frac{1}{P_n}\sum_{i=1}^{n} p_i x_i\Bigg]\Bigg)^{1/P_n^2},
\end{aligned}
$$

$$\vdots$$

$$
\begin{aligned}
G_{m-1}^{[m]}(t) = \Bigg(\prod_{i_1,\ldots,i_{m-1}=1}^{n} f^{p_{i_1}\cdots p_{i_{m-1}}}\Bigg[&\alpha_1(t)x_{i_1} + \cdots + \alpha_{m-1}(t)x_{i_{m-1}} \\
&+ \alpha_m(t)\frac{1}{P_n}\sum_{i=1}^{n} p_i x_i\Bigg]\Bigg)^{1/P_n^{m-1}}
\end{aligned}
$$

and

$$G^{[m]}(t)=\left(\prod_{i_1,\dots,i_m=1}^{n} f^{p_{i_1}\cdots p_{i_m}}\left[\alpha_1(t)x_{i_1}+\cdots+\alpha_m(t)x_{i_m}\right]\right)^{1/P_n^m},$$

where t is in T.

PROOF. The argument of the second part of (1.2.20) follows by (1.2.15) for the convex mapping–log f and we omit the details. □

REMARK 1.2.3. If in the above corollary we choose $f:(0,\infty)\to(0,\infty)$, $f(x)=x$, we obtain the following improvement of the arithmetic mean and geometric mean inequality

$$\begin{aligned}
&\frac{1}{P_n}\sum_{i=1}^{n} p_i x_i\\
&\geqslant\left(\prod_{i_1=1}^{n}\left[\alpha_i(t)x_{i_1}+\left(\alpha_2(t)+\cdots+\alpha_m(t)\right)\frac{1}{P_n}\sum_{i=1}^{n} p_i x_i\right]^{p_{i_1}}\right)^{1/P_n}\\
&\geqslant\left(\prod_{i_1,i_2=1}^{n}\left[\alpha_1(t)x_{i_1}+\alpha_2(t)x_{i_2}\right.\right.\\
&\qquad\left.\left.+\left(\alpha_3(t)+\cdots+\alpha_m(t)\right)\frac{1}{P_n}\sum_{i=1}^{n} p_i x_i\right]^{p_{i_1}p_{i_2}}\right)^{1/P_n^2}\\
&\geqslant\cdots\geqslant\\
&\geqslant\left(\prod_{i_1,\dots,i_{m-1}=1}^{n}\left[\alpha_1(t)x_{i_1}+\cdots+\alpha_m(t)\frac{1}{P_n}\sum_{i=1}^{n} p_i x_i\right]^{p_{i_1}\cdots p_{i_{m-1}}}\right)^{1/P_n^{m-1}}\\
&\geqslant\left(\prod_{i_1,\dots,i_m=1}^{n}\left[\alpha_1(t)x_{i_1}+\cdots+\alpha_m(t)x_{i_m}\right]^{p_{i_1}\cdots p_{i_m}}\right)^{1/P_n^m}\\
&\geqslant\left(\prod_{i=1}^{n} x_i^{p_i}\right)^{1/P_n}
\end{aligned}$$

for all t in T.

In view of the important role played by Jensen's inequality in analysis, many mathematicians have tried not only to establish (1.2.3) or (1.2.7) in a variety of ways but also to find different extensions, refinements and counterparts, see [90, 92,207,218,219,370,373] where further references are given.

The corresponding integral analogues of the well-known Jensen's inequality are also widely used in the mathematical analysis and applications.

The following integral analogue of Jensen's inequality is adapted from [174, p. 133].

THEOREM 1.2.5. *Let* $f:I=[a,b]\to\mathbb{R}$ *be a convex function. Let* $h:I\to(0,\infty)$ *and* $u:I\to\mathbb{R}_+=[0,\infty)$ *are integrable functions. Then*

$$f\left(\frac{\int_a^b h(t)u(t)\,\mathrm{d}t}{\int_a^b h(t)\,\mathrm{d}t}\right)\leqslant\frac{\int_a^b h(t)f(u(t))\,\mathrm{d}t}{\int_a^b h(t)\,\mathrm{d}t}\tag{1.2.21}$$

provided that all the integrals in (1.2.21) *are meaningful.*

PROOF. Let $\gamma>0$ be fixed. From the convexity of f it follows that there exists a $k\in\mathbb{R}$ such that

$$f(t)-f(\gamma)\geqslant k(t-\gamma)\quad\text{for all } t\geqslant 0.$$

Putting $t=u(t)$ and multiplying the resulting inequality by $h(t)$ we obtain after integration over $[a,b]$ that

$$\begin{aligned}\int_a^b h(t)f\big(u(t)\big)\,\mathrm{d}t-f(\gamma)\int_a^b h(t)\,\mathrm{d}t\\ \geqslant k\left\{\int_a^b h(t)u(t)\,\mathrm{d}t-\gamma\int_a^b h(t)\,\mathrm{d}t\right\}.\end{aligned}\tag{1.2.22}$$

Inequality (1.2.21) now follows by putting

$$\gamma=\frac{\int_a^b h(t)u(t)\,\mathrm{d}t}{\int_a^b h(t)\,\mathrm{d}t}.$$

The proof is complete. □

In [411] Steffensen uses his inequality which is now known in the literature as Steffensen's inequality (see [211, p. 107]) to derive a generalization of Jensen's inequality for convex functions. A corresponding inequality for integrals is also given in [411], see [211, p. 109]. For another generalization of Jensen's inequality and its integral analogue we refer the interested readers to Ciesieski [63] where analogous results are given for functions of two variables. In [31] Boas has also obtained some interesting results concerning Jensen's inequality and its integral analogues.

Let $f : I \to \mathbb{R}$ be a continuous convex function, where I is the range of the continuous function $g : [a, b] \to \mathbb{R}$. The following results are valid.

Jensen inequality. The inequality

$$f\left(\frac{\int_a^b g(x)\,\mathrm{d}\lambda(x)}{\int_a^b \mathrm{d}\lambda(x)}\right) \leqslant \frac{\int_a^b f(g(x))\,\mathrm{d}\lambda(x)}{\int_a^b \mathrm{d}\lambda(x)} \tag{1.2.23}$$

holds if f is continuous, provided that λ is nondecreasing, bounded, and $\lambda(a) \neq \lambda(b)$.

Jensen–Steffensen inequality. Inequality (1.2.23) holds if f is continuous and monotonic (in either sense) provided that λ is either continuous or of bounded variation, and it satisfies

$$\lambda(a) \leqslant \lambda(x) \leqslant \lambda(b), \quad x \in [a, b]; \qquad \lambda(b) > \lambda(a).$$

Jensen–Boas inequality. Inequality (1.2.23) holds if λ is continuous or of bounded variation and satisfies

$$\lambda(a) \leqslant \lambda(x_1) \leqslant \lambda(y_1) \leqslant \lambda(x_2) \leqslant \cdots \leqslant \lambda(y_{n-1}) \leqslant \lambda(x_n) \leqslant \lambda(b)$$

for all x_k in (y_{k-1}, y_k), $y_0 = a$, $y_n = b$, and $\lambda(b) > \lambda(a)$, provided that f is continuous and monotonic (in either sense) in each of the $n-1$ intervals (y_{k-1}, y_k).

For $n = 1$, we obtain the Jensen–Steffensen inequality from the Jensen–Boas inequality, and in the limit as $n \to \infty$, λ would increase and f would be required to be continuous, thus Jensen's inequality is a limiting case of the Jensen–Boas inequality.

In 1982, Pečarić [367] (see also [369]) has given an interesting and short proof of the Jensen–Boas inequality. In his proof he used only Jensen's inequality for sums, that is,

$$f\left(\frac{1}{P_n}\sum_{i=1}^{n} p_i x_i\right) \leqslant \frac{1}{P_n}\sum_{i=1}^{n} p_i f(x_i), \tag{1.2.24}$$

where $p_i \geqslant 0$ with $P_n = \sum_{i=1}^{n} p_i > 0$, $x_i \in I$ for $i = 1, \ldots, n$, and the Jensen–Steffensen inequality. Inequality (1.2.24) can easily be obtained from (1.2.23).

If $\lambda(a) < \lambda(y_1) < \lambda(y_2) < \cdots < \lambda(y_{n-1}) < \lambda(b)$, then from the Jensen–Steffensen inequality we have the inequalities

$$f\left(\frac{\int_{y_{k-1}}^{y_k} g(x)\,\mathrm{d}\lambda(x)}{\int_{y_{k-1}}^{y_k} \mathrm{d}\lambda(x)}\right) \leqslant \frac{\int_{y_{k-1}}^{y_k} f(g(x))\,\mathrm{d}\lambda(x)}{\int_{y_{k-1}}^{y_k} \mathrm{d}\lambda(x)}, \qquad k = 1, \ldots, n,$$

that is,

$$f(t_k) \leqslant \frac{1}{p_k} \int_{y_{k-1}}^{y_k} f\big(g(x)\big)\,\mathrm{d}\lambda(x), \qquad k = 1, \ldots, n,$$

with the notation

$$p_k = \int_{y_{k-1}}^{y_k} \mathrm{d}\lambda(x), \qquad t_k = \frac{\int_{y_{k-1}}^{y_k} g(x)\,\mathrm{d}\lambda(x)}{\int_{y_{k-1}}^{y_k} \mathrm{d}\lambda(x)}, \qquad k = 1, \ldots, n.$$

Since $p_k > 0$ and $t_k \in I$, $k = 1, \ldots, n$, from Jensen's inequality (1.2.24) we have

$$\begin{aligned} f\left(\frac{\int_a^b g(x)\,\mathrm{d}\lambda(x)}{\int_a^b \mathrm{d}\lambda(x)}\right) &= f\left(\frac{\sum_{k=1}^n p_k t_k}{\sum_{k=1}^n p_k}\right) \leqslant \frac{\sum_{k=1}^n p_k f(t_k)}{\sum_{k=1}^n p_k} \\ &\leqslant \frac{\sum_{k=1}^n p_k (1/p_k) \int_{y_{k-1}}^{y_k} f(g(x))\,\mathrm{d}\lambda(x)}{\sum_{k=1}^n p_k} \\ &= \frac{\int_a^b f(g(x))\,\mathrm{d}\lambda(x)}{\int_a^b \mathrm{d}\lambda(x)}. \end{aligned}$$

If $\lambda(y_{j-1}) = \lambda(y_j)$ for some j, then $\mathrm{d}\lambda(x) = 0$ on $[y_{j-1}, y_j]$ and

$$\int_a^b g(x)\,\mathrm{d}\lambda(x) - \sum_{k=1, k\neq j}^{n} p_k t_k, \qquad \int_a^b \mathrm{d}\lambda(x) = \sum_{k=1, k\neq j}^{n} p_k,$$

so, using (1.2.24), we can also easily prove that the Jensen–Boas inequality is valid.

For some interesting variants and generalizations of Jensen's integral inequality, see [211] and the references cited therein.

In 1975, Mitrinović and Vasić [214] use the so-called "centroid method" to obtain two new inequalities which are complementary to (the discrete version of) Jensen's inequality for convex functions. In [19] Beesack presents a general version of such inequalities using the same geometric ideas used in [214] but not using the centroid method itself. The results given in [19] extend the domain of the inequalities (even in the discrete case) and clarify the value of the constant appearing in the inequality.

The main results in [19] are given in the following theorems.

THEOREM 1.2.6. *Let ν be a nonnegative measure on σ-algebra of subsets of a set D and let q, f be real ν-measurable functions on D such that $q(x) > 0$, $-\infty < x_1 \leqslant f(x) \leqslant x_2 < \infty$ for all $x \in D$ and $\int_D q\, \mathrm{d}\nu = 1$. Let ϕ be a convex function on $I = [x_1, x_2]$ such that $\phi''(x) \geqslant 0$ with equality for most isolated points of I (so ϕ is strictly convex on I). If either*

(i) *$\phi(x) > 0$ for all $x \in I$ or*

(i′) *$\phi(x) > 0$ for $x_1 < x < x_2$, with either $\phi(x_1) = 0$, $\phi'(x_1) \neq 0$ or $\phi(x_2) = 0$, $\phi'(x_2) \neq 0$, or*

(ii) *$\phi(x) < 0$ for all $x \in I$ or*

(ii′) *$\phi(x) < 0$ for $x_1 < x < x_2$, with precisely one of $\phi(x_1) = 0$, $\phi(x_2) = 0$, then*

$$\int_D q\phi(f)\, \mathrm{d}\nu \leqslant \lambda\phi\left(\int_D qf\, \mathrm{d}\nu\right) \tag{1.2.25}$$

holds for some $\lambda > 1$ in cases (i) *and* (i′) *or $\lambda \in (0, 1)$ in cases* (ii) *and* (ii′). *More precisely, a value of λ (depending on x_1, x_2, ϕ) for* (1.2.25) *may be determined as follows. Set $\mu = [\phi(x_2) - \phi(x_1)]/(x_2 - x_1)$. If $\mu = 0$, let $x = \bar{x}$ be the unique solution of the equation $\phi'(x) = 0$, $x_1 < \bar{x} < x_2$, then $\lambda = \phi(x_1)/\phi(\bar{x})$ suffices for* (1.2.25). *In case $\mu \neq 0$, let $x = \bar{x}$ be the unique solution in $[x_1, x_2]$ of the equation*

$$g(x) = \mu\phi(x) - \phi'(x)\big[\phi(x_1) + \mu(x - x_1)\big] = 0, \tag{1.2.26}$$

then $\lambda = \mu/\phi'(\bar{x})$ suffices for (1.2.25). *Moreover, we have $x_1 < \bar{x} < x_2$ in the cases* (i) *and* (ii). *Moreover, equality holds in* (1.2.25) *if and only if $f(x) = x_i$ for $x \in D_i$ where D_1, D_2 are ν-measurable subsets of D such that $D = D_1 \cup D_2$ and $\bar{x} = x_1 \int_{D_1} q\, \mathrm{d}\nu + x_2 \int_{D_2} q\, \mathrm{d}\nu$.*

PROOF. We note that both integrals in (1.2.25) exist since both f and $\phi(f)$ are bounded measurable functions. In all cases, $\phi'(x)$ is continuous and strictly increasing on I so that, by the mean value theorem applied to μ, we have

$$\phi'(x_1) < \mu < \phi'(x_2). \tag{1.2.27}$$

Consider the pairs $\mathrm{A}(x_1, \phi(x_1))$, $\mathrm{B}(x_2, \phi(x_2))$ on the convex curve $y = \phi(x)$. The equation of the chord AB is

$$y = \phi(x_1) + \mu(x - x_1) \equiv m(x).$$

We also consider the family of convex curves with equations $y = \lambda\phi(x)$, $\lambda > 0$, and we show there is a unique $\lambda > 0$ such that the curve will be tangent to the line AB at a point $\overline{\mathrm{P}}(\bar{x}, \lambda\phi(\bar{x}))$ with $\bar{x} \in I$. (In fact, $x_1 < \bar{x} < x_2$ in cases (i), (ii).) This case holds if and only if the pair of equations,

$$\lambda\phi'(x) = \mu, \tag{1.2.28}$$

$$\lambda\phi(x) = m(x), \tag{1.2.29}$$

have a unique solution $(\bar{x}, \lambda)$ with $\bar{x} \in I$, $\lambda > 0$. In case $\mu = 0$, equations (1.2.28) and (1.2.29) reduce to $\lambda\phi'(x) = 0$, $\lambda\phi(x) = \phi(x_1)$. If $\phi(x_1) \neq 0$, these equations have the unique solution $(\bar{x}, \lambda)$ determined by $\phi'(x) = 0$, $\lambda = \phi(x_1)/\phi(\bar{x})$ where we observe that $x_1 < \bar{x} < x_2$, by the mean value theorem applied to μ. The case $\phi(x_1) = 0$ is impossible when $\mu \neq 0$ since then $\phi(x_2) = 0$ also, which is not the case. Note that $\lambda > 0$, when $\phi(x_1) \neq 0$.

When $\mu \neq 0$ we shall first consider only the cases (i) and (ii). By (1.2.28), $\lambda \neq 0$ and eliminating λ from the pair of equations (1.2.28), (1.2.29), we see that $x = \bar{x}$ must be a solution of equation (1.2.26). We now show that this equation has a unique solution on (x_1, x_2). First note that

$$g(x_1) = \phi(x_1)\big(\mu - \phi'(x_1)\big), \qquad g(x_2) = \phi(x_2)\big(\mu - \phi'(x_2)\big).$$

Since $\phi(x_1)$, $\phi(x_2)$ have the same sign, it follows from (1.2.27) that $g(x_1)$, $g(x_2)$ have opposite sign. Thus g has at least one zero on (x_1, x_2). Moreover,

$$g'(x) = -m(x)\phi''(x)$$

does not change sign on $[x_1, x_2]$. For, the linear function $m(x)$ has $m(x_i) = \phi(x_i)$ for $i = 1, 2$ and hence is either always positive in case (i) or always negative in case (ii), on $[x_1, x_2]$. It follows that g is strictly a monotonic function on $[x_1, x_2]$, and thus equation (1.2.26) has a unique solution $x = \bar{x} \in (x_1, x_2)$. Moreover, $\phi'(\bar{x}) \neq 0$ since if it were then setting $x = \bar{x}$ in (1.2.26) it would imply $0 = \mu\phi(\bar{x})$, which is impossible when $\mu \neq 0$ because $\phi(x) \neq 0$ on (x_1, x_2). If we now take $\lambda = \mu/\phi'(\bar{x})$ then it is easy to see that the pair $(\bar{x}, \lambda)$ satisfies equations (1.2.28), (1.2.29) and is the only such pair, with $x_1 < \bar{x} < x_2$,

$$\lambda\phi(\bar{x}) = \phi(x_1) + \mu(\bar{x} - x_1) = \left(1 - \frac{\bar{x} - x_1}{x_2 - x_1}\right)\phi(x_1) + \frac{\bar{x} - x_1}{x_2 - x_1}\phi(x_2)$$

or

$$\lambda\phi(\bar{x}) > \phi(\bar{x}). \tag{1.2.30}$$

From this idea, it follows that $\lambda > 1$ in case (i) and $\lambda < 1$ in case (ii). It remains to show that $\lambda > 0$ in case (ii) when $\mu \neq 0$. This result follows from (1.2.29) since $\lambda\phi(\bar{x}) = m(\bar{x})$ and, as noted above, ϕ and m have the same sign on I.

As for the cases (i′) and (ii′), which are relaxed versions of (i) and (ii), respectively, we omit details but note that in case (i′), if $\phi(x_1) = 0$, $\phi'(x_1) \neq 0$ then we necessarily have $\phi'(x_1) > 0$ and $\mu > 0$, while if $\phi(x_2) = 0$, $\phi'(x_2) \neq 0$ we must have $\mu < \phi'(x_2) < 0$. For the first of these conditions, we have $g(x) = \mu\phi(x) - \mu(x - x_1)\phi'(x) = \mu(x - x_1)[\phi'(X) - \phi'(x)]$ for $x_1 < X < x \leqslant x_2$, where $g(x) < 0$ for $x_1 < x \leqslant x_2$, so $\bar{x} = x_1$ is the unique solution of (1.2.26) on $[x_1, x_2]$, and equations (1.2.28), (1.2.29) clearly have the unique solution $x = x_1$ on $[x_1, x_2]$ with $\lambda = \mu/\phi'(x_1) > 1$. A similar analysis applies to the second case of (i′), where we now find $\bar{x} = x_2$ and $\lambda = \mu/\phi'(x_2) > 1$. For the two cases of (ii′), we observe in the first that $\phi'(x_1) < 0$ must hold, that $\bar{x} = x_1$, $0 < \lambda = \mu/\phi'(x_1) < 1$, and in the second that $\phi'(x_2) > 0$ must hold and $\bar{x} = x_2$, $0 < \lambda = \mu/\phi'(x_2) < 1$.

It only remains to prove inequality (1.2.25) with the value of λ determined above. To prove this point we note that since the line AB is tangent to the graph of the strictly convex (since $\lambda > 0$) function $\lambda\phi(x)$ at the point $\overline{\mathrm{P}}$, we have for all $x \in I$,

$$\lambda\phi(x) \geqslant m(x) = \phi(x_1) + \frac{\phi(x_2) - \phi(x_1)}{x_2 - x_1}(x - x_1),$$

with equality only for $x = \bar{x}$. We may take $x = \int_D qf\,\mathrm{d}\nu$ since this $x \in I$. This gives

$$\begin{aligned}\lambda\phi\left(\int_D qf\,\mathrm{d}\nu\right) &\geqslant \phi(x_1) + \frac{\phi(x_2) - \phi(x_1)}{x_2 - x_1}\left(\int_D qf\,\mathrm{d}\nu - x_1\right) \\ &= \int_D \left\{\phi(x_1) + \frac{\phi(x_2) - \phi(x_1)}{x_2 - x_1}(f - x_1)\right\}\mathrm{d}\nu \\ &\geqslant \int_D q\phi(f)\,\mathrm{d}\nu,\end{aligned}$$

precisely as at (1.2.30). Equality holds for the last step if and only if $f(x) = x_1$ or x_2 on ν-measurable subsets D_1 or D_2 of D. Hence equality holds in (1.2.25) precisely for such f where, in addition,

$$\bar{x} = \int_D qf\,\mathrm{d}\nu = x_1\int_{D_1} q\,\mathrm{d}\nu + x_2\int_{D_2} q\,\mathrm{d}\nu.$$

In case the measure $Q(\mathrm{A}) \equiv \int_{\mathrm{A}} q\,\mathrm{d}\nu$ is atomless, we observe that given any $\bar{x} \in [x_1, x_2]$ such sets D_1, D_2 exist but are not in general unique. □

COROLLARY 1.2.3. *Let all the hypotheses of Theorem* 1.2.6 *hold except that now* ϕ *is concave on* I *with* $\phi''(x) \leqslant 0$ *with equality for at most isolated points of* I. *Then*

$$\int_D q\phi(f)\,\mathrm{d}\nu \geqslant \lambda\phi\left(\int_D qf\,\mathrm{d}\nu\right), \tag{1.2.31}$$

where λ *is determined precisely as before. Now,* $\lambda > 1$ *holds if* $\phi(x) < 0$ *on* (x_1, x_2) *and* $0 < \lambda < 1$ *if* $\phi(x) > 0$ *on* (x_1, x_2). *Equality holds in* (i$'$) *for precisely the same* f *(if any) as in Theorem* 1.2.6.

This point follows from the Theorem 1.2.6 applied to the convex function $\phi_1 = -\phi$.

THEOREM 1.2.7. *Let* ν, D, q, f, x_1, x_2 *be as in Theorem* 1.2.6 *and let* $\phi(x)$ *be any differentiable function on* $I = [x_1, x_2]$ *such that* $\phi'(x)$ *exists and is strictly increasing on* I. *Then we have*

$$\int_D q\phi(f)\,\mathrm{d}\nu \leqslant \lambda + \phi\left(\int_D qf\,\mathrm{d}\nu\right) \tag{1.2.32}$$

for some λ, *satisfying* $0 < \lambda < (x_2 - x_1)[\mu - \phi'(x_1)]$, *where*

$$\mu = \frac{[\phi(x_2) - \phi(x_1)]}{x_2 - x_1}.$$

More precisely, λ *may be determined for* (1.2.32) *as follows. Let* $x = \bar{x}$ *be the unique solution of the equation* $\phi'(x) = \mu$, $x_1 < \bar{x} < x_2$, *then*

$$\lambda = \phi(x_1) - \phi(\bar{x}) + \mu(\bar{x} - x_1)$$

suffices in (1.2.32). *Equality holds in* (1.2.32) *only for* $f(x) = x_i$, $x \in D_i$, *where* D_1, D_2 *are* ν*-measurable subsets of* D *such that* $D = D_1 \cup D_2$ *and*

$$\bar{x} = x_1 \int_{D_1} q\,\mathrm{d}\nu + x_2 \int_{D_2} q\,\mathrm{d}\nu,$$

when such sets exist.

PROOF. The proof is similar to that given for Theorem 1.2.6 but is much easier. Using the same notation as before, we again have (1.2.27) and now look for the convex curve with the equation $y = \lambda + \phi(x)$ which is tangent to the chord AB at a point $(\bar{x}, \bar{y})$ with $x_1 < \bar{x} < x_2$. This result will occur if and only if $\bar{x}$, λ now satisfy the pair of equations,

$$\phi'(x) = \mu, \tag{1.2.33}$$

$$\lambda + \phi(x) = m(x). \tag{1.2.34}$$

Since ϕ' is strictly increasing on I it follows from the mean value theorem that equation (1.2.33) has a unique solution $\bar{x} \in (x_1, x_2)$, and then λ is uniquely determined from (1.2.34) as

$$\begin{aligned}\lambda &= m(\bar{x}) - \phi(\bar{x}) \\ &= \phi(x_1) - \phi(\bar{x}) + \mu(\bar{x} - x_1) \\ &= (\bar{x} - x_1)\big[\mu - \phi'(X)\big], \quad \text{where } x_1 < X < \bar{x}.\end{aligned}$$

From this we also obtain

$$0 < \lambda < (x_2 - x_1)\big[\mu - \phi'(x_1)\big].$$

The proof of inequality (1.2.32) is just as before since we now have, with this value of λ,

$$\lambda + \phi(x) \geqslant m(x) = \phi(x_1) + \frac{\phi(x_2) - \phi(x_1)}{x_2 - x_1}(x - x_1)$$

for all $x \in I$. Again, we set $x = \int_D q \,\mathrm{d}\nu$ and use the strict convexity of ϕ on I to obtain (1.2.32). The equality conditions follow precisely as in Theorem 1.2.6. □

COROLLARY 1.2.4. *Let all the hypotheses of Theorem* 1.2.7 *be satisfied except that* $\phi'(x)$ *is strictly decreasing on* I. *Then*

$$\phi\left(\int_D qf \,\mathrm{d}\nu\right) \leqslant \lambda + \int_D q\phi(f)\,\mathrm{d}\nu, \tag{1.2.35}$$

where

$$0 < \lambda < (x_2 - x_1)\big[\phi'(x_1) - \mu\big]$$

with

$$\mu = \frac{[\phi(x_2) - \phi(x_1)]}{x_2 - x_1}.$$

In fact, we may take $\lambda = \phi(\bar{x}) - \phi(x_1) - \mu(\bar{x} - x_1)$, *where* $x = \bar{x}$ *is the unique solution of the equation* $\phi'(x) = \mu$, $x_1 < \bar{x} < x_2$. *Equality holds in* (1.2.35) *under precisely the same condition as in* (1.2.32).

To prove this result we need to apply Theorem 1.2.7 to the function $\phi_1 = -\phi$ for which $\mu_1 = -\mu$, etc.

For interesting remarks and applications of Theorems 1.2.6 and 1.2.7 and their comparisons with other inequalities, we refer the reader to Beesack [19] and some of the references cited therein.

1.3 Jessen's and Related Inequalities

This section deals with a generalization of Jensen's inequality for convex functions due to Jessen [166] involving isotonic linear functionals. Some general inequalities complementary to Jessen's inequality given by Beesack and Pečarić [20] and Pečarić and Beesack [376] for convex functions involving isotonic linear functionals are also given.

Let E be a nonempty set and L be a linear class of real-valued functions $g: E \to \mathbb{R}$ having the following properties:

(L_1) $f, g \in L \to (af + bg) \in L$ for all $a, b \in \mathbb{R}$ and
(L_2) $1 \in L$, that is, if $f(t) = 1$, $t \in E$, then $f \in L$.

We also consider isotonic linear functionals $A: L \to \mathbb{R}$. That is, we suppose

(A_1) $A(af + bg) = aA(f) + bA(g)$ for $f, g \in L$, $a, b \in \mathbb{R}$ and
(A_2) $f \in L$, $f(t) \geqslant 0$ on $E \to A(f) \geqslant 0$ (A is isotonic).

We note that common examples of such isotonic linear functionals A are given by:

$$A(g) = \int_E g\,\mathrm{d}\mu \quad \text{or} \quad A(g) = \sum_{k \in E} p_k g_k,$$

where μ is a positive measure on E in the first case and E is a subset of natural numbers with $p_k > 0$ in the second case.

In 1931, Jessen [166] gives the following generalization of the Jensen inequality for convex functions (see [20, p. 537], also [390, p. 33]).

THEOREM 1.3.1. *Let L satisfy properties* (L_1), (L_2) *on a nonempty set E, and suppose ϕ is a convex function on an interval $I \subset \mathbb{R}$. If A is any isotonic linear functional with $A(1) = 1$ then, for all $g \in L$ such that $\phi(g) \in L$, we have $A(g) \in I$ and*

$$\phi\big(A(g)\big) \leqslant A\big(\phi(g)\big). \tag{1.3.1}$$

PROOF. First observe that if $I = [\alpha, \beta]$ and $g \in L$ with $\phi(g) \in L$, then we must have $\alpha \leqslant g(t) \leqslant \beta$, $t \in E$, where $\alpha = A(\alpha - 1) \leqslant A(g) \leqslant \beta$, so $A(g) \in I$. Since

ϕ is convex on I, for any $x_0 \in I$ there is a constant $m = m(x_0)$ such that

$$\phi(x) \geqslant \phi(x_0) + m(x - x_0), \quad \alpha \in I.$$

In this inequality, set $x = g(t)$, $x_0 = A(g)$, and apply the functional A to obtain

$$A(\phi(g)) \geqslant \phi(A(g)) + m(A(g) - A(g)),$$

proving (1.3.1). □

We now give three basic lemmas established in [20] which are complements of Jessen's inequality (1.3.1), in that they are inequalities of the form $A(\phi(g)) \leqslant \chi(A(g))$ for appropriate χ.

LEMMA 1.3.1. *Let ϕ be convex on $I = [m, M]$, $-\infty < m < M < \infty$, let L satisfy conditions* (L_1), (L_2) *and let A be any isotonic linear functional on L with $A(1) = 1$. Then, for all $g \in L$ such that $\phi(g) \in L$ (so $m \leqslant g(t) \leqslant M$ for all $t \in E$), we have*

$$A(\phi(g)) \leqslant \frac{(M - A(g))\phi(m) + (A(g) - m)\phi(M)}{M - m}. \tag{1.3.2}$$

PROOF. From the definition of convex function,

$$\phi(v) \leqslant \frac{w - v}{w - u}\phi(u) + \frac{v - u}{w - u}\phi(w), \quad u \leqslant v \leqslant w, u < w.$$

Now, let $u = m$, $v = g(t)$, $w = M$ to obtain

$$\phi(g(t)) \leqslant \frac{M - g(t)}{M - m}\phi(m) + \frac{g(t) - m}{M - m}\phi(M), \quad t \in E.$$

Since A satisfies (A_1), (A_2) and $A(k) = k$ holds for all constants k, (1.3.2) follows. □

LEMMA 1.3.2. (a) *Let L satisfy conditions* (L_1), (L_2) *and A satisfy conditions* (A_1), (A_2), *and $A(t) = 1$. Suppose ϕ is convex on $[m, M]$, $-\infty < m < M < \infty$, such that $\phi''(x) \geqslant 0$ with equality for at most isolated points of I (so that ϕ is strictly convex on I). Suppose that either* (i) *$\phi(x) > 0$ for all $x \in I$ or* (i′) *$\phi(x) > 0$ for $m < x < M$ with either $\phi(m) = 0$, $\phi'(m) \neq 0$ or $\phi(M) = 0$, $\phi'(M) \neq 0$; or* (ii) *$\phi(x) < 0$ for all $x \in I$ or* (ii′) *$\phi(x) < 0$ for $m < x < M$ with precisely one of $\phi(m) = 0$, $\phi(M) = 0$. Then, for all $g \in L$ such that $\phi(g) \in L$ (so $m \leqslant g(t) \leqslant M$ for all $t \in E$),*

$$A(\phi(g)) \leqslant \lambda\phi(A(g)) \tag{1.3.3}$$

holds for some $\lambda > 1$ *in cases* (i) *and* (i′) *or* $\lambda \in (0, 1)$ *in cases* (ii) *and* (ii′). *More precisely, a value of* λ *(depending only on* m, M, ϕ*) for* (1.3.3) *may be determined as follows. Set* $\mu = (\phi(M) - \phi(m))/(M - m)$. *If* $\mu = 0$, *let* $x = \bar{x}$ *be the unique solution of the equation* $\phi'(x) = 0$, $m < \bar{x} < M$, *then* $\lambda = \phi(m)/\phi(\bar{x})$ *suffices for* (1.3.3). *If* $\mu \neq 0$, *let* $x = \bar{x}$ *be the unique solution in* $[m, M]$ *of the equation*

$$\mu\phi(x) - \phi'(x)\{\phi(m) + \mu(x - m)\} = 0,$$

then $\lambda = \mu/\phi'(\bar{x})$ *suffices for* (1.3.3). *Moreover, we have* $m < \bar{x} < M$ *in the cases* (i) *and* (ii).

(b) *Let all the hypotheses of* (a) *hold except that now* ϕ *is concave on* I *with* $\phi''(x) \leqslant 0$, *with equality for at most isolated points of* I. *Then the reverse inequality to* (1.3.3) *holds, where* λ *is determined precisely as before. Now* $\lambda > 1$ *holds if* $\phi(x) < 0$ *on* (m, M) *and* $0 < \lambda < 1$ *if* $\phi(x) > 0$ *on* (m, M).

PROOF. (a) As in [212] (see also [19]) we consider the points $B(m, \phi(m))$ and $C(M, \phi(M))$ on the convex curve $y = \phi(x)$. The equation of the chord BC is

$$y = \phi(m) + \mu(x - m) \equiv h(x).$$

By Lemma 1.3.1, we obtain $A(\phi(g)) \leqslant h(A(g))$. If we consider the family of convex curves with equations $y = \lambda\phi(x)$, $\lambda > 0$, we can show as in [214] (or [19]) that there is a unique $\lambda > 0$ which satisfies the conditions stated in the lemma such that the curve will be tangent to the chord BC. Hence $h(y) \leqslant \lambda\phi(y)$ for all $y \in I$. Taking $y = A(g)$ it gives

$$A(\phi(g)) \leqslant h(A(g)) \leqslant \lambda\phi(A(g)),$$

proving (1.3.3).

(b) follows when (a) is applied to the convex function $\phi_1 = -\phi$. □

REMARK 1.3.1. It is clear that the last inequality in the above proof constitutes a refinement of (1.3.3).

LEMMA 1.3.3. (a) *Let* L, A *and* g *be as in Lemma* 1.3.2, *and let* $\phi(x)$ *be any differentiable function on* $I = [m, M]$ *such that* $\phi'(x)$ *exists and is strictly increasing on* I. *Then we have*

$$A(\phi(g)) \leqslant \lambda + \phi(A(g)) \tag{1.3.4}$$

for some λ *satisfying* $0 < \lambda < (M - m)\{\mu - \phi'(m)\}$, *where* $\mu = (\phi(M) - \phi(m))/(M - m)$. *More precisely,* λ *may be determined for* (1.3.4) *as follows. Let* $x = \bar{x}$

be the unique solution of the equation $\phi'(x) = \mu$, $m < \bar{x} < M$. *Then*

$$\lambda = \phi(m) - \phi(\bar{x}) + \mu(\bar{x} - m)$$

suffices in (1.3.4).

(b) *Let all the hypotheses of* (a) *be satisfied except that* $\phi'(x)$ *is strictly decreasing on* I. *Then*

$$\phi\big(A(g)\big) \leqslant \lambda + A\big(\phi(g)\big),$$

where $0 < \lambda < (M - m)\{\phi'(m) - \mu\}$ *with* μ *as in* (a). *In fact we may take* $\lambda = \phi(\bar{x}) - \phi(m) - \mu(\bar{x} - m)$ *with* $\bar{x}$ *as in* (a).

PROOF. (a) The proof is similar to that of Lemma 1.3.2. Using Lemma 1.3.1 we also have $A(\phi(g)) \leqslant h(A(g))$, where $y = h(x)$ is the equation of the chord joining $\mathrm{B}(m, \phi(m))$ to $\mathrm{C}(M, \phi(M))$. Now we consider the family of curves with equations $y = \lambda + \phi(x)$. We can show precisely as in [214] or [19], that there is a unique $\lambda > 0$ satisfying the stated conditions such that the curve will be tangent to the line BC. Therefore $h(A(g)) \leqslant \lambda + \phi(A(g))$, so

$$A\big(\phi(g)\big) \leqslant h\big(A(g)\big) \leqslant \lambda + \phi\big(A(g)\big),$$

proving (1.3.4).

(b) follows when (a) is applied to convex function $\phi_1 = -\phi$. □

REMARK 1.3.2. The last inequalities of the proof again constitute a refinement of (1.3.4).

REMARK 1.3.3. Lemmas 1.3.2 and 1.3.3 are generalizations of results from [214] and [19]. In [214] the special case,

$$A(g) = \frac{\sum_{i=1}^{n} p_i g_i}{\sum_{i=1}^{n} p_i}, \qquad p_i > 0,$$

was considered and conditions for equality to hold in (1.3.3) were given. The paper [19] deals with special case

$$A(g) = \int_D pg \,\mathrm{d}\nu, \quad \text{with} \int_D p \,\mathrm{d}\nu = 1,$$

and equality conditions were given for both (1.3.3) and (1.3.4) in this case.

The following results established by Beesack and Pečarić [20] deal with some theorems as applications or examples to the basic inequalities given in Theorem 1.3.1 and Lemmas 1.3.1–1.3.3, and include applications to generalized mean values with respect to the functional A.

THEOREM 1.3.2. *Let L satisfy conditions* (L_1), (L_2) *on E, and let A satisfy conditions* (A_1), (A_2) *with $A(1) = 1$. Suppose ϕ is convex on $I = [0, \infty)$, and $f : I \to \mathbb{R}$ satisfies the condition*

$$\phi(x) \leqslant f(x) \leqslant C\phi(Bx), \quad x \in I, \tag{1.3.5}$$

where $B, C > 0$ are constants. Then, for all $g \in L$ such that $g \geqslant 0$ on E and $f(Bg), \phi(Bg) \in L$, we have

$$f\big(A(g)\big) \leqslant CA\big(f(Bg)\big). \tag{1.3.6}$$

PROOF. Using both parts (1.3.5) and (1.3.1) we obtain

$$f\big(A(g)\big) \leqslant C\phi\big(BA(g)\big) = C\phi\big(A(Bg)\big) \leqslant CA\big(\phi(Bg)\big) \leqslant CA\big(f(Bg)\big).$$

The proof is complete. □

REMARK 1.3.4. Inequality (1.3.6) is a generalization of the sufficiency of a half of Theorem 1 of Mulholland [223]. When f satisfies (1.3.5) for some convex ϕ, Mulholland calls f quasiconvex on I.

THEOREM 1.3.3. *Let L, A be as in Theorem 1.3.1. Suppose ϕ is concave on an interval $I \subset \mathbb{R}$ and that $\psi(x) \equiv x\phi(x)$ is convex on I. Then, for all $g \in L$ such that $g^2, \phi(g), \psi(g) \in L$ and $A(g) > 0$, we have*

$$A\big(\phi(g)\big) \leqslant \phi\big(A(g)\big) \leqslant \frac{A(g\phi(g))}{A(g)} \leqslant \phi\left(\frac{A(g^2)}{A(g)}\right). \tag{1.3.7}$$

PROOF. The first and second inequalities of (1.3.7) are consequences of (1.3.1) applied to the convex functions $-\phi$ and $x\phi$. Since the operator $A_1(f) = A(gf)/A(g)$ is a linear, isotonic functional with $A_1(1) = 1$ and the last inequality of (1.3.7) also follows from (1.3.1). □

Let $I = (a, b)$, $-\infty \leqslant a < b \leqslant \infty$, and let $\psi, \chi : I \to \mathbb{R}$ be continuous and strictly monotonic. Suppose L and A satisfy the conditions (L_1), (L_2) and (A_1),

(A_2) with $A(1) = 1$ on a base set E, and that $\psi(g), \chi(g) \in L$ for some $g \in L$. We define the generalized mean with respect, to the operators A and ψ by

$$M_\psi(g; A) = \psi^{-1}\{A(\psi(g))\}, \quad g \in L. \tag{1.3.8}$$

Observe that if $\alpha \leqslant \psi(g(x)) \leqslant \beta$ for $x \in E$, then by the isotonic character of A, we have $\alpha \leqslant A(\psi(g)) \leqslant \beta$ so that M_ψ is well defined by (1.3.8). We note also that the above assumptions imply that $g(x) \in I$ for $x \in E$. In the following theorems given by Beesack and Pečarić [20] we assume that $g \in L$ satisfies the above conditions so that the theorems hold for such g.

THEOREM 1.3.4. *Under the above hypotheses we have*

$$M_\psi(g; A) \leqslant M_\chi(g; A), \tag{1.3.9}$$

provided either χ is increasing and $\phi = \chi \circ \psi^{-1}$ is convex, or χ is decreasing and ϕ is convex.

PROOF. For $g \in L$, we have both $\psi(g) \in L$ and $\chi(g) \in L$ by assumption. Hence $\phi(\psi(g)) = \chi(g) \in L$ for $g \in L$, so if ϕ is convex it follows from Jessen's inequality (1.3.1) that

$$\phi(A(\psi(g))) \leqslant A(\chi(g)).$$

Hence, if χ is increasing, so χ^{-1} is also increasing, we obtain

$$\chi^{-1}[\phi(A(\psi(g)))] \leqslant \chi^{-1}(A(\chi(g)))$$

which is (1.3.9). In case ϕ is concave, so $-\phi$ is convex, we obtain the first inequality above with the direction reversed. Since now χ^{-1} is decreasing with χ, we again obtain (1.3.9). □

REMARK 1.3.5. Theorem 1.3.4 is a generalization to functionals of the general mean value inequality given in [141, Theorem 92, p. 75].

THEOREM 1.3.5. *Let L, A, ψ and χ be as in Theorem* 1.3.4, *but with $I = [m, M]$ and $-\infty < m < M < \infty$. Then, for all $g \in L$ such that $m \leqslant g(t) \leqslant M$ for $t \in E$, we have*

$$\begin{aligned}&(\psi(M) - \psi(m))A(\chi(g)) - (\chi(M) - \chi(m))A(\psi(g)) \\ &\quad \leqslant \psi(M)\chi(m) - \chi(M)\psi(m),\end{aligned} \tag{1.3.10}$$

provided $\phi = \chi \circ \psi^{-1}$ is convex. The opposite inequality to (1.3.10) *holds when ϕ is concave.*

PROOF. In case ψ is increasing on I we have $m_1 = \psi(m) \leqslant \psi(g(t)) \leqslant \psi(M) = M_1$ for all $t \in E$. So, by Lemma 1.3.1 with m, M replaced by m_1, M_1, we have

$$A\big(\phi\big(\psi(g)\big)\big) \leqslant \big\{\psi(M) - \big(A\big(\psi(g)\big)\big)\chi(m) + \big(A\big(\psi(g)\big) - \psi(m)\big)\chi(M)\big\} \\ \times \big[\psi(M) - \psi(m)\big]^{-1}$$

which reduces to (1.3.10). If ψ is decreasing on I, we have $M_1 \leqslant \psi(g(t)) \leqslant m_1$ for $t \in E$ and, with an obvious modification of proof, the result follows as before. □

The following lemma given in [20] includes the corresponding versions of Jessen's inequality (1.3.1) and Lemmas 1.3.1–1.3.3.

LEMMA 1.3.4. *Let L satisfy conditions* (L_1), (L_2) *and A satisfy conditions* (A_1), (A_2) *on a base set E. Suppose $k \in L$ with $k \geqslant 0$ on E and $A(k) > 0$, and that ϕ is a convex function on an interval $I \subset \mathbb{R}$. For any function $g_1 : E \to \mathbb{R}$ such that $kg_1 \in L$ and $k\phi(g_1) \in L$, we have*

$$\phi\left(\frac{A(kg_1)}{A(k)}\right) \leqslant \frac{A(k\phi(g_1))}{A(k)}. \tag{1.3.11}$$

If in addition, $I = [m, M]$ where $-\infty < m < M < \infty$, then

$$A\big(k\phi(g_1)\big) \leqslant \frac{[MA(k) - A(kg_1)]\phi(m) + [A(kg_1) - mA(k)]\phi(M)}{M - m}. \tag{1.3.12}$$

Moreover, when ϕ satisfies the strict convexity conditions of Lemmas 1.3.2 *or* 1.3.3, *then*

$$A\big(k\phi(g_1)\big) \leqslant \lambda A(k)\psi\left(\frac{A(kg_1)}{A(k)}\right) \tag{1.3.13}$$

or

$$A\big(k\phi(g_1)\big) \leqslant A(k)\left\{\lambda + \phi\left(\frac{A(kg_1)}{A(k)}\right)\right\}, \tag{1.3.14}$$

where λ is determined as in Lemmas 1.3.2 *or* 1.3.3, *respectively.*

PROOF. In case $g_1 \in L$ and $\phi(g_1) \in L$, and k is such that $kh \in L$ for all $h \in L$, the functional $F : L \to \mathbb{R}$ defined by

$$F(h) = \frac{A(kh)}{A(k)}, \quad h \in L,$$

is an isotonic linear functional satisfying $F(1) = 1$. In this case, (1.3.11)–(1.3.14) follow from (1.3.1)–(1.3.4).

Under the weaker hypotheses stated above on k, g_1, we must proceed somewhat differently, essentially by giving a direct proof of (1.3.11)–(1.3.14) along the same lines used proving (1.3.1)–(1.3.4). It suffices to deal with (1.3.11) since similar modifications handle the other proofs. As before, if $I = [\alpha, \beta]$ then $k\phi(g_1) \in L$ implies $\alpha \leqslant g_1(t) \leqslant \beta$ for $t \in E$, whence $\alpha k(t) \leqslant k(t)g_1(t) \leqslant \beta k(t)$, so it follows that $x_0 = A(kg_1)/A(k) \in I$. The convexity of ϕ on I again yields

$$\phi\big(g_1(t)\big) \geqslant \phi(x_0) + m\big[g_1(t) - x_0\big], \quad t \in E,$$

so

$$k(t)\phi\big(g_1(t)\big) \geqslant \phi(x_0)k(t) + m\big[k(t)g_1(t) - x_0 k(t)\big], \quad t \in E,$$

for an appropriate constant m. Application of the linear isotonic functional A now gives (1.3.11). □

The following theorems given in [20] deal with Hölder's and Minkowski's inequalities respectively for isotonic functionals.

THEOREM 1.3.6. *Let L satisfy conditions* (L_1), (L_2) *and A satisfy conditions* (A_1), (A_2) *on a base set E. If $p > 1$ and $q = p/(p-1)$ so that $p^{-1} + q^{-1} = 1$, then if w, f, $g \geqslant 0$ on E and wf^p, wg^q, $wfg \in L$, we have*

$$A(wfg) \leqslant A^{1/p}\big(wf^p\big)A^{1/q}\big(wg^q\big). \tag{1.3.15}$$

In case $0 < p < 1$ (or $p < 0$) and $A(wg^q) > 0$ (or $A(wf^p) > 0$), the opposite inequality to (1.3.15) *holds.*

PROOF. Suppose first that $A(wg^q) > 0$ and $p > 1$. Then (1.3.15) follows from (1.3.11) by the substitutions

$$\phi(x) = x^p, \qquad g_1 = fg^{-q/p}, \qquad k = wg^q, \tag{1.3.16}$$

since then $k \in L$, $kg_1 = wfg \in L$ and $k\phi(g_1) = wf^p \in L$. Thus (1.3.15) holds in this case. In case $A(wf^p) > 0$, we may apply (1.3.15) with p, q, f, g replaced by q, p, g, f to obtain (1.3.15) again. Finally, suppose both $A(wg^q) = 0$ and $A(wf^p) = 0$. Since

$$0 \leqslant wfg \leqslant \frac{1}{p}wf^p + \frac{1}{q}wg^q \quad \text{on } E,$$

it follows that $A(wfg) = 0$ also, so again (1.3.15) holds. This proof completes the case $p > 1$.

For the case $0 < p < 1$, we have $P = 1/p > 1$ and so may apply (1.3.15) with p, q, f, g replaced by P, $Q = (1-p)^{-1}$, $f_1 = (fg)^p$, $g_1 = g^{-p}$ for which $wf_1^P = wfg$, $wg_1^Q = wg^q$ and $wf_1g_1 = wf^p$ which all belong to L. We obtain

$$A(wf^p) \leqslant A^p(wfg)A^{1-p}(wg^q),$$

which reduces to the opposite of (1.3.15) provided $A(wg^q) > 0$. Finally, if $p < 0$ then $0 < q < 1$, and we may apply the case just considered with p, q, f, g replaced by q, p, g, f, provided $A(wf^p) > 0$. □

THEOREM 1.3.7. *Let L and A be as in Theorem* 1.3.6. *If $p > 1$ and if w, f, $g \geqslant 0$ on E with $wf^p, wg^p, w(f+g)^p \in L$, then*

$$A^{1/p}\big(w(f+g)^p\big) \leqslant A^{1/p}\big(wf^p\big) + A^{1/p}\big(wg^p\big). \tag{1.3.17}$$

The opposite inequality to (1.3.17) *holds if $0 < p < 1$, and also if $p < 0$ provided $A(wf^p) > 0$, $A(wg^p) > 0$ in this case.*

PROOF. As in the proof of the ordinary Minkowski inequality, we write

$$w(f+g)^p = wf(f+g)^{p-1} + wg(f+g)^{p-1}.$$

Applying A to this, (1.3.15) then yields, in case $p > 1$,

$$A\big(w(f+g)^p\big) \leqslant \big\{A^{1/p}\big(wf^p\big) + A^{1/p}\big(wg^p\big)\big\}A^{1/q}\big(w(f+g)^p\big),$$

where $q = p/(p-1)$. Hence (1.3.17) follows if $A(w(f+g)^p) > 0$. However, if $A(w(f+g)^p) = 0$, then since $0 \leqslant wf^p$, $wg^p \leqslant w(f+g)^p$, we see that $A(wf^p) = A(wg^p) = 0$, and (1.3.17) still holds.

If $0 < p < 1$, the opposite of (1.3.15) yields the opposite of the last displayed inequality provided $A(w(f+g)^p) > 0$, and hence also the opposite of (1.3.17) if $A(w(f+g)^p) > 0$. As above, if $A(w(f+g)^p) = 0$ then $A(wf^p) = A(wg^p) = 0$, so the opposite of (1.3.17) still holds. Finally, if $p < 0$, we again obtain the opposite of the last displayed inequality provided both $A(wf^p) > 0$ and $A(wg^p) > 0$. If $A(w(f+g)^p) > 0$, the opposite of (1.3.17) follows. If $A(w(f+g)^p) = 0$, then the opposite of (1.3.17) clearly holds since then $A^{1/p}(w(f+g)^p) = +\infty$. □

We also observe that in the case $p < 0$, we have $0 \leqslant w(f+g)^p \leqslant wf^p, wg^p$, so that $A(w(f+g)^p) = 0$ if either $A(wf^p) = 0$ or $A(wg^p) = 0$. Thus, the opposite inequality to (1.3.17) holds ($\infty \geqslant \infty$) even in this degenerate case.

The following lemma given in [376] shows that the requirement ϕ be convex in (1.3.1) can be relaxed.

LEMMA 1.3.5. *Let L satisfy properties* (L_1), (L_2) *on a nonempty set E, and suppose $\phi : I \to \mathbb{R}$ is a function such that there exists a constant k for which*

$$\phi(y) - \phi(y_0) \geqslant k(y - y_0) \quad \text{for all } y \in I, \tag{1.3.18}$$

where y_0 is a fixed point in I. If $A : L \to \mathbb{R}$ is any isotonic linear functional with $A(1) = 1$ then, for all $g \in L$ such that $\phi(g) \in L$ and $A(g) = y_0$, inequality (1.3.1) *holds.*

The proof is similar to the proof of Jessen's inequality given in Theorem 1.3.1. If ϕ is convex on I then, for each $y_0 \in I$, a constant $k = k(y_0)$ exists such that (1.3.18) holds. Clearly not all ϕ satisfying (1.3.18) for some $y_0 \in I$, $k \in \mathbb{R}$ are convex.

For the next refinement of Jessen's inequality, we shall assume that Ω is an algebra of subsets of E, and that the linear class of functions $g : E \to \mathbb{R}$ satisfies not only (L_1), (L_2) but also

(L_3) $g \in L,\ E_1 \in \Omega \Rightarrow gC_{E_1} \in L,$

where C_{E_1} is the characteristic function of E_1. That is, $C_{E_1}(t) = 1$ for $t \in E_1$ and $C_{E_1}(t) = 0$ for $t \in E - E_1$. It follows from (L_2), (L_3) that $C_{E_1} \in L$ for all $E_1 \in \Omega$. Also, L contains constant functions by (L_1), (L_2). Observe that

$$A(C_{E_1}) + A(C_{E-E_1}) = 1, \qquad A(g) = A(gC_{E_1}) + A(gC_{E-E_1}). \tag{1.3.19}$$

LEMMA 1.3.6. *Let L satisfy properties* (L_1), (L_2) *and* (L_3) *on a nonempty set E, and suppose ϕ is a convex function on an interval $I \subset \mathbb{R}$. If A is any isotonic functional on L with $A(1) = 1$ then, for all $g \in L$ such that $\phi(g) \in L$, we have*

$$F_g(E) \geqslant F_g(E_1) + F_g(E - E_1) \geqslant F_g(E_1) \geqslant 0 \tag{1.3.20}$$

for all $E_1 \in \Omega$ such that $0 < A(C_{E_1}) < 1$, where

$$F_g(E_1) = A\big(\phi(g)C_{E_1}\big) - A(C_{E_1})\phi\left(\frac{A(gC_{E_1})}{A(C_{E_1})}\right). \tag{1.3.21}$$

Equality holds in the first part of (1.3.20) *for strictly convex ϕ if and only if*

$$\frac{A(gC_{E_1})}{A(C_{E_1})} = \frac{A(gC_{E-E_1})}{A(C_{E-E_1})}.$$

PROOF. To prove the first inequality of (1.3.20), set $p = A(C_{E_1})$, $q = A(C_{E-E_1})$ (so $p + q = 1$ by the first equation of (1.3.19)), $x = A(gC_{E_1})/A(C_{E_1})$, $y = A(gC_{E-E_1})/A(C_{E-E_1})$ in the convexity condition

$$p\phi(x) + q\phi(y) \geqslant \phi(px + qy), \quad p + q = 1, \tag{1.3.22}$$

and use the fact that $px + qy = A(g)$ by the second equation of (1.3.19) and similarly use

$$A(\phi(g)) = A(\phi(g)C_{E_1}) + A(\phi(g)C_{E-E_1}).$$

To prove the last inequality of (1.3.20), and hence also the second one, observe that inequality (1.3.1) applies to the isotonic linear functional $A_1 : L \to \mathbb{R}$ as defined by

$$A_1(g) = \frac{A(gC_{E_1})}{A(C_{E_1})}, \quad g \in L,$$

and having $A_1(1) = 1$. Equality holds in the first part of (1.3.20), and for strictly convex ϕ, this equality is the case if and only if $x = y$, since $0 < p, q < 1$. □

REMARK 1.3.6. The inequalities in (1.3.20) are a substantial refinement of the Jessen inequality (1.3.1) since (1.3.20) gives a lower bound ($\geqslant 0$) on the difference $A(\phi(g)) - \phi(A(g)) = F_g(E)$.

Next we give a generalization of Lemma 1.3.6, followed by a related generalization of an inequality of Knopp as established by Pečarić and Beesack in [376].

COROLLARY 1.3.1. *Let L satisfy properties* (L_1), (L_2) *and* (L_3) *on a nonempty set E, and set ψ, χ be strictly monotonic functions on an interval I such that $\phi = \chi \circ \psi^{-1}$ is convex on I. If A is any isotonic linear functional on L with $A(1) = 1$ then, for all $g \in L$ such that $\chi(g)$, $\psi(g) \in L$, we have*

$$H_g(E) \geqslant H_g(E_1) + H_g(E - E_1) \geqslant H_g(E_1) \geqslant 0 \tag{1.3.23}$$

for all $E_1 \in \Omega$ such that $0 < A(C_{E_1}) < 1$, where

$$H_g(E_1) = A(\chi(g)C_{E_1}) - A(C_{E_1})\phi\left(\frac{A(\psi(g)C_{E_1})}{A(C_{E_1})}\right). \tag{1.3.24}$$

Equality holds in the first part of (1.3.23) *for strictly convex ϕ if and only if*

$$\frac{A(\psi(g)C_{E_1})}{A(C_{E_1})} = \frac{A(\psi(g)C_{E-E_1})}{A(C_{E-E_1})}.$$

PROOF. Let $g \in L$ with $\chi(g) \in L$, $\psi(g) \in L$, and set $g_1 = \psi(g)$ so $g_1 \in L$ and $\phi(g_1) = \chi(g) \in L$. Since

$$F_{g_1}(E_1) \equiv H_g(E_1) \quad \text{for } E_1 \in \Omega,$$

inequality (1.3.23) follows from (1.3.20). The above equality condition for g likewise follows from that of Lemma 1.3.6 for $g_1 = \psi(g)$. Lemma 1.3.6 is, of course, just a special case ($\psi(x) \equiv x$, $\phi = \chi$) of Corollary 1.3.1, but only in the case ϕ is strictly monotonic. □

THEOREM 1.3.8. *Let L satisfy conditions* (L_1), (L_2) *on a nonempty set E and let A be an isotonic linear functional on L with $A(1) = 1$. Let ψ, χ be strictly monotonic functions on $I = [m, M]$, $-\infty < m < M < \infty$, such that $\phi = \chi \circ \psi^{-1}$ is convex on I. For all $g \in L$ such that $\chi(g) \in L$, $\psi(g) \in L$ (so $m \leqslant g(t) \leqslant M$ for all $t \in E$), we have*

$$\begin{aligned}0 &\leqslant A(\chi(g)) - \phi(A(\psi(g))) \\ &\leqslant \max_{0 \leqslant \theta \leqslant 1} \{\theta\chi(m) + (1-\theta)\chi(M) - \phi[\theta\psi(m) + (1-\theta)\psi(M)]\}. \qquad (1.3.25)\end{aligned}$$

PROOF. As in the proof of Corollary 1.3.1, set $g_1 = \psi(g)$. Then $g_1 \in L$, $\phi(g_1) \in L$ and $m_1 \leqslant g_1(t) \leqslant M_1$, where $m_1 = \psi(m)$, $M_1 = \psi(M)$ if ψ is increasing or $m_1 = \psi(M)$, $M_1 = \psi(m)$ if ψ is decreasing. By Lemma 1.3.1,

$$\begin{aligned}&A(\chi(g)) - \phi(A(\psi(g))) \\ &\quad = A(\phi(g_1)) - \phi(A(g_1)) \\ &\quad \leqslant \frac{(M_1 - A(g_1))\phi(m_1) + (A(g_1) - m_1)\phi(M_1)}{M_1 - m_1} - \phi(A(g_1)) \\ &\quad \leqslant \max_{0 \leqslant \theta \leqslant 1} \{\theta\phi(m_1) + (1-\theta)\phi(M_1) - \phi[\theta m_1 + (1-\theta)M_1]\}. \qquad (1.3.26)\end{aligned}$$

If $[m_1, M_1] = [\psi(m), \psi(M)]$ the right-hand inequality of (1.3.25) is reduced. If $[m_1, M_1] = [\psi(M), \psi(m)]$ the same result follows by making the substitution $\theta \to 1 - \theta$ in (1.3.25). The left-hand inequality is equivalent to $\phi(A(g_1)) \leqslant A(\phi(g_1))$ and so follows by (1.3.1). □

The following theorem given by Pečarić and Beesack in [376] deals with reformulation of inequalities (1.3.1) and (1.3.2) and presents an application which will illustrate the power of this reformulation.

THEOREM 1.3.9. *Let ϕ be a convex on an interval $I \supset [m, M]$, where $-\infty < m < M < \infty$. Suppose $g : E \to \mathbb{R}$ satisfies $m \leqslant g(t) \leqslant M$ for all $t \in E$, that $g \in L$ and $\phi(g) \in L$. Let $A : L \to \mathbb{R}$ be an isotonic linear functional with $A(1) = 1$ and let $p = p_g$, $q = q_g$ be nonnegative numbers (with $p + q > 0$) for which*

$$A(g) = \frac{pm + qM}{p + q}. \tag{1.3.27}$$

Then

$$\phi\left(\frac{pm + qM}{p + q}\right) \leqslant A\big(\phi(g)\big) \leqslant \frac{p\phi(m) + q\phi(M)}{p + q}. \tag{1.3.28}$$

PROOF. Observe first that since $m \leqslant A(g) \leqslant M$, there always exist $p \geqslant 0$, $q \geqslant 0$, $p + q > 0$ satisfying (1.3.27). The first inequality in (1.3.28) is (1.3.1), while the second of (1.3.28) is (1.3.2). □

REMARK 1.3.7. Observe that for the given p, q (determined by g) in (1.3.27), this inequality can be regarded as a refinement of the inequality obtained from the definition of convexity for the function ϕ. In case $A(g) = m$, it follows that $p_g > 0$, $q_g = 0$ and (1.3.28) reduces to $\phi(m) \leqslant A(\phi(g)) \leqslant \phi(m)$ so that no refinement is possible. Similarly, if $A(g) = M$ no genuine refinement of (1.3.28) is possible. We now show that whenever $m < A(g) < M$, a refinement of

$$\phi\left(\frac{pm + qM}{p + q}\right) \leqslant \frac{p\phi(m) + q\phi(M)}{p + q}$$

is possible for arbitrary $p > 0$, $q > 0$, one of the form of (1.3.28) but for an appropriately chosen function $g_1 \in L$, $g_1 = g_{1,p,q}$, namely

$$\phi\left(\frac{pm + qM}{p + q}\right) \leqslant A\big(\phi(g_1)\big) \leqslant \frac{p\phi(m) + q\phi(M)}{p + q},$$

provided only $\phi(g_1) \in L$. Hence, of course, as in (1.3.27), we have $m \leqslant g_1(t) \leqslant M$ for $t \in E$ and $A(g_1) = (pm + qM)/(p + q)$. To prove this idea we consider two cases: (a) $q/(p + q) \leqslant [A(g) - m]/(M - m)$ and (b) $q/(p + q) > [A(g) - m]/(M - m)$. If (a) holds, it suffices to take $g_1(t) = \alpha g(t) + m(1 + \alpha)$ for $t \in E$, with $\alpha = q(M - m)/\{(p + q)[A(g) - m]\}$, so that $0 < \alpha \leqslant 1$. If (b) holds, one can likewise verify that it suffices to take $g_1(t) = \beta g(t) + M(1 - \beta)$ for $t \in E$, with $\beta = p(M - m)/\{(p + q)[M - A(g)]\}$ while (b) implies $0 < \beta < 1$.

For various other applications and remarks on classical inequalities and means we refer the reader to [20,376] and the references cited therein.

1.4 Some General Inequalities Involving Convex Functions

In this section we shall give some inequalities involving convex functions established by various investigators in the past few years related to the well-know Jensen–Steffensen inequality

$$f\left(\frac{1}{P_n}\sum_{i=1}^{n} p_i x_i\right) \leqslant \frac{1}{P_n}\sum_{i=1}^{n} p_i f(x_i), \tag{1.4.1}$$

where x and p are two n-tuples of real numbers such that $x_i \in I$, $1 \leqslant i \leqslant n$, and I is an interval from $\mathbb{R}$, $P_n = \sum_{i=1}^{n} p_i > 0$, $f : I \to \mathbb{R}$ is convex, and for every monotonic n-tuple x if and only if

$$0 \leqslant P_k \leqslant P_n, \quad k = 1, 2, \ldots, n-1. \tag{M}$$

In a 1981 paper, Pečarić [366] has obtained necessary and sufficient conditions for the validity of reverse inequality, that is,

$$f\left(\frac{1}{P_n}\sum_{i=1}^{n} p_i x_i\right) \geqslant \frac{1}{P_n}\sum_{i=1}^{n} p_i f(x_i). \tag{1.4.2}$$

In [366], the following Fuch's generalization of the majorization theorem (see [122]) is used to establish the main results.

LEMMA 1.4.1. *Let $a_1 \geqslant \cdots \geqslant a_s$, $b_1 \geqslant \cdots \geqslant b_s$ and $q_1, \ldots, q_s$ be real numbers such that*

$$\sum_{i=1}^{k} q_i a_i \leqslant \sum_{i=1}^{k} q_i b_i, \quad 1 \leqslant k \leqslant s-1,$$

and

$$\sum_{i=1}^{s} q_i a_i = \sum_{i=1}^{s} q_i b_i.$$

Then for every convex function f,

$$\sum_{i=1}^{s} q_i f(a_i) \leqslant \sum_{i=1}^{s} q_i f(b_i). \tag{1.4.3}$$

In [366] Pečarić has given the following two theorems.

THEOREM 1.4.1. *Let x be a nonincreasing tuple of real numbers, $x_i \in I$, $1 \leqslant i \leqslant n$, p real n-tuple, and there exists x_j, $j \in (1, \ldots, n)$, such that*

$$\begin{aligned} &\sum_{i=1}^{k} p_i(x_i - x_j) \leqslant 0 \quad \text{for every } k \text{ such that } x_k \geqslant \bar{x} = \frac{1}{P_n}\sum_{i=1}^{n} p_i x_i, \\ &\sum_{i=k}^{n} p_i(x_i - x_j) \geqslant 0 \quad \text{for every } k \text{ such that } x_k \leqslant \bar{x}. \end{aligned} \tag{1.4.4}$$

(If $x_1 \leqslant \bar{x}$ the first condition in (1.4.4) *is taken to be vacuous and if $x_n \geqslant \bar{x}$ the second condition in* (1.4.4) *is taken to be vacuous.) If $\bar{x} \in I$ then, for every convex function $f : I \to \mathbb{R}$,* (1.4.2) *holds.*

If the reverse inequalities hold in (1.4.4), *then* (1.4.1) *holds.*

PROOF. Let $\bar{x} \in [x_{r+1}, x_r]$. By substitutions of

(i) $s = n+1$; $q_i = p_i$, $a_i = x_i$, $1 \leqslant i \leqslant r$; $q_{r+1} = -P_n$, $a_{r+1} = \bar{x}$; $q_i = p_{i-1}$, $a_i = x_{i-1}$, $r+2 \leqslant i \leqslant n+1$; $b_i = x_j$, $1 \leqslant i \leqslant n+1$, and

(ii) $s = n+1$; $a_i = x_j$, $1 \leqslant i \leqslant n+1$; $q_i = p_i$, $b_i = x_i$, $1 \leqslant i \leqslant r$; $q_{r+1} = -P_n$, $b_{r+1} = \bar{x}$; $q_i = p_{i-1}$, $b_i = x_{i-1}$, $r+2 \leqslant i \leqslant n+1$,

from Lemma 1.4.1, we get Theorem 1.4.1. It can easily be shown that the case $x_1 \leqslant \bar{x}$ and $x_n \geqslant \bar{x}$ can exist only for inequality (1.4.2). □

THEOREM 1.4.2. *Let x and p be two n-tuples of real numbers such that $x_i \in I$, $1 \leqslant i \leqslant n$, $\bar{x} \in I$ and $P_n > 0$. Inequality* (1.4.2) *holds for every convex function $f : I \to \mathbb{R}$ and for every monotonic n-tuple x if and only if there exists $m \in (1, \ldots, n)$ such that*

$$P_k \leqslant 0, \quad k < m, \quad \text{and} \quad \overline{P_k} \leqslant 0, \quad k > m, \tag{1.4.5}$$

where $\overline{P_k} = P_n - P_{k-1}$.

PROOF. Suppose that (1.4.5) holds. Using the identities

$$\begin{aligned} &\sum_{i=1}^{k} p_i(x_i - x_m) = (x_k - x_m)P_k + \sum_{i=1}^{k-1} P_i(x_i - x_{i+1}) \quad \text{and} \\ &\sum_{i=k}^{n} p_i(x_i - x_m) = (x_k - x_m)\overline{P_k} + \sum_{i=k+1}^{n} \overline{P_i}(x_i - x_{i-1}), \end{aligned} \tag{1.4.6}$$

we get, in the case $x_1 \geqslant \cdots \geqslant x_n$,

$$\sum_{i=1}^{k} p_i(x_i - x_m) \leqslant 0, \quad 1 \leqslant k \leqslant m, \quad \text{and}$$
$$\sum_{i=k}^{n} p_i(x_i - x_m) \geqslant 0, \quad m \leqslant k \leqslant n. \tag{1.4.7}$$

Let $\bar{x} \in [x_{r+1}, x_r]$ and let $m \leqslant r$. Then the conditions (1.4.4), for $j = m$ are obviously satisfied if $1 \leqslant k \leqslant m$ and $r \leqslant k \leqslant n$. Suppose that for k_1 and $m < k_1 < r$, the condition (1.4.4) is invalid, that is, $\sum_{i=1}^{k_1} p_i(x_i - x_m) \geqslant 0$. Since, from (1.4.7), we have $\sum_{i=k_1+1}^{n} p_i(x_i - x_m) \geqslant 0$, we get $\sum_{i=1}^{n} p_i(x_i - x_m) \geqslant 0$, that is, $\bar{x} \geqslant x_m$, what is evidently a contradiction. Analogously, in the case $m > r$ or $x_1 < \bar{x}$, $\bar{x} < x_n$, we get that (1.4.2) holds. If $x_1 \leqslant \cdots \leqslant x_n$, we can also prove that (1.4.2) is valid.

Next, suppose that (1.4.2) holds. Let $f(x) = x^2$, $x_i = 0$, $i = 1, \ldots, k-1$, and $x_i = 1$, $i = k, \ldots, n$. Then (1.4.2) becomes $(\overline{P}_k/P_n)^2 \geqslant \overline{P}_k/P_n$. Hence, $\overline{P}_k \leqslant 0$ or $P_{k-1} \leqslant 0$, $k = 2, \ldots, n$.

Now let $k < m$ and suppose that $\overline{P}_k \leqslant 0$. Let $x_i = 0$, $1 \leqslant i \leqslant k-1$, $x_i = 1$, $k \leqslant i \leqslant m-1$, and $x_i = 1+\varepsilon$, $m \leqslant i \leqslant n$. Then $\bar{x} = (\overline{P}_k + \overline{P}_m)/P_n$. Since $\overline{P}_k \leqslant 0$, we can choose ε sufficiently small that $\bar{x} < 1$. Let $f(z) = z - 1$ if $z \geqslant 1$ and $f(z) = 0$ for $z < 1$. Then (1.4.2) becomes $(1/P_n)\sum_{i=m}^{n} \varepsilon p_i \leqslant 0$, that is, $\overline{P}_m \leqslant 0$. Similarly we can conclude that $P_m \leqslant 0$ implies $P_k \leqslant 0$. So (1.4.5) for some $m \in (1, \ldots, n)$ must be satisfied. □

REMARK 1.4.1. Analogously we can prove (1.4.1). Indeed suppose that condition (M) holds. Using the identities (1.4.6) we get, in the case $x_1 \geqslant \cdots \geqslant x_n$, that for every $m = 1, \ldots, n$, (1.4.7) holds with the reverse inequalities. Since $\bar{x} \in [x_{r+1}, x_r]$, we can suppose that $\bar{x} \in [x_{r+1}, x_r]$. Then (1.4.4) with the reverse inequalities holds for $j = r$ and for $j = r+1$, that is, (1.4.1) is valid.

Next suppose that (1.4.1) holds. Let $f(x) = x^2$, $x_i = 1$, $1 \leqslant i \leqslant k$, and $x_i = 0$, $k+1 \leqslant i \leqslant n$. Then (1.4.1) becomes $(P_k/P_n)^2 \leqslant P_k/P_n$, that is, $0 \leqslant P_k \leqslant P_n$, $1 \leqslant k \leqslant n$.

Now we shall give the following corollaries from [366].

COROLLARY 1.4.1. *Let $x_1 \leqslant \cdots \leqslant x_m \leqslant 0 \leqslant x_{m+1} \leqslant \cdots \leqslant x_n$, $m \in (0, 1, \ldots, n)$, $x_i \in I$, $1 \leqslant i \leqslant n$, $0 \in I$, and p is real n-tuple.*

(i) *Inequality*

$$\sum_{i=1}^{n} p_i f(x_i) \geqslant f\left(\sum_{i=1}^{n} p_i x_i\right) + \sum_{i=1}^{n} (p_i - 1) f(0) \tag{1.4.8}$$

holds for every convex function $f : I \to \mathbb{R}$ *if and only if* $0 \leqslant P_k \leqslant 1$, $1 \leqslant k \leqslant m$, $0 \leqslant \overline{P}_k \leqslant 1$, $m+1 \leqslant k \leqslant n$.

(ii) *Let* $\sum_{i=1}^{n} p_i x_i \in I$. *The reverse inequality holds in* (1.4.8) *if and only if there exists* $j \leqslant m$ *such that*

$$P_i \leqslant 0, \quad i < j, \qquad P_i \geqslant 1, \quad j \leqslant i \leqslant m, \qquad \overline{P}_i \leqslant 0, \quad i \leqslant m+1,$$

or there exists $j \geqslant m$ *such that*

$$P_i \leqslant 0, \quad i \leqslant m, \qquad \overline{P}_i \geqslant 1, \quad m+1 \leqslant i \leqslant j, \quad \text{and} \quad \overline{P}_i \leqslant 0, \quad i > j.$$

PROOF. Let Theorems 1.4.1 and 1.4.2 hold for $n+1$ with $x_i = \hat{x}_i$ and $p_i = \hat{p}_i$, $1 \leqslant i \leqslant n+1$. By substitution, $\hat{x}_i = x_i$, $\hat{p}_i = p_i$, $1 \leqslant i \leqslant m$, $\hat{x}_{m+1} = 0$, $\hat{p}_{m+1} = 1 - P_n$, $\hat{x}_i = x_i$, $\hat{p}_i = p_{i-1}$, $m+2 \leqslant i \leqslant n+1$, we get Corollary 1.4.1. □

As a simple consequence of Theorem 1.4.2 we have the following corollary.

COROLLARY 1.4.2. *Let* x *and* p *be two* n*-tuples of real numbers such that* $x_i \in I$, $1 \leqslant i \leqslant n$, $\bar{x} \in I$ *and*

$$p_m > 0, \qquad p_i \leqslant 0, \quad i \neq m, \qquad P_n > 0.$$

Then (1.4.2) *holds for every convex function* $f : I \to \mathbb{R}$.

Next we shall give some inequalities for convex-dominated functions given by Dragomir and Ionescu in [88].

We shall introduce the following class of functions (see [88]).

DEFINITION 1.4.1. Let $g : I \to \mathbb{R}$ be a given convex function on interval I from $\mathbb{R}$. The real function $f : I \to \mathbb{R}$ is called g-convex dominated on I if the following condition is satisfied

$$\begin{aligned} &\left|\lambda f(x) + (1-\lambda) f(y) - f\big(\lambda x + (1-\lambda) y\big)\right| \\ &\quad \leqslant \lambda g(x) + (1-\lambda) g(y) - g\big(\lambda x + (1-\lambda) y\big) \end{aligned} \tag{1.4.9}$$

for all x, y in I and $\lambda \in [0, 1]$.

The following simple characterization of convex-dominated functions is valid.

LEMMA 1.4.2. *Let g be a convex function on I and $f : I \to \mathbb{R}$. Then the following statements are equivalent*:

(i) *f is g-convex dominated on I*;
(ii) *$g - f$ and $g + f$ are convex on I; and*
(iii) *there exist two convex mappings h, l on I such that $f = (h - l)/2$ and $g = (h + l)/2$.*

PROOF. (i) $\Leftrightarrow$ (ii). Condition (1.4.9) is equivalent to

$$\begin{aligned}
&\lambda\big(g(x) - f(x)\big) + (1 - \lambda)\big(g(y) - f(y)\big) \\
&\quad \geqslant g\big(\lambda x + (1 - \lambda)y\big) - f\big(\lambda x + (1 - \lambda)y\big), \\
&\lambda\big(g(x) + f(x)\big) + (1 - \lambda)\big(g(y) + f(y)\big) \\
&\quad \geqslant g\big(\lambda x + (1 - \lambda)y\big) + f\big(\lambda x + (1 - \lambda)y\big)
\end{aligned}$$

for all $x, y \in I$ and $\lambda \in [0, 1]$, that is, $g - f$ and $g + f$ are convex on I if and only if (1.4.9) holds.

(ii) $\Leftrightarrow$ (iii). It is obvious. □

Let $F(I)$ be the linear space of all real-valued functions defined on I and $J : F(I) \to \mathbb{R}$ be a functional satisfying the properties:

(J_1) $J(\alpha f + \beta g) = \alpha J(f) + \beta J(g)$ for all $\alpha, \beta \in \mathbb{R}$ and $f, g \in F(I)$, and
(J_2) $J(f) \geqslant 0$ for all convex functions f on I.

The following lemma plays an important role in the sequel.

LEMMA 1.4.3. *Let J be a functional satisfying conditions* (J_1), (J_2). *Then, for every convex function g and for every g-convex dominated function f on I, the following inequality holds*:

$$\big|g(f)\big| \leqslant J(g). \tag{1.4.10}$$

PROOF. Let g be a convex function and f be g-convex dominated on I. By Lemma 1.4.2, it follows that $g - f$ and $g + f$ are convex on I. Then

$$0 \leqslant J(g - f) = J(g) - J(f) \quad \text{and} \quad 0 \leqslant J(g + f) = J(g) + J(f)$$

which gives

$$-J(g) \leqslant J(f) \leqslant J(g).$$

Since $J(g) \geqslant 0$, inequality (1.4.10) is proven. □

In [88] Dragomir and Ionescu have obtained the following improvement of the Jensen inequality.

THEOREM 1.4.3. *Let g be a given convex function on I and $f : J \to \mathbb{R}$ be g-convex dominated. Then, for every $x_i \in I$, $p_i \geqslant 0$, $1 \leqslant i \leqslant n$, such that $P_n = \sum_{i=1}^n p_i > 0$, we have the inequality*

$$\left|\frac{1}{P_n}\sum_{i=1}^n p_i f(x_i) - f\left(\frac{1}{P_n}\sum_{i=1}^n p_i x_i\right)\right| \leqslant \frac{1}{P_n}\sum_{i=1}^n p_i g(x_i) - g\left(\frac{1}{P_n}\sum_{i=1}^n p_i x_i\right). \tag{1.4.11}$$

PROOF. Consider the functional

$$J(f) = \frac{1}{P_n}\sum_{i=1}^n p_i f(x_i) - f\left(\frac{1}{P_n}\sum_{i=1}^n p_i x_i\right), \quad f \in F(I).$$

Then J satisfies conditions (J_1) and (J_2) (by Jensen's inequality). Applying Lemma 1.4.3, we obtain inequality (1.4.11). The proof is finished. □

In [88] the following improvement of Fuch's generalization of the majorization theorem (see [122]) is given.

THEOREM 1.4.4. *Let $a_1 \geqslant \cdots \geqslant a_s$, $b_1 \geqslant \cdots \geqslant b_s$ and $q_1, \ldots, q_s$ be real numbers such that*

$$\sum_{i=1}^k q_i a_i \leqslant \sum_{i=1}^k q_i b_i, \quad 1 \leqslant k \leqslant s-1, \qquad \sum_{i=1}^s q_i a_i = \sum_{i=1}^s q_i b_i.$$

If g is convex on I and f is g-convex dominated on I, then the following inequality holds:

$$\left|\sum_{i=1}^s q_i\big(f(b_i) - f(a_i)\big)\right| \leqslant \sum_{i=1}^s \big(g(b_i) - g(a_i)\big). \tag{1.4.12}$$

PROOF. Consider the functional

$$J(f) = \sum_{i=1}^{s} q_i\big(f(b_i) - f(a_i)\big), \quad f \in F(I).$$

Then J satisfies conditions (J_1) and (J_2), by Fuchs' inequality, see also Lemma 1.4.1. Applying Lemma 1.4.3, we deduce inequality (1.4.12). □

The following theorem given in [88] deals with an improvement of the Jensen–Steffensen inequality.

THEOREM 1.4.5. *Let x and p be two n-tuples of real numbers such that $x_i \in I$, $1 \leqslant i \leqslant n$, and I is an interval from $\mathbb{R}$ and $P_n > 0$. Then the following statements are equivalent:*

(i) *For every convex function $g : I \to \mathbb{R}$, for every g-convex dominated function f and for all monotonic n-tuple x, inequality* (1.4.11) *hold.*

(ii) $0 \leqslant P_k \leqslant P_n$ *for all* $k = 1, 2, \ldots, n-1$.

PROOF. (i) ⇒ (ii) is obviously the Jensen–Steffensen inequality. (ii) ⇒(i) considers the functional

$$J(f) = \frac{1}{P_n}\sum_{i=1}^{n} p_i f(x_i) - f\left(\frac{1}{P_n}\sum_{i=1}^{n} p_i x_i\right), \quad f \in F(I).$$

Then J verifies conditions (J_1) and (J_2), by the Jensen–Steffensen inequality, see Theorem 1.2.2. Applying Lemma 1.4.3, we obtain (1.4.11). □

The next theorem given in [88] improves Pečarić's theorem given above in Theorem 1.4.1.

THEOREM 1.4.6. *Let x be a nonincreasing n-tuple of real numbers, $x_i \in I$, $1 \leqslant i \leqslant n$, p real n-tuple, and there exists x_j, $j \in (1, 2, \ldots, n)$, such that*

$$\begin{aligned}
&\sum_{i=1}^{k} p_i(x_i - x_j) \leqslant 0 \quad \textit{for every } k \textit{ such that } x_k \geqslant \bar{x} = \frac{1}{P_n}\sum_{i=1}^{n} p_i x_i,\\
&\sum_{i=1}^{n} p_i(x_i - x_j) \geqslant 0 \quad \textit{for every } k \textit{ such that } x_k \leqslant \bar{x}.
\end{aligned} \tag{1.4.13}$$

(If $x_1 < \bar{x}$ then the first condition in (1.4.13) *is taken to be vacuous, if $x_n \geqslant \bar{x}$ the second condition in* (1.4.13) *is taken to be vacuous.) If $\bar{x} \in I$, then for every*

convex function $g : I \to \mathbb{R}$ and for every g-convex dominated function $f : I \to \mathbb{R}$, we have

$$g\left(\frac{1}{P_n}\sum_{i=1}^{n} p_i x_i\right) - \frac{1}{P_n}\sum_{i=1}^{n} p_i g(x_i)$$
$$\geqslant \left| f\left(\frac{1}{P_n}\sum_{i=1}^{n} p_i x_i\right) - \frac{1}{P_n}\sum_{i=1}^{n} p_i f(x_i)\right|. \tag{1.4.14}$$

If the reverse inequalities in (1.4.13) *hold, then* (1.4.11) *holds.*

The proof follows by a similar argument by using Theorem 1.4.1 given above.

1.5 Hadamard's Inequalities

In 1893, J. Hadamard [134] investigates one of the fundamental inequalities in analysis, which is now known in the literature as Hadamard's inequality. Over the years many authors have developed various extensions, variants and refinements of Hadamard's inequality. In this section we shall deal with Hadamard's and related inequalities as established by various investigators during the past few years.

The following theorem deals with Hadamard's inequality involving convex functions.

THEOREM 1.5.1. *If $f : I \to \mathbb{R}$ is a convex function, where $I = [a, b]$ and $\mathbb{R}$ are a set of real numbers, then the inequalities*

$$f\left(\frac{a+b}{2}\right) \leqslant \frac{1}{b-a}\int_a^b f(x)\,\mathrm{d}x \leqslant \frac{f(a)+f(b)}{2} \tag{1.5.1}$$

are valid.

PROOF. Since f is convex on I, then for $t \in [0, 1]$, we have

$$f\bigl(ta + (1-t)b\bigr) \leqslant tf(a) + (1-t)f(b). \tag{1.5.2}$$

Integrating (1.5.2) with respect to t on $[0, 1]$ we get

$$\int_0^1 f\bigl(ta + (1-t)b\bigr)\,\mathrm{d}t \leqslant \frac{f(a)+f(b)}{2}. \tag{1.5.3}$$

On the other hand, since f is convex on I, then for $t \in [0, 1]$ we have

$$f\left(\frac{a+b}{2}\right) = f\left(\frac{ta+(1-t)b}{2} + \frac{(1-t)a+tb}{2}\right)$$
$$\leqslant \frac{1}{2}\big[f\big(ta+(1-t)b\big) + f\big((1-t)a+tb\big)\big]. \tag{1.5.4}$$

Integrating inequality (1.5.4) with respect to t on $[0, 1]$ we get

$$f\left(\frac{a+b}{2}\right) \leqslant \frac{1}{2}\int_0^1 \big[f\big(ta+(1-t)b\big) + f\big((1-t)a+tb\big)\big]\,\mathrm{d}t$$
$$= \frac{1}{2}\left[\int_0^1 f\big(ta+(1-t)b\big)\,\mathrm{d}t + \int_0^1 f\big((1-t)a+tb\big)\,\mathrm{d}t\right]. \tag{1.5.5}$$

By putting $1-t=s$ in the second integral on the right-hand side of (1.5.5), we have

$$f\left(\frac{a+b}{2}\right) \leqslant \frac{1}{2}\left[\int_0^1 f\big(ta+(1-t)b\big)\,\mathrm{d}t + \int_0^1 f\big(sa+(1-s)b\big)\,\mathrm{d}s\right]$$
$$= \int_0^1 f\big(ta+(1-t)b\big)\,\mathrm{d}t. \tag{1.5.6}$$

From (1.5.3) and (1.5.6), we get

$$f\left(\frac{a+b}{2}\right) \leqslant \int_0^1 f\big(ta+(1-t)b\big)\,\mathrm{d}t \leqslant \frac{f(a)+f(b)}{2}. \tag{1.5.7}$$

By putting $ta+(1-t)b = x$ in the integral involved in (1.5.7), it is easy to observe that

$$\int_0^1 f\big(ta+(1-t)b\big)\,\mathrm{d}t = \frac{1}{b-a}\int_a^b f(x)\,\mathrm{d}x. \tag{1.5.8}$$

Using (1.5.8) in (1.5.7) we get the required inequalities in (1.5.1) and the proof is complete. □

In 1976, A. Lupas [200] proves that if $p, q > 0$, f is convex on $I \supset [a, b]$ and $\nu = (pa+qb)/(p+q)$, then

$$f\left(\frac{pa+qb}{p+q}\right) \leqslant \frac{1}{2y}\int_{\nu-y}^{\nu+y} f(t)\,\mathrm{d}t \leqslant \frac{pf(a)+qf(b)}{p+q}, \tag{1.5.9}$$

provided $0 < y \leqslant [(b-a)/(p+q)]\min(p,q)$. The case $p = q = 1$, $y = (b-a)/2$ of (1.5.9) is Hadamard's inequality. In [376] it is shown that under the same hypotheses Hadamard's inequality yields the following refinement of (1.5.9).

THEOREM 1.5.2. *If $p, q > 0$, f is convex on $I \supset [a,b]$ and $\nu = (pa+qb)/(p+q)$, then*

$$f\left(\frac{pa+qb}{p+q}\right) \leqslant \frac{1}{2y}\int_{\nu-y}^{\nu+y} f(t)\,dt \leqslant \frac{1}{2}\big[f(\nu-y)+f(\nu+y)\big] \leqslant \frac{pf(a)+qf(b)}{p+q}. \tag{1.5.10}$$

PROOF. First observe that if $0 < y \leqslant [(b-a)/(p+q)]\min(p,q)$ then, by considering two cases ($0 < p \leqslant q$, $0 < q < p$), one easily verifies that $a \leqslant \nu - y < \nu + y \leqslant b$, so f is defined on $[\nu-y, \nu+y]$. By Hadamard's inequality (1.5.1) with a, b replaced by $\nu-y$, $\nu+y$, we obtain

$$f(\nu) \leqslant \frac{1}{2y}\int_{\nu-y}^{\nu+y} f(t)\,dt \leqslant \frac{1}{2}\big[f(\nu-y)+f(\nu+y)\big]. \tag{1.5.11}$$

By the definition of convexity, we have, for $a \leqslant x_1 < x_2 < x_3 \leqslant b$,

$$f(x_2) \leqslant \frac{x_3-x_2}{x_3-x_1}f(x_1) + \frac{x_2-x_1}{x_3-x_1}f(x_3).$$

Hence, taking $x_1 = a$, $x_3 = b$ we obtain

$$f(\nu-y) \leqslant \frac{b-(\nu-y)}{b-a}f(a) + \frac{\nu-y-a}{b-a}f(b), \tag{1.5.12}$$

$$f(\nu+y) \leqslant \frac{b-(\nu+y)}{b-a}f(a) + \frac{\nu+y-a}{b-a}f(b). \tag{1.5.13}$$

From (1.5.11)–(1.5.13), we now have

$$f(\nu) \leqslant \frac{1}{2y}\int_{\nu-y}^{\nu+y} f(t)\,dt \leqslant \frac{1}{2}\big[f(\nu-y)+f(\nu+y)\big]$$

$$\leqslant \frac{1}{2}\left[\frac{b-v}{b-a}f(a)+\frac{v-a}{b-a}f(b)\right]$$
$$=\frac{pf(a)+qf(b)}{p+q},$$

proving (1.5.10). □

In 1990, S.S. Dragomir [82] has given the following refinements of Hadamard's inequalities.

THEOREM 1.5.3. *Let* $f:[a,b]\to\mathbb{R}$ *be a convex mapping. Then, for all* $t\in[a,b]$, *we have the following inequalities*

$$f\left(\frac{a+b}{2}\right)\leqslant\frac{1}{(b-a)^2}\int_a^b\int_a^b f\big(tx+(1-t)y\big)\,\mathrm{d}x\,\mathrm{d}y$$
$$\leqslant\frac{1}{b-a}\int_a^b f(x)\,\mathrm{d}x$$
$$\leqslant\frac{f(a)+f(b)}{2}. \tag{1.5.14}$$

PROOF. Since f is convex on $[a,b]$, then for all $x,y\in[a,b]$ and $t\in[0,1]$, we have

$$f\big(tx+(1-t)y\big)\leqslant tf(x)+(1-t)f(y).$$

Integrating this inequality on $[a,b]\times[a,b]$ we get

$$\int_a^b\int_a^b f\big(tx+(1-t)y\big)\,\mathrm{d}x\,\mathrm{d}y\leqslant\int_a^b\int_a^b\big[tf(x)+(1-t)f(y)\big]\,\mathrm{d}x\,\mathrm{d}y$$
$$=(b-a)\int_a^b f(x)\,\mathrm{d}x,$$

which proves the second part of (1.5.14) by using the right half of Hadamard's inequality.

On the other hand, by Jensen's inequality for double integrals, we have

$$f\left(\frac{1}{(b-a)^2}\int_a^b\int_a^b\big(tx+(1-t)y\big)\,\mathrm{d}x\,\mathrm{d}y\right)$$
$$\leqslant\frac{1}{(b-a)^2}\int_a^b\int_a^b f\big(tx+(1-t)y\big)\,\mathrm{d}x\,\mathrm{d}y,$$

and since

$$\frac{1}{(b-a)^2}\int_a^b\int_a^b \big(tx+(1-t)y\big)\,\mathrm{d}x\,\mathrm{d}y=\frac{a+b}{2},$$

the proof is complete. □

COROLLARY 1.5.1. *Let f be as in Theorem 1.5.3. Then*

$$\begin{aligned}f\left(\frac{a+b}{2}\right)&\leqslant\frac{1}{(b-a)^2}\int_a^b\int_a^b f\left(\frac{x+y}{2}\right)\mathrm{d}x\,\mathrm{d}y\\&\leqslant\frac{1}{b-a}\int_a^b f(x)\,\mathrm{d}x\\&\leqslant\frac{f(a)+f(b)}{2}.\end{aligned}\tag{1.5.15}$$

THEOREM 1.5.4. *Let $f:[a,b]\to\mathbb{R}$ be a convex mapping on $[a,b]$. Then*

$$\begin{aligned}&\frac{1}{(b-a)^2}\int_a^b\int_a^b f\left(\frac{x+y}{2}\right)\mathrm{d}x\,\mathrm{d}y\\&\quad\leqslant\frac{1}{(b-a)^2}\int_a^b\int_a^b\int_0^1 f\big(tx+(1-t)y\big)\,\mathrm{d}x\,\mathrm{d}y\,\mathrm{d}t\\&\quad\leqslant\frac{1}{b-a}\int_a^b f(x)\,\mathrm{d}x.\end{aligned}\tag{1.5.16}$$

PROOF. Consider the mapping $g:[a,b]\to\mathbb{R}$ given by

$$g(t)=\frac{1}{(b-a)^2}\int_a^b\int_a^b f\big(tx+(1-t)y\big)\,\mathrm{d}x\,\mathrm{d}y.$$

For all $t_1,t_2\in[0,1]$ and $\alpha,\beta\geqslant 0$ with $\alpha+\beta=1$, we have

$$\begin{aligned}g(\alpha t_1+\beta t_2)&=\frac{1}{(b-a)^2}\int_a^b\int_a^b f\big((\alpha t_1+\beta t_2)x+(1-\alpha t_1-\beta t_2)y\big)\,\mathrm{d}x\,\mathrm{d}y\\&\leqslant\frac{\alpha}{(b-a)^2}\int_a^b\int_a^b f\big(t_1x+(1-t_1)y\big)\,\mathrm{d}x\,\mathrm{d}y\\&\quad+\frac{\beta}{(b-a)^2}\int_a^b\int_a^b f\big(t_2x+(1-t_2)y\big)\,\mathrm{d}x\,\mathrm{d}y\\&=\alpha g(t_1)+\beta g(t_2),\end{aligned}$$

which proves that g is convex on $[0, 1]$.

By means of Hadamard's inequalities for the convex mapping g and using Fubini's theorem for multiple integrals, we have

$$\begin{aligned}&\frac{1}{(b-a)^2}\int_a^b\int_a^b f\left(\frac{x+y}{2}\right)\mathrm{d}x\,\mathrm{d}y\\&= g\left(\frac{1}{2}\right) \leqslant \int_0^1 g(t)\,\mathrm{d}t\\&= \frac{1}{(b-a)^2}\int_a^b\int_a^b\int_0^1 f\big(tx+(1-t)y\big)\,\mathrm{d}x\,\mathrm{d}y\,\mathrm{d}t\\&\leqslant \frac{g(0)+g(1)}{2} = \frac{1}{b-a}\int_a^b f(x)\,\mathrm{d}x.\end{aligned}$$

The proof is complete. □

The next theorem given by Dragomir [86], in one sense, is an improvement of the "right" inequality in (1.5.1).

THEOREM 1.5.5. *Let $f:[a,b]\to\mathbb{R}$ be a differentiable convex function. Then the following inequalities*

$$\begin{aligned}0 &\leqslant \frac{1}{b-a}\int_a^b f(x)\,\mathrm{d}x - \frac{1}{(b-a)^2}\int_a^b\int_a^b f\big(tx+(1-t)y\big)\,\mathrm{d}x\,\mathrm{d}y\\&\leqslant t\left(\frac{f(a)+f(b)}{2} - \frac{1}{b-a}\int_a^b f(x)\,\mathrm{d}x\right)\end{aligned} \tag{1.5.17}$$

are valid for all t in $[0, 1]$.

PROOF. Since f is convex on $[a, b]$, we have

$$f\big(tx+(1-t)y\big) \leqslant tf(x)+(1-t)f(y)$$

for all x, y in $[a, b]$ and t in $[0, 1]$. Integrating the above inequality on $[a, b]^2$ we obtain

$$\begin{aligned}\int_a^b\int_a^b f\big(tx+(1-t)y\big)\,\mathrm{d}x\,\mathrm{d}y &\leqslant \int_a^b\int_a^b \big[tf(x)+(1-t)f(y)\big]\,\mathrm{d}x\,\mathrm{d}y\\&= (b-a)\int_a^b f(x)\,\mathrm{d}x,\end{aligned}$$

from which the first part of the proof of the inequalities in (1.5.17) follows.

On the other hand, since f is convex and derivable on $[a, b]$, we have

$$f\big(tx + (1-t)y\big) - f(y) \geqslant t(x-y)f'(y)$$

for all x, y in $[a, b]$ and t in $[0, 1]$. Integrating both sides of the above inequality on $[a, b]^2$ we get

$$\int_a^b\int_a^b f\big(tx + (1-t)y\big)\,\mathrm{d}x\,\mathrm{d}y - (b-a)\int_a^b f(x)\,\mathrm{d}x \geqslant t\int_a^b\int_a^b (x-y)f'(y)\,\mathrm{d}x\,\mathrm{d}y. \tag{1.5.18}$$

Since a simple calculation yields that

$$\int_a^b\int_a^b (x-y)f'(y)\,\mathrm{d}x\,\mathrm{d}y = (b-a)\int_a^b f(x)\,\mathrm{d}x - (b-a)^2\frac{f(a)+f(b)}{2},$$

by using this formula in (1.5.18), we obtain

$$(b-a)\int_a^b f(x)\,\mathrm{d}x - \int_a^b\int_a^b f\big(tx + (1-t)y\big)\,\mathrm{d}x\,\mathrm{d}y \leqslant t\left[(b-a)^2\frac{f(a)+f(b)}{2} - (b-a)\int_a^b f(x)\,\mathrm{d}x\right]$$

for all t in $[0, 1]$, which is the second inequality in (1.5.17). □

COROLLARY 1.5.2. *Let f be as in Theorem* 1.5.5. *Then we have*

$$0 \leqslant \frac{1}{b-a}\int_a^b f(x)\,\mathrm{d}x - \frac{1}{(b-a)^2}\int_a^b\int_a^b f\left(\frac{x+y}{2}\right)\mathrm{d}x\,\mathrm{d}y \leqslant \frac{1}{2}\left[\frac{f(a)+f(b)}{2} - \frac{1}{b-a}\int_a^b f(x)\,\mathrm{d}x\right].$$

The following theorem also is given in [86].

THEOREM 1.5.6. *Let f be as in Theorem* 1.5.5. *Then for all t in* $[0, 1]$, *we have the inequality*

$$0 \leqslant \frac{1}{b-a}\int_a^b f(x)\,\mathrm{d}x - \frac{1}{b-a}\int_a^b f\left(tx+(1-t)\frac{a+b}{2}\right)\mathrm{d}x$$
$$\leqslant (1-t)\left[\frac{f(a)+f(b)}{2} - \frac{1}{b-a}\int_a^b f(x)\,\mathrm{d}x\right]. \tag{1.5.19}$$

PROOF. First, we observe that

$$f\left(tx+(1-t)\frac{a+b}{2}\right) \leqslant tf(x)+(1-t)f\left(\frac{a+b}{2}\right)$$

for all x in $[a, b]$ and t in $[0, 1]$. Integrating the above inequality with respect to x over $[a, b]$ and using the left half of inequality (1.5.1) we have

$$\begin{aligned}&\frac{1}{b-a}\int_a^b f\left(tx+(1-t)\frac{a+b}{2}\right)\mathrm{d}x\\ &\quad\leqslant \frac{t}{b-a}\int_a^b f(x)\,\mathrm{d}x + (1-t)f\left(\frac{a+b}{2}\right)\\ &\quad\leqslant \frac{t}{b-a}\int_a^b f(x)\,\mathrm{d}x + \frac{1-t}{b-a}\int_a^b f(x)\,\mathrm{d}x\\ &\quad= \frac{1}{b-a}\int_a^b f(x)\,\mathrm{d}x.\end{aligned}$$

On the other hand, the function f being differentiate convex on $[a, b]$, we get

$$f\left(tx+(1-t)\frac{a+b}{2}\right) - f(x) \geqslant (1-t)\left(\frac{a+b}{2}-x\right)f'(x)$$

for all t in $[0, 1]$ and x in $[a, b]$. Integrating the above inequality with respect to x on $[a, b]$ we get

$$\frac{1}{b-a}\int_a^b f\left(tx+(1-t)\frac{a+b}{2}\right)\mathrm{d}x - \frac{1}{b-a}\int_a^b f(x)\,\mathrm{d}x$$
$$\geqslant (1-t)\int_a^b \left(\frac{a+b}{2}-x\right)f'(x)\,\mathrm{d}x \tag{1.5.20}$$

for all t in $[0, 1]$. A simple computation shows that

$$\int_a^b \left(\frac{a+b}{2} - x\right) f'(x)\,\mathrm{d}x = \int_a^b f(x)\,\mathrm{d}x - (b-a)\frac{f(a)+f(b)}{2}.$$

Using this equality in (1.5.20) we get the required inequality in (1.5.19). □

COROLLARY 1.5.3. *Let f be as in Theorem 1.5.6. Then we have*

$$\begin{aligned} 0 &\leqslant \frac{1}{b-a}\int_a^b f(x)\,\mathrm{d}x - \frac{2}{b-a}\int_{(3a+b)/4}^{(a+3b)/4} f(x)\,\mathrm{d}x \\ &\leqslant \frac{1}{2}\left[\frac{f(a)+f(b)}{2} - \frac{1}{b-a}\int_a^b f(x)\,\mathrm{d}x\right]. \end{aligned}$$

In [381] Pečarić and Dragomir and in [87] Dragomir obtain a generalization and refinement of Hadamard's inequality for isotonic linear functional. In the following theorems we give the results of [87,381]. We need the following lemmas given in [87,381].

LEMMA 1.5.1. *Let X be a linear space and C be its convex subset. Then the following statements are equivalent for a mapping $f: X \to \mathbb{R}$*

(i) *f is convex on C and*
(ii) *for all $x, y \in C$, the mapping $g_{x,y}: [0, 1] \to \mathbb{R}$, $g_{x,y} = f(tx + (1-t)y)$ is convex on* $[0, 1]$.

PROOF. (i) ⇒ (ii). Suppose $x, y \in C$ and let $t_1, t_2 \in [0, 1]$, $\lambda_1, \lambda_2 \geqslant 0$ with $\lambda_1 + \lambda_2 = 1$. Then

$$\begin{aligned} g_{x,y}(\lambda_1 t_1 + \lambda_2 t_2) &= f\big((\lambda_1 t_1 + \lambda_2 t_2)x + (1 - \lambda_1 t_1 - \lambda_2 t_2)y\big) \\ &= f\big((\lambda_1 t_1 + \lambda_2 t_2)x + \big(\lambda_1(1-t_1) + \lambda_2(1-t_2)\big)y\big) \\ &\leqslant \lambda_1 f\big(t_1 x + (1-t_1)y\big) + \lambda_2 f\big(t_2 x + (1-t_2)y\big) \\ &= \lambda_1 g_{x,y}(t_1) + \lambda_2 g_{x,y}(t_2), \end{aligned}$$

that is, $g_{x,y}$ is convex on $[0, 1]$.

(ii) ⇒ (i). Let $x, y \in C$ and $\lambda_1, \lambda_2 \geqslant 0$ with $\lambda_1 + \lambda_2 = 1$. Then

$$\begin{aligned} f(\lambda_1 x + \lambda_2 y) &= f\big(\lambda_1 x + (1-\lambda_1)y\big) \\ &= g_{x,y}(\lambda_2 \cdot 1 + \lambda_2 \cdot 0) \end{aligned}$$

$$\leqslant \lambda_1 g_{x,y}(1) + \lambda_2 g_{x,y}(0)$$
$$= \lambda_1 f(x) + \lambda_2 f(y),$$

that is, f is convex on C and the lemma is proved. □

LEMMA 1.5.2. *Let X be a real linear space and C be a convex subset. If $f: C \to \mathbb{R}$ is convex on C, then for all x, y in C, the mapping $g_{x,y}: [0,1] \to \mathbb{R}$ given by*

$$g_{x,y}(t) = \frac{1}{2}\big[f\big(tx + (1-t)y\big) + f\big((1-t)x + ty\big)\big]$$

is also convex on $[0,1]$. In addition, we have the inequality

$$f\left(\frac{x+y}{2}\right) \leqslant g_{x,y}(t) \leqslant \frac{f(a) + f(b)}{2}$$

for all x, y in C and $t \in [0,1]$.

PROOF. Suppose $x, y \in C$ and let $t_1, t_2 \in [0,1]$ and $\alpha, \beta \geqslant 0$ with $\alpha + \beta = 1$. Then

$$\begin{aligned}
g_{x,y}(\alpha t_1 + \beta t_2) &= \frac{1}{2}\big[f\big((\alpha t_1 + \beta t_2)x + (1 - \alpha t_1 - \beta t_2)y\big) \\
&\quad + f\big((1 - \alpha t_1 - \beta t_2)x + (\alpha t_1 + \beta t_2)y\big)\big] \\
&= \frac{1}{2}\big[f\big(\alpha\big(t_1 x + (1-t_1)y\big) + \beta\big(t_2 x + (1-t_2)y\big)\big) \\
&\quad + f\big(\alpha\big((1-t_1)x + t_1 y\big) + \beta\big((1-t_2)x + t_2 y\big)\big)\big] \\
&\leqslant \frac{1}{2}\big[\alpha f\big(t_1 x + (1-t_1)y\big) + \beta f\big(t_2 x + (1-t_2)y\big) \\
&\quad + \alpha f\big((1-t_1)x + t_1 y\big) + \beta f\big((1-t_2)x + t_2 y\big)\big] \\
&= \alpha g_{x,y}(t_1) + \beta g_{x,y}(t_2),
\end{aligned}$$

which shows that $g_{x,y}$ is convex on $[0,1]$.

By the convexity of f we can state

$$g_{x,y}(t) \geqslant f\left(\frac{1}{2}\big(tx + (1-t)y + (1-t)x + ty\big)\right) = f\left(\frac{x+y}{2}\right)$$

and also

$$g_{x,y}(t) \leqslant \frac{1}{2}\big[tf(x) + (1-t)f(y) + (1-t)f(x) + tf(y)\big] = \frac{f(x) + f(y)}{2},$$

for all t in $[0, 1]$, which completes the proof. □

In [381] Pečarić and Dragomir have given the following generalization of Hadamard's inequality for isotonic linear functionals.

THEOREM 1.5.7. *Let $f : C \subseteq X \to \mathbb{R}$ be a convex function on C, L and A satisfy conditions* (L_1), (L_2) *and* (A_1), (A_2) *in Section* 1.3, *and $h : E \to \mathbb{R}$, $0 \leqslant h(t) \leqslant 1$, $h \in L$ is such that $g_{x,y} \circ h \in L$ for x, y given in C, where E be a nonempty set. If $A(1) = 1$, then we have the inequality*

$$\begin{aligned} f\big(A(h)x + \big(1 - A(h)\big)y\big) &\leqslant A\big[f\big(hx + (1-h)y\big)\big] \\ &\leqslant A(h)f(x) + \big(1 - A(h)\big)f(y). \qquad (1.5.21) \end{aligned}$$

PROOF. Consider the mapping $g_{x,y} : [0, 1] \to \mathbb{R}$, $g_{x,y}(s) = f(sx + (1-s)y)$. Then, by the Lemma 1.5.1, we have $g_{x,y}$ is convex on $[0, 1]$. For all $t \in E$, we have

$$g_{x,y}\big(h(t) \cdot 1 + \big(1 - h(t)\big) \cdot 0\big) \leqslant h(t)g_{x,y}(1) + \big(1 - h(t)\big)g_{x,y}(0)$$

which implies

$$A\big(g_{x,y}(h)\big) \leqslant A(h)g_{x,y}(1) + \big(1 - A(h)\big)g_{x,y}(0),$$

that is,

$$A\big[f\big(hx + (1-h)y\big)\big] \leqslant A(h)f(x) + \big(1 - A(h)\big)f(y).$$

On the other hand, by using Jessen's inequality in Theorem 1.3.1 to $g_{x,y}$, we have

$$g_{x,y}\big(A(h)\big) \leqslant A\big(g_{x,y}(h)\big),$$

which gives

$$f\big(A(h)x + \big(1 - A(h)\big)y\big) \leqslant A\big[f\big(hx + (1-h)y\big)\big],$$

and the proof is complete. □

In [87] Dragomir has given the following refinement of Hadamard's inequality for isotonic linear functionals.

THEOREM 1.5.8. *Let $f : C \subseteq X \to \mathbb{R}$ be a convex function on convex set C, L and A satisfy conditions* (L_1), (L_2) *and* (A_1), (A_2) *in Section* 1.3, *and $h : E \to \mathbb{R}$, $0 \leqslant h(t) \leqslant 1$, $t \in E$, $h \in L$ is such that $f(hx + (1-h)y)$ and*

$f((1-h)x+hy)$ belong to L for x, y fixed in C. If $A(1)=1$, then we have the inequality

$$\begin{aligned}
f\left(\frac{x+y}{2}\right) &\leqslant \frac{1}{2}\big[f\big(A(h)x+\big(1-A(h)\big)y\big)+f\big(\big(1-A(h)\big)x+A(h)y\big)\big] \\
&\leqslant \frac{1}{2}\big(A\big[f\big(hx+(1-h)y\big)\big]+A\big[f\big((1-h)x+hy\big)\big]\big) \\
&\leqslant \frac{f(x)+f(y)}{2}.
\end{aligned} \tag{1.5.22}$$

PROOF. Consider the mapping $g_{x,y}:[0,1]\to\mathbb{R}$ given in Lemma 1.5.2. Then $g_{x,y}$ is convex on $[0,1]$. Applying Jessen's inequality in Theorem 1.3.1 for the mapping $g_{x,y}$, we get

$$g_{x,y}\big(A(h)\big)\leqslant A\big(g_{x,y}(h)\big).$$

But

$$g_{x,y}\big(A(h)\big)=\frac{1}{2}\big[f\big(A(h)x+\big(1-A(h)\big)y\big)+f\big(\big(1-A(h)\big)x+A(h)y\big)\big]$$

and

$$A\big(g_{x,y}(h)\big)=\frac{1}{2}\big(A\big[f\big(hx+(1-h)y\big)\big]+A\big[f\big((1-h)x+hy\big)\big]\big),$$

and the second inequality in (1.5.22) is proved.

To prove the first inequality in (1.5.22), we observe that by Lemma 1.5.2

$$f\left(\frac{x+y}{2}\right)\leqslant g_{x,y}\big(A(h)\big)$$

which is the desired statement.

Finally, we observe that by the convexity of f, we get

$$\frac{1}{2}\big[f\big(hx+(1-h)y\big)+f\big((1-h)x+hy\big)\big]\leqslant\frac{f(x)+f(y)}{2}\quad\text{on } E.$$

Applying to this inequality the functional A and since $A(1)=1$, we obtain the last part of (1.5.22). □

1.6 Inequalities of Hadamard Type I

During the past few years several papers have appeared which deals with Hadamard-type inequalities involving various classes of functions. In this section, we offer some basic inequalities of Hadamard type that have recently been published.

In 1985, Godunova and Levin [130] introduce the following class of functions.

A mapping $f : I \to \mathbb{R}$ is said to belong to the class $Q(I)$ if it is nonnegative and, for all $x, y \in I$ and $\lambda \in (0, 1)$, satisfies the inequality

$$f\big(\lambda x + (1-\lambda)y\big) \leqslant \frac{f(x)}{\lambda} + \frac{f(y)}{1-\lambda}, \tag{1.6.1}$$

where I is an interval from $\mathbb{R}$.

In [130] it is noted that all nonnegative monotone and nonnegative convex functions belong to this class and also proves the following result:

If $f \in Q(I)$ and $x, y, z \in I$, then

$$f(x)(x-y)(x-z) + f(y)(y-x)(y-z) + f(z)(z-x)(z-y) \geqslant 0. \tag{1.6.1$'$}$$

In fact, (1.6.1′) is equivalent to (1.6.1) so it can be used alternatively in the definition of the class $Q(I)$. For the case $f(x) = x^r$, $r \in \mathbb{R}$, inequality (1.6.1′) obviously coincides with the well-known Schur inequality.

The following result deals with an inequality of Hadamard type recently established in [93] for a class of functions $Q(I)$.

THEOREM 1.6.1. *Let $f \in Q(I)$, $a, b \in I$, with $a < b$ and $f \in L_1[a, b]$. Then*

$$f\left(\frac{a+b}{2}\right) \leqslant \frac{4}{b-a}\int_a^b f(x)\,\mathrm{d}x \tag{1.6.2}$$

and

$$\frac{1}{b-a}\int_a^b p(x) f(x)\,\mathrm{d}x \leqslant \frac{f(a)+f(b)}{2}, \tag{1.6.3}$$

where

$$p(x) = \frac{(b-x)(x-a)}{(b-a)^2}, \qquad x \in I.$$

The constant equal to 4 *in* (1.6.2) *is the best possible.*

PROOF. Since $f \in Q(I)$, we have for all $x, y \in I$ (with $\lambda = 1/2$ in (1.6.1)),

$$2\big(f(x) + f(y)\big) \geqslant f\left(\frac{x+y}{2}\right),$$

that is, with $x = ta + (1-t)b$, $y = (1-t)a + tb$, $t \in [0, 1]$,

$$2\big(f\big(ta + (1-t)b\big) + f\big((1-t)a + tb\big)\big) \geqslant f\left(\frac{a+b}{2}\right).$$

Integrating the above inequality over $[0, 1]$ we have

$$2\left(\int_0^1 f\big(ta + (1-t)b\big)\,\mathrm{d}t + \int_0^1 f\big((1-t)a + tb\big)\,\mathrm{d}t\right) \geqslant f\left(\frac{a+b}{2}\right). \tag{1.6.4}$$

Since

$$\int_0^1 f\big(ta + (1-t)b\big)\,\mathrm{d}t = \int_0^1 f\big((1-t)a + tb\big)\,\mathrm{d}t = \frac{1}{b-a}\int_a^b f(x)\,\mathrm{d}x,$$

we get inequality (1.6.2) from (1.6.4).

For the proof of (1.6.3), we first note that if $f \in Q(I)$ then, for all $a, b \in I$ and $\lambda \in [0, 1]$, it yields

$$\lambda(1-\lambda) f\big(\lambda a + (1-\lambda)b\big) \leqslant (1-\lambda) f(a) + \lambda f(b)$$

and

$$\lambda(1-\lambda) f\big((1-\lambda)a + \lambda b\big) \leqslant \lambda f(a) + (1-\lambda) f(b).$$

By adding these inequalities and integrating the resulting inequality on $[0, 1]$, we find that

$$\begin{aligned}&\int_0^1 \lambda(1-\lambda)\big(f\big(\lambda a + (1-\lambda)b\big) + f\big((1-\lambda)a + \lambda b\big)\big)\,\mathrm{d}\lambda \\ &\quad \leqslant f(a) + f(b).\end{aligned} \tag{1.6.5}$$

Moreover,

$$\begin{aligned}&\int_0^1 \lambda(1-\lambda) f\big(\lambda a + (1-\lambda)b\big)\,\mathrm{d}\lambda \\ &\quad = \int_0^1 \lambda(1-\lambda) f\big((1-\lambda)a + \lambda b\big)\,\mathrm{d}\lambda \\ &\quad = \frac{1}{b-a}\int_a^b \frac{(b-x)(x-a)}{(b-a)^2} f(x)\,\mathrm{d}x.\end{aligned} \tag{1.6.6}$$

We get (1.6.3) by combining (1.6.5) with (1.6.6).

The constant equal to 4 in (1.6.2) is the best possible because this inequality reduces to an equality for the function

$$f(x) = \begin{cases} 1, & a \leqslant x < \frac{a+b}{2}, \\ 4, & x = \frac{a+b}{2}, \\ 1, & \frac{a+b}{2} < x \leqslant b. \end{cases}$$

Moreover, this function is of the class $Q(I)$ because

$$\begin{aligned} \frac{f(x)}{\lambda} + \frac{f(y)}{1-\lambda} &\geqslant \frac{1}{\lambda} + \frac{1}{1-\lambda} \\ &= g(\lambda) \geqslant \min_{0<\lambda<1} g(\lambda) \\ &= g\left(\frac{1}{2}\right) = 4 \geqslant f\big(\lambda x + (1-\lambda)y\big). \end{aligned}$$

The proof is complete. □

In [93] the following class of functions is introduced.

A mapping $f : I \to \mathbb{R}$, belongs to the class $P(I)$ (I is an interval from $\mathbb{R}$) if it is nonnegative and, for all $x, y \in I$ and $\lambda \in [0, 1]$, satisfies the inequality

$$f\big(\lambda x + (1-\lambda)y\big) \leqslant f(x) + f(y). \tag{1.6.7}$$

Obviously, $Q(I) \supset P(I)$ and for applications it is important to note also that $P(I)$ contains all monotone, convex and quasi-convex functions, that is, functions satisfying $f(\lambda x + (1-\lambda)y) \leqslant \max(f(x), f(y))$.

In [93] the following version of Hadamard's inequality in the restricted class of functions is given.

THEOREM 1.6.2. *Let $f \in P(I)$, $a, b \in I$, with $a < b$ and $f \in L_1[a, b]$. Then*

$$f\left(\frac{a+b}{2}\right) \leqslant \frac{2}{b-a}\int_a^b f(x)\,\mathrm{d}x \leqslant 2\big(f(a) + f(b)\big). \tag{1.6.8}$$

Both inequalities are the best possible.

PROOF. According to (1.6.7) with $x = ta + (1-t)b$, $y = (1-t)a + tb$, $t \in [0, 1]$, and $\lambda = 1/2$ we find that

$$f\left(\frac{a+b}{2}\right) \leqslant f\big(ta + (1-t)b\big) + f\big((1-t)a + tb\big).$$

Integrating the above inequality with respect to t over $[0, 1]$, we obtain

$$f\left(\frac{a+b}{2}\right) \leqslant \int_0^1 \big(f\big(ta+(1-t)b\big)+f\big((1-t)a+tb\big)\big)\,\mathrm{d}t$$
$$= \frac{2}{b-a}\int_a^b f(x)\,\mathrm{d}x,$$

and the first inequality is proved. The proof of the second inequality follows by using (1.6.7) with $x=a$, $y=b$ and integrating with respect to λ over $[0, 1]$.

The first inequality in (1.6.8) reduces to an equality for the (nondecreasing) function

$$f(x) = \begin{cases} 0, & a \leqslant x < \frac{a+b}{2}, \\ 1, & \frac{a+b}{2} \leqslant x \leqslant b, \end{cases}$$

and the second inequality reduces to an equality for the (nondecreasing) function

$$f(x) = \begin{cases} 0, & x = a, \\ 1, & a < x \leqslant b. \end{cases}$$

The proof is complete. □

In [128] versions of the upper Hadamard inequality are developed for r-convex and r-concave functions.

Recall that a positive function f is log-convex on a real interval $[a,b]$ if, for all x, $y \in [a,b]$ and $\lambda \in [0, 1]$, we have

$$f\big(\lambda x+(1-\lambda)y\big) \leqslant f(x)^{\lambda} f(y)^{1-\lambda}. \tag{1.6.9}$$

If the reverse inequality holds, f is termed log-concave.

Also the power mean $M_r(x, y; \lambda)$ of order r of positive numbers x, y is denoted by

$$M_r(x, y; \lambda) = \begin{cases} \big(\lambda x^r+(1-\lambda)y^r\big)^{1/r} & \text{if } r \neq 0, \\ x^{\lambda} y^{1-\lambda} & \text{if } r = 0. \end{cases}$$

In the special case $\lambda = 1/2$, we contract this notation to $M_r(x, y)$.

In view of the above, a natural generalizing concept is that of r-convexity. A positive function f is r-convex on $[a,b]$ if, for all x, $y \in [a,b]$ and $\lambda \in [0, 1]$,

$$f\big(\lambda x+(1-\lambda)y\big) \leqslant M_r\big(f(x), f(y); \lambda\big). \tag{1.6.10}$$

The definition of r-convexity naturally complements the concept of r-concavity, in which the inequality is reversed. We have that 0-convex functions are simply log-convex functions and 1-convex functions are ordinary convex functions. For the latter the requirement that r-convex function be positive can clearly be relaxed.

Again, in all the above, we may take a real linear space X in place of the real line. The condition $x, y \in [a, b]$ then becomes $x, y \in U$ for U a convex set in X. In [128] the authors have developed Hadamard-type inequalities for log-convex functions and more generally for r-convex functions. It is convenient to separate the proof of the former special case as the functional representations differ in detail from those of the general case.

It will be convenient to invoke the logarithmic mean $L(x, y)$ of two positive numbers x, y, which is given by

$$L(x, y) = \begin{cases} \frac{x-y}{\ln x - \ln y}, & x \neq y, \\ x, & x = y, \end{cases}$$

and the generalized logarithmic means of order r of positive numbers x, y, defined by

$$F_r(x, y) = \begin{cases} \frac{r}{r+1} \frac{x^{r+1} - y^{r+1}}{x^r - y^r}, & r \neq 0, -1, x \neq y, \\ \frac{x-y}{\ln x - \ln y}, & r = 0, x \neq y, \\ xy \frac{\ln x - \ln y}{x-y}, & r = -1, x \neq y, \\ x, & x = y \end{cases}$$

(see [194]).

The following theorem proved in [128] deals with a version of the upper Hadamard inequality for log-convex functions.

THEOREM 1.6.3. *Let f be a positive, log-convex function on $[a, b]$. Then*

$$\frac{1}{b-a} \int_a^b f(t)\,\mathrm{d}t \leqslant L\big(f(a), f(b)\big).$$

For f a positive log-concave function, the inequality is reversed.

PROOF. First suppose that $f(a) \neq f(b)$. By (1.6.9) we have

$$\begin{aligned} \int_a^b f(t)\,\mathrm{d}t &= (b-a) \int_0^1 f\big(sb + (1-s)a\big)\,\mathrm{d}s \\ &\leqslant (b-a) \int_0^1 f(b)^s f(a)^{1-s}\,\mathrm{d}s \\ &= (b-a) f(a) \int_0^1 \left\{\frac{f(b)}{f(a)}\right\}^s \mathrm{d}s \end{aligned}$$

$$= (b-a)f(a)\left[\left(\frac{f(b)}{f(a)}\right)^s \Big/ \ln\left(\frac{f(b)}{f(a)}\right)\right]_0^1$$
$$= (b-a)\frac{f(b)-f(a)}{\ln f(b)-\ln f(a)}$$
$$= (b-a)L\big(f(a), f(b)\big).$$

For $f(a) = f(b)$, we have with the same development

$$\begin{aligned}\int_a^b f(t)\,\mathrm{d}t &= (b-a)\int_0^1 f\big(sb+(1-s)a\big)\,\mathrm{d}s \\ &\leqslant (b-a)\int_0^1 f(b)^s f(a)^{1-s}\,\mathrm{d}s \\ &= (b-a)\int_0^1 f(a)\,\mathrm{d}s \\ &= (b-a)f(a) \\ &= (b-a)L\big(f(a), f(b)\big).\end{aligned}$$

The proof is complete. □

A similar proof gives the following generalization established in [128].

THEOREM 1.6.4. *Let f be a positive, log-convex function on a convex set $U \subset X$, where X is a linear vector space. Then for $a, b \in U$,*

$$\int_0^1 f\big(sa+(1-s)b\big)\,\mathrm{d}s \leqslant L\big(f(a), f(b)\big).$$

The following inequalities may be derived by way of corollaries to Theorems 1.6.3 and 1.6.4.

COROLLARY 1.6.1. *Let f be a positive log-convex function on $[a, b]$. Then*

$$\begin{aligned}&\frac{1}{b-a}\int_a^b f(t)\,\mathrm{d}t \\ &\quad\leqslant \min_{x\in[a,b]} \frac{(x-a)L(f(a), f(x)) + (b-x)L(f(x), f(b))}{b-a}. \qquad (1.6.11)\end{aligned}$$

If f is a positive log-concave function, then

$$\frac{1}{b-a}\int_a^b f(t)\,\mathrm{d}t \geqslant \max_{x\in[a,b]} \frac{(x-a)L(f(a),f(x))+(b-x)L(f(x),f(b))}{b-a}. \tag{1.6.12}$$

PROOF. Let f be a positive log-convex function. Then by Theorem 1.6.3, we have that

$$\int_a^b f(t)\,\mathrm{d}t = \int_a^x f(t)\,\mathrm{d}t + \int_x^b f(t)\,\mathrm{d}t \leqslant (x-a)L\big(f(a),f(x)\big)+(b-x)L\big(f(x),f(b)\big)$$

for all $x \in [a,b]$, whence (1.6.11). Similarly we can prove (1.6.12). □

COROLLARY 1.6.2. *Let f be a positive log-convex function on $[a,b]$. Then*

$$\frac{1}{b-a}\int_a^b f(t)\,\mathrm{d}t \leqslant \frac{1}{n}\sum_{i=1}^{n} L\left(f\left(a+\frac{i-1}{n}(b-a)\right), f\left(a+\frac{i}{n}(b-a)\right)\right).$$

If f is a positive log-concave function, then inequality is reversed.

PROOF. The result follows by applying Theorem 1.6.3 to the integrals on the right-hand side in

$$\int_a^b f(t)\,\mathrm{d}t = \sum_{i=1}^{n}\int_{a+(i-1)(b-a)/n}^{a+i(b-a)/n} f(t)\,\mathrm{d}t.$$

□

COROLLARY 1.6.3.

(a) *Let f be a positive log-convex function on $[a,b]$. Then*

$$\frac{1}{b-a}\int_a^b f(t)\,\mathrm{d}t \leqslant M_{1/3}\big(f(a),f(b)\big),$$

while if f is log-concave, then

$$\frac{1}{b-a}\int_a^b f(t)\,\mathrm{d}t \geqslant \sqrt{f(a)f(b)}. \tag{1.6.13}$$

(b) *Let f be a positive, log-convex function on a convex set $U \subset X$, where X is a linear vector space. Then for $a, b \in U$,*

$$\int_0^1 f\big(sa + (1-s)b\big)\,\mathrm{d}s \leqslant M_{1/3}\big(f(a), f(b)\big),$$

while if f is log-concave, then

$$\int_0^1 f\big(sa + (1-s)b\big)\,\mathrm{d}s \geqslant \sqrt{f(a)f(b)}.$$

PROOF. Part (a) follows from Theorem 1.6.3 and the inequalities

$$G(a,b) \leqslant L(a,b) \leqslant M_{1/3}(a,b),$$

where $L(a,b)$, $M_r(a,b)$ are as defined above and $G(a,b)$ is the geometric mean (see also [194]). Part (b) follows similarly from Theorem 1.6.4. □

In [128] the following Hadamard-type inequality for r-convex functions is proved.

THEOREM 1.6.5. *Suppose f is a positive r-convex function on $[a,b]$. Then*

$$\frac{1}{b-a}\int_a^b f(t)\,\mathrm{d}t \leqslant F_r\big(f(a), f(b)\big).$$

If f is a positive r-concave function, then the inequality is reversed.

PROOF. The case $r = 0$ has been dealt with Theorem 1.6.3. Suppose that $r \neq 0, -1$. First assume that $f(a) \neq f(b)$. By (1.6.10) we have

$$\begin{aligned}
\int_a^b f(t)\,\mathrm{d}t &= (b-a)\int_0^1 f\big(sb + (1-s)a\big)\,\mathrm{d}s \\
&\leqslant (b-a)\int_0^1 \big\{sf^r(b) + (1-s)f^r(a)\big\}^{1/r}\,\mathrm{d}s \\
&= (b-a)\int_{f^r(a)}^{f^r(b)} \frac{t^{1/r}\,\mathrm{d}t}{f^r(b) - f^r(a)} \\
&= (b-a)\frac{r}{r+1}\frac{f^{r+1}(b) - f^{r+1}(a)}{f^r(b) - f^r(a)} \\
&= (b-a)F_r\big(f(a), f(b)\big).
\end{aligned}$$

For $f(a) = f(b)$, we have similarly

$$\begin{aligned}\int_a^b f(t)\,\mathrm{d}t &\leqslant (b-a)\int_0^1 \{sf^r(a) + (1-s)f^r(a)\}^{1/r}\,\mathrm{d}s \\ &= (b-a)f(a) \\ &= (b-a)F_r\big(f(a), f(a)\big).\end{aligned}$$

Finally, let $r = -1$. For $f(a) \neq f(b)$, we have again

$$\begin{aligned}\int_a^b f(t)\,\mathrm{d}t &\leqslant (b-a)\int_0^1 \{sf^{-1}(b) + (1-s)f^{-1}(a)\}^{-1}\,\mathrm{d}s \\ &= \frac{b-a}{1/f(b) - 1/f(a)}\int_{1/f(a)}^{1/f(b)} t^{-1}\,\mathrm{d}t \\ &= \frac{b-a}{1/f(b) - 1/f(a)}\left(\ln\frac{1}{f(b)} - \ln\frac{1}{f(a)}\right) \\ &= (b-a)f(a)f(b)\frac{\ln f(a) - \ln f(b)}{f(a) - f(b)} \\ &= (b-a)F_{-1}\big(f(a), f(b)\big).\end{aligned}$$

When $f(a) = f(b)$ the proof is similar. □

In concluding this section, we note that in [365] the authors have obtained some generalizations of the extensions of Hadamard's inequality to r-convex functions involving Stolarsky means. For inequalities of Hadamard's type, see also [118, 220].

1.7 Inequalities of Hadamard Type II

The classical inequality of the Hadamard type has been generalized and extended in several directions. In this section we shall give some results on refinements and variants of Hadamard's inequality given by various investigators.

In 1992, Dragomir [84] has obtained some refinements of Hadamard's inequalities (1.5.1) by considering the following two mappings involving convex functions

$$H(t) = \frac{1}{b-a}\int_a^b f\left(tx + (1-t)\frac{a+b}{2}\right)\mathrm{d}x \tag{1.7.1}$$

and

$$F(t) = \frac{1}{(b-a)^2} \int_a^b \int_a^b f\bigl(tx + (1-t)y\bigr)\,\mathrm{d}x\,\mathrm{d}y, \tag{1.7.2}$$

where $f:[a,b] \to \mathbb{R}$ is a convex function and H and F are real-valued functions defined on $[a,b]$.

The main results given in [84] are given in the following two theorems.

THEOREM 1.7.1. *Let $f:[a,b] \to \mathbb{R}$ be a convex function. Then*

(i) *H is convex on* $[0, 1]$.

(ii) *We have*

$$\inf_{t\in[0,1]} H(t) = H(0) = f\left(\frac{a+b}{2}\right)$$

and

$$\sup_{t\in[0,1]} H(t) = H(1) = \frac{1}{b-a}\int_a^b f(x)\,\mathrm{d}x.$$

(iii) *H increases monotonically on* $[0, 1]$.

PROOF. (i) Let $\alpha, \beta \geqslant 0$ with $\alpha + \beta = 1$ and $t_1, t_2 \in [0, 1]$. Then

$$\begin{aligned}
&H(\alpha t_1 + \beta t_2) \\
&\quad = \frac{1}{b-a}\int_a^b f\left(\alpha\left(t_1 x + (1-t_1)\frac{a+b}{2}\right) + \beta\left(t_2 x + (1-t_2)\frac{a+b}{2}\right)\right)\mathrm{d}x \\
&\quad \leqslant \alpha\frac{1}{b-a}\int_a^b f\left(t_1 x + (1-t_1)\frac{a+b}{2}\right)\mathrm{d}x \\
&\qquad + \beta\frac{1}{b-a}\int_a^b f\left(t_2 x + (1-t_2)\frac{a+b}{2}\right)\mathrm{d}x \\
&\quad = \alpha H(t_1) + \beta H(t_2)
\end{aligned}$$

which shows that H is convex in $[0, 1]$.

(ii) We shall prove the following inequalities

$$\begin{aligned}
f\left(\frac{a+b}{2}\right) &\leqslant H(t) \\
&\leqslant t\frac{1}{b-a}\int_a^b f(x)\,\mathrm{d}x + (1-t)f\left(\frac{a+b}{2}\right)
\end{aligned}$$

$$\leqslant \frac{1}{b-a}\int_a^b f(x)\,\mathrm{d}x \tag{1.7.3}$$

for all t in $[0, 1]$.

By Jensen's integral inequality, we have

$$H(t) \geqslant f\left(\frac{1}{b-a}\int_a^b \left[tx + (1-t)\frac{a+b}{2}\right]\mathrm{d}x\right)$$
$$= f\left(\frac{a+b}{2}\right).$$

Now, using the convexity of f, we get

$$H(t) \leqslant \frac{1}{b-a}\int_a^b \left[tf(x) + (1-t)f\left(\frac{a+b}{2}\right)\right]\mathrm{d}x$$
$$= t\frac{1}{b-a}\int_a^b f(x)\,\mathrm{d}x + (1-t)f\left(\frac{a+b}{2}\right),$$

and the second inequality in (1.7.3) is proven.

The last inequality is obvious because the mapping

$$g(t) = t\frac{1}{b-a}\int_a^b f(x)\,\mathrm{d}x + (1-t)f\left(\frac{a+b}{2}\right)$$

is increasing monotonically on $[0, 1]$.

(iii) Let $t_1, t_2 \in (0, 1)$ with $t_2 > t_1$. Then, H being convex on $(0, 1)$,

$$\frac{H(t_2) - H(t_1)}{t_2 - t_1} \geqslant H'_+(t_1)$$
$$= \frac{1}{b-a}\int_a^b f'_+\left(t_1 x + (1-t_1)\frac{a+b}{2}\right)\left(x - \frac{a+b}{2}\right)\mathrm{d}x.$$

Since f is convex on $[a, b]$, we deduce that

$$f\left(\frac{a+b}{2}\right) - f\left(t_1 x + (1-t_1)\frac{a+b}{2}\right)$$
$$\geqslant t_1 f'_+\left(t_1 x + (1-t_1)\frac{a+b}{2}\right)\left(\frac{a+b}{2} - x\right)$$

for every x in $[a, b]$. Thus

$$\begin{aligned}
&\frac{1}{b-a}\int_a^b f'_+\left(t_1x+(1-t_1)\frac{a+b}{2}\right)\left(x-\frac{a+b}{2}\right)\mathrm{d}x\\
&\geqslant \frac{1}{t_1}\left[\frac{1}{b-a}\int_a^b f\left(t_1x+(1-t_1)\frac{a+b}{2}\right)\mathrm{d}x - f\left(\frac{a+b}{2}\right)\right]\\
&=\frac{1}{t_1}\left[H(t_1)-f\left(\frac{a+b}{2}\right)\right]\geqslant 0.
\end{aligned}$$

Consequently $H(t_2) - H(t_1) \geqslant 0$ for $1 \geqslant t_2 > t_1 \geqslant 0$ which shows that H increases monotonically on $[0, 1]$. The proof is complete. □

THEOREM 1.7.2. *Let $f:[a,b]\to\mathbb{R}$ be a convex function. Then*

(i) $F(\sigma+\frac{1}{2}) = F(\frac{1}{2}-\sigma)$ *for all* $\sigma \in [0, \frac{1}{2}]$.
(ii) F *is convex on* $[0, 1]$.
(iii) *We have*

$$\sup_{t\in[0,1]} F(t) = F(0) = F(1) = \frac{1}{b-a}\int_a^b f(x)\,\mathrm{d}x$$

and

$$\inf_{t\in[0,1]} F(t) = F\left(\frac{1}{2}\right) = \frac{1}{(b-a)^2}\int_a^b\int_a^b f\left(\frac{x+y}{2}\right)\mathrm{d}x\,\mathrm{d}y.$$

(iv) *The following inequality is valid*

$$f\left(\frac{a+b}{2}\right) \leqslant F\left(\frac{1}{2}\right).$$

(v) F *decreases monotonically on* $[0, \frac{1}{2}]$ *and increases monotonically on* $[\frac{1}{2}, 1]$.
(vi) *We have the inequality*

$$H(t) \leqslant F(t) \quad \textit{for all } t \in [0, 1].$$

PROOF. (i) Let $\sigma \in [0, \frac{1}{2}]$. We have

$$F\left(\sigma+\frac{1}{2}\right) = \frac{1}{(b-a)^2}\int_a^b\int_a^b f\left(\left(\sigma+\frac{1}{2}\right)x + \left(\frac{1}{2}-\sigma\right)y\right)\mathrm{d}x\,\mathrm{d}y$$

$$= \frac{1}{(b-a)^2} \int_a^b \int_a^b f\left(\left(\frac{1}{2} - \sigma\right)x + \left(\sigma + \frac{1}{2}\right)y\right) \mathrm{d}x\,\mathrm{d}y$$

$$= F\left(\frac{1}{2} - \sigma\right).$$

(ii) The argument is similar to that in the proof of Theorem 1.7.1(i).

(iii) For all x, y in $[a, b]$ and t in $(0, 1]$, we have

$$f\big(tx + (1-t)y\big) \leqslant tf(x) + (1-t)f(y).$$

Integrating this inequality in $[a, b] \times [a, b]$ we get

$$\int_a^b \int_a^b f\big(tx + (1-t)y\big) \mathrm{d}x\,\mathrm{d}y \leqslant \int_a^b \int_a^b \big[tf(x) + (1-t)f(y)\big] \mathrm{d}x\,\mathrm{d}y$$
$$= (b-a) \int_a^b f(x)\,\mathrm{d}x$$

which shows that $F(t) \leqslant F(0) = F(1)$ for all t in $[0, 1]$. Since f is convex on $[a, b]$ for all $t \in [0, 1]$ and x, y in $[a, b]$, we have

$$\frac{1}{2}\big[f\big(tx + (1-t)y\big) + f\big(ty + (1-t)x\big)\big] \geqslant f\left(\frac{x+y}{2}\right).$$

Integrating this inequality in $[a, b] \times [a, b]$ we deduce

$$\int_a^b \int_a^b f\left(\frac{x+y}{2}\right) \mathrm{d}x\,\mathrm{d}y$$
$$\leqslant \frac{1}{2} \int_a^b \int_a^b \big[f\big(tx + (1-t)y\big) + f\big(ty + (1-t)x\big)\big] \mathrm{d}x\,\mathrm{d}y$$
$$= \int_a^b \int_a^b f\big(tx + (1-t)y\big) \mathrm{d}x\,\mathrm{d}y$$

which implies that $F(1/2) \leqslant F(t)$ for all t in $[0, 1]$, and the statement is proven.

(iv) Using Jensen's inequality for double integrals we have

$$\frac{1}{(b-a)^2} \int_a^b \int_a^b f\left(\frac{x+y}{2}\right) \mathrm{d}x\,\mathrm{d}y \geqslant f\left(\frac{1}{(b-a)^2} \int_a^b \int_a^b \left(\frac{x+y}{2}\right) \mathrm{d}x\,\mathrm{d}y\right).$$

Since a simple computation shows that

$$\frac{1}{(b-a)^2} \int_a^b \int_a^b \left(\frac{x+y}{2}\right) \mathrm{d}x\,\mathrm{d}y = \frac{a+b}{2},$$

the proof is complete.

(v) Since f is convex on $(0, 1)$, we have for $t_2 > t_1$, $t_1, t_2 \in (\frac{1}{2}, 1)$,

$$\frac{F(t_2) - F(t_1)}{t_2 - t_1} \geqslant F'_+(t_1) = \frac{1}{(b-a)^2} \int_a^b \int_a^b f'_+\big(t_1 x + (1-t_1)y\big)(x-y)\,\mathrm{d}x\,\mathrm{d}y.$$

By the convexity of f on $[a, b]$, we deduce

$$f\left(\frac{x+y}{2}\right) - f\big(t_1 x + (1-t_1)y\big) \geqslant f'_+\big(t_1 x + (1-t_1)y\big)\frac{(x-y)(1-2t_1)}{2}$$

for all x, y in $[a, b]$ and $t_1 \in (\frac{1}{2}, 1)$, which is equivalent to

$$(x-y) f'_+\big(t_1 x + (1-t_1)y\big) \geqslant \frac{2}{2t_1 - 1}\left[f\big(t_1 x + (1-t_1)y\big) - f\left(\frac{x+y}{2}\right)\right].$$

Integrating on $[a, b] \times [a, b]$ we obtain

$$F'_+(t_1) \geqslant \frac{2}{2t_1 - 1}\left(F(t_1) - F\left(\frac{1}{2}\right)\right) \geqslant 0, \quad t_1 \in \left(\frac{1}{2}, 1\right),$$

which shows that F increases monotonically on $[\frac{1}{2}, 1]$.

The fact that F increases monotonically on $[0, \frac{1}{2}]$ follows from the above conclusion using statement (i).

(vi) A simple computation shows that

$$H(t) = \frac{1}{b-a} \int_a^b f\left(\frac{1}{b-a} \int_a^b \big[tx + (1-t)y\big]\,\mathrm{d}y\right)\mathrm{d}x.$$

Using Jensen's integral inequality we derive

$$H(t) \leqslant \frac{1}{(b-a)^2} \int_a^b f\left(\int_a^b \big[tx + (1-t)y\big]\,\mathrm{d}y\right)\mathrm{d}x$$

for all t in $[0, 1]$, and the proof is complete. □

In [89] Dragomir and Ionescu have given the following theorem.

THEOREM 1.7.3. *Let $f : C \subseteq X \to \mathbb{R}$ be a convex mapping on a convex subset C of a linear space X. For a, b two given elements in C, define the mapping $F(a, b) : [0, 1] \to \mathbb{R}$ by*

$$F(a, b)(t) = \frac{1}{2}\big[f\big(ta + (1-t)b\big) + f\big((1-t)a + tb\big)\big]$$

for all t *in* $[0, 1]$. *Then*

(i) $F(a,b)(\sigma + \frac{1}{2}) = F(a,b)(\frac{1}{2} - \sigma)$ *for all* σ *in* $[0, \frac{1}{2}]$.
(ii) $\sup_{t\in[0,1]} F(a,b)(t) = F(a,b)(0) = F(a,b)(1) = (f(a) + f(b))/2$.
(iii) $\inf_{t\in[0,1]} F(a,b)(t) = F(a,b)(\frac{1}{2}) = f(\frac{a+b}{2})$.
(iv) $F(a,b)$ *is convex on* $[0, 1]$.
(v) *We have the generalized Hadamard inequalities*

$$f\left(\frac{a+b}{2}\right) \leqslant \int_0^1 f\big(ta + (1-t)b\big)\,\mathrm{d}t \leqslant \frac{f(a) + f(b)}{2}. \tag{1.7.4}$$

(vi) *Let* $p_i \geqslant 0$ *with* $P_n = \sum_{i=1}^n p_i > 0$ *and* t_i *are in* $[0, 1]$ *for all* $i = 1, \ldots, n$. *Then we have the following inequalities*

$$\begin{aligned} f\left(\frac{a+b}{2}\right) &\leqslant F(a,b)\left(\frac{1}{P_n}\sum_{i=1}^n p_i t_i\right) \\ &\leqslant \frac{1}{P_n}\sum_{i=1}^n p_i F(a,b)(t_i) \\ &\leqslant \frac{f(a) + f(b)}{2}, \end{aligned} \tag{1.7.5}$$

which is the discrete variant of Hadamard's result.

Moreover, if we assume that $X = \mathbb{R}$ *and* a, b *in* C, $a < b$, *here* C *is an interval of real numbers, we also have*

(vii) $F(a,b)$ *is monotone decreasing on* $[0, \frac{1}{2}]$ *and monotone increasing on* $[\frac{1}{2}, 1]$.

(viii) *We have the identity*

$$\int_a^b F(a,b)(t)\,\mathrm{d}t = \frac{1}{b-a}\int_a^b f(x)\,\mathrm{d}x.$$

(ix) *Hadamard's inequalities hold, that is,*

$$f\left(\frac{a+b}{2}\right) \leqslant \frac{1}{b-a}\int_a^b f(x)\,\mathrm{d}x \leqslant \frac{f(a) + f(b)}{2}.$$

(x) *If* f *is differentiable on* $[a, b]$, *then*

$$f\left(\frac{a+b}{2}\right) \geqslant \frac{f(a) + f(b)}{2} - \frac{b-a}{2}\big(f'(b) - f'(a)\big).$$

PROOF. (i) A simple computation shows that

$$\begin{aligned}
&F(a,b)\left(\sigma+\frac{1}{2}\right)\\
&=\frac{1}{2}\left[f\left(\left(\sigma+\frac{1}{2}\right)a+\left(\frac{1}{2}-\sigma\right)b\right)+f\left(\left(\frac{1}{2}-\sigma\right)a+\left(\sigma+\frac{1}{2}\right)b\right)\right]\\
&=F(a,b)\left(\frac{1}{2}-\sigma\right)
\end{aligned}$$

for all σ in $[0,\frac{1}{2}]$, which proves the statement.

(ii) Using the convexity of f we get

$$F(a,b)(t)\leqslant\frac{1}{2}\big[tf(a)+(1-t)f(b)+(1-t)f(a)+tf(b)\big]=\frac{f(a)+f(b)}{2}$$

for all t in $[0,1]$ and

$$F(a,b)(0)=F(a,b)(1)=\frac{f(a)+f(b)}{2},$$

which proves the assertion.

(iii) By the convexity of f, we also have

$$F(a,b)(t)\geqslant f\left[\frac{ta+(1-t)b+(1-t)a+tb}{2}\right]=f\left(\frac{a+b}{2}\right)$$

for all t in $[0,1]$ and

$$F(a,b)\left(\frac{1}{2}\right)=f\left(\frac{a+b}{2}\right),$$

which shows the statement.

(iv) Let $\alpha,\beta\geqslant 0$ with $\alpha+\beta=1$ and $t_1,t_2\in[0,1]$. Then

$$\begin{aligned}
F(a,b)(\alpha t_1+\beta t_2)&=\frac{1}{2}\big[f\big(\alpha\big[t_1a+(1-t_1)b\big]+\beta\big[t_2a+(1-t_2)b\big]\big)\\
&\qquad+f\big(\alpha\big[(1-t_1)a+t_1b\big]+\beta\big[(1-t_2)a+t_2b\big]\big)\big]\\
&\leqslant\frac{1}{2}\big[\alpha f\big(t_1a+(1-t_1)b\big)+\beta f\big(t_2a+(1-t_2)b\big)\\
&\qquad+\alpha f\big((1-t_1)a+t_1b\big)+\beta f\big((1-t_2)a+t_2b\big)\big]\\
&=\alpha F(a,b)(t_1)+\beta F(a,b)(t_2),
\end{aligned}$$

which shows that $F(a,b)$ is convex on $[0,1]$.

(v) $F(a,b)$ being convex on $[0,1]$, it is integrable on $[0,1]$ and by (ii) and (iii), we get

$$f\left(\frac{a+b}{2}\right) \leqslant \int_0^1 F(a,b)(t)\,\mathrm{d}t \leqslant \frac{f(a)+f(b)}{2}.$$

Since a simple calculation shows that

$$\int_0^1 F(a,b)(t)\,\mathrm{d}t = \int_0^1 f\big(ta+(1-t)b\big)\,\mathrm{d}t,$$

the proof of inequality (1.7.4) is complete.

(vi) The first inequality in (1.7.5) is obvious from (iii). The second inequality follows by Jensen's inequality applied for the convex mapping $F(a,b)$.

To prove the last inequality in (1.7.5), by (ii) we observe that

$$F(a,b)(t_i) \leqslant \frac{f(a)+f(b)}{2}$$

for all $i=1,\ldots,n$. By multiplying with $p_i \geqslant 0$ and summing these inequalities over i from 1 to n, we obtain the desired inequality.

(vii) $F(a,b)$ being convex on $(0,1)$ for all $t_2 > t_1$, with $t_1, t_2 \in [\frac{1}{2},1)$, we have

$$\begin{aligned}
&\frac{F(a,b)(t_2)-F(a,b)(t_1)}{t_2-t_1}\\
&\quad\geqslant F'(a,b)(t_1)\\
&\quad= \frac{b-a}{2}\big[f'_+\big((1-t_1)a+t_1b\big) - f'_+\big(t_1a+(1-t_1)b\big)\big].
\end{aligned}$$

Since $t_1 \in [\frac{1}{2},1)$, we have $(1-t_1)a+t_1b \geqslant t_1a+(1-t_1)b$ and because f'_+ is monotone increasing on (a,b), we deduce that

$$f'_+\big((1-t_1)a+t_1b\big) \geqslant f'_+\big(t_1a+(1-t_1)b\big),$$

that is, $F(a,b)$ is monotone increasing on $[\frac{1}{2},1)$ and by (ii) also in $[\frac{1}{2},1]$.

The fact that $F(a,b)$ is monotone decreasing on $[0,\frac{1}{2}]$ goes likewise.

(viii) It is easy to observe that

$$\int_0^1 f\big(ta+(1-t)b\big)\,\mathrm{d}t = \frac{1}{b-a}\int_a^b f(x)\,\mathrm{d}x.$$

(ix) Follows by (v) and (viii).

(x) Since f is differentiable on $[a, b]$, we have

$$f\big(ta + (1-t)b\big) \geqslant f(a) + (1-t)(b-a)f'(a),$$
$$f\big((1-t)a + tb\big) \geqslant f(a) + t(b-a)f'(a),$$

for all t in $[0, 1]$. Summing these inequalities we get

$$F(a,b)(t) \geqslant f(a) + \frac{b-a}{2} f'(a).$$

The fact that

$$F(a,b)(t) \geqslant f(b) - \frac{b-a}{2} f'(b)$$

for $t \in [0, 1]$, goes likewise. □

The next theorem deals with the inequalities of Hadamard's type for Lipschitzian mappings given in [92].

THEOREM 1.7.4. *Let $f : I \subseteq \mathbb{R} \to \mathbb{R}$ be an M-Lipschitzian mapping on I and $a, b \in I$ with $a < b$. Then we have the inequalities*

$$\left| f\left(\frac{a+b}{2}\right) - \frac{1}{b-a}\int_a^b f(x)\,\mathrm{d}x \right| \leqslant \frac{M}{4}(b-a) \tag{1.7.6}$$

and

$$\left| \frac{f(a)+f(b)}{2} - \frac{1}{b-a}\int_a^b f(x)\,\mathrm{d}x \right| \leqslant \frac{M}{3}(b-a), \tag{1.7.7}$$

where $M > 0$ is a Lipschitzian constant.

PROOF. Let $t \in [0, 1]$. Then we have for all $a, b \in I$,

$$\begin{aligned}
&\left| tf(a) + (1-t)f(b) - f\big(ta + (1-t)b\big) \right| \\
&\quad = \left| t\big(f(a) - f\big(ta + (1-t)b\big)\big) + (1-t)\big(f(b) - f\big(ta + (1-t)b\big)\big) \right| \\
&\quad \leqslant t\left| f(a) - f\big(ta + (1-t)b\big) \right| + (1-t)\left| f(b) - f\big(ta + (1-t)b\big) \right| \\
&\quad \leqslant tM\left| a - \big(ta + (1-t)b\big) \right| + (1-t)M\left| b - \big(ta + (1-t)b\big) \right| \\
&\quad = 2t(1-t)M|b-a|.
\end{aligned} \tag{1.7.8}$$

If we choose $t = 1/2$, we have also

$$\left| \frac{f(a)+f(b)}{2} - f\left(\frac{a+b}{2}\right) \right| \leqslant \frac{M}{2}|b-a|. \tag{1.7.9}$$

If we put $ta+(1-t)b$ instead of a and $(1-t)a+tb$ instead of b in (1.7.9), respectively, then we have

$$\left|\frac{f(ta+(1-t)b)+f((1-t)a+tb)}{2}-f\left(\frac{a+b}{2}\right)\right| \leqslant \frac{M|2t-1|}{2}|b-a| \tag{1.7.10}$$

for all $t\in[0,1]$. If we integrate inequality (1.7.10) on $[0,1]$, we have

$$\left|\frac{1}{2}\left[\int_0^1 f\big(ta+(1-t)b\big)\,\mathrm{d}t+\int_0^1 f\big((1-t)a+tb\big)\,\mathrm{d}t\right]-f\left(\frac{a+b}{2}\right)\right| \leqslant \frac{M|b-a|}{2}\int_0^1 |2t-1|\,\mathrm{d}t.$$

Thus, from

$$\int_0^1 f\big(ta+(1-t)b\big)\,\mathrm{d}t=\int_0^1 f\big((1-t)a+tb\big)\,\mathrm{d}t = \frac{1}{b-a}\int_a^b f(x)\,\mathrm{d}x$$

and

$$\int_0^1 |2t-1|\,\mathrm{d}t=\frac{1}{2},$$

we obtain inequality (1.7.6).

Note that, by inequality (1.7.8), we have

$$\big|tf(a)+(1-t)f(b)-f\big(ta+(1-t)b\big)\big|\leqslant 2t(1-t)M(b-a)$$

for all $t\in[0,1]$ and $a,b\in I$ with $a<b$. Integrating on $[0,1]$, we have

$$\left|f(a)\int_0^1 t\,\mathrm{d}t+f(b)\int_0^1(1-t)\,\mathrm{d}t-\int_0^1 f\big(ta+(1-t)b\big)\,\mathrm{d}t\right| \leqslant 2M(b-a)\int_0^1 t(1-t)\,\mathrm{d}t.$$

Hence from

$$\int_0^1 t\,\mathrm{d}t=\int_0^1(1-t)\,\mathrm{d}t=\frac{1}{2}\quad\text{and}\quad\int_0^1 t(1-t)\,\mathrm{d}t=\frac{1}{6},$$

we have

$$\left|\frac{f(a)+f(b)}{2}-\frac{1}{b-a}\int_a^b f(x)\,\mathrm{d}x\right| \leqslant \frac{M}{3}(b-a)$$

and so we have inequality (1.7.7). The proof is complete. □

For various other inequalities of Hadamard's type, see [47,135,349,356,357] and the references cited therein.

1.8 Some Inequalities Involving Concave Functions

There exists a vast literature on inequalities involving concave functions. In [203, 204] Maligranda, Pečarić and Persson and in [364] Pearce and Pečarić have given generalizations of well-known inequalities of Grüss [133], Barnes [15], Borell [37–39], Favard [111] and Berwald [28] which involve concave functions. This section deals with inequalities given in the above mentioned papers.

The following inequalities are well known.

Let f and g denote nonnegative concave functions on $[0, 1]$.

(i) If $p, q \geqslant 1$, then

$$\int_0^1 f(x)g(x)\,\mathrm{d}x \geqslant \frac{(p+1)^{1/p}(q+1)^{1/q}}{6}\|f\|_p\|g\|_q. \tag{1.8.1}$$

(ii) If $0 < p, q \leqslant 1$, then

$$\int_0^1 f(x)g(x)\,\mathrm{d}x \leqslant \frac{(p+1)^{1/p}(q+1)^{1/q}}{3}\|f\|_p\|g\|_q. \tag{1.8.2}$$

For $p > 0$ and $f \geqslant 0$, we use the usual notation $\|f\|_p = (\int_0^1 |f|^p\,\mathrm{d}x)^{1/p}$.

Inequalities (1.8.1) and (1.8.2) in general were proved by Barnes [15]. In the case $p = q = 1$, inequalities (1.8.1) and (1.8.2) were proved by Grüss [133]. We note that from inequality (1.8.1) it follows, in the special case $q = 1$, $g(x) = 1$, the Favard inequality [111]

$$\int_0^1 f(x)\,\mathrm{d}x \geqslant \frac{(p+1)^{1/p}}{2}\|f\|_p, \quad p \geqslant 1, \tag{1.8.3}$$

and similarly with (1.8.2). Therefore we quote (1.8.1) and (1.8.2) as Grüss–Barnes inequalities.

In [37] Borell observes the following inequality.

Let f and g be nonnegative concave functions on $[a,b]$. If $p,q \geqslant 1$, then

$$\int_0^1 f(x)g(x)\,\mathrm{d}x \geqslant \frac{(p+1)^{1/p}(q+1)^{1/q}}{6}\|f\|_p\|g\|_q + \frac{f(0)g(0)+f(1)g(1)}{6}. \tag{1.8.4}$$

Inequality (1.8.4) is here referred to as a Borell inequality.

Favard [111] proves the following result.

Let f be nonnegative and not identically zero, continuous and a concave function on $[a,b]$, and let ϕ be a convex function on $[0,2\bar{f}]$, where $\bar{f} = \frac{1}{b-a}\int_a^b f(t)\,\mathrm{d}t$. Then

$$\frac{1}{b-a}\int_a^b \phi[f(t)]\,\mathrm{d}t \leqslant \frac{1}{2\bar{f}}\int_0^{2\bar{f}} \phi(y)\,\mathrm{d}y = \int_0^1 \phi(2s\bar{f})\,\mathrm{d}s. \tag{1.8.5}$$

In [28] Berwald proves the following generalization of Favard's inequality.

Let f be nonnegative and not identically zero, continuous and a concave function on $[a,b]$, and let ψ be a strictly increasing function on $[0,\infty)$. Assume that ϕ is a convex function with respect to ψ, that is, $\phi \circ \psi^{-1}$ is convex function on $[0,\infty)$. If $\bar{z}$ is a positive (i.e., nonnegative and not identically zero) root of the equation

$$\frac{1}{\bar{z}}\int_0^{\bar{z}} \psi(y)\,\mathrm{d}y = \frac{1}{b-a}\psi[f(t)]\,\mathrm{d}t, \tag{1.8.6}$$

then

$$\frac{1}{b-a}\int_a^b \phi[f(t)]\,\mathrm{d}t \leqslant \frac{1}{\bar{z}}\int_0^{\bar{z}} \phi(y)\,\mathrm{d}y = \int_0^1 \phi(s\bar{z})\,\mathrm{d}s. \tag{1.8.7}$$

The rearrangement function is important and useful in inequalities of different type in the theory of symmetric spaces and in the theory of interpolation of operators. Therefore many properties of rearrangement are known. In [202] Maligranda has given an important property concerning the concavity and convexity of rearrangement.

For a measurable function f on an interval $[0,a]$, $0 < a < \infty$, the distribution function d_f is defined on $[0,\infty)$ by

$$d_f(\lambda) = m\big(\{s \in I\colon |f(s)| > \lambda\}\big), \quad \lambda \geqslant 0,$$

and the decreasing rearrangement f^* on $[0,\infty)$ by

$$f^*(t) = \inf\{\lambda \geqslant 0\colon d_f(\lambda) \leqslant t\}, \quad t \geqslant 0.$$

The functions d_f and f^* are decreasing and right continuous. The function d_f is bounded by $mI = a$ and so $f^*(t) = 0$ for $t \geqslant a$. Moreover, if $a_f = \operatorname{ess\,inf}_{x\in I} |f(x)| > 0$ then $d_f(\lambda) = a$ for $0 \leqslant \lambda < a_f$, and if $b_f = \operatorname{ess\,sup}_{x\in I} |f(x)| < \infty$ then $d_f(\lambda) = 0$ for $\lambda \geqslant b_f$.

Furthermore, $f^*(d_f(\lambda)) \leqslant \lambda$, $d_f(f^*(t)) \leqslant t$ and $f^*(d_f(\lambda) - \delta) > \lambda$ provides that $d_f(\lambda) \geqslant \delta > 0$, $d_f(f^*(t) - \varepsilon) > t$ provides that $f^*(t) \geqslant \varepsilon > 0$, and if d_f happens to be continuous and strictly decreasing, then f^* is simply the inverse function of d_f on the appropriate interval.

A function f defined on an interval $J \subset [0, \infty)$ is said to be concave on J if for all $x, y \in J$ and $0 \leqslant \alpha \leqslant 1$, we have

$$f\big(\alpha x + (1-\alpha)y\big) \geqslant \alpha f(x) + (1-\alpha) f(y).$$

In [202] the following theorem is given.

THEOREM 1.8.1. *Let f be a positive measurable function on $I = [0, a]$.*

(a) *If f is concave on I, then d_f is concave on $[0, b_f)$ and f^* is concave on I.*

(b) *If f is convex on I, then d_f is convex on $[a_f, \infty)$ and f^* is convex on $[0, \infty)$.*

PROOF. (a) Suppose that f is a positive concave function on I. Set $A_f(\lambda) = \{x \in I\colon\ f(x) > \lambda\}$, $\lambda \geqslant 0$. Then $d_f(\lambda) = m(A_f(\lambda))$ and for $0 \leqslant \lambda_1, \lambda_2 < b_f$, $0 < \alpha < 1$,

$$A_f\big(\alpha\lambda_1 + (1-\alpha)\lambda_2\big) \supset \alpha A_f(\lambda_1) + (1-\alpha) A_f(\lambda_2).$$

The sets $A_f(\lambda_1)$ and $A_f(\lambda_2)$ are convex. Therefore there are intervals I_1 and I_2 where the following equality is true (for measurable subsets I the well-known Brunn–Minkowski theorem gives only inequality $\geqslant$):

$$m(I_1 + I_2) = m(I_1) + m(I_2).$$

Thus

$$\begin{aligned} d_f\big(\alpha\lambda_1 + (1-\alpha)\lambda_2\big) &= m\big[A_f\big(\alpha\lambda_1 + (1-\alpha)\lambda_2\big)\big] \\ &\geqslant m\big[\alpha A_f(\lambda_1) + (1-\alpha) A_f(\lambda_2)\big] \\ &= \alpha m\big(A_f(\lambda_1)\big) + (1-\alpha) m\big(A_f(\lambda_2)\big) \\ &= \alpha d_f(\lambda_1) + (1-\alpha) d_f(\lambda_2), \end{aligned}$$

that is, d_f is a concave function on $[0, b_f)$.

Now, let for any fixed $0 \leqslant t_1, t_2 < a$ and sufficiently small $\varepsilon_1 > 0$, $\varepsilon_2 > 0$, be $\lambda_1 = f^*(t_1) - \varepsilon_1$, $\lambda_2 = f^*(t_2) - \varepsilon_2$. Then

$$\begin{aligned} d_f\big(\alpha\lambda_1 + (1-\alpha)\lambda_2\big) &\geqslant \alpha d_f(\lambda_1) + (1-\alpha)d_f(\lambda_2) \\ &= \alpha d_f\big(f^*(t_1) - \varepsilon_1\big) + (1-\alpha)d_f\big(f^*(t_2) - \varepsilon_2\big) \\ &> \alpha t_1 + (1-\alpha)t_2, \end{aligned}$$

that is,

$$\begin{aligned} f^*\big(\alpha t_1 + (1-\alpha)t_2\big) &> \alpha\lambda_1 + (1-\alpha)\lambda_2 \\ &= \alpha f^*(t_1) + (1-\alpha)f^*(t_2) - \alpha\varepsilon_1 - (1-\alpha)\varepsilon_2. \end{aligned}$$

Also, from the concavity of f we have $d_f(0) = a$ and so

$$\begin{aligned} d_f(\alpha\lambda_1) &\geqslant \alpha d_f(\lambda_1) + (1-\alpha)d_f(0) \\ &= \alpha d_f(\lambda_1) + (1-\alpha)a \\ &= \alpha d_f\big(f^*(t_1) - \varepsilon_1\big) + (1-\alpha)a \\ &> \alpha t_1 + (1-\alpha)a, \end{aligned}$$

which means that

$$\begin{aligned} f^*\big(\alpha t_1 + (1-\alpha)a\big) &> a\lambda_1 \\ &= \alpha f^*(t_1) - \alpha\varepsilon_1 \\ &= \alpha f^*(t_1) + (1-\alpha)f^*(a) - \alpha\varepsilon_1. \end{aligned}$$

But ε_1, ε_2 were arbitrary chosen, therefore f^* is concave on I.

(b) Suppose that f is a positive convex function on I. Set $B_f(\lambda) = I - A_f(\lambda)$, $\lambda \geqslant 0$. Then for $a_f \leqslant \lambda_1, \lambda_2 < \infty$, $0 < \alpha < 1$,

$$\alpha B_f(\lambda_1) + (1-\alpha)B_f(\lambda_2) \subset B_f\big(\alpha\lambda_1 + (1-\alpha)\lambda_2\big)$$

and so

$$\begin{aligned} &A_f\big(\alpha\lambda_1 + (1-\alpha)\lambda_2\big) \\ &\quad = I - B_f\big(\alpha\lambda_1 + (1-\alpha)\lambda_2\big) \subset I - \big[\alpha B_f(\lambda_1) + (1-\alpha)B_f(\lambda_2)\big]. \end{aligned}$$

The sets $B_f(\lambda_1)$ and $B_f(\lambda_2)$ are convex subsets of I, that is, intervals in I.

Therefore

$$\begin{aligned} d_f\big(\alpha\lambda_1+(1-\alpha)\lambda_2\big) &= m\big[A_f\big(\alpha\lambda_1+(1-\alpha)\lambda_2\big)\big] \\ &\leqslant a-m\big[\alpha B_f(\lambda_1)+(1-\alpha)B_f(\lambda_2)\big] \\ &= a-\alpha\big[a-d_f(\lambda_1)\big]-(1-\alpha)\big[a-d_f(\lambda_2)\big] \\ &= \alpha d_f(\lambda_1)+(1-\alpha)d_f(\lambda_2). \end{aligned}$$

Thus d_f is a convex function on $[a_f,\infty)$.

Now, let $t_1, t_2 \in [0,\infty)$, $0 \leqslant \alpha \leqslant 1$ and $\lambda_1 = f^*(t_1)$, $\lambda_2 = f^*(t_2)$. Then

$$\begin{aligned} d_f\big(\alpha\lambda_1+(1-\alpha)\lambda_2\big) &\leqslant \alpha d_f(\lambda_1)+(1-\alpha)d_f(\lambda_2) \\ &\leqslant \alpha t_1+(1-\alpha)t_2 \end{aligned}$$

and so

$$\begin{aligned} f^*\big(\alpha t_1+(1-\alpha)t_2\big) &\leqslant \alpha\lambda_1+(1-\alpha)\lambda_2 \\ &= \alpha f^*(t_1)+(1-\alpha)f^*(t_2); \end{aligned}$$

that is, f^* is convex on $[0,\infty)$. □

The following inequality given in [203] deals with a generalization of inequality (1.8.3).

THEOREM 1.8.2. *Let f and g be nonnegative and concave functions on $[0,1]$ and let $p, q \geqslant 1$. Then*

$$\int_0^1 (1-x)f(x)g(x)\,\mathrm{d}x \geqslant \frac{(p+1)^{1/p}(q+1)^{1/q}}{12}\|f\|_p\|g\|_q + \frac{f(0)g(0)}{6} \tag{1.8.8}$$

and

$$\int_0^1 xf(x)g(x)\,\mathrm{d}x \geqslant \frac{(p+1)^{1/p}(q+1)^{1/q}}{12}\|f\|_p\|g\|_q + \frac{f(1)g(1)}{6}. \tag{1.8.9}$$

Equality in (1.8.8) *and* (1.8.9) *occurs if either* (1) $f(x)=1-x$, $g(x)=x$ *(or* $f(x)=x$, $g(x)=1-x$*) or* (2) $f(x)=g(x)=x$ *or* (3) $f(x)=g(x)=1-x$.

PROOF. The assumptions that f and g are nonnegative and concave imply that we may assume $f, g \in C^1$ and that we have the following estimates:

$$f(x)-f(y) \leqslant (x-y)f'(y) \quad \text{and} \quad g(x)-g(y) \leqslant (x-y)g'(y),$$

for all $x, y \in (0, 1)$. Therefore, by multiplying the first inequality by $g(y)$ and the second by $f(y)$, and then adding, we obtain

$$f(x)g(y) + f(y)g(x) \leqslant 2f(y)g(y) + (x-y)\big[f'(y)g(y) + f(y)g'(y)\big].$$

Moreover,

$$\begin{aligned}(x-y)\big(f'(y)g(y) + f(y)g'(y)\big) &= (x-y)(fg)'(y)\\ &= f(x)g(y) - \frac{\mathrm{d}}{\mathrm{d}y}\big((y-x)f(y)g(y)\big),\end{aligned}$$

and we conclude that

$$f(x)g(y) + f(y)g(x) + \frac{\mathrm{d}}{\mathrm{d}y}\big((y-x)f(y)g(y)\big) \leqslant 3f(y)g(y).$$

By integrating over y from 0 to x, we obtain

$$f(x)\int_0^x g(y)\,\mathrm{d}y + g(x)\int_0^x f(y)\,\mathrm{d}y + xf(0)g(0) \leqslant 3\int_0^x f(y)g(y)\,\mathrm{d}y,$$

that is,

$$\frac{\mathrm{d}}{\mathrm{d}x}\left(\int_0^x f(y)\,\mathrm{d}y\int_0^x g(y)\,\mathrm{d}y\right) + xf(0)g(0) \leqslant 3\int_0^x f(y)g(y)\,\mathrm{d}y.$$

Now, an integration with respect to x from 0 to 1 gives

$$\int_0^1 f(y)\,\mathrm{d}y\int_0^1 g(y)\,\mathrm{d}y + \frac{f(0)g(0)}{2} \leqslant 3\int_0^1\left[\int_0^x f(y)g(y)\,\mathrm{d}y\right]\mathrm{d}x,$$

that is,

$$\frac{1}{3}\int_0^1 f(x)\,\mathrm{d}x\int_0^1 g(x)\,\mathrm{d}x + \frac{f(0)g(0)}{6} \leqslant \int_0^1 (1-x)f(x)g(x)\,\mathrm{d}x. \qquad (1.8.10)$$

Then using twice the Favard inequality (1.8.3) we obtain (1.8.8).

Similarly, by first integrating over y from x to 1 and after that over x from 0 to 1, we obtain

$$\frac{1}{3}\int_0^1 f(x)\,\mathrm{d}x\int_0^1 g(x)\,\mathrm{d}x + \frac{f(1)g(1)}{6} \leqslant \int_0^1 xf(x)g(x)\,\mathrm{d}x. \qquad (1.8.11)$$

Again using twice the Favard inequality (1.8.3) we obtain (1.8.9). Moreover, it is straightforward to check that we have equality both in (1.8.8) and (1.8.9) for all the cases (1)–(3), and the proof is complete. □

As an application of Theorem 1.8.1, Maligranda [202] has obtained the following well-known Favard inequality.

THEOREM 1.8.3. *Let f be a positive concave function on $I = [0, a]$ not identically zero and let ϕ be a convex function on $[0, 2\bar{f}]$, where $\bar{f} = \frac{1}{a}\int_0^a f(x)\,\mathrm{d}x$. Then*

$$\frac{1}{a}\int_0^a \phi\big[f(x)\big]\,\mathrm{d}x \leqslant \frac{1}{2\bar{f}}\int_0^{2\bar{f}} \phi(u)\,\mathrm{d}u. \tag{1.8.12}$$

PROOF. First, we will prove as in [147] that

$$\int_0^a f^*(t)\,\mathrm{d}t \leqslant \frac{x}{a}\left(2-\frac{x}{a}\right)\int_0^a f^*(t)\,\mathrm{d}t \quad \text{for all } x \in I. \tag{1.8.13}$$

In fact,

$$\begin{aligned}
&x\left(a-\frac{x}{2}\right)\int_0^a f^*(t)\,\mathrm{d}t - \frac{a^2}{2}\int_0^x f^*(t)\,\mathrm{d}t\\
&= \int_0^x (a-t)\,\mathrm{d}t \int_0^a f^*(t)\,\mathrm{d}t - \int_0^a (a-t)\,\mathrm{d}t \int_0^x f^*(t)\,\mathrm{d}t\\
&= \int_0^x (a-t)\,\mathrm{d}t \int_x^a f^*(t)\,\mathrm{d}t + \left[\int_0^x (a-t)\,\mathrm{d}t - \int_0^a (a-t)\,\mathrm{d}t\right]\int_0^x f^*(t)\,\mathrm{d}t\\
&= \int_0^x (a-t)\,\mathrm{d}t \int_x^a f^*(t)\,\mathrm{d}t - \int_x^a (a-t)\,\mathrm{d}t \int_0^x f^*(t)\,\mathrm{d}t\\
&= \int_0^x \int_x^a \big[(a-t)f^*(s) - (a-s)f^*(t)\big]\,\mathrm{d}s\,\mathrm{d}t,
\end{aligned}$$

and for $0 \leqslant t \leqslant x < a$, we have $s = (a-s)/(a-t)t + (1 + (a-s)/(a-t))a$. Then, from the concavity of f^*,

$$f^*(s) \geqslant \frac{a-s}{a-t}f^*(t) + \left(1-\frac{a-s}{a-t}\right)f^*(a) = \frac{a-s}{a-t}f^*(t),$$

that is, expression in the last integral is positive and inequality (1.8.13) holds. Let for $t \in I$,

$$g(t) = 2a^{-2} t \int_0^a f(x)\,dx.$$

Then $g^*(t) = 2a^{-2}(a-t)\int_0^a f(x)\,dx$ and (1.8.13) means that

$$\int_0^x f^*(t)\,dt \leqslant \int_0^x g^*(t)\,dt \quad \text{for all } x \in I. \tag{1.8.14}$$

Second, we shall need the well-known majorization theorem proved by Hardy, Littlewood and Pólya [141, Theorem 250].

If (1.8.14) holds and ϕ is a convex function on an interval which contains $f^*(I) \cup g^*(I)$, then

$$\int_0^a \phi\bigl[f^*(t)\,dt\bigr] \leqslant \int_0^a \phi\bigl[g^*(t)\bigr]\,dt. \tag{1.8.15}$$

In fact, for any $u, v \geqslant 0$, we have

$$\phi(u) - \phi(v) \leqslant \phi'_+(u)(u-v),$$

and, using the Stieltjes integral with $F(x) = \int_0^x [f^*(t) - g^*(t)]\,dt$, we get

$$\begin{aligned}
\int_0^a \phi'_+\bigl(f^*(t)\bigr)\bigl[f^*(t) - g^*(t)\bigr]\,dt &= \int_0^a \phi'_+\bigl(f^*(t)\bigr)\,dF(t) \\
&= \phi'_+\bigl(f^*(a)\bigr)F(a) - \int_0^a F(t)\,d\phi'_+\bigl(f^*(t)\bigr) \\
&\leqslant 0,
\end{aligned}$$

so (1.8.15) holds.

Now, since the function f and its rearrangement are equimeasurable, it follows that

$$\begin{aligned}
\int_0^a \phi\bigl(f(t)\bigr)\,dt &= \int_0^a \phi\bigl(f^*(t)\bigr)\,dt \leqslant \int_0^a \phi\bigl(g^*(t)\bigr)\,dt \\
&= \int_0^a \phi\bigl(g(t)\bigr)\,dt = \int_0^a \phi\Bigl(\frac{2}{a} t \bar{f}\Bigr)\,dt \\
&= \frac{a}{2\bar{f}} \int_0^{2\bar{f}} \phi(u)\,du.
\end{aligned}$$

The proof is complete. □

The following weighted versions of the majorization lemmas given in [204] will be needed in the proofs of further results.

LEMMA 1.8.1. *Let v be a weight function. If h is an increasing function on (a,b), then*

$$\int_a^x h(t)v(t)\,dt\int_a^b v(t)\,dt \leqslant \int_a^b h(t)v(t)\,dt\int_a^x v(t)\,dt \quad \text{for all } x\in[a,b].$$

If h is a decreasing function on (a,b), then the reverse inequality holds.

PROOF. If $\int_a^x v(t)\,dt=0$ then $v(t)=0$ a.e. on $[a,x]$, and we obtain the equality. Now assume that $\int_a^x v(t)\,dt>0$. If h is increasing, then

$$\begin{aligned}
&\int_a^x h(t)v(t)\,dt\int_a^b v(t)\,dt\\
&\quad=\int_a^x h(t)v(t)\,dt\left[\int_a^x v(t)\,dt+\int_x^b v(t)\,dt\right]\\
&\quad=\left[\int_a^b h(t)v(t)\,dt-\int_x^b h(t)v(t)\,dt\right]\int_a^x v(t)\,dt+\int_a^x h(t)v(t)\,dt\int_x^b v(t)\,dt\\
&\quad=\int_a^b h(t)v(t)\,dt\int_a^x v(t)\,dt-\int_x^b h(t)v(t)\,dt\int_a^x v(t)\,dt\\
&\qquad+\int_a^x h(t)v(t)\,dt\int_x^b v(t)\,dt\\
&\quad\leqslant\int_a^b h(t)v(t)\,dt\int_a^x v(t)\,dt-h(x)\int_x^b v(t)\,dt\int_a^x v(t)\,dt\\
&\qquad+h(x)\int_a^x v(t)\,dt\int_x^b v(t)\,dt\\
&\quad=\int_a^b h(t)v(t)\,dt\int_a^x v(t)\,dt.
\end{aligned}$$

The proof of the case with decreasing function h is similar. □

LEMMA 1.8.2. *Let w be a weight function and let f and g be positive integrable functions on $[a,b]$. Suppose that $\phi:[0,\infty)\to\mathbb{R}$ is a convex function and that*

$$\int_a^x f(t)w(t)\,\mathrm{d}t \leqslant \int_a^x g(t)w(t)\,\mathrm{d}t \quad \text{for all } x\in[a,b]$$

and

$$\int_a^b f(t)w(t)\,\mathrm{d}t = \int_a^b g(t)w(t)\,\mathrm{d}t.$$

(i) *If f is decreasing on $[a,b]$, then*

$$\int_a^b \phi[f(t)]w(t)\,\mathrm{d}t \leqslant \int_a^b \phi[g(t)]w(t)\,\mathrm{d}t.$$

(ii) *If g is increasing on $[a,b]$, then*

$$\int_a^b \phi[g(t)]w(t)\,\mathrm{d}t \leqslant \int_a^b \phi[f(t)]w(t)\,\mathrm{d}t.$$

PROOF. If we prove the inequalities for $\phi\in C^1(0,\infty)$, then the general case follows from the pointwise approximation of ϕ by smooth functions.

Since ϕ is a convex function on $[0,\infty)$, it follows that

$$\phi(u_1)-\phi(u_2)\leqslant\phi'(u_1)(u_1-u_2) \quad \text{for } u_1,u_2\geqslant 0.$$

If we set $F(x)=\int_a^x[f(t)-g(t)]w(t)\,\mathrm{d}t$, then $F(x)\leqslant 0$ for all $x\in[a,b]$, and $F(a)=F(b)=0$.

If f is decreasing on $[a,b]$, then

$$\begin{aligned}
&\int_a^b \{\phi[f(t)]-\phi[g(t)]\}w(t)\,\mathrm{d}t\\
&\quad\leqslant \int_a^b \phi'[f(t)]\{f(t)-g(t)\}w(t)\,\mathrm{d}t\\
&\quad= \int_a^b \phi'[f(t)]\,\mathrm{d}F(t) = \big[\phi'[f(t)]F(t)\big]_a^b - \int_a^b F(t)\,\mathrm{d}\{\phi'[f(t)]\}\\
&\quad= -\int_a^b F(t)\,\mathrm{d}\{\phi'[f(t)]\}\leqslant 0.
\end{aligned}$$

Similarly, if g is increasing, then

$$\begin{aligned}\int_a^b \{\phi[g(t)] - \phi[f(t)]\} w(t)\,\mathrm{d}t &\leqslant \int_a^b \phi'[g(t)]\{g(t) - f(t)\} w(t)\,\mathrm{d}t \\ &= \int_a^b \phi'[g(t)]\,\mathrm{d}[-F(t)] = -[\phi'[g(t)]F(t)]_a^b + \int_a^b F(t)\,\mathrm{d}\{\phi'[g(t)]\} \\ &= \int_a^b F(t)\,\mathrm{d}\{\phi'[g(t)]\} \leqslant 0.\end{aligned}$$

The proof is complete. □

Our next theorem deals with a weighted version of the Favard inequality established in [204].

In order to obtain the classical Favard result we need to define a number $\bar{f}$ connected with a positive concave function f. In the weighted situation the numbers $\bar{f}_i$ for an increasing function f, and $\bar{f}_d$ of a decreasing function f will in general be different. Of course, in the case when $w \equiv 1$ these numbers coincide and they are equal to the number $\bar{f}$ in the Favard result.

THEOREM 1.8.4. (i) *Let f be a positive increasing concave function on $[a, b]$. Assume that ϕ is a convex function on $[0, 2\bar{f}_i]$, where*

$$\bar{f}_i = \frac{(b-a)\int_a^b f(t)w(t)\,\mathrm{d}t}{[2\int_a^b (t-a)w(t)\,\mathrm{d}t]}. \tag{1.8.16}$$

Then

$$\frac{1}{b-a}\int_a^b \phi[f(t)]w(t)\,\mathrm{d}t \leqslant \int_0^1 \phi(2s\bar{f}_i)w[a(1-s)+bs]\,\mathrm{d}s. \tag{1.8.17}$$

If f is increasing convex function on $[a, b]$ and $f(a) = 0$, then the reverse inequality in (1.8.17) *holds.*

(ii) *Let f be a positive decreasing concave function on $[a, b]$. Assume that ϕ is a convex function on $[0, 2\bar{f}_d]$, where*

$$\bar{f}_d = \frac{(b-a)\int_a^b f(t)w(t)\,\mathrm{d}t}{[2\int_a^b (b-t)w(t)\,\mathrm{d}t]}. \tag{1.8.18}$$

Then

$$\frac{1}{b-a}\int_a^b \phi[f(t)]w(t)\,\mathrm{d}t \leqslant \int_0^1 \phi(2s\bar{f}_d)w[as+b(1-s)]\,\mathrm{d}s. \tag{1.8.19}$$

If f is a decreasing convex function on $[a,b]$ and $f(b)=0$, then the reverse inequality in (1.8.19) *holds.*

PROOF. (i) For the positive concave function f, the function h, defined by $h(t)=f(t)/(t-a)$, is decreasing on $(a,b]$. In fact, for $a<t_1\leqslant t_2\leqslant b$, we have

$$\begin{aligned} f(t_1) &= f\left(\frac{t_1-a}{t_2-a}t_2+\frac{t_2-a-(t_1-a)}{t_2-a}a\right)\\ &\geqslant \frac{t_1-a}{t_2-a}f(t_2)+\left(1-\frac{t_1-a}{t_2-a}\right)f(a)\\ &\geqslant \frac{t_1-a}{t_2-a}f(t_2). \end{aligned}$$

Using Lemma 1.8.1 with the weight $v(t)=(t-a)w(t)$ and with the decreasing function $h(t)=f(t)/(t-a)$ we obtain

$$\int_a^x (t-a)w(t)\,\mathrm{d}t\int_a^b f(t)w(t)\,\mathrm{d}t \leqslant \int_a^x f(t)w(t)\,\mathrm{d}t\int_a^b (t-a)w(t)\,\mathrm{d}t \tag{1.8.20}$$

for all $x\in[a,b]$. According to (1.8.16), inequality (1.8.20) can be written in the form

$$\int_a^x \frac{t-a}{b-a}2\bar{f}_i w(t)\,\mathrm{d}t \leqslant \int_a^x f(t)w(t)\,\mathrm{d}t \quad \text{for all } x\in[a,b].$$

Then using the majorization lemma (Lemma 1.8.2(ii)) we have (only here we are using the assumption that f is increasing)

$$\int_a^b \phi[f(t)]w(t)\,\mathrm{d}t \leqslant \int_a^b \phi\left(\frac{t-a}{b-a}2\bar{f}_i\right)w(t)\,\mathrm{d}t.$$

But

$$\begin{aligned} \frac{1}{b-a}\int_a^b \phi\left(\frac{t-a}{b-a}2\bar{f}_i\right)w(t)\,\mathrm{d}t &= \frac{1}{2\bar{f}_i}\int_0^{2\bar{f}_i}\phi(y)w\left(a+y\frac{b-a}{2\bar{f}_i}\right)\mathrm{d}y\\ &= \int_0^1 \phi(2s\bar{f}_i)w[a(1-s)+bs]\,\mathrm{d}s, \end{aligned}$$

and inequality (1.8.17) is proved.

(ii) For the positive concave function f, the function h, defined by $h(t) = f(t)/(b-t)$, is increasing on $[a,b]$. In fact, for $a \leqslant t_1 \leqslant t_2 < b$, we have

$$\begin{aligned} f(t_2) &= f\left(\frac{b-t_2}{b-t_1}t_1 + \frac{b-t_1-(b-t_2)}{b-t_1}b\right) \\ &\geqslant \frac{b-t_2}{b-t_1}f(t_1) + \left(1 - \frac{b-t_2}{b-t_1}\right)f(b) \\ &\geqslant \frac{b-t_2}{b-t_1}f(t_1). \end{aligned}$$

Using Lemma 1.8.1 with weight $v(t) = (b-t)w(t)$ and with the increasing function $h(t) = f(t)/(b-t)$ we obtain

$$\begin{aligned} &\int_a^x f(t)w(t)\,\mathrm{d}t \int_a^b (b-t)w(t)\,\mathrm{d}t \\ &\quad \leqslant \int_a^x (b-t)w(t)\,\mathrm{d}t \int_a^b f(t)w(t)\,\mathrm{d}t \end{aligned} \tag{1.8.21}$$

for all $x \in [a,b]$. In view of (1.8.18), inequality (1.8.21) can be written in the form

$$\int_a^x f(t)w(t)\,\mathrm{d}t \leqslant \int_a^x \frac{b-t}{b-a}2\bar{f}_d w(t)\,\mathrm{d}t \quad \text{for all } x \in [a,b].$$

Then using the majorization lemma (Lemma 1.8.2(i)) we have (assuming that f is increasing)

$$\int_a^b \phi[f(t)]w(t)\,\mathrm{d}t \leqslant \int_a^b \phi\left(\frac{b-t}{b-a}2\bar{f}_d\right)w(t)\,\mathrm{d}t.$$

But

$$\begin{aligned} \frac{1}{b-a}\int_a^b \phi\left(\frac{b-t}{b-a}2\bar{f}_d\right)w(t)\,\mathrm{d}t &= \frac{1}{2\bar{f}_d}\int_0^{2\bar{f}_d} \phi(y)w\left(b - y\frac{b-a}{2\bar{f}_d}\right)\mathrm{d}y \\ &= \int_0^1 \phi(2s\bar{f}_d)w\big[as + b(1-s)\big]\,\mathrm{d}s, \end{aligned}$$

and inequality (1.8.19) follows. The proof of the convex case is similar. □

In the following theorem we present a generalization of the Berwald inequality to the weighted case given in [204].

THEOREM 1.8.5. *Let ϕ be a convex function with respect to the strictly increasing function ψ, that is, $\phi \circ \psi^{-1}$ is convex.*

(i) *If f is a positive increasing concave function on $[a, b]$ and $\bar{z}_i$ is a positive root of the equation*

$$\frac{1}{\bar{z}_i}\int_0^{\bar{z}_i} \psi(y) w\left(a + \frac{b-a}{\bar{z}_i} y\right) \mathrm{d}y = \frac{1}{b-a}\int_a^b \psi[f(t)] w(t)\, \mathrm{d}t, \tag{1.8.22}$$

then

$$\frac{1}{b-a}\int_a^b \phi[f(t)] w(t)\, \mathrm{d}t \leqslant \int_0^1 \phi(s\bar{z}_i) w[a(1-s) + bs]\, \mathrm{d}s. \tag{1.8.23}$$

If f is an increasing convex function on $[a, b]$ with $f(a) = 0$, then the reverse inequality in (1.8.23) *holds.*

(ii) *If f is a positive decreasing concave function on $[a, b]$ and if $\bar{z}_d$ is a positive root of the equation*

$$\frac{1}{\bar{z}_d}\int_0^{\bar{z}_d} \psi(y) w\left(b - \frac{b-a}{\bar{z}_d} y\right) \mathrm{d}y = \frac{1}{b-a}\int_a^b \psi[f(t)] w(t)\, \mathrm{d}t, \tag{1.8.24}$$

then

$$\frac{1}{b-a}\int_a^b \phi[f(t)] w(t)\, \mathrm{d}t \leqslant \int_0^1 \phi(s\bar{z}_d) w[as + b(1-s)]\, \mathrm{d}s. \tag{1.8.25}$$

If f is a decreasing convex function on $[a, b]$ with $f(b) = 0$, then the reverse inequality in (1.8.25) *holds.*

PROOF. (i) If f is a positive increasing concave function on $[a, b]$, then there exists $t_0 \in [a, b]$ such that $[(t_0 - a)/(b-a)]\bar{z}_i = f(t_0)$ and

$$\frac{t-a}{b-a}\bar{z}_i \leqslant f(t) \quad \text{for all } t \in [a, t_0] \quad \text{and} \quad \frac{t-a}{b-a}\bar{z}_i \geqslant f(t) \quad \text{for all } t \in [t_0, b]. \tag{1.8.26}$$

Equality (1.8.22) can be written in the form

$$\int_a^b \psi\left(\frac{t-a}{b-a}\bar{z}_i\right) w(t)\, \mathrm{d}t = \int_a^b \psi[f(t)] w(t)\, \mathrm{d}t.$$

We prove that

$$\int_a^x \psi\left(\frac{t-a}{b-a}\bar{z}_i\right) w(t)\, \mathrm{d}t \leqslant \int_a^x \psi[f(t)] w(t)\, \mathrm{d}t \quad \text{for all } x \in (a, b). \tag{1.8.27}$$

If $a < x \leqslant t_0$, then inequality (1.8.27) follows immediately from (1.8.26). If $t_0 \leqslant x < b$, then, by using equality (1.8.22) and the second inequality in (1.8.26), we obtain

$$\begin{aligned}
&\int_a^x \psi\left(\frac{t-a}{b-a}\bar{z}_i\right) w(t)\,\mathrm{d}t \\
&= \int_a^b \psi\left(\frac{t-a}{b-a}\bar{z}_i\right) w(t)\,\mathrm{d}t - \int_x^b \psi\left(\frac{t-a}{b-a}\bar{z}_i\right) w(t)\,\mathrm{d}t \\
&= \int_a^b \psi\big[f(t)\big] w(t)\,\mathrm{d}t - \int_x^b \psi\left(\frac{t-a}{b-a}\bar{z}_i\right) w(t)\,\mathrm{d}t \\
&\leqslant \int_a^b \psi\big[f(t)\big] w(t)\,\mathrm{d}t - \int_x^b \psi\big[f(t)\big] w(t)\,\mathrm{d}t \\
&= \int_a^x \psi\big[f(t)\big] w(t)\,\mathrm{d}t. \qquad (1.8.28)
\end{aligned}$$

According to inequalities (1.8.27) and (1.8.28), the assumption that $\phi \circ \psi^{-1}$ is convex and Lemma 1.8.2(ii), we find that

$$\int_a^b \phi\big[f(t)\big] w(t)\,\mathrm{d}t \leqslant \int_a^b \phi\left(\frac{t-a}{b-a}\bar{z}_i\right) w(t)\,\mathrm{d}t.$$

Moreover,

$$\begin{aligned}
\frac{1}{b-a}\int_a^b \phi\left(\frac{t-a}{b-a}\bar{z}_i\right) w(t)\,\mathrm{d}t &= \frac{1}{\bar{z}_i}\int_0^{\bar{z}_i} \phi(y) w\left(a + \frac{b-a}{\bar{z}_i} y\right) \mathrm{d}y \\
&= \int_0^1 \phi(s\bar{z}_i) w\big[a(1-s) + bs\big]\,\mathrm{d}s,
\end{aligned}$$

and inequality (1.8.23) is proved.

(ii) If f is a positive decreasing concave function on $[a,b]$, then there exists $t_1 \in [a,b]$ such that $[(b-t_1)/(b-a)]\bar{z}_d = f(t_1)$ and

$$\begin{aligned}
f(t) &\leqslant \frac{b-t}{b-a}\bar{z}_d \quad \text{for all } t \in [a, t_1] \quad \text{and} \\
f(t) &\geqslant \frac{b-t}{b-a}\bar{z}_d \quad \text{for all } t \in [t_1, b].
\end{aligned} \qquad (1.8.29)$$

Equality (1.8.24) can be written in the following form

$$\int_a^b \psi\left(\frac{b-t}{b-a}\bar{z}_d\right) w(t)\,\mathrm{d}t = \int_a^b \psi\big[f(t)\big] w(t)\,\mathrm{d}t.$$

We prove that

$$\int_a^x \psi\big[f(t)\big] w(t)\,\mathrm{d}t \leqslant \int_a^x \psi\left(\frac{b-t}{b-a}\bar{z}_d\right) w(t)\,\mathrm{d}t \quad \text{for all } x \in (a,b). \tag{1.8.30}$$

If $a < x \leqslant t_1$, then inequality (1.8.30) follows immediately from (1.8.29). If $t_1 \leqslant x < b$, then, by using equality (1.8.24) and the second inequality in (1.8.29), we obtain

$$\begin{aligned}
&\int_a^x \psi\big[f(t)\big] w(t)\,\mathrm{d}t \\
&\quad = \int_a^b \psi\big[f(t)\big] w(t)\,\mathrm{d}t - \int_x^b \psi\big[f(t)\big] w(t)\,\mathrm{d}t \\
&\quad = \int_a^b \psi\left(\frac{b-t}{b-a}\bar{z}_d\right) w(t)\,\mathrm{d}t - \int_x^b \psi\big[f(t)\big] w(t)\,\mathrm{d}t \\
&\quad \leqslant \int_a^b \psi\left(\frac{b-t}{b-a}\bar{z}_d\right) w(t)\,\mathrm{d}t - \int_x^b \psi\left(\frac{b-t}{b-a}\bar{z}_d\right) w(t)\,\mathrm{d}t \\
&\quad = \int_a^x \psi\left(\frac{b-t}{b-a}\bar{z}_d\right) w(t)\,\mathrm{d}t.
\end{aligned}$$

By using inequality (1.8.30), equality (1.8.24), the assumption that $\phi \circ \psi^{-1}$ is convex and Lemma 1.8.2(i), we obtain

$$\int_a^b \phi\big[f(t)\big] w(t)\,\mathrm{d}t \leqslant \int_a^b \phi\left(\frac{b-t}{b-a}\bar{z}_d\right) w(t)\,\mathrm{d}t.$$

Furthermore,

$$\begin{aligned}
\frac{1}{b-a}\int_a^b \phi\left(\frac{b-t}{b-a}\bar{z}_d\right) w(t)\,\mathrm{d}t &= \frac{1}{\bar{z}_d}\int_a^{\bar{z}_d} \phi(y) w\left(b - \frac{b-a}{\bar{z}_d} y\right)\mathrm{d}y \\
&= \int_0^1 \phi(s\bar{z}_d) w\big[as + b(1-s)\big]\,\mathrm{d}s,
\end{aligned}$$

and inequality (1.8.25) is proved. □

1.9 Miscellaneous Inequalities

1.9.1 Pečarić and Dragomir [379]

Let $f: C \to \mathbb{R}$ be a convex (concave) function on C, $p_i \geqslant 0$, $i = 1, \ldots, n$, and $P_n = \sum_{i=1}^n p_i > 0$. Then the following inequality holds

$$\begin{aligned} f\left(\frac{1}{P_n}\sum_{i=1}^n p_i x_i\right) &\leqslant (\geqslant) \frac{1}{P_n^{k+1}} \sum_{i_1,\ldots,i_{k+1}=1}^n p_{i_1}\cdots p_{i_{k+1}} f\left(\frac{1}{k+1}\sum_{j=1}^{k+1} x_{i_j}\right) \\ &\leqslant (\geqslant) \frac{1}{P_n^k} \sum_{i_1,\ldots,i_k=1}^n p_{i_1}\cdots p_{i_k} f\left(\frac{1}{k}\sum_{j=1}^{n} x_{i_j}\right) \\ &\leqslant (\geqslant) \cdots \leqslant (\geqslant) \frac{1}{P_n^2} \sum_{i_1,i_2=1}^n p_{i_1} p_{i_2} f\left(\frac{x_{i_1}+x_{i_2}}{2}\right) \\ &\leqslant (\geqslant) \frac{1}{P_n}\sum_{i=1}^n p_i f(x_i), \end{aligned}$$

where k is a positive integer such that $1 \leqslant k \leqslant n-1$.

1.9.2 Dragomir [85]

Let $f: C \to \mathbb{R}$ be a convex mapping, $x_i \in C$, $p_i \geqslant 0$ and $P_n = \sum_{i=1}^n p_i > 0$, where C is a convex subset of real linear space X. Then the following inequality holds

$$\begin{aligned} &\frac{1}{P_n}\sum_{i=1}^n p_i f(x_i) - f\left(\frac{1}{P_n}\sum_{i=1}^n p_i x_i\right) \\ &\geqslant \frac{1}{P_n}\sum_{i=1}^n p_i f(x_i) - \frac{1}{P_n^k}\sum_{i_1,\ldots,i_k=1}^n p_{i_1}\cdots p_{i_k} f\left(\frac{1}{k}\sum_{j=1}^k x_{i_j}\right) \\ &\geqslant \left|\frac{1}{kP_n^k}\sum_{i_1,\ldots,i_k=1}^n p_{i_1}\cdots p_{i_k}\left|\sum_{j=1}^k f(x_{i_j})\right|\right. \\ &\qquad \left. - \frac{1}{P_n^k}\sum_{i_1,\ldots,i_k=1}^n p_{i_1}\cdots p_{i_k}\left|f\left(\frac{1}{k}\sum_{j=1}^k x_{i_j}\right)\right|\right| \geqslant 0 \end{aligned}$$

for every positive integer k such that $1 \leqslant k \leqslant n$.

1.9.3 Dragomir [83]

Let $f:[0, A] \to \mathbb{R}$ be a function such that the mapping $g(x) = f(x) - f(A-x)$ is (convex) concave on $[0, \overline{A}]$, $\overline{A} \leqslant A$. If $x_i \in [0, \overline{A}]$, $p_i \geqslant 0$, $i = 1, \ldots, n$, $x_{i_j} \in \{x_i\}_{i=1,\ldots,n}$, $p_{i_j} \in \{p_i\}_{i=1,\ldots,n}$ and $P_n = \sum_{i=1}^n p_i > 0$, then

$$\begin{aligned}
& f\left(\frac{1}{P_n}\sum_{i=1}^n p_i x_i\right) - f\left(\frac{1}{P_n}\sum_{i=1}^n (A - x_i) p_i\right) \\
& \geqslant (\leqslant) \frac{1}{P_n^{k+1}} \sum_{i_1,\ldots,i_{k-1}=1}^n p_{i_1}\cdots p_{i_{k-1}} f\left(\frac{1}{k+1}\sum_{j=1}^{k+1} x_{i_j}\right) \\
& \qquad - \frac{1}{P_n^{k+1}} \sum_{i_1,\ldots,i_{k-1}=1}^n p_{i_1}\cdots p_{i_{k-1}} f\left(\frac{1}{k+1}\sum_{j=1}^{k+1} (A - x_{i_j})\right) \\
& \geqslant (\leqslant) \frac{1}{P_n^{k}} \sum_{i_1,\ldots,i_{k}=1}^n p_{i_1}\cdots p_{i_{k}} f\left(\frac{1}{k}\sum_{j=1}^{k} x_{i_j}\right) \\
& \qquad - \frac{1}{P_n^{k}} \sum_{i_1,\ldots,i_{k}=1}^n p_{i_1}\cdots p_{i_{k}} f\left(\frac{1}{k}\sum_{j=1}^{k} (A - x_{i_j})\right) \\
& \quad \vdots \\
& \geqslant (\leqslant) \frac{1}{P_n}\sum_{i=1}^n p_i f(x_i) - \frac{1}{P_n}\sum_{i=1}^n p_i (A - x_i).
\end{aligned}$$

1.9.4 Pečarić [373]

Let the linear spaces X and Y be endowed with partial orders which are compatible with the linear structures of X and Y, respectively. Let $D \subset X$ be a convex set. A function $f: D \to Y$ is called 1-convex (convex of first order) if and only if the relation

$$f\left(\frac{x+y}{2}\right) \leqslant \frac{f(x)+f(y)}{2} \tag{A}$$

holds for all comparable $x, y \in D$ (i.e., such that either $x \leqslant y$ or $y \leqslant x$). Usually a partial order in X is generated by a cone $C \subset X$ and the functions fulfilling (A) are called $(C\text{–}J)$-convex. Let $f : D \to Y$ be a $(C\text{–}J)$-convex function. If either $x_1 \leqslant \cdots \leqslant x_n$ or $x_1 \geqslant \cdots \geqslant x_n$ hold for n points $x_1, \ldots, x_n$ from D, then

$$f\left(\frac{1}{n}\sum_{i=1}^{n} x_i\right) \leqslant \frac{1}{n}\sum_{i=1}^{n} f(x_i).$$

1.9.5 Pečarić [373]

Let $f : D \to Y$ be a $(C\text{–}J)$-convex function and let x_i, $i = 1, \ldots, n$, be points in D chosen so that $x_1 \leqslant \cdots \leqslant x_n$. Define

$$f_{m,n} = \binom{n}{m}^{-1} \sum_{1 \leqslant i_1 < \cdots < i_m \leqslant n} f\left(\frac{1}{m}(x_{i_1} + \cdots + x_{i_m})\right)$$

for $1 \leqslant m \leqslant n$. Then

$$f_{n,n} \leqslant \cdots \leqslant f_{m,n} \leqslant \cdots \leqslant f_{1,n}, \quad 1 \leqslant m \leqslant n,$$

where D and Y are as in Section 1.9.4.

1.9.6 Mond and Pečarić [219]

A mapping f defined on the set T of rectangular matrices with values in the set of (rectangular) matrices is said to be increasing if $A \leqslant B$ implies $f(A) \leqslant f(B)$. It is decreasing, by definition, if the mapping $A \to -f(A)$ is increasing. A mapping f is said to be semi-convex on T if $X, Y \in T$; $X \leqslant Y$ implies that, for any $0 < \lambda < 1$, the convex combination $\lambda X + (1 - \lambda)Y$ is in T, that is, T is a convex set, and

$$f\big(\lambda X + (1 - \lambda)Y\big) \leqslant \lambda f(X) + (1 - \lambda) f(Y).$$

It is semi-concave, by definition, if the mapping $X \to -F(X)$ is semi-convex. Let f be semi-convex on T, let p_i, $i = 1, \ldots, n$, be nonnegative numbers with $P_n = \sum_{i=1}^{n} p_i > 0$ and let $X_i \in T$, $i = 1, \ldots, n$, satisfy

$$X_1 \leqslant X_2 \leqslant \cdots \leqslant X_n \quad \text{or} \quad X_1 \geqslant X_2 \geqslant \cdots \geqslant X_n.$$

Then

$$f\left(\frac{1}{P_n}\sum_{i=1}^{n} p_i X_i\right) \leqslant \frac{1}{P_n}\sum_{i=1}^{n} p_i f(X_i).$$

1.9.7 Dragomir and Sándor [91]

Let $f:C\subseteq X\to\mathbb{R}$ be a uniformly-convex function defined on a convex subset C of a linear space X, $p_i\geqslant 0$, $P_n=\sum_{i=1}^n p_i>0$ and $x_i\in C$, $i=1,\dots,n$. Then

$$\frac{1}{P_n}\sum_{i=1}^n p_i f(x_i)-f\left(\frac{1}{P_n}\sum_{i=1}^n p_i x_i\right)\geqslant\frac{1}{P_n}\sum_{i=1}^n p_i\|x_i\|^2-\left\|\frac{1}{P_n}\sum_{i=1}^n p_i x_i\right\|^2\geqslant 0.$$

1.9.8 Dragomir and Sándor [91]

Let $a_1\geqslant\cdots\geqslant a_s$, $b_1\geqslant\cdots\geqslant b_s$ and $q_1,\dots,q_s$ be real numbers such that

$$\sum_{i=1}^k q_i a_i\leqslant\sum_{i=1}^k q_i b_i,\quad 1\leqslant k\leqslant s-1,\qquad\sum_{i=1}^s q_i a_i=\sum_{i=1}^s q_i b_i.$$

If f is uniformly-convex on the interval I (I contains a_i, b_i for $i=1,\dots,n$), then

$$\sum_{i=1}^s q_i\big(f(b_i)-f(a_i)\big)\geqslant\sum_{i=1}^s q_i\big(b_i^2-a_i^2\big)\geqslant 0.$$

1.9.9 Dragomir and Sándor [91]

Let x and p be two n-tuples of real numbers such that $P_n=\sum_{i=1}^n p_i>0$ and $0\leqslant P_k\leqslant P_n$, $k=1,\dots,n-1$, and x is a monotonic n-tuple. Then, for all uniformly-convex functions $f:I\to\mathbb{R}$, $x_i\in I$, $i=1,\dots,n$, we have

$$\frac{1}{P_n}\sum_{i=1}^n p_i f(x_i)-f\left(\frac{1}{P_n}\sum_{i=1}^n p_i x_i\right)\geqslant\frac{1}{P_n}\sum_{i=1}^n p_i x_i^2-\left(\frac{1}{P_n}\sum_{i=1}^n p_i x_i^2\right)^2\geqslant 0.$$

1.9.10 Fejér [114]

Let $f:[a,b]\to\mathbb{R}$ and $g:[a,b]\to\mathbb{R}_+$ be integrable and symmetric with respect to the line $x=(a+b)/2$, that is, $g((a+b)/2+t)=g((a+b)/2-t)$. Then

$$\frac{f(a)+f(b)}{2}\int_a^b g(t)\,\mathrm{d}t\leqslant\int_a^b f(t)g(t)\,\mathrm{d}t\leqslant f\left(\frac{a+b}{2}\right)\int_a^b g(t)\,\mathrm{d}t.$$

1.9.11 Pečarić [371]

If $f:[a,b]\to\mathbb{R}$ is a convex function and if $t\in[0,1]$, then

$$0\leqslant \frac{1}{b-a}\int_a^b f(x)\,\mathrm{d}x-\frac{1}{(b-a)^2}\int_a^b\int_a^b f\big(tx+(1-t)y\big)\,\mathrm{d}x\,\mathrm{d}y$$
$$\leqslant \min(t,1-t)\left(\frac{f(a)+f(b)}{2}-\frac{1}{b-a}\int_a^b f(x)\,\mathrm{d}x\right).$$

1.9.12 Pečarić [371]

Let $p,q>0$, f be convex on $I\supset[a,b]$, $A=(pa+qb)/(p+q)$ and c be a real number such that

$$0<c\leqslant\frac{b-a}{p+q}\min(p,q).$$

If $t\in[0,1]$, then

$$0\leqslant\frac{1}{2c}\int_{A-c}^{A+c}f(x)\,\mathrm{d}x-\frac{1}{4c^2}\int_{A-c}^{A+c}\int_{A-c}^{A+c}f\big(tx+(1-t)y\big)\,\mathrm{d}x\,\mathrm{d}y$$
$$\leqslant\min(t,1-t)\left(\frac{pf(a)+qf(b)}{p+q}-\frac{1}{2c}\int_{A-c}^{A+c}f(x)\,\mathrm{d}x\right).$$

1.9.13 Brenner and Alzer [41]

Suppose $p,q\in\mathbb{R}^+$, let $f:[a,b]\to\mathbb{R}$ be concave and let $g:[a,b]\to\mathbb{R}_0^+$ (where $\mathbb{R}^+$ and $\mathbb{R}_0^+$ denote the positive and nonnegative real numbers) be integrable and symmetric with respect to the line $x=A=(pa+qb)/(p+q)$, that is, $g(A+t)=g(A-t)$. If

$$0\leqslant y\leqslant\frac{b-a}{p+q}\min(p,q),$$

then

$$\frac{pf(a)+qf(b)}{p+q}\int_{A-y}^{A+y}g(t)\,\mathrm{d}t\leqslant\int_{A-y}^{A+y}f(t)g(t)\,\mathrm{d}t$$
$$\leqslant f\left(\frac{pa+qb}{p+q}\right)\int_{A-y}^{A+y}g(t)\,\mathrm{d}t.$$

1.9.14 Dragomir, Pečarić and Sándor [94]

Let $f : I \to \mathbb{R}$ ($I \subset \mathbb{R}$ is interval) be a continuous convex function and let $a, b \in I$, $n \in \mathbb{N} = \{1, 2, \dots\}$. Then

$$f\left(\frac{a+b}{2}\right) \leqslant \frac{1}{n}\sum_{i=1}^{n}\left(\left(\frac{i}{n+1}\right)a + \left(1 - \frac{i}{n+1}\right)b\right) \leqslant \frac{f(a)+f(b)}{2}.$$

1.9.15 Dragomir, Pečarić and Sándor [94]

Let $f : I \to \mathbb{R}$ be a continuous convex function and $a, b \in I$, $a < b$, $n \in \mathbb{N}$ ($\mathbb{N}$ is the set of natural numbers). Then

$$\begin{aligned} f\left(\frac{a+b}{2}\right) &\leqslant \frac{1}{(b-a)^{n+1}} \int_a^b \cdots \int_a^b f\left(\sum_{i=1}^{n+1} \frac{x_i}{n+1}\right) \mathrm{d}x_1 \cdots \mathrm{d}x_{n+1} \\ &\leqslant \frac{1}{(b-a)^{n}} \int_a^b \cdots \int_a^b f\left(\sum_{i=1}^{n} \frac{x_i}{n}\right) \mathrm{d}x_1 \cdots \mathrm{d}x_{n} \\ &\leqslant \\ &\vdots \\ &\leqslant \frac{1}{(b-a)^2} \int_a^b \int_a^b f\left(\frac{x_1+x_2}{2}\right) \mathrm{d}x_1\, \mathrm{d}x_2 \\ &\leqslant \frac{1}{b-a} \int_a^b f(x)\, \mathrm{d}x \leqslant \frac{f(a)+f(b)}{2}. \end{aligned}$$

1.9.16 Buşe, Dragomir and Barbu [49]

Let I be an interval with $a, b \in I^0$ (I^0 is an interior of I) and $a < b$. If $f : I \to \mathbb{R}$ is a convex function on I and $q_i(m) \geqslant 0$ for all $i, m \in \mathbb{N}$ ($\mathbb{N}$ is the set of natural numbers) then

$$\begin{aligned} f\left(\frac{a+b}{2}\right) &\leqslant \frac{1}{(b-a)^m} \int_a^b \cdots \int_a^b f\left(\frac{q_1(m)x_1 + \cdots + q_m(m)x_m}{Q_m}\right) \mathrm{d}x_1 \cdots \mathrm{d}x_m \\ &\leqslant \frac{1}{b-a} \int_a^b f(x)\, \mathrm{d}x, \end{aligned}$$

where $Q_m = q_1(m) + \cdots + q_m(m) > 0$, $m \in \mathbb{N}$.

1.9.17 Buşe, Dragomir and Barbu [49]

Let $f : I \subset \mathbb{R} \to \mathbb{R}$ be a convex function on I, $a, b \in I^0 = (a, b)$ with $a < b$ and $q_i(m) \geqslant 0$ for all i, $m \in \mathbb{N}$ ($\mathbb{N}$ is the set of natural numbers). If $Q_m = q_1(m) + \cdots + q_m(m) > 0$ and

$$\lim_{m\to\infty} \frac{q_1^2(m) + \cdots + q_m^2(m)}{Q_m^2} = 0,$$

then

$$\lim_{m\to\infty} \frac{1}{(b-a)^m} \int_a^b \cdots \int_a^b f\left(\frac{q_1(m)x_1 + \cdots + q_m(m)x_m}{Q_m}\right) \mathrm{d}x_1 \cdots \mathrm{d}x_m = f\left(\frac{a+b}{2}\right).$$

1.9.18 Pearce and Pečarić [364]

If $f : [a, b] \to [0, \infty)$ is continuous and concave, then

$$\frac{\alpha+\beta}{\alpha+2\beta} \max_{a\leqslant x\leqslant b} f^{\beta}(x) \int_a^b f^{\alpha}(t)\,\mathrm{d}t \leqslant \int_a^b f^{\alpha+\beta}(t)\,\mathrm{d}t,$$

holds for all real numbers α and β with $\alpha + \beta > 0$ and $0 < \beta \leqslant 1$.

1.9.19 Brenner and Alzer [41]

If $f : [a, b] \to (0, \infty)$ is continuous and concave, then

$$1 \leqslant \frac{1}{(b-a)^2} \int_a^b f(t)\,\mathrm{d}t \int_a^b \frac{\mathrm{d}t}{f(t)} \leqslant 1 + \log\left[f\left(\frac{a+b}{2}\right) \Big/ \sqrt{f(a)f(b)}\right].$$

1.9.20 Maligranda, Pečarić and Persson [203]

Assume that f and g are nonnegative functions on $[0, 1]$ such that the functions $f^{1/a}$ and $g^{1/b}$ are concave on $[0, 1]$ for some $a, b > 0$.

(i) If $p, q \geqslant 1$, then

$$\int_0^1 f(x)g(x)\,\mathrm{d}x \geqslant B(a+1, b+1)(pa+1)^{1/p}(qb+1)^{1/q}\|f\|_p\|g\|_q.$$

Equality occurs if $f(x) = cx^a$ and $g(x) = d(1-x)^b$ with c, d positive constants. Here B denotes the usual beta function $B(p,q) = \int_0^1 x^{p-1}(1-x)^{q-1}\,dx$.

(ii) If $0 < p, q \leqslant 1$, then

$$\int_0^1 f(x)g(x)\,dx \leqslant \frac{(pa+1)^{1/p}(qb+1)^{1/q}}{a+b+1}\|f\|_p\|g\|_q.$$

Equality occurs if either $f(x) = cx^a$ and $g(x) = dx^b$ or $f(x) = c(1-x)^a$ and $g(x) = d(1-x)^b$ with c, d positive constants.

1.9.21 Brenner and Alzer [41]

For a continuous and concave function $f:[a,b] \to \mathbb{R}$,

$$\frac{1}{2}\max_{a\leqslant x\leqslant b}\left[f(x) + \frac{(x-a)f(a)+(b-x)f(b)}{b-a}\right] \leqslant \frac{1}{b-a}\int_a^b f(t)\,dt.$$

Further, if f is strictly concave on a nondegenerate interval, then the inequality is strict.

1.9.22 Brenner and Alzer [41]

If $f:[a,b] \to \mathbb{R}$ is continuous and concave, then

$$\frac{1}{b-a}\int_a^b f(t)\,dt \leqslant \min_{a\leqslant x\leqslant b}\left[\frac{x-a}{b-a}f\left(\frac{a+x}{2}\right) + \frac{b-x}{b-a}f\left(\frac{x+b}{2}\right)\right].$$

Strict inequality holds if and only if there is a nondegenerate subinterval on which f is strictly concave.

1.9.23 Pečarić, Perić and Persson [383]

Let f, g be real-valued functions defined on the interval (a,b), $-\infty \leqslant a < b < \infty$. We say that f is C-decreasing (C-increasing), $C \geqslant 1$, if $f(t) \leqslant Cf(s)$ ($f(s) \leqslant Cf(t)$) whenever $s \leqslant t$, $t, s \in (a,b)$.

Let $\phi:[0,\infty) \to \mathbb{R}$ be a concave, nonnegative and differentiable function such that $\phi(0) = 0$ and let $-\infty \leqslant a < b < \infty$.

(a) If f is C-decreasing and g is increasing, differentiable and such that $g(a+0) = 0$, then

$$\phi\left(C\int_a^b f(x)\,dg(x)\right) \leqslant C\int_a^b \phi'\big(f(x)g(x)\big)f(x)\,dg(x).$$

(b) If f is C-increasing and g is increasing, differentiable and such that $g(a+0)=0$, then

$$\phi\left(\frac{1}{C}\int_a^b f(x)\,\mathrm{d}g(x)\right) \geqslant \frac{1}{C}\int_a^b \phi'\big(f(x)g(x)\big)f(x)\,\mathrm{d}g(x).$$

(c) If f is C-increasing and g is decreasing, differentiable and such that $g(b-0)=0$, then

$$\phi\left(C\int_a^b f(x)\,\mathrm{d}\big[-g(x)\big]\right) \leqslant C\int_a^b \phi'\big(f(x)g(x)\big)f(x)\,\mathrm{d}\big[-g(x)\big].$$

(d) If f is C-decreasing and g is decreasing, differentiable and such that $g(b-0)=0$, then

$$\phi\left(\frac{1}{C}\int_a^b f(x)\,\mathrm{d}\big[-g(x)\big]\right) \geqslant \frac{1}{C}\int_a^b \phi'\big(f(x)g(x)\big)f(x)\,\mathrm{d}\big[-g(x)\big].$$

(e) If the condition "ϕ is concave" is replaced by "ϕ is convex", then all the above inequalities in (a)–(d) hold in the reversed direction.

1.9.24 Farwing and Zwick [110]

Let $x_0, \ldots, x_n$ be given real numbers, where $a \leqslant x_0 \leqslant \cdots \leqslant x_n \leqslant b$. Let f be a real-valued function defined on $[a,b]$ and let $[x_0, \ldots, x_n]f$ denote the nth divided difference of f at the points $x_0, \ldots, x_n$. Let $f^{(n)}$ be a convex function on (a,b). Then

$$f^{(n)}\left[\frac{1}{n+1}\sum_{i=0}^{n} x_i\right] \leqslant n![x_0, \ldots, x_n]f \leqslant \frac{1}{n+1}\sum_{i=0}^{n} f^{(n)}(x_i).$$

If $x_0 \neq x_n$, then strict inequalities hold if and only if $f \notin P_{n+1}$ (the space of all polynomials of degree $\leqslant n+1$).

1.9.25 Abramovich, Mond and Pečarić [1]

Let $F(x_1, \ldots, x_n)$ be a complex function in n complex variables and let

$$\big|F(x_1, \ldots, x_n)\big| \leqslant \big|F\big(|x_1|, \ldots, |x_n|\big)\big|.$$

Let also $|F(x_1, \ldots, x_n)|$ be a concave function for $x=(x_1, \ldots, x_n) \in \mathbb{R}^n$. If $f_i(t)$, $i=1, \ldots, n$, $w(t)$ are complex functions of real variables and $f_i(t)w(t)$, $w(t)$ are integrable on $[a, b]$, then

$$A(c) = \left| \int_a^c w(t) F\big(f_1(t), \ldots, f_n(t)\big)\,\mathrm{d}t \right| + \int_c^b |w(t)|\,\mathrm{d}t \left| F\left(\frac{\int_c^b |w(t) f_1(t)|\,\mathrm{d}t}{\int_c^b |w(t)|\,\mathrm{d}t}, \ldots, \frac{\int_c^b |w(t) f_n(t)|\,\mathrm{d}t}{\int_c^b |w(t)|\,\mathrm{d}t} \right) \right|$$

is a decreasing function in c, $a \leqslant c \leqslant b$.

1.9.26 Grüss [133]

Let f, g be positive concave integrable functions on $I = [0, a]$. Then

$$\int_0^a f(s)\,\mathrm{d}s \int_0^a g(s)\,\mathrm{d}s \leqslant \frac{3}{2} a \int_0^a f(s) g(s)\,\mathrm{d}s.$$

1.9.27 Bergh [26]

Let f be a positive and quasi-concave function on $\mathbb{R}_+$, that is, $f(s) \leqslant \max(1, \frac{s}{t}) f(t)$. Assume that $0 < p \leqslant q \leqslant \infty$ and $0 < \alpha < 1$. Then

$$\left(\int_0^\infty \big(t^{-\alpha} f(t)\big)^q \frac{\mathrm{d}t}{t} \right)^{1/q} \leqslant p^{1/p} q^{-1/q} \big(\alpha(1-\alpha)\big)^{1/p - 1/q} \left(\int_0^\infty \big(t^{-\alpha} f(t)\big)^p \frac{\mathrm{d}t}{t} \right)^{1/p},$$

where equality holds for $f(t) = \min(1, t)$.

1.9.28 Maligranda, Pečarić and Persson [204]

Let $1 \leqslant p < \infty$ and let v and w be weight functions. Then the inequality

$$\left[\int_a^b f^p(t) w(t)\,\mathrm{d}t \right]^{1/p} \leqslant C \int_a^b f(t) v(t)\,\mathrm{d}t \qquad (*)$$

holds for all positive concave functions f on $[a, b]$ if and only if

$$\sup_{s \in (a,b)} \frac{\left[\int_a^b K(t,s)^p w(t)\,\mathrm{d}t\right]^{1/p}}{\int_a^b K(t,s) v(t)\,\mathrm{d}t} \leqslant C < \infty,$$

where K is the kernel given by

$$K(t,s) = \begin{cases} (s-a)(b-t) & \text{if } a \leqslant s \leqslant t \leqslant b, \\ (t-a)(b-s) & \text{if } a \leqslant t \leqslant s \leqslant b. \end{cases}$$

If $0 < p \leqslant 1$, then the reverse inequality in $(*)$ is valid if and only if

$$\inf_{s\in(a,b)} \frac{\left[\int_a^b K(t,s)^p w(t)\,\mathrm{d}t\right]^{1/p}}{\int_a^b K(t,s)v(t)\,\mathrm{d}t} \geqslant C > 0.$$

1.9.29 Borell [38]

Let $f_1, f_2, \dots, f_n$ be nonnegative concave functions on $[0,1]$ and let $p_k \geqslant 1$, $k = 1, \dots, n$. Then

$$C_n \int_0^1 \prod_{k=1}^n f_k(x)\,\mathrm{d}x \geqslant \prod_{k=1}^n (p_k+1)^{1/p_k} \|f_k\|_{p_k},$$

where for $p > 0$ and $f \geqslant 0$ the usual notation $\|f\|_p = (\int_0^1 f(x)^p\,\mathrm{d}x)^{1/p}$, $C_n = (n+1)!/([\frac{n}{2}]![\frac{n+1}{2}]!)$. Equality occurs if $f_k(x) = x$, $k \in I$, and $f_k(x) = 1-x$, $k \in I'$, for the sets of indices such that $I \cup I' = \{1,2,\dots,n\}$ and one of them contains $[\frac{n}{2}]$ elements.

1.9.30 Brenner and Alzer [41]

Let $f_1, f_2, \dots, f_n$ be nonnegative concave functions on $[0,1]$ and let $p_k \geqslant 1$, $k = 1, 2, \dots, n$. Then

$$\int_0^1 \prod_{k=1}^n f_k(x)\,\mathrm{d}x$$
$$\geqslant K_n \prod_{k=1}^n (1+p_k)^{1/p_k} \|f_k\|_{p_k} + \frac{1}{2(n+1)}\left(\prod_{k=1}^n f_k(0) + \prod_{k=1}^n f_k(1)\right),$$

where for $p > 0$ and $f \geqslant 0$ the usual notation $\|f\|_p = (\int_0^1 f^p(x)\,\mathrm{d}x)^{1/p}$ and $K_n = [\frac{n-1}{2}]![\frac{n}{2}]!/(2(n+1)(n-1)!)$.

1.10 Notes

One of the most fundamental inequalities for convex functions is that associated with the name of Jensen. Theorem 1.2.1 deals with a well-known Jensen inequality [164,165] which finds important applications in various branches of mathematics. Theorem 1.2.2 deals with a Jensen–Steffensen inequality, and its proof is due to Pečarić [367]. Theorem 1.2.3 is due to Pečarić [375] and Theorem 1.2.4 is due to Dragomir and Milošević [90]. Theorem 1.2.5 is taken from Beesack [19]. Theorem 1.3.1 is a generalization of Jensen's inequality established by Jessen [166] in 1931. The remaining results in Section 1.3 are taken from Beesack and Pečarić [20,376]. The results in Theorems 1.4.1 and 1.4.2 are due to Pečarić [366] and the results in Theorems 1.4.3–1.4.6 are taken from Dragomir and Ionescu [88].

Theorem 1.5.1 deals with the famous Hadamard inequality discovered in [134]. Theorem 1.5.2 is taken from Beesack and Pečarić [20]. Theorems 1.5.3–1.5.6 are due to Dragomir [82,86]. Theorems 1.6.1 and 1.6.2 are taken from Dragomir, Pečarić and Persson [93]. Theorems 1.6.3–1.6.5 are due to Gill, Pearce and Pečarić [128]. Theorems 1.7.1 and 1.7.2 are due to Dragomir [84] and Theorem 1.7.3 is taken from Dragomir and Ionescu [89] while Theorem 1.7.4 is taken from Dragomir, Cho and Kim [92]. Theorem 1.8.1 is taken from Maligranda [202] which deals with an important property concerning the concavity and convexity of rearrangement. Theorem 1.8.2 is taken from Maligranda, Pečarić and Persson [203] and Theorem 1.8.3 is due to Maligranda [202]. Theorems 1.8.4 and 1.8.5 are taken from Maligranda, Pečarić and Persson [204].

Chapter 2

Inequalities Related to Hardy's Inequality

2.1 Introduction

In the course of attempts to simplify the proof of Hilbert's double series theorem, G.H. Hardy [136] first proved in 1920 the most famous inequality which is now known in the literature as Hardy's inequality. Hardy's inequality is remarkable in terms of its simplicity, the large number of results to which it deals, and the variety of applications which can be related to it. Since from its discovery Hardy's inequality has evoked the interest of many mathematicians, and large number of papers have appeared which deal with new proofs, various extensions, refinements, generalizations and series analogues. In the past few years, various investigators have discovered many useful and new inequalities related to well-known Hardy's inequality. This chapter presents a number of new and basic inequalities related to Hardy's inequality recently investigated in order to achieve a diversity of desired goals.

2.2 Hardy's Series Inequality and Its Generalizations

There is a vast and growing literature related to the series inequalities. In this section we will give some basic inequalities involving series of terms, which find important applications in analysis.

In an attempt to give a simple proof of Hilbert's inequality, Hardy [136] (see also [141, Theorem 315]) establishes the following most fundamental inequality.

THEOREM 2.2.1. *If* $p > 1$, $a_n \geqslant 0$ *and* $A_n = a_1 + a_2 + \cdots + a_n$, *then*

$$\sum_{n=1}^{\infty}\left(\frac{A_n}{n}\right)^p < \left(\frac{p}{p-1}\right)^p \sum_{n=1}^{\infty} a_n^p, \tag{2.2.1}$$

unless all the a are zero. The constant is the best possible.

PROOF. The proof given here is due to Elliott [105] and is also given in [141]. By relabeling, if necessary, we may assume that $a_1 > 0$ and hence that each $A_n > 0$. We write α_n for A_n/n and agree that any number with suffix 0 is equal to 0. Now, by making use of the elementary inequality

$$x^{n+1} + ny^{n+1} \geqslant (n+1)xy^n, \tag{2.2.2}$$

$x, y \geqslant 0$ reals, we observe that

$$\begin{aligned}
\alpha_n^p - \frac{p}{p-1}\alpha_n^{p-1}a_n &= \alpha_n^p - \frac{p}{p-1}\{n\alpha_n - (n-1)\alpha_{n-1}\}\alpha_n^{p-1} \\
&= \alpha_n^p\left(1 - \frac{np}{p-1}\right) + \frac{(n-1)p}{p-1}\alpha_n^{p-1}\alpha_{n-1} \\
&\leqslant \alpha_n^p\left(1 - \frac{np}{p-1}\right) + \frac{n-1}{p-1}\{(p-1)\alpha_n^p + \alpha_{n-1}^p\} \\
&= \frac{1}{p-1}\{(n-1)\alpha_{n-1}^p - n\alpha_n^p\}.
\end{aligned} \tag{2.2.3}$$

By substituting $n = 1, 2, \ldots, N$ in (2.2.3) and adding the inequalities we have

$$\sum_{n=1}^N \alpha_n^p - \frac{p}{p-1}\sum_{n=1}^N \alpha_n^{p-1}a_n \leqslant -\frac{N\alpha_N^p}{p-1} \leqslant 0. \tag{2.2.4}$$

From (2.2.4) we observe that

$$\sum_{n=1}^N \alpha_n^p \leqslant \frac{p}{p-1}\sum_{n=1}^N \alpha_n^{p-1}a_n. \tag{2.2.5}$$

Using Hölder's inequality with indices $p, p/(p-1)$ on the right-hand side of (2.2.5) we have

$$\sum_{n=1}^N \alpha_n^p \leqslant \frac{p}{p-1}\left(\sum_{n=1}^N a_n^p\right)^{1/p}\left(\sum_{n=1}^N \alpha_n^p\right)^{(p-1)/p}. \tag{2.2.6}$$

Dividing by the last factor on the right-hand side (which is certainly positive) and raising the result to the pth power, we get

$$\sum_{n=1}^N \alpha_n^p \leqslant \left(\frac{p}{p-1}\right)^p \sum_{n=1}^N a_n^p. \tag{2.2.7}$$

When we make N tend to infinity we obtain (2.2.1), except that we have "less than or equal to" in place of "less than". In particular, we see that $\sum_{n=1}^{\infty} \alpha_n^p$ is finite.

Returning to (2.2.5), and replacing N by ∞, we obtain

$$\begin{aligned}\sum_{n=1}^{\infty} \alpha_n^p &\leqslant \frac{p}{p-1} \sum_{n=1}^{\infty} \alpha_n^{p-1} a_n \\ &\leqslant \frac{p}{p-1} \left(\sum_{n=1}^{\infty} a_n^p\right)^{1/p} \left(\sum_{n=1}^{\infty} \alpha_n^p\right)^{(p-1)/p}. \end{aligned} \tag{2.2.8}$$

There is an inequality in the second place unless (a_n^p) and (α_n^p) are proportion, that is, unless $a_n = C\alpha_n$, where C is independent of n. If this is so then $(a_1 = \alpha_1 > 0)$ C must be 1, and then $A_n = na_n$ for all n. This idea is inconsistent with the convergence of $\sum_{n=1}^{\infty} a_n^p$. Hence

$$\sum_{n=1}^{\infty} \alpha_n^p < \frac{p}{p-1} \left(\sum_{n=1}^{\infty} a_n^p\right)^{1/p} \left(\sum_{n=1}^{\infty} \alpha_n^p\right)^{(p-1)/p}, \tag{2.2.9}$$

and (2.2.1) follows from (2.2.9) as (2.2.7) followed from (2.2.6).

To prove the constant factor the best possible, we take

$$a_n = n^{-1/p}, \quad n \leqslant N, \qquad a_n = 0, \quad n > N.$$

Then

$$\sum_{n=1}^{\infty} a_n^p = \sum_{n=1}^{N} \frac{1}{n},$$

$$A_n = \sum_{1}^{n} v^{-1/p} > \int_1^n x^{-1/p}\, dx = \frac{p}{p-1} \left\{ n^{(p-1)/p} - 1 \right\}, \quad n \leqslant N,$$

$$\left(\frac{A_n}{n}\right)^p > \left(\frac{p}{p-1}\right)^p \left(\frac{1-\varepsilon_n}{n}\right), \quad n \leqslant N,$$

where $\varepsilon_n \to 0$ when $n \to \infty$. It follows that

$$\sum_{n=1}^{\infty} \left(\frac{A_n}{n}\right)^p > \sum_{n=1}^{N} \left(\frac{A_n}{n}\right)^p > \left(\frac{p}{p-1}\right)^p (1 - \eta_N) \sum_{n=1}^{\infty} a_n^p,$$

where $\eta_N \to 0$ when $N \to \infty$. Hence, any inequality of the type

$$\sum_{n=1}^{\infty}\left(\frac{A_n}{n}\right)^p < \left(\frac{p}{p-1}\right)^p (1-\varepsilon)\sum_{n=1}^{\infty} a_n^p$$

is false if a_n is chosen as above and N is sufficiently large. □

The above theorem states the relationship between the arithmetic means of a sequence and the sequence itself. This theorem along with its integral analogue was first proved by Hardy, which later went by the name "Hardy's inequality". The constant at the right-hand side of (2.2.1) is determined by Landu in [182], who showed that it is the best possible for each p.

There are many generalizations and extensions of Theorem 2.2.1, which have been proved by different writers in different ways; and we give some of these results here in the following theorems.

In 1926, Copson [69] generalizes Theorem 2.2.1 by replacing the arithmetic mean of a sequence by a weighted arithmetic mean. We shall first consider the following version of Copson's generalization of Hardy's inequality.

THEOREM 2.2.2. *Let* $p > 1, \lambda_n > 0, a_n > 0, n = 1, 2, \ldots, \sum_{n=1}^{\infty} \lambda_n a_n^p$ *converge, and further let* $\Lambda_n = \sum_{i=1}^{n} \lambda_i$, $A_n = \sum_{i=1}^{n} \lambda_i a_i$. *Then*

$$\sum_{n=1}^{\infty} \lambda_n \left(\frac{A_n}{\Lambda_n}\right)^p \leqslant \left(\frac{p}{p-1}\right)^p \sum_{n=1}^{\infty} \lambda_n a_n^p. \tag{2.2.10}$$

PROOF. We write $\alpha_n = A_n \Lambda_n^{-1}$ and agree that any number with suffix 0 is equal to 0. Now, by making use of the elementary inequality (2.2.2), we observe that

$$\begin{aligned}
&\lambda_n \alpha_n^p - \frac{p}{p-1}\lambda_n a_n \alpha_n^{p-1} \\
&= \lambda_n \alpha_n^p - \frac{p}{p-1}\alpha_n^{p-1}[\Lambda_n \alpha_n - \Lambda_{n-1}\alpha_{n-1}] \\
&= \left(\lambda_n - \Lambda_n \frac{p}{p-1}\right)\alpha_n^p + \frac{p}{p-1}\Lambda_{n-1}\alpha_{n-1}\alpha_n^{p-1} \\
&\leqslant \left(\lambda_n - \Lambda_n \frac{p}{p-1}\right)\alpha_n^p + \frac{\Lambda_{n-1}}{p-1}\left[\alpha_{n-1}^p + (p-1)\alpha_n^p\right] \\
&= \frac{1}{p-1}\left[(p\lambda_n - \lambda_n - p\Lambda_n + p\Lambda_{n-1} - \Lambda_{n-1})\alpha_n^p + \Lambda_{n-1}\alpha_{n-1}^p\right]
\end{aligned}$$

$$
\begin{aligned}
&= \frac{1}{p-1}\left[(p\lambda_n - \lambda_n - p(\Lambda_n - \Lambda_{n-1}) - \Lambda_{n-1})\alpha_n^p + \Lambda_{n-1}\alpha_{n-1}^p\right] \\
&= \frac{1}{p-1}\left[(p\lambda_n - \lambda_n - p\lambda_n - \Lambda_{n-1})\alpha_n^p + \Lambda_{n-1}\alpha_{n-1}^p\right] \\
&= \frac{1}{p-1}\left[-\Lambda_n\alpha_n^p + \Lambda_{n-1}\alpha_{n-1}^p\right] \\
&= \frac{1}{p-1}\left[\Lambda_{n-1}\alpha_{n-1}^p - \Lambda_n\alpha_n^p\right]. \qquad (2.2.11)
\end{aligned}
$$

By substituting $n = 1, \ldots, N$ in (2.2.11) and adding the inequalities, we have

$$
\sum_{n=1}^{N} \lambda_n \alpha_n^p - \frac{p}{p-1}\sum_{n=1}^{N} \lambda_n a_n \alpha_n^{p-1} \leqslant -\frac{1}{p-1}\Lambda_n \alpha_N^p \leqslant 0. \qquad (2.2.12)
$$

From (2.2.12) we observe that

$$
\sum_{n=1}^{N} \lambda_n \alpha_n^p \leqslant \frac{p}{p-1}\sum_{n=1}^{N} \lambda_n a_n \alpha_n^{p-1}. \qquad (2.2.13)
$$

Using Hölder's inequality with indices p, $p/(p-1)$ on the right-hand side of (2.2.13) we have

$$
\sum_{n-1}^{N} \lambda_n \alpha_n^p \leqslant \frac{p}{p-1}\left(\sum_{n=1}^{N} \lambda_n a_n^p\right)^{1/p}\left(\sum_{n=1}^{N} \lambda_n \alpha_n^p\right)^{(p-1)/p}.
$$

Dividing the above inequality by the last factor on the right-hand side and raising the result to the pth power, we obtain

$$
\sum_{n=1}^{N} \lambda_n \alpha_n^p \leqslant \left(\frac{p}{p-1}\right)^p \sum_{n=1}^{N} \lambda_n a_n^p. \qquad (2.2.14)
$$

When we make N tend to infinity we obtain (2.2.10). □

A version of the companion inequality proved by Copson [69] can be stated as follows.

THEOREM 2.2.3. *Let $p > 1$, $\lambda_n > 0$, $a_n > 0$ for $n = 1, 2, \ldots$, $\sum_{n=1}^{\infty} \lambda_n a_n^p$ converge, and further let*

$$\Lambda_n = \sum_{i=1}^{n} \lambda_i, \qquad A_n = \sum_{i=n}^{\infty} \frac{\lambda_i a_i}{\Lambda_i}.$$

Then

$$\sum_{n=1}^{\infty} \lambda_n A_n^p \leqslant p^p \sum_{n=1}^{\infty} \lambda_n a_n^p. \tag{2.2.15}$$

As in the proof of Theorem 2.2.1, see also Copson [69, p. 12], the constants involved in (2.2.10) and (2.2.15) are best possible. In 1928, Hardy in his paper [137] notes that the inequality given in Theorem 2.2.3 does not require a separate proof but can be derived from Copson's first inequality given in Theorem 2.2.2. In view of this remark, here we omit the proof of Theorem 2.2.3. For an independent proof of Theorem 2.2.3, see Copson [69].

In [139] Hardy and Littlewood generalizes Hardy's inequality in Theorem 2.2.1 as follows.

THEOREM 2.2.4. *Suppose $p > 0$, c is a real (but not necessarily positive) constant and $\sum_{n=1}^{\infty} a_n$ is a series of positive terms. Set*

$$A_{1n} = \sum_{k=1}^{n} a_k \quad \textit{and} \quad A_{n\infty} = \sum_{k=n}^{\infty} a_k.$$

If $p > 1$ we have

$$\sum_{n=1}^{\infty} n^{-c} A_{1n}^p \leqslant K \sum_{n=1}^{\infty} n^{-c} (na_n)^p \quad \textit{with } c > 1, \tag{2.2.16}$$

$$\sum_{n=1}^{\infty} n^{-c} A_{n\infty}^p \leqslant K \sum_{n=1}^{\infty} n^{-c} (na_n)^p \quad \textit{with } c < 1, \tag{2.2.17}$$

and if $p < 1$ we have

$$\sum_{n=1}^{\infty} n^{-c} A_{1n}^p \geqslant K \sum_{n=1}^{\infty} n^{-c} (na_n)^p \quad \textit{with } c > 1, \tag{2.2.18}$$

$$\sum_{n=1}^{\infty} n^{-c} A_{n\infty}^p \geqslant K \sum_{n=1}^{\infty} n^{-c} (na_n)^p \quad \textit{with } c < 1, \tag{2.2.19}$$

where K denotes a positive constant, not necessarily the same at each occurrence.

Theorem 2.2.4 was generalized by Leindler in [186], who replaced in (2.2.16)–(2.2.19) the sequence $\{n^{-c}\}$ by an arbitrary sequence $\{\lambda_n\}$: for instance, he proved the inequality

$$\sum_{n=1}^{\infty} \lambda_n A_{1n}^p \leqslant p^p \sum_{n=1}^{\infty} \lambda_n^{1-p} \left(\sum_{m=n}^{\infty} \lambda_m \right)^p a_n^p \tag{2.2.20}$$

with $p \geqslant 1$ and $\lambda_n > 0$.

In [226] Nemeth gives further generalizations by combing Hardy's inequality in Theorem 2.2.1 and the Hardy and Littlewood inequality in Theorem 2.2.4. In the following theorem we present the results given in [226]. We use the following definitions given in [226].

(i) $C \in M_1$ denotes that the matrix $C = (c_{m,v})$ satisfies the conditions:

$$c_{m,v} > 0, \quad v \leqslant m, \qquad c_{m,v} = 0, \quad v > m, m, v = 1, 2, \ldots, \quad \text{and}$$

$$0 < \frac{c_{m,v}}{c_{n,v}} \leqslant N_1, \quad 0 \leqslant v \leqslant n \leqslant m. \tag{2.2.21}$$

(ii) $C \in M_2$ denotes that $c_{m,v} > 0$ $(v \geqslant m)$ and $c_{m,v} = 0$ $(v < m, m, v = 1, 2, \ldots)$,

$$\frac{c_{m,v}}{c_{n,v}} \geqslant N_2, \quad 0 \leqslant n \leqslant m \leqslant v. \tag{2.2.22}$$

(iii) $C \in M_3$ denotes that $c_{v,m} > 0$ $(v \geqslant m)$ and $c_{v,m} = 0$ $(v < m, v, m = 1, 2, \ldots)$,

$$0 < \frac{c_{v,m}}{c_{v,n}} \leqslant N_3, \quad v \geqslant n \geqslant m \geqslant 0. \tag{2.2.23}$$

(iv) $C \in M_4$ denotes that $c_{v,m} > 0$ $(v \leqslant m)$ and $c_{v,m} = 0$ $(v > m, v, m = 1, 2, \ldots)$,

$$\frac{c_{v,m}}{c_{v,n}} \geqslant N_4, \quad 0 \leqslant v \leqslant m \leqslant n, \tag{2.2.24}$$

where N_i denote positive absolute constants for $i = 1, 2, 3, 4$.

The main result given by Nemeth in [226] follows.

THEOREM 2.2.5. *Let $a_n \geqslant 0$ and $\lambda_n > 0$, $n = 1, 2, \ldots$, be given, and let $C = (c_{m,k})$ be a triangular matrix.*

(a) *If $C \in M_1$ and $p \geqslant 1$, then*

$$\sum_{n=1}^{\infty} \lambda_n \left(\sum_{m=1}^{n} c_{n,m} a_m \right)^p \leqslant N_1^{p(p-1)} p^p \sum_{n=1}^{\infty} \lambda_n^{1-p} \left(\sum_{m=n}^{\infty} \lambda_m c_{m,n} \right)^p a_n^p. \tag{2.2.25}$$

(b) *If $C \in M_3$ and $p \geqslant 1$, then*

$$\sum_{m=1}^{\infty} \lambda_m \left(\sum_{n=m}^{\infty} c_{n,m} a_n \right)^p \leqslant N_3^{p(p-1)} p^p \sum_{m=1}^{\infty} \lambda_m^{1-p} \left(\sum_{n=1}^{m} \lambda_n c_{m,n} \right)^p a_m^p. \tag{2.2.26}$$

(c) *If $C \in M_2$ and $0 < p \leqslant 1$, then*

$$\sum_{n=1}^{\infty} \lambda_n \left(\sum_{v=n}^{\infty} c_{n,v} a_v \right)^p \geqslant N_2^{(1-p)p} p^p \sum_{n=1}^{\infty} \lambda_n^{1-p} \left(\sum_{k=1}^{n} c_{k,n} \lambda_k \right)^p a_n^p. \tag{2.2.27}$$

(d) *If $C \in M_4$ and $0 < p \leqslant 1$, then*

$$\sum_{m=1}^{\infty} \lambda_m \left(\sum_{n=1}^{m} c_{n,m} a_n \right)^p \geqslant N_4^{(1-p)p} p^p \sum_{m=1}^{\infty} \lambda_m^{1-p} \left(\sum_{n=m}^{\infty} \lambda_n c_{m,n} \right)^p a_m^p. \tag{2.2.28}$$

We note that Theorem 2.2.5 implies Leindler's theorem in [186], further if $\lambda_m = c_{m,m} f_{(m)}^{1-p}$ and we write $c_{m,n} f_{(m)}$ instead of elements of the matrix C, then assertion (a) includes Theorem 3 of Izumi, Izumi and Petersen [163], and in the case $\lambda_n = f_{(n)}^{-p}$ and $c_{k,n} = f(k) a_{k,n}$, assertion (d) reduces to Theorem 5 of Davis and Petersen [78].

In the proof of the above theorem, we require the following lemmas.

LEMMA 2.2.1 [78, Lemma 1]. *If $p > 1$ and $z_n \geqslant 0$, $n = 1, 2, \ldots$, then*

$$\left(\sum_{k=1}^{n} z_k \right)^p \leqslant p \sum_{k=1}^{n} z_k \left(\sum_{v=1}^{k} z_v \right)^{p-1}.$$

PROOF. Let $\lambda_r = z_1 + \cdots + z_r$, $r = 1, 2, \ldots, n$. Then

$$\begin{aligned}
(\lambda_n)^p &= p \int_0^{\lambda_n} x^{p-1} \, \mathrm{d}x \\
&= p \left(\int_0^{\lambda_1} + \int_{\lambda_1}^{\lambda_2} + \cdots + \int_{\lambda_{n-1}}^{\lambda_n} \right) x^{p-1} \, \mathrm{d}x
\end{aligned}$$

$$\leqslant p\{\lambda_1\lambda_1^{p-1} + (\lambda_2 - \lambda_1)\lambda_2^{p-1} + \cdots + (\lambda_n - \lambda_{n-1})\lambda_n^{p-1}\}$$
$$= p\{z_1^p + z_2(z_1 + z_2)^{p-1} + \cdots + z_n(z_1 + \cdots + z_n)^{p-1}\}.$$

The proof is complete. □

The proofs of the following lemmas are similar to that of Lemma 2.2.1 (see [226]).

LEMMA 2.2.2. *If* $0 < p < 1$ *and* $z_1 > 0$, $z_n \geqslant 0$, $n = 2, 3, \ldots$, *then*

$$\left(\sum_{k=1}^{n} z_k\right)^p \geqslant p \sum_{k=1}^{n} z_k \left(\sum_{v=1}^{k} z_v\right)^{p-1}.$$

LEMMA 2.2.3. *If* $0 < p < 1$ *and* $z_n \geqslant 0$, $n = 1, 2, \ldots$, *then for every natural number* N, *for which* $z_N > 0$,

$$\left(\sum_{k=n}^{N} z_k\right)^p \geqslant p \sum_{k=n}^{N} z_k \left(\sum_{v=k}^{N} z_v\right)^{p-1}.$$

LEMMA 2.2.4. *If* $p > 1$ *and* $z_n \geqslant 0$, $n = 1, 2, \ldots$, *then for every natural number* N,

$$\left(\sum_{k=n}^{N} z_k\right)^p \leqslant p \sum_{k=n}^{N} z_k \left(\sum_{v=k}^{N} z_v\right)^{p-1}.$$

PROOF OF THEOREM 2.2.5. For $p = 1$ the assertions are obvious; we have only to interchange the order of summations. Further we may assume that not all a_n vanish (otherwise the theorem is evident).

(a) By Lemma 2.2.1 we obtain, for $C = (c_{m,k}) \in M_1$,

$$\sum_{n=1}^{N} \lambda_n \left(\sum_{m=1}^{n} c_{n,m} a_m\right)^p \leqslant p \sum_{n=1}^{N} \lambda_n \sum_{m=1}^{n} c_{n,m} a_m \left(\sum_{k=1}^{m} c_{n,k} a_k\right)^{p-1}$$
$$\leqslant N_1^{p-1} p \sum_{n=1}^{N} \lambda_n \sum_{m=1}^{n} c_{n,m} a_m \left(\sum_{k=1}^{m} c_{m,k} a_k\right)^{p-1}$$
$$= N_1^{p-1} p \sum_{m=1}^{N} \left(\sum_{k=1}^{m} c_{m,k} a_k\right)^{p-1} a_m \sum_{n=m}^{N} \lambda_n c_{n,m}.$$

Hence, using Hölder's inequality, we have

$$\sum_{n=1}^{N}\lambda_n\left(\sum_{m=1}^{n}c_{n,m}a_m\right)^p$$

$$\leqslant N_1^{p-1}p\left\{\sum_{m=1}^{N}\lambda_m\left(\sum_{k=1}^{m}c_{m,k}a_k\right)^p\right\}^{1/q}\left\{\sum_{m=1}^{N}\lambda_m^{1-p}\left(\sum_{n=m}^{N}\lambda_n c_{n,m}\right)^p a_m^p\right\}^{1/p}$$

with $q = p/(p-1)$, which by standard computation gives assertion (a).

(b) By Lemma 2.2.4 we have, for $C = (c_{m,k}) \in M_3$,

$$\begin{aligned}\sum_{m=1}^{N}\lambda_m\left(\sum_{n=m}^{N}c_{n,m}a_n\right)^p &\leqslant p\sum_{m=1}^{N}\lambda_m\sum_{n=m}^{N}c_{n,m}a_n\left(\sum_{v=n}^{N}c_{v,m}a_v\right)^{p-1}\\ &\leqslant N_3^{p-1}p\sum_{m=1}^{N}\lambda_m\sum_{n=m}^{N}c_{n,m}a_n\left(\sum_{v=n}^{N}c_{v,n}a_v\right)^{p-1}\\ &= N_3^{p-1}p\sum_{n=1}^{N}\left(\sum_{v=n}^{N}c_{v,n}a_v\right)^{p-1}a_n\sum_{m=1}^{n}c_{n,m}\lambda_m.\end{aligned}$$

Hence, using Hölder's inequality, we have

$$\sum_{m=1}^{N}\lambda_m\left(\sum_{n=m}^{N}c_{n,m}a_n\right)^p$$

$$\leqslant N_3^{p-1}p\left\{\sum_{n=1}^{N}\lambda_n\left(\sum_{v=n}^{N}c_{v,n}a_v\right)^p\right\}^{1/q}\left\{\sum_{n=1}^{N}\lambda_n^{1-p}\left(\sum_{m=1}^{n}c_{n,m}\lambda_m\right)^p a_m^p\right\}^{1/p},$$

where $q = p/(p-1)$ which by standard computation gives assertion (b).

(c) Using Lemma 2.2.3 with an index n for which $a_N > 0$, we obtain

$$\begin{aligned}\sum_{n=1}^{N}\lambda_n\left(\sum_{v=n}^{N}c_{n,v}a_v\right)^p &\geqslant p\sum_{n=1}^{N}\lambda_n\sum_{v=n}^{N}c_{n,v}a_v\left(\sum_{k=v}^{N}c_{n,k}a_k\right)^{p-1}\\ &\geqslant N_2^{1-p}p\sum_{n=1}^{N}\lambda_n\sum_{v=n}^{N}c_{n,v}a_v\left(\sum_{k=v}^{N}c_{v,k}a_k\right)^{p-1}\end{aligned}$$

$$= N_2^{1-p} p \sum_{v=1}^{N} \left(\sum_{k=v}^{N} c_{v,k} a_k \right)^{p-1} a_v \sum_{n=1}^{v} \lambda_n c_{n,v}.$$

Hence, using Hölder's inequality [16, p. 19], we have

$$\sum_{n=1}^{N} \lambda_n \left(\sum_{v=n}^{N} c_{n,v} a_v \right)^p$$

$$\geqslant N_2^{1-p} p \left\{ \sum_{v=1}^{N} \lambda_v \left(\sum_{k=v}^{N} c_{v,k} a_k \right)^p \right\}^{1/q} \left\{ \sum_{v=1}^{N} \lambda_v^{1-p} \left(\sum_{n=1}^{v} \lambda_n c_{n,v} \right)^p a_v^p \right\}^{1/p}.$$

This result gives assertion (c) by standard computation.

(d) We may assume that $a_1 \neq 0$. Using Lemma 2.2.2, Hölder's inequality with indices $p, q = p/(p-1)$, we have

$$\sum_{m=1}^{N} \lambda_m \left(\sum_{n=1}^{m} c_{n,m} a_n \right)^p$$

$$\geqslant p \sum_{m=1}^{N} \lambda_m \sum_{n=1}^{m} c_{n,m} a_n \left(\sum_{k=1}^{n} c_{k,m} a_k \right)^{p-1}$$

$$\geqslant N_4^{1-p} p \sum_{m=1}^{N} \lambda_m \sum_{n=1}^{m} c_{n,m} a_n \left(\sum_{k=1}^{n} c_{k,n} a_k \right)^{p-1}$$

$$= N_4^{1-p} p \sum_{n=1}^{N} \left(\sum_{k=1}^{n} c_{k,n} a_k \right)^{p-1} a_n \sum_{m-n}^{N} \lambda_m c_{n,m}$$

$$\geqslant N_4^{1-p} p \left\{ \sum_{n=1}^{N} \lambda_n \left(\sum_{k=1}^{n} c_{k,n} a_k \right)^p \right\}^{1/q} \left\{ \sum_{n=1}^{N} \lambda_n^{1-p} \left(\sum_{m=n}^{N} \lambda_m c_{n,m} \right)^p a_n^p \right\}^{1/p}.$$

By standard computation this result gives assertion (d), and the proof is complete. □

In [196] Love has established generalizations of Hardy's and Copson's series inequalities by replacing means by more general linear transforms. The results in [196] are based on the following lemma.

LEMMA 2.2.5. *If g is a decreasing (equimeasurable) rearrangement of a nonnegative measurable function f on $(0, c)$, h is nonnegative and decreasing* *on* $(0, b)$, $0 < b < c$ *and* $p > 0$, *then*

$$\int_0^b f(u)h(u)\,\mathrm{d}u \leqslant \int_0^b g(u)h(u)\,\mathrm{d}u, \qquad \int_0^c f(u)^p\,\mathrm{d}u = \int_0^c g(u)^p\,\mathrm{d}u.$$

PROOF. The second result is immediate. For the first, let k be a decreasing rearrangement of f on $(0, b)$ and let $k(u) = f(u)$ for $b \leqslant u < c$. Also, let $h(u) = 0$ for $b \leqslant u < c$, so that h is decreasing on $(0, c)$. Observing that g is a decreasing rearrangement of k on $(0, c)$, two applications of Theorem 378 in [141], one on $(0, b)$ and the other on $(0, c)$, give

$$\int_0^b fh\,\mathrm{d}u \leqslant \int_0^b kh\,\mathrm{d}u = \int_0^c kh\,\mathrm{d}u \leqslant \int_0^c gh\,\mathrm{d}u = \int_0^b gh\,\mathrm{d}u. \qquad \square$$

The following theorems given in [196] generalize Copson's inequalities in Theorems 2.2.2 and 2.2.3 (see also [70]), restated perhaps more neatly.

THEOREM 2.2.6. *If* $p > 1$, $\alpha(t)$ *is nonnegative and decreasing in* $(0, 1]$,

$$A = \int_0^1 \alpha(t)t^{-1/p}\,\mathrm{d}t < \infty, \qquad \lambda_n > 0, \qquad \Lambda_m = \sum_{n=1}^{m} \lambda_n$$

and

$$|a_{mn}| \leqslant \frac{\lambda_n}{\Lambda_m}\alpha\left(\frac{\Lambda_n}{\Lambda_m}\right) \quad \text{for } 0 < n \leqslant m.$$

Then

$$\left(\sum_{m=1}^{\infty} \lambda_m \left|\sum_{n=1}^{m} a_{mn}x_n\right|^p\right)^{1/p} \leqslant A\left(\sum_{m=1}^{\infty} \lambda_m |x_m|^p\right)^{1/p},$$

where (x_n) *is a fixed sequence.*

PROOF. It will be enough to prove the inequality with upper terminal ∞ of the outer summations replaced by any positive integer M. Fix such M and a sequence (x_n).

Let $f(u) = |x_n|$ for $\Lambda_{n-1} < u \leqslant \Lambda_n$ and $0 < n \leqslant M$, where $\Lambda_0 = 0$. Let $g(u)$ be a decreasing rearrangement of $f(u)$ on $(0, \Lambda_M]$. For m such that $0 < m \leqslant M$, let

$$z_m = \lambda_m^{1/p} \sum_{n=1}^{m} |a_{mn}x_n| \quad \text{and} \quad \Lambda_{m-1} < s \leqslant \Lambda_m.$$

Then

$$\begin{aligned}
\frac{z_m}{\lambda_m^{1/p}} &\leqslant \sum_{n=1}^{m} \frac{\lambda_n}{\Lambda_m} \alpha\left(\frac{\Lambda_n}{\Lambda_m}\right) |x_n| \\
&= \frac{1}{\Lambda_m} \sum_{n=1}^{m} \int_{\Lambda_{n-1}}^{\Lambda_n} \alpha\left(\frac{\Lambda_n}{\Lambda_m}\right) f(u)\, \mathrm{d}u \\
&\leqslant \frac{1}{\Lambda_m} \sum_{n=1}^{m} \int_{\Lambda_{n-1}}^{\Lambda_n} \alpha\left(\frac{u}{\Lambda_m}\right) f(u)\, \mathrm{d}u \\
&= \frac{1}{\Lambda_m} \int_{0}^{\Lambda_m} \alpha\left(\frac{u}{\Lambda_m}\right) f(u)\, \mathrm{d}u \qquad (2.2.29) \\
&\leqslant \frac{1}{\Lambda_m} \int_{0}^{\Lambda_m} \alpha\left(\frac{u}{\Lambda_m}\right) g(u)\, \mathrm{d}u \\
&= \int_0^1 \alpha(t) g(\Lambda_m t)\, \mathrm{d}t \\
&\leqslant \int_0^1 \alpha(t) g(st)\, \mathrm{d}t. \qquad (2.2.30)
\end{aligned}$$

Lemma 2.2.5 has been used after (2.2.29). From (2.2.30),

$$\begin{aligned}
\left(\sum_{m=1}^{M} z_m^p\right)^{1/p} &= \left(\sum_{m=1}^{M} \frac{1}{\lambda_m} \int_{\Lambda_{m-1}}^{\Lambda_m} z_m^p\, \mathrm{d}s\right)^{1/p} \\
&\leqslant \left(\sum_{m=1}^{M} \int_{\Lambda_{m-1}}^{\Lambda_m} \left(\int_0^1 \alpha(t) g(st)\, \mathrm{d}t\right)^p \mathrm{d}s\right)^{1/p} \\
&= \left(\int_0^{\Lambda_M} \left(\int_0^1 \alpha(t) g(st)\, \mathrm{d}t\right)^p \mathrm{d}s\right)^{1/p} \\
&\leqslant \int_0^1 \left(\int_0^{\Lambda_M} \alpha(t)^p g(st)^p\, \mathrm{d}s\right)^{1/p} \mathrm{d}t \qquad (2.2.31) \\
&= \int_0^1 \alpha(t) t^{-1/p} \left(\int_0^{\Lambda_M} g(st)^p t\, \mathrm{d}s\right)^{1/p} \mathrm{d}t \\
&= \int_0^1 \alpha(t) t^{-1/p} \left(\int_0^{\Lambda_M t} g(u)^p\, \mathrm{d}u\right)^{1/p} \mathrm{d}t
\end{aligned}$$

$$\leqslant A\left(\int_0^{\Lambda_M} g(u)^p \, \mathrm{d}u\right)^{1/p} \tag{2.2.32}$$

$$= A\left(\int_0^{\Lambda_M} f(u)^p \, \mathrm{d}u\right)^{1/p} \tag{2.2.33}$$

$$= A\left(\sum_{m=1}^{M} \int_{\Lambda_{m-1}}^{\Lambda_m} f(u)^p \, \mathrm{d}u\right)^{1/p}$$

$$= A\left(\sum_{m=1}^{M} \lambda_m |x_m|^p\right)^{1/p}.$$

The "double integral" version of Minkowski's inequality [141, Theorem 202] has been used at (2.2.31), and Lemma 2.2.5 at (2.2.33). Making $M \to \infty$ the inequality follows. □

COROLLARY 2.2.1 (Copson's inequality, Theorem 2.2.2). *If* $p \geqslant c > 1$, (x_n), λ_n *and* Λ_n *are as in Theorem* 2.2.6, *and* $x_n \geqslant 0$, *then*

$$\left(\sum_{m=1}^{\infty} \lambda_m \Lambda_m^{-c} \left(\sum_{n=1}^{m} \lambda_n x_n\right)^p\right)^{1/p} \leqslant \frac{p}{c-1}\left(\sum_{m=1}^{\infty} \lambda_m \Lambda_m^{p-c} x_m^p\right)^{1/p}.$$

PROOF. For $0 < t \leqslant 1$ and $0 < n \leqslant m$, let

$$\alpha(t) = t^{c/p-1} \quad \text{and} \quad a_{mn} = \frac{\lambda_n \Lambda_n^{c/p-1}}{\Lambda_m^{c/p}} = \frac{\lambda_n}{\Lambda_m}\alpha\left(\frac{\Lambda_n}{\Lambda_m}\right). \tag{2.2.34}$$

Theorem 2.2.6 now applies with $A = p/(c-1)$ which after replacing x_n by $\Lambda_n^{1-c/p} x_n$ gives Copson's inequality. □

THEOREM 2.2.7. *If* $p > 1$, $t\alpha(t)$ *is nonnegative and increasing in* $[1, \infty)$,

$$B = \int_1^{\infty} \alpha(t) t^{-1/p} \, \mathrm{d}t < \infty, \qquad \lambda_n > 0, \qquad \Lambda_m = \sum_{n=1}^{m} \lambda_n$$

and

$$|a_{mn}| \leqslant \frac{\lambda_n}{\Lambda_m}\alpha\left(\frac{\Lambda_n}{\Lambda_m}\right) \quad \textit{for } 0 < m \leqslant n,$$

then

$$\left(\sum_{m=1}^{\infty}\lambda_m\left|\sum_{n=m}^{\infty}a_{mn}x_n\right|^p\right)^{1/p}\leqslant B\left(\sum_{m=1}^{\infty}\lambda_m|x_n|^p\right)^{1/p},$$

where (x_n) *is a fixed sequence.*

PROOF. Let $q=p'$, $\beta(t)=t^{-1}\alpha(t^{-1})$ and $b_{mn}=\lambda_n a_{mn}\lambda_m^{-1}$. Then $q>1$, $\beta(t)$ is nonnegative and decreasing in $(0,1]$,

$$\int_0^1\beta(u)u^{-1/q}\,\mathrm{d}u=B<\infty\quad\text{and}\quad|b_{mn}|\leqslant\frac{\lambda_n}{\Lambda_m}\beta\left(\frac{\Lambda_n}{\Lambda_m}\right),$$

for $0<n\leqslant m$. So Theorem 2.2.6 applies with p, $\alpha(t)$ and a_{mn} replaced by q, $\beta(t)$ and b_{mn}. Replacing x_n by $\lambda_n^{-1/q}y_n$, and defining $b_{mn}=0$ for $0<m<n$, the strong form of that theorem gives

$$\left(\sum_{m=1}^{\infty}\left(\sum_{n=1}^{\infty}\left|\lambda_m^{1/q}b_{mn}\lambda_n^{-1/q}y_n\right|\right)^q\right)^{1/q}\leqslant B\left(\sum_{m=1}^{\infty}|y_m|^q\right)^{1/q}$$

for all y_n. The converse of Hölder's inequality [141, Theorems 15 and 167] leads to the conjugate inequality

$$\left(\sum_{m=1}^{\infty}\left(\sum_{n=1}^{\infty}\left|\lambda_n^{1/q}b_{mn}\lambda_m^{-1/q}y_n\right|\right)^p\right)^{1/p}\leqslant B\left(\sum_{m=1}^{\infty}|y_m|^p\right)^{1/p}$$

for all y_n. Noting the definitions of b_{mn}, and putting $y_n=\lambda_n^{1/p}x_n$, this result reduces to (the strong form of) the stated conclusion. □

COROLLARY 2.2.2 (Copson's inequality, Theorem 2.2.3). *If* $p>1>c\geqslant 0$, λ_n *and* Λ_n *are as in Theorem* 2.2.7, $x_n\geqslant 0$ *and* $\sum_{n=1}^{\infty}\lambda_n x_n$ *is convergent, then*

$$\left(\sum_{m=1}^{\infty}\lambda_m\Lambda_m^{-c}\left(\sum_{n=m}^{\infty}\lambda_n x_n\right)^p\right)^{1/p}\leqslant\frac{p}{1-c}\left(\sum_{m=1}^{\infty}\lambda_m\Lambda_m^{p-c}x_m^p\right)^{1/p}.$$

PROOF. This proof follows from Corollary 2.2.1, but with $t\geqslant 1$ and $0<m\leqslant n$ in (2.2.34) and Theorem 2.2.7 used instead of Theorem 2.2.6. □

2.3 Series Inequalities Related to Those of Hardy, Copson and Littlewood

Hardy's inequality concerning the series of terms given in Theorem 2.2.1 has received wide attention from the book "*Inequalities*" written in 1934 by Hardy, Littlewood and Pólya. In this section we present some basic inequalities due to Copson and related to those of Hardy, Copson and Littlewood. In what follows, we assume that all the sums exist on the respective domains of definitions and agree that the value of any function $u(m,n)$ or $u(n)$ for $m=0$ or $n=0$ is zero.

In 1979, Copson [71] proves two series inequalities which in fact are the discrete analogues of the integral inequalities established earlier in 1932 by Hardy and Littlewood [140].

The first result established by Copson in [71] is given in the following theorem.

THEOREM 2.3.1. *Let $\{a_n\}$ be a sequence of real numbers such that $\sum_{-\infty}^{\infty} a_n^2$, $\sum_{-\infty}^{\infty}(\Delta^2 a_n)^2$ are convergent. Then*

$$\left\{\sum_{-\infty}^{\infty}(\Delta a_n)^2\right\} \leqslant \sum_{-\infty}^{\infty} a_n^2 \sum_{-\infty}^{\infty}(\Delta^2 a_n)^2.$$

Equality occurs if and only if $a_n = 0$ for all n, where $\Delta a_n = a_{n+1} - a_n$ and $\Delta^2 a_n = \Delta(\Delta a_n)$.

PROOF. Since

$$\Delta(a_n \Delta a_n) = a_{n+1}\Delta^2 a_n + (\Delta a_n)^2,$$

we have

$$\sum_{-M}^{N} a_{n+1}\Delta^2 a_n + \sum_{-M}^{N}(\Delta a_n)^2 = a_{N+1}\Delta^2 a_{N+1} - a_{-M}\Delta^2 a_{-M}.$$

But since $\sum_{-\infty}^{\infty} a_n^2$ convergent, a_n tends to zero as $n \to \infty$ and as $n \to -\infty$. Therefore

$$\lim_{\substack{M\to\infty \\ N\to\infty}} \left\{\sum_{-M}^{N} a_{n+1}\Delta^2 a_n + \sum_{-M}^{N}(\Delta a_n)^2\right\} = 0.$$

By Cauchy's inequality,

$$\sum_{-\infty}^{\infty} a_{n+1}\Delta^2 a_n$$

is absolutely convergent. Hence $\sum_{-\infty}^{\infty}(\Delta a_n)^2$ is also convergent, and

$$\sum_{-\infty}^{\infty}(\Delta a_n)^2 = -\sum_{-\infty}^{\infty} a_{n+1}\Delta^2 a_n.$$

Therefore

$$\left\{\sum_{-\infty}^{\infty}(\Delta a_n)^2\right\}^2 \leqslant \sum_{-\infty}^{\infty} a_{n+1}^2 \sum_{-\infty}^{\infty}(\Delta^2 a_n)^2. \tag{2.3.1}$$

Equality occurs if and only if there exists a real constant λ such that $\Delta^2 a_n = \lambda a_{n+1}$ for all values of n. The solution of this difference equation is

$$a_n = Ak_1^n + Bk_2^n,$$

where k_1 and k_2 are the roots of the equation

$$(k-1)^2 = \lambda k$$

if $\lambda \neq 0$, but it is

$$a_n = A + Bn$$

if $\lambda = 0$. The latter case is impossible since the series $\sum_{-\infty}^{\infty} a_n^2$ would diverge. If $\lambda \neq 0$, k_1 and k_2 are unequal and $k_1 k_2 = 1$. If k_1 and k_2 are real, one is numerically greater than unity, the other less, and $\sum_{-\infty}^{\infty} a_n^2$ diverges. If k_1 and k_2 are complex, $a_n = C\cos(n\alpha + \beta)$, and $\sum_{-\infty}^{\infty} a_n^2$ diverges again. Hence equality occurs in (2.3.1) if and only if $a_n = 0$ for all values of n. □

The second result established by Copson in [71] is embodied in the following theorem.

THEOREM 2.3.2. *Let $\{a_n\}$ be a sequence of real numbers such that $\sum_0^{\infty} a_n^2$, $\sum_0^{\infty}(\Delta^2 a_n)^2$ are convergent. Then*

$$\left\{\sum_{0}^{\infty}(\Delta a_n)^2\right\}^2 \leqslant 4\sum_{0}^{\infty} a_n^2 \sum_{0}^{\infty}(\Delta^2 a_n)^2.$$

Equality occurs if and only if $a_n = 0$ for all n.

PROOF. The proof depends on a generalization of Cauchy's inequality, that

$$D=\begin{vmatrix}\sum a_n^2 & \sum a_nb_n & \sum a_nc_n\\ \sum a_nb_n & \sum b_n^2 & \sum b_nc_n\\ \sum a_nc_n & \sum b_nc_n & \sum c_n^2\end{vmatrix}\geqslant 0,$$

a result stated in [141, p. 16]. Equality occurs if and only if the sequences $\{a_n\}, \{b_n\}, \{c_n\}$ are linearly independent. If we put $b_n = \Delta a_n$, $c_n = \Delta^2 a_n$, we obtain

$$D=\begin{vmatrix}A & H & G\\ H & B & F\\ G & F & C\end{vmatrix}\geqslant 0,$$

where

$$A=\sum_0^\infty a_n^2,\qquad B=\sum_0^\infty(\Delta a_n)^2,\qquad C=\sum_0^\infty(\Delta^2 a_n)^2,$$

$$F=\sum_0^\infty \Delta a_n\Delta^2 a_n,\qquad G=\sum_0^\infty a_n\Delta^2 a_n,\qquad H=\sum_0^\infty a_n\Delta a_n.$$

The series A and C are, by hypotheses, convergent. The series G is absolutely convergent by Cauchy's inequality. If we put $a_n = 0$ for all $n < 0$, we see that the series B is convergent; see the proof of Theorem 2.3.1. Then, again by Cauchy's inequality, the series F and H are absolutely convergent. The determinant D vanishes if and only if there exist real constants α, β, γ, not all zero, such that

$$\alpha a_n + \beta\Delta a_n + \gamma\Delta^2 a_n = 0$$

for all n. This implies, either that $a_n = 0$ for all n or that

$$a_n = r^n\cos(pr+q),$$

where p, q, r are real and $0 < r < 1$ since $\sum a_n^2$ convergent.

Now

$$\Delta a_n^2 = 2a_n\Delta a_n + (\Delta a_n)^2.$$

Summing from $n = 0$ to $n = \infty$, we find that

$$2H + B = -a_0^2,$$

using the fact that $a_n \to 0$ as $n \to \infty$. Similarly

$$2F + C = -b_0^2,$$

where $b_0 = \Delta a_0$. Lastly, from

$$\Delta(a_n \Delta a_n) = a_n \Delta^2 a_n + \Delta a_n \Delta^2 a_n + (\Delta a_n)^2,$$

we obtain

$$G + B + F = -a_0 b_0.$$

Hence

$$D = \begin{vmatrix} A & -\frac{1}{2}(a_0^2 + B) & G \\ -\frac{1}{2}(a_0^2 + B) & B & -\frac{1}{2}(b_0^2 + C) \\ G & -\frac{1}{2}(b_0^2 + C) & C \end{vmatrix}$$
$$= ABC + \frac{1}{2}(a_0^2 + B)(b_0^2 + C)G - \frac{1}{4}A(b_0^2 + C)^2 - BG^2 - \frac{1}{4}C(a_0^2 + B)^2.$$

Therefore

$$\begin{aligned} 2ABC + (a_0^2 + B)(b_0^2 + C)G - 2BG^2 &\geqslant \frac{1}{2}A(b_0^2 + C)^2 + \frac{1}{2}C(a_0^2 + B)^2 \\ &\geqslant (a_0^2 + B)(b_0^2 + C)\sqrt{(AC)}, \end{aligned}$$

by the arithmetic mean–geometric mean inequality. Equality at the last step occurs only when $A(b_0^2 + C)^2 = C(a_0^2 + B)^2$. Hence

$$2ABC - 2BG^2 \geqslant \{\sqrt{(AC)} - G\}(a_0^2 + B)(b_0^2 + C).$$

By Cauchy's inequality, $G^2 \leqslant AC$. If $G^2 < AC$, we can divide through by $\sqrt{(AC)} - G$, to obtain

$$2B\{\sqrt{(AC)} + G\} \geqslant (a_0^2 + B)(b_0^2 + C).$$

Therefore

$$\begin{aligned} 2B\sqrt{(AC)} &\geqslant (a_0^2 + B)(b_0^2 + C) - 2BG \\ &= (a_0^2 + B)(b_0^2 + C) - 2B(B + G) + 2B^2. \end{aligned}$$

But

$$B + G = -a_0 b_0 - F = -a_0 b_0 + \frac{1}{2}b_0^2 + \frac{1}{2}C.$$

Therefore

$$\begin{aligned}
2B\sqrt{(AC)} &\geqslant (a_0^2 + B)(b_0^2 + C) - 2B\left(\frac{1}{2}b_0^2 - a_0 b_0 + \frac{1}{2}C\right) + 2B^2 \\
&= a_0^2 b_0^2 + 2a_0 b_0 B + 2B^2 + a_0^2 C \\
&= (a_0 b_0 + B)^2 + B^2 + a_0^2 C.
\end{aligned}$$

This result gives

$$2B\sqrt{(AC)} - B^2 \geqslant (a_0 b_0 + B)^2 + a_0^2 C.$$

Therefore, if $G^2 < AC$,

$$B \leqslant 2\sqrt{(AC)},$$

the required inequality.

The relation $G^2 = AC$, that is,

$$\left\{\sum_0^\infty a_n \Delta^2 a_n\right\}^2 = \sum_0^\infty a_n^2 \sum_0^\infty (\Delta^2 a_n)^2$$

holds if and only if there exists a constant λ such that $\Delta^2 a_n = \lambda a_n$. If $\lambda = 0$, Δa_n is a constant and so is zero by the convergence condition which again implies that a_n is constant and so is zero. If $\lambda = 0$, $\{a_n\}$ is the null sequence. If $\lambda = 0$,

$$a_n = \alpha k_1^n + \beta k_2^n,$$

where k_1 and k_2 are the roots of $(k - 1)^2 = \lambda$. The roots are different. If $\lambda = -\mu^2 < 0$,

$$a_n = \alpha(1 + \mathrm{i}\mu)^n + \beta(1 - \mathrm{i}\mu)^n$$

which is impossible by the convergence condition for $|1 \pm \mathrm{i}\mu| > 1$, and a_n does not then tend to zero as $n \to \infty$. If $\lambda = v^2 > 0$, where $v > 0$,

$$a_n = \alpha(1 + v)^n + \beta(1 - v)^n$$

which tends to zero as $n \to \infty$ if and only if $\alpha = 0$ and $0 < v < 2$. Dropping the factor β, the only case when $G^2 = AC$ is $a_n = r^n$, where $-1 < r < 1$. This gives

$$A = \frac{1}{1 - r^2}, \qquad B = (r - 1)^2 A, \qquad C = (r - 1)^4 A,$$

and so

$$b = \sqrt{(AC)} < 2\sqrt{AC}.$$

The inequality also holds when $G^2 = AC$.

It remains to consider the conditions under which $B = 2\sqrt{(AC)}$. Going over the proof, we see that they are following:

(i) $\{a_n\}, \{\Delta a_n\}, \{\Delta^2 a_n\}$ are linearly dependent sequences,
(ii) $A(b_0^2 + C)^2 = C(a_0^2 + B)^2$,
(iii) $a_0 b_0 + B = 0$, $a_0^2 C = 0$.

In fact, (iii) implies that $\{a_n\}$ is the null-sequence, and (i) and (ii) follow. If $a_0 = 0$ then $B = 0$, hence $\Delta a_n = 0$ and a_n is zero for all n. If $C = 0$ then $\Delta^2 a_n = 0$, hence Δa_n is a constant and so is zero by the convergence condition. This result implies that a_n is zero for all n.

We have thus covered all the cases. We have proved that

$$\left\{\sum_0^\infty (\Delta a_n)^2\right\}^2 \leqslant 4\sum_0^\infty a_n^2 \sum_0^\infty (\Delta^2 a_n)^2,$$

with equality if and only if a_n is zero for all values of n. □

REMARK 2.3.1. As observed by Copson in [71] the constants in inequalities in Theorems 2.3.1 and 2.3.2 are best possible. For detailed discussion, see [71, pp. 110 and 114].

In [277] Pachpatte establishes some generalizations of Copson's inequality given in Theorem 2.2.2. The main result established in [277] is given in the following theorem.

THEOREM 2.3.3. *Let $f(u)$ be a real-valued positive convex function defined for $u > 0$. Let $p > 1$ be a constant, $\lambda_n > 0$, $a_n > 0$, $\sum_{n=1}^\infty \lambda_n f^p(a_n)$ converge, and further let $\Lambda_n = \sum_{i=1}^n \lambda_i$, $A_n = \sum_{i=1}^n \lambda_i a_i$. Then*

$$\sum_{n=1}^\infty \lambda_n f^p\left(\frac{A_n}{\Lambda_n}\right) \leqslant \left(\frac{p}{p-1}\right)^p \sum_{n=1}^\infty \lambda_n f^p(a_n). \tag{2.3.2}$$

PROOF. Since f is convex, by Jensen's inequality (see [174, p. 133]), we have

$$f\left(\frac{A_n}{\Lambda_n}\right) \leqslant \frac{F_n}{\Lambda_n}, \tag{2.3.3}$$

where $F_n = \sum_{i=1}^{n} \lambda_i f(a_i)$. We write $\alpha_n = F_n \Lambda_n^{-1}$ and agree that any number with suffix 0 is equal to 0. Now, by making use of the elementary inequality

$$u^{k+1} + kv^{k+1} \geqslant (k+1)uv^k, \quad u, v \geqslant 0, k \geqslant 1, \tag{2.3.4}$$

we observe that

$$\begin{aligned}
&\lambda_n \alpha_n^p - \frac{p}{p-1}\lambda_n f(a_n)\alpha_n^{p-1} \\
&= \lambda_n \alpha_n^p - \frac{p}{p-1}\alpha_n^{p-1}[\alpha_n \Lambda_n - \alpha_{n-1}\Lambda_{n-1}] \\
&= \left(\lambda_n - \frac{p}{p-1}\Lambda_n\right)\alpha_n^p + \frac{p}{p-1}\Lambda_{n-1}\alpha_{n-1}\alpha_n^{p-1} \\
&\leqslant \left(\lambda_n - \frac{p}{p-1}\Lambda_n\right)\alpha_n^p + \frac{\Lambda_{n-1}}{p-1}\left[\alpha_{n-1}^p + (p-1)\alpha_n^p\right] \\
&= \frac{1}{p-1}\left[\Lambda_{n-1}\alpha_{n-1}^p - \Lambda_n \alpha_n^p\right].
\end{aligned} \tag{2.3.5}$$

By substituting $n = 1, \ldots, m$ in (2.3.5) and adding the inequalities, we see that

$$\sum_{n=1}^{m}\left[\lambda_n \alpha_n^p - \frac{p}{p-1}\lambda_n f(a_n)\alpha_n^{p-1}\right] \leqslant -\frac{1}{p-1}\Lambda_m \alpha_m^p \leqslant 0. \tag{2.3.6}$$

From (2.3.6) we observe that

$$\sum_{n=1}^{m}\lambda_n \alpha_n^p \leqslant \frac{p}{p-1}\sum_{n=1}^{m}\left\{\lambda_n^{1/p} f(a_n)\right\}\left\{\lambda_n^{(p-1)/p}\alpha_n^{p-1}\right\}. \tag{2.3.7}$$

Using Hölder's inequality with indices $p, p/(p-1)$ on the right-hand side of (2.3.7) we have

$$\sum_{n=1}^{m}\lambda_n \alpha_n^p \leqslant \frac{p}{p-1}\left\{\sum_{n=1}^{m}\lambda_n f^p(a_n)\right\}^{1/p}\left\{\sum_{n=1}^{m}\lambda_n \alpha_n^p\right\}^{(p-1)/p}.$$

Dividing the above inequality by the last factor on the right-hand side and raising the result to the pth power, we obtain

$$\sum_{n=1}^{m}\lambda_n\left(\frac{F_n}{\Lambda_n}\right)^p \leqslant \left(\frac{p}{p-1}\right)^p \sum_{n=1}^{m}\lambda_n f^p(a_n). \tag{2.3.8}$$

Now, from (2.3.3) and (2.3.8), we have

$$\sum_{n=1}^{m} \lambda_n f^p\left(\frac{A_n}{\Lambda_n}\right) \leqslant \sum_{n=1}^{m} \lambda_n \left(\frac{F_n}{\Lambda_n}\right)^p \leqslant \left(\frac{p}{p-1}\right)^p \sum_{n=1}^{m} \lambda_n f^p(a_n). \tag{2.3.9}$$

By letting m tend to infinity in (2.3.9), we obtain the desired inequality in (2.3.2). The proof is complete. □

The next result established by Pachpatte in [307] deals with the Hardy-type series inequality in two independent variables.

THEOREM 2.3.4. *If $p > 1$ is a constant, $b(m, n) \geqslant 0$ for $m, n \in N$ (the set of natural numbers) and*

$$B(m, n) = \frac{1}{mn} \sum_{s=1}^{m} \sum_{t=1}^{n} \frac{1}{st} \sum_{x=1}^{s} \sum_{y=1}^{t} b(x, y) \tag{2.3.10}$$

for $m, n \in N$, then

$$\sum_{m=1}^{\infty} \sum_{n=1}^{\infty} B^p(m, n) \leqslant \left(\frac{p}{p-1}\right)^{4p} \sum_{m=1}^{\infty} \sum_{n=1}^{\infty} b^p(m, n). \tag{2.3.11}$$

The equality holds in (2.3.11) if $b(m, n) = 0$ for $m, n \in N$.

PROOF. If $b(m, n)$ is null, then (2.3.11) is trivially true. Let us suppose that $b(m, n) > 0$ for all $m, n \in N$. Let $M \geqslant 1$, $L \geqslant 1$ be any integers, and define

$$S_{ML} = \sum_{m=1}^{M} \sum_{n=1}^{L} B^p(m, n). \tag{2.3.12}$$

From (2.3.12) we observe that

$$S_{ML} = \sum_{m=1}^{M} m^{-p} \sum_{n=1}^{L} \alpha_1^p(m, n), \tag{2.3.13}$$

where

$$\alpha_1(m, n) = \frac{1}{n} \sum_{t=1}^{n} \frac{1}{t} \sum_{s=1}^{m} \frac{1}{s} \sum_{x=1}^{s} \sum_{y=1}^{t} b(x, y). \tag{2.3.14}$$

From (2.3.14) and using inequality (2.3.4), we observe that

$$\alpha_1^p(m,n) - \left(\frac{p}{p-1}\right)\frac{1}{n}\left\{\sum_{s=1}^{m}\frac{1}{s}\sum_{x=1}^{s}\sum_{y=1}^{t} b(x,y)\right\}\alpha_1^p(m,n)$$

$$= \alpha_1^p(m,n) - \left(\frac{p}{p-1}\right)\left\{n\alpha_1(m,n) - (n-1)\alpha_1(m,n-1)\right\}\alpha_1^{p-1}(m,n)$$

$$= \left\{1 - \left(\frac{p}{p-1}\right)n\right\}\alpha_1^p(m,n) + \left(\frac{p}{p-1}\right)(n-1)\alpha_1(m,n-1)\alpha_1^{p-1}(m,n)$$

$$\leqslant \left\{1 - \left(\frac{p}{p-1}\right)n\right\}\alpha_1^p(m,n)$$

$$+ \left(\frac{p}{p-1}\right)(n-1)\frac{1}{p}\left\{\alpha_1^p(m,n-1) + (p-1)\alpha_1^p(m,n)\right\}$$

$$= \left(\frac{1}{p-1}\right)\left\{(n-1)\alpha_1^p(m,n-1) - n\alpha_1^p(m,n)\right\}. \tag{2.3.15}$$

Now, keeping m fixed in (2.3.15) and letting $n = 1, \ldots, L$, and adding the inequalities, we have

$$\sum_{n=1}^{L}\alpha_1^p(m,n) - \left(\frac{p}{p-1}\right)\sum_{n=1}^{L}\frac{1}{n}\left\{\sum_{s=1}^{m}\frac{1}{s}\sum_{x=1}^{s}\sum_{y=1}^{n} b(x,y)\right\}\alpha_1^p(m,n)$$

$$\leqslant \left(\frac{1}{p-1}\right)\sum_{n=1}^{L}\left\{(n-1)\alpha_1^p(m,n-1) - n\alpha_1^p(m,n)\right\}$$

$$= -\left(\frac{1}{p-1}\right)L\alpha_1^p(m,L)$$

$$\leqslant 0. \tag{2.3.16}$$

From (2.3.16) and using Hölder's inequality with indices p, $p/(p-1)$, we observe that

$$\sum_{n=1}^{L}\alpha_1^p(m,n) \leqslant \left(\frac{p}{p-1}\right)\sum_{n=1}^{L}\frac{1}{n}\left\{\sum_{s=1}^{m}\frac{1}{s}\sum_{x=1}^{s}\sum_{y=1}^{n} b(x,y)\right\}\alpha_1^{p-1}(m,n)$$

$$\leqslant \left(\frac{p}{p-1}\right)\left\{\sum_{n=1}^{L}\left\{\frac{1}{n}\left\{\sum_{s=1}^{m}\frac{1}{s}\sum_{x=1}^{s}\sum_{y=1}^{n}b(x,y)\right\}\right\}^{p}\right\}^{1/p}$$

$$\times\left\{\sum_{n=1}^{L}\alpha_1^p(m,n)\right\}^{(p-1)/p}. \tag{2.3.17}$$

Dividing both sides of (2.3.17) by the last factor on the right-hand side and raising the result to the pth power we get

$$\sum_{n=1}^{L}\alpha_1^p(m,n)\leqslant\left(\frac{p}{p-1}\right)^p\sum_{n=1}^{L}\left\{\frac{1}{n}\left\{\sum_{s=1}^{m}\frac{1}{s}\sum_{x=1}^{s}\sum_{y=1}^{n}b(x,y)\right\}\right\}^p. \tag{2.3.18}$$

From (2.3.13) and (2.3.18), we observe that

$$S_{ML}\leqslant\left(\frac{p}{p-1}\right)^p\sum_{n=1}^{L}n^{-p}\sum_{s=1}^{M}\alpha_2^p(m,n), \tag{2.3.19}$$

where

$$\alpha_2(m,n)=\frac{1}{m}\sum_{s=1}^{m}\frac{1}{s}\sum_{x=1}^{s}\sum_{y=1}^{n}b(x,y). \tag{2.3.20}$$

From (2.3.20) and using inequality (2.3.4), it is easy to observe that

$$\alpha_2^p(m,n)-\left(\frac{p}{p-1}\right)\frac{1}{m}\left\{\sum_{x=1}^{m}\sum_{y=1}^{n}b(x,y)\right\}\alpha_2^{p-1}(m,n)$$

$$\leqslant\left(\frac{1}{p-1}\right)\left\{(m-1)\alpha_2^p(m-1,n)-m\alpha_2^p(m,n)\right\}. \tag{2.3.21}$$

Keeping n fixed in (2.3.21) and letting $m=1,\ldots,M$, and adding the inequalities, we have

$$\sum_{m=1}^{M}\alpha_2^p(m,n)-\left(\frac{p}{p-1}\right)\sum_{m=1}^{M}\frac{1}{m}\left\{\sum_{x=1}^{m}\sum_{y=1}^{n}b(x,y)\right\}\alpha_2^{p-1}(m,n)$$

$$\leqslant\left(\frac{1}{p-1}\right)\sum_{m=1}^{M}\left\{(m-1)\alpha_2^p(m-1,n)-m\alpha_2^p(m,n)\right\}$$

$$=-\left(\frac{1}{p-1}\right)M\alpha_2^p(M,n)\leqslant 0. \tag{2.3.22}$$

From (2.3.22) and by following the same procedure below (2.3.16) up to (2.3.18), we get

$$\sum_{m=1}^{M} \alpha_2^p(m,n) \leqslant \left(\frac{p}{p-1}\right)^p \sum_{m=1}^{M} \left\{ \frac{1}{m} \sum_{x=1}^{m} \sum_{y=1}^{n} b(x,y) \right\}^p. \tag{2.3.23}$$

From (2.3.19) and (2.3.23), we observe that

$$S_{ML} \leqslant \left(\frac{p}{p-1}\right)^{2p} \sum_{m=1}^{M} m^{-p} \sum_{n=1}^{L} \alpha_3^p(m,n), \tag{2.3.24}$$

where

$$\alpha_3(m,n) = \frac{1}{n} \sum_{y=1}^{n} \sum_{x=1}^{m} b(x,y). \tag{2.3.25}$$

From (2.3.25) and using (2.3.4), we observe that

$$\alpha_3^p(m,n) - \left(\frac{p}{p-1}\right) \left\{ \sum_{x=1}^{m} b(x,n) \right\} \alpha_3^{p-1}(m,n)$$
$$\leqslant \left(\frac{1}{p-1}\right) \left\{ (n-1)\alpha_3^p(m,n-1) - n\alpha_3^p(m,n) \right\}. \tag{2.3.26}$$

Now, by following the same procedure below (2.3.15) up to (2.3.18), we get

$$\sum_{n=1}^{L} \alpha_3^p(m,n) \leqslant \left(\frac{p}{p-1}\right)^p \sum_{n=1}^{L} \left\{ \sum_{x=1}^{m} b(x,y) \right\}^p. \tag{2.3.27}$$

From (2.3.24) and (2.3.27), we observe that

$$S_{ML} \leqslant \left(\frac{p}{p-1}\right)^{3p} \sum_{n=1}^{L} \sum_{m=1}^{M} \alpha_4^p(m,n), \tag{2.3.28}$$

where

$$\alpha_4(m,n) = \frac{1}{m} \sum_{x=1}^{m} b(x,n). \tag{2.3.29}$$

From (2.3.29) and using inequality (2.3.4), we observe that

$$\alpha_4^p(m,n) - \left(\frac{p}{p-1}\right) b(m,n) \alpha_4^{p-1}(m,n)$$

$$\leqslant \left(\frac{1}{p-1}\right)\left\{(m-1)\alpha_4^p(m-1,n) - m\alpha_4^p(m,n)\right\}. \tag{2.3.30}$$

Now, following the same procedure below (2.3.21) up to (2.3.23), we get

$$\sum_{m=1}^{M} \alpha_4^p(m,n) \leqslant \left(\frac{p}{p-1}\right)^p \sum_{m=1}^{M} b(m,n). \tag{2.3.31}$$

From (2.3.28) and (2.3.31), we observe that

$$S_{ML} \leqslant \left(\frac{p}{p-1}\right)^{4p} \sum_{m=1}^{M}\sum_{n=1}^{L} b(m,n). \tag{2.3.32}$$

By letting M and L tend to infinity in (2.3.32), we get the desired inequality in (2.3.11). The proof is complete. □

REMARK 2.3.2. If we define $B(m,n)$ in (2.3.10) by

$$B(m,n) = \frac{1}{mn}\sum_{s_1=1}^{m}\sum_{t_1=1}^{n}\frac{1}{s_1t_1}\sum_{s_2=1}^{s_1}\sum_{t_2=1}^{t_1}\cdots\sum_{s_{r-1}=1}^{s_{r-2}}\sum_{t_{r-1}=1}^{t_{r-2}}\frac{1}{s_{r-1}t_{r-1}}\sum_{s_r=1}^{s_{r-1}}\sum_{t_r=1}^{t_{r-1}} b(s_r,t_r) \tag{2.3.33}$$

for $m,n \in N$, then in place of inequality (2.3.11) we get

$$\sum_{m=1}^{\infty}\sum_{n=1}^{\infty} B^p(m,n) \leqslant \left(\frac{p}{p-1}\right)^{2rp}\sum_{m=1}^{\infty}\sum_{n=1}^{\infty} b^p(m,n). \tag{2.3.34}$$

The proof of inequality (2.3.34) is a natural extension of the proof of Theorem 2.3.4 given above. Further, we note that the inequality obtained in (2.3.34) can be extended to the functions of several independent variables.

In 1967, J.E. Littlewood [195] presents several open problems concerning elementary inequalities for infinite series which have their roots in the theory of orthogonal series. One of his simplest problem is to decide whether a constant k exists for which

$$\sum_n a_n^3 \sum_{m=1}^{n} a_m^2 A_m \leqslant k \sum_n a_n^4 A_n^2, \tag{2.3.35}$$

where $A_n = a_1 + \cdots + a_n$, and the inequality is to hold for all nonnegative numbers $a_1, a_2, \ldots,$ and k is an absolute constant. An answer to Littlewood's question was published in 1987 by G. Bennett [22–24] who shows that (2.3.35) is valid with $k=4$. Actually, Bennett proves the following more general result.

THEOREM 2.3.5. *Let $p,q \geqslant 1$. Then*

$$\sum_n a_n^p A_n^q \left[\sum_{m\geqslant n} a_m^{1+p/q}\right]^q \leqslant \left[\frac{2p-1}{p}q\right]^q \sum_n [a_n^p A_n^q]^2, \tag{2.3.36}$$

where a's are arbitrary nonnegative numbers with partial sum $A_n = a_1+\cdots+a_n$.

PROOF. The proof involves just two applications of Hölder's inequality. We may assume that only finitely many of the a_n's are positive, say $a_n = 0$ whenever $n > N$. To keep the notation manageable, we set $b_n = a_n A_n^{q/p}$ and $c_n = \sum_{m\geqslant n} a_m^{1+p/q}$. Elementary estimates give

$$A_n^{1+q/p} \geqslant b_1+\cdots+b_n \quad (= B_n,\ \text{say}) \tag{2.3.37}$$

and

$$c_n^r - c_{n+1}^r \leqslant r c_n^{r-1} a_n^{1+p/q}, \tag{2.3.38}$$

where r ($\geqslant 1$) is to be chosen later.

Letting $\theta = (2p-1)^{-1}$ so that $0<\theta\leqslant 1$, the left-hand side of (2.3.36) may be rewritten as

$$L = \sum_n b_n^{p(1-\theta)} b_n^{p\theta} c_n^q.$$

Applying Hölder's inequality with indices $2/(1-\theta)$ and $2/(1+\theta)$, we have

$$\begin{aligned} L &\leqslant \left[\sum_n b_n^{2p}\right]^{(1-\theta)/2} \left[\sum_n b_n^{2p\theta/(1+\theta)} c_n^{2q/(1+\theta)}\right]^{(1+\theta)/2} \\ &= \left[\sum_n b_n^{2p}\right]^{(1-\theta)/2} \left[\sum_n b_n c_n^r\right]^{(1+\theta)/2}, \end{aligned}$$

where we have set

$$r = \frac{2q}{1+\theta} = \frac{(2p-1)q}{p} \geqslant 1. \tag{2.3.39}$$

Thus, to prove (2.3.36), it suffices to show that

$$\sum_n b_n c_n^r \leqslant r^r \sum_n b_n^{2p}. \tag{2.3.40}$$

Summing by parts, and then applying (2.3.37) and (2.3.38), we see that

$$\sum_{n=1}^{N} b_n c_n^r = \sum_{n=1}^{N-1} B_n\left[c_n^r - c_{n+1}^r\right] + B_N c_N^r$$

$$\leqslant r\sum_{n=1}^{N} a_n^{1+p/q} A_n^{1+q/p} c_n^{r-1}$$

$$= r\sum_{n=1}^{N} b_n^{1/r^*} c_n^{r-1} b_n^{2p/r},$$

where $r^* = r/(r-1)$ is the conjugate of r (see (2.3.39)). Applying Hölder's inequality once more gives

$$\sum_n b_n c_n^r \leqslant r\left[\sum_n b_n c_n^r\right]^{1/r^*}\left[\sum_n b_n^{2p}\right]^{1/r},$$

which is equivalent to (2.3.40). The proof is complete. □

REMARK 2.3.3. Setting $p = 2$ and $q = 1$ in the theorem above, and interchanging the order of summation on the left-hand side of (2.3.36), it shows that (2.3.35) holds with $k = 3/2$. Furthermore, setting $p = 1$ and $q = 2$ leads to

$$\sum_n a_n A_n^2\left[\sum_{m\geqslant n} a_m^{3/2}\right]^2 \leqslant 4\sum_n a_n^2 A_n^4. \tag{2.3.41}$$

For further results related to Littlewood's problem, see [195].

In [326] Pachpatte has established the inequalities in the following theorem which are similar to that of Littlewood's inequality given in (2.3.35).

THEOREM 2.3.6. *Let $p \geqslant 1$, $q \geqslant 1$, $r \geqslant 1$ be real constants. If $a_n \geqslant 0$, $n = 1, 2, \ldots$, and $A_n = \sum_{m=1}^{n} a_m$, then*

$$\sum_{n=1}^{N} a_n A_n \leqslant \frac{N+1}{2}\sum_{n=1}^{N} a_n^2, \tag{2.3.42}$$

$$\sum_{n=1}^{N} A_n^{p+q} \leqslant \left[(p+q)(N+1)\right]^q \sum_{n=1}^{N} A_n^p a_n^q \tag{2.3.43}$$

and

$$\sum_{n=1}^{N} A_n^{p+q} a_n^r \leqslant \left[(p+q+r)(N+1)\right]^q \sum_{n=1}^{N} A_n^p a_n^{q+r}. \tag{2.3.44}$$

PROOF. Rewriting the left-hand side of (2.3.42), and using the Schwarz inequality, interchanging the order of summations, and using the elementary inequality $a^{1/2}b^{1/2} \leqslant (a+b)/2$, $a \geqslant 0, b \geqslant 0$ reals, we observe that

$$\begin{aligned}
\sum_{n=1}^{N} a_n A_n &= \sum_{n=1}^{N} (\sqrt{n}a_n)\left(\frac{1}{\sqrt{n}} \sum_{m=1}^{n} a_m\right) \\
&\leqslant \left[\sum_{n=1}^{N} n a_n^2\right]^{1/2} \left[\sum_{n=1}^{N} \frac{1}{n}\left(\sum_{m=1}^{n} a_m\right)^2\right]^{1/2} \\
&\leqslant \left[\sum_{n=1}^{N} n a_n^2\right]^{1/2} \left[\sum_{n=1}^{N} \frac{1}{n}\left(\sum_{m=1}^{n} 1\right)\left(\sum_{m=1}^{n} a_m^2\right)\right]^{1/2} \\
&= \left[\sum_{n=1}^{N} n a_n^2\right]^{1/2} \left[\sum_{n=1}^{N} \left(\sum_{m=1}^{n} a_m^2\right)\right]^{1/2} \\
&= \left[\sum_{n=1}^{N} n a_n^2\right]^{1/2} \left[\sum_{m=1}^{N} a_m^2\left(\sum_{n=m}^{N} 1\right)\right]^{1/2} \\
&= \left[\sum_{n=1}^{N} n a_n^2\right]^{1/2} \left[\sum_{m=1}^{N} (N-m+1) a_m^2\right]^{1/2} \\
&\leqslant \frac{1}{2}\left[\sum_{n=1}^{N} n a_n^2 + \sum_{n=1}^{N} (N-n+1) a_n^2\right] \\
&= \frac{N+1}{2} \sum_{n=1}^{N} a_n^2.
\end{aligned}$$

This proves the required inequality in (2.3.42).

By taking $z_m = a_m$ and $\alpha = p+q$ in the following inequality (see [226])

$$\left(\sum_{m=1}^{n} z_m\right)^{\alpha} \leqslant \alpha \sum_{m=1}^{n} z_m \left(\sum_{k=1}^{m} z_k\right)^{\alpha-1},$$

where $\alpha > 1$ is a constant and $z_m \geqslant 0$, $m = 1, 2, \ldots,$ we have

$$A_n^{p+q} \leqslant (p+q) \sum_{m=1}^{n} a_m A_m^{p+q-1}. \tag{2.3.45}$$

By taking the sum on both sides of (2.3.45) from 1 to N and interchanging the order of the summation, we observe that

$$\begin{aligned}
\sum_{n=1}^{N} A_n^{p+q} &\leqslant (p+q) \sum_{n=1}^{N} \left(\sum_{m=1}^{n} a_m A_m^{p+q-1} \right) \\
&= (p+q) \sum_{m=1}^{N} a_m A_m^{p+q-1} (N - m + 1) \\
&\leqslant (p+q)(N+1) \sum_{n=1}^{N} \left(a_n A_n^{p+q-1} \right) \\
&= (p+q)(N+1) \sum_{n=1}^{N} \left(A_n^{p/q} a_n \right) \left(A_n^{p+q-1-p/q} \right).
\end{aligned} \tag{2.3.46}$$

By using Hölder's inequality with indices $q, q/(q-1)$ on the right-hand side of (2.3.46), we have

$$\sum_{n=1}^{N} A_n^{p+q} \leqslant (p+q)(N+1) \left[\sum_{n=1}^{N} A_n^p a_n^q \right]^{1/q} \left[\sum_{n=1}^{N} A_n^{p+q} \right]^{(q-1)/q}. \tag{2.3.47}$$

Dividing by the last factor on the right-hand side of (2.3.47) and raising to the qth power of the resulting inequality, we get the desired inequality in (2.3.43).

By rewriting the left-hand side of (2.3.44) and using Hölder's inequality with indices $(q+r)/r$, $(q+r)/q$ and the inequality (2.3.43), we observe that

$$\begin{aligned}
\sum_{n=1}^{N} A_n^{p+q} a_n^r &= \sum_{n=1}^{N} \left(A_n^{pr/(q+r)} a_n^p \right) \left(A_n^{p+q-pr/(q+r)} \right) \\
&\leqslant \left[\sum_{n=1}^{N} A_n^p a_n^{q+r} \right]^{r/(q+r)} \left[\sum_{n=1}^{N} A_n^{p+q+r} \right]^{q/(q+r)}
\end{aligned}$$

$$\begin{aligned}&\leqslant \left[\sum_{n=1}^{N} A_n^p a_n^{q+r}\right]^{r/(q+r)}\\&\quad\times\left[\left[(p+q+r)(N+1)\right]^{q+r}\sum_{n=1}^{N} A_n^p a_n^{q+r}\right]^{q/(q+r)}\\&=\left[(p+q+r)(N+1)\right]^{q}\sum_{n=1}^{N} A_n^p a_n^{q+r}.\end{aligned}$$

This result is the required inequality in (2.3.44), and the proof is complete. □

REMARK 2.3.4. By taking $p = 1$, $q = 1$ in (2.3.43), we get the lower bound on the left-hand side of the inequality given in (2.3.42).

2.4 Hardy's Integral Inequality and Its Generalizations

One of the many fundamental mathematical discoveries of G.H. Hardy is the following integral inequality [141, Theorem 327] discovered in 1920 in the course of attempts to simplify the proof of Hilbert's double series theorem.

THEOREM 2.4.1. *If $p > 1$, $f(x) \geqslant 0$ and $F(x) = \int_0^x f(t)\,\mathrm{d}t$, then*

$$\int_0^\infty \left(\frac{F}{x}\right)^p \mathrm{d}x < \left(\frac{p}{p-1}\right)^p \int_0^\infty f^p\,\mathrm{d}x, \tag{2.4.1}$$

unless $f \equiv 0$. The constant is the best possible.

PROOF. The proof given here is due to Hardy [136] and is also given in [141, pp. 242–243]. We may suppose f is not null.

Let $n > 0$, $f_n = \min(f, n)$, $F_n = \int_0^x f_n\,\mathrm{d}x$ and let X_0 be so large that f, and so f_n, F_n are not null in $(0, X)$ when $X > X_0$. We have

$$\begin{aligned}\int_0^X \left(\frac{F_n}{x}\right)^p \mathrm{d}x &= -\frac{1}{p-1}\int_0^X F_n^p \frac{\mathrm{d}}{\mathrm{d}x}\left(x^{1-p}\right)\mathrm{d}x\\&=\left[-\frac{x^{1-p}F_n^p(x)}{p-1}\right]_0^X + \frac{p}{p-1}\int_0^X \left(\frac{F_n}{x}\right)^{p-1} f_n\,\mathrm{d}x\\&\leqslant \frac{p}{p-1}\int_0^X \left(\frac{F_n}{x}\right)^{p-1} f_n\,\mathrm{d}x\end{aligned}$$

since the integral term vanishes at $x = 0$ in virtue of $F_n = o(x)$. Using Hölder's inequality with indices $p, p/(p-1)$ we have

$$\int_0^X \left(\frac{F_n}{x}\right)^p \mathrm{d}x \leqslant \frac{p}{p-1}\left(\int_0^X \left(\frac{F_n}{x}\right)^p \mathrm{d}x\right)^{1/p'} \left(\int_0^X f_n^p\right)^{1/p}, \tag{2.4.2}$$

where $p' = p/(p-1)$. The left-hand side being positive (and finite), this inequality gives

$$\int_0^X \left(\frac{F_n}{x}\right)^p \mathrm{d}x \leqslant \left(\frac{p}{p-1}\right)^p \int_0^X f_n^p \,\mathrm{d}x.$$

We make $n \to \infty$ in this inequality, the result being to suppress the two suffixes n. Making $X \to \infty$ we have

$$\int_0^\infty \left(\frac{F}{x}\right)^p \mathrm{d}x \leqslant \left(\frac{p}{p-1}\right)^p \int_0^\infty f^p \,\mathrm{d}x,$$

the desired result with "$\leqslant$" for "$<$". Making $n \to \infty$ and then $X \to \infty$ in (2.4.2) we have

$$\int_0^\infty \left(\frac{F}{x}\right)^p \mathrm{d}x \leqslant \frac{p}{p-1}\left(\int_0^\infty \left(\frac{F}{x}\right)^p\right)^{1/p} \left(\int_0^\infty f^p \,\mathrm{d}x\right)^{1/p}. \tag{2.4.3}$$

The integrals in this inequality being now known to be all finite and positive, (2.4.3) gives

$$\int_0^\infty \left(\frac{F}{x}\right)^p \mathrm{d}x < \left(\frac{p}{p-1}\right)^p \int_0^\infty f^p \,\mathrm{d}x,$$

unless $x^{-p}F^p$ and f^p are effectively proportional, which is impossible since it would make f a power of x, and $\int_0^\infty f^p$ is divergent. □

The proof that the constant is the best possible follows the same lines as before in Theorem 2.2.1: take $f(x) = 0$ for $x < 1$, $f(x) = x^{-1/p-\varepsilon}$ for $x \geqslant 1$, where $\varepsilon > 0$ is a constant.

The above inequality is now known in the literature as Hardy's integral inequality. In the past few years, a number of generalizations, variants and extensions of the above inequality have been given by several investigators. Here we give some of these results in the following theorems.

In a paper [137] published in 1928, Hardy himself proved the following generalization of the inequality given in Theorem 2.4.1.

THEOREM 2.4.2. *If $p > 1$, $m \neq 1$, $f(x) \geqslant 0$ and $R(x)$ is defined by*

$$R(x) = \begin{cases} \int_0^x f(t)\,\mathrm{d}t, & m > 1, \\ \int_x^\infty f(t)\,\mathrm{d}t, & m < 1, \end{cases} \tag{2.4.4}$$

then

$$\int_0^\infty x^{-m} R^p(x)\,\mathrm{d}x < \left\{\frac{p}{|m-1|}\right\}^p \int_0^\infty x^{-m}\bigl(xf(x)\bigr)^p\,\mathrm{d}x, \tag{2.4.5}$$

unless $f \equiv 0$. The constant is the best possible.

The proof of this theorem follows by the same arguments as in the proof of Theorem 2.4.1 with suitable modifications. Here we leave the details to the reader.

In the following two theorems we present the main results given by Copson in [70] whose proofs are based on the ideas of the proofs of similar results given by Levinson in [190] and by Pachpatte in [254].

THEOREM 2.4.3. *Let $p \geqslant 1$, $m > 1$ be constants. Let $f(x)$ be a nonnegative function on $(0, \infty)$ and let $r(t)$ be a positive function on $(0, \infty)$ and let*

$$R(x) = \int_0^x r(t)\,\mathrm{d}t, \qquad F(x) = \int_0^x r(t) f(t)\,\mathrm{d}t. \tag{2.4.6}$$

Then

$$\int_0^\infty R^{-m}(x) r(x) F^p(x)\,\mathrm{d}x \leqslant \left(\frac{p}{m-1}\right)^p \int_0^\infty R^{p-m}(x) r(x) f^p(x)\,\mathrm{d}x. \tag{2.4.7}$$

PROOF. Let $0 < a < b < \infty$ and define, for $m > 1$,

$$F_a(x) = \int_a^x r(t) f(t)\,\mathrm{d}t$$

for $x \in (a, b)$ with $F_0(x) = F(x)$. Integrating by parts gives

$$\int_a^b R^{-m}(x) r(x) F_a^p(x)\,\mathrm{d}x = \left[\frac{R^{-m+1}(x)}{-m+1} F_a^p(x)\right]_a^b - \int_a^b \frac{R^{-m+1}(x)}{-m+1} p F_a^{p-1}(x) r(x) f(x)\,\mathrm{d}x. \tag{2.4.8}$$

Since $m > 1$, from (2.4.8) we observe that

$$\begin{aligned}\int_a^b & R^{-m}(x)r(x)F_a^p(x)\,\mathrm{d}x \\ &\leqslant \frac{p}{m-1}\int_a^b R^{-m+1}(x)r(x)f(x)F_a^{p-1}(x)\,\mathrm{d}x \\ &= \frac{p}{m-1}\int_a^b \left\{R^{(p-m)/p}(x)r^{1/p}(x)f(x)\right\} \\ &\qquad \times \left\{R^{-m(p-1)/p}(x)r^{(p-1)/p}(x)F_a^{p-1}(x)\right\}\mathrm{d}x. \qquad (2.4.9)\end{aligned}$$

Using Hölder's inequality with indices p, $p/(p-1)$ on the right-hand side of (2.4.9) we obtain

$$\begin{aligned}\int_a^b & R^{-m}(x)r(x)F_a^p(x)\,\mathrm{d}x \\ &\leqslant \left(\frac{p}{m-1}\right)\left\{\int_a^b R^{p-m}(x)r(x)f^p(x)\,\mathrm{d}x\right\}^{1/p} \\ &\quad \times \left\{\int_a^b R^{-m}(x)r(x)F_a^p(x)\,\mathrm{d}x\right\}^{(p-1)/p}. \qquad (2.4.10)\end{aligned}$$

Dividing both sides of (2.4.10) by the second integral factor on the right-hand side of (2.4.10), and raising both sides to the pth power, we obtain

$$\int_a^b R^{-m}(x)r(x)F_a^p(x)\,\mathrm{d}x \leqslant \left(\frac{p}{m-1}\right)^p \int_a^b R^{p-m}(x)r(x)f^p(x)\,\mathrm{d}x. \qquad (2.4.11)$$

From (2.4.11) we have

$$\int_a^b R^{-m}(x)r(x)F_a^p(x)\,\mathrm{d}x \leqslant \left(\frac{p}{m-1}\right)^p \int_0^\infty R^{p-m}(x)r(x)f^p(x)\,\mathrm{d}x. \qquad (2.4.12)$$

Let $a < c < b$. Then from (2.4.12) we have

$$\int_c^b R^{-m}(x)r(x)F_a^p(x)\,\mathrm{d}x \leqslant \left(\frac{p}{m-1}\right)^p \int_0^\infty R^{p-m}(x)r(x)f^p(x)\,\mathrm{d}x. \qquad (2.4.13)$$

Letting $a \to 0$ in (2.4.13) we have

$$\int_c^b R^{-m}(x)r(x)F^p(x)\,\mathrm{d}x \leqslant \left(\frac{p}{m-1}\right)^p \int_0^\infty R^{p-m}(x)r(x)f^p(x)\,\mathrm{d}x.$$

Since this inequality holds for arbitrary $0 < c < b$, it follows that

$$\int_0^\infty R^{-m}(x)r(x)F^p(x)\,dx \leqslant \left(\frac{p}{m-1}\right)^p \int_0^\infty R^{p-m}(x)r(x)f^p(x)\,dx.$$

The proof is complete. □

THEOREM 2.4.4. *Let $p \geqslant 1$, $m < 1$ be constants. Let $f(x)$, $r(x)$ and $R(x)$ be as defined in Theorem 2.4.3. If $F(x)$ is defined by*

$$F(x) = \int_x^\infty r(t)f(t)\,dt \tag{2.4.14}$$

for $x \in (0, \infty)$, then

$$\int_0^\infty R^{-m}(x)r(x)F^p(x)\,dx \leqslant \left(\frac{p}{1-m}\right)^p \int_0^\infty R^{p-m}(x)r(x)f^p(x)\,dx. \tag{2.4.15}$$

PROOF. Let $0 < a < b < \infty$ and define for $m < 1$,

$$F_b(x) = \int_x^b r(t)f(t)\,dt$$

for $x \in (a, b)$ with $F_\infty(x) = F(x)$. Integrating by parts gives

$$\int_a^b R^{-m}(x)r(x)F_b^p(x)\,dx = \left[\frac{R^{-m+1}}{-m+1}F_b^p(x)\right]_a^b - \int_a^b \frac{R^{-m+1}(x)}{-m+1}pF_b^{p-1}(x)\big(-r(x)f(x)\big)\,dx. \tag{2.4.16}$$

Since $m < 1$, from (2.4.16) we observe that

$$\begin{aligned}\int_a^b R^{-m}(x)r(x)F_b^p(x)\,dx &\leqslant \frac{p}{1-m}\int_a^b R^{-m+1}(x)r(x)f(x)F_b^{p-1}(x)\,dx \\ &= \frac{p}{1-m}\int_a^b \left\{R^{(p-m)/p}(x)r^{1/p}(x)f(x)\right\} \\ &\quad \times \left\{R^{-m(p-1)/p}(x)r^{(p-1)/p}(x)F_b^{p-1}(x)\right\}dx.\end{aligned} \tag{2.4.17}$$

Using Hölder's inequality with indices p, $p/(p-1)$ on the right-hand side of (2.4.17) we obtain

$$\int_a^b R^{-m}(x)r(x)F_b^p(x)\,\mathrm{d}x \leqslant \frac{p}{1-m}\left\{\int_a^b R^{p-m}(x)r(x)f^p(x)\,\mathrm{d}x\right\}^{1/p} \times \left\{\int_a^b R^{-m}(x)r(x)F_b^p(x)\,\mathrm{d}x\right\}^{(p-1)/p}. \tag{2.4.18}$$

Dividing both sides of (2.4.18) by the second integral factor on the right-hand side of (2.4.18), and raising both sides to the pth power, we obtain

$$\int_a^b R^{-m}(x)r(x)F_b^p(x)\,\mathrm{d}x \leqslant \left(\frac{p}{1-m}\right)^p \int_a^b R^{p-m}(x)r(x)f^p(x)\,\mathrm{d}x. \tag{2.4.19}$$

From (2.4.19) we have

$$\int_a^b R^{-m}(x)r(x)F_b^p(x)\,\mathrm{d}x \leqslant \left(\frac{p}{1-m}\right)^p \int_0^\infty R^{p-m}(x)r(x)f^p(x)\,\mathrm{d}x. \tag{2.4.20}$$

Let $a < c < b$. Then from (2.4.20) we have

$$\int_a^c R^{-m}(x)r(x)F_b^p(x)\,\mathrm{d}x \leqslant \left(\frac{p}{1-m}\right)^p \int_0^\infty R^{p-m}(x)r(x)f^p(x)\,\mathrm{d}x. \tag{2.4.21}$$

Letting $b \to \infty$ in (2.4.21) gives

$$\int_a^c R^{-m}(x)r(x)F^p(x)\,\mathrm{d}x \leqslant \left(\frac{p}{1-m}\right)^p \int_0^\infty R^{p-m}(x)r(x)f^p(x)\,\mathrm{d}x.$$

Since this holds for any $a, c, 0 < a < c$, it follows that

$$\int_0^\infty R^{-m}(x)r(x)F^p(x)\,\mathrm{d}x \leqslant \left(\frac{p}{1-m}\right)^p \int_0^\infty R^{p-m}(x)r(x)f^p(x)\,\mathrm{d}x.$$

The proof is complete. □

REMARK 2.4.1. In [70] Copson has given the two companion results corresponding to Theorems 2.4.3 and 2.4.4 when $0 < p \leqslant 1$ and he also has given two more companion results corresponding to Theorems 2.4.3 and 2.4.4 when $m = 1$. For more details, we refer the interested readers to [70].

In the next two theorems we give the generalizations of Hardy's inequality established by Love in [199]. In what follows, the functions involved have domains in $(0,\infty)$ or $(0,\infty)^2$ and ranges in $[0,\infty]$. The word "increasing" is used for "nondecreasing" and similarly for "decreasing". A function $\sigma(x)$ is called submultiplicative if its values are positive and satisfy $\sigma(xy)\leqslant\sigma(x)\sigma(y)$ for all x and y in $(0,\infty)$. The integrals are Lebesgue integrals, and they are said to exist even if their values are infinite since their integrals are measurable and nonnegative. The words "measurable" and "Measurable" are used to denote linear and plane measurability respectively. The qth power of the number $f(x)$ is denoted by $f(x)^q$, not by $f^q(x)$.

THEOREM 2.4.5. *Let* $1\leqslant q<\infty$, $0<b\leqslant\infty$ *and* $b_1=\max\{b,1\}$. *Let* $\sigma(x)$ *be submultiplicative and measurable on* $(0,\infty)$, $\tau(x)$ *be decreasing and positive on* $(0,b)$, $H(x,y)$ *be Measurable, nonnegative and homogeneous of degree* $h-1$ *on* $0<y\leqslant x\leqslant b_1$, *and*

$$A=\int_0^1 H(1,t)\{t^{-1}\sigma(t^{-1})\}^{1/q}\,\mathrm{d}t<\infty.$$

If $f(x)$ *is measurable and nonnegative on* $(0,b)$ *and*

$$\|f\|=\left(\int_0^b f(x)^q\sigma(x)\tau(x)\,\mathrm{d}x\right)^{1/q}<\infty,$$

then

$$Hf(x)=x^{1/h}\int_0^x H(x,y)f(y)\,\mathrm{d}y$$

exists finitely for almost all x *in* $(0,b)$, *and*

$$\|Hf\|\leqslant A\|f\|.$$

PROOF. By Fubini's theorem, there is $\xi\in(0,b)$ such that $H(\xi,y)$ is measurable on $0<y<\xi$. Therefore $H(\xi,\xi t)$ is measurable on $0<t<1$, hence so is

$$H(x,xt)=\left(\frac{x}{\xi}\right)^{h-1}H(\xi,\xi t)$$

for each $x\in(0,b_1)$. This result with (2.4.22) below ensures the existence, finite or infinite, of $Hf(x)$ for each $x\in(0,b]$ and also of A.

For $0<x<b$,

$$Hf(x)=\frac{1}{x^h}\int_0^1 H(x,xt)f(xt)x\,\mathrm{d}t=\int_0^1 H(1,t)f(xt)\,\mathrm{d}t. \tag{2.4.22}$$

Using the form [141, Theorem 202] of Minkowski's inequality at (2.4.23),

$$\begin{aligned}
\|Hf\| &= \left(\int_0^b \left(\int_0^1 H(1,t) f(xt)\,\mathrm{d}t\right)^q \sigma(x)\tau(x)\,\mathrm{d}x\right)^{1/q} \\
&\leqslant \int_0^1 \left(\int_0^b H(1,t)^q f(xt)^q \sigma(x)\tau(x)\,\mathrm{d}x\right)^{1/q} \mathrm{d}t \qquad (2.4.23) \\
&= \int_0^1 H(1,t)\left(\int_0^{bt} f(y)^q \sigma\big(yt^{-1}\big)\tau\big(yt^{-1}\big)t^{-1}\,\mathrm{d}y\right)^{1/q} \mathrm{d}t \\
&\leqslant \int_0^1 H(1,t)\left(\int_0^{b} f(y)^q \sigma(y)\sigma\big(t^{-1}\big)\tau(y)t^{-1}\,\mathrm{d}y\right)^{1/q} \mathrm{d}t \\
&= A\|f\|.
\end{aligned}$$

In particular, $\|Hf\| < \infty$ and since σ and τ are positive in $(0, b)$, it follows that $Hf(x)$, which exists for all x in $(0, b)$ as already shown, and also is finite for almost all x in $(0, b)$. □

REMARK 2.4.2. The first half of Hardy's inequality in the form of Theorem 330 in [141] is the case of Theorem 2.4.5 in which $b = \infty$, $r > 1$, $\sigma(x) = x^{q-r}$, $\tau(x) = 1$, $H(x, y) = x^{h-1}$ and $A = q/(r-1)$. Hardy's inequality in its original form of Theorem 327 in [141] is the case $r = q$ and thus $\sigma(x) = 1$.

THEOREM 2.4.6. *Let $1 \leqslant q < \infty$, $0 \leqslant a < \infty$ and $a_1 = \min\{a, 1\}$. Let $\sigma(x)$ be submultiplicative and measurable on $(0, \infty)$, $\tau(x)$ be increasing and positive on (a, ∞), $H(x, y)$ be Measurable, nonnegative and homogeneous of degree $h - 1$ on $a_1 \leqslant x \leqslant y < \infty$, and*

$$B = \int_1^\infty H(1,t)\big\{t^{-1}\sigma\big(t^{-1}\big)\big\}^{1/q}\,\mathrm{d}t < \infty.$$

If $f(x)$ is measurable and nonnegative on $(0, \infty)$ and

$$\|f\| = \left(\int_a^\infty f(x)^q \sigma(x)\tau(x)\,\mathrm{d}x\right)^{1/q} < \infty,$$

then

$$Hf(x) = \frac{1}{x^h}\int_x^\infty H(x,y) f(y)\,\mathrm{d}y$$

exists finitely for almost all x in (a, ∞), and

$$\|Hf\| \leqslant B\|f\|.$$

The proof of this theorem is formally the same as the proof of Theorem 2.4.5 with suitable changes and thus we omit the details.

REMARK 2.4.3. Theorems 2.4.5 and 2.4.6 are generalizations of Theorems 1.1 and 1.3 in [198]. In [199] there are theorems about the best possible constants and discrete analogues of Theorems 2.4.5 and 2.4.6.

In [221] Muckenhoupt has given the following more general versions of Hardy's inequality with weights. In which $0 \cdot \infty$ is taken to be 0 and the usual convention is used for the integrals if p or p' is ∞.

THEOREM 2.4.7. *Let $1 \leqslant p \leqslant \infty$, there is a finite C for which*

$$\left[\int_0^\infty \left|U(x)\int_0^x f(t)\,\mathrm{d}t\right|^p \mathrm{d}x\right]^{1/p} \leqslant C\left[\int_0^\infty |V(x)f(x)|^p\,\mathrm{d}x\right]^{1/p} \tag{2.4.24}$$

is true for real f if and only if

$$B = \sup_{r>0}\left[\int_r^\infty |U(x)|^p\,\mathrm{d}x\right]^{1/p}\left[\int_0^r |V(x)|^{-p'}\,\mathrm{d}x\right]^{1/p'} < \infty, \tag{2.4.25}$$

where $1/p + 1/p' = 1$ and $U(x)$, $V(x)$ are weight functions. Furthermore, if C is the least constant for which (2.4.24) *holds, then $B \leqslant C \leqslant p^{1/p}(p')^{1/p'}B$ for $1 < p < \infty$ and $B = C$ if $p = 1$ or ∞.*

THEOREM 2.4.8. *If $1 \leqslant p \leqslant \infty$, there is a finite C such that*

$$\left[\int_0^\infty \left|U(x)\int_x^\infty f(t)\,\mathrm{d}t\right|^p \mathrm{d}x\right]^{1/p} \leqslant C\left[\int_0^\infty |V(x)f(x)|^p\,\mathrm{d}x\right]^{1/p} \tag{2.4.26}$$

is true for real f if and only if

$$B = \sup_{r>0}\left[\int_0^r |U(x)|^p\,\mathrm{d}x\right]^{1/p}\left[\int_r^\infty |V(x)|^{-p'}\,\mathrm{d}x\right]^{1/p'} < \infty,$$

where $1/p + 1/p' = 1$ and $U(x)$, $V(x)$ are weight functions. Furthermore, if C is the least constant for which (2.4.26) *is true, then $B \leqslant C \leqslant p^{1/p}(p')^{1/p'}B$.*

PROOFS OF THEOREMS 2.4.7 AND 2.4.8. For Theorem 2.4.7 it is sufficient to prove the asserted inequalities between B and C. The new proof is the proof that $C \leqslant Bp^{1/p}(p')^{1/p'}$. The proof given here that $B \leqslant C$ is standard.

To prove that $C \leqslant Bp^{1/p}(p')^{1/p'}$ for $1 < p < \infty$, it will be shown that

$$\left[\int_0^\infty \left|U(x)\int_0^x f(t)\,\mathrm{d}t\right|^p \mathrm{d}x\right]^{1/p} \leqslant Bp^{1/p}(p')^{1/p'}\left[\int_0^\infty |V(x)f(x)|^p\,\mathrm{d}x\right]^{1/p}. \tag{2.4.27}$$

To do this, let $h(x) = [\int_0^x |V(t)|^{-p'}\,\mathrm{d}t]^{1/(pp')}$. By Hölder's inequality, the pth power of the left-hand side of (2.4.27) is bounded by

$$\int_0^\infty |U(x)|^p\left[\int_0^x |f(t)V(t)h(t)|^p\,\mathrm{d}t\right]\left[\int_0^x |V(u)h(u)|^{-p'}\,\mathrm{d}u\right]^{p/p'}\mathrm{d}x.$$

Simple special arguments justify this expression even if $V(t)h(t)$ is 0 or ∞ on a set of positive measure provided the right-hand side of (2.4.27) is finite. Fubini's theorem shows that this expression equals to

$$\int_0^\infty |f(t)V(t)h(t)|^p\left(\int_t^\infty |U(x)|^p\left[\int_0^x |V(u)h(u)^{-p'}|\,\mathrm{d}t\right]^{p-1}\mathrm{d}u\right)\mathrm{d}t. \tag{2.4.28}$$

Now, by performing the inner integration, it is apparent that

$$\int_t^\infty |U(x)|^p\left[\int_0^x |V(u)h(u)|^{-p'}\,\mathrm{d}u\right]^{p-1}\mathrm{d}u \tag{2.4.29}$$

equals to

$$(p')^{p-1}\int_t^\infty |U(x)|^p\left[\int_0^x |V(u)|^{-p'}\,\mathrm{d}u\right]^{(p-1)/p'}\mathrm{d}x.$$

By the definition of B, this expression is bounded above by

$$(Bp')^{p-1}\int_t^\infty |U(x)|^p\left[\int_x^\infty |U(u)|^p\,\mathrm{d}u\right]^{-1/p'}\mathrm{d}x. \tag{2.4.30}$$

Performing the outer integration shows that this expression equals to

$$p(Bp')^{p-1}\left[\int_t^\infty |U(x)|^p\,\mathrm{d}x\right]^{1/p}. \tag{2.4.31}$$

By the definition of B, this expression is bounded by

$$pB^p(p')^{p-1}|h(t)|^{-p}. \tag{2.4.32}$$

Now in (2.4.28) we use the fact that (2.4.29) is bounded above by (2.4.32). This shows that (2.4.28) is bounded above by the pth power of the right-hand side of (2.4.27) and completes the proof of (2.4.27) for $1 < p < \infty$.

For $p = 1$ and $p = \infty$, the fact that $C \leqslant B$ is proved by showing that

$$\left[\int_0^\infty \left|U(x)\int_0^x f(t)\,\mathrm{d}t\right|^p \mathrm{d}x\right]^{1/p} \leqslant B\left[\int_0^\infty |V(x)f(x)|^p\,\mathrm{d}x\right]^{1/p}. \qquad (2.4.33)$$

For $p = 1$, inequality (2.4.33) follows just by interchanging the order of integration on the left-hand side of the inequality.

If $p = \infty$,

$$\left|U(x)\int_0^x f(t)\,\mathrm{d}t\right| \leqslant \Big[\operatorname*{ess\,sup}_{0\leqslant t\leqslant x}|f(t)V(t)|\Big]|U(x)|\int_0^x |V(t)|^{-1}\,\mathrm{d}t,$$

and (2.4.33) follows immediately.

To prove that $B \leqslant C$, observe that for a nonnegative f, a reduction of the intervals of integration in (2.4.24) shows that for $r > 0$,

$$\left[\int_r^\infty |U(x)|^p\,\mathrm{d}x\right]^{1/p}\left|\int_0^r f(t)\,\mathrm{d}t\right| \leqslant C\left[\int_0^r |V(x)f(x)|^p\,\mathrm{d}x\right]^{1/p}. \qquad (2.4.34)$$

It is sufficient to show that

$$\left[\int_r^\infty |U(x)|^p\,\mathrm{d}x\right]^{1/p}\left[\int_0^r |V(x)|^{-p'}\,\mathrm{d}x\right]^{1/p'} \leqslant C. \qquad (2.4.35)$$

If $p \neq 1$ and $0 < \int_0^r |V(x)|^{-p'}\,\mathrm{d}x < \infty$, inequality (2.4.35) follows immediately from (2.4.34) by taking $f(x) = |V(x)|^{-p'}$. If $p = 1$ and $0 < \operatorname{ess\,sup}_{0<x<r} 1/|V(x)| < \infty$, (2.4.35) follows from (2.4.34) by letting f be the characteristic function of the set where $1/|V(x)| \geqslant -1/n + \operatorname{ess\,sup}_{0<x<r} 1/|V(x)|$ and then letting $n \to \infty$. If $[\int_0^r |V(x)|^{-p'}\,\mathrm{d}x]^{1/p'} = 0$, (2.4.35) is immediate. If $[\int_0^r |1/V(x)|^{p'}\,\mathrm{d}x]^{1/p'} = \infty$, there exists an $f(x)$ such that $[\int_0^r |f(x) \times V(x)|^p\,\mathrm{d}x]^{1/p} < \infty$ and $\int_0^r f(x)\,\mathrm{d}x = \infty$. Then if $C < \infty$, (2.4.34) with this f shows that $[\int_r^\infty |U(x)|^p\,\mathrm{d}x]^{1/p} = 0$ so (2.4.35) holds. If $C = \infty$, (2.4.35) is obviously true.

To prove Theorem 2.4.8, assume first that $0 < U(x) < \infty$ and $0 < V(x) < \infty$ almost everywhere on $[0, \infty)$. Let g be a function in $L^{p'}$. By Fubini's theorem,

$$\int_0^\infty \left[U(x)\int_x^\infty f(t)\,\mathrm{d}t\right]g(x)\,\mathrm{d}x \qquad (2.4.36)$$

equals to

$$\int_0^\infty V(t)f(t)\left[\frac{1}{V(t)}\int_0^t g(x)U(x)\,\mathrm{d}x\right]\mathrm{d}t.$$

By Hölder's inequality, this expression is bounded above by

$$\left[\int_0^\infty |f(t)V(t)|^p\,\mathrm{d}t\right]^{1/p}\left[\int_0^\infty\left|\frac{1}{V(t)}\int_0^t g(x)U(x)\,\mathrm{d}x\right|^{p'}\mathrm{d}t\right]^{1/p'}.$$

By Theorem 2.4.7, this expression is bounded by

$$p^{1/p}\left(p'\right)^{1/p'}B\left[\int_0^\infty |f(t)V(t)|^p\,\mathrm{d}t\right]^{1/p}\left[\int_0^\infty |g(x)|^{p'}\right]^{1/p'},$$

and the converse of Hölder's inequality shows that $C \leqslant p^{1/p}(p')^{1/p'}B$. Simple limiting arguments take care of the cases where U and V are 0 or ∞ on a set of positive measure. The proof that $B \leqslant C$ is most easily done by imitating the corresponding proof of Theorem 2.4.7. □

In [221] another necessary and sufficient condition is given for (2.4.24) to hold with a finite C and also the question of general measures is also considered. For various other inequalities related to Hardy's integral inequality, see [17,18,40,65, 146,148,167,168,199,217,221].

2.5 Further Generalizations of Hardy's Integral Inequality

In the past few years a large number of papers has appeared in the literature which deal with various generalizations of the Hardy's integral inequality. In this section we shall deal with some generalizations of Hardy's integral inequality established by Levinson in [190] and Pachpatte in [295,344].

In 1964, Levinson [190] proved the following generalizations of Hardy's integral inequality.

THEOREM 2.5.1. *On an open interval, finite or infinite, let $\phi(u) \geqslant 0$ be defined and have a second derivative $\phi'' \geqslant 0$. For some $p > 1$, let*

$$\phi\phi'' \geqslant \left(1-\frac{1}{p}\right)(\phi')^2. \tag{2.5.1}$$

At the ends of the interval, let ϕ take its limiting values, finite or infinite. Then, if for $0 < x < \infty$, the range of values of $f(x)$ lie in the closed interval of definition *of ϕ and if G is defined by*

$$G(x) = \frac{1}{x}\int_0^x f(t)\,\mathrm{d}t \tag{2.5.2}$$

for $0 < x < \infty$, then

$$\int_0^\infty \phi\big(G(x)\big)\,\mathrm{d}x < \left(\frac{p}{p-1}\right)^p \int_0^\infty \phi\big(f(x)\big)\,\mathrm{d}x, \tag{2.5.3}$$

unless $\phi \equiv 0$.

PROOF. Let

$$\psi(u) = \big(\phi(u)\big)^{1/p} \geqslant 0.$$

Then by (2.5.1), $\psi''(u) \geqslant 0$ where $\psi(u) > 0$. Hence $\psi(u)$ is convex. Then by Jensen's inequality,

$$\psi\left(\frac{1}{x}\int_0^x f(t)\,\mathrm{d}t\right) \leqslant \frac{1}{x}\int_0^x \psi\big(f(t)\big)\,\mathrm{d}t. \tag{2.5.4}$$

Hardy's inequality in Theorem 2.4.1 applied to $\psi(f(x))$ gives

$$\int_0^\infty \left(\frac{1}{x}\int_0^x \psi\big(f(t)\big)\,\mathrm{d}t\right)^p \mathrm{d}x < \left(\frac{p}{p-1}\right)^p \int_0^\infty \big(\psi\big(f(x)\big)\big)^p\,\mathrm{d}x.$$

Using (2.5.4) and $\psi^p = \phi$ the proof of the theorem is complete. □

REMARK 2.5.1. We note that in the special case when $\phi(u) = u^p$, $u \geqslant 0$, it shows the constant is the best possible.

Theorem 2.5.1 is essentially a special case of the following theorem given by Levinson in [190, Theorem 2].

THEOREM 2.5.2. *Let ϕ and f be as in Theorem 2.5.1. For $x > 0$ let $r(x) > 0$, be continuous and monotone nondecreasing and set*

$$R(x) = \int_0^x r(t)\,\mathrm{d}t. \tag{2.5.5}$$

Then

$$\int_0^\infty \phi\left(\frac{1}{R(x)}\int_0^x r(t)f(t)\,\mathrm{d}t\right)\mathrm{d}x \leqslant \left(\frac{p}{p-1}\right)^p \int_0^\infty \phi\big(f(x)\big)\,\mathrm{d}x. \tag{2.5.6}$$

PROOF. First we prove the special case corresponding to $\phi(u) = u^p$, $p > 1$. The general form then follows by using Jensen's inequality and the special case. Let $\phi = u^p$, $u \geqslant 0$, and in this part of the proof let $f \geqslant 0$ and $f \in L^p(0, \infty)$. Let

$$F_a(x) = \frac{1}{R(x)} \int_a^x r(t) f(t)\, dt$$

and let

$$g_a(x) = r^{-1/p}(x) \int_a^x r(t) f(t)\, dt.$$

For $0 < a < b < \infty$, integrating by parts gives

$$\begin{aligned}\int_a^b F_a^p(x)\, dx &= \int_a^b r(x) R^{-p}(x) g_a^p(x)\, dx \\ &= \left[-\frac{1}{p-1} R^{-p+1}(x) g_a^p(x) \right]_a^b + \frac{p}{p-1} \int_a^b F_a^{p-1}(x) f(x)\, dx \\ &\quad - \frac{1}{p-1} \int_a^b R^{-p+1}(x) r^{-1}(x) g_a^p(x) r'(x)\, dx. \qquad (2.5.7)\end{aligned}$$

From (2.5.7) it is easy to observe that

$$(p-1) \int_a^b F_a^p(x)\, dx + \int_a^b \frac{R(x)}{r^2(x)} r'(x) F_a^p(x)\, dx \leqslant p \int_a^b F_a^{p-1}(x) f(x)\, dx. \qquad (2.5.8)$$

Since $r(x)$ is nondecreasing,

$$(p-1) \int_a^b F_a^p(x)\, dx \leqslant p \int_a^b F_a^{p-1}(x) f(x)\, dx.$$

Applying Hölder's inequality with indices p, $p/(p-1)$ to the right member and dividing through gives

$$(p-1)\left(\int_a^b F_a^p(x)\, dx \right)^{1/p} \leqslant p \left(\int_a^b f^p(x)\, dx \right)^{1/p}.$$

Or letting

$$\left(\frac{p}{p-1} \right)^p = k \qquad (2.5.9)$$

gives

$$\int_a^b F_a^p(x)\,\mathrm{d}x \leqslant k\int_a^b f^p(x)\,\mathrm{d}x.$$

Let $c > a$. Then

$$\int_c^b F_a^p(x)\,\mathrm{d}x \leqslant k\int_0^\infty f^p(x)\,\mathrm{d}x.$$

Now let $a \to 0$ on the left. Then

$$\int_c^b \left(\frac{1}{R(x)}\int_0^x r(t)f(t)\,\mathrm{d}t\right)^p \mathrm{d}x \leqslant k\int_0^\infty f^p(x)\,\mathrm{d}x.$$

Since this inequality holds for arbitrary $b > c > 0$, it follows that

$$\int_0^\infty \left(\frac{1}{R(x)}\int_0^x r(t)f(t)\,\mathrm{d}t\right)^p \mathrm{d}x \leqslant k\int_0^\infty f^p(x)\,\mathrm{d}x.$$

Next, let $\psi = \phi^{1/p}$ as in the proof of Theorem 2.5.1. By Jensen's inequality,

$$\psi\left(\frac{1}{R(x)}\int_0^x r(t)f(t)\,\mathrm{d}t\right) \leqslant \frac{1}{R(x)}\int_0^x r(t)\psi\big(f(t)\big)\,\mathrm{d}t. \tag{2.5.10}$$

Using (2.5.9) and (2.5.8) we get

$$\int_0^\infty \left(\frac{1}{R(x)}\int_0^x r(t)\psi\big(f(t)\big)\,\mathrm{d}t\right)^p \mathrm{d}x \leqslant \left(\frac{p}{p-1}\right)^p \int_0^\infty \psi\big(f(x)\big)^p\,\mathrm{d}x.$$

Combining the above inequality with (2.5.10) and using $\psi^p = \phi$ proves (2.5.6) and the proof is complete. □

REMARK 2.5.2. In the case $r \equiv 1$, Theorem 2.5.2 is a slightly weaker version of Theorem 2.5.1. If $\phi(u) = \mathrm{e}^u$ and $f(x)$ is replaced by $\log f(x)$, where now $f \geqslant 0$, then (2.5.6) becomes, on letting $p \to \infty$,

$$\int_0^\infty \exp\left(\frac{1}{R(x)}\int_0^x r(t)\log f(t)\,\mathrm{d}t\right)\mathrm{d}x \leqslant \mathrm{e}\int_0^\infty f(x)\,\mathrm{d}x,$$

which for $r \equiv 1$ is a consequence of a result of Knopp [141, p. 250]. For any ϕ for which $\phi\phi'' \geqslant \phi'^2$, one can let $p \to \infty$ in (2.5.6) to get for such a case

$$\int_0^\infty \frac{1}{R(x)}\phi\left(\int_0^x r(t)f(t)\,\mathrm{d}t\right)\mathrm{d}x \leqslant \mathrm{e}\int_0^\infty \phi\big(f(x)\big)\,\mathrm{d}x.$$

If ϕ is monotone and hence has an inverse, the analogue of introducing the logarithm can be used here.

The following theorem also given by Levinson in [190] is valid and does not require r to be monotone.

THEOREM 2.5.3. *Let ϕ and f be as in Theorem 2.5.1. Let $r(x) > 0$, $x > 0$, be absolutely continuous and let $R(x)$ as in (2.5.5) exist. Let there exist $\lambda > 0$ such that for almost all $x > 0$,*

$$\frac{r'R}{r^2} + p - 1 \geqslant \frac{p}{\lambda}. \tag{2.5.11}$$

Then

$$\int_0^\infty \frac{1}{R(x)}\phi\left(\int_0^x r(t)f(t)\,\mathrm{d}t\right)\mathrm{d}x \leqslant \lambda^p \int_0^\infty \phi\bigl(f(x)\bigr)\,\mathrm{d}x. \tag{2.5.12}$$

PROOF. The proof proceeds as for Theorem 2.5.2 up to (2.5.8). Using (2.5.11) in (2.5.8) gives

$$\int_a^b F_a^p(x)\,\mathrm{d}x \leqslant \lambda \int_a^b F_a^{p-1}(x)f(x)\,\mathrm{d}x.$$

Proceeding now much as in the proof of Theorem 2.5.2 gives

$$\int_0^\infty \frac{1}{R(x)}\left(\int_0^x r(t)f(t)\,\mathrm{d}t\right)^p \mathrm{d}x \leqslant \lambda^p \int_0^\infty f^p(x)\,\mathrm{d}x. \tag{2.5.13}$$

Applying (2.5.13) to $\psi(f)$ instead of f and using (2.5.10) completes the proof. □

REMARK 2.5.3. The case $r \equiv 1$, $\phi = u^p$ with $\lambda = p/(p-1)$ shows the constant in (2.5.12) to be the best possible.

In 1992, Pachpatte [295] has established the following generalizations of the certain extensions of well-known Hardy's inequality given by Chan in [52].

THEOREM 2.5.4. *Let $p > 1$ be a constant and r be a positive and absolutely continuous function on $(1, b)$ where $1 < b \leqslant \infty$. Let f be a nonnegative function on $(1, b)$ and $r(t)f(t)/t$ is integrable on $(1, b)$. Let*

$$1 - px(\log x)\frac{r'(x)}{r(x)} \geqslant \frac{1}{\alpha} \tag{2.5.14}$$

for almost all $x \in (1, b)$ and for some constant $\alpha > 0$. If $F(x)$ is defined by

$$F(x) = \frac{1}{r(x)} \int_x^b \frac{r(t)f(t)}{t}\,\mathrm{d}t$$

for $x \in (1, b)$, then

$$\int_1^b x^{-1}F^p(x)\,\mathrm{d}x \leqslant (\alpha p)^p \int_1^b x^{-1}\big[(\log x)f(x)\big]^p\,\mathrm{d}x. \tag{2.5.15}$$

THEOREM 2.5.5. *Let $p > 1$ be a constant and r be a positive and absolutely continuous function on $(0, 1)$. Let f be a nonnegative function on $(0, 1)$ and $r(t)f(t)/t$ is integrable on $(0, 1)$. Let*

$$1 - px(\log x)\frac{r'(x)}{r(x)} \geqslant \frac{1}{\beta} \tag{2.5.16}$$

for almost all $x \in (0, 1)$ and for some constant $\beta > 0$. If $F(x)$ is defined by

$$F(x) = \frac{1}{r(x)} \int_0^x \frac{r(t)f(t)}{t}\,\mathrm{d}t$$

for $x \in (0, 1)$, then

$$\int_0^1 x^{-1}F^p(x)\,\mathrm{d}x \leqslant (\beta p)^p \int_0^1 x^{-1}\big[|\log x| f(x)\big]^p\,\mathrm{d}x. \tag{2.5.17}$$

REMARK 2.5.4. We note that in the special cases when $r(x) = 1$ and $\alpha = \beta = 1$, and setting $f(t) = th(t)$, the inequalities established in Theorems 2.5.4 and 2.5.5 reduces respectively to inequalities (1a) and (2a) given in Theorems 1 and 2 by Chan in [52]. If we replace the function $r(x)$ by $1/r(x)$ in Theorems 2.5.4 and 2.5.5, then we get the corresponding variants of the inequalities in Theorems 2.5.4 and 2.5.5 which may be of interest in certain situations.

PROOFS OF THEOREMS 2.5.4 AND 2.5.5. If f is null, then inequality (2.5.15) is trivially true. We assume that f is not null. Integrating the left-hand side in (2.5.15) by parts we have

$$\int_1^b x^{-1}F^p(x)\,\mathrm{d}x = p\int_1^b x^{-1}(\log x)f(x)F^{p-1}(x)\,\mathrm{d}x$$
$$+ p\int_1^b (\log x)\frac{r'(x)}{r(x)}F^p(x)\,\mathrm{d}x. \tag{2.5.18}$$

From (2.5.18) we observe that

$$\int_1^b \left[1 - px(\log x)\frac{r'(x)}{r(x)}\right] x^{-1} F^p(x)\,\mathrm{d}x$$
$$= p\int_1^b x^{-1}(\log x) f(x) F^{p-1}(x)\,\mathrm{d}x. \tag{2.5.19}$$

By using (2.5.14) in (2.5.19) and then using Hölder's inequality with indices p, $p/(p-1)$ on the right-hand side, we obtain

$$\int_1^b x^{-1} F^p(x)\,\mathrm{d}x$$
$$\leqslant \alpha p \int_1^b \left[\left(x^{-1}\right)^{-(p-1)/p} x^{-1}(\log x) f(x)\right]\left[\left(x^{-1}\right)^{(p-1)/p} F^{p-1}(x)\right]\mathrm{d}x$$
$$\leqslant \alpha p \left\{\int_1^b \left[\left(x^{-1}\right)^{-(p-1)}\left(x^{-1}\right)^p\left((\log x) f(x)\right)^p\right]\mathrm{d}x\right\}^{1/p}$$
$$\times \left\{\int_1^b x^{-1} F^p(x)\,\mathrm{d}x\right\}^{(p-1)/p}. \tag{2.5.20}$$

Dividing both sides of (2.5.20) by the second integral factor on the right-hand side of (2.5.20), and then taking the pth power of both sides of the resulting inequality, we get the desired inequality (2.5.15). This proof completes Theorem 2.5.4.

In order to prove Theorem 2.5.5, we may suppose f is not null. Integrating the left-hand side in (2.5.17) by parts we have

$$\int_0^1 x^{-1} F^p(x)\,\mathrm{d}x = -p\int_0^1 x^{-1}(\log x) f(x) F^{p-1}(x)\,\mathrm{d}x$$
$$+ p\int_0^1 (\log x)\frac{r'(x)}{r(x)} F^p(x)\,\mathrm{d}x. \tag{2.5.21}$$

From (2.5.21) we observe that

$$\int_0^1 \left[1 - px(\log x)\frac{r'(x)}{r(x)}\right] x^{-1} F^p(x)\,\mathrm{d}x$$
$$\leqslant p\int_0^1 x^{-1}|\log x| f(x) F^{p-1}(x)\,\mathrm{d}x. \tag{2.5.22}$$

Now, by using (2.5.16) in (2.5.22) and by following the arguments as in the proof of Theorem 2.5.4 given below inequality (2.5.19), we get the required inequality in (2.5.17) and the proof of Theorem 2.5.5 is complete. □

In the following theorems, we present the results established by Pachpatte in [344] by using a uniform treatment based on the idea used first by Levinson [190] and then by Pachpatte in [252–255,268,295], Pachpatte and Love [278] and Pečarić and Love [382]. These results yield in special cases some of the earlier as well as recent generalizations of Hardy's inequality given in [70,162,198,382].

THEOREM 2.5.6. *Let $p>1$, $m\neq 1$ be constants. Let $f(x)$ be a nonnegative and integrable function on $(0,\infty)$ and let $h(x)$ be a positive continuous function on $(0,\infty)$, and let $H(x)=\int_0^x h(t)\,\mathrm{d}t$. Let $w(x)$ and $r(x)$ be positive and absolutely continuous functions on $(0,\infty)$. Let*

$$1-\frac{1}{m-1}\frac{H(x)}{h(x)}\frac{w'(x)}{w(x)}+\frac{p}{m-1}\frac{H(x)}{h(x)}\frac{r'(x)}{r(x)}\geqslant\frac{1}{\alpha_1}\quad \textit{for } m>1, \tag{2.5.23}$$

$$1-\frac{1}{1-m}\frac{H(x)}{h(x)}\frac{w'(x)}{w(x)}-\frac{p}{1-m}\frac{H(x)}{h(x)}\frac{r'(x)}{r(x)}\geqslant\frac{1}{\alpha_2}\quad \textit{for } m<1, \tag{2.5.24}$$

for almost all $x\in(0,\infty)$ and for some positive constants α_1,α_2. If $F(x)$ is defined by

$$F(x)=\begin{cases}\frac{1}{r(x)}\int_0^x r(t)h(t)f(t)\,\mathrm{d}t & \textit{for } m>1,\\ \frac{1}{r(x)}\int_x^\infty r(t)h(t)f(t)\,\mathrm{d}t & \textit{for } m<1,\end{cases} \tag{2.5.25}$$

for $x\in(0,\infty)$, then

$$\int_0^\infty w(x)H(x)^{-m}h(x)F^p(x)\,\mathrm{d}x \leqslant\left[\alpha\left(\frac{p}{|m-1|}\right)\right]^p\int_0^\infty w(x)H(x)^{p-m}h(x)f^p(x)\,\mathrm{d}x, \tag{2.5.26}$$

where $\alpha=\max\{\alpha_1,\alpha_2\}$. Equality holds in (2.5.26) if $f(x)\equiv 0$.

THEOREM 2.5.7. *Let $0\leqslant a<b<\infty$, $p>0$ and $q>0$, $\alpha>0$ be constants. Let $w(x)$, $r(x)$ be positive and locally absolutely continuous on (a,b) and $f(x)$ be almost everywhere nonnegative and measurable on (a,b). Let*

$$F(x)=\frac{1}{r(x)}\int_0^x\frac{r(t)f(t)}{t\log(b/t)}\,\mathrm{d}t \tag{2.5.27}$$

for all x *in* $[a, b)$, *and*

$$F(x) = \mathrm{o}\big((b-x)^{-q/p}\big) \quad \text{as } x \to b^{-}. \tag{2.5.28}$$

If $p > 1$ *and*

$$1 - \frac{1}{q} x \frac{w'(x)}{w(x)} \log\frac{b}{x} + \frac{p}{q} x \frac{r'(x)}{r(x)} \log\frac{b}{x} \geqslant \frac{1}{\alpha} \tag{2.5.29}$$

for almost all x *in* (a, b), *then*

$$\int_a^b w(x)\frac{1}{x}\left(\log\frac{b}{x}\right)^{q-1} F^p(x)\,\mathrm{d}x \leqslant \left(\frac{\alpha p}{q}\right)^p \int_a^b w(x)\frac{1}{x}\left(\log\frac{b}{x}\right)^{q-1} f^p(x)\,\mathrm{d}x. \tag{2.5.30}$$

If $0 < p < 1$ *and the reverse inequality in* (2.5.29) *holds, then the reverse inequality in* (2.5.30) *also holds.*

THEOREM 2.5.8. *Let* $a < b < R$, $p > 1$, $q < 1$, $\alpha > 0$ *be constants. Let* $w(x), r(x)$ *be positive and locally absolutely continuous in* (a, b). *Let* $h(x)$ *be a positive continuous function and let* $H(x) = \int_a^x h(t)\,\mathrm{d}t$, *for* $x \in (a, b)$. *Let* $f(x)$ *be nonnegative and measurable on* (a, b). *Let*

$$1 - \frac{1}{1-q}\frac{H(x)}{h(x)}\frac{w'(x)}{w(x)}\log\left(\frac{H(R)}{H(x)}\right) + \frac{p}{1-q}\frac{H(x)}{h(x)}\frac{r'(x)}{r(x)}\log\left(\frac{H(R)}{H(x)}\right) \geqslant \frac{1}{\alpha} \tag{2.5.31}$$

for almost all x *in* (a, b). *If* $F(x)$ *is defined by*

$$F(x) = \frac{1}{r(x)}\int_a^x r(t)h(t)f(t)\,\mathrm{d}t \tag{2.5.32}$$

for all $x \in (a, b)$, *then*

$$\begin{aligned}
&\int_a^b w(x)H^{-1}(x)h(x)\left(\log\left(\frac{H(R)}{H(x)}\right)\right)^{-q} F^p(x)\,\mathrm{d}x \\
&\quad \leqslant \left(\frac{\alpha p}{1-q}\right)^p \int_a^b w(x)\big(H(x)\big)^{p-1} h(x)\left(\log\left(\frac{H(R)}{H(x)}\right)\right)^{p-q} f^p(x)\,\mathrm{d}x.
\end{aligned} \tag{2.5.33}$$

PROOFS OF THEOREMS 2.5.6–2.5.8. For the proof of Theorem 2.5.6, let $0 < a < b < \infty$ and define, for $m > 1$,

$$F_a(x) = \frac{1}{r(x)}\int_a^x r(t)h(t)f(t)\,\mathrm{d}t,$$

with $F_0(x) = F(x)$. Integrating by parts gives

$$\begin{aligned}
&\int_a^b w(x)H(x)^{-m}h(x)F_a^p(x)\,\mathrm{d}x \\
&= \left[w(x)F_a^p(x)\frac{H(x)^{-m+1}}{-m+1}\right]_a^b \\
&\quad - \int_a^b \frac{H(x)^{-m+1}}{-m+1}\Bigg[w'(x)F_a^p(x) + w(x)pF_a^{p-1}(x) \\
&\qquad \times \left\{\frac{r(x)(r(x)h(x)f(x)) - r'(x)\int_a^x r(t)h(t)f(t)\,\mathrm{d}t}{r^2(t)}\right\}\Bigg]\mathrm{d}x.
\end{aligned} \tag{2.5.34}$$

Since $m > 1$, from (2.5.34) we observe that

$$\begin{aligned}
&\int_a^b \left[1 - \frac{1}{m-1}\frac{H(x)}{h(x)}\frac{w'(x)}{w(x)} + \frac{p}{m-1}\frac{H(x)}{h(x)}\frac{r'(x)}{r(x)}\right]w(x)H(x)^{-m}h(x)F_a^p(x)\,\mathrm{d}x \\
&\leqslant \frac{p}{m-1}\int_a^b w(x)H(x)^{-m+1}h(x)f(x)F_a^{p-1}(x)\,\mathrm{d}x \\
&= \frac{p}{m-1}\int_a^b \left[w(x)^{1/p}H(x)^{(p-m)/p}h(x)^{1/p}f(x)\right] \\
&\qquad \times \left[w(x)^{(p-1)/p}H(x)^{-m(p-1)/p}h(x)^{(p-1)/p}F_a^{p-1}(x)\right]\mathrm{d}x.
\end{aligned} \tag{2.5.35}$$

Using (2.5.23) and applying Hölder's inequality with indices p, $p/(p-1)$ on the right-hand side of (2.5.35), we obtain

$$\begin{aligned}
&\int_a^b w(x)H(x)^{-m}h(x)F_a^p(x)\,\mathrm{d}x \\
&\leqslant \alpha_1\left(\frac{p}{m-1}\right)\left\{\int_a^b w(x)H(x)^{p-m}h(x)f^p(x)\,\mathrm{d}x\right\}^{1/p} \\
&\qquad \times \left\{\int_a^b w(x)H^{-m}(x)h(x)F_a^p(x)\,\mathrm{d}x\right\}^{(p-1)/p}.
\end{aligned} \tag{2.5.36}$$

Dividing both sides of (2.5.36) by the second integral factor on the right-hand side of (2.5.36), and raising both sides to the pth power, we obtain

$$\int_a^b w(x)H^{-m}(x)h(x)F_a^p(x)\,\mathrm{d}x \leqslant \left[\alpha_1\left(\frac{p}{m-1}\right)\right]^p \int_a^b w(x)H^{p-m}(x)h(x)f^p(x)\,\mathrm{d}x. \tag{2.5.37}$$

From (2.5.37) we have

$$\int_a^b w(x)H^{-m}(x)h(x)F_a^p(x)\,\mathrm{d}x \leqslant \left[\alpha_1\left(\frac{p}{m-1}\right)\right]^p \int_0^\infty w(x)H^{p-m}(x)h(x)f^p(x)\,\mathrm{d}x. \tag{2.5.38}$$

Let $a < c < b$. Then, from (2.5.38), we have

$$\int_c^b w(x)H^{-m}(x)h(x)F_a^p(x)\,\mathrm{d}x \leqslant \left[\alpha_1\left(\frac{p}{m-1}\right)\right]^p \int_0^\infty w(x)H^{p-m}(x)h(x)f^p(x)\,\mathrm{d}x. \tag{2.5.39}$$

Letting $a \to 0$ in (2.5.39) we have

$$\int_c^b w(x)H^{-m}(x)h(x)F^p(x)\,\mathrm{d}x \leqslant \left[\alpha_1\left(\frac{p}{m-1}\right)\right]^p \int_0^\infty w(x)H^{p-m}(x)h(x)f^p(x)\,\mathrm{d}x. \tag{2.5.40}$$

Since this inequality holds for arbitrary $0 < c < b$, it follows that

$$\int_0^\infty w(x)H^{-m}(x)h(x)F^p(x)\,\mathrm{d}x \leqslant \left[\alpha_1\left(\frac{p}{m-1}\right)\right]^p \int_0^\infty w(x)H^{p-m}(x)h(x)f^p(x)\,\mathrm{d}x. \tag{2.5.41}$$

Let $0 < a < b < \infty$ and define, for $m < 1$,

$$F_b(x) = \frac{1}{r(x)} \int_x^b r(t)h(t)f(t)\,\mathrm{d}t,$$

with $F_{\infty}(x) = F(x)$. Now, by following the same steps as in the proof of inequality (2.5.41) with suitable modifications, we obtain

$$\int_0^{\infty} w(x)H^{-m}(x)h(x)F^p(x)\,\mathrm{d}x \leqslant \left[\alpha_2\left(\frac{p}{1-m}\right)\right]^p \int_0^{\infty} w(x)H^{p-m}(x)h(x)f^p(x)\,\mathrm{d}x. \tag{2.5.42}$$

From (2.5.41) and (2.5.42), we obtain

$$\int_0^{\infty} w(x)H^{-m}(x)h(x)F^p(x)\,\mathrm{d}x \leqslant \left[\alpha\left(\frac{p}{|m-1|}\right)\right]^p \int_0^{\infty} w(x)H^{p-m}(x)h(x)f^p(x)\,\mathrm{d}x.$$

The proof of Theorem 2.5.6 is complete.

To prove Theorem 2.5.7, let $p > 1$ and suppose that $a > 0$. Integrating by parts we have

$$\begin{aligned}
&\int_a^b w(x)\frac{1}{x}\left(\log\left(\frac{b}{x}\right)\right)^{q-1} F^p(x)\,\mathrm{d}x \\
&= \left[-w(x)F^p(x)\frac{1}{q}\left(\log\left(\frac{b}{x}\right)\right)^q\right]_a^b \\
&\quad + \int_a^b \frac{1}{q}\left(\log\left(\frac{b}{x}\right)\right)^q \Bigg\{w'(x)F^p(x) + w(x)pF^{p-1}(x) \\
&\qquad \times \left[r(x)\frac{r(x)f(x)}{x\log(1/x)} - r'(x)\int_a^x \frac{r(t)f(t)}{t\log(b/t)}\,\mathrm{d}t\right]\frac{1}{r^2(x)}\Bigg\}\,\mathrm{d}x.
\end{aligned} \tag{2.5.43}$$

From (2.5.43) we observe that

$$\begin{aligned}
&\int_a^b \left[1 - \frac{1}{q}x\frac{w'(x)}{w(x)}\log\left(\frac{b}{x}\right) + \frac{p}{q}x\frac{r'(x)}{r(x)}\log\left(\frac{b}{x}\right)\right] \\
&\qquad \times w(x)\frac{1}{x}\left(\log\left(\frac{b}{x}\right)\right)^{q-1} F^p(x)\,\mathrm{d}x
\end{aligned}$$

$$= \frac{p}{q} \int_a^b \left[w^{1/p}(x) \frac{x^{(p-1)/p} (\log(b/x))^{q-(q-1)(p-1)/p}}{x \log(b/x)} f(x) \right] \times \left[w^{(p-1)/p} x^{-(p-1)/p} \left(\log\left(\frac{b}{x}\right)\right)^{(q-1)(p-1)/p} F^{p-1}(x) \right] \mathrm{d}x. \tag{2.5.44}$$

Using (2.5.29) and applying Hölder's inequality with indices p, $p/(p-1)$ we obtain

$$\int_a^b w(x) \frac{1}{x} \left(\log\left(\frac{b}{x}\right)\right)^{q-1} F^p(x)\,\mathrm{d}x \leqslant \alpha \frac{p}{q} \left\{ \int_a^b \frac{w(x) x^{p-1} (\log(b/x))^{q+p-1}}{x^p (\log(b/x))^p} f^p(x)\,\mathrm{d}x \right\}^{1/p} \times \left\{ \int_a^b w(x) x^{-1} \left(\log\left(\frac{b}{x}\right)\right)^{q-1} F^p(x)\,\mathrm{d}x \right\}^{(p-1)/p}. \tag{2.5.45}$$

Dividing both sides of (2.5.45) by the second integral factor on the right-hand side of (2.5.45) and taking the pth power on both sides of the resulting inequality we get the desired inequality in (2.5.30). Similarly, if $0 < p < 1$ and (2.5.29) is reversed, the reverse inequality in (2.5.30) is obtained.

Suppose instead that $a = 0$. If $0 < a' < b$, all the hypotheses hold with a replaced by a', under their respective conditions on p and in (2.5.29). Call these inequalities (2.5.30′). As $a' \downarrow a^+$, the modified $F(x)$ increases toward the value given in (2.5.27), and both sides of (2.5.30′) tend to the corresponding sides of (2.5.30), using the monotonic convergence theorem for the left-hand sides. This fact limits processes thus produces inequality (2.5.30) as required. This result completes the proof of Theorem 2.5.7.

We next establish the inequality (2.5.33) in Theorem 2.5.8. Integrating by parts we obtain

$$\int_a^b w(x) H(x)^{-1} h(x) \left(\log\left(\frac{H(R)}{H(x)}\right)\right)^{-q} F^p(x)\,\mathrm{d}x = -\left\{ \left[\frac{w(x) F^p(x) (\log(H(R)/H(x)))^{-q+1}}{-q+1} \right]_a^b - \int_a^b \frac{(\log(H(R)/H(x)))^{-q+1}}{-q+1} \right.$$

$$\times \left\{ w'(x) F^p(x) + w(x) p F^{p-1}(x) \right.$$

$$\left. \times \frac{1}{r^2(x)} \left[r(x) \big(r(x) h(x) f(x) \big) - r'(x) \int_a^x r(t) h(t) f(t) \, dt \right] \right\} dx \Bigg\}. \tag{2.5.46}$$

From (2.5.46) we observe that

$$\begin{aligned} &\int_a^b \left[1 - \frac{1}{1-q} \frac{H(x)}{h(x)} \frac{w'(x)}{w(x)} \log\left(\frac{H(R)}{H(x)} \right) + \frac{p}{1-q} \frac{H(x)}{h(x)} \frac{r'(x)}{r(x)} \log\left(\frac{H(R)}{H(x)} \right) \right] \\ &\quad \times w(x) H(x)^{-1} h(x) \left(\log\left(\frac{H(R)}{H(x)} \right) \right)^{-q} F^p(x) \, dx \\ &\leqslant \frac{p}{1-q} \int_a^b w(x) h(x) f(x) \left(\log\left(\frac{H(R)}{H(x)} \right) \right)^{-q+1} F^{p-1}(x) \, dx \\ &= \frac{p}{1-q} \int_a^b \left\{ \left(\log\left(\frac{H(R)}{H(x)} \right) \right)^{-q} \right\}^{1/p} \log\left(\frac{H(R)}{H(x)} \right) \\ &\qquad \times w(x)^{1/p} \big(H(x)^{-1} \big)^{-(p-1)/p} \big(h(x) \big)^{1/p} f(x) \\ &\qquad \times \left\{ \left(\log\left(\frac{H(R)}{H(x)} \right) \right)^{-q} \right\}^{(p-1)/p} \\ &\qquad \times w(x)^{(p-1)/p} F^{p-1}(x) \big(H(x)^{-1} h(x) \big)^{(p-1)/p} \, dx. \end{aligned} \tag{2.5.47}$$

Using (2.5.31) and applying Hölder's inequality with indices p, $p/(p-1)$ we obtain

$$\begin{aligned} &\int_a^b w(x) H(x)^{-1} h(x) \left(\log\left(\frac{H(R)}{H(x)} \right) \right)^{-q} F^p(x) \, dx \\ &\quad \leqslant \frac{p}{1-q} \left\{ \int_a^b \left(\log\left(\frac{H(R)}{H(x)} \right) \right)^{-q} \left(\log\left(\frac{H(R)}{H(x)} \right) \right)^p \right. \\ &\qquad\qquad \left. \times w(x) \big(H(x)^{-1} \big)^{-(p-1)} h(x) f^p(x) \, dx \right\}^{1/p} \\ &\qquad \times \left\{ \int_a^b w(x) H(x)^{-1} h(x) \left(\log\left(\frac{H(R)}{H(x)} \right) \right)^{-q} F^p(x) \, dx \right\}^{(p-1)/p}. \end{aligned} \tag{2.5.48}$$

Dividing both sides of (2.5.48) by the second integral factor on the right-hand side of (2.5.48) and taking the pth power on both sides of the resulting inequality we get the required inequality in (2.5.33). The proof of Theorem 2.5.8 is complete. □

2.6 Hardy-Type Integral Inequalities

There are many extensions and variants of Hardy's integral inequality (2.4.1) established by different writers in different ways. In this section we give some of these results established in the literature during the past few years.

In 1968, Izurai and Izumi [162] obtained the following variant of Hardy's inequality in Theorem 2.4.1.

THEOREM 2.6.1. *Let $p > 1$ and $s < -1$ and let f be nonnegative and integrable on $(0, b)$. If $x^s f(x)^p$ is integrable, then*

$$\int_0^b x^s G(x)^p \, \mathrm{d}x \leqslant \left(\frac{p}{-s-1}\right)^p \int_0^b x^s \left| f\left(\frac{x}{2}\right) - f(x)\right|^p \mathrm{d}x, \tag{2.6.1}$$

where

$$G(x) = \int_{x/2}^x t^{-1} f(t) \, \mathrm{d}t. \tag{2.6.2}$$

PROOF. The proof is based on the idea used in [434, p. 20]. First of all we have

$$\begin{aligned} G(x) &= \int_{x/2}^x t^{-1} f(t) \, \mathrm{d}t = \int_{x/2}^x t^{-s/(p-1)} t^{s/p} f(t) \, \mathrm{d}t \\ &\leqslant \left(\int_{x/2}^x t^s f(t)^p \, \mathrm{d}t\right)^{1/p} \left(\int_{x/2}^x t^{-(p+s)q/p} \, \mathrm{d}t\right)^{1/q} \\ &= \mathrm{O}\big(x^{-(s+1)/p}\big) \quad \text{as } x \to 0, \end{aligned}$$

where $\frac{1}{p} + \frac{1}{q} = 1$, and then, if $s \neq -1$,

$$\begin{aligned} &\int_0^b x^s G(x)^p \, \mathrm{d}x \\ &= \left[\frac{x^{s+1}}{s+1} G(x)^p\right]_0^b - \frac{p}{s+1} \int_0^b x^{s+1} G(x)^{p-1} \left(\frac{f(x)}{x} - \frac{1}{2}\frac{f(x/2)}{x/2}\right) \mathrm{d}x. \end{aligned}$$

If $s < -1$, then the first term on the right-hand side is negative and hence the second term is positive, and

$$
\begin{aligned}
&\int_0^b x^s G(x)^p \,\mathrm{d}x \\
&\quad\leqslant \frac{p}{-s-1} \int_0^b x^s G(x)^{p-1} \left(f(x) - f\left(\frac{x}{2}\right)\right) \mathrm{d}x \\
&\quad\leqslant \frac{p}{-s-1} \int_0^b x^{s(p-1)/p} G(x)^{p-1} x^{s-s(p-1)/p} \left| f(x) - f\left(\frac{x}{2}\right)\right| \mathrm{d}x \\
&\quad\leqslant \frac{p}{-s-1} \left(\int_0^b x^s G(x)^p \,\mathrm{d}x \right)^{1/q} \left(\int_0^b x^s \left| f(x) - f\left(\frac{x}{2}\right)\right|^p \mathrm{d}x \right)^{1/p}.
\end{aligned}
$$

Dividing both sides of the above inequality by the first integral factor on the right-hand side and taking the pth power of both sides we get inequality (2.6.1). □

REMARK 2.6.1. We note that in [162] Izumi and Izumi also have given some other similar variants of Hardy's inequality so as to use the same in order to establish some inequalities for Fourier series. For other results, we refer the interested readers to [162].

In [255] and [337] Pachpatte has established the following variants of Hardy's inequality in Theorem 2.4.1.

THEOREM 2.6.2. *Let $p > 1$, $m > 1$ be constants. Let f be a nonnegative and integrable function on $(0, b)$, $b > 0$ is a constant. Let r be a positive and absolutely continuous function on $(0, b)$. Let*

$$1 + \frac{p}{m-1} x \frac{r'(x)}{r(x)} \geqslant \frac{1}{\alpha} \tag{2.6.3}$$

for almost all $x \in (0, b)$ and for some constant $\alpha > 0$. Then

$$
\begin{aligned}
&\int_0^b x^{-m} F^p(x) \,\mathrm{d}x \\
&\quad\leqslant \left\{ \alpha \left(\frac{p}{m-1} \right) \right\}^p \int_0^b x^{-m} \left\{ \frac{1}{r(x)} \left| r(x) f(x) - r\left(\frac{x}{2}\right) f\left(\frac{x}{2}\right)\right| \right\}^p \mathrm{d}x, \qquad (2.6.4)
\end{aligned}
$$

where

$$F(x) = \frac{1}{r(x)} \int_{x/2}^x \frac{r(t) f(t)}{t} \,\mathrm{d}t.$$

PROOF. Integrating by parts the left-hand side in (2.6.4) we obtain

$$\begin{aligned}
&\int_0^b x^{-m} F^p(x)\,\mathrm{d}x \\
&\quad = \left[\frac{x^{-m+1}}{-m+1} F^p(x)\right]_0^b \\
&\qquad + \frac{p}{m-1}\int_0^b x^{-m+1}\frac{1}{xr(x)}\left\{r(x)f(x) - r\left(\frac{x}{2}\right)f\left(\frac{x}{2}\right)\right\}F^{p-1}(x)\,\mathrm{d}x \\
&\qquad - \frac{p}{m-1}\int_0^b x^{-m+1}\frac{r'(x)}{r(x)}F^p(x)\,\mathrm{d}x. \qquad (2.6.5)
\end{aligned}$$

From (2.6.5) we observe that

$$\begin{aligned}
&\int_0^b \left[1 + \frac{p}{m-1} x \frac{r'(x)}{r(x)}\right] x^{-m} F^p(x)\,\mathrm{d}x \\
&\quad \leqslant \left(\frac{p}{m-1}\right)\int_0^b \left[\left(x^{-m}\right)^{(p-1)/p} F^{p-1}(x)\right] \\
&\qquad\qquad \times \left[\left(x^{-m}\right)^{-(p-1)/p} x^{-m}\frac{1}{r(x)}\left|r(x)f(x) - r\left(\frac{x}{2}\right)f\left(\frac{x}{2}\right)\right|\right]\mathrm{d}x. \qquad (2.6.6)
\end{aligned}$$

From (2.6.3) and using Hölder's inequality with indices p, $p/(p-1)$ on the right-hand side of (2.6.6), we obtain

$$\begin{aligned}
&\int_0^b x^{-m} F^p(x)\,\mathrm{d}x \\
&\quad \leqslant \alpha\left(\frac{p}{m-1}\right)\left\{\int_0^b x^{-m} F^p(x)\,\mathrm{d}x\right\}^{(p-1)/p} \\
&\qquad \times \left\{\int_0^b x^{-m}\left[\frac{1}{r(x)}\left|r(x)f(x) - r\left(\frac{x}{2}\right)f\left(\frac{x}{2}\right)\right|\right]^p \mathrm{d}x\right\}^{1/p}. \qquad (2.6.7)
\end{aligned}$$

Dividing both sides of (2.6.7) by the first integral factor on the right-hand side in (2.6.7) and taking the pth power on both sides we get the desired inequality in (2.6.4). The proof is complete. □

REMARK 2.6.2. If we take in Theorem 2.6.2, $r(x) = 1$, $\alpha = 1$ and $-m = s$ where $s < -1$ is a constant, then we get the inequality in Theorem 2.6.1 given in [162, Theorem 2].

THEOREM 2.6.3. *Let $p > 1$, $m > 1$ be constants. Let f be a nonnegative and integrable function on $(0, b)$, $0 < b < \infty$. If $F(x)$ is defined by*

$$F(x) = \int_{x/2}^{x} \frac{1}{t}\left(\int_{t/2}^{t} \frac{f(s)}{s}\,\mathrm{d}s\right)\mathrm{d}t \tag{2.6.8}$$

for $x \in (0, b)$, then

$$\int_0^b x^{-m} F^p(x)\,\mathrm{d}x \leqslant \left(\frac{p}{m-1}\right)^{2p} \int_0^b x^{-m}\left|f(x) - f\left(\frac{x}{4}\right)\right|^p \mathrm{d}x. \tag{2.6.9}$$

PROOF. Integrating the left-hand side in (2.6.9) by parts we have

$$\begin{aligned}
&\int_0^b x^{-m} F^p(x)\,\mathrm{d}x \\
&\quad = -\frac{b^{-m+1}}{m-1} F^p(b) + \left(\frac{p}{m-1}\right) \\
&\qquad \times \int_0^b x^{-m+1} F^{p-1}(x)\left[\frac{1}{x}\int_{x/2}^{x} \frac{f(s)}{s}\,\mathrm{d}s - \frac{1}{2}\frac{1}{x/2}\int_{x/4}^{x/2} \frac{f(s)}{s}\,\mathrm{d}s\right]\mathrm{d}x.
\end{aligned} \tag{2.6.10}$$

From (2.6.10) and using Hölder's inequality with indices p, $p/(p-1)$ we observe that

$$\begin{aligned}
&\int_0^b x^{-m} F^p(x)\,\mathrm{d}x \\
&\quad \leqslant \left(\frac{p}{m-1}\right)\int_0^b x^{-m} F^{p-1}(x)\left|\int_{x/2}^{x} \frac{f(s)}{s}\,\mathrm{d}s - \int_{x/4}^{x/2} \frac{f(s)}{s}\,\mathrm{d}s\right|\mathrm{d}x \\
&\quad \leqslant \left(\frac{p}{m-1}\right)\int_0^b x^{-m} F^{p-1}(x)\left\{\int_{x/4}^{x} \frac{|f(s)|}{s}\,\mathrm{d}s\right\}\mathrm{d}x \\
&\quad = \left(\frac{p}{m-1}\right) \\
&\qquad \times \int_0^b \left[\left(x^{-m}\right)^{(p-1)/p} F^{p-1}(x)\right]\left[\left(x^{-m}\right)^{-(p-1)/p+1}\int_{x/4}^{x} \frac{|f(s)|}{s}\,\mathrm{d}s\right]\mathrm{d}x
\end{aligned}$$

$$\leqslant \left(\frac{p}{m-1}\right)\left[\int_0^b x^{-m} F^p(x)\,\mathrm{d}x\right]^{-(p-1)/p}$$
$$\times \left[\int_0^b x^{-m}\left\{\int_{x/4}^x \frac{|f(s)|}{s}\,\mathrm{d}s\right\}^p \mathrm{d}x\right]^{1/p}. \tag{2.6.11}$$

Dividing both sides of (2.6.11) by the first integral factor on the right-hand side in (2.6.11) and taking the pth power of both sides we get

$$\int_0^b x^{-m} F^p(x)\,\mathrm{d}x \leqslant \left(\frac{p}{m-1}\right)^p \int_0^b x^{-m}\left\{\int_{x/4}^x \frac{|f(s)|}{s}\,\mathrm{d}s\right\}^p \mathrm{d}x. \tag{2.6.12}$$

Now, integrating by parts the integral on the right-hand side in (2.6.12), we have

$$\int_0^b x^{-m}\left\{\int_{x/4}^x \frac{|f(s)|}{s}\,\mathrm{d}s\right\}^p \mathrm{d}x$$
$$= -\frac{b^{-m+1}}{m-1}\left\{\int_{b/4}^b \frac{|f(s)|}{s}\,\mathrm{d}s\right\} + \left(\frac{p}{m-1}\right)$$
$$\times \int_0^b x^{-m+1}\left\{\int_{x/4}^x \frac{|f(s)|}{s}\,\mathrm{d}s\right\}^{p-1}\left[\frac{1}{x}|f(x)| - \frac{1}{4}\frac{1}{x/4}\left|f\left(\frac{x}{4}\right)\right|\right]\mathrm{d}x. \tag{2.6.13}$$

From (2.6.13) and using Hölder's inequality with indices p, $p/(p-1)$ we observe that

$$\int_0^b x^{-m}\left\{\int_{x/4}^x \frac{|f(s)|}{s}\,\mathrm{d}s\right\}^p \mathrm{d}x$$
$$\leqslant \left(\frac{p}{m-1}\right)\int_0^b \left[\left(x^{-m}\right)^{(p-1)/p}\left\{\int_{x/4}^x \frac{|f(s)|}{s}\,\mathrm{d}s\right\}^{p-1}\right]$$
$$\times \left[\left(x^{-m}\right)^{-(p-1)/p+1}\left|f(x) - f\left(\frac{x}{4}\right)\right|\right]\mathrm{d}x$$
$$\leqslant \left(\frac{p}{m-1}\right)\left[\int_0^b x^{-m}\left\{\int_{x/4}^x \frac{|f(s)|}{s}\,\mathrm{d}s\right\}^p \mathrm{d}x\right]^{(p-1)/p}$$
$$\times \left[\int_0^b x^{-m}\left|f(x) - f\left(\frac{x}{4}\right)\right|^p \mathrm{d}x\right]^{1/p}. \tag{2.6.14}$$

Dividing both sides of (2.6.14) by the first integral factor on the right-hand side in (2.6.14) and taking the pth power of both sides we get

$$\int_0^{b} x^{-m}\left(\int_{x/4}^{x} \frac{|f(s)|}{s}\,ds\right)^p dx \leqslant \left(\frac{p}{m-1}\right)^p \int_0^b x^{-m}\left|f(x) - f\left(\frac{x}{4}\right)\right|^p dx. \tag{2.6.15}$$

Using (2.6.15) in (2.6.12) we get the desired inequality in (2.6.9). The proof is complete. □

Motivated by Hardy's inequality, in 1987, Pachpatte [254] establishes the following inequalities.

THEOREM 2.6.4. *Let $m \neq 1$ and $p_j > 1$, $j = 1,2,3$, be constants. Let $f_j(x)$, $j = 1,2,3$, be nonnegative and integrable functions on $(0,\infty)$ and let $r_j(x)$, $j = 1,2,3$, be positive and absolutely continuous functions on $(0,\infty)$. Let*

$$1 + \left(\frac{2p_j}{m-1}\right)x\frac{r_j'(x)}{r_j(x)} \geqslant \frac{1}{\alpha_j} \quad \textit{for } m > 1, \tag{2.6.16}$$

$$1 - \left(\frac{2p_j}{1-m}\right)x\frac{r_j'(x)}{r_j(x)} \geqslant \frac{1}{\beta_j} \quad \textit{for } m < 1, \tag{2.6.17}$$

for almost all $x > 0$ and for some positive constants α_j, β_j, $j = 1,2,3$. If $F_j(x)$, $j = 1,2,3$, are defined by

$$F_j(x) = \begin{cases} \frac{1}{r_j(x)}\int_0^x \frac{r_j(t)f_j(t)}{t}\,dt & \textit{for } m > 1, \\ \frac{1}{r_j(x)}\int_x^\infty \frac{r_j(t)f_j(t)}{t}\,dt & \textit{for } m < 1, \end{cases} \tag{2.6.18}$$

then

$$\int_0^\infty x^{-m}\left[F_1^{p_1}(x)F_2^{p_2}(x) + F_2^{p_2}(x)F_3^{p_3}(x) + F_3^{p_3}(x)F_1^{p_1}(x)\right]dx \leqslant \sum_{j=1}^{3}\left[\left\{\lambda\left(\frac{2p_j}{|m-1|}\right)\right\}^{2p_j}\int_0^\infty x^{-m} f_j^{2p_j}(x)\,dx\right], \tag{2.6.19}$$

where $\lambda = \max\{\alpha_j, \beta_j\}$ for $j = 1,2,3$. Equality holds in (2.6.19) if $f_j(x) \equiv 0$, $j = 1,2,3$.

THEOREM 2.6.5. *Let m, p_j, f_j, r_j be as in Theorem 2.6.4. Let*

$$1 + \left(\frac{4p_j}{m-1}\right) x \frac{r_j'(x)}{r_j(x)} \geqslant \frac{1}{\alpha_j^{\cdot}} \quad \textit{for } m > 1, \tag{2.6.20}$$

$$1 - \left(\frac{4p_j}{1-m}\right) x \frac{r_j'(x)}{r_j(x)} \geqslant \frac{1}{\beta_j^{\cdot}} \quad \textit{for } m < 1, \tag{2.6.21}$$

for almost all $x > 0$ and for some positive constants $\alpha_j^{\cdot}$, $\beta_j^{\cdot}$, $j = 1, 2, 3$. If $F_j(x)$, $j = 1, 2, 3$, are defined by (2.6.18), *then*

$$\int_0^\infty x^{-m} F_1^{p_1}(x) F_2^{p_2}(x) F_3^{p_3}(x) \left[F_1^{p_1}(x) + F_2^{p_2}(x) + F_3^{p_3}(x)\right] \mathrm{d}x$$

$$\leqslant \sum_{j=1}^{3} \left[\left\{\lambda^{\cdot}\left(\frac{4p_j}{|m-1|}\right)\right\}^{4p_j} \int_0^\infty x^{-m} f_j^{4p_j}(x)\, \mathrm{d}x\right], \tag{2.6.22}$$

where $\lambda^{\cdot} = \max\{\alpha_j^{\cdot}, \beta_j^{\cdot}\}$ for $j = 1, 2, 3$. Equality holds in (2.6.22) *if $f_j(x) \equiv 0$, $j = 1, 2, 3$.*

REMARK 2.6.3. In the special cases when $p_j = p > 1$, $f_j = f$, $F_j = F$, $r_j = 1$, $\lambda = 1$ and $\lambda^{\cdot} = 1$, the inequalities established in (2.6.19) and (2.6.22) reduce, respectively, to the following inequalities

$$\int_0^\infty x^{-m} F^{2p}(x)\, \mathrm{d}x \leqslant \left(\frac{2p}{|m-1|}\right)^{2p} \int_0^\infty x^{-m} f^{2p}(x)\, \mathrm{d}x \tag{2.6.23}$$

and

$$\int_0^\infty x^{-m} F^{4p}(x)\, \mathrm{d}x \leqslant \left(\frac{4p}{|m-1|}\right)^{4p} \int_0^\infty x^{-m} f^{4p}(x)\, \mathrm{d}x. \tag{2.6.24}$$

We note that the inequalities obtained in (2.6.23) and (2.6.24) are the slight variants of Hardy's inequality in Theorem 2.4.2.

THEOREM 2.6.6. *Let $m \neq 1$ and $p_i > 1$, $i = 1, \ldots, n$, be constants. Let $g_i(x)$, $i = 1, \ldots, n$, be nonnegative and integrable functions on $(0, \infty)$ and let $k_i(x)$, $i = 1, \ldots, n$, be positive and absolutely continuous functions on $(0, \infty)$. Let*

$$1 + \left(\frac{np_i}{m-1}\right) x \frac{k_i'(x)}{k_i(x)} \geqslant \frac{1}{\gamma_i} \quad \textit{for } m > 1, \tag{2.6.25}$$

$$1 - \left(\frac{np_i}{1-m}\right) x \frac{k_i'(x)}{k_i(x)} \geqslant \frac{1}{\delta_i} \quad \textit{for } m < 1, \tag{2.6.26}$$

for almost all $x > 0$ and for some positive constants γ_i, δ_i, $i = 1, \ldots, n$. If $G_i(x)$, $i = 1, \ldots, n$, are defined by

$$G_i(x) = \begin{cases} \frac{1}{k_i(x)} \int_0^x \frac{k_i(t) g_i(t)}{t} \, \mathrm{d}t & \text{for } m > 1, \\ \frac{1}{k_i(x)} \int_x^\infty \frac{k_i(t) g_i(t)}{t} \, \mathrm{d}t & \text{for } m < 1, \end{cases} \tag{2.6.27}$$

then

$$\int_0^\infty x^{-m} \prod_{i=1}^n G_i^{p_i}(x) \, \mathrm{d}x \leqslant \frac{1}{n} \sum_{i=1}^n \left[\left\{ \mu \left(\frac{np_i}{|m-1|} \right) \right\}^{np_i} \int_0^\infty x^{-m} g_i^{np_i}(x) \, \mathrm{d}x \right], \tag{2.6.28}$$

where $\mu = \max\{\gamma_i, \delta_i\}$ for $i = 1, \ldots, n$. Equality holds in (2.6.28) if $g_i(x) \equiv 0$, $i = 1, \ldots, n$.

REMARK 2.6.4. In the special case when $p_i = p > 1$, $g_i = g$, $G_i = G$, $k_i = 1$ and $\mu = 1$, the inequality established in (2.6.28) reduces to the following inequality

$$\int_0^\infty x^{-m} G^{np}(x) \, \mathrm{d}x \leqslant \left(\frac{np}{|m-1|} \right)^{np} \int_0^\infty x^{-m} g^{np}(x) \, \mathrm{d}x,$$

which in turn is a variant of Hardy's inequality in Theorem 2.4.2.

The following inequalities established by Pachpatte in [254] have their origins in the variant of Hardy's inequality given by Izumi and Izumi in Theorem 2.6.1.

THEOREM 2.6.7. *Let $m > 1$ and $p_j > 1$, $j = 1, 2, 3$, be constants. Let $h_j(x)$, $j = 1, 2, 3$, be nonnegative and integrable functions on $(0, b)$ and let $z_j(x)$, $j = 1, 2, 3$, be positive and absolutely continuous functions on $(0, b)$. Let*

$$1 + \left(\frac{2p_j}{m-1} \right) x \frac{z_j'(x)}{z_j(x)} \geqslant \frac{1}{\eta_j} \tag{2.6.29}$$

for almost all $x \in (0, b)$ and for some positive constants η_j, $j = 1, 2, 3$. If $H_j(x)$, $j = 1, 2, 3$, are defined by

$$H_j(x) = \frac{1}{z_j(x)} \int_{x/2}^x \frac{z_j(t) h_j(t)}{t} \, \mathrm{d}t, \tag{2.6.30}$$

then

$$\int_0^b x^{-m}\left[H_1^{p_1}(x)H_2^{p_2}(x)+H_2^{p_2}(x)H_3^{p_3}(x)+H_3^{p_3}(x)H_1^{p_1}(x)\right]\mathrm{d}x$$

$$\leqslant \sum_{j=1}^{3}\left[\left\{\eta_j\left(\frac{2p_j}{m-1}\right)\right\}^{2p_j}\right.$$

$$\left.\times\int_0^b x^{-m}\left\{\frac{1}{z_j(x)}\left|z_j(x)h_j(x)-z_j\left(\frac{x}{2}\right)h_j\left(\frac{x}{2}\right)\right|\right\}^{2p_j}\mathrm{d}x\right]. \tag{2.6.31}$$

THEOREM 2.6.8. *Let m, p_j, h_j, z_j be as defined in Theorem 2.6.7. Let*

$$1+\left(\frac{4p_j}{m-1}\right)x\frac{z_j'(x)}{z_j(x)}\geqslant\frac{1}{\xi_j} \tag{2.6.32}$$

for almost all $x \in (0,b)$ and for some positive constants ξ_j, $j=1,2,3$. If $H_j(x)$, $j=1,2,3$, defined by (2.6.30), *then*

$$\int_0^b x^{-m}H_1^{p_1}(x)H_2^{p_2}(x)H_3^{p_3}(x)\left[H_1^{p_1}(x)+H_2^{p_2}(x)+H_3^{p_3}(x)\right]\mathrm{d}x$$

$$\leqslant \sum_{j=1}^{3}\left[\left\{\xi_j\left(\frac{4p_j}{m-1}\right)\right\}^{4p_j}\right.$$

$$\left.\times\int_0^b x^{-m}\left\{\frac{1}{z_j(x)}\left|z_j(x)h_j(x)-z_j\left(\frac{x}{2}\right)h_j\left(\frac{x}{2}\right)\right|\right\}^{4p_j}\mathrm{d}x\right]. \tag{2.6.33}$$

REMARK 2.6.5. In the special cases when $p_j = p > 1, h_j = h, H_j = H$, $z_j = 1, \eta_j = 1$ and $\xi_j = 1$, the inequalities established in (2.6.31) and (2.6.33) reduce, respectively, to the inequalities

$$\int_0^b x^{-m}H^{2p}(x)\,\mathrm{d}x \leqslant \left(\frac{2p}{m-1}\right)^{2p}\int_0^b x^{-m}\left|h(x)-h\left(\frac{x}{2}\right)\right|^{2p}\mathrm{d}x, \tag{2.6.34}$$

and

$$\int_0^b x^{-m}H^{4p}(x)\,\mathrm{d}x \leqslant \left(\frac{4p}{m-1}\right)^{4p}\int_0^b x^{-m}\left|h(x)-h\left(\frac{x}{2}\right)\right|^{4p}\mathrm{d}x. \tag{2.6.35}$$

We note that, by setting $-m = s$ and $2p = p^{\cdot} > 1$ in (2.6.34) and $4p = p^{\cdot} > 1$ in (2.6.35), the inequalities obtained in (2.6.34) and (2.6.35) reduce to the inequality established by Izumi and Izumi in Theorem 2.6.1.

THEOREM 2.6.9. *Let $m > 1$ and $p_i > 1$, $i = 1, \dots, n$, be constants. Let $q_i(t)$, $i = 1, \dots, n$, be nonnegative and integrable functions on $(0, b)$ and let $w_i(x)$, $i = 1, \dots, n$, be positive and absolutely continuous functions on $(0, b)$. Let*

$$1 + \left(\frac{np_i}{m-1}\right) x \frac{w_i'(x)}{w_i(x)} \geqslant \frac{1}{\gamma_i} \tag{2.6.36}$$

for almost all $x \in (0, b)$ and for some positive constants γ_i, $i = 1, \dots, n$. If $Q_i(x)$, $i = 1, \dots, n$, are defined by

$$Q_i(x) = \frac{1}{w_i(x)} \int_{x/2}^{x} \frac{w_i(t) q_i(t)}{t} \, dt, \tag{2.6.37}$$

then

$$\begin{aligned} &\int_0^b x^{-m} \prod_{i=1}^{n} Q_i^{p_i}(x) \, dx \\ &\quad \leqslant \frac{1}{n} \sum_{i=1}^{n} \Bigg[\left\{ \gamma_i \left(\frac{np_i}{m-1} \right) \right\}^{np_i} \\ &\qquad \times \int_0^b x^{-m} \left\{ \frac{1}{w_i(x)} \left| w_i(x) q_i(x) - w_i\left(\frac{x}{2}\right) q_i\left(\frac{x}{2}\right) \right| \right\}^{np_i} dx \Bigg]. \end{aligned} \tag{2.6.38}$$

REMARK 2.6.6. We note that in the special case when $p_i = p > 1$, $q_i = q$, $Q_i = Q$, $w_i = 1$ and $\gamma_i = 1$, the inequality established in (2.6.38) reduces to the following inequality

$$\int_0^b x^{-m} Q^{np}(x) \, dx \leqslant \left(\frac{np}{m-1}\right)^{np} \int_0^b x^{-m} \left| q(x) - q\left(\frac{x}{2}\right) \right|^{np} dx,$$

which in turn, by setting $-m = s$ and $np = p^{\cdot} > 1$, reduces to the inequality of Izumi and Izumi given in Theorem 2.6.1.

PROOFS OF THEOREMS 2.6.4–2.6.9. Let $0 < a < b < \infty$ and define, for $m > 1$,

$$F_{ja}(x) = \frac{1}{r_j(x)} \int_a^x \frac{r_j(t) f_j(t)}{t} \, dt, \tag{2.6.39}$$

with $F_{j0}(x) = F_j(x)$ for $j = 1, 2, 3$. From (2.6.39) and using the elementary inequalities $c_1c_2 + c_2c_3 + c_3c_1 \leqslant c_1^2 + c_2^2 + c_3^2$ (for c_1, c_2, c_3 real), we observe that

$$\begin{aligned} &F_{1a}^{p_1}(x)F_{2a}^{p_2}(x) + F_{2a}^{p_2}(x)F_{3a}^{p_3}(x) + F_{3a}^{p_3}(x)F_{1a}^{p_1}(x) \\ &\quad \leqslant F_{1a}^{2p_1}(x) + F_{2a}^{2p_2}(x) + F_{3a}^{2p_3}(x). \end{aligned} \tag{2.6.40}$$

Multiplying both sides of (2.6.40) by x^{-m} and integrating from a to b we have

$$\begin{aligned} &\int_a^b x^{-m}\left[F_{1a}^{p_1}(x)F_{2a}^{p_2}(x) + F_{2a}^{p_2}(x)F_{3a}^{p_3}(x) + F_{3a}^{p_3}(x)F_{1a}^{p_1}(x)\right]\mathrm{d}x \\ &\quad \leqslant \int_a^b x^{-m}F_{1a}^{2p_1}(x)\,\mathrm{d}x + \int_a^b x^{-m}F_{2a}^{2p_2}(x)\,\mathrm{d}x + \int_a^b x^{-m}F_{3a}^{2p_3}(x)\,\mathrm{d}x. \end{aligned} \tag{2.6.41}$$

Integrating by parts we have

$$\begin{aligned} &\int_a^b x^{-m}F_{ja}^{2p_j}(x)\,\mathrm{d}x \\ &\quad = \left[\frac{x^{-m+1}}{-m+1}F_{ja}^{2p_j}(x)\right]_a^b \\ &\qquad - \int_a^b \frac{x^{-m+1}}{-m+1}2p_jF_{ja}^{2p_j-1}(x) \\ &\qquad\quad \times \left\{\frac{1}{r_j^2(x)}\left[r_j(x)\left(\frac{r_j(x)f_j(x)}{x}\right) - r_j'(x)\left(\int_a^x \frac{r_j(t)f_j(t)}{t}\,\mathrm{d}t\right)\right]\right\}\mathrm{d}x \end{aligned} \tag{2.6.42}$$

for $j = 1, 2, 3$. Since $m > 1$, from (2.6.42) we observe that

$$\begin{aligned} &\int_a^b \left[1 + \left(\frac{2p_j}{m-1}\right)x\frac{r_j'(x)}{r_j(x)}\right]x^{-m}F_{ja}^{2p_j}(x)\,\mathrm{d}x \\ &\quad \leqslant \left(\frac{2p_j}{m-1}\right)\int_a^b \left[(x^m)^{(2p_j-1)/(2p_j)}f_j(x)\right]\left[(x^m)^{-(2p_j-1)/(2p_j)}F_{ja}^{2p_j-1}(x)\right]\mathrm{d}x. \end{aligned} \tag{2.6.43}$$

From (2.6.16) and applying Hölder's inequality with indices $2p_j, 2p_j/(2p_j-1)$

on the right-hand side of (2.6.43), we obtain

$$\begin{aligned}\int_a^b & x^{-m} F_{ja}^{2p_j}(x)\,\mathrm{d}x \\ & \leqslant \alpha_j\left(\frac{2p_j}{m-1}\right)\left\{\int_a^b x^{-m} f_j^{2p_j}(x)\,\mathrm{d}x\right\}^{1/(2p_j)} \\ & \quad\times\left\{\int_a^b x^{-m} F_{ja}^{2p_j}(x)\,\mathrm{d}x\right\}^{(2p_j-1)/(2p_j)}. \end{aligned} \tag{2.6.44}$$

Dividing both sides of (2.6.44) by the second integral factor on the right-hand side of (2.6.44) and raising both sides to the $2p_j$th power we obtain

$$\int_a^b x^{-m} F_{ja}^{2p_j}(x)\,\mathrm{d}x \leqslant \left\{\alpha_j\left(\frac{2p_j}{m-1}\right)\right\}^{2p_j} \int_a^b x^{-m} f_j^{2p_j}(x)\,\mathrm{d}x \tag{2.6.45}$$

for $j = 1, 2, 3$. Using (2.6.45) in (2.6.41) we have

$$\begin{aligned}\int_a^b & x^{-m}\left[F_{1a}^{p_1}(x)F_{2a}^{p_2}(x) + F_{2a}^{p_2}(x)F_{3a}^{p_3}(x) + F_{3a}^{p_3}(x)F_{1a}^{p_1}(x)\right]\mathrm{d}x \\ & \leqslant \sum_{j=1}^{3}\left[\left\{\alpha_j\left(\frac{2p_j}{m-1}\right)\right\}^{2p_j} \int_a^b x^{-m} f_j^{2p_j}(x)\,\mathrm{d}x\right]. \end{aligned} \tag{2.6.46}$$

From (2.6.46) we have

$$\begin{aligned}\int_a^b & x^{-m}\left[F_{1a}^{p_1}(x)F_{2a}^{p_2}(x) + F_{2a}^{p_2}(x)F_{3a}^{p_3}(x) + F_{3a}^{p_3}(x)F_{1a}^{p_1}(x)\right]\mathrm{d}x \\ & \leqslant \sum_{j=1}^{3}\left[\left\{\alpha_j\left(\frac{2p_j}{m-1}\right)\right\}^{2p_j} \int_0^\infty x^{-m} f_j^{2p_j}(x)\,\mathrm{d}x\right]. \end{aligned} \tag{2.6.47}$$

Let $a < c < b$. Then, from (2.6.47), we have

$$\begin{aligned}\int_c^b & x^{-m}\left[F_{1a}^{p_1}(x)F_{2a}^{p_2}(x) + F_{2a}^{p_2}(x)F_{3a}^{p_3}(x) + F_{3a}^{p_3}(x)F_{1a}^{p_1}(x)\right]\mathrm{d}x \\ & \leqslant \sum_{j=1}^{3}\left[\left\{\alpha_j\left(\frac{2p_j}{m-1}\right)\right\}^{2p_j} \int_0^\infty x^{-m} f_j^{2p_j}(x)\,\mathrm{d}x\right]. \end{aligned} \tag{2.6.48}$$

Letting $a \to 0$ on the left-hand side of (2.6.48) we have

$$\int_c^b x^{-m}\left[F_1^{p_1}(x)F_2^{p_2}(x)+F_2^{p_2}(x)F_3^{p_3}(x)+F_3^{p_3}(x)F_1^{p_1}(x)\right]\mathrm{d}x \leqslant \sum_{j=1}^{3}\left[\left\{\alpha_j\left(\frac{2p_j}{m-1}\right)\right\}^{2p_j}\int_0^\infty x^{-m}f_j^{2p_j}(x)\,\mathrm{d}x\right].$$

Since this inequality holds for arbitrary $0 < c < b$, it follows that

$$\int_0^\infty x^{-m}\left[F_1^{p_1}(x)F_2^{p_2}(x)+F_2^{p_2}(x)F_3^{p_3}(x)+F_3^{p_3}(x)F_1^{p_1}(x)\right]\mathrm{d}x \leqslant \sum_{j=1}^{3}\left[\left\{\alpha_j\left(\frac{2p_j}{m-1}\right)\right\}^{2p_j}\int_0^\infty x^{-m}f_j^{2p_j}(x)\,\mathrm{d}x\right]. \tag{2.6.49}$$

Let $0 < a < b < \infty$ and define, for $m < 1$,

$$F_{jb}(x)=\frac{1}{r_j(x)}\int_x^b \frac{r_j(t)f_j(t)}{t}\,\mathrm{d}t, \tag{2.6.50}$$

with $F_{j\infty}(x)=F_j(x)$ for $j=1,2,3$. Now, by following the same steps as in the proof of inequality (2.6.49) with suitable modifications, we obtain

$$\int_0^\infty x^{-m}\left[F_1^{p_1}(x)F_2^{p_2}(x)+F_2^{p_2}(x)F_3^{p_3}(x)+F_3^{p_3}(x)F_1^{p_1}(x)\right]\mathrm{d}x \leqslant \sum_{j=1}^{3}\left[\left\{\beta_j\left(\frac{2p_j}{m-1}\right)\right\}^{2p_j}\int_0^\infty x^{-m}f_j^{2p_j}(x)\,\mathrm{d}x\right]. \tag{2.6.51}$$

From (2.6.49) and (2.6.51), we obtain

$$\int_0^\infty x^{-m}\left[F_1^{p_1}(x)F_2^{p_2}(x)+F_2^{p_2}(x)F_3^{p_3}(x)+F_3^{p_3}(x)F_1^{p_1}(x)\right]\mathrm{d}x \leqslant \sum_{j=1}^{3}\left[\left\{\lambda\left(\frac{2p_j}{|m-1|}\right)\right\}^{2p_j}\int_0^\infty x^{-m}f_j^{2p_j}(x)\,\mathrm{d}x\right].$$

The proof of Theorem 2.6.4 is complete.

In order to establish inequality (2.6.22) in Theorem 2.6.5, let $0 < a < b < \infty$ and define $F_{ja}(x)$ by (2.6.39). From (2.6.39) and using the elementary inequalities $c_1c_2c_3(c_1+c_2+c_3) \leqslant \frac{1}{3}(c_1c_2+c_2c_3+c_3c_1)^2$, $(c_1c_2+c_2c_3+c_3c_1) \leqslant$

$c_1^2+c_2^2+c_3^2$ and $(c_1+c_2+c_3)^2 \leqslant 3(c_1^2+c_2^2+c_3^2)$ (for c_1, c_2, c_3 real) (see [211, pp. 201, 203]), we observe that

$$F_{1a}^{p_1}(x)F_{2a}^{p_2}(x)F_{3a}^{p_3}(x)\big[F_{1a}^{p_1}(x)+F_{2a}^{p_2}(x)+F_{3a}^{p_3}(x)\big] \leqslant F_{1a}^{4p_1}(x)+F_{2a}^{4p_2}(x)+F_{3a}^{4p_3}(x). \tag{2.6.52}$$

Multiplying both sides of (2.6.52) by x^{-m} and integrating from a to b we have

$$\int_a^b x^{-m}F_{1a}^{p_1}(x)F_{2a}^{p_2}(x)F_{3a}^{p_3}(x)\big[F_{1a}^{p_1}(x)+F_{2a}^{p_3}(x)+F_{3a}^{p_3}(x)\big]\,dx \leqslant \int_a^b x^{-m}F_{1a}^{4p_1}(x)\,dx+\int_a^b x^{-m}F_{2a}^{4p_2}(x)\,dx+\int_a^b x^{-m}F_{3a}^{4p_3}(x)\,dx. \tag{2.6.53}$$

The rest of the proof of Theorem 2.6.5 follows exactly the same steps as in the proof of Theorem 2.6.4 below inequality (2.6.41) with suitable changes, and hence we omit the further details.

Let $0<a<b<\infty$ and define, for $m>1$,

$$G_{ia}(x)=\frac{1}{k_i(x)}\int_a^x \frac{k_i(t)g_i(t)}{t}\,dt, \tag{2.6.54}$$

with $G_{i0}(x)=G_i(x), i=1,\dots,n$. From (2.6.54) and using the elementary inequalities $(\prod_{i=1}^n c_i)^{1/n} \leqslant \frac{1}{n}\sum_{i=1}^n c_i$ and $(\sum_{i=1}^n c_i)^n \leqslant n^{n-1}\sum_{i=1}^n c_i^n$ (for $c_1,\dots,c_n \geqslant 0$ real for $n \geqslant 1$) (see [211]), we observe that

$$\prod_{i=1}^n G_{ia}^{p_i}(x)=\left[\left\{\prod_{i=1}^n G_{ia}^{p_i}(x)\right\}^{1/n}\right]^n \leqslant \frac{1}{n}\sum_{i=1}^n G_{ia}^{np_i}(x). \tag{2.6.55}$$

Multiplying both sides of (2.6.55) by x^{-m} and integrating from a to b we have

$$\int_a^b x^{-m}\prod_{i=1}^n G_{ia}^{p_i}(x)\,dx \leqslant \frac{1}{n}\int_a^b x^{-m}\left\{\sum_{i=1}^n G_{ia}^{np_i}(x)\right\}dx.$$

Now, by following the same steps as in the proof of Theorem 2.6.4 below inequality (2.6.41) with suitable modifications, we obtain the desired inequality in (2.6.28). The proof of Theorem 2.6.6 is complete.

From (2.6.30) and using the elementary inequality $c_1c_2 + c_2c_3 + c_3c_1 \leqslant c_1^2 + c_2^2 + c_3^2$ (for c_1, c_2, c_3 real), we observe that

$$H_1^{p_1}(x)H_2^{p_2}(x) + H_2^{p_2}(x)H_3^{p_3}(x) + H_3^{p_3}(x)H_1^{p_1}(x) \leqslant H_1^{2p_1}(x) + H_2^{p_2}(x) + H_3^{p_3}(x). \tag{2.6.56}$$

Multiplying both sides of (2.6.56) by x^{-m} and integrating from 0 to b we have

$$\int_0^b x^{-m}\left[H_1^{p_1}(x)H_2^{p_2}(x) + H_2^{p_2}(x)H_3^{p_3}(x) + H_3^{p_3}(x)H_1^{p_1}(x)\right]\mathrm{d}x \leqslant \int_0^b x^{-m}H_1^{2p_1}(x)\,\mathrm{d}x + \int_0^b x^{-m}H_2^{2p_2}(x)\,\mathrm{d}x + \int_0^b x^{-m}H_3^{2p_3}(x)\,\mathrm{d}x. \tag{2.6.57}$$

Integrating by parts we obtain

$$\begin{aligned}\int_0^b x^{-m}H_j^{2p_j}(x)\,\mathrm{d}x &= \left[\frac{x^{-m+1}}{-m+1}H_j^{2p_j}(x)\right]_0^b \\ &\quad + \left(\frac{2p_j}{m-1}\right)\int_0^b x^{-m}\frac{1}{z_j(x)}\left\{z_j(x)h_j(x) - z_j\left(\frac{x}{2}\right)h_j\left(\frac{x}{2}\right)\right\}H_j^{2p_j-1}(x)\,\mathrm{d}x \\ &\quad - \left(\frac{2p_j}{m-1}\right)\int_0^b x^{-m+1}\frac{z_j'(x)}{z_j(x)}H_j^{2p_j}(x)\,\mathrm{d}x.\end{aligned} \tag{2.6.58}$$

From (2.6.58) we observe that

$$\begin{aligned}\int_0^b \left[1 + \left(\frac{2p_j}{m-1}\right)x\frac{z_j'(x)}{z_j(x)}\right]x^{-m}H_j^{2p_j}(x)\,\mathrm{d}x &\leqslant \left(\frac{2p_j}{m-1}\right)\int_0^b \left[\left(x^{-m}\right)^{-(2p_j-1)/2p_j}x^{-m}\frac{1}{z_j(x)}\right. \\ &\quad \times \left.\left|z_j(x)h_j(x) - z_j\left(\frac{x}{2}\right)h_j\left(\frac{x}{2}\right)\right|\right] \\ &\quad \times \left[\left(x^{-m}\right)^{(2p_j-1)/(2p_j)}H_j^{2p_j-1}(x)\right]\mathrm{d}x.\end{aligned} \tag{2.6.59}$$

From (2.6.29) and using Hölder's inequality with indices $2p_j, 2p_j/(2p_j - 1)$ we obtain

$$\begin{aligned}&\int_0^b x^{-m} H_j^{2p_j}(x)\,\mathrm{d}x\\&\leqslant \eta_j\left(\frac{2p_j}{m-1}\right)\\&\quad\times\left\{\int_0^b x^{-m}\left\{\frac{1}{z_j(x)}\left|z_j(x)h_j(x) - z_j\left(\frac{x}{2}\right)h_j\left(\frac{x}{2}\right)\right|\right\}^{2p_j}\mathrm{d}x\right\}^{1/(2p_j)}\\&\quad\times\left\{\int_0^b x^{-m} H_j^{2p_j}(x)\,\mathrm{d}x\right\}^{(2p_j-1)/(2p_j)}.\end{aligned}\tag{2.6.60}$$

Dividing both sides of (2.6.60) by the second integral factor on the right-hand side of (2.6.60) and taking the $2p_j$th power on both sides of the resulting inequality, we obtain

$$\begin{aligned}&\int_0^b x^{-m} H_j^{2p_j}(x)\,\mathrm{d}x\\&\leqslant \left\{\eta_j\left(\frac{2p_j}{m-1}\right)\right\}^{2p_j}\\&\quad\times\int_0^b x^{-m}\left\{\frac{1}{z_j(x)}\left|z_j(x)h_j(x) - z_j\left(\frac{x}{2}\right)h_j\left(\frac{x}{2}\right)\right|\right\}^{2p_j}\mathrm{d}x\end{aligned}\tag{2.6.61}$$

for $j = 1, 2, 3$. Using (2.6.61) in (2.6.57) we obtain the desired inequality in (2.6.31). The proof of Theorem 2.6.7 is complete. □

The proofs of Theorems 2.6.8 and 2.6.9 follow from the proof of Theorem 2.6.7 and by similar arguments as in the proofs of Theorems 2.6.5 and 2.6.6 and with suitable modification. Here we omit the details.

2.7 Multidimensional Hardy-Type Inequalities

In view of wider applications, Hardy's integral inequality has been generalized in various directions. The present section is devoted to the multidimensional Hardy-type inequalities investigated by Pachpatte in [293,315,333,341]. The analysis used in the proofs is based on the applications of well-known Fubini's theorem and Jensen's integral inequality.

In [315] Pachpatte established the following Hardy-type integral inequalities involving functions of two independent variables.

THEOREM 2.7.1. *Let f be a nonnegative integrable function on $\Delta = (0, a) \times (0, b)$, where a, b are positive constants, and r_1, r_2 be positive and absolutely continuous functions on $(0, \infty)$ such that*

$$1 + \frac{p}{p-1} x \frac{r_1'(x)}{r_1(x)} \geqslant \frac{1}{\alpha}, \tag{2.7.1}$$

$$1 + \frac{p}{p-1} y \frac{r_2'(y)}{r_2(y)} \geqslant \frac{1}{\beta}, \tag{2.7.2}$$

for almost all $x, y \in (0, \infty)$ and some constants $\alpha > 0, \beta > 0, p > 1$. If F is defined by

$$F(x, y) = \frac{1}{r_1(x) r_2(y)} \int_{x/2}^{x} \int_{y/2}^{y} \frac{r_1(s) r_2(t) f(s, t)}{st} \, dt \, ds \tag{2.7.3}$$

for $(x, y) \in \Delta$, then

$$\begin{aligned} &\int_0^a \int_0^b \left\{ \frac{F(x, y)}{xy} \right\}^p dy \, dx \\ &\leqslant (\alpha\beta)^p \left(\frac{p}{p-1} \right)^{2p} \\ &\quad \times \int_0^a \int_0^b \left\{ \frac{1}{xy r_1(x) r_2(y)} \left| r_2(y) \left(r_1(x) f(x, y) - r_2\left(\frac{x}{2}\right) f\left(\frac{x}{2}, y\right) \right) \right. \right. \\ &\qquad \left. \left. - r_2\left(\frac{y}{2}\right) \left(r_1(x) f\left(x, \frac{y}{2}\right) - r_1\left(\frac{x}{2}\right) f\left(\frac{x}{2}, \frac{y}{2}\right) \right) \right| \right\}^p dy \, dx. \end{aligned} \tag{2.7.4}$$

THEOREM 2.7.2. *Let g be a nonnegative integrable function on Δ and the functions r_1, r_2 and the constant p be as defined in Theorem 2.7.1 satisfying the conditions (2.7.1) and (2.7.2). If G is defined by*

$$G(x, y) = \frac{1}{r_1(x) r_2(y)} \int_{x/2}^{x} \int_{0}^{y} \frac{r_1(s) r_2(t) g(s, t)}{st} \, dt \, ds \tag{2.7.5}$$

for $(x, y) \in \Delta$, *then*

$$\int_0^a \int_0^b \left\{ \frac{G(x,y)}{xy} \right\}^p \mathrm{d}y\,\mathrm{d}x \leqslant (\alpha\beta)^p \left(\frac{p}{p-1} \right)^{2p} \times \int_0^a \int_0^b \left\{ \frac{1}{xyr_1(x)} \left| r_1(x) g(x,y) - r_1\left(\frac{x}{2}\right) g\left(\frac{x}{2}, y\right) \right| \right\}^p \mathrm{d}y\,\mathrm{d}x. \tag{2.7.6}$$

THEOREM 2.7.3. *Let h be a nonnegative integrable function on Δ and the functions r_1, r_2 and the constant p be as in Theorem* 2.7.1 *satisfying* (2.7.1) *and* (2.7.2). *If H is defined by*

$$H(x,y) = \frac{1}{r_1(x) r_2(y)} \int_0^x \int_{y/2}^y \frac{r_1(s) r_2(t) h(s,t)}{st} \,\mathrm{d}t\,\mathrm{d}s \tag{2.7.7}$$

for $(x, y) \in \Delta$, *then*

$$\int_0^a \int_0^b \left\{ \frac{H(x,y)}{xy} \right\}^p \mathrm{d}y\,\mathrm{d}x \leqslant (\alpha\beta)^p \left(\frac{p}{p-1} \right)^{2p} \times \int_0^a \int_0^b \left\{ \frac{1}{xyr_2(y)} \left| r_2(y) h(x,y) - r_2\left(\frac{y}{2}\right) h\left(x, \frac{y}{2}\right) \right| \right\}^p \mathrm{d}y\,\mathrm{d}x. \tag{2.7.8}$$

REMARK 2.7.1. The inequalities (2.7.4), (2.7.6) and (2.7.8) extend the result provided in [255, Theorem 1, with $p = m$ therein] to the case of two independent variables. In the special case of $r_1(x) = r_2(y) = 1$, $\alpha = \beta = 1$, inequalities (2.7.4), (2.7.6) and (2.7.8) reduce to

$$\int_0^a \int_0^b \left\{ \frac{F(x,y)}{xy} \right\}^p \mathrm{d}y\,\mathrm{d}x \leqslant \left(\frac{p}{p-1} \right)^{2p} \times \int_0^a \int_0^b \left\{ \frac{1}{xy} \left| f(x,y) - f\left(\frac{x}{2}, y\right) - f\left(x, \frac{y}{2}\right) + f\left(\frac{x}{2}, \frac{y}{2}\right) \right| \right\}^p \mathrm{d}y\,\mathrm{d}x, \tag{2.7.9}$$

$$\int_0^a \int_0^b \left\{ \frac{G(x,y)}{xy} \right\}^p \mathrm{d}y\,\mathrm{d}x$$
$$\leqslant \left(\frac{p}{p-1}\right)^{2p} \int_0^a \int_0^b \left\{ \frac{1}{xy} \left| g(x,y) - g\left(\frac{x}{2}, y\right) \right| \right\}^p \mathrm{d}y\,\mathrm{d}x, \tag{2.7.10}$$

$$\int_0^a \int_0^b \left\{ \frac{H(x,y)}{xy} \right\}^p \mathrm{d}y\,\mathrm{d}x$$
$$\leqslant \left(\frac{p}{p-1}\right)^{2p} \int_0^a \int_0^b \left\{ \frac{1}{xy} \left| h(x,y) - h\left(x, \frac{y}{2}\right) \right| \right\}^p \mathrm{d}y\,\mathrm{d}x, \tag{2.7.11}$$

respectively. Inequalities (2.7.9)–(2.7.11) can be considered as the two independent variables generalizations of the slight variant of Hardy's inequality given in Theorem 2.6.1.

PROOFS OF THEOREMS 2.7.1–2.7.3. For $f = 0$ inequality (2.7.4) is trivially true. We assume that f is positive and denote by I the integral on the left-hand side of (2.7.4). By using Fubini's theorem (see [3, p. 18]), we observe that

$$I = \int_0^a \big(x r_1(x)\big)^{-p} I_1(x)\,\mathrm{d}x, \tag{2.7.12}$$

where

$$I_1(x) = \int_0^b y^{-p} \left\{ \frac{1}{r_2(y)} \int_{y/2}^{y} \frac{r_2(t)}{t} \left(\int_{x/2}^{x} \frac{r_1(s) f(s,t)}{s} \mathrm{d}s \right) \mathrm{d}t \right\}^p \mathrm{d}y. \tag{2.7.13}$$

By keeping x fixed in (2.7.13) and integrating by parts, we have the relation

$$\begin{aligned}
I_1(x) = {} & -\frac{b^{-p+1}}{p-1} \left\{ \frac{1}{r_2(b)} \int_{b/2}^{b} \frac{r_2(t)}{t} \left[\int_{x/2}^{x} \frac{r_1(s) f(s,t)}{s} \mathrm{d}s \right] \mathrm{d}t \right\}^p \\
& + \frac{p}{p-1} \int_0^b y^{-p+1} \left\{ \frac{1}{r_2(y)} \int_{y/2}^{y} \frac{r_2(t)}{t} \left[\int_{x/2}^{x} \frac{r_1(s) f(s,t)}{s} \mathrm{d}s \right] \mathrm{d}t \right\}^{p-1} \\
& \times \frac{1}{r_2^2(y)} \left[r_2(y) \left(\frac{r_2(y)}{y} \int_{x/2}^{x} \frac{r_1(s) f(s,y)}{s} \mathrm{d}s \right.\right. \\
& \left. - \frac{1}{2} \frac{r_2(y/2)}{y/2} \int_{x/2}^{x} \frac{r_1(s) f(s,y/2)}{s} \mathrm{d}s \right) \\
& \left. - r_2'(y) \int_{y/2}^{y} \frac{r_2(t)}{t} \left(\int_{x/2}^{x} \frac{r_1(s) f(s,t)}{s} \mathrm{d}s \right) \mathrm{d}t \right] \mathrm{d}y,
\end{aligned} \tag{2.7.14}$$

implying

$$\int_0^b \left(1+\frac{p}{p-1} y \frac{r_2'(y)}{r_2(y)}\right) y^{-p} \left\{\frac{1}{r_2(y)} \int_{y/2}^{y} \frac{r_2(t)}{t} \left[\int_{x/2}^{x} \frac{r_1(s) f(s,t)}{s} \mathrm{d}s\right] \mathrm{d}t\right\}^p \mathrm{d}y$$

$$\leqslant \frac{p}{p-1} \int_0^b \left\{ y^{-p+1} \left[\frac{1}{r_2(y)} \int_{y/2}^{y} \frac{r_2(t)}{t} \left(\int_{x/2}^{x} \frac{r_1(s) f(s,t)}{s} \mathrm{d}s\right) \mathrm{d}t\right]^{p-1} \right.$$
$$\times \frac{1}{y r_2(y)} \left| r_2(y) \int_{x/2}^{x} \frac{r_1(s) f(s,y)}{s} \mathrm{d}s \right.$$
$$\left. \left. - r_2\left(\frac{y}{2}\right) \int_{x/2}^{x} \frac{r_1(s) f(s,y/2)}{s} \mathrm{d}s \right| \right\} \mathrm{d}y. \tag{2.7.15}$$

From (2.7.1), (2.7.15), using Hölder's integral inequality with indices p, $p/(p-1)$ we get

$$I_1(x) \leqslant \frac{\alpha p}{p-1} I_1^{(p-1)/p}(x)$$
$$\times \left[\int_0^b \left\{\frac{1}{y r_2(y)} \left| r_2(y) \int_{x/2}^{x} \frac{r_1(s) f(s,y)}{s} \mathrm{d}s \right.\right.\right.$$
$$\left.\left.\left. - r_2\left(\frac{y}{2}\right) \int_{x/2}^{x} \frac{r_1(s) f(s,y/2)}{s} \mathrm{d}s \right| \right\}^p\right]^{1/p}. \tag{2.7.16}$$

Dividing both sides of (2.7.16) by $I_1^{(p-1)/p}(x)$ and raising the result to the pth power we get

$$I_1(x) \leqslant \left(\frac{\alpha p}{p-1}\right)^p \int_0^b y^{-p} \left\{\frac{1}{r_2(y)} \left| r_2(y) \int_{x/2}^{x} \frac{r_1(s) f(s,y)}{s} \mathrm{d}s \right.\right.$$
$$\left.\left. - r_2\left(\frac{y}{2}\right) \int_{x/2}^{x} \frac{r_1(s) f(s,y/2)}{s} \mathrm{d}s \right| \right\}^p \mathrm{d}y. \tag{2.7.17}$$

Substituting (2.7.17) in (2.7.12) and using Fubini's theorem, we obtain

$$I \leqslant \left(\frac{\alpha p}{p-1}\right)^p \int_0^b y^{-p} I_2(y) \,\mathrm{d}y, \tag{2.7.18}$$

where

$$I_2(y) = \int_0^a x^{-p} \left\{ \frac{|m(x,y)|}{r_1(x) r_2(y)} \right\}^p \mathrm{d}x, \tag{2.7.19}$$

$$m(x,y) = r_2(y) \int_{x/2}^x \frac{r_1(s) f(s,y)}{s} \,\mathrm{d}s - r_2\left(\frac{y}{2}\right) \int_{x/2}^x \frac{r_1(s) f(s,y/2)}{s} \,\mathrm{d}s. \tag{2.7.20}$$

Now, keeping y fixed in (2.7.19), using the relation $|m(x,y)| = m(x,y) \times \operatorname{sgn} m(x,y)$ and integrating by parts, we have the equality

$$\begin{aligned} I_2(y) = {} & \frac{a^{-p+1}}{-p+1} \left\{ \frac{|m(a,y)|}{r_1(a) r_2(y)} \right\}^p \\ & + \frac{p}{p-1} \int_0^a x^{-p+1} \left\{ \frac{|m(x,y)|}{r_1(x) r_2(y)} \right\}^{p-1} \\ & \qquad \times \frac{\operatorname{sgn} m(x,y)}{x r_1(x) r_2(y)} \left[r_2(y) \left(r_1(x) f(x,y) - r_1\left(\frac{x}{2}\right) f\left(\frac{x}{2}, y\right) \right) \right. \\ & \qquad \left. - r_2\left(\frac{y}{2}\right) \left(r_1(x) f\left(x, \frac{y}{2}\right) - r_1\left(\frac{x}{2}\right) f\left(\frac{x}{2}, \frac{y}{2}\right) \right) \right] \mathrm{d}x \\ & - \frac{p}{p-1} \int_0^a x^{-p+1} \left\{ \frac{|m(x,y)|}{r_1(x) r_2(y)} \right\}^{p-1} \frac{r_1'(x)}{r_1(x)} \frac{m(x,y) \operatorname{sgn} m(x,y)}{r_1(x) r_2(y)} \,\mathrm{d}x, \end{aligned} \tag{2.7.21}$$

which implies

$$\begin{aligned} & \int_0^a \left(1 + \frac{p}{p-1} x \frac{r_1'(x)}{r_1(x)} \right) x^{-p} \left\{ \frac{|m(x,y)|}{r_1(x) r_2(y)} \right\}^p \mathrm{d}x \\ & \leqslant \frac{p}{p-1} \int_0^a \left[x^{-p+1} \left\{ \frac{|m(x,y)|}{r_1(x) r_2(y)} \right\}^{p-1} \right] \\ & \qquad \times \left[\frac{1}{x r_1(x) r_2(y)} \left| r_2(y) \left(r_1(x) f(x,y) - r_1\left(\frac{x}{2}\right) f\left(\frac{x}{2}, y\right) \right) \right. \right. \\ & \qquad \left. \left. - r_2\left(\frac{y}{2}\right) \left(r_1(x) f\left(x, \frac{y}{2}\right) - r_1\left(\frac{x}{2}\right) f\left(\frac{x}{2}, \frac{y}{2}\right) \right) \right| \right] \mathrm{d}x. \end{aligned} \tag{2.7.22}$$

From (2.7.2), (2.7.22), using Hölder's inequality with indices p, $p/(p-1)$, we get

$$\begin{aligned} I_2(y) \leqslant{} & \frac{\beta p}{p-1} I_2^{(p-1)/p}(y) \\ & \times \Bigg[\int_0^a x^{-p} \Bigg\{ \frac{1}{r_1(x) r_2(y)} \Bigg| r_2(y) \bigg(r_1(x) f(x,y) - r_1\Big(\frac{x}{2}\Big) f\Big(\frac{x}{2}, y\Big) \bigg) \\ & - r_2\Big(\frac{y}{2}\Big) \bigg(r_1(x) f\Big(x, \frac{y}{2}\Big) - r_1\Big(\frac{x}{2}\Big) f\Big(\frac{x}{2}, \frac{y}{2}\Big) \bigg) \Bigg| \Bigg\}^p \mathrm{d}x \Bigg]^{1/p}. \end{aligned} \tag{2.7.23}$$

Dividing both sides of (2.7.23) by $I_2^{(p-1)/p}(y)$, and raising to the pth power, we get

$$\begin{aligned} I_2(y) \leqslant{} & \Big(\frac{\beta p}{p-1} \Big)^p \\ & \times \int_0^a x^{-p} \Bigg\{ \frac{1}{r_1(x) r_2(y)} \Bigg| r_2(y) \bigg(r_1(x) f(x,y) - r_1\Big(\frac{x}{2}\Big) f\Big(\frac{x}{2}, y\Big) \bigg) \\ & - r_2\Big(\frac{y}{2}\Big) \bigg(r_1(x) f\Big(x, \frac{y}{2}\Big) - r_1\Big(\frac{x}{2}\Big) f\Big(\frac{x}{2}, \frac{y}{2}\Big) \bigg) \Bigg| \Bigg\}^p \mathrm{d}x. \end{aligned} \tag{2.7.24}$$

Substituting (2.7.24) in (2.7.18) and using Fubini's theorem we get (2.7.4), which completes the proof of Theorem 2.7.1.

The proofs of Theorems 2.7.2 and 2.7.3 follow by the similar arguments as in the proof of Theorem 2.7.1 with suitable modifications. Here we omit the details. □

In the following theorems we present Hardy-type integral inequalities involving functions of several independent variables. In what follows, let B and H be subsets of the n-dimensional Euclidean space $\mathbb{R}^n$ defined by $B = \{x \in \mathbb{R}^n\colon 0 < x < \infty\}$ and $H = \{x \in \mathbb{R}^n\colon a < x < b\}$, where $0, a, b \in \mathbb{R}^n$ and $a > 0$. For the functions $u(z)$ and $v(z)$ defined on B and H, respectively, we denote by $\int_B u(z)\,\mathrm{d}z$, $\int_{B_{x,y}} u(z)\,\mathrm{d}z$ and $\int_H v(z)\,\mathrm{d}z$, $\int_{H_{x,y}} v(z)\,\mathrm{d}z$ the n-fold integrals

$$\int_0^\infty \cdots \int_0^\infty u(z_1, \ldots, z_n)\,\mathrm{d}z_n \cdots \mathrm{d}z_1, \qquad \int_{x_1}^{y_1} \cdots \int_{x_n}^{y_n} u(z_1, \ldots, z_n)\,\mathrm{d}z_n \cdots \mathrm{d}z_1$$

and

$$\int_{a_1}^{b_1}\cdots\int_{a_n}^{b_n} v(z_1,\ldots,z_n)\,\mathrm{d}z_n\cdots\mathrm{d}z_1, \qquad \int_{x_1}^{y_1}\cdots\int_{x_n}^{y_n} v(z_1,\ldots,z_n)\,\mathrm{d}z_n\cdots\mathrm{d}z_1,$$

respectively, where $x_i < y_i$ and $a_i < b_i$. Hereafter, in this section without further mention, we assume that all inequalities between vectors are componentwise and all the integrals exists on the respective domains of their definitions.

In 1992, Pachpatte [293] has established the following multivariate version of Hardy's integral inequality given in Theorem 2.4.1.

THEOREM 2.7.4. *Let $p > 1$ be a constant. Let $f(x)$ be a nonnegative and integrable function on B, and define*

$$F(x) = \int_{B_{0,x}} f(y)\,\mathrm{d}y, \quad x \in B, \tag{2.7.25}$$

then

$$\int_B \left(\prod_{i=1}^n x_i\right)^{-p} F^p(x)\,\mathrm{d}x \leqslant \left(\frac{p}{p-1}\right)^{np} \int_B f^p(x)\,\mathrm{d}x. \tag{2.7.26}$$

Equality holds in (2.7.26) *if $f(x) \equiv 0$.*

PROOF. Let $a = (a_1,\ldots,a_n) \in B$, $b = (b_1,\ldots,b_n) \in B$, $0 < a < b < \infty$, and define

$$F_a(x) = \int_{B_{a,x}} f(y)\,\mathrm{d}y, \quad x \in B. \tag{2.7.27}$$

From (2.7.27) and by Fubini's theorem [3, p. 18], we have

$$\begin{aligned}
&\int_{B_{a,b}} \left(\prod_{i=1}^n x_i\right)^{-p} F_a^p(x)\,\mathrm{d}x \\
&= \int_{a_1}^{b_1}\cdots\int_{a_{n-1}}^{b_{n-1}} (x_1x_2\cdots x_{n-1})^{-p} \\
&\quad\times \left\{\int_{a_n}^{b_n} x_n^{-p}\left(\int_{a_n}^{x_n}\left(\int_{a_1}^{x_1}\cdots\right.\right.\right. \\
&\quad\left.\left.\left.\cdots\int_{a_{n-1}}^{x_{n-1}} f(s_1,\ldots,s_{n-1},s_n)\,\mathrm{d}s_{n-1}\cdots\mathrm{d}s_1\right)\mathrm{d}s_n\right)^p \mathrm{d}x_n\right\}\mathrm{d}x_{n-1}\cdots\mathrm{d}x_1.
\end{aligned} \tag{2.7.28}$$

By keeping $x_1, \ldots, x_{n-1}$ fixed and integrating by parts, and using the fact that $p > 1$ and Hölder's inequality with indices p, $p/(p-1)$, we see that

$$\begin{aligned}
&\int_{a_n}^{b_n} x_n^{-p} \left\{ \int_{a_n}^{x_n} \left(\int_{a_1}^{x_1} \cdots \int_{a_{n-1}}^{x_{n-1}} f(s_1, \ldots, s_{n-1}, s_n)\, \mathrm{d}s_{n-1} \cdots \mathrm{d}s_1 \right) \mathrm{d}s_n \right\}^p \mathrm{d}x_n \\
&= \frac{b_n^{-p+1}}{-p+1} \left\{ \int_{a_n}^{b_n} \left(\int_{a_1}^{x_1} \cdots \int_{a_{n-1}}^{x_{n-1}} f(s_1, \ldots, s_{n-1}, s_n)\, \mathrm{d}s_{n-1} \cdots \mathrm{d}s_1 \right) \mathrm{d}s_n \right\}^{p-1} \\
&\quad + \frac{p}{p-1} \int_{a_n}^{b_n} x_n^{-p+1} \left\{ \int_{a_n}^{x_n} \left(\int_{a_1}^{x_1} \cdots \int_{a_{n-1}}^{x_{n-1}} f(s_1, \ldots, s_{n-1}, s_n) \right.\right. \\
&\qquad\qquad \left.\left. \times\, \mathrm{d}s_{n-1} \cdots \mathrm{d}s_1 \right) \mathrm{d}s_n \right\}^{p-1} \\
&\qquad \times \left\{ \int_{a_1}^{x_1} \cdots \int_{a_{n-1}}^{x_{n-1}} f(s_1, \ldots, s_{n-1}, x_n)\, \mathrm{d}s_{n-1} \cdots \mathrm{d}s_1 \right\} \mathrm{d}x_n \\
&\leqslant \frac{p}{p-1} \int_{a_n}^{b_n} x_n^{-p+1} \left\{ \int_{a_n}^{x_n} \left(\int_{a_1}^{x_1} \cdots \int_{a_{n-1}}^{x_{n-1}} f(s_1, \ldots, s_{n-1}, s_n) \right.\right. \\
&\qquad\qquad \left.\left. \times\, \mathrm{d}s_{n-1} \cdots \mathrm{d}s_1 \right) \mathrm{d}s_n \right\}^{p-1} \\
&\qquad \times \left\{ \int_{a_1}^{x_1} \cdots \int_{a_{n-1}}^{x_{n-1}} f(s_1, \ldots, s_{n-1}, x_n)\, \mathrm{d}s_{n-1} \cdots \mathrm{d}s_1 \right\} \mathrm{d}x_n \\
&\leqslant \frac{p}{p-1} \left\{ \int_{a_n}^{b_n} x_n^{-p} \left\{ \int_{a_n}^{x_n} \left(\int_{a_1}^{x_1} \cdots \int_{a_{n-1}}^{x_{n-1}} f(s_1, \ldots, s_{n-1}, s_n) \right.\right.\right. \\
&\qquad\qquad \left.\left.\left. \times\, \mathrm{d}s_{n-1} \cdots \mathrm{d}s_1 \right) \mathrm{d}s_n \right\}^p \mathrm{d}x_n \right\}^{(p-1)/p} \\
&\quad \times \left\{ \int_{a_n}^{b_n} \left\{ \int_{a_1}^{x_1} \cdots \int_{a_{n-1}}^{x_{n-1}} f(s_1, \ldots, s_{n-1}, x_n)\, \mathrm{d}s_{n-1} \cdots \mathrm{d}s_1 \right\}^p \mathrm{d}x_n \right\}^{1/p}.
\end{aligned} \tag{2.7.29}$$

Dividing both sides of (2.7.29) by the first integral factor on the right-hand side of (2.7.29), and raising both sides to the pth power, we get

$$\int_{a_n}^{b_n} x_n^{-p} \left\{ \int_{a_n}^{x_n} \left(\int_{a_1}^{x_1} \cdots \int_{a_{n-1}}^{x_{n-1}} f(s_1, \ldots, s_{n-1}, s_n)\, \mathrm{d}s_{n-1} \cdots \mathrm{d}s_1 \right) \mathrm{d}s_n \right\}^p \mathrm{d}x_n$$

$$\leqslant \left(\frac{p}{p-1}\right)^p \int_{a_n}^{b_n} \left\{ \int_{a_1}^{x_1} \cdots \int_{a_{n-1}}^{x_{n-1}} f(s_1, \ldots, s_{n-1}, x_n)\, \mathrm{d}s_{n-1} \cdots \mathrm{d}s_1 \right\}^p \mathrm{d}x_n. \tag{2.7.30}$$

Substituting (2.7.30) in (2.7.28) and using Fubini's theorem we have

$$\begin{aligned}
&\int_{B_{a,b}} \left(\prod_{i=1}^{n} x_i\right)^{-p} F_a^p(x)\, \mathrm{d}x \\
&\leqslant \left(\frac{p}{p-1}\right)^p \int_{a_1}^{b_1} \cdots \int_{a_{n-1}}^{b_{n-1}} (x_1 x_2 \cdots x_{n-1})^{-p} \\
&\quad \times \left\{ \int_{a_n}^{b_n} \left\{ \int_{a_1}^{x_1} \cdots \int_{a_{n-1}}^{x_{n-1}} f(s_1, \ldots, s_{n-1}, x_n)\, \mathrm{d}s_{n-1} \cdots \mathrm{d}s_1 \right\}^p \mathrm{d}x_n \right\} \\
&\qquad \times \mathrm{d}x_{n-1} \cdots \mathrm{d}x_1 \\
&= \left(\frac{p}{p-1}\right)^p \int_{a_1}^{b_1} \cdots \int_{a_{n-2}}^{b_{n-2}} \int_{a_n}^{b_n} (x_1 x_2 \cdots x_{n-2})^{-p} \\
&\quad \times \left\{ \int_{a_{n-1}}^{b_{n-1}} x_{n-1}^{-p} \left\{ \int_{a_{n-1}}^{x_{n-1}} \left\{ \int_{a_1}^{x_1} \cdots \right.\right.\right. \\
&\qquad \left.\left.\left. \cdots \int_{a_{n-2}}^{x_{n-2}} f(s_1, \ldots, s_{n-2}, s_{n-1}, x_n)\, \mathrm{d}s_{n-2} \cdots \mathrm{d}s_1 \right\} \mathrm{d}s_{n-1} \right\}^p \mathrm{d}x_{n-1} \right\} \\
&\qquad \times \mathrm{d}x_n\, \mathrm{d}x_{n-2} \cdots \mathrm{d}x_1.
\end{aligned} \tag{2.7.31}$$

Now, by following exactly the same arguments as above, we obtain

$$\begin{aligned}
&\int_{a_{n-1}}^{b_{n-1}} x_{n-1}^{-p} \left\{ \int_{a_{n-1}}^{x_{n-1}} \left(\int_{a_1}^{x_1} \cdots \right.\right. \\
&\qquad \left.\left. \cdots \int_{a_{n-2}}^{x_{n-2}} f(s_1, \ldots, s_{n-2}, s_{n-1}, x_n)\, \mathrm{d}s_{n-2} \cdots \mathrm{d}s_1 \right) \mathrm{d}s_{n-1} \right\}^p \mathrm{d}x_{n-1} \\
&\leqslant \left(\frac{p}{p-1}\right)^p \int_{a_{n-1}}^{x_{n-1}} \left\{ \int_{a_1}^{x_1} \cdots \right. \\
&\qquad \left. \cdots \int_{a_{n-2}}^{x_{n-2}} f(s_1, \ldots, s_{n-2}, x_{n-1}, x_n)\, \mathrm{d}s_{n-2} \cdots \mathrm{d}s_1 \right\}^p \mathrm{d}x_{n-1}.
\end{aligned} \tag{2.7.32}$$

Substituting (2.7.32) in (2.7.31) we have

$$\begin{aligned}
&\int_{B_{a,b}} \left(\prod_{i=1}^{n} x_i\right)^{-p} F_a^p(x)\,\mathrm{d}x \\
&\quad \leqslant \left(\frac{p}{p-1}\right)^{2p} \\
&\quad \times \int_{a_1}^{b_1} \cdots \int_{a_{n-2}}^{b_{n-2}} \int_{a_n}^{b_n} (x_1 \cdots x_{n-2})^{-p} \\
&\quad \times \left\{ \int_{a_{n-1}}^{b_{n-1}} \left\{ \int_{a_1}^{x_1} \cdots \int_{a_{n-2}}^{x_{n-2}} f(s_1, \ldots, s_{n-2}, x_{n-1}, x_n)\,\mathrm{d}s_{n-2} \cdots \mathrm{d}s_1 \right\}^p \mathrm{d}x_{n-1} \right\} \\
&\qquad\qquad \times \mathrm{d}x_n\,\mathrm{d}x_{n-2} \cdots \mathrm{d}x_1. \qquad (2.7.33)
\end{aligned}$$

Continuing in this way we finally get

$$\int_{B_{a,b}} \left(\prod_{i=1}^{n} x_i\right)^{-p} F_a^p(x)\,\mathrm{d}x \leqslant \left(\frac{p}{p-1}\right)^{np} \int_{B_{a,b}} f^p(x)\,\mathrm{d}x. \qquad (2.7.34)$$

Let $c = (c_1, \ldots, c_n) \in B$ and $a < c < b$. Then, from (2.7.34), we have

$$\int_{B_{c,b}} \left(\prod_{i=1}^{n} x_i\right)^{-p} F_a^p(x)\,\mathrm{d}x \leqslant \left(\frac{p}{p-1}\right)^{np} \int_{B} f^p(x)\,\mathrm{d}x. \qquad (2.7.35)$$

Letting $a \to 0$, that is, $a_i \to 0$ on the left-hand side of (2.7.35) we have

$$\int_{B_{c,b}} \left(\prod_{i=1}^{n} x_i\right)^{-p} F^p(x)\,\mathrm{d}x \leqslant \left(\frac{p}{p-1}\right)^{np} \int_{B} f^p(x)\,\mathrm{d}x. \qquad (2.7.36)$$

Since this inequality holds for arbitrary $0 < c < b$, it follows that

$$\int_{B} \left(\prod_{i=1}^{n} x_i\right)^{-p} F^p(x)\,\mathrm{d}x \leqslant \left(\frac{p}{p-1}\right)^{np} \int_{B} f^p(x)\,\mathrm{d}x. \qquad (2.7.37)$$

The proof is complete. □

The following theorem established by Pachpatte [293] is a multivariate version of a slight variant of Levinson's inequality given in [190, Theorem 1].

THEOREM 2.7.5. *Let $\phi(u) \geqslant 0$ be defined on an open interval, finite or infinite, and at the ends of the intervals, let ϕ take its limiting values, finite or infinite. For some $p > 1$, let $\phi^{1/p}(u)$ be convex. If, for $x \in B$, the range of values of $f(x)$ lies in the closed interval of definition of ϕ and $\phi(f(x))$ is integrable on B and if $F(x)$ is as defined in Theorem 2.7.4, then*

$$\int_B \phi\left(\left(\prod_{i=1}^n x_i\right)^{-1} F(x)\right) \mathrm{d}x \leqslant \left(\frac{p}{p-1}\right)^{np} \int_B \phi(f(x))\, \mathrm{d}x. \tag{2.7.38}$$

Equality holds in (2.7.38) if $\phi(f(x)) \equiv 0$.

PROOF. Let $\psi(u) = \{\phi(u)\}^{1/p} \geqslant 0$. Then $\psi(u)$ is convex. By Jensen's inequality (see [174, p. 133]), we have

$$\psi\left(\left(\prod_{i=1}^n x_i\right)^{-1} F(x)\right) \leqslant \left(\prod_{i=1}^n x_i\right)^{-1} \int_{B_{0,x}} \psi(f(y))\, \mathrm{d}y. \tag{2.7.39}$$

Applying (2.7.26) with $f(x)$ replaced by $\psi(f(x))$ we have

$$\int_B \left\{\left(\prod_{i=1}^n x_i\right)^{-1} \int_{B_{0,x}} \psi(f(y))\, \mathrm{d}y\right\}^p \leqslant \left(\frac{p}{p-1}\right)^{np} \int_B \{\psi(f(x))\}^p\, \mathrm{d}x. \tag{2.7.40}$$

Using $\phi = \psi^p$ and (2.7.39) we have

$$\begin{aligned}\phi\left(\left(\prod_{i=1}^n x_i\right)^{-1} F(x)\right) &= \left\{\psi\left(\left(\prod_{i=1}^n x_i\right)^{-1}\right) \int_{B_{0,x}} f(y)\, \mathrm{d}y\right\}^p \\ &\leqslant \left\{\left(\prod_{i=1}^n x_i\right)^{-1} \int_{B_{0,x}} \psi(f(y))\, \mathrm{d}y\right\}^p.\end{aligned} \tag{2.7.41}$$

From (2.7.41) and (2.7.40), we observe that

$$\begin{aligned}\int_B \phi\left(\left(\prod_{i=1}^n x_i\right)^{-1} F(x)\right) \mathrm{d}x &\leqslant \int_B \left\{\left(\prod_{i=1}^n x_i\right)^{-1} \int_{B_{0,x}} \psi(f(y))\, \mathrm{d}y\right\}^p \mathrm{d}x \\ &\leqslant \left(\frac{p}{p-1}\right)^{np} \int_B \{\psi(f(x))\}^p\, \mathrm{d}x \\ &= \left(\frac{p}{p-1}\right)^{np} \int_B \phi(f(x))\, \mathrm{d}x.\end{aligned}$$

The proof is complete. □

REMARK 2.7.2. We note that in [33, p. 465] Boas and Imoru have obtained a two-dimensional version of Hardy's inequality in Theorem 2.4.1 by using different analysis. Here the approach to obtain a multivariate version of a slight variant of Hardy's inequality is more direct and quite elementary.

In [341] Pachpatte has established the following multivariate variants of Hardy's inequality.

THEOREM 2.7.6. *Let $p > 1$ be a constant and $f(x)$ be a nonnegative and integrable function on $B_{1,a}$. If*

$$F(x) = \int_{B_{x,a}} f(y)\,\mathrm{d}y, \quad x \in B_{1,a}, \tag{2.7.42}$$

then

$$\int_{B_{1,a}} \left(\prod_{i=1}^{n} x_i^{-1}\right) F^p(x)\,\mathrm{d}x \leqslant p^{np} \int_{B_{1,a}} \left(\prod_{i=1}^{n} x_i^{-1}\right) \left[\left(\prod_{i=1}^{n} x_i \log x_i\right) f(x)\right]^p \mathrm{d}x. \tag{2.7.43}$$

PROOF. If f is null, inequality (2.7.43) is trivially true. We assume that f is not null. Denote by I the integral on the left-hand side of (2.7.43). By Fubini's theorem [3, p. 18] we observe that

$$I = \int_1^{a_1} \cdots \int_1^{a_{n-1}} \left(\prod_{i=1}^{n} x_i^{-1}\right) I_n\,\mathrm{d}x_{n-1} \cdots \mathrm{d}x_1, \tag{2.7.44}$$

where

$$I_n = \int_1^{a_n} x_n^{-1} \left(\int_{x_n}^{a_n} \left(\int_{x_1}^{a_1} \cdots \right.\right.$$

$$\left.\left.\cdots \int_{x_{n-1}}^{a_{n-1}} f(y_1, \ldots, y_{n-1}, y_n)\,\mathrm{d}y_{n-1} \cdots \mathrm{d}y_1\right) \mathrm{d}y_n\right)^p \mathrm{d}x_n. \tag{2.7.45}$$

By keeping $x_1, \ldots, x_{n-1}$ fixed in (2.7.45) and integrating by parts, we have

$$I_n = \left[(\log x_n)\left(\int_{x_n}^{a_n} \left(\int_{x_1}^{a_1} \cdots \right.\right.\right.$$

$$\cdots \int_{x_{n-1}}^{a_{n-1}} f(y_1, \ldots, y_{n-1}, y_n)\,\mathrm{d}y_{n-1} \cdots \mathrm{d}y_1 \Big) \mathrm{d}y_n \Big)^p \Big]_1^{a_n}$$

$$- \int_1^{a_n} \Big((\log x_n) p \Big[\int_{x_n}^{a_n} \Big(\int_{x_1}^{a_1} \cdots$$

$$\cdots \int_{x_{n-1}}^{a_{n-1}} f(y_1, \ldots, y_{n-1}, y_n)\,\mathrm{d}y_{n-1} \cdots \mathrm{d}y_1 \Big) \mathrm{d}y_n \Big]^{p-1}$$

$$\times \Big(- \int_{x_1}^{a_1} \cdots \int_{x_{n-1}}^{a_{n-1}} f(y_1, \ldots, y_{n-1}, x_n)\,\mathrm{d}y_{n-1} \cdots \mathrm{d}y_1 \Big) \mathrm{d}y_n \Big) \mathrm{d}x_n. \tag{2.7.46}$$

From (2.7.46) we observe that

$$I_n = p \int_1^{a_n} \Big[x_n^{(p-1)/p} (\log x_n)$$

$$\times \Big(\int_{x_1}^{a_1} \cdots \int_{x_{n-1}}^{a_{n-1}} f(y_1, \ldots, y_{n-1}, x_n)\,\mathrm{d}y_{n-1} \cdots \mathrm{d}y_1 \Big) \Big]$$

$$\times \Big[x_n^{-(p-1)/p} \int_{x_n}^{a_n} \Big(\int_{x_1}^{a_1} \cdots$$

$$\cdots \int_{x_{n-1}}^{a_{n-1}} f(y_1, \ldots, y_{n-1}, y_n)\,\mathrm{d}y_{n-1} \cdots \mathrm{d}y_1 \Big) \mathrm{d}y_n \Big]^{p-1} \mathrm{d}x_n. \tag{2.7.47}$$

Now, applying Hölder's inequality with indices p, $p/(p-1)$ on the right-hand side of (2.7.47), we obtain

$$I_n \leqslant p \Big[\int_1^{a_n} \Big(x_n^{(p-1)/p} (\log x_n) \Big(\int_{x_1}^{a_1} \cdots$$

$$\cdots \int_{x_{n-1}}^{a_{n-1}} f(y_1, \ldots, y_{n-1}, x_n)\,\mathrm{d}y_{n-1} \cdots \mathrm{d}y_1 \Big) \Big)^p \mathrm{d}x_n \Big]^{1/p} I_n^{(p-1)/p}. \tag{2.7.48}$$

Dividing both sides of (2.7.48) by the second factor on the right-hand side of (2.7.48), and then raising both sides to the pth power, we obtain

$$I_n \leqslant p^p \int_1^{a_n} x_n^{p-1} (\log x_n)^p$$

$$\times \Big(\int_{x_1}^{a_1} \cdots \int_{x_{n-1}}^{a_{n-1}} f(y_1, \ldots, y_{n-1}, x_n)\,\mathrm{d}y_{n-1} \cdots \mathrm{d}y_1 \Big)^p \mathrm{d}x_n. \tag{2.7.49}$$

Substituting (2.7.49) in (2.7.44) and using Fubini's theorem we observe that

$$I \leqslant p^p \int_1^{a_1} \cdots \int_1^{a_{n-2}} \left(\prod_{i=1}^{n-2} x_i^{-1} \right) \int_1^{a_n} x_n^{p-1} (\log x_n)^p I_{n-1} \, dx_n \, dx_{n-2} \cdots dx_1, \tag{2.7.50}$$

where

$$I_{n-1} = \int_1^{a_{n-1}} x_{n-1}^{-1} \left(\int_{x_{n-1}}^{a_{n-1}} \left(\int_{x_1}^{a_1} \cdots \right.\right.$$
$$\left.\left. \cdots \int_{x_{n-2}}^{a_{n-2}} f(y_1, \ldots, y_{n-2}, y_{n-1}, x_n) \, dy_{n-2} \cdots dy_1 \right) dy_{n-1} \right)^p dx_{n-1}. \tag{2.7.51}$$

Now, by following exactly the same arguments as above with suitable modifications, we obtain

$$I_{n-1} \leqslant p^p \int_1^{a_{n-1}} x_{n-1}^{p-1} (\log x_{n-1})^p \left(\int_{x_1}^{a_1} \cdots \right.$$
$$\left. \cdots \int_{x_{n-2}}^{a_{n-2}} f(y_1, \ldots, y_{n-2}, x_{n-1}, x_n) \, dy_{n-2} \cdots dy_1 \right)^p dx_{n-1}. \tag{2.7.52}$$

Substituting (2.7.52) in (5.7.50) and again using Fubini's theorem we have

$$I \leqslant p^{2p} \int_1^{a_1} \int_1^{a_{n-3}} \left(\prod_{i=1}^{n-3} x_i^{-1} \right)$$
$$\times \int_1^{a_{n-1}} x_{n-1}^{p-1} (\log x_{n-1})^p \int_1^{a_n} x_n^{p-1} (\log x_n)^p I_{n-2} \, dx_n \, dx_{n-1} \, dx_{n-3} \cdots dx_1,$$

where

$$I_{n-2} = \int_1^{a_{n-2}} x_{n-2}^{-1} \left(\int_{x_{n-2}}^{a_{n-2}} \left(\int_{x_1}^{a_1} \cdots \right.\right.$$
$$\left.\left. \cdots \int_{x_{n-3}}^{a_{n-3}} f(y_1, \ldots, y_{n-3}, y_{n-2}, x_{n-1}, x_n) \, dy_{n-3} \cdots dy_1 \right) dy_{n-2} \right)^p dx_{n-2}.$$

Continuing in this way we finally get

$$I \leqslant p^{np} \int_{B_{1,a}} \left(\prod_{i=1}^{n} x_i^{-1} \right) \left[\left(\prod_{i=1}^{n} x_i \log x_i \right) f(x) \right]^p \mathrm{d}x.$$

This is the required inequality in (2.7.43) and the proof is complete. □

THEOREM 2.7.7. *Let $p > 1$ be a constant and $f(x)$ be a nonnegative and integrable function on $B_{0,1}$. If*

$$F(x) = \int_{B_{0,x}} f(y)\,\mathrm{d}y, \quad x \in B_{0,1}, \tag{2.7.53}$$

then

$$\int_{B_{0,1}} \left(\prod_{i=1}^{n} x_i^{-1} \right) F^p(x)\,\mathrm{d}x \leqslant p^{np} \int_{B_{0,1}} \left(\prod_{i=1}^{n} x_i^{-1} \right) \left[\left(\prod_{i=1}^{n} x_i |\log x_i| \right) f(x) \right]^p \mathrm{d}x. \tag{2.7.54}$$

The proof of this theorem follows by the same arguments as in the proof of Theorem 2.7.6 given above with suitable modifications. We omit the details.

REMARK 2.7.3. In the special case when $n = 1$, the inequalities established in Theorems 2.7.6 and 2.7.7 reduce respectively to inequalities (1a) and (2a) given by Chan in [52] in Theorems 1 and 2, respectively. In [52] the results are obtained by using the method of Banson [25]. Here our proofs are more direct and elementary.

In [333] Pachpatte establishes the following Hardy-type inequalities involving functions of several variables.

THEOREM 2.7.8. *Let $p > 1$ be a constant. Let $f(x)$ be a nonnegative and integrable function on B and let $r_i(x_i)$, $i = 1, \ldots, n$, be positive and absolutely continuous functions on $(0, \infty)$ and let*

$$R_i(x_i) = \int_0^{x_i} r_i(y_i)\,\mathrm{d}y_i \tag{2.7.55}$$

exist. Let

$$1+\left(\frac{1}{p-1}\right)\frac{R_i(x_i)r_i'(x_i)}{r_i^2(x_i)} \geqslant \frac{1}{\alpha_i} \tag{2.7.56}$$

for all $x_i > 0$ *and for some positive constants* α_i, $i = 1, \ldots, n$. *If* $F(x)$ *is defined by*

$$F(x) = \frac{1}{\prod_{i=1}^n R_i(x_i)} \int_{B_{0,x}} \left(\prod_{i=1}^n r_i(y_i) f(y)\,\mathrm{d}y \right) \tag{2.7.57}$$

for $x \in B$, *then*

$$\int_B F^p(x)\,\mathrm{d}x \leqslant \prod_{i=1}^n \left(\frac{p\alpha_i}{p-1}\right)^p \int_B f^p(x)\,\mathrm{d}x. \tag{2.7.58}$$

Equality holds in (2.7.58) *if* $f(x) \equiv 0$.

THEOREM 2.7.9. *Let* p *and* f *be as in Theorem* 2.7.8. *Let* $r_i(x_i)$, $i = 1, \ldots, n$, *be positive, continuous and monotone nondecreasing functions on* $(0, \infty)$. *If* $R_i(x_i)$ *and* $F(x)$ *be as defined in* (2.7.55) *and* (2.7.57), *respectively, where* $r_i(x_i)$ *are as defined above, then*

$$\int_B F^p(x)\,\mathrm{d}x \leqslant \left(\frac{p}{p-1}\right)^{np} \int_B f^p(x)\,\mathrm{d}x. \tag{2.7.59}$$

Equality holds in (2.7.59) *if* $f(x) \equiv 0$.

THEOREM 2.7.10. *Let* p, f, r_i, $i = 1, \ldots, n$, *be as defined in Theorem* 2.7.8. *Let*

$$1+\left(\frac{p}{p-1}\right)\frac{x_i r_i'(x_i)}{r_i(x_i)} \geqslant \frac{1}{\beta_i} \tag{2.7.60}$$

for all $x_i > 0$ *and for some positive constants* β_i, $i = 1, \ldots, n$. *If* $G(x)$ *is defined by*

$$G(x) = \frac{1}{\prod_{i=1}^n x_i r_i(x_i)} \int_{B_{0,x}} \left(\prod_{i=1}^n r_i(y_i) \right) f(y)\,\mathrm{d}y \tag{2.7.61}$$

for $x \in B$, *then*

$$\int_B G^p(x)\,\mathrm{d}x \leqslant \prod_{i=1}^n \left(\frac{p\beta_i}{p-1}\right)^p \int_B f^p(x)\,\mathrm{d}x. \tag{2.7.62}$$

Equality holds in (2.7.62) *if* $f(x) \equiv 0$.

THEOREM 2.7.11. *Let* $p, f, r_i, i = 1, \dots, n$, *be as defined in Theorem* 2.7.8. *Let*

$$1 - \left(\frac{p}{p-1}\right)\frac{x_i r_i'(x_i)}{r_i(x_i)} \geqslant \frac{1}{\gamma_i} \tag{2.7.63}$$

for all $x_i > 0$ *and for some positive constants* $\gamma_i, i = 1, \dots, n$. *If* $H(x)$ *is defined by*

$$H(x) = \prod_{i=1}^{n}\left(\frac{r_i(x_i)}{x_i}\right)\int_{B_{0,x}}\left(\frac{1}{\prod_{i=1}^{n} r_i(y_i)}\right) f(y)\,\mathrm{d}y \tag{2.7.64}$$

for $x \in B$, *then*

$$\int_B H^p(x)\,\mathrm{d}x \leqslant \prod_{i=1}^{n}\left(\frac{p\gamma_i}{p-1}\right)^p \int_B f^p(x)\,\mathrm{d}x. \tag{2.7.65}$$

Equality holds in (2.7.65) *if* $f(x) \equiv 0$.

REMARK 2.7.4. We note that (i) in the special cases when $r_i(x_i) = 1$ and $\alpha_i = 1$ in (2.7.56) and $n = 1$, the inequalities established in Theorems 2.7.8 and 2.7.9 reduce to the slight variant of Hardy's inequality given in Theorem 2.4.1, (ii) in the special case when $n = 1$, the inequalities established in Theorems 2.7.10 and 2.7.11 reduce to the inequalities established by Levinson in [190, Theorems 4 and 5].

THEOREM 2.7.12. *Let* $\phi(u) \geqslant 0$ *be defined on an open interval, finite or infinite, and at the ends of the interval, let* ϕ *take its limiting values, finite or infinite. For some* $p > 1$, *let* $\phi^{1/p}(u)$ *be convex. Let* $r_i(x_i)$ *and* $R_i(x_i)$ *be as defined in Theorem* 2.7.8 *satisfying the condition* (2.7.56). *If, for* $x \in B$, *the range of values of* $f(x)$ *lie in the closed interval of definition of* ϕ *and* $\phi(f(x))$ *is integrable on* B, *and if* $F(x)$ *is defined by* (2.7.57), *then*

$$\int_B \phi\big(F(x)\big)\,\mathrm{d}x \leqslant \prod_{i=1}^{n}\left(\frac{p\alpha_i}{p-1}\right)^p \int_B \phi\big(f(x)\big)\,\mathrm{d}x. \tag{2.7.66}$$

Equality holds in (2.7.66) *if* $\phi(f(x)) \equiv 0$.

THEOREM 2.7.13. *Let* $\phi, p, \phi^{1/p}, f, \phi(f)$ *be as defined in Theorem* 2.7.12. *Let* $r_i(x_i)$ *and* $R_i(x_i)$ *be as defined in Theorem* 2.7.9. *If* $F(x)$ *is defined as in*

Theorem 2.7.9, *then*

$$\int_B \phi\big(F(x)\big)\,\mathrm{d}x \leqslant \left(\frac{p}{p-1}\right)^{np} \int_B \phi\big(f(x)\big)\,\mathrm{d}x. \tag{2.7.67}$$

Equality holds in (2.7.67) *if* $\phi(f(x)) \equiv 0$.

REMARK 2.7.5. The inequalities established in Theorems 2.7.12 and 2.7.13 can be considered as the multivariate versions of the inequalities given by Levinson in [190, Theorems 3 and 2]. However, our hypotheses on $\phi(u)$ differs slightly from those used by Levinson in [190]. Here the condition $\phi''(u) \geqslant 0$ used in [190, p. 389] is not needed and the condition

$$\phi\phi'' \geqslant \left(1 - \frac{1}{p}\right)(\phi')^2, \quad p > 1,$$

used in [190, p. 389] is replaced by $\phi^{1/p}$ being convex, since this fact is all required for the application of Jensen's inequality in the proofs.

PROOFS OF THEOREMS 2.7.8–2.7.13. Let $a = (a_1, \dots, a_n) \in B$, $b = (b_1, \dots, b_n) \in B$, $0 = (0, \dots, 0) \in \mathbb{R}^n$ be such that $0 < a < b < \infty$, and define

$$F_a(x) = \frac{1}{\prod_{i=1}^n R_i(x_i)} \int_{B_{a,x}} \left(\prod_{i=1}^n r_i(y_i)\right) f(y)\,\mathrm{d}y \tag{2.7.68}$$

for $x \in B$. From (2.7.68) and by Fubini's theorem (see [3, p. 18]), we have

$$\begin{aligned}
&\int_{B_{a,b}} F_a^p(x)\,\mathrm{d}x \\
&\quad = \int_{a_1}^{b_1} \cdots \int_{a_{n-1}}^{b_{n-1}} \frac{1}{(\prod_{i=1}^{n-1} R_i(x_i))^p} \\
&\qquad \times \left[\int_{a_n}^{b_n} R_n^{-p}(x_n) \left[\int_{a_n}^{x_n} r_n(y_n) \left(\int_{a_1}^{x_1} \cdots \right.\right.\right. \\
&\qquad \left.\left.\left. \cdots \int_{a_{n-1}}^{x_{n-1}} \left(\prod_{i=1}^{n-1} r_i(y_i)\right) f(y_1, \dots, y_{n-1}, y_n)\,\mathrm{d}y_{n-1} \cdots \mathrm{d}y_1\right)\mathrm{d}y_n\right]^p \mathrm{d}x_n\right] \\
&\qquad \times \mathrm{d}x_{n-1} \cdots \mathrm{d}x_1.
\end{aligned} \tag{2.7.69}$$

By keeping $x_1, \dots, x_{n-1}$ fixed and integrating by parts, we have

$$
\begin{aligned}
&\int_{a_n}^{b_n} R_n^{-p}(x_n) \\
&\quad \times \left[\int_{a_n}^{x_n} r_n(y_n) \left(\int_{a_1}^{x_1} \cdots \right.\right. \\
&\quad \left.\left. \cdots \int_{a_{n-1}}^{x_{n-1}} \left(\prod_{i=1}^{n-1} r_i(y_i) \right) f(y_1, \dots, y_{n-1}, y_n)\, \mathrm{d}y_{n-1} \cdots \mathrm{d}y_1 \right) \mathrm{d}y_n \right]^p \mathrm{d}x_n \\
&= \int_{a_n}^{b_n} R_n^{-p}(x_n) r_n(x_n) \frac{1}{r_n(x_n)} \\
&\qquad \times \left[\int_{a_n}^{x_n} r_n(y_n) \left(\int_{a_1}^{x_1} \cdots \right.\right. \\
&\qquad \left.\left. \cdots \int_{a_{n-1}}^{x_{n-1}} \left(\prod_{i=1}^{n-1} r_i(y_i) \right) f(y_1, \dots, y_{n-1}, x_n)\, \mathrm{d}y_{n-1} \cdots \mathrm{d}y_1 \right) \mathrm{d}y_n \right]^p \mathrm{d}x_n \\
&= \frac{R_n^{-p+1}(b_n)}{-p+1} \frac{1}{r_n(b_n)} \left[\int_{a_n}^{b_n} r_n(y_n) \left(\int_{a_1}^{x_1} \cdots \right.\right. \\
&\qquad \left.\left. \cdots \int_{a_{n-1}}^{x_{n-1}} \left(\prod_{i=1}^{n-1} r_i(y_i) \right) f(y_1, \dots, y_{n-1}, y_n)\, \mathrm{d}y_{n-1} \cdots \mathrm{d}y_1 \right) \mathrm{d}y_n \right]^p \\
&\qquad + \left(\frac{p}{p-1} \right) \int_{a_n}^{b_n} R_n^{-p+1}(x_n) \left[\int_{a_n}^{x_n} r_n(y_n) \left(\int_{a_1}^{x_1} \cdots \right.\right. \\
&\qquad \left.\left. \cdots \int_{a_{n-1}}^{x_{n-1}} \left(\prod_{i=1}^{n-1} r_i(y_i) \right) f(y_1, \dots, y_{n-1}, y_n)\, \mathrm{d}y_{n-1} \cdots \mathrm{d}y_1 \right) \mathrm{d}y_n \right]^{p-1} \\
&\qquad \times \left(\int_{a_1}^{x_1} \cdots \int_{a_{n-1}}^{x_{n-1}} \left(\prod_{i=1}^{n-1} r_i(y_i) \right) f(y_1, \dots, y_{n-1}, x_n)\, \mathrm{d}y_{n-1} \cdots \mathrm{d}y_1 \right) \mathrm{d}x_n \\
&\qquad - \left(\frac{1}{p-1} \right) \int_{a_n}^{b_n} \frac{R_n^{-p+1}(x_n) r_n'(x_n)}{r_n^2(x_n)} \left[\int_{a_n}^{x_n} r_n(y_n) \left(\int_{a_1}^{x_1} \cdots \right.\right. \\
&\qquad \left.\left. \cdots \int_{a_{n-1}}^{x_{n-1}} \left(\prod_{i=1}^{n-1} r_i(y_i) \right) f(y_1, \dots, y_{n-1}, y_n)\, \mathrm{d}y_{n-1} \cdots \mathrm{d}y_1 \right) \mathrm{d}y_n \right]^p \mathrm{d}x_n .
\end{aligned}
\tag{2.7.70}
$$

Since $p > 1$, from (2.7.70) we observe that

$$\int_{a_n}^{b_n}\left[1+\left(\frac{1}{p-1}\right)\frac{R_n(x_n)r_n'(x_n)}{r_n^2(x_n)}\right]R_n^{-p}(x_n)\left[\int_{a_n}^{x_n} r_n(y_n)\left(\int_{a_1}^{x_1}\cdots\right.\right.$$

$$\left.\left.\cdots\int_{a_{n-1}}^{x_{n-1}}\left(\prod_{i=1}^{n-1} r_i(y_i)\right) f(y_1,\ldots,y_{n-1},y_n)\,\mathrm{d}y_{n-1}\cdots\mathrm{d}y_1\right)\mathrm{d}y_n\right]^p \mathrm{d}x_n$$

$$\leqslant\left(\frac{p}{p-1}\right)\int_{a_n}^{b_n} R_n^{-p+1}(x_n)\left[\int_{a_n}^{x_n} r_n(y_n)\left(\int_{a_1}^{x_1}\cdots\right.\right.$$

$$\left.\left.\cdots\int_{a_{n-1}}^{x_{n-1}}\left(\prod_{i=1}^{n-1} r_i(y_i)\right) f(y_1,\ldots,y_{n-1},y_n)\,\mathrm{d}y_{n-1}\cdots\mathrm{d}y_1\right)\mathrm{d}y_n\right]^{p-1}$$

$$\times\left(\int_{a_1}^{x_1}\cdots\int_{a_{n-1}}^{x_{n-1}}\left(\prod_{i=1}^{n-1} r_i(y_i)\right) f(y_1,\ldots,y_{n-1},x_n)\,\mathrm{d}y_{n-1}\cdots\mathrm{d}y_1\right)\mathrm{d}x_n. \tag{2.7.71}$$

From (2.7.56) and applying Hölder's inequality with indices p, $p/(p-1)$ on the right-hand side of (2.7.71), we obtain

$$\int_{a_n}^{b_n} R_n^{-p}(x_n)\left[\int_{a_n}^{x_n} r_n(y_n)\left(\int_{a_1}^{x_1}\cdots\int_{a_{n-1}}^{x_{n-1}}\left(\prod_{i=1}^{n-1} r_i(y_i)\right)\right.\right.$$

$$\left.\left.\times f(y_1,\ldots,y_{n-1},y_n)\,\mathrm{d}y_{n-1}\cdots\mathrm{d}y_1\right)\mathrm{d}y_n\right]^p \mathrm{d}x_n$$

$$\leqslant\left(\frac{p\alpha_n}{p-1}\right)\left[\int_{a_n}^{b_n} R_n^{-p}(x_n)\left[\int_{a_n}^{x_n} r_n(y_n)\left(\int_{a_1}^{x_1}\cdots\int_{a_{n-1}}^{x_{n-1}}\left(\prod_{i=1}^{n-1} r_i(y_i)\right)\right.\right.\right.$$

$$\left.\left.\left.\times f(y_1,\ldots,y_{n-1},y_n)\,\mathrm{d}y_{n-1}\cdots\mathrm{d}y_1\right)\mathrm{d}y_n\right]^p \mathrm{d}x_n\right]^{(p-1)/p}$$

$$\times\left[\int_{a_n}^{b_n}\left[\int_{a_1}^{x_1}\cdots\int_{a_{n-1}}^{x_{n-1}}\left(\prod_{i=1}^{n-1} r_i(y_i)\right)\right.\right.$$

$$\left.\left.\times f(y_1,\ldots,y_{n-1},y_n)\,\mathrm{d}y_{n-1}\cdots\mathrm{d}y_1\right]^p \mathrm{d}x_n\right]^{1/p}. \tag{2.7.72}$$

Dividing both sides of (2.7.72) by the first integral factor on the right-hand side of (2.7.72) and then raising both sides to the pth power we obtain

$$\int_{a_n}^{b_n} R_n^{-p}(x_n)\left[\int_{a_n}^{x_n} r_n(y_n)\left(\int_{a_1}^{x_1}\cdots\int_{a_{n-1}}^{x_{n-1}}\left(\prod_{i=1}^{n-1} r_i(y_i)\right)\right.\right.$$
$$\left.\left.\times f(y_1,\dots,y_{n-1},y_n)\,\mathrm{d}y_{n-1}\cdots\mathrm{d}y_1\right)\mathrm{d}y_n\right]^p \mathrm{d}x_n$$
$$\leqslant\left(\frac{p\alpha_n}{p-1}\right)^p\int_{a_n}^{b_n}\left[\int_{a_1}^{x_1}\cdots\int_{a_{n-1}}^{x_{n-1}}\left(\prod_{i=1}^{n-1} r_i(y_i)\right)\right.$$
$$\left.\times f(y_1,\dots,y_{n-1},x_n)\,\mathrm{d}y_{n-1}\cdots\mathrm{d}y_1\right]^p \mathrm{d}x_n. \tag{2.7.73}$$

Substituting (2.7.73) in (2.7.69) and using Fubini's theorem we have

$$\int_{B_{a,b}} F_a^p(x)\,\mathrm{d}x$$
$$\leqslant\left(\frac{p\alpha_n}{p-1}\right)\int_{a_1}^{b_1}\cdots\int_{a_{n-1}}^{b_{n-1}}\frac{1}{(\prod_{i=1}^{n-1}R_i(x_i))^p}\left[\int_{a_n}^{b_n}\left[\int_{a_1}^{x_1}\cdots\right.\right.$$
$$\left.\left.\cdots\int_{a_{n-1}}^{x_{n-1}}\left(\prod_{i=1}^{n-1} r_i(y_i)\right)f(y_1,\dots,y_{n-1},x_n)\,\mathrm{d}y_{n-1}\cdots\mathrm{d}y_1\right]^p\mathrm{d}x_n\right]$$
$$\times\,\mathrm{d}x_{n-1}\cdots\mathrm{d}x_1$$
$$=\left(\frac{p\alpha_n}{p-1}\right)^p\int_{a_1}^{b_1}\cdots\int_{a_{n-2}}^{b_{n-2}}\int_{a_n}^{b_n}\frac{1}{(\prod_{i=1}^{n-2}R_i(x_i))^p}$$
$$\times\left[\int_{a_{n-1}}^{b_{n-1}}R_{n-1}^{-p}(x_{n-1})\left[\int_{a_{n-1}}^{x_{n-1}}r_{n-1}(y_{n-1})\left(\int_{a_1}^{x_1}\cdots\int_{a_{n-2}}^{x_{n-2}}\left(\prod_{i=1}^{n-2} r_i(y_i)\right)\right.\right.\right.$$
$$\left.\left.\left.\times f(y_1,\dots,y_{n-2},y_{n-1},x_n)\,\mathrm{d}y_{n-2}\cdots\mathrm{d}y_1\right)\mathrm{d}y_{n-1}\right]^p\mathrm{d}x_{n-1}\right]$$
$$\times\,\mathrm{d}x_n\,\mathrm{d}x_{n-2}\cdots\mathrm{d}x_1. \tag{2.7.74}$$

Now, by following exactly the same arguments as above, we obtain

$$\begin{aligned}
&\int_{a_{n-1}}^{b_{n-1}} R_{n-1}^{-p}(x_{n-1}) \\
&\quad \times \left[\int_{a_{n-1}}^{x_{n-1}} r_{n-1}(y_{n-1}) \left(\int_{a_1}^{x_1} \cdots \int_{a_{n-2}}^{x_{n-2}} \left(\prod_{i=1}^{n-2} r_i(y_i)\right)\right.\right. \\
&\qquad\qquad \left.\left. \times f(y_1, \ldots, y_{n-2}, y_{n-1}, x_n)\,\mathrm{d}y_{n-2} \cdots \mathrm{d}y_1\right) \mathrm{d}y_{n-1}\right]^p \mathrm{d}x_{n-1} \\
&\leqslant \left(\frac{p\alpha_{n-1}}{p-1}\right)^p \int_{a_{n-1}}^{b_{n-1}} \left[\int_{a_1}^{x_1} \cdots \int_{a_{n-2}}^{x_{n-2}} \left(\prod_{i=1}^{n-2} r_i(y_i)\right)\right. \\
&\qquad\qquad \left. \times f(y_1, \ldots, y_{n-2}, x_{n-1}, x_n)\,\mathrm{d}y_{n-2} \cdots \mathrm{d}y_1\right]^p \mathrm{d}x_{n-1}.
\end{aligned} \tag{2.7.75}$$

Substituting (2.7.75) in (2.7.74) we have

$$\begin{aligned}
&\int_{B_{a,b}} F_a^p(x)\,\mathrm{d}x \\
&\quad \leqslant \left(\frac{p\alpha_n}{p-1}\right)^p \left(\frac{p\alpha_{n-1}}{p-1}\right)^p \\
&\qquad \times \int_{a_1}^{b_1} \cdots \int_{a_{n-2}}^{b_{n-2}} \int_{a_n}^{b_n} \frac{1}{(\prod_{i=1}^{n-2} R_i(x_i))^p} \left[\int_{a_{n-1}}^{b_{n-1}} \left[\int_{a_1}^{x_1} \cdots\right.\right. \\
&\qquad \cdots \left.\left. \int_{a_{n-2}}^{x_{n-2}} \left(\prod_{i=1}^{n-2} r_i(y_i)\right) f(y_1, \ldots, y_{n-2}, x_{n-1}, x_n)\,\mathrm{d}y_{n-2} \cdots \mathrm{d}y_1\right]^p \mathrm{d}x_{n-1}\right] \\
&\qquad\qquad \times \mathrm{d}x_n\,\mathrm{d}x_{n-2} \cdots \mathrm{d}x_1.
\end{aligned} \tag{2.7.76}$$

Continuing in this way we finally get

$$\int_{B_{a,b}} F_a^p(x)\,\mathrm{d}x \leqslant \prod_{i=1}^{n} \left(\frac{p\alpha_i}{p-1}\right)^p \int_{B_{a,b}} f^p(x)\,\mathrm{d}x. \tag{2.7.77}$$

Let $c = (c_1, \dots, c_n) \in B$ and $a < c < b$. Then, from (2.7.77), we have

$$\int_{B_{c,b}} F_a^p(x)\,\mathrm{d}x \leqslant \prod_{i=1}^{n}\left(\frac{p\alpha_i}{p-1}\right)^p \int_B f^p(x)\,\mathrm{d}x. \tag{2.7.78}$$

Letting $a \to 0$, that is, $a_i \to 0$ on the left-hand side of (2.7.78) we have

$$\int_{B_{c,b}} F^p(x)\,\mathrm{d}x \leqslant \prod_{i=1}^{n}\left(\frac{p\alpha_i}{p-1}\right)^p \int_B f^p(x)\,\mathrm{d}x. \tag{2.7.79}$$

Since this holds for arbitrary $0 < c < b$, it follows that

$$\int_{B} F^p(x)\,\mathrm{d}x \leqslant \prod_{i=1}^{n}\left(\frac{p\alpha_i}{p-1}\right)^p \int_B f^p(x)\,\mathrm{d}x. \tag{2.7.80}$$

The proof of Theorem 2.7.8 is complete.

The proof of Theorem 2.7.9 proceeds in the same way as the proof of Theorem 2.7.8. By following the same arguments as in Theorem 2.7.8 we obtain (2.7.70). □

Since $r_i(x_i)$ are monotone nondecreasing, from (2.7.70) we observe that (see [190, p. 391])

$$\begin{aligned}
&\int_{a_n}^{b_n} R_n^{-p}(x_n) \\
&\quad\times\left[\int_{a_n}^{x_n} r_n(y_n)\left(\int_{a_1}^{x_1}\cdots\right.\right. \\
&\qquad\left.\left.\cdots\int_{a_{n-1}}^{x_{n-1}}\left(\prod_{i=1}^{n-1} r_i(y_i)\right) f(y_1,\dots,y_{n-1},y_n)\,\mathrm{d}y_{n-1}\cdots\mathrm{d}y_1\right)\mathrm{d}y_n\right]^p \mathrm{d}x_n \\
&\leqslant \left(\frac{p}{p-1}\right)\int_{a_n}^{b_n} R_n^{-p+1}(x_n)\left[\int_{a_n}^{x_n} r_n(y_n)\left(\int_{a_1}^{x_1}\cdots\right.\right. \\
&\qquad\left.\left.\cdots\int_{a_{n-1}}^{x_{n-1}}\left(\prod_{i=1}^{n-1} r_i(y_i)\right) f(y_1,\dots,y_{n-1},y_n)\,\mathrm{d}y_{n-1}\cdots\mathrm{d}y_1\right)\mathrm{d}y_n\right]^{p-1} \\
&\quad\times\left(\int_{a_1}^{x_1}\cdots\int_{a_{n-1}}^{x_{n-1}}\left(\prod_{i=1}^{n-1} r_i(y_i)\right) f(y_1,\dots,y_{n-1},x_n)\,\mathrm{d}y_{n-1}\cdots\mathrm{d}y_1\right)\mathrm{d}x_n.
\end{aligned}$$

The rest of the proof of Theorem 2.7.9 follows the same steps as in the proof of Theorem 2.7.8 below inequality (2.7.71) with suitable changes and hence we omit the further details.

The proofs of Theorems 2.7.10 and 2.7.11 can be completed by following the similar arguments as in the proof of Theorem 2.7.8 (see also [190, p. 393]) with suitable modifications. We leave the details to the reader.

Let $\psi(u) = (\phi(u))^{1/p} \geqslant 0$. Then $\psi(u)$ is convex. By repeated application of Jensen's inequality (see [174, p. 133]), we have

$$\psi\big(F(x)\big) \leqslant \frac{1}{\prod_{i=1}^{n} R_i(x_i)} \int_{B_{0,x}} \prod_{i=1}^{n} r_i(y_i)\psi\big(f(y)\big)\,\mathrm{d}y \tag{2.7.81}$$

for $x \in B$. Applying (2.7.58) to $\psi(f(x))$ instead of $f(x)$ we have

$$\begin{aligned}
&\int_B \left[\frac{1}{\prod_{i=1}^{n} R_i(x_i)} \int_{B_{0,x}} \left(\prod_{i=1}^{n} r_i(y_i)\right)\psi\big(f(y)\big)\,\mathrm{d}y\right]^p \mathrm{d}x \\
&\quad\leqslant \prod_{i=1}^{n}\left(\frac{p\alpha_i}{p-1}\right)^p \int_B \big(\psi\big(f(x)\big)\big)^p\,\mathrm{d}x.
\end{aligned} \tag{2.7.82}$$

Using $\phi(u) = \psi^p(u)$ and (2.7.81) we have

$$\begin{aligned}
\phi\big(F(x)\big) &= \big(\psi\big(F(x)\big)\big)^p \\
&\leqslant \left[\frac{1}{\prod_{i=1}^{n} R_i(x_i)} \int_{B_{0,x}} \left(\prod_{i=1}^{n} r_i(y_i)\right)\psi\big(f(y)\big)\,\mathrm{d}y\right]^p.
\end{aligned} \tag{2.7.83}$$

From (2.7.83) and (2.7.82), we observe that

$$\begin{aligned}
\int_B \phi\big(F(x)\big)\,\mathrm{d}x &\leqslant \int_B \left[\frac{1}{\prod_{i=1}^{n} R_i(x_i)} \int_{B_{0,x}} \left(\prod_{i=1}^{n} r_i(y_i)\right)\psi\big(f(y)\big)\,\mathrm{d}y\right]^p \mathrm{d}x \\
&\leqslant \prod_{i=1}^{n}\left(\frac{p\alpha_i}{p-1}\right)^p \int_B \big(\psi\big(f(x)\big)\big)^p\,\mathrm{d}x \\
&= \prod_{i=1}^{n}\left(\frac{p\alpha_i}{p-1}\right)^p \int_B \phi\big(f(x)\big)\,\mathrm{d}x.
\end{aligned}$$

The proof of Theorem 2.7.12 is complete.

The proof of Theorem 2.7.13 proceeds in the same way as in the proof of Theorem 2.7.12 with suitable changes and hence we omit the details. □

REMARK 2.7.6. The multidimensional variants of Hardy's inequality are recently given by other investigators by using different techniques. Here we note that the above results are established by using elementary analysis and the inequalities obtained in Theorems 2.7.8–2.7.13 are of independent interest.

2.8 Inequalities Similar to Hilbert's Inequality

The well-known inequality due to Hilbert and its integral analogue can be stated as follows (see [141, p. 226]).

THEOREM A. *If $p > 1$, $p' = p/(p-1)$ and $\sum a_m^p \leqslant A$, $\sum b_n^{p'} \leqslant B$, the summations running from* 1 *to* ∞, *then*

$$\sum\sum \frac{a_m b_n}{m+n} < \frac{\pi}{\sin(\pi/p)} A^{1/p} B^{1/p'},$$

unless the sequence $\{a_m\}$ or $\{b_n\}$ is null.

THEOREM B. *If $p > 1$, $p' = p/(p-1)$ and $\int_0^\infty f^p(x)\,\mathrm{d}x \leqslant F$, $\int_0^\infty g^{p'}(y) \times \mathrm{d}y \leqslant G$, then*

$$\int_0^\infty \int_0^\infty \frac{f(x)g(y)}{x+y}\,\mathrm{d}x\,\mathrm{d}y < \frac{\pi}{\sin(\pi/p)} F^{1/p} G^{1/p'},$$

unless $f \equiv 0$ or $g \equiv 0$.

The inequalities in Theorems A and B were studied extensively and numerous variants, generalizations and extensions appear in the literature, see [141,210,213, 424] and the references cited therein. Recently in a series of papers [334,335,342, 343,350,352,353] Pachpatte has established a number of new inequalities similar to the inequalities given in Theorems A and B. In this section we present some of these results.

In [334] Pachpatte has given the following inequality similar to that of Hilbert's inequality in Theorem A.

THEOREM 2.8.1. *Let $p \geqslant 1$, $q \geqslant 1$ be constants and $\{a_m\}$ and $\{b_n\}$ be two nonnegative sequences of real numbers defined for $m = 1, 2, \ldots, k$ and $n = 1, 2, \ldots, r$, where k, r are natural numbers and define $A_m = \sum_{s=1}^m a_s$ and*

$B_n = \sum_{t=1}^{n} b_t$. Then

$$\sum_{m=1}^{k}\sum_{n=1}^{r}\frac{A_m^p B_n^q}{m+n} \leqslant C(p,q,k,r)\left(\sum_{m=1}^{k}(k-m+1)\left(A_m^{p-1}a_m\right)^2\right)^{1/2} \times\left(\sum_{n=1}^{r}(r-n+1)\left(B_n^{q-1}b_n\right)^2\right)^{1/2}, \tag{2.8.1}$$

unless $\{a_m\}$ or $\{b_n\}$ is null, where

$$C(p,q,k,r)=\frac{1}{2}pq\sqrt{kr}. \tag{2.8.2}$$

PROOF. By using the following inequality (see [78,226])

$$\left(\sum_{m=1}^{n} z_m\right)^{\alpha} \leqslant \alpha\sum_{m=1}^{n} z_m\left(\sum_{k=1}^{m} z_k\right)^{\alpha-1},$$

where $\alpha \geqslant 1$ is a constant and $z_m \geqslant 0$, $m = 1, 2, \ldots$, it is easy to observe that

$$A_m^p \leqslant p\sum_{s=1}^{m} a_s A_s^{p-1}, \quad m = 1, 2, \ldots, k, \tag{2.8.3}$$

$$B_n^q \leqslant q\sum_{t=1}^{n} b_t B_t^{q-1}, \quad n = 1, 2, \ldots, r. \tag{2.8.4}$$

From (2.8.3), (2.8.4) and using the Schwarz inequality and the elementary inequality $c^{1/2}d^{1/2} \leqslant (c+d)/2$ (for c, d nonnegative reals), we observe that

$$\begin{aligned} A_m^p B_n^q &\leqslant pq\left(\sum_{s=1}^{m} a_s A_s^{p-1}\right)\left(\sum_{t=1}^{n} b_t B_t^{q-1}\right)\\ &\leqslant pq(m)^{1/2}\left(\sum_{s=1}^{m}\left(a_s A_s^{p-1}\right)^2\right)^{1/2}(n)^{1/2}\left(\sum_{t=1}^{n}\left(b_t B_t^{q-1}\right)^2\right)^{1/2}\\ &\leqslant \frac{1}{2}pq(m+n)\left(\sum_{s=1}^{m}\left(a_s A_s^{p-1}\right)^2\right)^{1/2}\left(\sum_{t=1}^{n}\left(b_t B_t^{q-1}\right)^2\right)^{1/2}. \end{aligned} \tag{2.8.5}$$

Dividing both sides of (2.8.5) by $m+n$ and then taking the sum over n from 1 to r, first, and then the sum over n from 1 to k, and using the Schwarz inequality and then interchanging the order of the summations (see [226,326]) we observe that

$$
\begin{aligned}
&\sum_{m=1}^{k}\sum_{n=1}^{r}\frac{A_m^p B_n^q}{m+n}\\
&\quad\leqslant \frac{1}{2}pq\left\{\sum_{m=1}^{k}\left(\sum_{s=1}^{m}(a_s A_s^{p-1})^2\right)^{1/2}\right\}\left\{\sum_{n=1}^{r}\left(\sum_{t=1}^{n}(b_t B_t^{q-1})^2\right)^{1/2}\right\}\\
&\quad\leqslant \frac{1}{2}pq(k)^{1/2}\left\{\sum_{m=1}^{k}\left(\sum_{s=1}^{m}(a_s A_s^{p-1})^2\right)\right\}^{1/2}(r)^{1/2}\left\{\sum_{n=1}^{r}\left(\sum_{t=1}^{n}(b_t B_t^{q-1})^2\right)\right\}^{1/2}\\
&\quad= pq\sqrt{kr}\left\{\sum_{s=1}^{k}(a_s A_s^{p-1})^2\left(\sum_{m=s}^{k}1\right)\right\}^{1/2}\left\{\sum_{t=1}^{r}(b_t B_t^{q-1})^2\left(\sum_{n=t}^{r}1\right)\right\}^{1/2}\\
&\quad= C(p,q,k,r)\left(\sum_{s=1}^{k}(a_s A_s^{p-1})^2(k-s-1)\right)^{1/2}\\
&\qquad\times\left(\sum_{t=1}^{r}(b_t B_t^{q-1})^2(r-t+1)\right)^{1/2}\\
&\quad= C(p,q,k,r)\left(\sum_{m=1}^{k}(k-m+1)\big(a_m A_m^{p-1}\big)^2\right)^{1/2}\\
&\qquad\times\left(\sum_{n=1}^{r}(r-n+1)\big(b_n B_n^{q-1}\big)^2\right)^{1/2}.
\end{aligned}
$$

The proof is complete. □

REMARK 2.8.1. If we take $p=q=1$ in Theorem 2.8.1, then inequality (2.8.1) reduces to the following inequality

$$
\begin{aligned}
&\sum_{m=1}^{k}\sum_{n=1}^{r}\frac{A_m B_n}{m+n}\\
&\quad\leqslant C(1,1,k,r)\left(\sum_{m=1}^{k}(k-m+1)(a_m)^2\right)^{1/2}\left(\sum_{n=1}^{r}(r-n+1)(b_n)^2\right)^{1/2},
\end{aligned}
\tag{2.8.6}
$$

where $C(1,1,k,r)$ is obtained by taking $p=q=1$ in (2.8.2).

The next result established by Pachpatte in [334] deals with the further generalization of the inequality obtained in (2.8.6).

THEOREM 2.8.2. *Let $\{a_m\}$, $\{b_n\}$, A_m, B_n be as in Theorem 2.8.1. Let $\{p_m\}$ and $\{q_n\}$ be two positive sequences for $m=1,2,\dots k$ and $n=1,2,\dots,r$, and define $P_m=\sum_{s=1}^{m} p_s$ and $Q_n=\sum_{t=1}^{n} q_t$. Let ϕ and ψ be two real-valued, nonnegative, convex and submultiplicative functions defined on $\mathbb{R}_+=[0,\infty)$. Then*

$$\sum_{m=1}^{k}\sum_{n=1}^{r}\frac{\phi(A_m)\psi(B_n)}{m+n} \leqslant M(k,r)\left(\sum_{m=1}^{k}(k-m+1)\left[p_m\phi\left(\frac{a_m}{p_m}\right)\right]^2\right)^{1/2} \times\left(\sum_{n=1}^{r}(r-n+1)\left[q_n\psi\left(\frac{b_n}{q_n}\right)\right]^2\right)^{1/2}, \tag{2.8.7}$$

where

$$M(k,r)=\frac{1}{2}\left(\sum_{m=1}^{k}\left[\frac{\phi(P_m)}{P_m}\right]^2\right)^{1/2}\left(\sum_{n=1}^{r}\left[\frac{\psi(Q_n)}{Q_n}\right]^2\right)^{1/2}. \tag{2.8.8}$$

PROOF. From the hypotheses and by using Jensen's inequality and Schwarz inequality (see [211]), it is easy to observe that

$$\begin{aligned}\phi(A_m) &= \phi\left(P_m\sum_{s=1}^{m}\frac{p_s a_s}{p_s}\Big/\sum_{s=1}^{m}p_s\right)\\ &\leqslant \phi(P_m)\phi\left(\sum_{s=1}^{m}\frac{p_s a_s}{p_s}\Big/\sum_{s=1}^{m}p_s\right)\\ &\leqslant \frac{\phi(P_m)}{P_m}\sum_{s=1}^{m}p_s\phi\left(\frac{a_s}{p_s}\right)\\ &\leqslant \frac{\phi(P_m)}{P_m}(m)^{1/2}\left\{\sum_{s=1}^{m}\left[p_s\phi\left(\frac{a_s}{p_s}\right)\right]^2\right\}^{1/2}\end{aligned} \tag{2.8.9}$$

and similarly,

$$\psi(B_n) \leqslant \frac{\psi(Q_n)}{Q_n}(n)^{1/2}\left\{\sum_{t=1}^{n}\left[q_t\psi\left(\frac{b_t}{q_t}\right)\right]^2\right\}^{1/2}. \tag{2.8.10}$$

From (2.8.9) and (2.8.10) and using the elementary inequality $c^{1/2}d^{1/2} \leqslant (c+d)/2$ (c, d nonnegative reals), we observe that

$$\phi(A_m)\psi(B_n) \leqslant \frac{1}{2}(m+n)\left[\frac{\phi(P_m)}{P_m}\left\{\sum_{s=1}^{m}\left[p_s\phi\left(\frac{a_s}{p_s}\right)\right]^2\right\}^{1/2}\right]$$
$$\times\left[\frac{\psi(Q_n)}{Q_n}\left\{\sum_{t=1}^{n}\left[q_t\psi\left(\frac{b_t}{q_t}\right)\right]^2\right\}^{1/2}\right]. \tag{2.8.11}$$

Dividing both sides of (2.8.11) by $m+n$ and then taking the sum over n from 1 to r, first, and then the sum over m from 1 to k, and using the Schwarz inequality and then interchanging the order of the summations we observe that

$$\sum_{m=1}^{k}\sum_{n=1}^{r}\frac{\phi(A_m)\psi(B_n)}{m+n}$$
$$\leqslant \frac{1}{2}\left\{\sum_{m=1}^{k}\left[\frac{\phi(P_m)}{P_m}\left\{\sum_{s=1}^{m}\left[p_s\phi\left(\frac{a_s}{p_s}\right)\right]^2\right\}^{1/2}\right]\right\}$$
$$\times\left\{\sum_{n=1}^{r}\left[\frac{\psi(Q_n)}{Q_n}\left\{\sum_{t=1}^{n}\left[q_t\psi\left(\frac{b_t}{q_t}\right)\right]^2\right\}^{1/2}\right]\right\}$$
$$\leqslant \frac{1}{2}\left(\sum_{m=1}^{k}\left[\frac{\phi(P_m)}{P_m}\right]^2\right)^{1/2}\left(\sum_{m=1}^{k}\left(\sum_{s=1}^{m}\left[p_s\phi\left(\frac{a_s}{p_s}\right)\right]^2\right)\right)^{1/2}$$
$$\times\left(\sum_{n=1}^{r}\left[\frac{\psi(Q_n)}{Q_n}\right]^2\right)^{1/2}\left(\sum_{n=1}^{r}\left(\sum_{t=1}^{n}\left[q_t\psi\left(\frac{b_t}{q_t}\right)\right]^2\right)\right)^{1/2}$$
$$= M(k,r)\left(\sum_{s=1}^{k}\left[p_s\phi\left(\frac{a_s}{p_s}\right)\right]^2\left(\sum_{m=s}^{k}1\right)\right)^{1/2}$$
$$\times\left(\sum_{t=1}^{r}\left[q_t\psi\left(\frac{b_t}{q_t}\right)\right]^2\left(\sum_{n=t}^{r}1\right)\right)^{1/2}$$

$$
\begin{aligned}
&= M(k,r)\left(\sum_{s=1}^{k}\left[p_s\phi\left(\frac{a_s}{p_s}\right)\right]^2(k-s+1)\right)^{1/2}\\
&\quad\times\left(\sum_{t=1}^{r}\left[q_t\psi\left(\frac{b_t}{q_t}\right)\right]^2(r-t+1)\right)^{1/2}\\
&= M(k,r)\left(\sum_{m=1}^{k}(k-m+1)\left[p_m\phi\left(\frac{a_m}{p_m}\right)\right]^2\right)^{1/2}\\
&\quad\times\left(\sum_{n=1}^{r}(r-n+1)\left[q_n\psi\left(\frac{b_n}{q_n}\right)\right]^2\right)^{1/2}.
\end{aligned}
$$

The proof is complete. □

REMARK 2.8.2. By applying the elementary inequality $c^{1/2}d^{1/2} \leqslant (c+d)/2$ (for c, d nonnegative reals) on the right-hand sides of (2.8.1) and (2.8.7), we get respectively the following inequalities

$$
\begin{aligned}
\sum_{m=1}^{k}\sum_{n=1}^{r}\frac{A_m^p B_n^q}{m+n} \leqslant \frac{1}{2}C(p,q,k,r)\Bigg[&\sum_{m=1}^{k}(k-m+1)\left(A_m^{p-1}a_m\right)^2\\
&+\sum_{n=1}^{r}(r-n+1)\left(B_n^{q-1}b_n\right)^2\Bigg]
\end{aligned} \tag{2.8.12}
$$

and

$$
\begin{aligned}
\sum_{m=1}^{k}\sum_{n=1}^{r}\frac{\phi(A_m)\psi(B_n)}{m+n} \leqslant \frac{1}{2}M(k,r)\Bigg[&\sum_{m=1}^{k}(k-m+1)\left[p_m\phi\left(\frac{a_m}{p_m}\right)\right]^2\\
&+\sum_{n=1}^{r}(r-n+1)\left[q_n\psi\left(\frac{b_n}{q_n}\right)\right]^2\Bigg].
\end{aligned} \tag{2.8.13}
$$

The following two theorems established by Pachpatte in [334] deal with slight variants of the inequality given in Theorem 2.8.2.

THEOREM 2.8.3. *Let $\{a_m\}$ and $\{b_n\}$ be as in Theorem 2.8.1, and define $A_m = \frac{1}{m}\sum_{s=1}^{m} a_s$ and $B_n = \frac{1}{n}\sum_{t=1}^{n} b_t$, for $m = 1, 2, \ldots, k$ and $n = 1, 2, \ldots, r$, where k, r are natural numbers. Let ϕ and ψ be two real-valued, nonnegative and convex functions defined on $\mathbb{R}_+$. Then*

$$\sum_{m=1}^{k}\sum_{n=1}^{r}\frac{mn}{m+n}\phi(A_m)\psi(B_n) \leqslant C(1,1,k,r)\left(\sum_{m=1}^{k}(k-m+1)\left[\phi(a_m)\right]^2\right)^{1/2} \times \left(\sum_{n=1}^{r}(r-n+1)\left[\psi(b_n)\right]^2\right)^{1/2}, \tag{2.8.14}$$

where $C(1, 1, k, r)$ is defined by taking $p = q = 1$ in (2.8.2).

PROOF. From the hypotheses and by using Jensen's inequality and Schwarz inequality, it is easy to observe that

$$\begin{aligned}\phi(A_m) &= \phi\left(\frac{1}{m}\sum_{s=1}^{m} a_s\right)\\ &\leqslant \frac{1}{m}\sum_{s=1}^{m}\phi(a_s)\\ &\leqslant \frac{1}{m}(m)^{1/2}\left\{\sum_{s=1}^{m}\left[\phi(a_s)\right]^2\right\}^{1/2}\end{aligned} \tag{2.8.15}$$

and similarly,

$$\psi(B_n) \leqslant \frac{1}{n}(n)^{1/2}\left\{\sum_{t=1}^{n}\left[\psi(b_t)\right]^2\right\}^{1/2}. \tag{2.8.16}$$

The rest of the proof can be completed by following the same steps as in the proof of Theorems 2.8.1 and 2.8.2 with suitable changes, and hence we omit the details. □

THEOREM 2.8.4. *Let $\{a_m\}$, $\{b_n\}$, $\{p_m\}$, $\{q_n\}$, P_m, Q_n be as in Theorem 2.8.2 and define $A_m = \frac{1}{P_m}\sum_{s=1}^{m} p_s a_s$ and $B_n = \frac{1}{Q_n}\sum_{t=1}^{n} q_t b_t$ for $m = 1, 2, \ldots, k$ and $n = 1, 2, \ldots, r$, where k, r are the natural numbers. Let ϕ and ψ be as in Theo-*

rem 2.8.3. *Then*

$$
\begin{aligned}
&\sum_{m=1}^{k}\sum_{n=1}^{r} P_m Q_n \frac{\phi(A_m)\psi(B_n)}{m+n} \\
&\leqslant C(1,1,k,r)\left(\sum_{m=1}^{k}(k-m+1)\left[p_m\phi(a_m)\right]^2\right)^{1/2} \\
&\quad\times\left(\sum_{n=1}^{r}(r-n+1)\left[q_n\psi(b_n)\right]^2\right)^{1/2}, \qquad (2.8.17)
\end{aligned}
$$

where $C(1,1,k,r)$ *is defined by taking* $p=q=1$ *in* (2.8.2).

PROOF. From the hypotheses and using Jensen's inequality and Schwarz inequality, it is easy to observe that

$$
\begin{aligned}
\phi(A_m) &= \phi\left(\frac{1}{P_m}\sum_{s=1}^{m} p_s a_s\right) \\
&\leqslant \frac{1}{P_m}\sum_{s=1}^{m} p_s\phi(a_s) \\
&\leqslant \frac{1}{P_m}(m)^{1/2}\left\{\sum_{s=1}^{m}\left[p_s\phi(a_s)\right]^2\right\}^{1/2} \qquad (2.8.18)
\end{aligned}
$$

and similarly,

$$
\psi(B_n) \leqslant \frac{1}{Q_n}(n)^{1/2}\left\{\sum_{t=1}^{n}\left[q_t\psi(b_t)\right]^2\right\}^{1/2}. \qquad (2.8.19)
$$

Proceeding as in the proofs of Theorems 2.8.1 and 2.8.2 given above with suitable modifications we get the required inequality in (2.8.17). □

In [350] Pachpatte has established the following theorem.

THEOREM 2.8.5. *Let* $\{a_m\}$ *and* $\{b_n\}$ *be two nonnegative sequences of real numbers defined for* $m=0,1,2,\ldots,k$ *and* $n=0,1,2,\ldots,r$ *and* $a_0=b_0=0$, *where*

k, r are natural numbers. Then

$$\sum_{m=1}^{k}\sum_{n=1}^{r}\frac{a_m b_n}{m+n} \leqslant \frac{\sqrt{kr}}{2}\left(\sum_{m=1}^{k}(k-m+1)(\nabla a_m)^2\right)^{1/2}\left(\sum_{n=1}^{r}(r-n+1)(\nabla b_n)^2\right)^{1/2}, \tag{2.8.20}$$

where $\nabla a_m = a_m - a_{m-1}$ and $\nabla b_n = b_n - b_{n-1}$.

PROOF. From the hypotheses, it is easy to observe that the following identities hold

$$a_m = \sum_{s=1}^{m}\nabla a_s, \quad m = 1, 2, \ldots, k, \tag{2.8.21}$$

and

$$b_n = \sum_{t=1}^{n}\nabla b_t, \quad n = 1, 2, \ldots, r. \tag{2.8.22}$$

From (2.8.21) and (2.8.22) and using Schwarz inequality and the elementary inequality $c^{1/2}d^{1/2} \leqslant (c+d)/2$ (for c, d nonnegative reals), we observe that

$$\begin{aligned} a_m b_n &= \left(\sum_{s=1}^{m}\nabla a_s\right)\left(\sum_{t=1}^{n}\nabla b_t\right) \\ &\leqslant (m)^{1/2}\left(\sum_{s=1}^{m}(\nabla a_s)^2\right)^{1/2}(n)^{1/2}\left(\sum_{t=1}^{n}(\nabla b_t)^2\right)^{1/2} \\ &\leqslant \frac{1}{2}(m+n)\left(\sum_{s=1}^{m}(\nabla a_s)^2\right)^{1/2}\left(\sum_{t=1}^{n}(\nabla b_t)^2\right)^{1/2}. \end{aligned}$$

Rewriting the above inequality and taking the sum over n from 1 to r, first, and then the sum over m of the resulting inequality from 1 to k and using Schwarz inequality and then interchanging the order of the summations we observe that

$$
\begin{aligned}
&\sum_{m=1}^{k}\sum_{n=1}^{r}\frac{a_m b_n}{m+n}\\
&\quad\leqslant \frac{1}{2}\sum_{m=1}^{k}\left(\sum_{s=1}^{m}(\nabla a_s)^2\right)^{1/2}\sum_{n=1}^{r}\left(\sum_{t=1}^{n}(\nabla b_t)^2\right)^{1/2}\\
&\quad\leqslant \frac{1}{2}(k)^{1/2}\left(\sum_{m=1}^{k}\left(\sum_{s=1}^{m}(\nabla a_s)^2\right)\right)^{1/2}(r)^{1/2}\left(\sum_{n=1}^{r}\left(\sum_{t=1}^{n}(\nabla b_t)^2\right)\right)^{1/2}\\
&\quad= \frac{1}{2}\sqrt{kr}\left(\sum_{s=1}^{k}(\nabla a_s)^2\left(\sum_{m=s}^{k}1\right)\right)^{1/2}\left(\sum_{t=1}^{r}(\nabla b_t)^2\left(\sum_{n=t}^{r}1\right)\right)^{1/2}\\
&\quad= \frac{1}{2}\sqrt{kr}\left(\sum_{s=1}^{k}(\nabla a_s)^2(k-s+1)\right)^{1/2}\left(\sum_{t=1}^{r}(\nabla b_t)^2(r-t+1)\right)^{1/2}\\
&\quad= \frac{1}{2}\sqrt{kr}\left(\sum_{m=1}^{k}(k-m+1)(\nabla a_m)^2\right)^{1/2}\left(\sum_{n=1}^{r}(r-n+1)(\nabla b_n)^2\right)^{1/2}.
\end{aligned}
$$

The proof is complete. □

The next theorem deals with the further generalization of Theorem 2.8.5 and is established by Pachpatte in [335].

THEOREM 2.8.6. *Let $\{a_m\}$ and $\{b_n\}$ be two nonnegative sequences of real numbers defined for $m = 1, 2, \ldots, k$ and $n = 1, 2, \ldots, r$ with $a_0 = b_0 = 0$ and let $\{p_m\}$ and $\{q_n\}$ be two positive sequences of real numbers defined for $m = 1, 2, \ldots, k$ and $n = 1, 2, \ldots, r$, where k, r are natural numbers, and define $P_m = \sum_{s=1}^{m} p_s$ and $Q_n = \sum_{t=1}^{n} q_t$. Let ϕ and ψ be two real-valued nonnegative, convex and submultiplicative functions defined on $\mathbb{R}_+ = [0, \infty)$. Then*

$$
\begin{aligned}
\sum_{m=1}^{k}\sum_{n=1}^{r}\frac{\phi(a_m)\psi(b_n)}{m+n} &\leqslant M(k,r)\left(\sum_{m=1}^{k}(k-m+1)\left[p_m\phi\left(\frac{\nabla a_m}{p_m}\right)\right]^2\right)^{1/2}\\
&\quad\times\left(\sum_{n=1}^{r}(r-n+1)\left[q_n\psi\left(\frac{\nabla b_n}{q_n}\right)\right]^2\right)^{1/2}, \qquad (2.8.23)
\end{aligned}
$$

where

$$M(k,r)=\frac{1}{2}\left(\sum_{m=1}^{k}\left[\frac{\phi(P_m)}{P_m}\right]^2\right)^{1/2}\left(\sum_{n=1}^{r}\left[\frac{\psi(Q_n)}{Q_n}\right]^2\right)^{1/2} \tag{2.8.24}$$

and $\nabla a_m = a_m - a_{m-1}$, $\nabla b_n = b_n - b_{n-1}$.

PROOF. From the hypotheses, it is easy to observe that the following identities hold

$$a_m=\sum_{s=1}^{m}\nabla a_s, \quad m=1,2,\ldots,k, \tag{2.8.25}$$

$$b_n=\sum_{t=1}^{n}\nabla b_t, \quad n=1,2,\ldots,r. \tag{2.8.26}$$

From (2.8.25) and (2.8.26) and using Jensen's inequality (see [211]), we observe that

$$\begin{aligned}\phi(a_m)&=\phi\left(P_m\sum_{s=1}^{m}p_s\frac{\nabla a_s}{p_s}\Big/\sum_{s=1}^{m}p_s\right)\\&\leqslant\phi(P_m)\phi\left(\sum_{s=1}^{m}p_s\frac{\nabla a_s}{p_s}\Big/\sum_{s=1}^{m}p_s\right)\\&\leqslant\phi(P_m)\sum_{s=1}^{m}p_s\phi\left(\frac{\nabla a_s}{p_s}\right)\Big/P_m\end{aligned} \tag{2.8.27}$$

and similarly,

$$\psi(b_n)\leqslant\psi(Q_n)\sum_{t=1}^{n}q_t\psi\left(\frac{\nabla b_t}{q_t}\right)\Big/Q_n. \tag{2.8.28}$$

The rest of the proof can be completed by following the same arguments as in the proof of Theorem 2.8.2 with suitable modifications and here we omit the further details. □

REMARK 2.8.3. If we apply the elementary inequality $c^{1/2}d^{1/2}\leqslant(c+d)/2$ (for c,d nonnegative reals) on the right-hand sides of (2.8.20) and (2.8.23) then we get respectively the following inequalities

$$\sum_{m=1}^{k}\sum_{n=1}^{r}\frac{a_m b_n}{m+n} \leqslant \frac{1}{4}\sqrt{kr}\left[\sum_{m=1}^{k}(k-m+1)(\nabla a_m)^2+\sum_{n=1}^{r}(r-n+1)(\nabla b_n)^2\right] \quad (2.8.29)$$

and

$$\sum_{m=1}^{k}\sum_{n=1}^{r}\frac{\phi(a_m)\psi(b_n)}{m+n} \leqslant \frac{1}{2}M(k,r)\left[\sum_{m=1}^{k}(k-m+1)\left[p_m\phi\left(\frac{\nabla a_m}{p_m}\right)\right]^2 + \sum_{n=1}^{r}(r-n+1)\left[q_n\psi\left(\frac{\nabla b_n}{q_n}\right)\right]^2\right]. \quad (2.8.30)$$

In [353] Pachpatte has established the following inequality similar to that of the extension of Hilbert's inequality given in [141, p. 253].

THEOREM 2.8.7. *Let $p > 1$, $q > 1$ be constants and $\frac{1}{p}+\frac{1}{q}=1$. Let $\{a_m\}$ and $\{b_n\}$ be two sequences of real numbers defined for $m = 1, 2, \ldots, k$ and $n = 1, 2, \ldots, r$, where k, r are natural numbers with $a_0 = b_0 = 0$. Then*

$$\sum_{m=1}^{k}\sum_{n=1}^{r}\frac{|a_m||b_n|}{qm^{p-1}+pn^{q-1}} \leqslant M(p,q,k,r)\left(\sum_{m=1}^{k}(k-m+1)|\nabla a_m|^p\right)^{1/p}\left(\sum_{n=1}^{r}(r-n+1)|\nabla b_n|^q\right)^{1/q}, \quad (2.8.31)$$

where

$$M(p,q,k,r)=\frac{1}{pq}k^{(p-1)/p}r^{(q-1)/q} \quad (2.8.32)$$

and $\nabla a_m = a_m - a_{m-1}$, $\nabla b_n = b_n - b_{n-1}$.

PROOF. From the hypotheses, it is easy to observe that the following identities hold:

$$a_m = \sum_{s=1}^{m} \nabla a_s, \tag{2.8.33}$$

$$b_n = \sum_{t=1}^{n} \nabla b_t, \tag{2.8.34}$$

for $m = 1, 2, \ldots, k$, $n = 1, 2, \ldots, r$. From (2.8.33) and (2.8,34) and using Hölder's inequality with indices $p, p/(p-1)$ and $q, q/(q-1)$, respectively, we have

$$|a_m| \leqslant (m)^{(p-1)/p} \left(\sum_{s=1}^{m} |\nabla a_s|^p \right)^{1/p}, \tag{2.8.35}$$

$$|b_n| \leqslant (n)^{(q-1)/q} \left(\sum_{t=1}^{n} |\nabla b_t|^q \right)^{1/q}, \tag{2.8.36}$$

for $m = 1, 2, \ldots, k, n = 1, 2, \ldots, r$. From (2.8.35), (2.8.36) and using the elementary inequality

$$z_1 z_2 \leqslant \frac{z_1^p}{p} + \frac{z_2^q}{q}, \quad z_1 \geqslant 0, z_2 \geqslant 0, \frac{1}{p} + \frac{1}{q} = 1, p > 1, \tag{2.8.37}$$

we observe that

$$\begin{aligned} |a_m||b_n| &\leqslant (m)^{(p-1)/p}(n)^{(q-1)/q} \left(\sum_{s=1}^{m} |\nabla a_s|^p \right)^{1/p} \left(\sum_{t=1}^{n} |\nabla b_t|^q \right)^{1/q} \\ &\leqslant \left[\frac{m^{p-1}}{p} + \frac{n^{q-1}}{q} \right] \left(\sum_{s=1}^{m} |\nabla a_s|^p \right)^{1/p} \left(\sum_{t=1}^{n} |\nabla b_t|^q \right)^{1/q} \end{aligned} \tag{2.8.38}$$

for $m = 1, 2, \ldots, k, n = 1, 2, \ldots, r$. From (2.8.38) we observe that

$$\frac{|a_m||b_n|}{qm^{p-1} + pn^{q-1}} \leqslant \frac{1}{pq} \left(\sum_{s=1}^{m} |\nabla a_s|^p \right)^{1/p} \left(\sum_{t=1}^{n} |\nabla b_t|^q \right)^{1/q} \tag{2.8.39}$$

for $m = 1, 2, \ldots, k, n = 1, 2, \ldots, r$. Taking the sum on both sides of (2.8.39) first over n from 1 to r and then the sum over m from 1 to k of the resulting inequality and using Hölder's inequality with indices $p, p/(p-1)$ and $q, q/(q-1)$ and interchanging the order of summations we observe that

$$\begin{aligned}
&\sum_{m=1}^{k}\sum_{n=1}^{r}\frac{|a_m||b_n|}{qm^{p-1}+pn^{q-1}}\\
&\quad\leqslant \frac{1}{pq}\left\{\sum_{m=1}^{k}\left(\sum_{s=1}^{m}|\nabla a_s|^p\right)^{1/p}\right\}\left\{\sum_{n=1}^{r}\left(\sum_{t=1}^{n}|\nabla b_t|^q\right)^{1/q}\right\}\\
&\quad\leqslant \frac{1}{pq}(k)^{(p-1)/p}\left\{\sum_{m=1}^{k}\left(\sum_{s=1}^{m}|\nabla a_s|^p\right)\right\}^{1/p}(r)^{(q-1)/q}\left\{\sum_{n=1}^{r}\left(\sum_{t=1}^{n}|\nabla b_t|^q\right)\right\}^{1/q}\\
&\quad= M(p,q,k,r)\left(\sum_{m=1}^{k}(k-m+1)|\nabla a_m|^p\right)^{1/p}\left(\sum_{n=1}^{r}(r-n+1)|\nabla b_n|^q\right)^{1/q}.
\end{aligned}$$

The proof is complete. □

The following two independent variable version of the inequality given in Theorem 2.8.7 is also established by Pachpatte in [353]. In what follows, we denote by $\mathbb{R}$ the set of real numbers. Let $\mathbb{N} = \{1, 2, \ldots\}$, $\mathbb{N}_0 = \{0, 1, 2, \ldots\}$, $\mathbb{N}_\alpha = \{0, 1, 2, \ldots, \alpha\}$, $\alpha \in \mathbb{N}$. For a function $v(s,t): \mathbb{N}_0 \times \mathbb{N}_0 \to \mathbb{R}$, we define the operators $\nabla_1 v(s,t) = v(s,t) - v(s-1,t)$, $\nabla_2 v(s,t) = v(s,t) - v(s,t-1)$ and $\nabla_2\nabla_1 v(s,t) = \nabla_2(\nabla_1 v(s,t)) = \nabla_1(\nabla_2 v(s,t))$.

THEOREM 2.8.8. *Let $p > 1, q > 1$ be constants and $\frac{1}{p} + \frac{1}{q} = 1$. Let $a(s,t): \mathbb{N}_x \times \mathbb{N}_y \to \mathbb{R}$, $b(k,r): \mathbb{N}_z \times \mathbb{N}_w \to \mathbb{R}$ and $a(0,t) = b(0,t) = 0$, $a(s,0) = b(s,0) = 0$. Then*

$$\begin{aligned}
&\sum_{s=1}^{x}\sum_{t=1}^{y}\left(\sum_{k=1}^{z}\sum_{r=1}^{w}\frac{|a(s,t)||b(k,r)|}{q(st)^{p-1}+p(kr)^{q-1}}\right)\\
&\quad\leqslant L(p,q,x,y,z,w)\left(\sum_{s=1}^{x}\sum_{t=1}^{y}(x-s+1)(y-t+1)\left|\nabla_2\nabla_1 a(s,t)\right|^p\right)^{1/p}\\
&\qquad\times\left(\sum_{k=1}^{z}\sum_{r=1}^{w}(z-k+1)(w-r+1)\left|\nabla_2\nabla_1 b(k,r)\right|^q\right)^{1/q} \qquad (2.8.40)
\end{aligned}$$

for x, y, z, w in $\mathbb{N}$, where

$$L(p,q,x,y,z,w) = \frac{1}{pq}(xy)^{(p-1)/p}(zw)^{(q-1)/q} \tag{2.8.41}$$

for x, y, z, w in $\mathbb{N}$.

PROOF. From the hypotheses, it is easy to observe that the following identities hold:

$$a(s,t) = \sum_{\xi=1}^{s}\sum_{\eta=1}^{t} \nabla_2\nabla_1 a(\xi,\eta), \tag{2.8.42}$$

$$b(k,r) = \sum_{\sigma=1}^{k}\sum_{\tau=1}^{r} \nabla_2\nabla_1 b(\sigma,\tau) \tag{2.8.43}$$

for $(s,t) \in \mathbb{N}_x \times \mathbb{N}_y$, $(k,r) \in \mathbb{N}_z \times \mathbb{N}_w$. From (2.8.42), (2.8.43) and using Hölder's inequality with indices p, $p/(p-1)$ and $q, q/(q-1)$, respectively, we have

$$\left|a(s,t)\right| \leqslant (st)^{(p-1)/p}\left(\sum_{\xi=1}^{s}\sum_{\eta=1}^{t}\left|\nabla_2\nabla_1 a(\xi,\eta)\right|^p\right)^{1/p}, \tag{2.8.44}$$

$$\left|b(k,r)\right| \leqslant (kr)^{(q-1)/q}\left(\sum_{\sigma=1}^{k}\sum_{\tau=1}^{r}\left|\nabla_2\nabla_1 b(\sigma,\tau)\right|^q\right)^{1/q}, \tag{2.8.45}$$

for $(s,t) \in \mathbb{N}_x \times \mathbb{N}_y$, $(k,r) \in \mathbb{N}_z \times \mathbb{N}_w$. From (2.8.44), (2.8.45) and using the elementary inequality (2.8.37), it is easy to observe that

$$\begin{aligned} &\frac{|a(s,t)||b(k,r)|}{q(st)^{p-1} + p(kr)^{q-1}} \\ &\quad \leqslant \frac{1}{pq}\left(\sum_{\xi=1}^{s}\sum_{\eta=1}^{t}\left|\nabla_2\nabla_1 a(\xi,\eta)\right|^p\right)^{1/p}\left(\sum_{\sigma=1}^{k}\sum_{\tau=1}^{r}\left|\nabla_2\nabla_1 b(\sigma,\tau)\right|^q\right)^{1/q} \end{aligned} \tag{2.8.46}$$

for $(s,t) \in \mathbb{N}_x \times \mathbb{N}_y$, $(k,r) \in \mathbb{N}_z \times \mathbb{N}_w$. Taking the sum on both sides of (2.8.46) first over r from 1 to w and over k from 1 to z and then taking the sum on both sides of the resulting inequality first over t from 1 to y and over s from 1 to x and then using Hölder's inequality with indices p, $p/(p-1)$ and $q, q/(q-1)$ and interchanging the order of the summation we observe that

$$\sum_{s=1}^{x}\sum_{t=1}^{y}\left(\sum_{k=1}^{z}\sum_{r=1}^{w}\frac{|a(s,t)||b(k,r)|}{q(st)^{p-1}+p(kr)^{q-1}}\right)$$

$$\leqslant \frac{1}{pq}\left\{\sum_{s=1}^{x}\sum_{t=1}^{y}\left(\sum_{\xi=1}^{s}\sum_{\eta=1}^{t}\left|\nabla_2\nabla_1 a(\xi,\eta)\right|^{p}\right)^{1/p}\right\}$$

$$\times\left\{\sum_{k=1}^{z}\sum_{r=1}^{w}\left(\sum_{\sigma=1}^{k}\sum_{\tau=1}^{r}\left|\nabla_2\nabla_1 b(\sigma,\tau)\right|^{q}\right)^{1/q}\right\}$$

$$\leqslant \frac{1}{pq}(xy)^{(p-1)/p}\left\{\sum_{s=1}^{x}\sum_{t=1}^{y}\left(\sum_{\xi=1}^{s}\sum_{\eta=1}^{t}\left|\nabla_2\nabla_1 a(\xi,\eta)\right|^{p}\right)\right\}^{1/p}$$

$$\times (zw)^{(q-1)/q}\left\{\sum_{k=1}^{z}\sum_{r=1}^{w}\left(\sum_{\sigma=1}^{k}\sum_{\tau=1}^{r}\left|\nabla_2\nabla_1 b(\sigma,\tau)\right|^{q}\right)\right\}^{1/q}$$

$$= L(p,q,x,y,z,w)\left(\sum_{s=1}^{x}\sum_{t=1}^{y}(x-s+1)(y-t+1)\left|\nabla_2\nabla_1 a(s,t)\right|^{p}\right)^{1/p}$$

$$\times\left(\sum_{k=1}^{z}\sum_{r=1}^{w}(z-k+1)(w-r+1)\left|\nabla_2\nabla_1 b(k,r)\right|^{q}\right)^{1/q}.$$

The proof is complete. □

REMARK 2.8.4. If we apply the elementary inequality (2.8.37) on the right-hand sides of (2.8.31) and (2.8.40), then we get respectively the following inequalities

$$\sum_{m=1}^{k}\sum_{n=1}^{r}\frac{|a_m||b_n|}{qm^{p-1}+pn^{q-1}}$$

$$\leqslant M(p,q,k,r)\left[\frac{1}{p}\sum_{m=1}^{k}(k-m+1)|\nabla a_m|^{p}+\frac{1}{q}\sum_{n=1}^{r}(r-n+1)|\nabla b_n|^{q}\right] \tag{2.8.47}$$

and

$$\sum_{s=1}^{x}\sum_{t=1}^{y}\left(\sum_{k=1}^{z}\sum_{r=1}^{w}\frac{|a(s,t)||b(k,r)|}{q(st)^{p-1}+p(kr)^{q-1}}\right)$$
$$\leqslant L(p,q,x,y,z,w)\left[\frac{1}{p}\sum_{s=1}^{x}\sum_{t=1}^{y}(x-s+1)(y-t+1)\left|\nabla_2\nabla_1 a(s,t)\right|^p\right.$$
$$\left.+\frac{1}{q}\sum_{k=1}^{z}\sum_{r=1}^{w}(z-k+1)(w-r+1)\left|\nabla_2\nabla_1 b(k,r)\right|^q\right]. \tag{2.8.48}$$

The following theorems deals with the integral analogues of the inequalities in Theorems 2.8.1–2.8.4 established by Pachpatte in [334].

THEOREM 2.8.9. *Let $p \geqslant 1, q \geqslant 1$ and $f(\sigma) \geqslant 0$, $g(\tau) \geqslant 0$ for $\sigma \in (0,x)$, $\tau \in (0,y)$, where x, y are positive real numbers, and define $F(s)=\int_0^s f(\sigma)\,\mathrm{d}\sigma$ and $G(t)=\int_0^t g(\tau)\,\mathrm{d}\tau$, for $s\in(0,x)$, $t\in(0,y)$. Then*

$$\int_0^x\int_0^y\frac{F^p(s)G^q(t)}{s+t}\,\mathrm{d}s\,\mathrm{d}t \leqslant D(p,q,x,y)\left(\int_0^x (x-s)\left(F^{p-1}(s)f(s)\right)^2\mathrm{d}s\right)^{1/2}$$
$$\times\left(\int_0^y (y-t)\left(G^{q-1}(t)\right)^2\mathrm{d}t\right)^{1/2}, \tag{2.8.49}$$

unless $f\equiv 0$ or $g\equiv 0$, where

$$D(p,q,x,y)=\frac{1}{2}pq\sqrt{xy}. \tag{2.8.50}$$

PROOF. From the hypotheses, it is easy to observe that

$$F^p(s)=p\int_0^s F^{p-1}(\sigma)f(\sigma)\,\mathrm{d}\sigma,\quad s\in(0,x), \tag{2.8.51}$$

$$G^q(t)=q\int_0^t G^{q-1}(\tau)g(\tau)\,\mathrm{d}\tau,\quad t\in(0,y). \tag{2.8.52}$$

From (2.8.51) and (2.8.52) and using Schwarz inequality and the elementary inequality $c^{1/2}d^{1/2} \leqslant (c+d)/2$ (for c, d nonnegative reals), we observe that

$$
\begin{aligned}
&F^p(s)G^q(t)\\
&= pq\left(\int_0^s F^{p-1}(\sigma)f(\sigma)\,\mathrm{d}\sigma\right)\left(\int_0^t G^{q-1}(\tau)g(\tau)\,\mathrm{d}\tau\right)\\
&\leqslant pq(s)^{1/2}\left(\int_0^s \left(F^{p-1}(\sigma)f(\sigma)\right)^2\mathrm{d}\sigma\right)^{1/2}(t)^{1/2}\left(\int_0^t \left(G^{q-1}(\tau)g(\tau)\right)^2\mathrm{d}\tau\right)^{1/2}\\
&\leqslant \frac{1}{2}pq(s+t)\left(\int_0^s \left(F^{p-1}(\sigma)f(\sigma)\right)^2\mathrm{d}\sigma\right)^{1/2}\left(\int_0^t \left(G^{q-1}(\tau)g(\tau)\right)^2\mathrm{d}\tau\right)^{1/2}.
\end{aligned}
\tag{2.8.53}
$$

Dividing both sides of (2.8.53) by $s+t$ and then integrating over t from 0 to y, first, and then integrating the resulting inequality over s from 0 to x and using Schwarz inequality we observe that

$$
\begin{aligned}
&\int_0^x\int_0^y \frac{F^p(s)G^q(t)}{s+t}\,\mathrm{d}s\,\mathrm{d}t\\
&\leqslant \frac{1}{2}pq\left\{\int_0^x\left(\int_0^s \left(F^{p-1}(\sigma)f(\sigma)\right)^2\mathrm{d}\sigma\right)^{1/2}\right\}\\
&\quad\times\left\{\int_0^y\left(\int_0^t \left(G^{q-1}(\tau)g(\tau)\right)^2\mathrm{d}\tau\right)^{1/2}\right\}\\
&\leqslant \frac{1}{2}pq(x)^{1/2}\left\{\int_0^x\left(\int_0^s \left(F^{p-1}(\sigma)f(\sigma)\right)^2\mathrm{d}\sigma\right)\mathrm{d}s\right\}^{1/2}\\
&\quad\times(y)^{1/2}\left\{\int_0^y\left(\int_0^t \left(G^{q-1}(\tau)g(\tau)\right)^2\mathrm{d}\tau\right)\mathrm{d}t\right\}^{1/2}\\
&= D(p,q,x,y)\left(\int_0^x (x-s)\left(F^{p-1}(s)f(s)\right)^2\mathrm{d}s\right)^{1/2}\\
&\quad\times\left(\int_0^y (y-t)\left(G^{q-1}(t)g(t)\right)^2\mathrm{d}t\right)^{1/2}.
\end{aligned}
$$

The proof is complete. □

REMARK 2.8.5. In the special case when $p = q = 1$, inequality (2.8.49) reduces to the following inequality

$$\int_0^x \int_0^y \frac{F(s)G(t)}{s+t}\,ds\,dt \leqslant D(1,1,x,y)\left(\int_0^x (x-s)f^2(s)\,ds\right)^{1/2}\left(\int_0^y (y-t)g^2(t)\,dt\right)^{1/2}, \tag{2.8.54}$$

where $D(1,1,x,y)$ is obtained by taking $p = q = 1$ in (2.8.50).

THEOREM 2.8.10. *Let f, g, F, G be as in Theorem 2.8.9. Let $p(\sigma)$ and $q(\tau)$ be two positive functions defined for $\sigma \in (0,x)$, $\tau \in (0,y)$, and define $P(s) = \int_0^s p(\sigma)\,d\sigma$ and $Q(t) = \int_0^t q(\tau)\,d\tau$, for $s \in (0,x)$, $t \in (0,y)$, where x, y are positive real numbers. Let ϕ and ψ be as in Theorem 2.8.2. Then*

$$\int_0^x \int_0^y \frac{\phi(F(s))\psi(G(t))}{s+t}\,ds\,dt \leqslant L(x,y)\left(\int_0^x (x-s)\left[p(s)\phi\left(\frac{f(s)}{p(s)}\right)\right]^2 ds\right)^{1/2} \times\left(\int_0^y (y-t)\left[q(t)\psi\left(\frac{g(t)}{q(t)}\right)\right]^2 dt\right)^{1/2}, \tag{2.8.55}$$

where

$$L(x,y) = \frac{1}{2}\left(\int_0^x \left[\frac{\phi(P(s))}{P(s)}\right]^2 ds\right)^{1/2}\left(\int_0^y \left[\frac{\psi(Q(t))}{Q(t)}\right]^2 dt\right)^{1/2}. \tag{2.8.56}$$

PROOF. From the hypotheses and by using Jensen's inequality and the Schwarz inequality, it is easy to observe that

$$\begin{aligned}\phi(F(s)) &= \phi\left(P(s)\int_0^s p(\sigma)\frac{f(\sigma)}{p(\sigma)}\,d\sigma \Big/ \int_0^s p(\sigma)\,d\sigma\right)\\ &\leqslant \frac{\phi(P(s))}{P(s)}\int_0^s p(\sigma)\phi\left(\frac{f(\sigma)}{p(\sigma)}\right) d\sigma\\ &\leqslant \left[\frac{\phi(P(s))}{P(s)}\right](s)^{1/2}\left\{\int_0^s \left[p(\sigma)\phi\left(\frac{f(\sigma)}{p(\sigma)}\right)\right]^2 d\sigma\right\}^{1/2}\end{aligned} \tag{2.8.57}$$

and similarly,

$$\psi\big(G(t)\big) \leqslant \left[\frac{\psi(Q(t))}{Q(t)}\right](t)^{1/2}\left\{\int_0^t \left[q(\tau)\psi\left(\frac{g(\tau)}{q(\tau)}\right)\right]^2 \mathrm{d}\tau\right\}^{1/2}. \tag{2.8.58}$$

From (2.8.57) and (2.8.58) and using the elementary inequality $c^{1/2}d^{1/2} \leqslant (c+d)/2$ (for c, d nonnegative reals), we observe that

$$\phi\big(F(s)\big)\psi\big(G(t)\big) \leqslant \frac{1}{2}(s+t)\left[\frac{\phi(P(s))}{P(s)}\left\{\int_0^s \left[p(\sigma)\phi\left(\frac{f(\sigma)}{p(\sigma)}\right)\right]^2 \mathrm{d}\sigma\right\}^{1/2}\right]$$
$$\times \left[\frac{\psi(Q(t))}{Q(t)}\left\{\int_0^t \left[q(\tau)\psi\left(\frac{g(\tau)}{q(\tau)}\right)\right]^2 \mathrm{d}\tau\right\}^{1/2}\right]. \tag{2.8.59}$$

The rest of the proof can be completed by following the same steps as in the proof of Theorem 2.8.9 and closely looking at the proof of Theorem 2.8.2, hence we omit the details. □

THEOREM 2.8.11. *Let f, g be as in Theorem 2.8.9, and define $F(s) = \frac{1}{s}\int_0^s f(\sigma)\,\mathrm{d}\sigma$ and $G(t) = \frac{1}{t}\int_0^t g(\tau)\,\mathrm{d}\tau$, for $s \in (0, x)$, $t \in (0, y)$, where x, y are positive real numbers. Let ϕ and ψ be as in Theorem 2.8.3. Then*

$$\int_0^x\int_0^y \frac{st}{s+t}\phi\big(F(s)\big)\psi\big(G(t)\big)\,\mathrm{d}s\,\mathrm{d}t$$
$$\leqslant D(1,1,x,y)\left(\int_0^x (x-s)\big[\phi\big(f(s)\big)\big]^2\,\mathrm{d}s\right)^{1/2}$$
$$\times \left(\int_0^y (y-t)\big[\psi\big(g(t)\big)\big]^2\,\mathrm{d}t\right)^{1/2}, \tag{2.8.60}$$

where $D(1,1,x,y)$ is obtained by taking $p = q = 1$ in (2.8.50).

THEOREM 2.8.12. *Let f, g, p, q, P, Q be as in Theorem 2.8.10, and define $F(s) = \frac{1}{P(s)}\int_0^s p(\sigma)f(\sigma)\,\mathrm{d}\sigma$ and $G(t) = \frac{1}{Q(t)}\int_0^t q(\tau)g(\tau)\,\mathrm{d}\tau$, for $s \in (0, x)$, $t \in (0, y)$, where x, y are positive real numbers. Let ϕ and ψ be as defined in Theorem 2.8.3. Then*

$$\int_0^x\int_0^y \frac{P(s)Q(t)\phi(F(s))\psi(G(t))}{s+t}\,\mathrm{d}s\,\mathrm{d}t$$
$$\leqslant D(1,1,x,y)\left(\int_0^x (x-s)\big[p(s)\phi\big(f(s)\big)\big]^2\,\mathrm{d}s\right)^{1/2}$$

$$\times\left(\int_0^y (y-t)\left[q(t)\psi\left(g(t)\right)\right]^2 \mathrm{d}t\right)^{1/2}, \tag{2.8.61}$$

where $D(1,1,x,y)$ *is defined by taking* $p=q=1$ *in* (2.8.50).

The proofs of Theorems 2.8.11 and 2.8.12 can be completed by following the proof of Theorem 2.8.10 and closely looking at the proofs of Theorems 2.8.3 and 2.8.4 and by making use of the integral versions of Jensen's and the Schwarz inequalities. Here we omit the details.

The integral analogues of Theorems 2.8.5 and 2.8.6 established by Pachpatte in [335,350] are given in the following theorems.

THEOREM 2.8.13. *Let* $f\in C^1([0,x),\mathbb{R}_+)$, $g\in C^1([0,y),\mathbb{R}_+)$ *and* $f(0)=g(0)=0$, *where* $\mathbb{R}_+=[0,\infty)$, $x,y\in\mathbb{R}$, *the set of real numbers. Then*

$$\int_0^x\int_0^y \frac{f(s)g(t)}{s+t}\,\mathrm{d}s\,\mathrm{d}t$$
$$\leqslant \frac{1}{2}\sqrt{xy}\left(\int_0^x (x-s)f'(s)^2\,\mathrm{d}s\right)^{1/2}\left(\int_0^y (y-t)g'(t)^2\,\mathrm{d}t\right)^{1/2}, \tag{2.8.62}$$

where "$'$" *denotes the derivative of a function.*

PROOF. From the hypotheses we have the following identities

$$f(s)=\int_0^s f'(\sigma)\,\mathrm{d}\sigma, \quad s\in[0,x), \tag{2.8.63}$$

$$g(t)=\int_0^t g'(\tau)\,\mathrm{d}\tau, \quad t\in[0,y). \tag{2.8.64}$$

From (2.8.63), (2.8.64) and using the Schwarz inequality and the elementary inequality $c^{1/2}d^{1/2}\leqslant(c+d)/2$ (for c,d nonnegative reals), we observe that

$$\begin{aligned} f(s)g(t) &= \left(\int_0^s f'(\sigma)\,\mathrm{d}\sigma\right)\left(\int_0^t g'(\tau)\,\mathrm{d}\tau\right) \\ &\leqslant (s)^{1/2}\left(\int_0^s f'^2(\sigma)\,\mathrm{d}\sigma\right)^{1/2}(t)^{1/2}\left(\int_0^t g'^2(\tau)\,\mathrm{d}\tau\right)^{1/2} \\ &\leqslant \frac{1}{2}(s+t)\left(\int_0^s f'^2(\sigma)\,\mathrm{d}\sigma\right)^{1/2}\left(\int_0^t g'^2(\tau)\,\mathrm{d}\tau\right)^{1/2}. \end{aligned} \tag{2.8.65}$$

Rewriting (2.8.65) and then integrating over t from 0 to y, first, and then integrating the resulting inequality over s from 0 to x and using the Schwarz inequality we observe that

$$\begin{aligned}
&\int_0^x \int_0^y \frac{f(s)g(t)}{s+t}\,\mathrm{d}s\,\mathrm{d}t \\
&\quad \leqslant \frac{1}{2}\left(\int_0^x \left(\int_0^s f'^2(\sigma)\,\mathrm{d}\sigma\right)^{1/2} \mathrm{d}s\right)\left(\int_0^y \left(\int_0^t g'^2(\tau)\,\mathrm{d}\tau\right)^{1/2} \mathrm{d}t\right) \\
&\quad \leqslant \frac{1}{2}(x)^{1/2}\left(\int_0^x \left(\int_0^s f'^2(\sigma)\,\mathrm{d}\sigma\right) \mathrm{d}s\right)^{1/2} \\
&\qquad \times (y)^{1/2}\left(\int_0^y \left(\int_0^t g'^2(\tau)\,\mathrm{d}\tau\right) \mathrm{d}t\right)^{1/2} \\
&\quad = \frac{1}{2}\sqrt{xy}\left(\int_0^x (x-s)f'^2(s)\,\mathrm{d}s\right)^{1/2}\left(\int_0^y (y-t)g'^2(t)\,\mathrm{d}t\right)^{1/2}.
\end{aligned}$$

The proof is complete. □

THEOREM 2.8.14. *Let $f \in C^1([0,x),\mathbb{R}_+)$, $g \in C^1([0,y),\mathbb{R}_+)$ with $f(0) = g(0) = 0$ and let $p(\sigma)$ and $q(\tau)$ be two positive functions defined for $\sigma \in [0,x)$ and $\tau \in [0,y)$, and $P(s) = \int_0^s p(\sigma)\,\mathrm{d}\sigma$ and $Q(t) = \int_0^t q(\tau)\,\mathrm{d}\tau$, for $s \in [0,x)$ and $t \in [0,y)$, where x, y are positive real numbers. Let ϕ and ψ be as in Theorem 2.8.6. Then*

$$\begin{aligned}
&\int_0^x \int_0^y \frac{\phi(f(s))\psi(g(t))}{s+t}\,\mathrm{d}s\,\mathrm{d}t \\
&\quad \leqslant L(x,y)\left(\int_0^x (x-s)\left[p(s)\phi\left(\frac{f'(s)}{p(s)}\right)\right]^2 \mathrm{d}s\right)^{1/2} \\
&\qquad \times \left(\int_0^y (y-t)\left[q(t)\psi\left(\frac{g'(t)}{q(t)}\right)\right]^2 \mathrm{d}t\right)^{1/2},
\end{aligned} \tag{2.8.66}$$

where

$$L(x,y) = \frac{1}{2}\left(\int_0^x \left[\frac{\phi(P(s))}{P(s)}\right]^2 \mathrm{d}s\right)^{1/2}\left(\int_0^y \left[\frac{\psi(Q(t))}{Q(t)}\right]^2 \mathrm{d}t\right)^{1/2} \tag{2.8.67}$$

and "$'$" denotes the derivative of a function.

PROOF. From the hypotheses we have identities (2.8.63) and (2.8.64). From (2.8.63) and (2.8.64) and using Jensen's integral inequality (see [211]), we observe that

$$\begin{aligned}
\phi\big(f(s)\big) &= \phi\left(P(s)\int_0^s p(\sigma)\frac{f'(\sigma)}{p(\sigma)}\,\mathrm{d}\sigma \Big/ \int_0^s p(\sigma)\,\mathrm{d}\sigma\right) \\
&\leqslant \phi\big(P(s)\big)\phi\left(\int_0^s p(\sigma)\frac{f'(\sigma)}{p(\sigma)}\,\mathrm{d}\sigma \Big/ \int_0^s p(\sigma)\,\mathrm{d}\sigma\right) \\
&\leqslant \left[\frac{\phi(P(s))}{P(s)}\right]\int_0^s p(\sigma)\phi\left(\frac{f'(\sigma)}{p(\sigma)}\right)\mathrm{d}\sigma
\end{aligned} \tag{2.8.68}$$

and similarly,

$$\psi\big(g(t)\big) \leqslant \left[\frac{\psi(Q(t))}{Q(t)}\right]\int_0^t q(\tau)\psi\left(\frac{g'(\tau)}{q(\tau)}\right)\mathrm{d}\tau. \tag{2.8.69}$$

From (2.8.68) and (2.8.69) and using the elementary inequality $c^{1/2}d^{1/2} \leqslant (c+d)/2$ (for c, d nonnegative reals), we observe that

$$\begin{aligned}
&\phi\big(f(s)\big)\psi\big(g(t)\big) \\
&\quad\leqslant \left[\left[\frac{\phi(P(s))}{P(s)}\right]\int_0^s p(\sigma)\phi\left(\frac{f'(\sigma)}{p(\sigma)}\right)\mathrm{d}\sigma\right] \\
&\qquad\times \left[\left[\frac{\psi(Q(t))}{Q(t)}\right]\int_0^t q(\tau)\psi\left(\frac{g'(\tau)}{q(\tau)}\right)\mathrm{d}\tau\right] \\
&\quad\leqslant \left[\left[\frac{\phi(P(s))}{P(s)}\right](s)^{1/2}\left\{\int_0^s \left[p(s)\phi\left(\frac{f'(\sigma)}{p(\sigma)}\right)\right]^2\mathrm{d}\sigma\right\}^{1/2}\right] \\
&\qquad\times \left[\left[\frac{\psi(Q(t))}{Q(t)}\right](t)^{1/2}\left\{\int_0^t \left[q(\tau)\psi\left(\frac{g'(\tau)}{q(\tau)}\right)\right]^2\mathrm{d}\tau\right\}^{1/2}\right] \\
&\quad\leqslant \frac{1}{2}(s+t)\left[\left[\frac{\phi(P(s))}{P(s)}\right]\left\{\int_0^s \left[p(\sigma)\phi\left(\frac{f'(\sigma)}{p(\sigma)}\right)\right]^2\mathrm{d}\sigma\right\}^{1/2}\right] \\
&\qquad\times \left[\left[\frac{\psi(Q(t))}{Q(t)}\right]\left\{\int_0^t \left[q(\tau)\psi\left(\frac{g'(\tau)}{q(\tau)}\right)\right]^2\mathrm{d}\tau\right\}^{1/2}\right].
\end{aligned} \tag{2.8.70}$$

The rest of the proof can be completed by following the same steps as in the proof of Theorem 2.8.13, and hence we omit it here. □

REMARK 2.8.6. If we apply the elementary inequality $c^{1/2}d^{1/2} \leqslant (c+d)/2$ (for c, d nonnegative reals) on the right-hand sides of (2.8.62) and (2.8.66), we get respectively the following inequalities

$$\int_0^x \int_0^y \frac{f(s)g(t)}{s+t}\,\mathrm{d}s\,\mathrm{d}t \leqslant \frac{1}{4}\sqrt{xy}\left[\int_0^x (x-s)f'^2(s)\,\mathrm{d}s + \int_0^y (y-t)g'^2(t)\,\mathrm{d}t\right] \tag{2.8.71}$$

and

$$\int_0^x \int_0^y \frac{\phi(f(s))\psi(g(t))}{s+t}\,\mathrm{d}s\,\mathrm{d}t \leqslant \frac{1}{2}L(x,y)\left[\int_0^x (x-s)\left[p(s)\phi\left(\frac{f'(s)}{p(s)}\right)\right]^2 \mathrm{d}s + \int_0^y (y-t)\left[q(t)\psi\left(\frac{g'(t)}{q(t)}\right)\right]^2 \mathrm{d}t\right]. \tag{2.8.72}$$

The integral analogue of Theorem 2.8.7 established by Pachpatte in [353] is given in the following theorem.

THEOREM 2.8.15. *Let $p > 1, q > 1$ be constants and $\frac{1}{p} + \frac{1}{q} = 1$. Let $f(s)$ and $g(t)$ be real-valued continuous functions defined on $I_x = [0, x)$ and $I_y = [0, y)$, respectively, and $f(0) = g(0) = 0$. Then*

$$\int_0^x \int_0^y \frac{|f(s)||g(t)|}{qs^{p-1} + pt^{q-1}}\,\mathrm{d}s\,\mathrm{d}t \leqslant K(p,q,x,y)\left(\int_0^x (x-s)\left|f'(s)\right|^p \mathrm{d}s\right)^{1/p}\left(\int_0^y (y-t)\left|g'(t)\right|^q \mathrm{d}t\right)^{1/q} \tag{2.8.73}$$

for $x, y \in I_0 = (0, \infty)$, where

$$K(p,q,x,y) = \frac{1}{pq}x^{(p-1)/p}y^{(q-1)/q} \tag{2.8.74}$$

for $x, y \in I_0$.

PROOF. From the hypotheses we have the following identities

$$f(s) = \int_0^s f'(\sigma)\,\mathrm{d}\sigma \tag{2.8.75}$$

and

$$g(t) = \int_0^t g'(\tau)\,\mathrm{d}\tau, \tag{2.8.76}$$

for $s \in I_x, t \in I_y$. From (2.8.75) and (2.8.76) and using Hölder's integral inequality with indices $p, p/(p-1)$ and $q, q/(q-1)$, respectively, we have

$$\left|f(s)\right| \leqslant (s)^{(p-1)/p}\left(\int_0^s \left|f'(\sigma)\right|^p \mathrm{d}\sigma\right)^{1/p}, \tag{2.8.77}$$

$$\left|g(t)\right| \leqslant (t)^{(q-1)/q}\left(\int_0^t \left|g'(\tau)\right| \mathrm{d}\tau\right)^{1/q} \tag{2.8.78}$$

for $s \in I_x, t \in I_y$. From (2.8.77), (2.8.78) and using the elementary inequality (2.8.37), we observe that

$$\begin{aligned}
\left|f(s)\right|\left|g(t)\right| &\leqslant (s)^{(p-1)/p}(t)^{(q-1)/q}\left(\int_0^s \left|f'(\sigma)\right|^p \mathrm{d}\sigma\right)^{1/p}\left(\int_0^t \left|g'(\tau)\right|^q \mathrm{d}\tau\right)^{1/q}\\
&\leqslant \left[\frac{s^{p-1}}{p} + \frac{t^{q-1}}{q}\right]\left(\int_0^s \left|f'(\sigma)\right|^p \mathrm{d}\sigma\right)^{1/p}\left(\int_0^t \left|g'(\tau)\right|^q \mathrm{d}\tau\right)^{1/q}
\end{aligned}$$

for $s \in I_x, t \in I_y$. From the above inequality we observe that

$$\frac{|f(s)||g(t)|}{qs^{p-1} + pt^{q-1}} \leqslant \frac{1}{pq}\left(\int_0^s \left|f'(\sigma)\right|^p \mathrm{d}\sigma\right)^{1/p}\left(\int_0^t \left|g'(\tau)\right|^q \mathrm{d}\tau\right)^{1/q} \tag{2.8.79}$$

for $s \in I_x, t \in I_y$. Integrating both sides of (2.8.79) over t from 0 to y, first, and then integrating the resulting inequality over s from 0 to x and using Hölder's integral inequality with indices $p, p/(p-1)$ and $q, q/(q-1)$ we observe that

$$\begin{aligned}
&\int_0^x\int_0^y \frac{|f(s)||g(t)|}{qs^{p-1} + pt^{q-1}}\,\mathrm{d}s\,\mathrm{d}t\\
&\quad\leqslant \frac{1}{pq}\left\{\int_0^x\left(\int_0^s \left|f'(\sigma)\right|^p\right)^{1/p}\mathrm{d}s\right\}\left\{\int_0^y\left(\int_0^t \left|g'(\tau)\right|^q\right)^{1/q}\mathrm{d}t\right\}\\
&\quad\leqslant \frac{1}{pq}(x)^{(p-1)/p}\left\{\int_0^x\left(\int_0^s \left|f'(\sigma)\right|^p \mathrm{d}\sigma\right)\mathrm{d}s\right\}^{1/p}\\
&\qquad\times (y)^{(q-1)/q}\left\{\int_0^y\left(\int_0^t \left|g'(\tau)\right|^q \mathrm{d}\tau\right)\mathrm{d}t\right\}^{1/q}\\
&\quad= K(p,q,x,y)\left(\int_0^x (x-s)\left|f'(s)\right|^p \mathrm{d}s\right)^{1/p}\left(\int_0^y (y-t)\left|g'(t)\right|^q \mathrm{d}t\right)^{1/q}.
\end{aligned}$$

The proof is complete. □

REMARK 2.8.7. If we apply the elementary inequality (2.8.37) on the right-hand side of (2.8.73), then we get the following inequality

$$\int_0^x \int_0^y \frac{|f(s)||g(t)|}{qs^{p-1} + pt^{q-1}}\,\mathrm{d}s\,\mathrm{d}t \leqslant K(p,q,x,y)\left[\frac{1}{p}\int_0^x (x-s)\left|f'(s)\right|^p \mathrm{d}s + \frac{1}{q}\int_0^y (y-t)\left|g'(t)\right|^q \mathrm{d}t\right]. \tag{2.8.80}$$

In [342] Pachpatte establishes the following inequality similar to the integral analogue of Hilbert's inequality.

THEOREM 2.8.16. *Let $n \geqslant 1$ be an integer. Let $u \in C^n(I_x, \mathbb{R})$, $v \in C^n(I_y, \mathbb{R})$ and $u^{(i)}(0) = v^{(i)}(0) = 0$ for $i = 0, 1, 2, \ldots, n-1$, where $I_x = [0, x)$, $I_y = [0, y)$. Then*

$$\int_0^x \int_0^y \frac{|u^{(k)}(s)||v^{(k)}(t)|}{s^{2n-2k-1} + t^{2n-2k-1}}\,\mathrm{d}s\,\mathrm{d}t \leqslant M_1(n,k,x,y)\left(\int_0^x (x-s)\left|u^{(n)}(s)\right|^2 \mathrm{d}s\right)^{1/2} \times \left(\int_0^y (y-t)\left|v^{(n)}(t)\right|^2 \mathrm{d}t\right)^{1/2}, \tag{2.8.81}$$

where

$$M_1(n,k,x,y) = \frac{1}{2}\frac{\sqrt{xy}}{[(n-k-1)!]^2(2n-2k-1)}. \tag{2.8.82}$$

PROOF. From the hypotheses and Taylor expansion, we have

$$u^{(k)}(s) = \frac{1}{(n-k-1)!}\int_0^s (s-\sigma)^{n-k-1} u^{(n)}(\sigma)\,\mathrm{d}\sigma, \tag{2.8.83}$$

$$v^{(k)}(t) = \frac{1}{(n-k-1)!}\int_0^t (t-\tau)^{n-k-1} v^{(n)}(\tau)\,\mathrm{d}\tau, \tag{2.8.84}$$

for $s \in I_x, t \in I_y$. From (2.8.83) and using the Schwarz integral inequality, we have

$$\begin{aligned}
\left|u^{(k)}(s)\right| &\leqslant \frac{1}{(n-k-1)!}\int_0^s (s-\sigma)^{n-k-1}\left|u^{(n)}(\sigma)\right| \mathrm{d}\sigma \\
&\leqslant \frac{1}{(n-k-1)!}\left(\int_0^s (s-\sigma)^{2(n-k-1)}\,\mathrm{d}\sigma\right)^{1/2}\left(\int_0^s \left|u^{(n)}(\sigma)\right|^2 \mathrm{d}\sigma\right)^{1/2} \\
&= \frac{1}{(n-k-1)!}\frac{s^{(2n-2k-1)/2}}{(2n-2k-1)^{1/2}}\left(\int_0^s \left|u^{(n)}(\sigma)\right|^2 \mathrm{d}\sigma\right)^{1/2} \qquad (2.8.85)
\end{aligned}$$

for $s \in I_x$. Similarly, from (2.8.84) and using the Schwarz integral inequality, we have

$$\left|v^{(k)}(t)\right| \leqslant \frac{1}{(n-k-1)!}\frac{t^{(2n-2k-1)/2}}{(2n-2k-1)^{1/2}}\left(\int_0^t \left|v^{(n)}(\tau)\right|^2 \mathrm{d}\tau\right)^{1/2} \qquad (2.8.86)$$

for $t \in I_y$. From (2.8.85), (2.8.86) and using the elementary inequality $c^{1/2}d^{1/2} \leqslant (c+d)/2$ (for c, d nonnegative reals), we have

$$\begin{aligned}
\left|u^{(k)}(s)\right|\left|v^{(k)}(t)\right| &\leqslant \frac{1}{2}\frac{1}{[(n-k-1)!]^2(2n-2k-1)}\left[s^{2n-2k-1}+t^{2n-2k-1}\right] \\
&\quad\times\left(\int_0^s \left|u^{(n)}(\sigma)\right|^2 \mathrm{d}\sigma\right)^{1/2}\left(\int_0^t \left|v^{(n)}(\tau)\right|^2 \mathrm{d}\tau\right)^{1/2} \qquad (2.8.87)
\end{aligned}$$

for $s \in I_x, t \in I_y$. Rewriting (2.8.87) and then integrating over t from 0 to y, first, and then integrating the resulting inequality over s from 0 to x and using the Schwarz integral inequality we have

$$\begin{aligned}
&\int_0^x\int_0^y \frac{|u^{(k)}(s)||v^{(k)}(t)|}{s^{2n-2k-1}+t^{2n-2k-1}}\,\mathrm{d}s\,\mathrm{d}t \\
&\quad\leqslant \frac{1}{2}\frac{1}{[(n-k-1)!]^2(2n-2k-1)} \\
&\quad\times\left(\int_0^x\left(\int_0^s \left|u^{(n)}(\sigma)\right|^2 \mathrm{d}\sigma\right)^{1/2}\mathrm{d}s\right)\left(\int_0^y\left(\int_0^t \left|v^{(n)}(\tau)\right|^2 \mathrm{d}\tau\right)^{1/2}\mathrm{d}t\right)
\end{aligned}$$

$$\leqslant \frac{1}{2}\frac{1}{[(n-k-1)!]^2(2n-2k-1)}\sqrt{x}\left(\int_0^x\left(\int_0^s\left|u^{(n)}(\sigma)\right|^2 \mathrm{d}\sigma\right)\mathrm{d}s\right)^{1/2}$$

$$\times\sqrt{y}\left(\int_0^y\left(\int_0^t\left|v^{(n)}(\tau)\right|^2 \mathrm{d}\tau\right)\mathrm{d}t\right)^{1/2}$$

$$= M_1(n,x,y)\left(\int_0^x (x-s)\left|u^{(n)}(s)\right|^2 \mathrm{d}s\right)^{1/2}\left(\int_0^y (y-t)\left|v^{(n)}(t)\right|^2 \mathrm{d}t\right)^{1/2}.$$

The proof is complete. □

REMARK 2.8.8. In the special case when $k=0$, inequality (2.8.81) reduces to the following inequality

$$\int_0^x\int_0^y \frac{|u(s)||v(t)|}{s^{2n-1}+t^{2n-1}}\,\mathrm{d}s\,\mathrm{d}t$$
$$\leqslant M_1(n,0,x,y)\left(\int_0^x (x-s)\left|u^{(n)}(s)\right|^2 \mathrm{d}s\right)^{1/2}\left(\int_0^y (y-t)\left|v^{(n)}(t)\right|^2 \mathrm{d}t\right)^{1/2}, \tag{2.8.88}$$

and, by taking $n=1$, inequality (2.8.88) reduces to the slight variant of the inequality given in Theorem 2.8.13. If we apply the elementary inequality $c^{1/2}d^{1/2} \leqslant (c+d)/2$ (for c,d nonnegative reals) on the right-hand side of (2.8.81), then we get the following inequality

$$\int_0^x\int_0^y \frac{|u^{(k)}(s)||v^{(k)}(t)|}{s^{2n-2k-1}+t^{2n-2k-1}}\,\mathrm{d}s\,\mathrm{d}t$$
$$\leqslant \frac{1}{2}M_1(n,k,x,y)\left[\int_0^x (x-s)\left|u^{(n)}(s)\right|^2 \mathrm{d}s+\int_0^y (y-t)\left|v^{(n)}(t)\right|^2 \mathrm{d}t\right]. \tag{2.8.89}$$

The integral analogue of the inequality in Theorem 2.8.8 established by Pachpatte in [353] is given in the following theorem.

In what follows, we use the notations $I=[0,\infty)$, $I_0=(0,\infty)$, $I_\beta=[0,\beta)$, $\beta\in I_0$, denotes the subintervals of $\mathbb{R}$. For any function $u: I\times I\to\mathbb{R}$ we denote the partial derivatives $\frac{\partial}{\partial s}u(s,t)$, $\frac{\partial}{\partial t}u(s,t)$ and $\frac{\partial^2}{\partial s\,\partial t}u(s,t)$ by $D_1u(s,t)$, $D_2u(s,t)$ and $D_2D_1u(s,t)=D_1D_2u(s,t)$, respectively.

THEOREM 2.8.17. *Let $p>1$, $q>1$ be constants and $\frac{1}{p}+\frac{1}{q}=1$. Let $f(s,t)$ and $g(s,t)$ be real-valued continuous functions defined on $I_x\times I_y$ and $I_z\times I_w$,* *respectively, and* $f(0,t) = g(0,t) = 0$, $f(s,0) = g(s,0) = 0$. *Then*

$$\int_0^x \int_0^y \left(\int_0^z \int_0^w \frac{|f(s,t)||g(k,r)|}{q(st)^{p-1} + p(kr)^{q-1}} \,dk\,dr \right) ds\,dt$$
$$\leqslant C(p,q,x,y,z,w) \left(\int_0^x \int_0^y (x-s)(y-t) \left| D_2 D_1 f(s,t) \right|^p ds\,dt \right)^{1/p}$$
$$\times \left(\int_0^z \int_0^w (z-k)(w-r) \left| D_2 D_1 g(k,r) \right|^q dk\,dr \right)^{1/q} \tag{2.8.90}$$

for $x, y, z, w \in I_0$, *where*

$$C(p,q,x,y,z,w) = \frac{1}{pq} (xy)^{(p-1)/p} (zw)^{(q-1)/q} \tag{2.8.91}$$

for $x, y, z, w \in I_0$.

PROOF. From the hypotheses we have the following identities

$$f(s,t) = \int_0^s \int_0^t D_2 D_1 f(\xi,\eta)\,d\xi\,d\eta, \tag{2.8.92}$$

$$g(k,r) = \int_0^k \int_0^r D_2 D_1 g(\sigma,\tau)\,d\sigma\,d\tau, \tag{2.8.93}$$

for $(s,t) \in I_x \times I_y$, $(k,r) \in I_z \times I_w$. From (2.8.92), (2.8.93) and using Hölder's integral inequality with indices $p, p/(p-1)$ and $q, q/(q-1)$, respectively, we observe that

$$\left| f(s,t) \right| \leqslant (st)^{(p-1)/p} \left(\int_0^s \int_0^t \left| D_2 D_1 f(\xi,\eta) \right|^p d\xi\,d\eta \right)^{1/p}, \tag{2.8.94}$$

$$\left| g(k,r) \right| \leqslant (kr)^{(q-1)/q} \left(\int_0^k \int_0^r \left| D_2 D_1 g(\sigma,\tau) \right|^q d\sigma\,d\tau \right)^{1/q}, \tag{2.8.95}$$

for $(s,t) \in I_x \times I_y$, $(k,r) \in I_z \times I_w$. From (2.8.94) and (2.8.95) and using the elementary inequality (2.8.37), it is easy to observe that

$$\frac{|f(s,t)||g(k,r)|}{q(st)^{p-1} + p(kr)^{q-1}} \leqslant \frac{1}{pq} \left(\int_0^s \int_0^t \left| D_2 D_1 f(\xi,\eta) \right|^p d\xi\,d\eta \right)^{1/p}$$
$$\times \left(\int_0^k \int_0^r \left| D_2 D_1 g(\sigma,\tau) \right|^q d\sigma\,d\tau \right)^{1/q} \tag{2.8.96}$$

for $(s,t) \in I_x \times I_y$, $(k,r) \in I_z \times I_w$. Integrating both sides of (2.8.96) first over r from 0 to w and over k from 0 to z and then integrating both sides of the resulting inequality over t from 0 to y and over s from 0 to x and using Hölder's inequality with indices $p, p/(p-1)$ and $q, q/(q-1)$ and Fubini's theorem we observe that

$$\begin{aligned}
&\int_0^x\int_0^y\left(\int_0^z\int_0^w \frac{|f(s,t)||g(k,r)|}{q(st)^{p-1}+p(kr)^{q-1}}\,\mathrm{d}k\,\mathrm{d}r\right)\mathrm{d}s\,\mathrm{d}t\\
&\leqslant \frac{1}{pq}\left[\int_0^x\int_0^y\left(\int_0^s\int_0^t \left|D_2D_1f(\xi,\eta)\right|^p\,\mathrm{d}\xi\,\mathrm{d}\eta\right)^{1/p}\mathrm{d}s\,\mathrm{d}t\right]\\
&\quad\times\left[\int_0^z\int_0^w\left(\int_0^k\int_0^r \left|D_2D_1g(\sigma,\tau)\right|^q\,\mathrm{d}\sigma\,\mathrm{d}\tau\right)^{1/q}\mathrm{d}k\,\mathrm{d}r\right]\\
&\leqslant \frac{(xy)^{(p-1)/p}}{pq}\left(\int_0^x\int_0^y\left(\int_0^s\int_0^t \left|D_2D_1f(\xi,\eta)\right|^p\,\mathrm{d}\xi\,\mathrm{d}\eta\right)\mathrm{d}s\,\mathrm{d}t\right)^{1/p}\\
&\quad\times(zw)^{(q-1)/q}\left(\int_0^z\int_0^w\left(\int_0^k\int_0^r \left|D_2D_1g(\sigma,\tau)\right|^q\,\mathrm{d}\sigma\,\mathrm{d}\tau\right)\mathrm{d}k\,\mathrm{d}r\right)^{1/q}\\
&= C(p,q,x,y,z,w)\left(\int_0^x\int_0^y (x-s)(y-t)\left|D_2D_1f(s,t)\right|^p\,\mathrm{d}s\,\mathrm{d}t\right)^{1/p}\\
&\quad\times\left(\int_0^z\int_0^w (z-k)(w-r)\left|D_2D_1g(k,r)\right|^q\,\mathrm{d}k\,\mathrm{d}r\right)^{1/q}.
\end{aligned}$$

The proof is complete. □

REMARK 2.8.9. By using the elementary inequality (2.8.37) on the right-hand side of (2.8.90), we get the following inequality

$$\begin{aligned}
&\int_0^x\int_0^y\left(\int_0^z\int_0^w \frac{|f(s,t)||g(k,r)|}{q(st)^{p-1}+p(kr)^{q-1}}\,\mathrm{d}k\,\mathrm{d}r\right)\mathrm{d}s\,\mathrm{d}t\\
&\leqslant C(p,q,x,y,z,w)\left[\frac{1}{p}\int_0^x\int_0^y (x-s)(y-t)\left|D_2D_1f(s,t)\right|^p\,\mathrm{d}s\,\mathrm{d}t\right.\\
&\qquad\left.+\frac{1}{q}\int_0^z\int_0^w (z-k)(w-r)\left|D_2D_1g(k,r)\right|^q\,\mathrm{d}k\,\mathrm{d}r\right].
\end{aligned}\tag{2.8.97}$$

2.9 Miscellaneous Inequalities

2.9.1 Hardy and Littlewood [139]

Suppose that $a_n \geqslant 0$, $n = 1, 2, \ldots$, and c is a real number. Set

$$A_{m,n} = \sum_{v=m}^{n} a_v.$$

If $p > 1$ we have

$$\sum_{n=1}^{\infty} n^{-c} A_{1,n}^p \leqslant K \sum_{n=1}^{\infty} n^{-c} (na_n)^p \quad \text{with } c > 1,$$

$$\sum_{n=1}^{\infty} n^{-c} A_{n,\infty}^p \leqslant K \sum_{n=1}^{\infty} n^{-c} (na_n)^p \quad \text{with } c < 1,$$

and if $0 < p < 1$ we have

$$\sum_{n=1}^{\infty} n^{-c} A_{1,n}^p \geqslant K \sum_{n=1}^{\infty} n^{-c} (na_n)^p \quad \text{with } c > 1,$$

$$\sum_{n=1}^{\infty} n^{-c} A_{n,\infty}^p \geqslant K \sum_{n=1}^{\infty} n^{-c} (na_n)^p \quad \text{with } c < 1,$$

where K denotes a positive absolute constant, not necessary the same at each occurrence.

2.9.2 Leindler [186]

Let $a_n \geqslant 0$ and $\lambda_n \geqslant 0$, $n = 1, 2, \ldots$, be given. Let $v_1 < \cdots < v_n < \cdots$ denote the indices for which $\lambda_{v_n} > 0$. Let N denote the number of the positive terms of the sequence λ_n provided this number is finite; in the contrary case, set $N = \infty$. Set $v_0 = 0$, and if $N < \infty$ then $v_{N+1} = \infty$. Using the notations

$$A_{m,n} = \sum_{i=m}^{n} a_i \quad \text{and} \quad \Lambda_{m,n} = \sum_{i=m}^{n} \lambda_i, \quad 1 \leqslant m \leqslant n \leqslant \infty,$$

we have the following inequalities

$$\sum_{n=1}^{\infty} \lambda_n A_{1,n}^p \leqslant p^p \sum_{n=1}^{N} \lambda_{v_n}^{1-p} \Lambda_{v_n,\infty}^p A_{v_{n-1}+1,v_n}^p,$$

$$\sum_{n=1}^{\infty} \lambda_n A_{n,\infty}^p \leqslant p^p \sum_{n=1}^{N} \lambda_{v_n}^{1-p} \Lambda_{1,v_n}^p A_{v_n,v_{n+1}-1}^p,$$

for $p \geqslant 1$ (the constant p^p being the best possible one), and

$$\sum_{n=1}^{N} \lambda_{v_n}^{1-p} \Lambda_{v_n,\infty}^p A_{v_{n-1}+1,v_n}^p \leqslant 8 \sum_{n=1}^{\infty} \lambda_n A_{1,n}^p,$$

$$\sum_{n=1}^{N} \lambda_{v_n}^{1-p} \Lambda_{1,v_n}^p A_{v_n,v_{n+1}-1}^p \leqslant 9 \sum_{n=1}^{\infty} \lambda_n A_{n,\infty}^p,$$

for $0 < p < 1$.

2.9.3 Izumi, Izumi and Petersen [163]

Let $p > 1$, $a_m \geqslant 0$, $m = 1, 2, \ldots$, $g(m) > 0$, $m = 1, 2, \ldots$, and $C = (c_{m,k})$ be a positive triangular matrix (i.e., $c_{m,k} = 0$ for $k > m$ and $c_{m,k} > 0$ for $k \leqslant m$, $m = 1, 2, \ldots$). If

$$\sum_{m=1}^{\infty} c_{m,m} < \infty$$

and

$$\sum_{m=n}^{\infty} g(m) c_{m,n}^p \leqslant A_1 g(n) c_{n,n}^{p-1} \quad \text{for all } n \geqslant 1,$$

then

$$\sum_{m=1}^{\infty} g(m) \left(\sum_{n=1}^{m} c_{m,n} a_n \right)^p \leqslant A_2 \sum_{m=1}^{\infty} g(m) a_m^p,$$

where A_1, A_2 are constants independent of the terms under the summation sign.

2.9.4 Izumi, Izumi and Petersen [163]

Let $a_m \geqslant 0$, $m = 1, 2, \ldots$, $g(m) > 0$, $m = 1, 2, \ldots$, and $C = (c_{m,k})$ be a positive triangular matrix which satisfies the conditions

$$\sum_{m=v}^{\infty} g(m) c_{m,\mu} c_{m,v} \leqslant A_1 g(v) c_{v,\mu} \quad \text{for all } \mu \geqslant v,$$

then

$$\sum_{m=1}^{\infty} g(m) \left(\sum_{v=1}^{m} c_{m,v} a_v \right)^2 \leqslant A_2 \sum_{m=1}^{\infty} g(m) a_m^2,$$

where A_1, A_2 are constants independent of the terms under the summation sign.

2.9.5 Love [196]

If $p > 1$, $\alpha(t) \geqslant 0$, $\alpha(t)$ is decreasing in $(0, 1]$, $t\alpha(t)$ is increasing in $[1, \infty)$,

$$c = \int_0^t \alpha(t) t^{-1/p}\,\mathrm{d}t < \infty, \qquad \lambda_n > 0, \qquad \Lambda_m = \sum_{n=1}^{m} \lambda_n,$$

$$|a_{mn}| \leqslant \frac{\lambda_n}{\Lambda_m} \alpha\left(\frac{\Lambda_n}{\Lambda_m}\right) \quad \text{for } m \neq n, \qquad |a_{nn}| \leqslant 2 \frac{\lambda_n}{\Lambda_n} \alpha(1),$$

then

$$\left(\sum_{m=1}^{\infty} \left| \sum_{n=1}^{\infty} a_{mn} x_n \right|^p \right)^{1/p} \leqslant c \left(\sum_{m=1}^{\infty} \lambda_m |x_m|^p \right)^{1/p},$$

where $\{x_n\}$ is an arbitrary fixed sequence.

2.9.6 Love [196]

If $q \geqslant p > 1$, $1 - (p^{-1} - q^{-1}) \leqslant r^{-1} \leqslant 1$, $\alpha(t)$ is nonnegative and decreasing in $(0, 1]$,

$$L = \left(\int_0^1 \alpha(t)^r t^{-r/q}\,\mathrm{d}t \right)^{1/r} < \infty, \qquad \lambda_n > 0, \qquad \Lambda_m = \sum_{n=1}^{m} \lambda_n$$

and

$$|a_{mn}| \leqslant \frac{\lambda_n^{1/p-1/q+1/r}}{\lambda_m^{1/r}} \alpha\left(\frac{\Lambda_n}{\Lambda_m}\right) \quad \text{for } 0 < n \leqslant m,$$

then

$$\left(\sum_{m=1}^{\infty}\lambda_m\left|\sum_{n=1}^{m}a_{mn}x_n\right|^q\right)^{1/q}\leqslant L\left(\sum_{n=1}^{\infty}\lambda_m|x_m|^p\right)^{1/p},$$

where $\{x_n\}$ is a fixed sequence.

2.9.7 Pachpatte [307]

If $p>1$ is a constant, $a(n)\geqslant 0$ for $n\in\mathbb{N}$, the set of natural numbers, and

$$A(n)=\frac{1}{n}\sum_{m_1=1}^{n}\frac{1}{m_1}\sum_{m_2=1}^{m_1}\cdots\sum_{m_{r-1}=1}^{m_{r-2}}\frac{1}{m_{r-1}}\sum_{m_r=1}^{m_{r-1}}a(m_r)$$

for $n\in\mathbb{N}$ with $m_0=n$, then

$$\sum_{n=1}^{\infty}A^p(n)\leqslant\left(\frac{p}{p-1}\right)^{rp}\sum_{n=1}^{\infty}a^p(n). \tag{2.9.1}$$

Equality holds in (2.9.1) if $a(n)=0$ for $n\in\mathbb{N}$.

2.9.8 Pachpatte [292]

Let $\mathbb{R}$ denote the set of real numbers and B be a subset of $\mathbb{R}^n$ defined by $B=\{x\in\mathbb{R}^n\colon \mathbf{1}\leqslant x<\infty\}$ where $\mathbf{1}=(1,\dots,1)\in\mathbb{R}^n$. For a function $u:B\to\mathbb{R}$ denote

$$\sum_B u(y)=\sum_{y_1=1}^{\infty}\cdots\sum_{y_n=1}^{\infty}u(y_1,\dots,y_n)$$

and

$$\sum_{B_{1,x}}u(y)=\sum_{y_1=1}^{x_1}\cdots\sum_{y_n=1}^{x_n}u(y_1,\dots,y_n),$$

where $\mathbf{1}=(1,\dots,1)\in B$, $x=(x_1,\dots,x_n)\in B$ such that $\mathbf{1}\leqslant x$, that is, $1\leqslant x_i$. Assume that all inequalities between vectors are componentwise and all the sums exist on the respective domains of their definitions and the value of any function $u(x_1,\dots,x_n)$ with any of its component zero is equal to 0. If $p>1$ is a constant, $f(x)\geqslant 0$ for $x\subset B$ and

$$A(x)=\sum_{B_{1,x}}f(y),\quad x\in B, \tag{2.9.2}$$

then

$$\sum_B \left(\frac{A(x)}{\prod_{i=1}^n x_i} \right) \leqslant \left(\frac{p}{p-1} \right)^{np} \sum_B f^p(x). \tag{2.9.3}$$

Equality holds in (2.9.3) if $f(x) = 0$ for all $x_i, i = 1, 2, \ldots, n$.

2.9.9 Pachpatte [277]

Let $f_j(u)$, $j = 1, 2$, be real-valued positive convex functions defined for $u > 0$. Let $p_j \geqslant 1$, $j = 1, 2$, be constants, $\lambda_n > 0$, $a_n^{(j)} > 0$, $j = 1, 2$, $\sum_{n=1}^{\infty} \lambda_n \times f_j^{p_1+p_2}(a_n^j)$, $j = 1, 2$, converge, and further let $\Lambda_n = \sum_{i=1}^n \lambda_i$, $A_n^{(j)} = \sum_{i=1}^n \lambda_i \times a_i^{(j)}$, $j = 1, 2$. Then

$$\begin{aligned}
&\sum_{n=1}^{\infty} \lambda_n f_1^{p_1}\left(\frac{A_n^{(1)}}{\Lambda_n} \right) f_2^{p_2}\left(\frac{A_n^{(2)}}{\Lambda_n} \right) \\
&\quad \leqslant \left(\frac{p_1 + p_2}{p_1 + p_2 - 1} \right)^{p_1+p_2} \\
&\qquad \times \left[\left(\frac{p_1}{p_1 + p_2} \right) \sum_{n=1}^{\infty} \lambda_n f_1^{p_1+p_2}(a_n^{(1)}) + \left(\frac{p_2}{p_1 + p_2} \right) \sum_{n=1}^{\infty} \lambda_n f_2^{p_1+p_2}(a_n^{(2)}) \right].
\end{aligned}$$

2.9.10 Pachpatte [277]

(i) Let $f_j(u)$, $j = 1, 2, 3$, be real-valued positive convex functions defined for $u > 0$. Let $p_j \geqslant 1$, $j = 1, 2, 3$, be constants, $\lambda_n > 0$, $a_n^{(j)} > 0$, $j = 1, 2, 3$, $\sum_{n=1}^{\infty} \lambda_n f_j^{2p_j}(a_n^{(j)})$, $j = 1, 2, 3$, converge, and further let $\Lambda_n = \sum_{i=1}^n \lambda_i$, $A_n^{(j)} = \sum_{i=1}^n \lambda_i a_i^{(j)}$, $j = 1, 2, 3$. Then

$$\begin{aligned}
&\sum_{n=1}^{\infty} \lambda_n \left[f_1^{p_1}\left(\frac{A_n^{(1)}}{\Lambda_n} \right) f_2^{p_2}\left(\frac{A_n^{(2)}}{\Lambda_n} \right) \right. \\
&\qquad \left. + f_2^{p_2}\left(\frac{A_n^{(2)}}{\Lambda_n} \right) f_3^{p_3}\left(\frac{A_n^{(3)}}{\Lambda_n} \right) + f_3^{p_3}\left(\frac{A_n^{(3)}}{\Lambda_n} \right) f_1^{p_1}\left(\frac{A_n^{(1)}}{\Lambda_n} \right) \right] \\
&\quad \leqslant \sum_{j=1}^{3} \left(\frac{2p_j}{2p_j - 1} \right)^{2p_j} \sum_{n=1}^{\infty} \lambda_n f_j^{2p_j}(a_n^{(j)}).
\end{aligned}$$

(ii) Let $f_j, p_j, \Lambda_n, A_n^{(j)}$ be as in (i) and $\sum_{n=1}^{\infty} \lambda_n f_j^{4p_j}(a_n^{(j)})$, $j = 1, 2, 3$, converge. Then

$$\sum_{n=1}^{\infty} \lambda_n f_1^{p_1}\left(\frac{A_n^{(1)}}{\Lambda_n}\right) f_2^{p_2}\left(\frac{A_n^{(2)}}{\Lambda_n}\right) f_3^{p_3}\left(\frac{A_n^{(3)}}{\Lambda_n}\right)$$
$$\times \left[f_1^{p_1}\left(\frac{A_n^{(1)}}{\Lambda_n}\right) + f_2^{p_2}\left(\frac{A_n^{(2)}}{\Lambda_n}\right) + f_3^{p_3}\left(\frac{A_n^{(3)}}{\Lambda_n}\right)\right]$$
$$\leqslant \sum_{j=1}^{3}\left(\frac{4p_j}{4p_j - 1}\right)^{4p_j} \sum_{n=1}^{\infty} \lambda_n f_j^{4p_j}(a_n^{(j)}).$$

2.9.11 Pachpatte [277]

Let $f_j(u)$, $j = 1, \ldots, m$, be real-valued positive convex functions defined for $u > 0$. Let $p_j > 1$, $j = 1, \ldots, m$, be constants, $\lambda_n > 0$, $a_n^{(j)} > 0$, $j = 1, \ldots, m$, $\sum_{n=1}^{\infty} \lambda_n f_j^{mp_j}(a_n^{(j)})$, $j = 1, \ldots, m$, converge, and further let $\Lambda_n = \sum_{i=1}^{n} \lambda_i$, $A_n^{(j)} = \sum_{i=1}^{n} \lambda_i a_i^{(j)}$, $j = 1, \ldots, m$. Then

$$\sum_{n=1}^{\infty} \lambda_n \prod_{j=1}^{m} f_j^{p_j}\left(\frac{A_n^{(j)}}{\Lambda_n}\right) \leqslant \frac{1}{m} \sum_{j=1}^{m}\left(\frac{mp_j}{mp_j - 1}\right)^{mp_j} \sum_{n=1}^{\infty} \lambda_n f_j^{mp_j}(a_n^{(j)}).$$

2.9.12 Pachpatte [305]

Let $H(u)$ be a real-valued nonnegative convex function defined for $u > 0$. Let $\lambda_n > 0$, $a_n \geqslant 0$ and

$$\Lambda_n = \sum_{i=1}^{n} \lambda_i, \qquad A_n = \sum_{i=1}^{n} \lambda_i a_i, \qquad Q_n = \sum_{i=1}^{n} \lambda_i H(a_i).$$

(i) If $p \geqslant 0, q > 1$ be real constants and $\sum_{n=1}^{\infty} \lambda_n [Q_n/\Lambda_n]^p [H(a_n)]^q < \infty$, then

$$\sum_{n=1}^{\infty} \lambda_n \left[H\left(\frac{A_n}{\Lambda_n}\right)\right]^{p+q} \leqslant M^q \sum_{n=1}^{\infty} \lambda_n \left[\frac{Q_n}{\Lambda_n}\right]^p [H(a_n)]^q,$$

where $M = (p+q)/(p+q-1)$.

(ii) If p, q, r be nonnegative real constants such that $q+r>1$ and $\sum_{n=1}^{\infty} \lambda_n \times [Q_n/\Lambda_n]^p [H(a_n)]^{q+r} < \infty$, then

$$\sum_{n=1}^{\infty} \lambda_n \left[H\left(\frac{A_n}{\Lambda_n}\right)\right]^{p+q} \left[H(a_n)\right]^r \leqslant B^q \sum_{n=1}^{\infty} \lambda_n \left[\frac{Q_n}{\Lambda_n}\right]^p \left[H(a_n)\right]^{q+r},$$

where $B = (p+q+r)/(p+q+r-1)$.

2.9.13 Bennett [23]

Let $p, q, r \geqslant 1$. If $\{a_n\}_{n=1}^{\infty}$ is a sequence of nonnegative numbers with partial sums $A_n = a_1 + a_2 + \cdots + a_n$, then

$$\sum_n a_n^p A_n^q \left(\sum_{m \geqslant n} a_m^{1+p/q}\right)^r \leqslant \left(\frac{p(q+r)-q}{p}\right)^r \sum_n \left(a_n^p A_n^q\right)^{1+r/q}.$$

2.9.14 Alzer [9]

Let $a_1, \ldots, a_N$ be nonnegative real numbers such that $a_1 \leqslant a_2 \leqslant \cdots \leqslant a_N$ and let $A_n = \sum_{i=1}^{n} a_i$. If $p \geqslant 1$, $q>0$, $r>0$ are real numbers such that

$$d = \frac{p(q+r)-q}{p} \geqslant k,$$

where $k \geqslant 1$ is an integer, then

$$\sum_{n=1}^{N} a_n^p A_n^q \left[\sum_{m=n}^{N} a_m^{1+p/q}\right]^r \leqslant \prod_{i=0}^{k-1} (d-i)^{r/k} \sum_{n-1}^{N} \left(a_n^p A_n^q\right)^{1+r/q}.$$

If $k=1$ then assumption $a_1 \leqslant a_2 \leqslant \cdots \leqslant a_N$ can be dropped.

2.9.15 Cochran and Lee [65]

If γ and p are constants with $\gamma \geqslant 0$ and $p \geqslant 1$, and $0 \leqslant x_n \leqslant 1$, then

$$\sum_{m=1}^{\infty} m^{\gamma} \left(\prod_{n=1}^{m} x_n^{n^{p-1}}\right)^{p/m^p} \leqslant \mathrm{e}^{(\gamma+1)p} \sum_{m=1}^{\infty} m^{\gamma} x_m.$$

2.9.16 Andersen and Heinig [14]

Let $\{K(m,n)\}$ be a nonnegative double sequence defined in $D = \{(m,n) \in \mathbb{Z} \times \mathbb{Z}\colon n \leqslant m\}$ such that $K(m,n)$ is nonincreasing in m and nondecreasing in n where $\mathbb{Z}$ denote the set of integers.

(i) If $1 \leqslant p \leqslant q < \infty$ and $\{u_n\}$, $\{v_n\}$ are nonnegative sequences such that, for some β, $0 \leqslant \beta \leqslant 1$, and all integers γ,

$$\left(\sum_{n=r}^{\infty}\{K(n,r)\}^{\beta q}u_n^q\right)^{1/q}\left(\sum_{n=-\infty}^{r}\{K(r,n)\}^{(1-\beta)p'}v_n^{-p'}\right)^{1/p'} \leqslant C < \infty, \tag{2.9.4}$$

where p' denotes the conjugate index of p, then for all sequences $\{a_n\}$,

$$\left(\sum_{n=-\infty}^{\beta}\left|u_n\sum_{m=-\infty}^{n}K(n,m)a_m\right|^q\right)^{1/q} \leqslant AC\left(\sum_{n=-\infty}^{\infty}|v_na_n|^p\right)^{1/p}.$$

In case $p = 1$, the second sum in (2.9.4) is replaced in the usual way by the supremum for $n \leqslant r$.

(ii) If $1 \leqslant p \leqslant q \leqslant \infty$ and $\{u_n\}$, $\{v_n\}$ are nonnegative sequences such that, for some β, $0 \leqslant \beta \leqslant 1$, and all integers r,

$$\left(\sum_{n=-\infty}^{r}\{K(r,n)\}^{\beta q}u_n^q\right)^{1/q}\left(\sum_{n=r}^{\infty}\{K(n,r)\}^{(1-\beta)p'}v_n^{-p'}\right)^{1/p'} \leqslant C^{\cdot} < \infty,$$

where p' denotes the conjugate index of p, then for all sequences $\{a_n\}$,

$$\left(\sum_{n=-\infty}^{\infty}\left|u_n\sum_{m=n}^{\infty}K(m,n)a_m\right|^q\right)^{1/q} \leqslant AC^{\cdot}\left(\sum_{n=-\infty}^{\infty}|v_na_n|^p\right)^{1/p}.$$

2.9.17 Pachpatte [309]

Let $\alpha \geqslant 0$, $p \geqslant 0$, $q \geqslant 1$ be real constants. Let $\mathbb{N} = \{1,2,\ldots\}$, $\mathbb{N}_0 = \{0,1,2,\ldots\}$, $\mathbb{N}_{0,m} = \{0,1,2,\ldots,m\}$ for some fixed $m \in \mathbb{N}$ and Δ is the forward difference operator. Let $\{u_n\}$, $n \in \mathbb{N}_{0,m}$, be a sequence of real numbers. Then

$$\sum_{n=0}^{m-1}n^{\alpha}|u_n|^{p+q} \leqslant M\left\{\sum_{n=0}^{m-1}(n+1)^{\alpha+q}|u_n|^p\left(\frac{|u_n|}{m}+|\Delta u_n|\right)^q\right\}^{1/q}$$

$$\times\left\{\sum_{n=0}^{m-1}(n+1)^{\alpha}|u_n|^{p+q}\right\}^{1/q'},$$

$$\sum_{n=0}^{m-1}n^{\alpha}|u_n|^{p+q}\leqslant M\left\{\sum_{n=0}^{m-1}(n+1)^{\alpha+1}|u_n|^{p}\left(\frac{|u_n|}{m}+|\Delta u_n|\right)^{q}\right\}^{1/q}$$

$$\times\left\{\sum_{n=0}^{m-1}(n+1)^{\alpha+1}|u_n|^{p+q}\right\}^{1/q'},$$

$$\sum_{n=0}^{m-1}n^{\alpha}|u_n|^{p+q}\leqslant M\left\{\sum_{n=0}^{m-1}(n+1)^{(\alpha+1)q}|u_n|^{p}\left(\frac{|u_n|}{m}+|\Delta u_n|\right)^{q}\right\}^{1/q}$$

$$\times\left\{\sum_{n=0}^{m-1}|u_n|^{p+q}\right\}^{1/q'},$$

$$\sum_{n=0}^{m-1}n^{\alpha}|u_n|^{p+q}\leqslant M\left\{\sum_{n=0}^{m-1}(n+1)^{\alpha(q-1)}|u_n|^{p}\left(\frac{|u_n|}{m}+|\Delta u_n|\right)^{q}\right\}^{1/q}$$

$$\times\left\{\sum_{n=0}^{m-1}(n+1)^{\alpha(q'-1)+q'}|u_n|^{p+q}\right\}^{1/q'},$$

where

$$M=\max\left\{\frac{\alpha+2}{\alpha+1},\frac{p+q}{\alpha+1}\right\}\quad\text{and}\quad q'=\frac{q}{q-1}.$$

2.9.18 Pachpatte [303]

Let $p>1$, $\lambda_n>0$, $a_n>0$, $n=1,2,\ldots$, $A_n=\sum_{i=1}^{n}\lambda_i a_i$, $\Lambda_n=\sum_{i=1}^{n}\lambda_i$ and that $\sum_{n=1}^{\infty}\lambda_n a_n^p$ converges. Then

$$\sum_{n=1}^{\infty}\lambda_n\left(\frac{\nabla A_n^p}{\Lambda_n^{p-1}}\right)\leqslant p\left(\frac{p}{p-1}\right)^{p-1}\sum_{n=1}^{\infty}\lambda_n a_n^p,$$

where $\nabla A_n^p=A_n^p-A_{n-1}^p$ and that any number with suffix zero is equal to 0.

2.9.19 Andersen [13]

Let $\lambda \geqslant 0$, $1 \leqslant p \leqslant q \leqslant \infty$ and suppose $\{u_n\}_1^\infty$, $\{v_m\}_1^\infty$ are nonnegative extended real-valued sequences. There is a constant C independent of $\{a_m\}_1^\infty$, $\{b_n\}_1^\infty$ such that

$$\left|\sum_{n=1}^{\infty}\sum_{m=1}^{\infty}\frac{a_m b_n}{(m+n)^\lambda}\right| \leqslant C\left(\sum_{m=1}^{\infty}|a_m|^p v_m\right)^{1/p}\left(\sum_{n=1}^{\infty}|b_n|^{q'} u_n\right)^{1/q'}$$

if and only if there is a finite constant K such that

$$r^\lambda\left(\sum_{n=1}^{\infty}\frac{u_n}{(n+r)^{\lambda q}}\right)^{1/q}\left(\sum_{m=1}^{\infty}\frac{v_m^{-1/(p-1)}}{(m+r)^{\lambda p'}}\right)^{1/p'} \leqslant K$$

for all positive integers r, where p' and q' denote the conjugate index of p and q, respectively.

2.9.20 Izumi and Izumi [162]

Let $p > 1$ and $s > -1$ and let f be a nonnegative, nonincreasing and integrable function on $(0, b)$. If $x^s f(x)^p$ is integrable, then

$$\int_0^b x^s G(x)^p \,\mathrm{d}x \leqslant A\int_0^b x^s\left(f\left(\frac{x}{2}\right) - f(x)\right)^p \mathrm{d}x + A\left(\int_{b/2}^b f(x)\,\mathrm{d}x\right)^p,$$

where $G(x) = \int_{x/2}^x \frac{f(t)}{t}\,\mathrm{d}t$ and A is a constant depending only on p and s.

2.9.21 Shum [407]

Let $p > 1, r \neq 1$ and $f(x) \in L[0,b]$ or $f(x) \in L[a,\infty]$ according as $r > 1$ or $r < 1$, where $a > 0$, $b > 0$. If $F(x)$ is defined by

$$F(x) = \begin{cases} \int_0^x f(t)\,\mathrm{d}t, & r > 1, \\ \int_x^\infty f(t)\,\mathrm{d}t, & r < 1, \end{cases}$$

and if $\int_0^b x^{-r}(xf)^p\,dx < \infty$ in (i) and $\int_a^\infty x^{-r}(xf)^p\,dx < \infty$ in (ii), then

$$\text{(i)}\quad \int_0^b x^{-r}F^p\,dx + \frac{p}{r-1}b^{1-r}F^p(b) \leqslant \left(\frac{p}{r-1}\right)^p \int_0^b x^{-r}(xf)^p\,dx \quad \text{for } r > 1,$$

$$\text{(ii)}\quad \int_a^\infty x^{-r}F^p\,dx + \frac{p}{1-r}F^p(a) \leqslant \left(\frac{p}{1-r}\right)^p \int_a^\infty x^{-r}(xf)^p\,dx \quad \text{for } r < 1,$$

with equality in (i) or (ii) only for $f \equiv 0$, where the constant $(p/(r-1))^p$ or $(p/(1-r))^p$ is the best possible when the left-hand side of (i) or (ii) is unchanged, respectively.

2.9.22 Mohapatra and Russell [217]

Suppose that $a(\cdot,\cdot)$ is defined on $\mathbb{R}_+ \times \mathbb{R}_+$, with $a(x,t) \geqslant 0$ for $0 < t < x$, $a(x,t) = 0$ for $t > x$, and suppose that, for some constant $K_1 \geqslant 1$,

$$a(x,t) \leqslant K_1 a(y,t) \quad \text{for } x > y > t. \tag{2.9.5}$$

Let $g(x) \geqslant 0$, $x \in \mathbb{R}_+$, and $g(\cdot)a(\cdot,t) \in L(0,t)$ for each $t > 0$, and write

$$G_2(t) = \int_0^t g(x)a(x,t)\,dx, \quad t > 0.$$

Let $f(t) \geqslant 0$, $t \in \mathbb{R}_+$, and $a(x,\cdot)f(\cdot) \in L(0,x)$ for each $x > 0$, and write

$$F_1(x) = \int_0^x a(x,t)f(t)\,dt, \quad x > 0.$$

(a$_1$) If $1 < p < \infty$, $0 < m \leqslant \infty$, $g(x) > 0$ on $(0,m)$, then

$$\int_0^m gF_1^p\,dx \leqslant \left(pK_1^{p-1}\right)^p \int_0^m g^{1-p}(G_2 f)^p\,dx. \tag{2.9.6}$$

(a$_2$) If $0 < p < 1$, $0 \leqslant r < \infty$, $F_1(x) > 0$ on $\mathbb{R}_+$, then

$$\int_r^\infty gF_1^p\,dx \geqslant \left(pK_1^{p-1}\right)^p \int_r^\infty g^{1-p}(G_2 f)^p\,dx. \tag{2.9.7}$$

(a_3) If $p = 1$ then hypothesis (2.9.5) is not required and (2.9.6) for $0 < m < \infty$ and (2.9.7) for $0 < r < \infty$ hold, with equality in (2.9.6) for $m = \infty$ and (2.9.7) for $r = 0$.

2.9.23 Mohapatra and Russell [217]

Assume that $a(\cdot,\cdot)$ is defined on $\mathbb{R}_+ \times \mathbb{R}_+$ with $a(x,t) \geqslant 0$ for $0 < x < t$, $a(x,t) = 0$ for $x > t$, and suppose that, for some constant $K_2 \geqslant 1$,

$$a(x,t) \leqslant K_2 a(y,t) \quad \text{for } x < y < t. \tag{2.9.8}$$

Let $g(x) \geqslant 0$, $x \in \mathbb{R}_+$, and $g(\cdot)a(\cdot,t) \in L(0,t)$ for each $t > 0$, and write

$$G_1(t) = \int_0^t g(x)a(x,t)\,\mathrm{d}x, \quad t > 0.$$

Let $f(t) \geqslant 0$, $t \in \mathbb{R}_+$, and $a(x,\cdot)f(\cdot) \in L(x,\infty)$ for each $x > 0$, and write

$$F_2(x) = \int_x^\infty a(x,t)f(t)\,\mathrm{d}t, \quad x > 0.$$

(b_1) If $1 < p < \infty$, $0 \leqslant r < \infty$, $g(x) \geqslant 0$ on (r,∞), then

$$\int_r^\infty gF_2^p\,\mathrm{d}x \leqslant \left(pK_2^{p-1}\right)^p \int_r^\infty g^{1-p}(G_1 f)^p\,\mathrm{d}x. \tag{2.9.9}$$

(b_2) If $0 < p < 1$, $0 < m \leqslant \infty$, $F_2(x) > 0$ on $\mathbb{R}_+$, then

$$\int_0^m gF_2^p\,\mathrm{d}x \geqslant \left(pK_2^{p-1}\right)^p \int_0^m g^{1-p}(G_1 f)^p\,\mathrm{d}x. \tag{2.9.10}$$

(b_3) If $p = 1$ then hypothesis (2.9.8) is not required and (2.9.9) for $0 < r < \infty$ and (2.9.10) for $0 < m < \infty$ hold, with equality in (2.9.9) for $r = 0$ and (2.9.10) for $m = \infty$.

2.9.24 Levinson [190]

Let $p > 1$, $f(x) \geqslant 0$ and $r(x) > 0$, $x > 0$, be absolutely continuous. Let

$$\frac{xr'}{r} - \frac{p-1}{p} \geqslant \frac{1}{\lambda}$$

for almost all $x > 0$ and for some $\lambda > 0$. If

$$J(x) = \frac{r(x)}{x} \int_x^\infty \frac{f(t)}{r(t)}\,\mathrm{d}t,$$

then

$$\int_0^\infty J^p(x)\,\mathrm{d}x \leqslant \lambda^p \int_0^\infty f^p(x)\,\mathrm{d}x.$$

2.9.25 Pachpatte [255]

Let $p > 1$, $m > 1$ be constants. Let f be a nonnegative and integrable function on $(0, b)$, $b > 0$ is a constant. Let r be a positive and absolutely continuous function on $(0, b)$. Let

$$1 - \frac{p}{m-1} x \frac{r'(x)}{r(x)} \geqslant \frac{1}{\alpha}$$

for almost all $x \in (0, b)$ and for some constant $\alpha > 0$. Then

$$\int_0^b x^{-m} F^p(x)\,\mathrm{d}x \leqslant \left(\alpha\left(\frac{p}{m-1}\right)\right)^p \int_0^b x^{-m}\left(r(x)\left|\frac{f(x)}{r(x)} - \frac{f(x/2)}{r(x/2)}\right|\right)^p \mathrm{d}x,$$

where

$$F(x) = r(x) \int_{x/2}^x \left(\frac{f(t)}{tr(t)}\right) \mathrm{d}t.$$

2.9.26 Pachpatte [255]

Let p, f, r be as in Section 2.9.25 and $R(x) = \int_0^x r(t)\,\mathrm{d}t$. Let q be a positive and absolutely continuous function on $(0, b)$. Let

$$1 - \frac{1}{p-1} \frac{R(x)}{r(x)} \frac{q'(x)}{q(x)} + \frac{1}{p-1} \frac{R(x)r'(x)}{r^2(x)} \geqslant \frac{1}{\alpha}$$

for almost all $x \in (0, b)$ for some constant $\alpha > 0$. Then

$$\int_0^b q(x) F^p(x)\,\mathrm{d}x$$
$$\leqslant \left(\alpha\left(\frac{p}{p-1}\right)\right)^p \int_0^b q(x)\left\{\frac{1}{xr(x)}\left|r(x)f(x) - r\left(\frac{x}{2}\right)f\left(\frac{x}{2}\right)\right|\right\}^p \mathrm{d}x,$$

where

$$F(x) = \frac{1}{R(x)} \int_{x/2}^{x} \frac{r(t) f(t)}{t} \, \mathrm{d}t.$$

2.9.27 Pachpatte [268]

Let $m \neq 1$ and $p_j > 1$, $j = 1, 2$, be constants such that $\frac{1}{p_1} + \frac{1}{p_2} = 1$. Let $f_j(x)$, $j = 1, 2$, be nonnegative and integrable functions on $(0, \infty)$ and let $r_j(x)$, $j = 1, 2$, be positive and absolutely continuous functions on $(0, \infty)$. Let

$$1 + \left(\frac{p_j}{m-1}\right) x \frac{r_j'(x)}{r_j(x)} \geqslant \frac{1}{\alpha_j} \quad \text{for } m > 1,$$

$$1 - \left(\frac{p_j}{1-m}\right) x \frac{r_j'(x)}{r_j(x)} \geqslant \frac{1}{\beta_j} \quad \text{for } m < 1,$$

for almost all $x > 0$ and for some positive constants α_j, β_j, $j = 1, 2$. If $F_j(x)$, $j = 1, 2$, are defined by

$$F_j(x) = \begin{cases} \frac{1}{r_j(x)} \int_0^x \frac{r_j(t) f_j(t)}{t} \, \mathrm{d}t & \text{for } m > 1, \\ \frac{1}{r_j(x)} \int_x^\infty \frac{r_j(t) f_j(t)}{t} \, \mathrm{d}t & \text{for } m < 1, \end{cases} \tag{2.9.11}$$

then

$$\int_0^\infty x^{-m} F_1(x) F_2(x) \, \mathrm{d}x \leqslant \frac{1}{p_1} \left(\lambda_1 \left(\frac{p_1}{|m-1|}\right)\right)^{p_1} \int_0^\infty x^{-m} f_1^{p_1}(x) \, \mathrm{d}x + \frac{1}{p_2} \left(\lambda_2 \left(\frac{p_2}{|m-1|}\right)\right)^{p_2} \int_0^\infty x^{-m} f_2^{p_2}(x) \, \mathrm{d}x, \tag{2.9.12}$$

where $\lambda_j = \max\{\alpha_j, \beta_j\}$ for $j = 1, 2$. Equality holds in (2.9.12) if $f_j(x) \equiv 0$.

2.9.28 Pachpatte [268]

Let m, f_j, r_j be as in Section 2.9.27 and $p_j \geqslant 1$, $j = 1, 2$, be constants. Let

$$1 + \left(\frac{p_1 + p_2}{m-1}\right) x \frac{r_j'(x)}{r_j(x)} \geqslant \frac{1}{\alpha_j^*} \quad \text{for } m > 1,$$

$$1 - \left(\frac{p_1 + p_2}{1-m}\right) x \frac{r_j'(x)}{r_j(x)} \geqslant \frac{1}{\beta_j^*} \quad \text{for } m > 1,$$

for almost all $x > 0$ and for some positive constants α_j^*, β_j^*, $j = 1, 2$. If $F_j(x)$, $j = 1, 2$, are defined by (2.9.11), then

$$\int_0^\infty x^{-m} F_1^{p_1}(x) F_2^{p_2}(x)\,dx \leqslant \left(\frac{p_1}{p_1+p_2}\right)\left(\lambda_1^*\left(\frac{p_1+p_2}{|m-1|}\right)\right)^{p_1+p_2} \int_0^\infty x^{-m} f_1^{p_1+p_2}(x)\,dx + \left(\frac{p_1}{p_1+p_2}\right)\left(\lambda_2^*\left(\frac{p_1+p_2}{|m-1|}\right)\right)^{p_1+p_2} \int_0^\infty x^{-m} f_2^{p_1+p_2}(x)\,dx, \tag{2.9.13}$$

where $\lambda_j^* = \max\{\alpha_j^*, \beta_j^*\}$ for $j = 1, 2$. Equality holds in (2.9.13) if $f(x) \equiv 0$.

2.9.29 Pachpatte [268]

Let $m > 1$ and $p_j > 1$, $j = 1, 2$, be constants such that $\frac{1}{p_1} + \frac{1}{p_2} = 1$. Let $h_j(x)$, $j = 1, 2$, be nonnegative and integrable functions on $(0, b)$ and let $z_j(x)$, $j = 1, 2$, be positive and absolutely continuous functions on $(0, b)$. Let

$$1 + \left(\frac{p_j}{m-1}\right) x \frac{z_j'(x)}{z_j(x)} \geqslant \frac{1}{\gamma_j}$$

for almost all $x \in (0, b)$ and for some positive constants γ_j, $j = 1, 2$. If $H_j(x)$, $j = 1, 2$, defined by

$$H_j(x) = \frac{1}{z_j(x)} \int_{x/2}^x \frac{z_j(t) h_j(t)}{t}\,dt, \tag{2.9.14}$$

then

$$\int_0^b x^{-m} H_1(x) H_2(x)\,dx \leqslant \frac{1}{p_1}\left(\gamma_1\left(\frac{p_1}{m-1}\right)\right)^{p_1} \int_0^b x^{-m}\left\{\frac{1}{z_1(x)}\left|z_1(x)h_1(x) - z_1\left(\frac{x}{2}\right)h_1\left(\frac{x}{2}\right)\right|\right\}^{p_1} dx + \frac{1}{p_2}\left(\gamma_2\left(\frac{p_2}{m-1}\right)\right)^{p_2} \times \int_0^b x^{-m}\left\{\frac{1}{z_2(x)}\left|z_2(x)h_2(x) - z_2\left(\frac{x}{2}\right)h_2\left(\frac{x}{2}\right)\right|\right\}^{p_2} dx.$$

2.9.30 Pachpatte [268]

Let m, h_j, z_j be as in Section 2.9.29 and $p_j \geqslant 1$, $j = 1, 2$, be constants. Let

$$1 + \left(\frac{p_1 + p_2}{m - 1}\right) x \frac{z_j'(x)}{z_j(x)} \geqslant \frac{1}{\delta_j}$$

for almost all $x \in (0, b)$ and for some positive constants δ_j, $j = 1, 2$. If $H_j(x)$, $j = 1, 2$, are defined by (2.9.14), then

$$\begin{aligned}
&\int_0^b x^{-m} H_1^{p_1}(x) H_2^{p_2}(x)\,\mathrm{d}x \\
&\leqslant \left(\frac{p_1}{p_1 + p_2}\right)\left(\delta_1\left(\frac{p_1 + p_2}{m - 1}\right)\right)^{p_1+p_2} \\
&\quad \times \int_0^b x^{-m}\left\{\frac{1}{z_1(x)}\left|z_1(x)h_1(x) - z_1\left(\frac{x}{2}\right)h_1\left(\frac{x}{2}\right)\right|\right\}^{p_1+p_2} \mathrm{d}x \\
&\quad + \left(\frac{p_2}{p_1 + p_2}\right)\left(\delta_2\left(\frac{p_1 + p_2}{m - 1}\right)\right)^{p_1+p_2} \\
&\quad \times \int_0^b x^{-m}\left\{\frac{1}{z_2(x)}\left|z_2(x)h_2(x) - z_2\left(\frac{x}{2}\right)h_2\left(\frac{x}{2}\right)\right|\right\}^{p_1+p_2} \mathrm{d}x.
\end{aligned}$$

2.9.31 Pachpatte [337]

Let $p > 1$, $m > 1$ be constants. Let f be a nonnegative and integrable function on $(0, b)$, $0 < b < \infty$. If $F(x)$ is defined by

$$F(x) = \int_0^x \frac{1}{t}\left(\int_{t/2}^t \frac{f(s)}{s}\,\mathrm{d}s\right)\mathrm{d}t$$

for $x \in (0, b)$, then

$$\int_0^b x^{-m} F^p(x)\,\mathrm{d}x \leqslant \left(\frac{p}{m - 1}\right)^{2p} \int_0^b x^{-m}\left|f(x) - f\left(\frac{x}{2}\right)\right|^p \mathrm{d}x.$$

2.9.32 Pachpatte [337]

Let p, m and f be as in Section 2.9.31. If $F(x)$ is defined by

$$F(x) = \int_{x/2}^x \frac{1}{t}\left(\int_0^t \frac{f(s)}{s}\,\mathrm{d}s\right)\mathrm{d}t$$

for $x \in (0, b)$, then

$$\int_0^b x^{-m} F^p(x)\,\mathrm{d}x \leqslant \left(\frac{p}{m-1}\right)^{2p} \int_0^b x^{-m} \left| f(x) - f\left(\frac{x}{2}\right)\right|^p \mathrm{d}x.$$

2.9.33 Pachpatte [255]

Let $p > 1$, $m > 1$ be constants. Let f be a nonnegative and integrable function on $(0, b)$. Let r be a positive and absolutely continuous function on $(0, b)$. Let

$$1 + \frac{p}{m-1} x \frac{r'(x)}{r(x)} + \frac{p}{m-1} \geqslant \frac{1}{\alpha}$$

for almost all $x \in (0, b)$ and for some constant $\alpha > 0$. Then

$$\int_0^b x^{-m} F^p(x)\,\mathrm{d}x$$
$$\leqslant \left\{\alpha\left(\frac{p}{m-1}\right)\right\}^p \int_0^b x^{-m} \left\{\frac{1}{r(x)}\left| r(x) f(x) - \frac{1}{2} r\left(\frac{x}{2}\right) f\left(\frac{x}{2}\right)\right|\right\}^p \mathrm{d}x,$$

where

$$F(x) = \frac{1}{x r(x)} \int_{x/2}^x r(t) f(t)\,\mathrm{d}t.$$

2.9.34 Pachpatte [255]

Let p, m, f and r be as in Section 2.9.33. Let

$$1 - \frac{p}{m-1} x \frac{r'(x)}{r(x)} + \frac{p}{m-1} \geqslant \frac{1}{\alpha}$$

for almost all $x \in (0, b)$ and for some constant $\alpha > 0$. Then

$$\int_0^b x^{-m} F^p(x)\,\mathrm{d}x \leqslant \left(\alpha\left(\frac{p}{m-1}\right)\right)^p \int_0^b x^{-m} \left\{ r(x) \left| \frac{f(x)}{r(x)} - \frac{1}{2}\frac{f(x/2)}{r(x/2)}\right|\right\}^p \mathrm{d}x,$$

where

$$F(x) = \frac{r(x)}{x} \int_{x/2}^x \frac{f(t)}{r(t)}\,\mathrm{d}t.$$

2.9.35 Pachpatte [253]

Assume that:

(H_1) Let $p > 1$ be a constant. Let $f(x)$ be a nonnegative and integrable function on $(0, \infty)$. Let $r_i(x)$, $i = 1, \ldots, n$, be positive and absolutely continuous functions on $(0, \infty)$.

(H_2) There exist positive constants α_i, β_i such that for almost all $x > 0$,

$$1 + \left(\frac{p}{m-1}\right) x \frac{r_i'(x)}{r_i(x)} \geqslant \frac{1}{\alpha_i} \quad \text{for } m > 1,$$

$$1 - \left(\frac{p}{1-m}\right) x \frac{r_i'(x)}{r_i(x)} \geqslant \frac{1}{\beta_i} \quad \text{for } m < 1,$$

for $i = 1, \ldots, n$ hold. If $m \neq 1$ and $\int_0^\infty x^{-m} f^p(x)\,\mathrm{d}x < \infty$ and $E(x)$ defined by

$$E(x) = \begin{cases} \frac{1}{r_1(x)} \int_0^x \frac{r_1(t_1)}{t_1} \frac{1}{r_2(t_1)} \int_0^{t_1} \frac{r_2(t_2)}{t_2} \frac{1}{r_3(t_2)} \cdots \int_0^{t_{n-2}} \frac{r_{n-1}(t_{n-1})}{t_{n-1}} \frac{1}{r_n(t_{n-1})} \\ \qquad \times \int_0^{t_{n-1}} \frac{r_n(t_n)}{t_n} f(t_n)\,\mathrm{d}t_n\,\mathrm{d}t_{n-1} \cdots \mathrm{d}t_2\,\mathrm{d}t_1, \quad m > 1, \\ \frac{1}{r_1(x)} \int_x^\infty \frac{r_1(t_1)}{t_1} \frac{1}{r_2(t_1)} \int_{t_1}^\infty \frac{r_2(t_2)}{t_2} \frac{1}{r_3(t_2)} \cdots \int_{t_{n-2}}^\infty \frac{r_{n-1}(t_{n-1})}{t_{n-1}} \frac{1}{r_{n-1}(t_{n-1})} \\ \qquad \times \int_{t_{n-1}}^\infty \frac{r_n(t_n)}{t_n} f(t_n)\,\mathrm{d}t_n\,\mathrm{d}t_{n-1} \cdots \mathrm{d}t_2\,\mathrm{d}t_1, \quad m < 1, \end{cases}$$

then

$$\int_0^\infty x^{-m} E^p(x)\,\mathrm{d}x \leqslant \left[\left(\prod_{i=1}^n \alpha_i\right)\left(\frac{p}{m-1}\right)^n\right]^p \int_0^\infty x^{-m} f^p(x)\,\mathrm{d}x \quad \text{for } m > 1,$$

$$\int_0^\infty x^{-m} E^p(x)\,\mathrm{d}x \leqslant \left[\left(\prod_{i=1}^n \beta_i\right)\left(\frac{p}{1-m}\right)^n\right]^p \int_0^\infty x^{-m} f^p(x)\,\mathrm{d}x \quad \text{for } m < 1.$$

2.9.36 Pachpatte [253]

Assume that (H_1) in Section 2.9.35 and following hypothesis (H_3) hold.

(H_3) There exist positive constants k_i such that, for almost all $x > 0$,

$$1 + \left(\frac{1}{p-1}\right) \frac{1}{r_i^2(x)} R_i(x) r_i'(x) \geqslant \frac{1}{k_i}$$

for $i = 1, \ldots, n$ and

$$R_i(x) = \int_0^x r_i(t)\,\mathrm{d}t, \quad i = 1, \ldots, n,$$

holds. Let $\phi(u) \geqslant 0$ be defined on an open interval, finite or infinite, have a second derivative $\phi'' \geqslant 0$ and

$$\phi\phi'' \geqslant \left(1 - \frac{1}{p}\right)\phi'^2,$$

and at the ends of the interval ϕ take its limiting values, finite or infinite, and for $0 < x < \infty$, the range of values of $f(x)$ lie in the closed interval of definition of ϕ. If $\int_0^\infty \phi(f(x))\,\mathrm{d}x < \infty$ and $J(x)$ is defined by

$$J(x) = \frac{1}{R_1(x)} \int_0^x r_1(t_1) \frac{1}{R_2(t_1)} \int_0^{t_1} r_2(t_2) \frac{1}{R_3(t_2)} \cdots \int_0^{t_{n-2}} r_{n-1}(t_{n-1}) \times \frac{1}{R_n(t_{n-1})} \int_0^{t_{n-1}} r_n(t_n) f(t_n)\,\mathrm{d}t_n\,\mathrm{d}t_{n-1} \cdots \mathrm{d}t_2\,\mathrm{d}t_1,$$

then

$$\int_0^\infty \phi(J(x))\,\mathrm{d}x \leqslant \left[\left(\prod_{i=1}^n k_i\right)\left(\frac{p}{p-1}\right)^n\right]^p \int_0^\infty \phi(f(x))\,\mathrm{d}x.$$

2.9.37 Pachpatte [310]

Let B be a subset of the n-dimensional Euclidean space $\mathbb{R}^n$ defined by $B = \{x \in \mathbb{R}^n\colon\ 0 < x < \infty\}$ for $0 \in \mathbb{R}^n$. For the function $u(z)$ defined on B, denote by $\int_B u(z)\,\mathrm{d}z$, $\int_{B_{x,y}} u(z)\,\mathrm{d}z$ the n-fold integrals

$$\int_0^\infty \cdots \int_0^\infty u(z_1, \ldots, z_n)\,\mathrm{d}z_n \cdots \mathrm{d}z_1,$$

$$\int_{x_1}^{y_1} \cdots \int_{x_n}^{y_n} u(z_1, \ldots, z_n)\,\mathrm{d}z_n \cdots \mathrm{d}z_1,$$

respectively, where $x_i < y_i$.

(i) Let $p \geqslant 1$ and $c > 1$ be constants. Let $f(x)$ be a nonnegative and integrable function on B and let $r_i(x_i)$, $i = 1, \ldots, n$, be nonnegative continuous functions on $(0, \infty)$ and let

$$R_i(x_i) = \int_0^{x_i} r_i(y_i)\,\mathrm{d}y_i, \quad 0 < x_i < \infty.$$

If $F(x)$ is defined by

$$F(x)=\int_{B_{0,x}}\left(\prod_{i=1}^{n}r_i(y_i)\right)f(y)\,\mathrm{d}y,\qquad x\in B,$$

then

$$\int_B\left(\prod_{i=1}^{n}R_i^{-c}(x_i)r_i(x_i)\right)F^p(x)\,\mathrm{d}x \leqslant\left(\frac{p}{c-1}\right)^{np}\int_B\left(\prod_{i=1}^{n}R_i^{p-c}(x_i)r_i(x_i)\right)f^p(x)\,\mathrm{d}x. \tag{2.9.15}$$

Equality holds in (2.9.15) if $f(x)\equiv 0$.

(ii) Let p, f, r_i, R_i be defined as in (i) and $c<1$ be a constant. If $F(x)$ is defined by

$$F(x)=\int_{B_{x,\infty}}\left(\prod_{i=1}^{n}r_i(y_i)\right)f(y)\,\mathrm{d}y,\qquad x\in B,$$

then

$$\int_B\left(\prod_{i=1}^{n}R_i^{-c}(x_i)r_i(x_i)\right)F^p(x)\,\mathrm{d}x \leqslant\left(\frac{p}{1-c}\right)^{np}\int_B\left(\prod_{i=1}^{n}R_i^{p-c}(x_i)r_i(x_i)\right)f^p(x)\,\mathrm{d}x. \tag{2.9.16}$$

Equality holds in (2.9.16) if $f(x)\equiv 0$.

2.9.38 Pachpatte [310]

Let H be a subset of the n-dimensional Euclidean space $\mathbb{R}^n$ defined by $H=\{x\in\mathbb{R}^n\colon a<x<b\}$, where $a,b\in\mathbb{R}^n$ and $a>0$. For a function $u(z)$ defined on H, we denote by $\int_H u(z)\,\mathrm{d}z$, $\int_{H_{x,y}} u(z)\,\mathrm{d}z$ the n-fold integrals

$$\int_{a_1}^{b_1}\cdots\int_{a_n}^{b_n}u(z_1,\ldots,z_n)\,\mathrm{d}z_n\cdots\mathrm{d}z_1,\qquad \int_{x_1}^{y_1}\cdots\int_{x_n}^{y_n}u(z_1,\ldots,z_n)\,\mathrm{d}z_n\cdots\mathrm{d}z_1,$$

respectively, where $a_i<b_i$, $x_i<y_i$.

(i) Let $p \geqslant 1$, $\mu < 1$ be constants and $0 < a < b \leqslant M$, where $0, a, b, M$ are in $\mathbb{R}^n$. Let $g(x)$ be a nonnegative and integrable function on H and let $r_i(x_i)$, $i = 1, \ldots, n$, be nonnegative continuous functions on $a_i < x_i \leqslant M_i$ and let

$$R_i(x_i) = \int_{a_i}^{x_i} r_i(y_i)\, \mathrm{d}y_i, \quad a_i < x_i \leqslant M_i.$$

If $G(x)$ is defined by

$$G(x) = \int_{H_{a,x}} \left(\prod_{i=1}^{n} r_i(y_i) \right) g(y)\, \mathrm{d}y, \quad a < x < b,$$

then

$$\int_H \left(\prod_{i=1}^{n} \left\{ \log\left(\frac{R_i(M_i)}{R_i(x_i)} \right) \right\}^{-\mu} R_i^{-1}(x_i) r_i(x_i) \right) G^p(x)\, \mathrm{d}x$$
$$\leqslant \left(\frac{p}{1-\mu} \right)^{np} \int_H \left(\prod_{i=1}^{n} \left\{ \log\left(\frac{R_i(M_i)}{R_i(x_i)} \right) \right\}^{p-\mu} R_i^{p-1}(x_i) r_i(x_i) \right) g^p(x)\, \mathrm{d}x. \tag{2.9.17}$$

Equality holds in (2.9.17) if $g(x) \equiv 0$.

(ii) Let p, a, b, M, g, r_i, R_i be as in part (i) above and $\mu > 1$ be a constant. If $G(x)$ is defined by

$$G(x) = \int_{H_{x,b}} \left(\prod_{i=1}^{n} r_i(y_i) \right) g(y)\, \mathrm{d}y, \quad a < x < b,$$

then

$$\int_H \left(\prod_{i=1}^{n} \left\{ \log\left(\frac{R_i(M_i)}{R_i(x_i)} \right) \right\}^{-\mu} R_i^{-1}(x_i) r_i(x_i) \right) G^p(x)\, \mathrm{d}x$$
$$\leqslant \left(\frac{p}{\mu-1} \right)^{np} \int_H \left(\prod_{i=1}^{n} \left\{ \log\left(\frac{R_i(M_i)}{R_i(x_i)} \right) \right\}^{p-\mu} R_i^{p-1}(x_i) r_i(x_i) \right) g^p(x)\, \mathrm{d}x. \tag{2.9.18}$$

Equality holds in (2.9.18) if $g(x) \equiv 0$.

2.9.39 Pachpatte [342]

Let $n \geqslant 2$, $0 \leqslant k \leqslant n-2$ be integers. Let $r \in C^1(I, \mathbb{R}_+)$, $u \in C^n(I_x, \mathbb{R})$, $v \in C^n(I_y, \mathbb{R})$ and $u^{(i-1)}(0) = v^{(i-1)}(0) = 0$, $i = 1, \ldots, n$, where $I_x = [0, x)$, $I_y = [0, y)$. Then

$$\int_0^x \int_0^y \frac{|u^{(k)}(s)||v^{(k)}(t)|}{s^{2n-2k-3} + t^{2n-2k-3}} \, \mathrm{d}s \, \mathrm{d}t$$

$$\leqslant M(n, k, x, y) \left(\int_0^x (x-s) \frac{s}{r^2(s)} \left(\int_0^s \left| \left(r(\sigma) u^{(n-1)}(\sigma) \right)' \right|^2 \mathrm{d}\sigma \right) \mathrm{d}s \right)^{1/2}$$

$$\times \left(\int_0^y (y-t) \frac{t}{r^2(t)} \left(\int_0^t \left| \left(r(\xi) v^{(n-1)}(\xi) \right)' \right|^2 \mathrm{d}\xi \right) \mathrm{d}t \right)^{1/2},$$

where

$$M(n, k, x, y) = \frac{1}{2} \frac{\sqrt{xy}}{[(n-k-2)!]^2 (2n-2k-3)}.$$

2.9.40 Pachpatte [342]

Let $n \geqslant 1$, $0 \leqslant k \leqslant n-1$ be integers. Let $r \in C^n(I, \mathbb{R}_+)$, $u \in C^{2n}(I_x, \mathbb{R})$, $v \in C^{2n}(I_y, \mathbb{R})$ and $u^{(i-1)}(0) = v^{(i-1)}(0) = 0$, $(r(0) u^{(n)}(0))^{(i-1)} = (r(0) \times v^{(n)}(0))^{(i-1)} = 0$, $i = 1, \ldots, n$, where $I_x = [0, x)$, $I_y = [0, y)$. Then

$$\int_0^x \int_0^y \frac{|u^{(k)}(s)||v^{(k)}(t)|}{s^{2n-2k-1} + t^{2n-2k-1}} \, \mathrm{d}s \, \mathrm{d}t$$

$$\leqslant M(n, k, x, y) \left(\int_0^x (x-s) \frac{s^{2n-1}}{r^2(s)} \left(\int_0^s \left| \left(r(\sigma) u^{(n)}(\sigma) \right)^{(n)} \right|^2 \mathrm{d}\sigma \right) \mathrm{d}s \right)^{1/2}$$

$$\times \left(\int_0^y (y-t) \frac{t^{2n-1}}{r^2(t)} \left(\int_0^t \left| \left(r(\xi) v^{(n)}(\xi) \right)^{(n)} \right|^2 \mathrm{d}\xi \right) \mathrm{d}t \right)^{1/2},$$

where

$$M(n, k, x, y) = \frac{1}{2} \frac{\sqrt{xy}}{[(n-1)!(n-k-1)!(2n-1)]^2 (2n-2k-1)}.$$

2.10 Notes

In 1920, G.H. Hardy proved the inequality given in Theorem 2.2.1. He deduced inequality (2.2.1) from the corresponding inequality for integrals. Theorems

2.2.2 and 2.2.3 are due to Copson [69]. Theorem 2.2.4 deals with a generalization of Hardy's inequality in Theorem 2.2.1 and is established by Hardy and Littlewood in [139]. Theorem 2.2.5 is due to Nemeth [226] which deals with generalizations of the Hardy–Littlewood inequality given in Theorem 2.2.4. Theorems 2.2.6 and 2.2.7 are taken from Love [196].

The results in Theorems 2.3.1 and 2.3.2 are due to Copson [71]. Theorems 2.3.3 and 2.3.4 are taken from Pachpatte [277] and [307]. The result in Theorem 2.3.5 is due to Bennett [22] which, in fact, is motivated by Littlewood's problem [195]. The inequalities in Theorem 2.3.6 are taken from Pachpatte [326]. Theorem 2.4.1 deals with the famous integral inequality first discovered by Hardy in [136]. Theorem 2.4.2 is a further generalization of the inequality in Theorem 2.4.1 and also established by Hardy in [137]. The inequalities in Theorems 2.4.3 and 2.4.4 are due to Copson [70]. Theorems 2.4.5 and 2.4.6 are taken from Love [199]. Theorems 2.4.7 and 2.4.8 deal with the general versions of Hardy's integral inequality with weights and are taken from Muckenhoupt [221].

Theorems 2.5.1–2.5.3 deal with certain generalizations of well-known Hardy's integral inequality established by Levinson in [190]. The results given in Theorems 2.5.4–2.5.8 deal with some basic generalizations of Hardy's integral inequality established by Pachpatte [295,344]. Theorem 2.6.1 deals with the variant of Hardy's integral inequality and is proved by Izumi and Izumi in [162]. The results in Theorems 2.6.2–2.6.9 deal with some extensions and variants of Hardy's integral inequality established by Pachpatte [254,255,337] (see also [266,305,306]). The results given in Section 2.7 deal with certain multidimensional generalizations and variants of Hardy's integral inequality established by Pachpatte in [293,315,333,341]. Section 2.8 contains some new results on inequalities similar to Hilbert's inequality established by Pachpatte in [334,335, 342,343,350,352,353]. Section 2.9 deals with miscellaneous inequalities which claim their origin in well-known Hardy's inequalities.

Chapter 3

Opial-Type Inequalities

3.1 Introduction

In 1960, Z. Opial [231] discovered one of the most fundamental integral inequalities involving a function and its derivative, which is now known in the literature as Opial's inequality. In the same year, C. Olech [230] published a note which addresses a simpler proof of Opial's inequality under weaker conditions. Starting from the pioneer papers [230,231], the result of Opial has received considerable attention and many papers have appeared, which provides with the simple proofs, various generalizations, extensions and discrete analogues of Opial's inequality and its generalizations. The importance of Opial's inequality and its generalizations and extensions lies in successful utilization to many interesting applications in the theory of differential equations. Good surveys of the work on such inequalities together with many references are contained in monographs [4,211,215]. In the past few years, numerous variants, generalizations and extensions of Opial's inequality which involves functions of one and many independent variables have been found in various directions. This chapter deals with important fundamental results on Opial-type inequalities recently investigated in the literature by various investigators.

3.2 Opial-Type Integral Inequalities

In [231] Opial established the following interesting integral inequality

$$\int_0^h \left|u(t)u'(t)\right| \mathrm{d}t \leqslant \frac{h}{4} \int_0^h \left|u'(t)\right|^2 \mathrm{d}t, \tag{3.2.1}$$

where $u(t) \in C^1[0,h]$, $u(t) > 0$ in $(0,h)$ such that $u(0) = u(h) = 0$. In (3.2.1) the constant $\frac{h}{4}$ is the best possible. The first simple proof of inequality (3.2.1) is given by Olech [230] in his paper published along with Opial's paper [231]. In the years thereafter, numerous variants, generalizations and extensions of inequality (3.2.1) have appeared in the literature; see [4,211,215] and the references given therein. In this section we present a weaker form of (3.2.1) and its simplified proof based on Olech [230] as well as variants established by various investigators during the past few years.

We begin with the following weaker form of Opial's inequality (3.2.1) which Olech establishes in [230].

THEOREM 3.2.1. *Let u be an absolutely continuous function on $[0,h]$ and let $u(0) = u(h) = 0$. Then inequality (3.2.1) holds. Equality holds in (3.2.1) if and only if*

$$u(t) = \begin{cases} cx & \text{for } 0 \leqslant x \leqslant \frac{h}{2}, \\ c(h-x) & \text{for } \frac{h}{2} \leqslant x \leqslant h, \end{cases} \tag{3.2.2}$$

where c is a constant.

PROOF. Let $y(t) = \int_0^t |u'(s)|\,\mathrm{d}s$ and $z(t) = \int_t^h |u'(s)|\,\mathrm{d}s$. Then we have the following relations

$$y'(t) = |u'(t)| = -z'(t) \tag{3.2.3}$$

and

$$|u(t)| \leqslant y(t), \qquad |u(t)| \leqslant z(t), \tag{3.2.4}$$

for $t \in [0,h]$. From (3.2.3) and (3.2.4), we get

$$\int_0^{h/2} |u(t)u'(t)|\,\mathrm{d}t \leqslant \int_0^{h/2} y(t)y'(t)\,\mathrm{d}t = \frac{1}{2}y^2\left(\frac{h}{2}\right) \tag{3.2.5}$$

and

$$\int_{h/2}^{h} |u(t)u'(t)|\,\mathrm{d}t \leqslant -\int_{h/2}^{h} z(t)z'(t)\,\mathrm{d}t = \frac{1}{2}z^2\left(\frac{h}{2}\right). \tag{3.2.6}$$

From (3.2.5) and (3.2.6), we find that

$$\int_0^{h} |u(t)u'(t)|\,\mathrm{d}t \leqslant \frac{1}{2}\left[y^2\left(\frac{h}{2}\right) + z^2\left(\frac{h}{2}\right)\right]. \tag{3.2.7}$$

On the other hand, using the Cauchy–Schwarz inequality, we have

$$y^2\left(\frac{h}{2}\right) = \left(\int_0^{h/2} \left|u'(t)\right| \mathrm{d}t\right)^2 \leqslant \frac{h}{2}\int_0^{h/2} \left|u'(t)\right|^2 \mathrm{d}t, \tag{3.2.8}$$

$$z^2\left(\frac{h}{2}\right) = \left(\int_{h/2}^{h} \left|u'(t)\right| \mathrm{d}t\right)^2 \leqslant \frac{h}{2}\int_{h/2}^{h} \left|u'(t)\right|^2 \mathrm{d}t. \tag{3.2.9}$$

Now the desired inequality in (3.2.1) follows from (3.2.7)–(3.2.9).

Now suppose that the equality holds in (3.2.1), that is,

$$\int_0^h \left|u(t)u'(t)\right| \mathrm{d}t = \frac{h}{4}\int_0^h \left|u'(t)\right|^2 \mathrm{d}t. \tag{3.2.10}$$

Then from (3.2.7)–(3.2.9), we get

$$\left(\int_0^{h/2} \left|u'(t)\right| \mathrm{d}t\right)^2 = \frac{h}{2}\int_0^{h/2} \left|u'(t)\right|^2 \mathrm{d}t, \tag{3.2.11}$$

$$\left(\int_{h/2}^{h} \left|u'(t)\right| \mathrm{d}t\right)^2 = \frac{h}{2}\int_{h/2}^{h} \left|u'(t)\right|^2 \mathrm{d}t. \tag{3.2.12}$$

It is easy to see that equalities (3.2.11) and (3.2.12) are possible if and only if $|u'(t)| = \text{constant}$ almost everywhere in $[0, \frac{h}{2}]$ and in $[\frac{h}{2}, h]$. Hence $y(t)$ and $z(t)$ are linear. Further, it follows from (3.2.10), (3.2.7), (3.2.11) and (3.2.12) that $|u(t)| = y(t)$ for $0 \leqslant t \leqslant \frac{h}{2}$ and $|u(t)| = z(t)$ for $\frac{h}{2} \leqslant t \leqslant h$. These facts imply (3.2.2). The proof is complete. □

In [419] Traple has given the inequalities in the following theorem.

THEOREM 3.2.2. *Let p be a nonnegative and continuous function on $[0, h]$. Let u be an absolutely continuous function on $[0, h]$ with $u(0) = u(h) = 0$. Then the following inequalities hold*

$$\int_0^h p(t)\left|u(t)\right|^2 \mathrm{d}t \leqslant \frac{h}{4}\left(\int_0^h p(t)\,\mathrm{d}t\right)\left(\int_0^h \left|u'(t)\right|^2 \mathrm{d}t\right), \tag{3.2.13}$$

$$\int_0^h p(t)\left|u(t)\right|\left|u'(t)\right| \mathrm{d}t \leqslant \left(\frac{h}{4}\int_0^h p^2(t)\,\mathrm{d}t\right)^{1/2}\left(\int_0^h \left|u'(t)\right|^2 \mathrm{d}t\right). \tag{3.2.14}$$

PROOF. For $t \in [0, h]$, we have

$$\left|u(t)\right| \leqslant \int_0^t \left|u'(s)\right| \mathrm{d}s, \qquad \left|u(t)\right| \leqslant \int_t^h \left|u'(s)\right| \mathrm{d}s. \tag{3.2.15}$$

From (3.2.15) we observe that

$$\left|u(t)\right| \leqslant \frac{1}{2}\int_0^h \left|u'(t)\right| \mathrm{d}t.$$

From the above inequality and using the Schwarz inequality, we have

$$\begin{aligned}\int_0^h p(t)\left|u(t)\right|^2 \mathrm{d}t &\leqslant \frac{1}{4}\int_0^h p(t)\left(\int_0^h \left|u'(t)\right|^2 \mathrm{d}t\right)^2 \\ &\leqslant \frac{h}{4}\left(\int_0^h p(t)\,\mathrm{d}t\right)\left(\int_0^h \left|u'(t)\right|^2 \mathrm{d}t\right),\end{aligned}$$

which is the required inequality in (3.2.13).

By using inequality (3.2.13) and the Schwarz inequality, we observe that

$$\begin{aligned}&\int_0^h p(t)\left|u(t)\right|\left|u'(t)\right| \mathrm{d}t \\ &\quad\leqslant \left(\int_0^h p^2(t)\left|u(t)\right|^2 \mathrm{d}t\right)^{1/2}\left(\int_0^h \left|u'(t)\right|^2 \mathrm{d}t\right)^{1/2} \\ &\quad\leqslant \left(\frac{h}{4}\int_0^h p^2(t)\,\mathrm{d}t\int_0^h \left|u'(t)\right|^2 \mathrm{d}t\right)^{1/2}\left(\int_0^h \left|u'(t)\right|^2 \mathrm{d}t\right)^{1/2} \\ &\quad= \left(\frac{h}{4}\int_0^h p^2(t)\,\mathrm{d}t\right)^{1/2}\left(\int_0^h \left|u'(t)\right|^2 \mathrm{d}t\right).\end{aligned}$$

The proof is complete. □

In the following two theorems we present the inequalities of the Opial type established by Pachpatte in [348].

THEOREM 3.2.3. *Let $p \geqslant 0$, $q \geqslant 1$, $m \geqslant 1$ be real numbers. If $u \in C^1([0,h],\mathbb{R})$ satisfies $u(0) = u(h) = 0$, then*

$$\int_0^h \left|u(t)\right|^{m(p+q)} \mathrm{d}t \leqslant \left[(p+q)^m I(m)\right]^q \int_0^h \left|u(t)\right|^{mp}\left|u'(t)\right|^{mq} \mathrm{d}t, \tag{3.2.16}$$

$$\int_0^h \left|u(t)\right|^{m(p+q)} \mathrm{d}t \leqslant \left[(p+q)^m I(m)\right]^{p+q} \int_0^h \left|u'(t)\right|^{m(p+q)} \mathrm{d}t, \tag{3.2.17}$$

where

$$I(m) = \int_0^h \left[t^{1-m} + (h-t)^{1-m}\right]^{-1} \mathrm{d}t. \tag{3.2.18}$$

PROOF. From the hypotheses, we have the following identities

$$u^{p+q}(t) = (p+q)\int_0^t u^{p+q-1}(s)u'(s)\,\mathrm{d}s, \tag{3.2.19}$$

$$u^{p+q}(t) = -(p+q)\int_t^h u^{p+q-1}(s)u'(s)\,\mathrm{d}s, \tag{3.2.20}$$

for $t \in [0,h]$. From (3.2.19), (3.2.20) and using Hölder's inequality with indices $m, m/(m-1)$, we obtain

$$\begin{aligned} &\left|u(t)\right|^{m(p+q)} \\ &\quad\leqslant (p+q)^m t^{m-1} \int_0^t \left|u(s)\right|^{p+q-1}\left|u'(s)\right|^m \mathrm{d}s, \end{aligned} \tag{3.2.21}$$

$$\begin{aligned} &\left|u(t)\right|^{m(p+q)} \\ &\quad\leqslant (p+q)^m (h-t)^{m-1} \int_t^h \left|u(s)\right|^{m(p+q-1)}\left|u'(s)\right|^m \mathrm{d}s, \end{aligned} \tag{3.2.22}$$

for $t \in [0,h]$. Multiplying (3.2.21) by t^{1-m} and (3.2.22) by $(h-t)^{1-m}$ and adding these inequalities we obtain

$$\begin{aligned} &\left[t^{1-m} + (h-t)^{1-m}\right]\left|u(t)\right|^{m(p+q)} \\ &\quad\leqslant (p+q)^m \int_0^h \left|u(s)\right|^{m(p+q-1)}\left|u'(s)\right|^m \mathrm{d}s. \end{aligned} \tag{3.2.23}$$

From (3.2.23) we get

$$\begin{aligned} &\left|u(t)\right|^{m(p+q)} \\ &\quad\leqslant (p+q)^m \left[t^{1-m} + (h-t)^{1-m}\right]^{-1} \\ &\qquad\times \int_0^h \left[\left|u(s)\right|^{mp/q}\left|u'(s)\right|^m\right]\left[\left|u(s)\right|^{m(p+q-1)-mp/q}\right] \mathrm{d}s \end{aligned} \tag{3.2.24}$$

for $t \in [0,h]$. Integrating (3.2.24) on $[0,h]$ and using Hölder's inequality with indices $q, q/(q-1)$, we obtain

$$\begin{aligned}
&\int_0^h |u(t)|^{m(p+q)}\,dt \\
&\quad\leqslant (p+q)^m I(m) \int_0^h \big[|u(t)|^{mp/q}|u'(t)|^m\big]\big[|u(t)|^{m(p+q-1)-mp/q}\big]\,dt \\
&\quad\leqslant (p+q)^m I(m) \\
&\qquad\times \left(\int_0^h |u(t)|^{mp}|u'(t)|^{mq}\,dt\right)^{1/q} \left(\int_0^h |u(t)|^{m(p+q)}\,dt\right)^{(q-1)/q}.
\end{aligned} \tag{3.2.25}$$

If $\int_0^h |u(t)|^{m(p+q)}\,dt = 0$ then (3.2.16) is trivially true, otherwise, dividing both sides of (3.2.25) by $(\int_0^h |u(t)|^{m(p+q)}\,dt)^{(q-1)/q}$ and then taking the qth power on both sides of the resulting inequality we get the required inequality in (3.2.16).

By using Hölder's inequality with indices $(p+q)/p$, $(p+q)/q$ to the integral on the right-hand side of (3.2.16) and following the arguments as in the last part of the proof of inequality (3.2.16) with suitable changes, we get the required inequality in (3.2.17). The proof is complete. □

THEOREM 3.2.4. *Let $p \geqslant 0$, $q \geqslant 1$, $r \geqslant 0$, $m \geqslant 1$ be real numbers. If $u \in C^1([0,h],\mathbb{R})$ satisfies $u(0) = u(h) = 0$, then*

$$\begin{aligned}
&\int_0^h |u(t)|^{m(p+q)}|u'(t)|^{mr}\,dt \\
&\quad\leqslant \big[(p+q+r)^m I(m)\big]^q \int_0^h |u(t)|^{mp}|u'(t)|^{m(q+r)}\,dt,
\end{aligned} \tag{3.2.26}$$

$$\begin{aligned}
&\int_0^h |u(t)|^{m(p+q)}|u'(t)|^{mr}\,dt \\
&\quad\leqslant \big[(p+q+r)^m I(m)\big]^{p+q} \int_0^h |u'(t)|^{m(p+q+r)}\,dt,
\end{aligned} \tag{3.2.27}$$

where $I(m)$ is defined by (3.2.18).

PROOF. Rewriting the integral on the left-hand side of (3.2.26) and using Hölder's inequality with indices $(q+r)/r$, $(q+r)/q$ and inequality (3.2.16),

we observe that

$$\begin{aligned}
&\int_0^h |u(t)|^{m(p+q)} |u'(t)|^{mr}\,\mathrm{d}t \\
&\quad = \int_0^h \big[|u(t)|^{m(pr/(q+r))} |u'(t)|^{mr}\big]\big[|u(t)|^{m(p+q)-m(pr/(q+r))}\big]\,\mathrm{d}t \\
&\quad \leqslant \left[\int_0^h |u(t)|^{mp} |u'(t)|^{m(q+r)}\,\mathrm{d}t\right]^{r/(q+r)} \left[\int_0^h |u(t)|^{m(p+q+r)}\,\mathrm{d}t\right]^{q/(q+r)} \\
&\quad \leqslant \left[\int_0^h |u(t)|^{mp} |u'(t)|^{m(q+r)}\,\mathrm{d}t\right]^{r/(q+r)} \\
&\qquad \times \left[\big[(p+q+r)^m I(m)\big]^{q+r} \int_0^h |u(t)|^{mp} |u'(t)|^{m(q+r)}\,\mathrm{d}t\right]^{q/(q+r)} \\
&\quad = \big[(p+q+r)^m I(m)\big]^q \int_0^h |u(t)|^{mp} |u'(t)|^{m(q+r)}\,\mathrm{d}t.
\end{aligned}$$

This result is the required inequality in (3.2.26).

From (3.2.26) and using Hölder's inequality with indices $(p+q)/p$, $(p+q)/q$, we observe that

$$\begin{aligned}
&\int_0^h |u(t)|^{m(p+q)} |u'(t)|^{mr}\,\mathrm{d}t \\
&\quad \leqslant \big[(p+q+r)^m I(m)\big]^q \\
&\qquad \times \int_0^h \big[|u(t)|^{mp} |u'(t)|^{m(rp/(p+q))}\big]\big[|u'(t)|^{m(q+r)-m(rp/(p+q))}\big]\,\mathrm{d}t \\
&\quad \leqslant \big[(p+q+r)^m I(m)\big]^q \\
&\qquad \times \left[\int_0^h |u(t)|^{m(p+q)} |u'(t)|^{mr}\,\mathrm{d}t\right]^{p/(p+q)} \left[\int_0^h |u'(t)|^{m(p+q+r)}\,\mathrm{d}t\right]^{q/(p+q)}.
\end{aligned}$$

Now, by following the arguments as in the last part of the proof of inequality (3.2.16) with suitable modifications, we get the required inequality in (3.2.27). The proof is complete. □

REMARK 3.2.1. We note that the inequalities in (3.2.16) and (3.2.27) are similar to that of Opial's inequality given in (3.2.1) which in turn yield respectively the lower and upper bounds on the integral of the form involved on the left-hand side of (3.2.1) while the inequalities obtained in (3.2.17) and (3.2.26) are different from those of (3.2.1).

In [304] Pachpatte has established the inequalities in the following theorems which can be considered as their origin to well-known Weyl's inequality [423], see also [141, p. 165] and Opial's inequality in (3.2.1).

THEOREM 3.2.5. *Let $\alpha \geqslant 0$, $p \geqslant 0$, $q \geqslant 1$ be real constants and f be a real-valued continuously differentiable function defined on $(0, b)$ for a fixed real number $b > 0$. Then the following inequalities hold*

$$\int_0^b t^\alpha |f(t)|^{p+q}\,dt \leqslant M\left(\int_0^b t^{\alpha+1}|f(t)|^p\left(\frac{|f(t)|}{b}+|f'(t)|\right)^q dt\right)^{1/q} \times\left(\int_0^b t^{\alpha+1}|f(t)|^{p+q}\,dt\right)^{(q-1)/q}, \tag{3.2.28}$$

$$\int_0^b t^\alpha |f(t)|^{p+q}\,dt \leqslant M\left(\int_0^b t^{(\alpha+1)q}|f(t)|^p\left(\frac{|f(t)|}{b}+|f'(t)|\right)^q dt\right)^{1/q} \times\left(\int_0^b |f(t)|^{p+q}\,dt\right)^{(q-1)/q}, \tag{3.2.29}$$

where $M = \{\frac{\alpha+2}{\alpha+1}, \frac{2(p+q)}{\alpha+1}\}$.

REMARK 3.2.2. It is interesting to note that, if the function f is continuously differentiable on $(0, \infty)$, then letting $b \to \infty$ in (3.2.28) and (3.2.29) we get respectively the following inequalities

$$\int_0^\infty t^\alpha |f(t)|^{p+q}\,dt \leqslant M\left(\int_0^\infty t^{\alpha+1}|f(t)|^p|f'(t)|^q\,dt\right)^{1/q} \times\left(\int_0^\infty t^{\alpha+1}|f(t)|^{p+q}\,dt\right)^{(q-1)/q}, \tag{3.2.30}$$

$$\int_0^\infty t^\alpha |f(t)|^{p+q}\,dt \leqslant M\left(\int_0^\infty t^{(\alpha+1)q}|f(t)|^p|f'(t)|^q\,dt\right)^{1/q} \times\left(\int_0^\infty |f(t)|^{p+q}\,dt\right)^{(q-1)/q}. \tag{3.2.31}$$

In particular, if we take $\alpha = 0$, $p = 0$, $q = 2$, then the inequalities obtained in (3.2.30), (3.2.31) reduce to the slight variants of Weyl's inequality given in [141, p. 165].

THEOREM 3.2.6. *Let α, p, q, f be as defined in Theorem 3.2.5. Then the following inequalities hold*

$$\int_0^b t^\alpha |f(t)|^{p+q}\,dt \leqslant M^q \int_0^b t^{\alpha+q}|f(t)|^p\left(\frac{|f(t)|}{b} + |f'(t)|\right)^q dt, \tag{3.2.32}$$

$$\int_0^b t^\alpha |f(t)|^{p+q}\,dt \leqslant M^{p+q} \int_0^b t^{\alpha+p+q}\left(\frac{|f(t)|}{b} + |f'(t)|\right)^{p+q} dt, \tag{3.2.33}$$

where M is as defined in Theorem 3.2.5.

REMARK 3.2.3. If the function f is continuously differentiable on $(0, \infty)$, then letting $b \to \infty$ in (3.2.32) and (3.2.33) we get respectively the following inequalities

$$\int_0^\infty t^\alpha |f(t)|^{p+q}\,dt \leqslant M^q \int_0^\infty t^{\alpha+q}|f(t)|^p|f'(t)|^q\,dt, \tag{3.2.34}$$

$$\int_0^\infty t^\alpha |f(t)|^{p+q}\,dt \leqslant M^{p+q} \int_0^\infty t^{\alpha+p+q}|f'(t)|^{p+q}\,dt. \tag{3.2.35}$$

Here we note that the inequalities obtained in (3.2.34) and (3.2.35) are similar to that of the Opial-type inequality (3.2.1) and those of well-known Hardy's inequality (2.4.1).

THEOREM 3.2.7. *Let α, p, q be as defined in Theorem 3.2.5 and let f be a real-valued continuously differentiable function defined on (a, b) for fixed real numbers $a < b$. Then the following inequalities hold*

$$\int_a^b |t|^\alpha |f(t)|^{p+q}\,dt \leqslant |H| + \left(\frac{p+q}{\alpha+1}\right)\left(\int_a^b |t|^{\alpha+1}|f(t)|^p|f'(t)|^q\,dt\right)^{1/q} \times \left(\int_a^b |t|^{\alpha+1}|f(t)|^{p+q}\,dt\right)^{(q-1)/q}, \tag{3.2.36}$$

$$\int_a^b |t|^\alpha |f(t)|^{p+q}\,\mathrm{d}t \leqslant |H| + \left(\frac{p+q}{\alpha+1}\right)\left(\int_a^b |t|^{(\alpha+1)q}|f(t)|^p|f'(t)|^q\,\mathrm{d}t\right)^{1/q} \times \left(\int_a^b |f(t)|^{p+q}\,\mathrm{d}t\right)^{(q-1)/q}, \tag{3.2.37}$$

where

$$H = \frac{1}{\alpha+1}\left[b^{\alpha+1}(\operatorname{sgn} b)^\alpha |f(b)|^{p+q} - a^{\alpha+1}(\operatorname{sgn} a)^\alpha |f(a)|^{p+q}\right]. \tag{3.2.38}$$

REMARK 3.2.4. We note that, in the special case when $q = 1$, the inequalities obtained in (3.2.36) and (3.2.37) reduces to the following inequality

$$\int_a^b |t|^\alpha |f(t)|^{p+1}\,\mathrm{d}t \leqslant |H_0| + \left(\frac{p+1}{\alpha+1}\right)\int_a^b |t|^{\alpha+1}|f(t)|^p|f'(t)|\,\mathrm{d}t, \tag{3.2.39}$$

where H_0 is defined by the right-hand side of (3.2.38) taking $q = 1$.

PROOFS OF THEOREMS 3.2.5–3.5.7. Integrating by parts we have the following identity

$$\int_0^b \left[t^{\alpha+1} - \frac{1}{b}t^{\alpha+2}\right]|f(t)|^{p+q-1}|f'(t)|\operatorname{sgn} f(t)\,\mathrm{d}t = -\int_0^b \left[(\alpha+1)t^\alpha - \frac{1}{b}(\alpha+2)t^{\alpha+1}\right]\frac{|f(t)|^{p+q}}{p+q}\,\mathrm{d}t. \tag{3.2.40}$$

From (3.2.40) we observe that

$$\begin{aligned}(\alpha+1)\int_0^b t^\alpha |f(t)|^{p+q}\,\mathrm{d}t &= (\alpha+2)\int_0^b t^{\alpha+1}\frac{1}{b}|f(t)|^{p+q}\,\mathrm{d}t \\ &\quad - (p+q)\int_0^b t^{\alpha+1}\left[1 - \frac{1}{b}t\right]|f(t)|^{p+q-1}|f'(t)|\operatorname{sgn} f(t)\,\mathrm{d}t\end{aligned}$$

$$
\begin{aligned}
&\leqslant (\alpha+2)\int_0^b t^{\alpha+1}\frac{1}{b}\left|f(t)\right|^{p+q}\,\mathrm{d}t \\
&\quad + (p+q)\int_0^b t^{\alpha+1}\left[1+\frac{1}{b}t\right]\left|f(t)\right|^{p+q-1}\left|f'(t)\right|\mathrm{d}t. \qquad (3.2.41)
\end{aligned}
$$

From (3.2.41) we observe that

$$
\begin{aligned}
&\int_0^b t^{\alpha}\left|f(t)\right|^{p+q}\,\mathrm{d}t \\
&\quad\leqslant \left(\frac{\alpha+2}{\alpha+1}\right)\int_0^b t^{\alpha+1}\frac{1}{b}\left|f(t)\right|^{p+q}\,\mathrm{d}t \\
&\qquad + \frac{2(p+q)}{\alpha+1}\int_0^b t^{\alpha+1}\left|f(t)\right|^{p+q-1}\left|f'(t)\right|\mathrm{d}t \\
&\quad\leqslant M\int_0^b t^{\alpha+1}\left|f(t)\right|^{p+q-1}\left[\frac{|f(t)|}{b}+\left|f'(t)\right|\right]\mathrm{d}t \\
&\quad= M\int_0^b\left[t^{(\alpha+1)/q}\left|f(t)\right|^{p/q}\left(\frac{|f(t)|}{b}+\left|f'(t)\right|\right)\right] \\
&\qquad\qquad \times\left[t^{(\alpha+1)((q-1)/q)}\left|f(t)\right|^{p+q-1-p/q}\right]\mathrm{d}t. \qquad (3.2.42)
\end{aligned}
$$

Now, by using Hölder's inequality with indices $q, q/(q-1)$, we get the required inequality in (3.2.28).

In order to establish inequality (3.2.29), we rewrite the last inequality in (3.2.42) in the following form

$$
\begin{aligned}
\int_0^b t^{\alpha}\left|f(t)\right|^{p+q}\,\mathrm{d}t &\leqslant M\int_0^b\left[t^{\alpha+1}\left|f(t)\right|^{p/q}\left(\frac{|f(t)|}{b}+\left|f'(t)\right|\right)\right] \\
&\qquad \times\left[\left|f(t)\right|^{p+q-1-p/q}\right]\mathrm{d}t. \qquad (3.2.43)
\end{aligned}
$$

By using Hölder's inequality with indices $q, q/(q-1)$ on the right-hand side of (3.2.43), we get the desired inequality in (3.2.29). The proof of Theorem 3.2.5 is complete.

By following the same arguments as in the proof of Theorem 3.2.5, we have the inequality (3.2.42). Rewriting the inequality (3.2.42) and using Hölder's inequality with indices $q, q/(q-1)$, we have

$$\int_0^b t^\alpha \big|f(t)\big|^{p+q}\,\mathrm{d}t \leqslant M \int_0^b \Big[t^{\alpha+1-\alpha((q-1)/q)}\big|f(t)\big|^{p/q}\Big(\frac{|f(t)|}{b}+\big|f'(t)\big|\Big)\Big]$$
$$\times \big[t^{\alpha((q-1)/q)}\big|f(t)\big|^{p+q-1-p/q}\big]\,\mathrm{d}t$$
$$\leqslant M\bigg(\int_0^b t^{\alpha+q}\big|f(t)\big|^p\Big(\frac{|f(t)|}{b}+\big|f'(t)\big|\Big)^q\,\mathrm{d}t\bigg)^{1/q}$$
$$\times\bigg(\int_0^b t^\alpha\big|f(t)\big|^{p+q}\,\mathrm{d}t\bigg)^{(q-1)/q}. \tag{3.2.44}$$

If $\int_0^b t^\alpha |f(t)|^{p+q}\,\mathrm{d}t = 0$ then (3.2.32) is trivially true; otherwise, dividing both sides of (3.2.44) by $(\int_0^b t^\alpha |f(t)|^{p+q}\,\mathrm{d}t)^{(q-1)/q}$ and then taking the qth power on both sides of the resulting inequality, we get the required inequality in (3.2.32).

Rewriting inequality (3.2.42) and using Hölder's inequality with indices $p+q$, $(p+q)/(p+q-1)$, we have

$$\int_0^b t^\alpha \big|f(t)\big|^{p+q}\,\mathrm{d}t \leqslant M \int_0^b \Big[t^{\alpha+1-\alpha((p+q-1)/(p+q))}\Big(\frac{|f(t)|}{b}+\big|f'(t)\big|\Big)\Big]$$
$$\times \big[t^{\alpha((p+q-1)/(p+q))}\big|f(t)\big|^{p+q-1}\big]\,\mathrm{d}t$$
$$\leqslant M\bigg(\int_0^b t^{\alpha+p+q}\Big(\frac{|f(t)|}{b}+\big|f'(t)\big|\Big)^{p+q}\,\mathrm{d}t\bigg)^{1/(p+q)}$$
$$\times\bigg(\int_0^b t^\alpha\big|f(t)\big|^{p+q}\,\mathrm{d}t\bigg)^{(p+q-1)/(p+q)}. \tag{3.2.45}$$

Now, by following the arguments as in the last part of the proof of inequality (3.2.32) with suitable modifications, we get the required inequality in (3.2.33). The proof of Theorem 3.2.6 is complete.

By rewriting and integrating by parts the integral on the left-hand side of (3.2.36), we have

$$\int_a^b |t|^\alpha \big|f(t)\big|^{p+q}\,\mathrm{d}t = \int_a^b t^\alpha(\operatorname{sgn} t)^\alpha\big|f(t)\big|^{p+q}\,\mathrm{d}t$$
$$= H - \Big(\frac{p+q}{\alpha+1}\Big)\int_a^b t^{\alpha+1}\big|f(t)\big|^{p+q-1}f'(t)(\operatorname{sgn} t)^\alpha\,\mathrm{d}t. \tag{3.2.46}$$

From (3.2.46) we observe that

$$\begin{aligned}
&\int_a^b |t|^\alpha \left|f(t)\right|^{p+q} \mathrm{d}t \\
&\quad \leqslant |H| + \left(\frac{p+q}{\alpha+1}\right) \int_a^b |t|^{\alpha+1} \left|f(t)\right|^{p+q-1} \left|f'(t)\right| \mathrm{d}t \\
&\quad \leqslant |H| + \left(\frac{p+q}{\alpha+1}\right) \int_a^b \left[|t|^{(\alpha+1)/q} \left|f(t)\right|^{p/q} \left|f'(t)\right|\right] \\
&\qquad \times \left[|t|^{(\alpha+1)((q-1)/q)} \left|f(t)\right|^{p+q-1-p/q}\right] \mathrm{d}t. \qquad (3.2.47)
\end{aligned}$$

Now, by using Hölder's inequality with indices $q, q/(q-1)$ to the integral on the right-hand side of (3.2.47), we get the required inequality in (3.2.36).

The proof of inequality (3.2.37) is similar to that of the proof of inequality (3.2.36) given above with suitable modifications, so we omit it here. The proof of Theorem 3.2.7 is complete. □

3.3 Wirtinger–Opial-Type Integral Inequalities

There is extensive literature on integral inequalities involving functions and their derivatives which claim their origin to the well-known Wirtinger- and Opial-type integral inequalities (see [141,211]). In this section we present some results established in [51,238,241,283].

In [238] Pachpatte has established the Wirtinger- and Opial-type integral inequalities in the following three theorems. In what follows, the symbol $D^k u(x)$ denotes the kth derivative of $u(x)$ with $D^0 u(x) = u(x)$ for $x \in [a,b] = I$ and we write $D^1 u(x) = Du(x)$ for $x \in I$.

THEOREM 3.3.1. *Let p_{i-1}, $i = 1, \ldots, n$, be real-valued nonnegative continuous functions defined on I. Let $f, g \in C^{(n-1)}(I)$ and $D^{n-1} f(x)$, $D^{n-1} g(x)$ are absolutely continuous on I with $D^k f(a) = D^k f(b) = 0$, $D^k g(a) = D^k g(b) = 0$ for $0 \leqslant k \leqslant n-1$. Then the following inequalities hold*

$$\begin{aligned}
&\int_a^b \sum_{i=1}^n p_{i-1}(t)\left|D^{i-1} f(t)\right|\left|D^{i-1} g(t)\right| \mathrm{d}t \\
&\quad \leqslant \frac{1}{2}\left(\frac{b-a}{4}\right) \\
&\qquad \times \sum_{i=1}^n \left[\left(\int_a^b p_{i-1}(t)\,\mathrm{d}t\right)\left(\int_a^b \left[\left|D^i f(t)\right|^2 + \left|D^i g(t)\right|^2\right] \mathrm{d}t\right)\right], \qquad (3.3.1)
\end{aligned}$$

$$\left(\int_a^b \sum_{i=1}^n p_{i-1}(t)\left|D^{i-1}f(t)\right|\left|D^{i-1}g(t)\right|\,\mathrm{d}t\right)^2$$
$$\leqslant \frac{n}{2}\left(\frac{1}{4}\right)^2(b-a)^4$$
$$\times \sum_{i=1}^n\left[\left(\int_a^b p_{i-1}^2(t)\,\mathrm{d}t\right)\left(\int_a^b\left[\left|D^i f(t)\right|^4+\left|D^i g(t)\right|^4\right]\mathrm{d}t\right)\right], \tag{3.3.2}$$

$$\prod_{i=1}^n\left(\int_a^b p_{i-1}(t)\left|D^{i-1}f(t)\right|\left|D^{i-1}g(t)\right|\,\mathrm{d}t\right)$$
$$\leqslant \left(\frac{1}{2}\right)^n\left(\frac{b-a}{4}\right)^n$$
$$\times \prod_{i=1}^n\left[\left(\int_a^b p_{i-1}(t)\,\mathrm{d}t\right)\left(\int_a^b\left[\left|D^i f(t)\right|^2+\left|D^i g(t)\right|^2\right]\mathrm{d}t\right)\right]. \tag{3.3.3}$$

REMARK 3.3.1. In the special case when $D^i f = D^i g$, the inequalities established in Theorem 3.3.1 reduce to the new integral inequalities of the Wirtinger type studied by many authors in the literature (see [211]). In this special case, it is easy to observe from the inequalities in (3.3.1) and (3.3.3) that the following inequality

$$\int_a^b p_{i-1}(t)\left|D^{i-1}f(t)\right|^2\mathrm{d}t$$
$$\leqslant \left(\frac{b-a}{4}\right)\left(\int_a^b p_{i-1}(t)\,\mathrm{d}t\right)\left(\int_a^b\left|D^i f(t)\right|^2\mathrm{d}t\right) \tag{3.3.4}$$

holds for $i = 1,\ldots,n$, which in turn is a further generalization of the integral inequality established by Traple in [419, p. 160]. Further if we take $D^i g = D^i f$ and $n = 2$, then inequality (3.3.2) reduces to the following integral inequality

$$\int_a^b\left[p_0(t)\left|f(t)\right|^2+p_1(t)\left|Df(t)\right|^2\right]\mathrm{d}t$$
$$\leqslant \left(\frac{1}{4}\right)^2(b-a)^4 2\left[\left(\int_a^b p_0^2(t)\,\mathrm{d}t\right)\left(\int_a^b\left|Df(t)\right|^4\mathrm{d}t\right)\right.$$
$$\left.+\left(\int_a^b p_1^2(t)\,\mathrm{d}t\right)\left(\int_a^b\left|D^2 f(t)\right|^4\mathrm{d}t\right)\right]. \tag{3.3.5}$$

The inequalities of the type (3.3.5) are considered by the authors in [18,211] by using a different technique. However, the bounds obtained on the right-hand side in (3.3.5) cannot be compared with the bound obtained in [18,211].

THEOREM 3.3.2. *Let p_{i-1}, f, g be as in Theorem* 3.3.1. *Then the following inequalities hold*

$$\int_a^b \sum_{i=1}^n p_{i-1}(t)\left|D^{i-1}f(t)\right|^2\left|D^{i-1}g(t)\right|^2 \mathrm{d}t \leqslant \frac{1}{2}\left(\frac{1}{4}\right)^2 (b-a)^3 \times \sum_{i=1}^n \left[\left(\int_a^b p_{i-1}(t)\,\mathrm{d}t\right)\left(\int_a^b \left[\left|D^i f(t)\right|^4 + \left|D^i g(t)\right|^4\right]\mathrm{d}t\right)\right], \tag{3.3.6}$$

$$\left(\int_a^b \sum_{i=1}^n p_{i-1}(t)\left|D^{i-1}f(t)\right|^2\left|D^{i-1}g(t)\right|^2 \mathrm{d}t\right)^2 \leqslant \frac{n}{2}\left(\frac{1}{4}\right)^4 (b-a)^8 \times \sum_{i=1}^n \left[\left(\int_a^b p_{i-1}^2(t)\,\mathrm{d}t\right)\left(\int_a^b \left[\left|D^i f(t)\right|^8 + \left|D^i g(t)\right|^8\right]\mathrm{d}t\right)\right], \tag{3.3.7}$$

$$\prod_{i=1}^n \left(\int_a^b p_{i-1}(t)\left|D^{i-1}f(t)\right|^2\left|D^{i-1}g(t)\right|^2 \mathrm{d}t\right) \leqslant \left(\frac{1}{2}\right)^n\left(\frac{1}{4}\right)^{2n} (b-a)^{3n} \times \prod_{i=1}^n \left[\left(\int_a^b p_{i-1}(t)\,\mathrm{d}t\right)\left(\int_a^b \left[\left|D^i f(t)\right|^4 + \left|D^i g(t)\right|^4\right]\mathrm{d}t\right)\right]. \tag{3.3.8}$$

REMARK 3.3.2. In the special case when $D^i g = D^i f$, the inequalities in (3.3.6)–(3.3.8) reduce to the new inequalities. In this special case, it is easy to observe from inequalities (3.3.6) and (3.3.8) that the following inequality

$$\int_a^b p_{i-1}(t)\left|D^{i-1}f(t)\right|^4 \mathrm{d}t \leqslant \left(\frac{1}{4}\right)^2 (b-a)^3$$

$$\times \left(\int_a^b p_{i-1}(t)\,\mathrm{d}t \right) \left(\int_a^b \left| D^i f(t) \right|^4 \mathrm{d}t \right) \quad (3.3.9)$$

holds for $i = 1, \ldots, n$.

THEOREM 3.3.3. *Let p_{i-1}, f, g be as in Theorem 3.3.1. Then the following inequalities hold*

$$\int_a^b \sum_{i=1}^n p_{i-1}(t) \left[\left| D^{i-1} f(t) \right| \left| D^i g(t) \right| + \left| D^{i-1} g(t) \right| \left| D^i f(t) \right| \right] \mathrm{d}t$$

$$\leqslant \sum_{i=1}^n \left[\left(\left(\frac{b-a}{4} \right) \int_a^b p_{i-1}^2(t)\,\mathrm{d}t \right)^{1/2} \right.$$

$$\left. \times \left(\int_a^b \left[\left| D^i f(t) \right|^2 + \left| D^i g(t) \right|^2 \right] \mathrm{d}t \right) \right], \quad (3.3.10)$$

$$\int_a^b \sum_{i=1}^n p_{i-1}(t) \left[\left| D^{i-1} f(t) \right|^2 \left| D^i g(t) \right|^2 + \left| D^{i-1} g(t) \right|^2 \left| D^i f(t) \right|^2 \right] \mathrm{d}t$$

$$\leqslant \sum_{i=1}^n \left[\left(\left(\frac{1}{4} \right)^2 (b-a)^3 \left(\int_a^b p_{i-1}^2(t)\,\mathrm{d}t \right) \right)^{1/2} \right.$$

$$\left. \times \left(\int_a^b \left[\left| D^i f(t) \right|^4 + \left| D^i g(t) \right|^4 \right] \mathrm{d}t \right) \right]. \quad (3.3.11)$$

REMARK 3.3.3. We note that, for $n = 1$, the inequality established in (3.3.10) reduces to the inequality established by Pachpatte in [243]. In the special case when $D^i f = D^i g$, the inequalities established in (3.3.10) and (3.3.11) reduce to the Opial-type integral inequalities.

PROOFS OF THEOREMS 3.3.1–3.3.3. From the hypotheses, for every $t \in I$ and $i = 1, \ldots, n$, we have

$$D^{i-1} f(t) = \int_a^t D^i f(s)\,\mathrm{d}s, \qquad D^{i-1} f(t) = -\int_t^b D^i f(s)\,\mathrm{d}s, \quad (3.3.12)$$

$$D^{i-1} g(t) = \int_a^t D^i g(s)\,\mathrm{d}s, \qquad D^{i-1} g(t) = -\int_t^b D^i g(s)\,\mathrm{d}s. \quad (3.3.13)$$

From (3.3.12) and (3.3.13), we observe that

$$|D^{i-1}f(t)| \leqslant \frac{1}{2}\int_a^b |D^i f(t)|\,\mathrm{d}t, \tag{3.3.14}$$

$$|D^{i-1}g(t)| \leqslant \frac{1}{2}\int_a^b |D^i g(t)|\,\mathrm{d}t, \tag{3.3.15}$$

for $i = 1, \ldots, n$. From (3.3.14) and (3.3.15), and using the elementary inequality $cd \leqslant \frac{1}{2}(c^2 + d^2)$ (for c, d reals) and Schwarz inequality, we obtain

$$\begin{aligned}
&\sum_{i=1}^{n} p_{i-1}(t)|D^{i-1}f(t)||D^{i-1}g(t)| \\
&\quad\leqslant \frac{1}{4}\sum_{i=1}^{n} p_{i-1}(t)\left(\int_a^b |D^i f(t)|\,\mathrm{d}t\right)\left(\int_a^b |D^i g(t)|\,\mathrm{d}t\right) \\
&\quad\leqslant \frac{1}{4}\sum_{i=1}^{n} p_{i-1}(t)\frac{1}{2}\left[\left(\int_a^b |D^i f(t)|\,\mathrm{d}t\right)^2 + \left(\int_a^b |D^i g(t)|\,\mathrm{d}t\right)^2\right] \\
&\quad\leqslant \frac{1}{2}\left(\frac{b-a}{4}\right) \\
&\qquad\times \sum_{i=1}^{n} p_{i-1}(t)\left(\int_a^b \left[|D^i f(t)|^2 + |D^i g(t)|^2\right]\mathrm{d}t\right).
\end{aligned} \tag{3.3.16}$$

Integrating both sides of (3.3.16) from a to b we obtain the desired inequality in (3.3.1).

Taking the square on both sides of inequality (3.3.1) and using the elementary inequality $(c_1 + \cdots + c_n)^2 \leqslant n(c_1^2 + \cdots + c_n^2)$ (for $c_1, \ldots, c_n$ reals), Schwarz inequality, and the elementary inequality $(c+d)^2 \leqslant 2(c^2 + d^2)$ (for c, d reals), we obtain

$$\begin{aligned}
&\left(\int_a^b \sum_{i=1}^{n} p_{i-1}(t)|D^{i-1}f(t)||D^{i-1}g(t)|\,\mathrm{d}t\right)^2 \\
&\quad\leqslant n\left(\frac{1}{2}\right)^2\left(\frac{1}{4}\right)^2(b-a)^2 \\
&\qquad\times \sum_{i=1}^{n}\left[\left(\int_a^b p_{i-1}(t)\,\mathrm{d}t\right)^2\left(\int_a^b \left[|D^i f(t)|^2 + |D^i g(t)|^2\right]\mathrm{d}t\right)^2\right]
\end{aligned}$$

$$
\begin{aligned}
&\leqslant n\left(\frac{1}{2}\right)^2\left(\frac{1}{4}\right)^2(b-a)^4 \\
&\quad\times\sum_{i=1}^{n}\left[\left(\int_a^b p_{i-1}^2(t)\,\mathrm{d}t\right)\left(\int_a^b \left[\left|D^i f(t)\right|^2+\left|D^i g(t)\right|^2\right]^2\mathrm{d}t\right)\right] \\
&\leqslant \frac{n}{2}\left(\frac{1}{4}\right)^2(b-a)^4 \\
&\quad\times\sum_{i=1}^{n}\left[\left(\int_a^b p_{i-1}^2(t)\,\mathrm{d}t\right)\left(\int_a^b \left[\left|D^i f(t)\right|^4+\left|D^i g(t)\right|^4\right]\mathrm{d}t\right)\right].
\end{aligned}
$$

The proof of inequality (3.3.20) is complete.

From (3.3.14) and (3.3.15), and using the elementary inequality $cd \leqslant \frac{1}{2}(c^2 + d^2)$ (for c, d reals) and Schwarz inequality, we obtain, for $i = 1, \ldots, n$,

$$
\begin{aligned}
&\int_a^b p_{i-1}(t)\left|D^{i-1}f(t)\right|\left|D^{i-1}g(t)\right|\mathrm{d}t \\
&\quad\leqslant \frac{1}{4}\left(\int_a^b p_{i-1}(t)\,\mathrm{d}t\right)\left(\int_a^b \left|D^i f(t)\right|\mathrm{d}t\right)\left(\int_a^b \left|D^i g(t)\right|\mathrm{d}t\right) \\
&\quad\leqslant \frac{1}{4}\left(\int_a^b p_{i-1}(t)\,\mathrm{d}t\right)\frac{1}{2}\left[\left(\int_a^b \left|D^i f(t)\right|\mathrm{d}t\right)^2+\left(\int_a^b \left|D^i g(t)\right|\mathrm{d}t\right)^2\right] \\
&\quad\leqslant \frac{1}{2}\left(\frac{b-a}{4}\right)\left(\int_a^b p_{i-1}(t)\,\mathrm{d}t\right)\left(\int_a^b \left[\left|D^i f(t)\right|^2+\left|D^i g(t)\right|^2\right]\mathrm{d}t\right).
\end{aligned}
\tag{3.3.17}
$$

From (3.3.17) we have

$$
\begin{aligned}
&\prod_{i=1}^{n}\left(\int_a^b p_{i-1}(t)\left|D^{i-1}f(t)\right|\left|D^{i-1}g(t)\right|\mathrm{d}t\right) \\
&\quad\leqslant \left(\frac{1}{2}\right)^n\left(\frac{b-a}{4}\right)^n \\
&\qquad\times\prod_{i=1}^{n}\left[\left(\int_a^b p_{i-1}(t)\,\mathrm{d}t\right)\left(\int_a^b \left[\left|D^i f(t)\right|^2+\left|D^i g(t)\right|^2\right]\mathrm{d}t\right)\right],
\end{aligned}
$$

which is the desired inequality in (3.3.3) and the proof of Theorem 3.3.1 is complete. □

From (3.3.14), (3.3.15) and using Schwarz inequality, we observe that

$$\left|D^{i-1}f(t)\right|^2 \leqslant \left(\frac{1}{4}\right)(b-a)\int_a^b \left|D^i f(t)\right|^2 \mathrm{d}t, \tag{3.3.18}$$

$$\left|D^{i-1}g(t)\right|^2 \leqslant \left(\frac{1}{4}\right)(b-a)\int_a^b \left|D^i g(t)\right|^2 \mathrm{d}t, \tag{3.3.19}$$

for $i = 1, \ldots, n$. Now, from (3.3.18) and (3.3.19), and following exactly the same arguments as in the proof of inequality (3.3.1) in Theorem 3.3.1, we obtain the desired inequality in (3.3.6).

The details of the proofs of inequalities (3.3.7) and (3.3.8) follow from (3.3.18) and (3.3.19), and following exactly the same arguments as in the proofs of inequalities (3.3.2) and (3.3.3) given in Theorem 3.3.1 and hence we omit further details.

By virtue of Schwarz inequality, the inequality (3.3.4) and the elementary inequality $c^{1/2}d^{1/2} \leqslant \frac{1}{2}(c+d)$ (for $c, d \geqslant 0$, reals), we observe that

$$\begin{aligned}
&\int_a^b \sum_{i=1}^n p_{i-1}(t)\left[\left|D^{i-1}f(t)\right|\left|D^i g(t)\right| + \left|D^{i-1}g(t)\right|\left|D^i f(t)\right|\right]\mathrm{d}t \\
&\leqslant \sum_{i=1}^n \left[\left(\int_a^b p_{i-1}^2(t)\left|D^{i-1}f(t)\right|^2 \mathrm{d}t\right)^{1/2}\left(\int_a^b \left|D^i g(t)\right|^2 \mathrm{d}t\right)^{1/2}\right] \\
&\quad + \sum_{i=1}^n \left[\left(\int_a^b p_{i-1}^2(t)\left|D^{i-1}g(t)\right|^2 \mathrm{d}t\right)^{1/2}\left(\int_a^b \left|D^i f(t)\right|^2 \mathrm{d}t\right)^{1/2}\right] \\
&\leqslant \sum_{i=1}^n \left[\left(\left(\frac{b-a}{4}\right)\left(\int_a^b p_{i-1}^2(t)\,\mathrm{d}t\right)\left(\int_a^b \left|D^i f(t)\right|^2 \mathrm{d}t\right)\right)^{1/2}\right. \\
&\qquad \left.\times \left(\int_a^b \left|D^i g(t)\right|^2 \mathrm{d}t\right)^{1/2}\right] \\
&\quad + \sum_{i=1}^n \left[\left(\left(\frac{b-a}{4}\right)\left(\int_a^b p_{i-1}^2(t)\,\mathrm{d}t\right)\left(\int_a^b \left|D^i g(t)\right|^2 \mathrm{d}t\right)\right)^{1/2}\right. \\
&\qquad \left.\times \left(\int_a^b \left|D^i f(t)\right|^2 \mathrm{d}t\right)^{1/2}\right]
\end{aligned}$$

$$= 2\sum_{i=1}^{n}\left[\left(\left(\frac{b-a}{4}\right)\int_a^b p_{i-1}^2(t)\,dt\right)^{1/2} \times \left(\int_a^b |D^i f(t)|^2\,dt\right)^{1/2}\left(\int_a^b |D^i g(t)|^2\,dt\right)^{1/2}\right]$$

$$\leqslant \sum_{i=1}^{n}\left[\left(\left(\frac{b-a}{4}\right)\int_a^b p_{i-1}^2(t)\,dt\right)^{1/2}\left(\int_a^b \left[|D^i f(t)|^2 + |D^i g(t)|^2\right] dt\right)\right],$$

which is the desired inequality in (3.3.10).

The details of the proof of inequality (3.3.11) follow by the same argument as in the proof of inequality (3.3.10) given above by using inequality (3.3.9) in place of inequality (3.3.4). We omit the details.

In [241] Pachpatte has established the inequalities in the following theorem.

THEOREM 3.3.4. *Let $p(t)$ be a real-valued nonnegative continuous function defined on $I = [0, b]$. Let $f \in C^{(n-1)}(I)$ with $D^{r-1} f(t)$ absolutely continuous for $t \in I$ and $D^{r-1} f(0) = D^{r-1} f(b) = 0$, for $r = 1, \ldots, n$. Then the following inequalities hold*

$$\int_0^b p(t)\left(\prod_{r=1}^{n} |D^{r-1} f(t)|\right)^{2/n} dt \leqslant \frac{b}{4n}\left(\int_0^b p(t)\,dt\right)\left(\int_0^b\left(\sum_{r=1}^{n} |D^r f(t)|^2\right) dt\right), \tag{3.3.20}$$

$$\int_0^b p(t)\left(\prod_{r=1}^{n} |D^{r-1} f(t)|^2\right)^{2/n} dt \leqslant \frac{1}{n}\left(\int_0^b p(t)\,dt\right) \times \left(\sum_{r=1}^{n}\left[\left(\int_0^b |D^{r-1} f(t)|^2\,dt\right)\left(\int_0^b |D^r f(t)|^2\,dt\right)\right]\right), \tag{3.3.21}$$

$$\int_0^b p(t)\left(\prod_{r=1}^{n} |D^{r-1} f(t)|\right)^{1/n}\left(\sum_{r=1}^{n} |D^r f(t)|\right) dt \leqslant \left(\frac{b}{4}\int_0^b p^2(t)\,dt\right)^{1/2}\left(\int_0^b\left(\sum_{r=1}^{n} |D^r f(t)|^2\right) dt\right). \tag{3.3.22}$$

REMARK 3.3.4. In the special cases when $n = 1$ and $n = 2$, the inequalities established in Theorem 3.3.4 reduces to Wirtinger- and Opial-type inequalities, see [241].

PROOF OF THEOREM 3.3.4. From the hypotheses we have the following identities

$$D^{r-1} f(t) = \int_0^t D^r f(s)\, ds, \tag{3.3.23}$$

$$D^{r-1} f(t) = -\int_t^b D^r f(s)\, ds, \tag{3.3.24}$$

for $t \in I$ and $r = 1, \ldots, n$. From (3.3.23) and (3.3.24), we obtain

$$\left|D^{r-1} f(t)\right| \leqslant \frac{1}{2}\int_0^b \left|D^r f(t)\right| dt \tag{3.3.25}$$

for $t \in I$ and $r = 1, \ldots, n$. From (3.3.25) and using the elementary inequalities

$$\left(\prod_{i=1}^n a_i\right)^{1/n} \leqslant \frac{1}{n}\sum_{i=1}^n a_i \tag{3.3.26}$$

(for $a_1, \ldots, a_n \geqslant 0$ reals and $n \geqslant 1$) and

$$\left(\sum_{i=1}^n a_i\right)^2 \leqslant n\sum_{i=1}^n a_i^2 \tag{3.3.27}$$

(for $a_1, \ldots, a_n$ reals) and Schwarz inequality, we obtain

$$\begin{aligned}
\left(\prod_{r=1}^n \left|D^{r-1} f(t)\right|\right)^{2/n} &\leqslant \left[\left(\frac{1}{2}\right)^n\right]^{2/n}\left[\left(\prod_{r=1}^n\left(\int_0^b \left|D^r f(t)\right| dt\right)\right)^{1/n}\right]^2 \\
&\leqslant \frac{1}{4}\left[\frac{1}{n}\left(\sum_{r=1}^n \int_0^b \left|D^r f(t)\right| dt\right)\right]^2 \\
&\leqslant \frac{1}{4n^2}\left[n\sum_{r=1}^n\left(\int_0^b \left|D^r f(t)\right| dt\right)^2\right] \\
&\leqslant \frac{b}{4n}\int_0^b\left(\sum_{r=1}^n \left|D^r f(t)\right|^2\right) dt.
\end{aligned} \tag{3.3.28}$$

Multiplying (3.3.28) by $p(t)$ and integrating the resulting inequality from 0 to b we obtain the desired inequality in (3.3.20).

From the hypotheses, we have the following identities

$$\left[D^{r-1} f(t)\right]^2 = 2\int_0^t D^{r-1} f(s) D^r f(s)\,\mathrm{d}s, \tag{3.3.29}$$

$$\left[D^{r-1} f(t)\right]^2 = -2\int_t^b D^{r-1} f(s) D^r f(s)\,\mathrm{d}s, \tag{3.3.30}$$

for $t \in I$ and $r = 1, \ldots, n$. From (3.3.29) and (3.3.30), we obtain

$$\left|D^{r-1} f(t)\right|^2 \leqslant \int_0^b \left|D^{r-1} f(t)\right|\left|D^r f(t)\right| \mathrm{d}t \tag{3.3.31}$$

for $t \in I$ and $r = 1, \ldots, n$. From (3.3.31) and using inequalities (3.3.26), (3.3.27) and Schwarz inequality, we obtain

$$\begin{aligned}
\left(\prod_{r=1}^{n} \left|D^{r-1} f(t)\right|^2\right)^{2/n} &\leqslant \left[\left(\prod_{r=1}^{n}\left(\int_0^b \left|D^{r-1} f(t)\right|\left|D^r f(t)\right| \mathrm{d}t\right)\right)^{1/n}\right]^2 \\
&\leqslant \left[\frac{1}{n}\sum_{r=1}^{n}\int_0^b \left|D^{r-1} f(t)\right|\left|D^r f(t)\right| \mathrm{d}t\right]^2 \\
&\leqslant \frac{1}{n^2}\left[n\sum_{r=1}^{n}\left(\int_0^b \left|D^{r-1} f(t)\right|\left|D^r f(t)\right| \mathrm{d}t\right)^2\right] \\
&\leqslant \frac{1}{n}\sum_{r=1}^{n}\left[\left(\int_0^b \left|D^{r-1} f(t)\right|^2 \mathrm{d}t\right)\left(\int_0^b \left|D^r f(t)\right|^2 \mathrm{d}t\right)\right].
\end{aligned} \tag{3.3.32}$$

Multiplying both sides of (3.3.32) by $p(t)$ and integrating the resulting inequality from 0 to b we obtain the desired inequality in (3.3.21).

By using Schwarz inequality and inequalities (3.3.20) and (3.3.27), we observe that

$$\begin{aligned}
&\int_0^b p(t)\left(\prod_{r=1}^{n}\left|D^{r-1} f(t)\right|\right)^{1/n}\left(\sum_{r=1}^{n}\left|D^r f(t)\right|\right)\mathrm{d}t \\
&\leqslant \left(\int_0^b p^2(t)\left(\prod_{r=1}^{n}\left|D^{r-1} f(t)\right|\right)^{2/n} \mathrm{d}t\right)^{1/2}\left(\int_0^b\left(\sum_{r=1}^{n}\left|D^r f(t)\right|\right)^2 \mathrm{d}t\right)^{1/2}
\end{aligned}$$

$$\leqslant \left(\frac{b}{4n}\left(\int_0^b p^2(t)\,\mathrm{d}t\right)\left(\int_0^b\left(\sum_{r=1}^n\left|D^r f(t)\right|^2\right)\mathrm{d}t\right)\right)^{1/2}$$

$$\times\left(\int_0^b n\left(\sum_{r=1}^n\left|D^r f(t)\right|^2\right)\mathrm{d}t\right)^{1/2}$$

$$=\left(\frac{b}{4}\int_0^b p^2(t)\,\mathrm{d}t\right)^{1/2}\left(\int_0^b\left(\sum_{r=1}^n\left|D^r f(t)\right|^2\right)\mathrm{d}t\right).$$

This result is the desired inequality in (3.3.22). The proof is complete. □

In [51] Calvert established the following inequalities by using the method of Olech [230].

THEOREM 3.3.5. *Let u be absolutely continuous on (a,b) with $u(a)=0$, where $-\infty \leqslant a < b < \infty$. Let $f(t)$ be a continuous complex-valued function defined for all t in the range of u and for all real t of the form $t(s)=\int_a^s |u'(s)|\,\mathrm{d}s$. Suppose that $|f(t)| \leqslant f(|t|)$ for all t, and that $f(t_1) \leqslant f(t_2)$ for $0 \leqslant t_1 \leqslant t_2$. Let r be positive, continuous, and $\int_a^b r^{1-q}(t)\,\mathrm{d}t < \infty$, where $1/p+1/q=1$, $p>1$. Let $F(s)=\int_0^s f(\sigma)\,\mathrm{d}\sigma$, $s>0$. Then the following inequality holds*

$$\int_a^b \left|f\big(u(t)\big)u'(t)\right|\mathrm{d}t$$

$$\leqslant F\left(\left(\int_a^b r^{1-q}(t)\,\mathrm{d}t\right)^{1/q}\left(\int_a^b r(t)\left|u'(t)\right|^p\mathrm{d}t\right)^{1/p}\right), \tag{3.3.33}$$

with equality if and only if $u(t)=A\int_a^t r^{1-q}(s)\,\mathrm{d}s$. The same result (but with equality for $u(t)=\int_t^b r^{1-q}(s)\,\mathrm{d}s$) holds if $u(b)=0$ and $-\infty < a < b \leqslant \infty$, where A is a constant.

PROOF. Let $z(t)=\int_a^t |u'(s)|\,\mathrm{d}s$, $t\in(a,b)$. Then $z'(t)=|u'(t)|$, and it follows that

$$\int_a^b \left|f\big(u(t)\big)u'(t)\right|\mathrm{d}t = \int_a^b\left|f\left(\int_a^t u'(s)\,\mathrm{d}s\right)u'(t)\right|\mathrm{d}t$$

$$\leqslant \int_a^b f\left(\left|\int_a^t u'(s)\,\mathrm{d}s\right|\right)\left|u'(t)\right|\mathrm{d}t$$

$$\leqslant \int_a^b f\left(\int_a^t |u'(s)|\,\mathrm{d}s\right)|u'(t)|\,\mathrm{d}t$$
$$= \int_a^b f(z(t))z'(t)\,\mathrm{d}t = F(z(b)). \tag{3.3.34}$$

Now, using Hölder's inequality with indices p, q, we get

$$\begin{aligned} z(b) &= \int_a^b |u'(t)|\,\mathrm{d}t \\ &= \int_a^b r^{-1/p}(t) r^{1/p}(t) z'(t)\,\mathrm{d}t \\ &\leqslant \left(\int_a^b r^{1-q}(t)\,\mathrm{d}t\right)^{1/q}\left(\int_a^b r(t) z'(t)^p\,\mathrm{d}t\right)^{1/p}. \end{aligned} \tag{3.3.35}$$

Inequality (3.3.33) now follows from (3.3.34) and (3.3.35) and the fact that F is nondecreasing. □

REMARK 3.3.5. Let $f(t) = t^{p-1}$, $p > 1$, $u(a) = 0$, $-\infty \leqslant a < b < \infty$. Then

$$\int_a^b |u^{p-1}(t)u'(t)|\,\mathrm{d}t \leqslant \frac{1}{p}\left(\int_a^b r^{1-q}(t)\,\mathrm{d}t\right)^{p-1}\int_a^b r(t)|u'(t)|^p\,\mathrm{d}t.$$

THEOREM 3.3.6. *Let u and v be absolutely continuous functions on (a, b) with $u(a) = v(a) = 0$, where $-\infty \leqslant a < b < \infty$. Let $r(t)$ and $s(t)$ be positive, continuous functions on (a, b) and $\int_a^b (r(t))^{-2}\,\mathrm{d}t < \infty$, $\int_a^b (s(t))^{-2}\,\mathrm{d}t < \infty$. Then the following inequality holds*

$$\int_a^b \left[|u(t)v'(t)| + |v(t)u'(t)|\right]\mathrm{d}t$$
$$\leqslant \left[\int_a^b (r(t))^{-2}\,\mathrm{d}t \int_a^b (s(t))^{-2}\,\mathrm{d}t \int_a^b r^2(t)|u'(t)|^2\,\mathrm{d}t \int_a^b s^2(t)|v'(t)|^2\,\mathrm{d}t\right]^{1/2}, \tag{3.3.36}$$

with equality if and only if $u(t) = A\int_a^b (r(t))^{-2}\,\mathrm{d}t$ and $v(t) = B\int_a^b (s(t))^{-2}\,\mathrm{d}t$, where A, B are constants.

PROOF. Let $x(t) = \int_a^t |u'(\sigma)|\,\mathrm{d}\sigma$ and $y(t) = \int_a^t |v'(\sigma)|\,\mathrm{d}\sigma$, then $x'(t) = |u'(t)|$ and $y'(t) = |v'(t)|$. It is easy to observe that $|u(t)| = |\int_a^t u'(\sigma)\,\mathrm{d}\sigma| \leqslant$

$\int_a^b |u'(\sigma)|\,\mathrm{d}\sigma = x(t)$ and $|v(t)| \leqslant y(t)$. Thus, it follows that

$$\begin{aligned}\int_a^b \big[\big|u(t)v'(t)\big| + \big|v(t)u'(t)\big|\big]\,\mathrm{d}t &\leqslant \int_a^b \big(x(t)y'(t) + y(t)x'(t)\big)\,\mathrm{d}t \\ &= \int_a^b \frac{\mathrm{d}}{\mathrm{d}t}\big(x(t)y(t)\big)\,\mathrm{d}t \\ &= x(b)y(b),\end{aligned}$$

where

$$\begin{aligned}x(b) &= \int_a^b \big(r(t)\big)^{-1} r(t)\big|u'(t)\big|\,\mathrm{d}t \\ &\leqslant \left(\int_a^b \big(r(t)\big)^{-2}\,\mathrm{d}t\right)^{1/2} \left(\int_a^b \big(r(t)\big)^2\big|u'(t)\big|^2\,\mathrm{d}t\right)^{1/2}\end{aligned}$$

and

$$\begin{aligned}y(b) &= \int_a^b \big(s(t)\big)^{-1} s(t)\big|v'(t)\big|\,\mathrm{d}t \\ &\leqslant \left(\int_a^b \big(s(t)\big)^{-2}\,\mathrm{d}t\right)^{1/2} \left(\int_a^b \big(s(t)\big)^2\big|v'(t)\big|^2\,\mathrm{d}t\right)^{1/2}.\end{aligned}$$

Inequality (3.3.36) now follows immediately from the above obtained inequalities. □

REMARK 3.3.6. If we take $v(t) = u(t)$ and $r(t) = s(t) = 1$ in (3.3.36), then we get the following inequality

$$\int_a^b \big|u(t)u'(t)\big|\,\mathrm{d}t \leqslant \frac{b-a}{2}\int_a^b \big|u'(t)\big|^2\,\mathrm{d}t.$$

In [283] Pachpatte has established the following inequality.

THEOREM 3.3.7. *Let u_r, $r = 1, \dots, m$, be absolutely continuous functions defined on $[a, b]$ with $u_r(a) = u_r(b) = 0$. Let $g_r(u)$, $r = 1, \dots, m$, be continuous functions defined for all u in the range of u_r and for all real t of the form $t(s) = \int_a^s |u_r'(\sigma)|\,\mathrm{d}\sigma$ or $t(s) = \int_b^s |u_r'(\sigma)|\,\mathrm{d}\sigma$; $|g_r(u)| \leqslant g_r(|u|)$ for all u and*

$g_r(u_1) \leqslant g_r(u_2)$ for $0 \leqslant u_1 \leqslant u_2$. Then for every $c \in (a,b)$, the following inequality holds

$$\int_a^b \left\{\prod_{r=1}^m \left|g_r\big(u_r(t)\big)u_r'(t)\right|\right\}^{1/m} \mathrm{d}t$$
$$\leqslant \frac{1}{m}\sum_{i=1}^m \left[G_r\left(\int_a^c \left|u_r'(t)\right| \mathrm{d}t\right) + G_r\left(\int_c^b \left|u_r'(t)\right| \mathrm{d}t\right)\right], \tag{3.3.37}$$

where $G_r(u) = \int_0^u g_r(t)\,\mathrm{d}t$ for $u > 0$ and $r = 1, \ldots, m$.

PROOF. Let $c \in [a,b]$ and define

$$z_r(t) = \int_0^t \left|u_r'(s)\right| \mathrm{d}s \tag{3.3.38}$$

for $a \leqslant t \leqslant c$ and $r = 1, \ldots, m$ and

$$v_r(t) = \int_t^b \left|u_r'(s)\right| \mathrm{d}s \tag{3.3.39}$$

for $c \leqslant t \leqslant b$ and $r = 1, \ldots, m$. From (3.3.38) and (3.3.39), we have

$$z_r'(t) = \left|u_r'(t)\right| \tag{3.3.40}$$

for $a \leqslant t \leqslant c$ and $r = 1, \ldots, m$ and

$$v_r'(t) = -\left|u_r'(t)\right| \tag{3.3.41}$$

for $c \leqslant t \leqslant b$ and $r = 1, \ldots, m$. We note that

$$u_r(t) = \int_a^t u_r'(s)\,\mathrm{d}s \tag{3.3.42}$$

for $a \leqslant t \leqslant c$ and $r = 1, \ldots, m$ and

$$u_r(t) = -\int_t^b u_r'(s)\,\mathrm{d}s \tag{3.3.43}$$

for $c \leqslant t \leqslant b$ and $r = 1, \ldots, m$. From the hypotheses, the arithmetic mean–geometric mean inequalities (3.3.26), (3.3.42), (3.3.40) and (3.3.38) we observe that

$$\begin{aligned}
\int_a^c &\left\{\prod_{r=1}^m \left|g_r\big(u_r(t)\big)u_r'(t)\right|\right\}^{1/m} \mathrm{d}t \\
&\leqslant \frac{1}{m}\sum_{r=1}^m \int_a^c \left|g_r\big(u_r(t)\big)u_r'(t)\right| \mathrm{d}t \\
&\leqslant \frac{1}{m}\sum_{r=1}^m \int_a^c g_r\left(\int_a^t \left|u_r'(s)\right| \mathrm{d}s\right)\left|u_r'(t)\right| \mathrm{d}t \\
&= \frac{1}{m}\sum_{r=1}^m \int_a^c g_r\big(z_r(t)\big)z_r'(t)\,\mathrm{d}t \\
&= \frac{1}{m}\sum_{r=1}^m \int_0^{z_r(c)} g_r(t)\,\mathrm{d}t \\
&= \frac{1}{m}\sum_{r=1}^m G_r\big(z_r(c)\big) \\
&= \frac{1}{m}\sum_{r=1}^m G_r\left(\int_a^c \left|u_r'(t)\right| \mathrm{d}t\right). \qquad (3.3.44)
\end{aligned}$$

Similarly, from the hypotheses and inequalities (3.3.26), (3.3.43), (3.3.41) and (3.3.39), we have

$$\int_c^b \left\{\prod_{r=1}^m \left|g_r\big(u_r(t)\big)u_r'(t)\right|\right\}^{1/m} \mathrm{d}t \leqslant \frac{1}{m}\sum_{r=1}^m G_r\left(\int_c^b \left|u_r'(t)\right| \mathrm{d}t\right). \qquad (3.3.45)$$

The desired inequality in (3.3.37) follows from (3.3.44) and (3.3.45) and the proof is complete. □

REMARK 3.3.7. In the special case when $m = 1$, inequality (3.3.37) reduces to

$$\int_a^b \left|g_1\big(u_1(t)\big)u_1'(t)\right| \mathrm{d}t \leqslant G_1\left(\int_a^c \left|u_1'(t)\right| \mathrm{d}t\right) + G_1\left(\int_c^b \left|u_1'(t)\right| \mathrm{d}t\right) \qquad (3.3.46)$$

for $c \in [a, b]$. Inequality (3.3.46) is a variant of the inequality due to Calvert given in Theorem 3.3.5. On taking $g_1(t) = t$ and hence $G_1(u) = \int_0^u \sigma\,\mathrm{d}\sigma = u^2/2$ and

$c = (a+b)/2$ in (3.3.46) and using Schwarz inequality on the right-hand side of the resulting inequality, we obtain Opial's inequality given in (3.2.1).

3.4 Inequalities Related to Opial's Inequality

Opial's inequality given in (3.2.1) or its generalizations and variants have many important applications in the theory of differential equations. In the past few years many authors have obtained various useful generalizations and extensions of this inequality. In this section we offer some basic inequalities established by Pachpatte in [239,303], which claim their origin in Opial's inequality.

In [239] Pachpatte has established the inequalities in the following theorems which deal with the Opial-type integral inequalities involving two functions and their first-order derivatives.

THEOREM 3.4.1. *Let $p(t)$ be positive and continuous function on a finite or infinite interval $a < t < b$ such that $\int_a^b p^{-1}(t)\,\mathrm{d}t < \infty$. If $u(t)$ and $v(t)$ are absolutely continuous functions on (a,b) and $u(a) = u(b) = 0$, $v(a) = v(b) = 0$, then*

$$\int_a^b \left[\left|u(t)v'(t)\right| + \left|v(t)u'(t)\right|\right]\mathrm{d}t \leqslant \frac{1}{2}A\int_a^b p(t)\left[\left|u'(t)\right|^2 + \left|v'(t)\right|^2\right]\mathrm{d}t, \quad (3.4.1)$$

where

$$A = \int_a^c p^{-1}(t)\,\mathrm{d}t = \int_c^b p^{-1}(t)\,\mathrm{d}t, \quad a \leqslant c \leqslant b. \quad (3.4.2)$$

Equality holds in (3.4.1) *if and only if*

$$u(t) = v(t) = M\int_a^t p^{-1}(s)\,\mathrm{d}s, \quad a \leqslant t \leqslant c,$$

$$u(t) = v(t) = M\int_t^b p^{-1}(s)\,\mathrm{d}s, \quad c \leqslant t \leqslant b,$$

where M is a constant.

REMARK 3.4.1. In the special case when $u(t) = v(t)$, Theorem 3.4.1 reduces to the inequality established by Yang [428, Theorem 1] which in turn contains as a special case Opial's inequality given in Theorem 3.2.1.

THEOREM 3.4.2. *Let $p(t)$ be positive and continuous function on an interval $a \leqslant t \leqslant c$ with $\int_a^c p^{-1}(t)\,\mathrm{d}t < \infty$, and let $q(t)$ be bounded, positive, continuous and* *nonincreasing function on $a \leqslant t \leqslant c$. If $u(t)$ and $v(t)$ are absolutely continuous functions on $a \leqslant t \leqslant c$ and $u(a) = v(a) = 0$, then*

$$\int_a^c q(t)\big[|u(t)v'(t)| + |v(t)u'(t)|\big]\,dt \leqslant \frac{1}{2}\int_a^c p^{-1}(t)\,dt \int_a^c p(t)q(t)\big[|u'(t)|^2 + |v'(t)|^2\big]\,dt, \tag{3.4.3}$$

with equality if and only if $q(t) =$ constant and $u(t) = v(t) = M\int_a^t p^{-1}(s)\,ds$ for $a \leqslant t \leqslant c$, where M is a constant.

THEOREM 3.4.3. *Let $p(t)$ be positive and continuous function on an interval $c \leqslant t \leqslant b$ with $\int_c^b p^{-1}(t)\,dt < \infty$, and let $q(t)$ be bounded, positive, continuous and nondecreasing function on $c \leqslant t \leqslant b$. If $u(t)$ and $v(t)$ are absolutely continuous functions on $c \leqslant t \leqslant b$ and $u(b) = v(b) = 0$, then*

$$\int_c^b q(t)\big[|u(t)v'(t)| + |v(t)u'(t)|\big]\,dt \leqslant \frac{1}{2}\int_c^b p^{-1}(t)\,dt \int_c^b p(t)q(t)\big[|u'(t)|^2 + |v'(t)|^2\big]\,dt, \tag{3.4.4}$$

with equality if and only if $q(t) =$ constant and $u(t) = v(t) = M\int_t^b p^{-1}(s)\,ds$ for $c \leqslant t \leqslant b$, where M is a constant.

REMARK 3.4.2. In the special case when $u(t) = v(t)$, Theorems 3.4.2 and 3.4.3 reduce to Theorems 3 and 3′ given in [428].

THEOREM 3.4.4. *If $u(t)$ and $v(t)$ are absolutely continuous functions on $a \leqslant t \leqslant b$ with $u(a) = u(b) = 0$, $v(a) = v(b) = 0$, then*

$$\int_a^b |u(t)v(t)|^m\big[|u(t)v'(t)| + |v(t)u'(t)|\big]\,dt \leqslant \frac{(b-a)^{2m+1}}{2^{2(m+1)}(m+1)}\int_a^b \big[|u'(t)|^{2(m+1)} + |v'(t)|^{2(m+1)}\big]\,dt, \tag{3.4.5}$$

where $m \geqslant 0$ is a constant. Equality holds in (3.4.5) if and only if

$$u(t) = v(t) = M(t-a), \quad a \leqslant t \leqslant c,$$
$$u(t) = v(t) = M(b-t), \quad c \leqslant t \leqslant b,$$

where M is a constant.

REMARK 3.4.3. It is interesting to note that, in the special case when $u(t) = v(t)$ and $2m + 1 = n$, Theorem 3.4.4 reduces to the inequality established by Yang [428, Theorem 4] which in itself contains as a special case Opial's inequality when $n = 1$, $a = 0$ and $b = h$.

PROOFS OF THEOREMS 3.4.1–3.4.4. Let $c \in [a, b]$ and define

$$y(t) = \int_a^t |u'(s)|\,\mathrm{d}s, \qquad z(t) = \int_a^t |v'(s)|\,\mathrm{d}s \tag{3.4.6}$$

for $a \leqslant t \leqslant c$ and

$$r(t) = -\int_t^b |u'(s)|\,\mathrm{d}s, \qquad w(t) = -\int_t^b |v'(s)|\,\mathrm{d}s \tag{3.4.7}$$

for $c \leqslant t \leqslant b$, then we have

$$y'(t) = |u'(t)|, \qquad z'(t) = |v'(t)| \tag{3.4.8}$$

for $a \leqslant t \leqslant c$ and

$$r'(t) = |u'(t)|, \qquad w'(t) = |v'(t)| \tag{3.4.9}$$

for $c \leqslant t \leqslant b$. We note that

$$u(t) = \int_a^t u'(s)\,\mathrm{d}s, \qquad v(t) = \int_a^t v'(s)\,\mathrm{d}s \tag{3.4.10}$$

for $a \leqslant t \leqslant c$ and

$$u(t) = -\int_t^b u'(s)\,\mathrm{d}s, \qquad v(t) = -\int_t^b v'(s)\,\mathrm{d}s \tag{3.4.11}$$

for $c \leqslant t \leqslant b$. From (3.4.10) and (3.4.6) and (3.4.11) and (3.4.7), we observe that

$$|u(t)| \leqslant y(t), \qquad |v(t)| \leqslant z(t) \tag{3.4.12}$$

for $a \leqslant t \leqslant c$ and

$$|u(t)| \leqslant -r(t), \qquad |v(t)| \leqslant -w(t) \tag{3.4.13}$$

for $c \leqslant t \leqslant b$. Now, from (3.4.12), (3.4.8), and upon using the elementary inequality

$$\alpha\beta \leqslant \frac{1}{2}\left[\alpha^2 + \beta^2\right] \quad \text{for } \alpha, \beta \text{ reals}, \tag{3.4.14}$$

the definitions of $y(t)$ and $z(t)$ given in (3.4.6) and Schwarz inequality, we have

$$\begin{aligned}
&\int_a^c \left[|u(t)v'(t)| + |v(t)u'(t)| \right] dt \\
&\leqslant \int_a^c \left[y(t)z'(t) + z(t)y'(t) \right] dt \\
&= \int_a^c \frac{d}{dt}\left[y(t)z(t) \right] dt \\
&= y(c)z(c) \\
&\leqslant \frac{1}{2}\left[y^2(c) + z^2(c) \right] \\
&= \frac{1}{2}\left[\left(\int_a^c \left(\frac{1}{\sqrt{p(t)}} \right) \sqrt{p(t)}|u'(t)|\, dt \right)^2 + \left(\int_a^c \left(\frac{1}{\sqrt{p(t)}} \right) \sqrt{p(t)}|v'(t)|\, dt \right)^2 \right] \\
&\leqslant \frac{1}{2} \int_a^c p^{-1}(t)\, dt \int_a^c p(t)\left[|u'(t)|^2 + |v'(t)|^2 \right] dt. \qquad (3.4.15)
\end{aligned}$$

Similarly, from (3.4.13), (3.4.9) and upon using the elementary inequality (3.4.14), the definitions of $r(t)$ and $w(t)$ given in (3.4.7) and Schwarz inequality, we have

$$\begin{aligned}
&\int_c^b \left[|u(t)v'(t)| + |v(t)u'(t)| \right] dt \\
&\quad \leqslant \frac{1}{2} \int_c^b p^{-1}(t)\, dt \int_c^b p(t)\left[|u'(t)|^2 + |v'(t)|^2 \right] dt. \qquad (3.4.16)
\end{aligned}$$

From (3.4.15), (3.4.16) and the definition of A given in (3.4.2), the desired inequality in (3.4.1) follows. The proof of Theorem 3.4.1 is complete.

Let $c \in [a, b]$ and define

$$y(t) = \int_a^t \sqrt{q(s)}|u'(s)|\, ds, \qquad z(t) = \int_a^t \sqrt{q(s)}|v'(s)|\, ds \qquad (3.4.17)$$

for $a \leqslant t \leqslant c$ and

$$r(t) = -\int_t^b \sqrt{q(s)}|u'(s)|\, ds, \qquad w(t) = -\int_t^b \sqrt{q(s)}|v'(s)|\, ds \qquad (3.4.18)$$

for $c \leqslant t \leqslant b$, then we have

$$y'(t) = \sqrt{q(t)}|u'(t)|, \qquad z'(t) = \sqrt{q(t)}|v'(t)| \qquad (3.4.19)$$

for $a \leqslant t \leqslant c$ and

$$r'(t) = \sqrt{q(t)}\,|u'(t)|, \qquad w'(t) = \sqrt{q(t)}\,|v'(t)| \tag{3.4.20}$$

for $c \leqslant t \leqslant b$. Now, from (3.4.10), (3.4.17), nonincreasing character of $q(t)$ on $a \leqslant t \leqslant c$, in Theorem 3.4.2 and (3.4.11), (3.4.18), nondecreasing character of $q(t)$ on $c \leqslant t \leqslant b$, in Theorem 3.4.3, we observe that

$$|u(t)| \leqslant \left(\frac{1}{\sqrt{q(t)}}\right)y(t), \qquad |v(t)| \leqslant \left(\frac{1}{\sqrt{q(t)}}\right)z(t) \tag{3.4.21}$$

for $a \leqslant t \leqslant c$ and

$$|u(t)| \leqslant -\left(\frac{1}{\sqrt{q(t)}}\right)r(t), \qquad |v(t)| \leqslant -\left(\frac{1}{\sqrt{q(t)}}\right)w(t) \tag{3.4.22}$$

for $c \leqslant t \leqslant b$, respectively. Now the proofs of Theorems 3.4.2 and 3.4.3 follow by closely looking at the proof of Theorem 3.4.1 given above with suitable modifications. We omit the details.

From (3.4.8) and (3.4.12) and using the elementary inequality (3.4.14), the definitions of $y(t)$ and $z(t)$ given in (3.4.6), Schwarz inequality and Hölder's inequality, we have

$$\begin{aligned}
&\int_a^c |u(t)v(t)|^m\big[|u(t)v'(t)| + |v(t)u'(t)|\big]\,\mathrm{d}t \\
&\quad \leqslant \int_a^c y^m(t)z^m(t)\big[y(t)z'(t) + z(t)y'(t)\big]\,\mathrm{d}t \\
&\quad = \int_a^c \frac{\mathrm{d}}{\mathrm{d}t}\left(\frac{1}{m+1}y^{m+1}(t)z^{m+1}(t)\,\mathrm{d}t\right) \\
&\quad = \frac{1}{m+1}y^{m+1}(c)z^{m+1}(c) \\
&\quad \leqslant \frac{1}{2(m+1)}\big[\big(y^{m+1}(c)\big)^2 + \big(z^{m+1}(c)\big)^2\big] \\
&\quad = \frac{1}{2(m+1)}\left[\left\{\left(\int_a^c |u'(t)|\,\mathrm{d}t\right)^2\right\}^{m+1} + \left\{\left(\int_a^c |v'(t)|\,\mathrm{d}t\right)^2\right\}^{m+1}\right] \\
&\quad \leqslant \frac{(c-a)^{2m+1}}{2(m+1)}\int_a^c \big[|u'(t)|^{2(m+1)} + |v'(t)|^{2(m+1)}\big]\,\mathrm{d}t.
\end{aligned} \tag{3.4.23}$$

Similarly, from (3.4.9), (3.4.13) and upon using the elementary inequality (3.4.14), the definitions of $r(t)$ and $w(t)$ given in (3.4.7), Schwarz inequality and Hölder's inequality, we obtain

$$\int_c^b |u(t)v(t)|^m \big[|u(t)v'(t)| + |v(t)u'(t)|\big]\,\mathrm{d}t$$

$$\leqslant \frac{(b-c)^{2m+1}}{2(m+1)} \int_c^b \big[|u'(t)|^{2(m+1)} + |v'(t)|^{2(m+1)}\big]\,\mathrm{d}t. \tag{3.4.24}$$

Now taking $c = (a+b)/2$, we obtain the desired inequality in (3.4.5) from (3.4.23) and (3.4.24).

In the following theorems we present some general inequalities similar to Opial's inequality established by Pachpatte in [303]. In what follows, for the sake of brevity we write f_i for $f_i(|u_i(t)|)$, f_i' for $f_i'(|u_i(t)|)$, u_i' for $u_i'(t)$, with $t \in [a,b]$ and use the notation

$$\begin{aligned} &L[f_1, \ldots, f_n, f_1', \ldots, f_n', u_1', \ldots, u_n'] \\ &\quad = f_1 \cdots f_{n-1} f_n' |u_n'| \\ &\qquad + f_1 \cdots f_{n-2} f_{n-1}' |u_{n-1}'| f_n + \cdots + f_1' |u_1'| f_2 \cdots f_n, \quad n \geqslant 2. \end{aligned}$$

THEOREM 3.4.5. *Let $u_i(t)$, $i = 1, \ldots, n$, be real-valued absolutely continuous functions on $[a,b]$ with $u_i(a) = 0$. Let $f_i(r)$, $i = 1, \ldots, n$, be real-valued nonnegative continuous nondecreasing functions for $r \geqslant 0$ and $f_i(0) = 0$ such that $f_i'(r)$ exist, nonnegative, continuous and nondecreasing for $r \geqslant 0$. Then, the following inequality holds*

$$\int_a^b L[f_1, \ldots, f_n, f_1', \ldots, f_n', u_1', \ldots, u_n']\,\mathrm{d}t \leqslant \prod_{i=1}^n f_i\left(\int_a^b |u_i'(t)|\,\mathrm{d}t\right). \tag{3.4.25}$$

Inequality (3.4.25) *also holds if we replace the condition $u_i(a) = 0$ by $u_i(b) = 0$.*

As an immediate consequence of Theorem 3.4.5 we have the following result.

THEOREM 3.4.6. *Assume that in the hypotheses of Theorem* 3.4.5 *we have $u_i = u$ and $f_i = f$. Then*

$$\int_a^b \{f(|u(t)|)\}^{n-1} f'(|u(t)|)|u'(t)|\,\mathrm{d}t \leqslant \frac{1}{n}\left\{f\left(\int_a^b |u'(t)|\,\mathrm{d}t\right)\right\}^n. \tag{3.4.26}$$

Inequality (3.4.26) *also holds if we replace the condition $u(a) = 0$ by $u(b) = 0$.*

REMARK 3.4.4. If we take $n=2$ in (3.4.26), then we get the following inequality

$$\int_a^b f(|u(t)|)f'(|u(t)|)|u'(t)|\,\mathrm{d}t \leqslant \frac{1}{2}\left\{f\left(\int_a^b |u'(t)|\,\mathrm{d}t\right)\right\}^2, \tag{3.4.27}$$

which is analogous to the inequality given in [130]. Further, by taking $f(r)=r^{m+1}$ in (3.4.27), $m\geqslant 0$ is a constant, and using Hölder's inequality with indices $2(m+1)$ and $2(m+1)/(2m+1)$ to the resulting integral on the right-hand side, we see that (3.4.27) reduces to the following inequality

$$\int_a^b |u(t)|^{2m+1}|u'(t)|\,\mathrm{d}t \leqslant \frac{(b-a)^{2m+1}}{2(m+1)}\int_a^b |u'(t)|^{2(m+1)}\,\mathrm{d}t, \tag{3.4.28}$$

which reduces to the form of Opial's inequality given in [211, Theorem 2′, p. 154] when $m=0$.

A slightly different version of the inequality given in Theorem 3.4.5 is embodied in the following theorem.

THEOREM 3.4.7. *Let u_i, f_i, f_i' be as in Theorem* 3.4.5. *Let $p_i(t)>0$ be defined on $[a,b]$ and $\int_a^b p_i(t)\,\mathrm{d}t=1$, $i=1,\dots,n$. If $h(r)$ is a positive, convex and increasing function for $r>0$, then*

$$\int_a^b L[f_1,\dots,f_n,f_1',\dots,f_n',u_1',\dots,u_n']\,\mathrm{d}t \leqslant \prod_{i=1}^n f_i\left(h^{-1}\left(\int_a^b p_i(t)h\left(\frac{|u_i'(t)|}{p_i(t)}\right)\mathrm{d}t\right)\right). \tag{3.4.29}$$

Inequality (3.4.29) *also holds if we replace the condition $u_i(a)=0$ by $u_i(b)=0$.*

The following result is an easy consequence of Theorem 3.4.7.

THEOREM 3.4.8. *Assume that in the hypotheses of Theorem* 3.4.7 *we have $u_i=u$, $f_i=f$ and $p_i=p$. Then*

$$\int_a^b \{f(|u(t)|)\}^{n-1}f'(|u(t)|)|u'(t)|\,\mathrm{d}t \leqslant \frac{1}{n}\left\{f\left(h^{-1}\left(\int_a^b p(t)h\left(\frac{|u'(t)|}{p(t)}\right)\mathrm{d}t\right)\right)\right\}^n. \tag{3.4.30}$$

Inequality (3.4.30) *also holds if we replace the condition $u(a)=0$ by $u(b)=0$.*

REMARK 3.4.5. If we take $n = 2$ in (3.4.30), then we get the following inequality

$$\int_a^b f(|u(t)|) f'(|u(t)|) |u'(t)| \,\mathrm{d}t \leqslant \frac{1}{2}\left\{ f\left(h^{-1}\left(\int_a^b p(t) h\left(\frac{|u'(t)|}{p(t)} \right) \mathrm{d}t \right) \right) \right\}^2. \tag{3.4.31}$$

We also note that in the special cases inequality (3.4.31) yields the various inequalities as discussed in Remark 3.4.4.

PROOFS OF THEOREMS 3.4.5–3.4.8. Let $t \in [a, b]$ and define

$$z_i(t) = \int_a^t |u'(s)| \,\mathrm{d}s, \quad i = 1, \ldots, n, \tag{3.4.32}$$

implying

$$z_i'(t) = |u_i(t)|, \quad t \in [a, b], i = 1, \ldots, n. \tag{3.4.33}$$

For $t \in [a, b]$ we have the following identities

$$u_i(t) = \int_a^t u_i'(s) \,\mathrm{d}s, \quad i = 1, \ldots, n. \tag{3.4.34}$$

From (3.4.34) and (3.4.32), we observe that

$$|u_i(t)| \leqslant z_i(t), \quad i = 1, \ldots, n. \tag{3.4.35}$$

Using (3.4.35), (3.4.33) and (3.4.32) we get

$$\begin{aligned}
&\int_a^b L[f_1, \ldots, f_n, f_1', \ldots, f_n', u_1', \ldots, u_n'] \,\mathrm{d}t \\
&\quad \leqslant \int_a^b \big[f_1(z_1(t)) \cdots f_{n-1}(z_{n-1}(t)) f_n'(z_n(t)) z_n'(t) \\
&\qquad + f_1(z_1(t)) \cdots f_{n-2}(z_{n-2}(t)) f_{n-1}'(z_{n-1}(t)) z_{n-1}'(t) f_n(z_n(t)) \\
&\qquad + \cdots + f_1'(z_1(t)) z_1'(t) f_2(z_2(t)) \cdots f_n(z_n(t)) \big] \,\mathrm{d}t \\
&\quad = \int_a^b \frac{\mathrm{d}}{\mathrm{d}t}\left[\prod_{i=1}^n f_i(z_i(t)) \right] \mathrm{d}t
\end{aligned}$$

$$= \prod_{i=1}^{n} f_i\big(z_i(b)\big)$$

$$= \prod_{i=1}^{n} f_i\left(\int_a^b \left|u_i'(t)\right| \mathrm{d}t\right),$$

being the required inequality in (3.4.25). Defining $z_i(t) = \int_t^b u_i'(s)\,\mathrm{d}s$ in case of $u_i(b) = 0$, then observing that $|u_i(t)| \leqslant z_i(t)$, similarly as above, we get (3.4.25). The proof of Theorem 3.4.5 is complete. □

From the hypotheses of Theorem 3.4.7, we observe that

$$\int_a^b \left|u_i'(t)\right| \mathrm{d}t = \left\{\int_a^b p_i(t) \frac{|u_i'(t)|}{p_i(t)} \,\mathrm{d}t\right\}\left\{\int_a^b p_i(t)\,\mathrm{d}t\right\}^{-1}, \quad i = 1, \ldots, n. \tag{3.4.36}$$

Since h is convex, from (3.4.36) and using Jensen's inequality [174, p. 113], we obtain

$$h\left(\int_a^b \left|u_i'(t)\right| \mathrm{d}t\right) \leqslant \int_a^b p_i(t) h\left(\frac{|u_i'(t)|}{p_i(t)}\right) \mathrm{d}t \tag{3.4.37}$$

which implies

$$\int_a^b \left|u_i'(t)\right| \mathrm{d}t \leqslant h^{-1}\left(\int_a^b p_i(t) h\left(\frac{|u_i'(t)|}{p_i(t)}\right) \mathrm{d}t\right). \tag{3.4.38}$$

All the hypotheses of Theorem 3.4.5 being satisfied we get (3.4.25). Using (3.4.38) in (3.4.25), we obtain the required inequality in (3.4.29). The proof of Theorem 3.4.7 is complete. □

We omit the proofs of Theorems 3.4.6 and 3.4.8 being immediate from those of Theorems 3.4.5 and 3.4.7.

3.5 General Opial-Type Integral Inequalities

Since its discovery in 1960, Opial's integral inequality has been generalized in various directions by several authors. In this section we shall give the inequalities established by Godunova and Levin [130], Rozanova [399] and Pachpatte [346] which contain as special cases many known generalizations of Opial's integral inequality given by other investigators.

In 1967, Godunova and Levin [130] established the following inequalities.

THEOREM 3.5.1. *Let $u(t)$ be real-valued absolutely continuous function defined on $[a,b]$ with $u(a)=0$. Let F be real-valued convex, increasing function on $[0,\infty)$ with $F(0)=0$. Then the following inequality holds*

$$\int_a^b F'(|u(t)|)|u'(t)|\,\mathrm{d}t \leqslant F\left(\int_a^b |u'(t)|\,\mathrm{d}t\right). \tag{3.5.1}$$

PROOF. Define $z(t)=\int_a^t |u'(s)|\,\mathrm{d}s$, $t\in[a,b]$. Then $z'(t)=|u'(t)|$ and $|u(t)|\leqslant z(t)$. Thus, it follows that

$$\begin{aligned}\int_a^b F'(|u(t)|)|u'(t)|\,\mathrm{d}t &\leqslant \int_a^b F'(z(t))z'(t)\,\mathrm{d}t\\ &= \int_a^b \frac{\mathrm{d}}{\mathrm{d}t}F(z(t))\,\mathrm{d}t\\ &= F(z(b))\\ &= F\left(\int_a^b |u'(t)|\,\mathrm{d}t\right).\end{aligned}$$

The proof is complete. □

REMARK 3.5.1. By taking $F(r)=r^2$ and hence $F'(r)=2r$ in (3.5.1) and using Hölder's inequality on the right-hand side of the resulting inequality, we get Opial's inequality given in [211, Theorem 2′, p. 154].

THEOREM 3.5.2. *Let $u(t)$ be real-valued absolutely continuous function defined on $[a,b]$ with $u(a)=u(b)=0$. Let F, g be real-valued convex and increasing functions on $[0,\infty)$ with $F(0)=0$. Further, let $p(t)$ be real-valued positive on $[a,b]$ and $\int_a^b p(t)\,\mathrm{d}t=1$. Then the following inequality holds*

$$\int_a^b F'(|u(t)|)|u'(t)|\,\mathrm{d}t \leqslant 2F\left(g^{-1}\left(\int_a^b p(t)g\left(\frac{|u'(t)|}{2p(t)}\right)\mathrm{d}t\right)\right). \tag{3.5.2}$$

PROOF. Let $c\in(a,b)$, so that in the interval $[a,c]$ the function $u(t)$ satisfies the hypotheses of Theorem 3.5.1 and the following inequality holds

$$\int_a^c F'(|u(t)|)|u'(t)|\,\mathrm{d}t \leqslant F\left(\int_a^c |u'(t)|\,\mathrm{d}t\right). \tag{3.5.3}$$

Next, in the interval $[c, b]$ the function $u(t)$ is absolutely continuous and $u(b) = 0$. By following the similar argument as in the proof of Theorem 3.5.1, we obtain

$$\int_c^b F'(|u(t)|)|u'(t)|\,\mathrm{d}t \leqslant F\left(\int_c^b |u'(t)|\,\mathrm{d}t\right). \tag{3.5.4}$$

If we choose c so that

$$\int_a^c |u'(t)|\,\mathrm{d}t = \int_c^b |u'(t)|\,\mathrm{d}t = \frac{1}{2}\int_a^b |u'(t)|\,\mathrm{d}t, \tag{3.5.5}$$

then a combination of (3.5.4) and (3.5.5) gives

$$\int_a^b F'(|u(t)|)|u'(t)|\,\mathrm{d}t \leqslant 2F\left(\frac{1}{2}\int_a^b |u'(t)|\,\mathrm{d}t\right). \tag{3.5.6}$$

Since $g(t)$ is convex, by using Jensen's inequality (see [174, p. 133]), we have

$$g\left(\int_a^b \left(\frac{|u'(t)|}{2p(t)}\right)p(t)\,\mathrm{d}t \Big/ \int_a^b p(t)\,\mathrm{d}t\right) \leqslant \int_a^b g\left(\frac{|u'(t)|}{2p(t)}\right)p(t)\,\mathrm{d}t \Big/ \int_a^b p(t)\,\mathrm{d}t$$

which in view of $\int_a^b p(t)\,\mathrm{d}t = 1$ and the increasing nature of g, implies

$$\frac{1}{2}\int_a^b |u'(t)|\,\mathrm{d}t \leqslant g^{-1}\left(\int_a^b p(t)g\left(\frac{|u'(t)|}{2p(t)}\right)\mathrm{d}t\right). \tag{3.5.7}$$

Using the increasing behavior of F and (3.5.7) in (3.5.6), inequality (3.5.2) follows. □

In 1972, Rozanova [399] obtained the following extension of the inequality given in Theorem 3.5.1.

THEOREM 3.5.3. *Let $u(t)$ be absolutely continuous on $[a, b]$ and $u(a) = 0$. Let F, g be as in Theorem* 3.5.2 *and let $p(t) \geqslant 0$, $p'(t) > 0$, $t \in [a, b]$, with $p(a) = 0$. Then*

$$\int_a^b p'(t)F'\left(p(t)g\left(\frac{|u(t)|}{p(t)}\right)\right)g\left(\frac{|u'(t)|}{p'(t)}\right)\mathrm{d}t \leqslant F\left(\int_a^b p'(t)g\left(\frac{|u'(t)|}{p'(t)}\right)\mathrm{d}t\right). \tag{3.5.8}$$

Moreover, equality holds in (3.5.8) *for the function $u(t) = cp(t)$, where c is a constant.*

PROOF. Define $z(t) = \int_a^t |u'(s)|\,\mathrm{d}s$, $t \in [a,b]$. Then $z'(t) = |u'(t)|$ and $|u(t)| \leqslant z(t)$. Let $t \in (a,b)$, then by using Jensen's inequality (see [174, p. 133]), it follows that

$$\begin{aligned} g\left(\frac{|u(t)|}{p(t)}\right) &\leqslant g\left(\frac{z(t)}{p(t)}\right) \\ &\leqslant g\left(\int_a^t p'(s)\frac{|u'(s)|}{p'(s)}\,\mathrm{d}s \Big/ \int_a^t p'(s)\,\mathrm{d}s\right) \\ &\leqslant \frac{1}{p(t)}\int_a^t p'(s) g\left(\frac{z'(s)}{p'(s)}\right)\mathrm{d}s. \end{aligned}$$

Using the above inequality we obtain

$$\begin{aligned} &\int_a^b p'(t) g\left(\frac{|u'(t)|}{p'(t)}\right) F'\left(p(t) g\left(\frac{|u(t)|}{p(t)}\right)\right)\mathrm{d}t \\ &\quad\leqslant \int_a^b p'(t) g\left(\frac{z'(t)}{p'(t)}\right) F'\left(\int_a^t p'(s) g\left(\frac{z'(s)}{p'(s)}\right)\mathrm{d}s\right)\mathrm{d}t \\ &\quad= \int_a^b \frac{\mathrm{d}}{\mathrm{d}t}\left[F\left(\int_a^t p'(s) g\left(\frac{z'(s)}{p'(s)}\right)\mathrm{d}s\right)\right]\mathrm{d}t \\ &\quad= F\left(\int_a^b p'(t) g\left(\frac{z'(t)}{p'(t)}\right)\mathrm{d}t\right) \\ &\quad= F\left(\int_a^b p'(t) g\left(\frac{|u'(t)|}{p'(t)}\right)\mathrm{d}t\right), \end{aligned}$$

which is the same as (3.5.8). The proof is complete. □

The following theorems deal with the generalized Opial-type integral inequalities established by Pachpatte in [346].

THEOREM 3.5.4. *Let u_i, v_i, $i = 1, \ldots, n$, be real-valued absolutely continuous functions on $I = [a,b]$, $a, b \in \mathbb{R}_+ = [0, \infty)$, with $u_i(a) = v_i(a) = 0$, $i = 1, \ldots, n$. Let F, G be real-valued nonnegative, continuous and nondecreasing functions on $\mathbb{R}_+^n$ with $F(0,\ldots,0) = 0$, $G(0,\ldots,0) = 0$ such that all their first-order partial derivatives F_i', G_i', $i = 1, \ldots, n$, are nonnegative, continuous and nondecreasing on $\mathbb{R}_+^n$. Let ϕ_i, ψ_i, $i = 1, \ldots, n$, be real-valued positive, convex and increasing functions on $(0, \infty)$. Let $r_i(t) \geqslant 0$, $r_i'(t) > 0$, $r_i(a) = 0$, $e_i(t) \geqslant 0$, $e_i'(t) > 0$,*

$e_i(a) = 0$, $i = 1, \ldots, n$. *Then the following integral inequality holds*

$$\int_a^b \Bigg[F\left(r_1(t)\phi_1\left(\frac{|u_1(t)|}{r_1(t)}\right), \ldots, r_n(t)\phi_n\left(\frac{|u_n(t)|}{r_n(t)}\right)\right)$$
$$\times \sum_{i=1}^n G_i'\left(e_1(t)\psi_1\left(\frac{|v_1(t)|}{e_1(t)}\right), \ldots, e_n(t)\psi_n\left(\frac{|v_n(t)|}{e_n(t)}\right)\right) e_i'(t)\psi_i\left(\frac{|v_i'(t)|}{e_i'(t)}\right)$$
$$+ G\left(e_1(t)\psi_1\left(\frac{|v_1(t)|}{e_1(t)}\right), \ldots, e_n(t)\psi_n\left(\frac{|v_n(t)|}{e_n(t)}\right)\right)$$
$$\times \sum_{i=1}^n F_i'\left(r_1(t)\phi_1\left(\frac{|u_1(t)|}{r_1(t)}\right), \ldots, r_n(t)\phi_n\left(\frac{|u_n(t)|}{r_n(t)}\right)\right) r_i'(t)\phi_i\left(\frac{|u_i'(t)|}{r_i'(t)}\right)\Bigg] \mathrm{d}t$$
$$\leqslant F\left(\int_a^b r_1'(t)\phi_1\left(\frac{|u_1'(t)|}{r_1'(t)}\right) \mathrm{d}t, \ldots, \int_a^b r_n'(t)\phi_n\left(\frac{|u_n'(t)|}{r_n'(t)}\right) \mathrm{d}t\right)$$
$$\times G\left(\int_a^b e_1'(t)\psi_1\left(\frac{|v_1'(t)|}{e_1'(t)}\right) \mathrm{d}t, \ldots, \int_a^b e_n'(t)\psi_n\left(\frac{|v_n'(t)|}{e_n'(t)}\right) \mathrm{d}t\right). \tag{3.5.9}$$

PROOF. From the hypotheses on u_i, v_i, r_i, e_i, $i = 1, \ldots, n$, we have

$$|u_i(t)| = \left| \int_a^t u_i'(s)\,\mathrm{d}s \right| \leqslant \int_a^t |u_i'(s)|\,\mathrm{d}s, \tag{3.5.10}$$

$$|v_i(t)| = \left| \int_a^t v_i'(s)\,\mathrm{d}s \right| \leqslant \int_a^t |v_i'(s)|\,\mathrm{d}s, \tag{3.5.11}$$

$$r_i(t) = \int_a^t r_i'(s)\,\mathrm{d}s, \tag{3.5.12}$$

$$e_i(t) = \int_a^t e_i'(s)\,\mathrm{d}s, \tag{3.5.13}$$

for $t \in I$. From (3.5.10)–(3.5.13) and using the hypotheses on ϕ_i, ψ_i, $i = 1, \ldots, n$, and Jensen's inequality [174, p. 133], we obtain

$$\phi_i\left(\frac{|u_i(t)|}{r_i(t)}\right) \leqslant \phi_i\left(\int_a^t \frac{r_i'(s)|u_i'(s)|}{r_i'(s)}\,\mathrm{d}s \Big/ \int_a^t r_i'(s)\,\mathrm{d}s \right)$$
$$\leqslant \frac{1}{r_i(t)} \int_a^t r_i'(s)\phi_i\left(\frac{|u_i'(s)|}{r_i'(s)}\right) \mathrm{d}s \tag{3.5.14}$$

and

$$\psi_i\left(\frac{|v_i(t)|}{e_i(t)}\right) \leqslant \psi_i\left(\int_a^t \frac{e_i'(s)|v_i'(s)|}{e_i'(s)}\,\mathrm{d}s \Big/ \int_a^t e_i'(s)\,\mathrm{d}s\right)$$

$$\leqslant \frac{1}{e_i(t)}\int_a^t e_i'(s)\psi_i\left(\frac{|v_i'(s)|}{e_i'(s)}\right)\mathrm{d}s, \tag{3.5.15}$$

for $t \in I$. From (3.5.14), (3.5.15) and using the hypotheses F, F_i', G, G_i', $i = 1, \ldots, n$, we observe that

$$\int_a^b \Bigg[F\left(r_1(t)\phi_1\left(\frac{|u_1(t)|}{r_1(t)}\right), \ldots, r_n(t)\phi_n\left(\frac{|u_n(t)|}{r_n(t)}\right)\right)$$

$$\times \sum_{i=1}^n G_i'\left(e_1(t)\psi_1\left(\frac{|v_1(t)|}{e_1(t)}\right), \ldots, e_n(t)\psi_n\left(\frac{|v_n(t)|}{e_n(t)}\right)\right)e_i'(t)\psi_i\left(\frac{|v_i'(t)|}{e_i'(t)}\right)$$

$$+ G\left(e_1(t)\psi_1\left(\frac{|v_1(t)|}{e_1(t)}\right), \ldots, e_n(t)\psi_n\left(\frac{|v_n(t)|}{e_n(t)}\right)\right)$$

$$\times \sum_{i=1}^n F_i'\left(r_1(t)\phi_1\left(\frac{|u_1(t)|}{r_1(t)}\right), \ldots, r_n(t)\phi_n\left(\frac{|u_n(t)|}{r_n(t)}\right)\right)r_i'(t)\phi_i\left(\frac{|u_i'(t)|}{r_i'(t)}\right)\Bigg]\mathrm{d}t$$

$$\leqslant \int_a^b \Bigg[F\left(\int_a^t r_1'(s)\phi_1\left(\frac{|u_1'(s)|}{r_1'(s)}\right)\mathrm{d}s, \ldots, \int_a^t r_n'(s)\phi_n\left(\frac{|u_n'(s)|}{r_n'(s)}\right)\mathrm{d}s\right)$$

$$\times \sum_{i=1}^n G_i'\left(\int_a^t e_1'(s)\psi_1\left(\frac{|v_1'(s)|}{e_1'(s)}\right)\mathrm{d}s, \ldots, \int_a^t e_n'(s)\psi_n\left(\frac{|v_n'(s)|}{e_n'(s)}\right)\mathrm{d}s\right)$$

$$\times e_i'(t)\psi_i\left(\frac{|v_i'(t)|}{e_i'(t)}\right)$$

$$+ G\left(\int_a^t e_1'(s)\psi_1\left(\frac{|v_1'(s)|}{e_1'(s)}\right)\mathrm{d}s, \ldots, \int_a^t e_n'(s)\psi_n\left(\frac{|v_n'(s)|}{e_n'(s)}\right)\mathrm{d}s\right)$$

$$\times \sum_{i=1}^n F_i'\left(\int_a^t r_1'(s)\phi_1\left(\frac{|u_1'(s)|}{r_1'(s)}\right)\mathrm{d}s, \ldots, \int_a^t r_n'(s)\phi_n\left(\frac{|u_n'(s)|}{r_n'(s)}\right)\mathrm{d}s\right)$$

$$\times r_i'(t)\phi_i\left(\frac{|u_i'(t)|}{r_i'(t)}\right)\Bigg]\mathrm{d}t$$

$$= \int_a^b \frac{\mathrm{d}}{\mathrm{d}t}\left[F\left(\int_a^t r_1'(s)\phi_1\left(\frac{|u_1'(s)|}{r_1'(s)}\right)\mathrm{d}s, \ldots, \int_a^t r_n'(s)\phi_n\left(\frac{|u_n'(s)|}{r_n'(s)}\right)\mathrm{d}s\right)\right.$$

$$\left.\times G\left(\int_a^t e_1'(s)\psi_1\left(\frac{|v_1'(s)|}{e_1'(s)}\right)\mathrm{d}s, \ldots, \int_a^t e_n'(s)\psi_n\left(\frac{|v_n'(s)|}{e_n'(s)}\right)\mathrm{d}s\right)\right]\mathrm{d}t$$

$$= F\left(\int_a^b r_1'(t)\phi_1\left(\frac{|u_1'(t)|}{r_1'(t)}\right)\mathrm{d}t, \ldots, \int_a^b r_n'(t)\phi_n\left(\frac{|u_n'(t)|}{r_n'(t)}\right)\mathrm{d}t\right)$$

$$\times G\left(\int_a^b e_1'(t)\psi_1\left(\frac{|v_1'(t)|}{e_1'(t)}\right)\mathrm{d}t, \ldots, \int_a^b e_n'(t)\psi_n\left(\frac{|v_n'(t)|}{e_n'(t)}\right)\mathrm{d}t\right).$$

The proof of inequality (3.5.9) is complete. □

THEOREM 3.5.5. *Let u_i, v_i, F, F_i', G, G_i', ϕ_i, ψ_i, $i = 1, \ldots, n$, be as in Theorem 3.5.4. Let p_i, q_i, $i = 1, \ldots, n$, be real-valued positive functions defined on I and $\int_a^b p_i(t)\,\mathrm{d}t = 1$, $\int_a^b q_i(t)\,\mathrm{d}t = 1$, $i = 1, \ldots, n$. Let h_i, w_i, $i = 1, \ldots, n$, be real-valued positive, convex, and increasing functions on $(0, \infty)$. Then the following integral inequality holds*

$$\int_a^b \left[F\left(r_1(t)\phi_1\left(\frac{|u_1(t)|}{r_1(t)}\right), \ldots, r_n(t)\phi_n\left(\frac{|u_n(t)|}{r_n(t)}\right)\right)\right.$$

$$\times \sum_{i=1}^n G_i'\left(e_1(t)\psi_1\left(\frac{|v_1(t)|}{e_1(t)}\right), \ldots, e_n(t)\psi_n\left(\frac{|v_n(t)|}{e_n(t)}\right)\right)e_i'(t)\psi_i\left(\frac{|v_i'(t)|}{e_i'(t)}\right)$$

$$+ G\left(e_1(t)\psi_1\left(\frac{|v_1(t)|}{e_1(t)}\right), \ldots, e_n(t)\psi_n\left(\frac{|v_n(t)|}{e_n(t)}\right)\right)$$

$$\left.\times \sum_{i=1}^n F_i'\left(r_1(t)\phi_1\left(\frac{|u_1(t)|}{r_1(t)}\right), \ldots, r_n(t)\phi_n\left(\frac{|u_n(t)|}{r_n(t)}\right)\right)r_i'(t)\phi_i\left(\frac{|u_i'(t)|}{r_i'(t)}\right)\right]\mathrm{d}t$$

$$\leqslant F\left(h_1^{-1}\left(\int_a^b p_1(t)h_1\left(r_1'(t)\phi_1\left(\frac{|u_1'(t)|}{r_1(t)}\right)\Big/p_1(t)\right)\mathrm{d}t\right),\right.$$

$$\left.\ldots, h_n^{-1}\left(\int_a^b p_n(t)h_n\left(r_n'(t)\phi_n\left(\frac{|u_n'(t)|}{r_n(t)}\right)\Big/p_n(t)\right)\mathrm{d}t\right)\right)$$

$$\times G\left(w_1^{-1}\left(\int_a^b q_1(t)w_1\left(e_1'(t)\psi_1\left(\frac{|v_1'(t)|}{e_1'(t)}\right)\Big/q_1(t)\right)\mathrm{d}t\right),\right.$$

$$\left.\ldots, w_n^{-1}\left(\int_a^b q_n(t)w_n\left(e_n'(t)\psi_n\left(\frac{|v_n'(t)|}{e_n'(t)}\right)\Big/q_n(t)\right)\mathrm{d}t\right)\right). \quad (3.5.16)$$

PROOF. From the assumptions, we write

$$\int_a^b r_i'(t)\phi_i\left(\frac{|u_i'(t)|}{r_i'(t)}\right)\mathrm{d}t$$
$$= \int_a^b \left(p_i(t) r_i'(t)\phi_i\left(\frac{|u_i'(t)|}{r_i'(t)}\right) \Big/ p_i(t)\right)\mathrm{d}t \Big/ \int_a^b p_i(t)\,\mathrm{d}t, \quad (3.5.17)$$

$$\int_a^b e_i'(t)\psi_i\left(\frac{|u_i'(t)|}{r_i'(t)}\right)\mathrm{d}t$$
$$= \int_a^b \left(q_i(t) e_i'(t)\psi_i\left(\frac{|v_i'(t)|}{e_i'(t)}\right) \Big/ q_i(t)\right)\mathrm{d}t \Big/ \int_a^b q_i(t)\,\mathrm{d}t, \quad (3.5.18)$$

for $i = 1, \ldots, n$. From (3.5.17), (3.5.18) and using the hypotheses on h_i, w_i, $i = 1, \ldots, n$, and Jensen's inequality [174, p. 133], we obtain

$$h_i\left(\int_a^b r_i'(t)\phi_i\left(\frac{|u_i'(t)|}{r_i'(t)}\right)\mathrm{d}t\right)$$
$$\leqslant \int_a^b p_i(t) h_i\left(r_i'(t)\phi_i\left(\frac{|u_i'(t)|}{r_i'(t)}\right) \Big/ p_i(t)\right)\mathrm{d}t, \quad (3.5.19)$$

$$w_i\left(\int_a^b e_i'(t)\psi_i\left(\frac{|v_i'(t)|}{e_i'(t)}\right)\mathrm{d}t\right)$$
$$\leqslant \int_a^b q_i(t) w_i\left(e_i'(t)\psi_i\left(\frac{|v_i'(t)|}{e_i'(t)}\right) \Big/ q_i(t)\right)\mathrm{d}t, \quad (3.5.20)$$

for $i = 1, \ldots, n$. From (3.5.19) and (3.5.20), we observe that

$$\int_a^b r_i'(t)\phi_i\left(\frac{|u_i'(t)|}{r_i'(t)}\right)\mathrm{d}t$$
$$\leqslant h_i^{-1}\left(\int_a^b p_i(t) h_i\left(r_i'(t)\phi_i\left(\frac{|u_i'(t)|}{r_i'(t)}\right) \Big/ p_i(t)\right)\mathrm{d}t\right), \quad (3.5.21)$$

$$\int_a^b e_i'(t)\psi_i\left(\frac{|v_i'(t)|}{e_i'(t)}\right)\mathrm{d}t$$
$$\leqslant w_i^{-1}\left(\int_a^b q_i(t) w_i\left(e_i'(t)\psi_i\left(\frac{|v_i'(t)|}{e_i'(t)}\right) \Big/ q_i(t)\right)\mathrm{d}t\right), \quad (3.5.22)$$

for $i = 1, \ldots, n$. Since all the hypotheses of Theorem 3.5.4 are among those of Theorem 3.5.5, we see that inequality (3.5.9) holds. Now using (3.5.21), (3.5.22) on the right-hand side of (3.5.9) we get the desired inequality in (3.5.16) and the proof is complete. □

REMARK 3.5.2. (i) In the special case when $n = 1$, the inequalities established in Theorems 3.5.4 and 3.5.5 reduce to the inequalities established by Pachpatte in [302]. (ii) If we take $G = 1$ and hence $G_i' = 0$, $i = 1, \ldots, n$, in (3.5.9) and (3.5.16), then we get the inequalities established by Pečarić and Brnetić in [377]. For a detailed discussion on the further special versions of the inequalities given in (3.5.9) and (3.5.16), see [302,346,399].

In [30] Bloom has established some Opial-type inequalities involving generalized Hardy operators. The results given in [30] are based on the observation made by Sinnamon in [409] and also as discussed by Bloom in [30, p. 28].

An operator T acting on $\mathbb{R}$ is called a Hardy operator, if T has the form

$$Tf(t) = \int_a^t f(s)\,\mathrm{d}s \tag{3.5.23}$$

for $t \in I = [a, b]$, $a, b \in \mathbb{R}_+ = [0, \infty)$, where $f(t)$ is real-valued continuous function defined on I. We say that the function $f(t)$ belongs to the class U if it can be represented in the form (3.5.23). We note that the results given below also hold, if the Hardy operator T has the form

$$Tf(t) = \int_t^b f(s)\,\mathrm{d}s \tag{3.5.24}$$

for $t \in I$, where $f(t)$ is continuous function on I.

The following theorems deal with the inequalities established by Pachpatte in [346].

THEOREM 3.5.6. *Let $f_i, g_i \in U$, $i = 1, \ldots, n$, and F, F_i', G, G_i', $i = 1, \ldots, n$, be as in Theorem 3.5.4. Then the following integral inequality holds*

$$\begin{aligned}\int_a^b \Bigg[& F\big(|Tf_1(t)|, \ldots, |Tf_n(t)|\big) \sum_{i=1}^n G_i'\big(|Tg_1(t)|, \ldots, |Tg_n(t)|\big)|g_i(t)| \\ & + G\big(|Tg_1(t)|, \ldots, |Tg_n(t)|\big) \sum_{i=1}^n F_i'\big(|Tf_1(t)|, \ldots, |Tf_n(t)|\big)|f_i(t)|\Bigg]\,\mathrm{d}t \\ \leqslant{} & F\bigg(\int_a^b |f_1(t)|\,\mathrm{d}t, \ldots, \int_a^b |f_n(t)|\,\mathrm{d}t\bigg) G\bigg(\int_a^b |g_1(t)|\,\mathrm{d}t, \ldots, \int_a^b |g_n(t)|\,\mathrm{d}t\bigg).\end{aligned} \tag{3.5.25}$$

PROOF. From the hypotheses, it is easy to observe that

$$\left|Tf_i(t)\right| = \left|\int_a^t f_i(s)\,\mathrm{d}s\right| \leqslant \int_a^t \left|f_i(s)\right|\mathrm{d}s, \tag{3.5.26}$$

$$\left|Tg_i(t)\right| = \left|\int_a^t g_i(s)\,\mathrm{d}s\right| \leqslant \int_a^t \left|g_i(s)\right|\mathrm{d}s, \tag{3.5.27}$$

for $t \in I$, $i = 1, \dots, n$. From (3.5.26), (3.5.27) and using the assumptions, we observe that

$$\begin{aligned}
&\int_a^b \Bigg[F\big(\left|Tf_1(t)\right|, \dots, \left|Tf_n(t)\right|\big) \sum_{i=1}^n G_i'\big(\left|Tg_1(t)\right|, \dots, \left|Tg_n(t)\right|\big)\left|g_i(t)\right| \\
&\qquad + G\big(\left|Tg_1(t)\right|, \dots, \left|Tg_n(t)\right|\big) \sum_{i=1}^n F_i'\big(\left|Tf_1(t)\right|, \dots, \left|Tf_n(t)\right|\big)\left|f_i(t)\right|\Bigg]\mathrm{d}t \\
&\quad \leqslant \int_a^b \Bigg[F\left(\int_a^t \left|f_1(s)\right|\mathrm{d}s, \dots, \int_a^t \left|f_n(s)\right|\mathrm{d}s\right) \\
&\qquad\quad \times \sum_{i=1}^n G_i'\left(\int_a^t \left|g_1(s)\right|\mathrm{d}s, \dots, \int_a^t \left|g_n(s)\right|\mathrm{d}s\right)\left|g_i(t)\right| \\
&\qquad\quad + G\left(\int_a^t \left|g_1(s)\right|\mathrm{d}s, \dots, \int_a^t \left|g_n(s)\right|\mathrm{d}s\right) \\
&\qquad\quad \times \sum_{i=1}^n F_i'\left(\int_a^t \left|f_1(s)\right|\mathrm{d}s, \dots, \int_a^t \left|f_n(s)\right|\mathrm{d}s\right)\left|f_i(t)\right|\Bigg]\mathrm{d}t \\
&\quad = \int_a^b \frac{\mathrm{d}}{\mathrm{d}t}\Bigg[F\left(\int_a^t \left|f_1(s)\right|\mathrm{d}s, \dots, \int_a^t \left|f_n(s)\right|\mathrm{d}s\right) \\
&\qquad\quad \times G\left(\int_a^t \left|g_1(s)\right|\mathrm{d}s, \dots, \int_a^t \left|g_n(s)\right|\mathrm{d}s\right)\Bigg]\mathrm{d}t \\
&\quad = F\left(\int_a^b \left|f_1(t)\right|\mathrm{d}t, \dots, \int_a^b \left|f_n(t)\right|\mathrm{d}t\right) \\
&\qquad \times G\left(\int_a^b \left|g_1(t)\right|\mathrm{d}t, \dots, \int_a^b \left|g_n(t)\right|\mathrm{d}t\right).
\end{aligned}$$

The proof is complete. □

THEOREM 3.5.7. *Let f_i, g_i, F, F_i', G, G_i' be as in Theorem 3.5.6 and p_i, q_i, h_i, w_i for $i = 1, \ldots, n$ be as in Theorem 3.5.5. Then, the following integral inequality holds*

$$\begin{aligned}
&\int_a^b \Bigg[F\big(|Tf_1(t)|, \ldots, |Tf_n(t)|\big) \sum_{i=1}^n G_i'\big(|Tg_1(t)|, \ldots, |Tg_n(t)|\big)|g_i(t)| \\
&\qquad + G\big(|Tg_1(t)|, \ldots, |Tg_n(t)|\big) \sum_{i=1}^n F_i'\big(|Tf_1(t)|, \ldots, |Tf_n(t)|\big)|f_i(t)| \Bigg] \mathrm{d}t \\
&\leqslant F\left(h_1^{-1}\left(\int_a^b p_1(t)h_1\left(\frac{|f_1(t)|}{p_1(t)}\right)\mathrm{d}t\right),\right. \\
&\qquad \left.\ldots, h_n^{-1}\left(\int_a^b p_n(t)h_n\left(\frac{|f_n(t)|}{p_n(t)}\right)\mathrm{d}t\right)\right) \\
&\quad \times G\left(w_1^{-1}\left(\int_a^b q_1(t)w_1\left(\frac{|g_1(t)|}{q_1(t)}\right)\mathrm{d}t\right),\right. \\
&\qquad \left.\ldots, w_n^{-1}\left(\int_a^b q_n(t)w_n\left(\frac{|g_n(t)|}{q_n(t)}\right)\mathrm{d}t\right)\right).
\end{aligned} \tag{3.5.28}$$

The proof can be completed by following the proof of Theorem 3.5.6 and closely looking at the proofs of Theorems 3.5.4 and 3.5.5 with suitable modifications. Here we omit the details.

REMARK 3.5.3. In the special cases, the inequalities given in Theorems 3.5.6 and 3.5.7 yield various new inequalities of the Opial type which are different from those of given by Bloom in [30]. For further generalizations of Theorems 3.5.6 and 3.5.7, see [346].

3.6 Opial-Type Inequalities Involving Higher-Order Derivatives

In 1968, Willett [425] established the following inequality

$$\int_a^x \left|u(t)u^{(n)}(t)\right| \mathrm{d}t \leqslant \frac{(x-a)^n}{2} \int_a^x \left|u^{(n)}(t)\right|^2 \mathrm{d}t, \tag{3.6.1}$$

where $x \in [a, b]$, $u \in C^{(n)}[a, b]$ with $u^{(i)}(a) = 0$ for $i = 0, 1, 2, \ldots, n-1$ and $n \geqslant 1$. Further results on some improvements, variants and generalizations of inequality (3.6.1) are given by a number of investigators, see [4] and the references given therein. In this section we shall present certain variants and extensions of inequality (3.6.1) investigated by Pachpatte in [239,296,312,317].

In the year 1986, Pachpatte [239] has proved the following Opial-type integral inequalities involving two functions and their nth-order derivatives.

THEOREM 3.6.1. *Let $u, v \in C^{(n-1)}[a,b]$ such that $u^{(k)}(a) = v^{(k)}(a) = 0$ for $k = 0, 1, 2, \dots, n-1$, where $n \geqslant 1$. Let $u^{(n-1)}$, $v^{(n-1)}$ be absolutely continuous and $\int_a^b |u^{(n)}(t)|^2\,dt < \infty$, $\int_a^b |v^{(n)}(t)|^2\,dt < \infty$. Then*

$$\int_a^b \left[\left|u(t)v^{(n)}(t)\right| + \left|v(t)u^{(n)}(t)\right|\right] dt \leqslant B(b-a)^n \int_a^b \left[\left|u^{(n)}(t)\right|^2 + \left|v^{(n)}(t)\right|^2\right] dt, \tag{3.6.2}$$

where

$$B = \frac{1}{2n!}\left(\frac{n}{2n-1}\right)^{1/2}. \tag{3.6.3}$$

Equality holds in (3.6.2) if and only if $n = 1$ and $u^{(n)}(t) = v^{(n)}(t) = M$, where M is a constant.

THEOREM 3.6.2. *Let $u, v \in C[a,b]$ and $u', \dots, u^{(n-1)}, v', \dots, v^{(n-1)}$ are piecewise continuous, $u^{(n-1)}$, $v^{(n-1)}$ are absolutely continuous with $\int_a^b |u^{(n)}(t)|^2\,dt < \infty$, $\int_a^b |v^{(n)}(t)|^2\,dt < \infty$, $u^{(k)}(a) = v^{(k)}(a) = 0$, $u^{(k)}(b) = v^{(k)}(b) = 0$ for $k = 0, 1, \dots, n-1$, where $n \geqslant 1$, then*

$$\int_a^b \left[\left|u(t)v^{(n)}(t)\right| + \left|v(t)u^{(n)}(t)\right|\right] dt \leqslant B\left(\frac{b-a}{2}\right)^n \int_a^b \left[\left|u^{(n)}(t)\right|^2 + \left|v^{(n)}(t)\right|^2\right] dt, \tag{3.6.4}$$

where B is given by (3.6.3). Equality holds in (3.6.4) if and only if $n = 1$ and $u(t) = v(t) = M(t-a)^n$, $a \leqslant t \leqslant (a+b)/2$; $u(t) = v(t) = M(b-t)^n$, $(a+b)/2 \leqslant t \leqslant b$, where M is a constant.

REMARK 3.6.1. We note that, in the special case when we take $u(t) = v(t)$ in Theorems 3.6.1 and 3.6.2, we get the integral inequalities established by Das in [77, Theorem 1 and Remark on p. 259] which in turn contains as a special case the Opial inequality (3.2.1) and the sharper version of the inequality established by Willett in (3.6.1). In order to avoid duplication, we omit the detailed discussion concerning the equalities in (3.6.2) and (3.6.4) and refer the reader to [77, p. 259] and [4, pp. 130–131].

PROOFS OF THEOREMS 3.6.1 AND 3.6.2. From the hypotheses of Theorem 3.6.1 we have

$$u(t)=\frac{1}{(n-1)!}\int_a^t (t-s)^{n-1}u^{(n)}(s)\,\mathrm{d}s, \tag{3.6.5}$$

$$v(t)=\frac{1}{(n-1)!}\int_a^t (t-s)^{n-1}v^{(n)}(s)\,\mathrm{d}s, \tag{3.6.6}$$

for $t\in[a,b]$. Now, multiplying (3.6.5) and (3.6.6) by $v^{(n)}(t)$ and $u^{(n)}(t)$, respectively, and upon using Schwarz inequality, we obtain

$$\left|u(t)v^{(n)}(t)\right| \leqslant \frac{|v^{(n)}(t)|(t-a)^{n-1/2}}{(n-1)!(2n-1)^{1/2}}\left(\int_a^t \left|u^{(n)}(s)\right|^2\mathrm{d}s\right)^{1/2}, \tag{3.6.7}$$

$$\left|v(t)u^{(n)}(t)\right| \leqslant \frac{|u^{(n)}(t)|(t-a)^{n-1/2}}{(n-1)!(2n-1)^{1/2}}\left(\int_a^t \left|v^{(n)}(s)\right|^2\mathrm{d}s\right)^{1/2}. \tag{3.6.8}$$

From (3.6.7) and (3.6.8), we obtain

$$\begin{aligned}
&\int_a^b \left[\left|u(t)v^{(n)}(t)\right|+\left|v(t)u^{(n)}(t)\right|\right]\mathrm{d}t\\
&\quad\leqslant \frac{1}{(n-1)!(2n-1)^{1/2}}\\
&\qquad\times\int_a^b (t-a)^{n-1/2}\left[\left|v^{(n)}(t)\right|\left(\int_a^t \left|u^{(n)}(s)\right|^2\mathrm{d}s\right)^{1/2}\right.\\
&\qquad\qquad\left.+\left|u^{(n)}(t)\right|\left(\int_a^t \left|v^{(n)}(t)\right|^2\mathrm{d}t\right)^{1/2}\right]\mathrm{d}t.
\end{aligned} \tag{3.6.9}$$

Now, first applying Schwarz inequality and then upon using the elementary inequalities $(\alpha+\beta)^2\leqslant 2(\alpha^2+\beta^2)$ and $\alpha^{1/2}\beta^{1/2}\leqslant\frac{1}{2}(\alpha+\beta)$, $\alpha,\beta\geqslant 0$ (for α,β reals) to the right-hand side of (3.6.9), we obtain

$$\begin{aligned}
&\int_a^b \left[\left|u(t)v^{(n)}(t)\right|+\left|v(t)u^{(n)}(t)\right|\right]\mathrm{d}t\\
&\quad\leqslant \frac{1}{(n-1)!(2n-1)^{1/2}}\left(\int_a^b (t-a)^{2(n-1/2)}\,\mathrm{d}t\right)^{1/2}
\end{aligned}$$

$$\times\left(\int_a^b\left[\left|v^{(n)}(t)\right|\left(\int_a^t\left|u^{(n)}(s)\right|^2\mathrm{d}s\right)^{1/2}\right.\right.$$
$$\left.\left.+\left|u^{(n)}(t)\right|\left(\int_a^t\left|v^{(n)}(s)\right|^2\mathrm{d}s\right)^{1/2}\right]^2\mathrm{d}t\right)^{1/2}$$
$$\leqslant\frac{(b-a)^n}{(n-1)!(2n-1)^{1/2}(2n)^{1/2}}$$
$$\times\left(2\int_a^b\left[\left|v^{(n)}(t)\right|^2\left(\int_a^t\left|u^{(n)}(s)\right|^2\mathrm{d}s\right)\right.\right.$$
$$\left.\left.+\left|u^{(n)}(t)\right|^2\left(\int_a^t\left|v^{(n)}(s)\right|^2\mathrm{d}s\right)\right]\mathrm{d}t\right)^{1/2}$$
$$=\frac{\sqrt{2}(b-a)^n}{(n-1)!(2n-1)^{1/2}(2n)^{1/2}}$$
$$\times\left(\int_a^b\frac{\mathrm{d}}{\mathrm{d}t}\left\{\left(\int_a^t\left|u^{(n)}(s)\right|^2\mathrm{d}s\right)\left(\int_a^t\left|v^{(n)}(s)\right|^2\mathrm{d}s\right)\right\}\mathrm{d}t\right)^{1/2}$$
$$=\frac{\sqrt{2}(b-a)^n}{(n-1)!(2n-1)^{1/2}(2n)^{1/2}}\left(\left(\int_a^b\left|u^{(n)}(s)\right|^2\mathrm{d}s\right)\left(\int_a^b\left|v^{(n)}(s)\right|^2\mathrm{d}s\right)\right)^{1/2}$$
$$\leqslant\frac{1}{2n!}\left(\frac{n}{2n-1}\right)^{1/2}(b-a)^n\int_a^b\left[\left|u^{(n)}(t)\right|^2+\left|v^{(n)}(t)\right|^2\right]\mathrm{d}t.$$

The proof of Theorem 3.6.1 is complete.

The proof of Theorem 3.6.2 follows immediately on using (3.6.2) once on $[a, \frac{a+b}{2}]$ and again on $[\frac{a+b}{2}, b]$, where on the latter interval, in view of the assumptions on u, v, we have

$$u(t)=\frac{(-1)^n}{(n-1)!}\int_t^b(s-t)^{n-1}u^{(n)}(s)\,\mathrm{d}s,$$
$$v(t)=\frac{(-1)^n}{(n-1)!}\int_t^b(s-t)^{n-1}v^{(n)}(s)\,\mathrm{d}s.$$

The details are omitted. □

In [296] Pachpatte has established the Opial-type inequalities in the following theorems, involving functions and their higher-order derivatives. In what follows, we let $I=[a,b]$ and $r_k(t)>0$, $k=1,\ldots,n-1$, and $z(t)$ be sufficiently smooth functions on $[a, b]$. The r-derivative of the function z is defined as follows.

$$\begin{cases} D_r^{(0)} z = z, \\ D_r^{(k)} z = r_k\big(D_r^{(k-1)} z\big)', \quad k = 1, \ldots, n-1 \ \big(\text{“}\,'\,\text{”} = \tfrac{\mathrm{d}}{\mathrm{d}t} = D\big), \\ D_r^{(n)} z = \big(D_r^{(n-1)} z\big)'. \end{cases} \tag{3.6.10}$$

Further, we set

$$R_k(a, t) = \begin{cases} 1, & \text{if } k = n-1, \\ \int_a^t \frac{\mathrm{d}s_{k+1}}{r_{k+1}(s_{k+1})} \int_a^{s_{k+1}} \frac{\mathrm{d}s_{k+2}}{r_{k+2}(s_{k+2})} \cdots \int_a^{s_{n-2}} \frac{\mathrm{d}s_{n-1}}{r_{n-1}(s_{n-1})}, & \text{if } 0 \leqslant k < n-1. \end{cases} \tag{3.6.11}$$

THEOREM 3.6.3. *Let $r_j > 0$, $j = 1, \ldots, n-1$, u, v be real-valued continuous functions defined on I and r-derivatives of u, v exist, be continuous on I and such that $D_r^{(i)} u(a) = D_r^{(i)} v(a) = 0$, $i = 0, 1, \ldots, n-1$, for $n \geqslant 1$ and $a \in I$. Then*

$$\int_a^b \big[\big|(D_r^{(k)} u)(t)\big|\big|(D_r^{(n)} v)(t)\big| + \big|(D_r^{(k)} v)(t)\big|\big|(D_r^{(n)} u)(t)\big|\big]\,\mathrm{d}t$$
$$\leqslant \begin{cases} \frac{(b-a)}{2} \int_a^b \big[\big|(D_r^{(n)} u)(t)\big|^2 + \big|(D_r^{(n)} v)(t)\big|^2\big]\,\mathrm{d}t, & \textit{if } k = n-1, \\ M \int_a^b \big[\big|(D_r^{(n)} u)(t)\big|^2 + \big|(D_r^{(n)} v)(t)\big|^2\big]\,\mathrm{d}t, & \textit{if } 0 \leqslant k < n-1, \end{cases} \tag{3.6.12}$$

where

$$M = \left[\frac{1}{2} \int_a^b (t-a) R_k^2(a, t)\,\mathrm{d}t\right]^{1/2}. \tag{3.6.13}$$

REMARK 3.6.2. (i) If we take $k = 0$ in inequality (3.6.12), then it reduces to the following inequality

$$\int_a^b \big[\big|u(t)\big|\big|(D_r^{(n)} v)(t)\big| + \big|v(t)\big|\big|(D_r^{(n)} u)(t)\big|\big]\,\mathrm{d}t$$
$$\leqslant M_0 \int_a^b \big[\big|(D_r^{(n)} u)(t)\big|^2 + \big|(D_r^{(n)} v)(t)\big|^2\big]\,\mathrm{d}t, \tag{3.6.14}$$

where

$$M_0 = \left[\frac{1}{2} \int_a^b (t-a) R_0^2(a, t)\,\mathrm{d}t\right]^{1/2}. \tag{3.6.15}$$

(ii) Putting $v = u$ in (3.6.12) and $r_j = 1$, $j = 1, \ldots, n-1$, we get

$$\int_a^b |u(t)||u^{(n)}(t)|\,dt \leqslant C_n \int_a^b |u^{(n)}(t)|^2\,dt, \tag{3.6.16}$$

where $C_n = \frac{1}{2}\frac{\sqrt{n}}{n!}(b-a)^n$. Inequality (3.6.16) contains Opial's inequality given in Theorem 2′ in [211, p. 154] with $n = 1$.

THEOREM 3.6.4. *Let p, q be positive constants satisfying $p + q > 1$ and r_j, u be as in Theorem* 3.6.3. *Then*

$$\int_a^b |(D_r^{(k)}u)(t)|^p |(D_r^{(n)}u)(t)|^q\,dt$$
$$\leqslant \begin{cases} q^{q/(p+q)}(p+q)^{-1}(b-a)^p \int_a^b |(D_r^{(n)}u)(t)|^{p+q}\,dt, & \textit{if } k = n-1, \\ N \int_a^b |(D_r^{(n)}u)(t)|^{p+q}\,dt, & \textit{if } 0 \leqslant k < n-1, \end{cases} \tag{3.6.17}$$

where

$$N = \left(\frac{q}{p+q}\right)^{q/(p+q)} \left[\int_a^b (t-a)^{p+q-1} R_k^{p+q}(a,t)\,dt\right]^{p/(p+q)}. \tag{3.6.18}$$

REMARK 3.6.3. If we take $k = 0$ in (3.6.17), we get

$$\int_a^b |u(t)|^p |(D_r^{(n)}u)(t)|^q\,dt \leqslant N_0 \int_a^b |(D_r^{(n)}u)(t)|^{p+q}\,dt, \tag{3.6.19}$$

where N_0 is obtained by the right-hand side of (3.6.18) by taking $k = 0$.

THEOREM 3.6.5. *Let r_j, u be as in Theorem* 3.6.3. *Then*

$$\int_a^b \prod_{i=0}^{n} |(D_r^{(i)}u)(t)|\,dt \leqslant Q\left[\int_a^b |(D_r^{(n)}u)(t)|^2\,dt\right]^{(n+1)/2}, \tag{3.6.20}$$

where

$$Q = \left[\frac{1}{n+1}\int_a^b (t-a)^n \prod_{i=0}^{n-2} R_i^2(a,t)\,dt\right]^{1/2}. \tag{3.6.21}$$

REMARK 3.6.4. By setting $r_j = 1$, $j = 1, \ldots, n-1$, then taking $n = 1$ and using the usual convention that $\prod_{i=n_1}^{n_2} R_i^2(a,t) = 1$ for $n_1 \geqslant n_2$, where n_1 and n_2 are integers, we see that inequality (3.6.20) reduces to Opial's inequality given in Theorem 2′ in [211, p. 154].

PROOFS OF THEOREMS 3.6.3–3.6.5. Under the assumptions of Theorem 3.6.3 for any $t \in I$, we have

$$(D_r^{(k)}u)(t) = \begin{cases} \int_a^t (D_r^{(n)}u)(s)\,\mathrm{d}s, & \text{if } k = n-1, \\ \int_a^t \frac{\mathrm{d}s_{k+1}}{r_{k+1}(s_{k+1})} \int_a^{s_{k+1}} \frac{\mathrm{d}s_{k+2}}{r_{k+2}(s_{k+2})} \cdots \int_a^{s_{n-2}} \frac{\mathrm{d}s_{n-1}}{r_{n-1}(s_{n-1})} \int_a^{s_{n-1}} (D_r^{(n)}u)(s)\,\mathrm{d}s & \\ \quad \leqslant R_k(a,t) \int_a^t \left|(D_r^{(n)}u)(s)\right| \mathrm{d}s, & \text{if } 0 \leqslant k < n-1. \end{cases} \tag{3.6.22}$$

Now, multiplying (3.6.22) by $|(D_r^{(n)}v)(t)|$ and (3.6.22) with $u = v$ by $|(D_r^{(n)}u)(t)|$, respectively, and making use of the properties of modulus and Schwarz inequality, we get the inequalities

$$\begin{aligned} &\left|(D_r^{(k)}u)(t)\right|\left|(D_r^{(n)}v)(t)\right| \\ &\quad \leqslant \begin{cases} \left|(D_r^{(n)}v)(t)\right| \int_a^t \left|(D_r^{(n)}u)(s)\right| \mathrm{d}s, \\ \quad \text{if } k = n-1, \\ (t-a)^{1/2} R_k(a,t) \left|(D_r^{(n)}v)(t)\right| \left(\int_a^t \left|(D_r^{(n)}u)(s)\right|^2 \mathrm{d}s\right)^{1/2}, \\ \quad \text{if } 0 \leqslant k < n-1, \end{cases} \end{aligned}$$

$$\begin{aligned} &\left|(D_r^{(k)}v)(t)\right|\left|(D_r^{(n)}u)(t)\right| \\ &\quad \leqslant \begin{cases} \left|(D_r^{(n)}u)(t)\right| \int_a^t \left|(D_r^{(n)}v)(s)\right| \mathrm{d}s, \\ \quad \text{if } k = n-1, \\ (t-a)^{1/2} R_k(a,t) \left|(D_r^{(n)}u)(t)\right| \left(\int_a^t \left|(D_r^{(n)}v)(s)\right|^2 \mathrm{d}s\right)^{1/2}, \\ \quad \text{if } 0 \leqslant k < n-1, \end{cases} \end{aligned}$$

which imply

$$\begin{aligned} &\int_a^b \left[\left|(D_r^{(k)}u)(t)\right|\left|(D_r^{(n)}v)(t)\right| + \left|(D_r^{(k)}v)(t)\right|\left|(D_r^{(n)}u)(t)\right|\right] \mathrm{d}t \\ &\quad \leqslant \begin{cases} \int_a^b \frac{\mathrm{d}}{\mathrm{d}t}\left[\left(\int_a^t \left|(D_r^{(n)}u)(s)\right| \mathrm{d}s\right)\left(\int_a^t \left|(D_r^{(n)}v)(s)\right| \mathrm{d}s\right)\right] \mathrm{d}t, & \text{if } k = n-1, \\ \int_a^b (t-a)^{1/2} R_k(a,t) & \\ \quad \times \left[\left|(D_r^{(n)}v)(t)\right| \left(\int_a^t \left|(D_r^{(n)}u)(s)\right|^2 \mathrm{d}s\right)^{1/2}\right. & \\ \quad\quad \left. + \left|(D_r^{(n)}u)(t)\right| \left(\int_a^t \left|(D_r^{(n)}v)(s)\right|^2 \mathrm{d}s\right)^{1/2}\right] \mathrm{d}t, & \text{if } 0 \leqslant k < n-1. \end{cases} \end{aligned} \tag{3.6.23}$$

In order to prove inequality (3.6.12), we consider the following two cases.

Case I. Let $k = n - 1$. From (3.6.23), using the elementary inequality $\alpha\beta \leqslant \frac{1}{2}(\alpha^2 + \beta^2)$, $\alpha, \beta \geqslant 0$ (for α, β reals) and the Schwarz inequality, we obtain

$$\begin{aligned}
&\int_a^b \left[\left|(D_r^{(k)}u)(t)\right|\left|(D_r^{(n)}v)(t)\right| + \left|(D_r^{(k)}v)(t)\right|\left|(D_r^{(n)}u)(t)\right|\right] \mathrm{d}t \\
&\quad \leqslant \left(\int_a^b \left|(D_r^{(n)}u)(t)\right| \mathrm{d}t\right)\left(\int_a^b \left|(D_r^{(n)}v)(t)\right| \mathrm{d}t\right) \\
&\quad \leqslant \frac{1}{2}(b-a)\int_a^b \left[\left|(D_r^{(n)}u)(t)\right|^2 + \left|(D_r^{(n)}v)(t)\right|^2\right] \mathrm{d}t,
\end{aligned} \tag{3.6.24}$$

being the required inequality in (3.6.12) for $k = n - 1$.

Case II. Let $0 \leqslant k < n - 1$. From (3.6.23), using Schwarz inequality, the elementary inequality $(\alpha + \beta)^2 \leqslant 2(\alpha^2 + \beta^2)$ and $\sqrt{\alpha\beta} \leqslant \frac{1}{2}(\alpha + \beta)$, $\alpha, \beta \geqslant 0$ (α, β reals), we get

$$\begin{aligned}
&\int_a^b \left[\left|(D_r^{(k)}u)(t)\right|\left|(D_r^{(n)}v)(t)\right| + \left|(D_r^{(k)}v)(t)\right|\left|(D_r^{(n)}u)(t)\right|\right] \mathrm{d}t \\
&\quad \leqslant \left(\int_a^b (t-a)R_k^2(a,t)\,\mathrm{d}t\right)^{1/2} \\
&\qquad \times \left\{\int_a^b \left[\left|(D_r^{(n)}v)(t)\right|\left(\int_a^t \left|(D_r^{(n)}u)(s)\right|^2 \mathrm{d}s\right)^{1/2}\right.\right. \\
&\qquad\qquad \left.\left. + \left|(D_r^{(n)}u)(t)\right|\left(\int_a^t \left|(D_r^{(n)}v)(s)\right|^2 \mathrm{d}s\right)^{1/2}\right]^2 \mathrm{d}t\right\}^{1/2} \\
&\quad \leqslant \left(\int_a^b (t-a)R_k^2(a,t)\,\mathrm{d}t\right)^{1/2} \\
&\qquad \times \left\{2\int_a^b \left[\left|(D_r^{(n)}v)(t)\right|^2\left(\int_a^t \left|(D_r^{(n)}u)(s)\right|^2 \mathrm{d}s\right)\right.\right. \\
&\qquad\qquad \left.\left. + \left|(D_r^{(n)}u)(t)\right|^2\left(\int_a^t \left|(D_r^{(n)}v)(s)\right|^2 \mathrm{d}s\right)\right] \mathrm{d}t\right\}^{1/2} \\
&\quad = \left(2\int_a^b (t-a)R_k^2(a,t)\,\mathrm{d}t\right)^{1/2}
\end{aligned}$$

$$\times \left\{\int_a^b \frac{\mathrm{d}}{\mathrm{d}t}\left[\left(\int_a^t \left|(D_r^{(n)}u)(s)\right|^2 \mathrm{d}s\right)\left(\int_a^t \left|(D_r^{(n)}v)(s)\right|^2 \mathrm{d}s\right)\right]\right\}^{1/2}$$

$$= \left(2\int_a^b (t-a)R_k^2(a,t)\,\mathrm{d}t\right)^{1/2}$$

$$\times \left(\int_a^b \left|(D_r^{(n)}u)(t)\right|^2 \mathrm{d}t\right)^{1/2}\left(\int_a^b \left|(D_r^{(n)}v)(t)\right|^2 \mathrm{d}t\right)^{1/2}$$

$$\leqslant \left(\frac{1}{2}\int_a^b (t-a)R_k^2(a,t)\,\mathrm{d}t\right)^{1/2}\int_a^b \left[\left|(D_r^{(n)}u)(t)\right|^2 + \left|(D_r^{(n)}v)(t)\right|^2\right]\mathrm{d}t, \tag{3.6.25}$$

being the required inequality in (3.6.12) for $0 \leqslant k < n-1$. Thus the proof of Theorem 3.6.3 is complete.

To prove Theorem 3.6.4, take modulo and pth power on both sides of (3.6.22) and use Hölder's inequality with indices $p+q$ and $(p+q)/(p+q-1)$, so we have

$$\left|(D_r^{(k)}u)(t)\right|^p$$
$$\leqslant \begin{cases} (t-a)^{p(p+q-1)/(p+q)}\left(\int_a^t \left|(D_r^{(n)}u)(s)\right|^{p+q} \mathrm{d}s\right)^{p/(p+q)}, \\ \quad \text{if } k = n-1, \\ (t-a)^{p(p+q-1)/(p+q)} R_k^p(a,t)\left(\int_a^t \left|(D_r^{(n)}u)(s)\right|^{p+q} \mathrm{d}s\right)^{p/(p+q)}, \\ \quad \text{if } 0 \leqslant k < n-1. \end{cases} \tag{3.6.26}$$

Now, multiplying both sides of (3.6.26) by $|(D_r^{(n)}u)(t)|^q$, then integrating from a to b and applying Hölder's inequality with indices $(p+q)/p$, $(p+q)/q$ to the integrals on the right-hand side, we find

$$\int_a^b \left|(D_r^{(k)}u)(t)\right|^p \left|(D_r^{(n)}u)(t)\right|^q \mathrm{d}t$$

$$\leqslant \begin{cases} \left(\int_a^b (t-a)^{p+q-1}\,\mathrm{d}t\right)^{p/(p+q)} \\ \quad \times \left(\int_a^b \left|(D_r^{(n)}u)(t)\right|^{p+q}\left(\int_a^t \left|(D_r^{(n)}u)(s)\right|^{p+q}\mathrm{d}s\right)^{p/q}\mathrm{d}t\right)^{q/(p+q)}, \\ \text{if } k = n-1, \\ \left(\int_a^b (t-a)^{p+q-1} R_k^{p+q}(a,t)\,\mathrm{d}t\right)^{p/(p+q)} \\ \quad \times \left(\int_a^b \left|(D_r^{(n)}u)(t)\right|^{p+q}\left(\int_a^t \left|(D_r^{(n)}u)(s)\right|^{p+q}\mathrm{d}s\right)^{p/q}\mathrm{d}t\right)^{q/(p+q)} \\ \text{if } 0 \leqslant k < n-1, \end{cases}$$

$$= \begin{cases} q^{q/(p+q)}(p+q)^{-1}(b-a)^p \int_a^b \left|(D_r^{(n)}u)(t)\right|^{p+q} \mathrm{d}t, & \text{if } k = n-1, \\ N \int_a^b \left|(D_r^{(n)}u)(t)\right|^{p+q} \mathrm{d}t, & \text{if } 0 \leqslant k < n-1. \end{cases} \tag{3.6.27}$$

The proof of Theorem 3.6.4 is complete.

By the hypotheses of Theorem 3.6.5, we have inequality (3.6.22). Taking $k = 0, 1, \ldots, n-1$ in (3.6.22) and modulo, we get

$$\left|(D_r^{(0)}u)(t)\right| \leqslant R_0(a,t) \int_a^t \left|(D_r^{(n)}u)(s)\right| \mathrm{d}s,$$

$$\left|(D_r^{(1)}u)(t)\right| \leqslant R_1(a,t) \int_a^t \left|(D_r^{(n)}u)(s)\right| \mathrm{d}s,$$

$$\vdots$$

$$\left|(D_r^{(n-2)}u)(t)\right| \leqslant R_{n-2}(a,t) \int_a^t \left|(D_r^{(n)}u)(s)\right| \mathrm{d}s,$$

$$\left|(D_r^{(n-1)}u)(t)\right| \leqslant \int_a^t \left|(D_r^{(n)}u)(s)\right| \mathrm{d}s.$$

From these inequalities and using Schwarz inequality, we obtain

$$\begin{aligned} \prod_{i=0}^{n} \left|(D_r^{(i)}u)(t)\right| &\leqslant \prod_{i=0}^{n-2} R_i(a,t)\left|(D_r^{(n)}u)(t)\right| \left(\int_a^t \left|(D_r^{(n)}u)(s)\right| \mathrm{d}s\right)^n \\ &\leqslant \prod_{i=0}^{n-2} R_i(a,t)\left|(D_r^{(n)}u)(t)\right|(t-a)^{n/2}\left(\int_a^t \left|(D_r^{(n)}u)(s)\right|^2 \mathrm{d}s\right)^{n/2}. \end{aligned} \tag{3.6.28}$$

Integrating both sides of (3.6.28) from a to b and using again Schwarz inequality, we find

$$\begin{aligned} &\int_a^b \prod_{i=0}^{n} \left|(D_r^{(i)}u)(t)\right| \mathrm{d}t \\ &\quad \leqslant \int_a^b (t-a)^{n/2} \prod_{i=0}^{n-2} R_i(a,t)\left|(D_r^{(n)}u)(t)\right| \left(\int_a^t \left|(D_r^{(n)}u)(s)\right|^2 \mathrm{d}s\right)^{n/2} \mathrm{d}t \end{aligned}$$

$$\leqslant \left(\int_a^b (t-a)^n \prod_{i=0}^{n-2} R_i^2(a,t)\,\mathrm{d}t \right)^{1/2}$$

$$\times \left(\int_a^b \left|\left(D_r^{(n)}u\right)(t)\right|^2 \left(\int_a^t \left|\left(D_r^{(n)}u\right)(s)\right|^2 \mathrm{d}s \right)^n \mathrm{d}t \right)^{1/2}$$

$$= Q\left(\int_a^b \left|\left(D_r^{(n)}u\right)(t)\right|^2 \mathrm{d}t \right)^{(n+1)/2}.$$

The proof of Theorem 3.6.5 is complete. □

REMARK 3.6.5. We note that, by following the proof of Theorem 3.6.3 with suitable modifications, we can establish inequality (3.6.14) for $u = v$ in the following useful variant

$$\int_a^t |u(s)|\left|\left(D_r^{(n-1)}u\right)(s)\right| \mathrm{d}s \leqslant M_0^*(t) \int_a^t \left|\left(D_r^{(n-1)}u\right)(s)\right|^2 \mathrm{d}s, \tag{3.6.29}$$

where $a, t \in I$ and r_j, u are as defined in Theorem 3.6.3 but with $j = 1, \ldots, n-2$, $i = 0, 1, \ldots, n-2$ for $n \geqslant 2$,

$$M_0^*(t) = \left[\frac{1}{2} \int_a^t (s-a) R_0^{*2}(a,s)\,\mathrm{d}s \right]^{1/2} \tag{3.6.30}$$

with

$$R_0^*(a,t) = \int_a^t \frac{\mathrm{d}s_1}{r_1(s_1)} \int_a^{s_1} \frac{\mathrm{d}s_2}{r_2(s_2)} \cdots \int_a^{s_{n-3}} \frac{\mathrm{d}s_{n-2}}{r_{n-2}(s_{n-2})}. \tag{3.6.31}$$

Inequality (3.6.29) is formulated in the framework of Willett's inequality (3.6.1) which is suitable in certain applications.

It is easy to observe that the constant obtained in (3.6.16) is better than that in (3.6.1).

Inequality (3.6.16) with sharper constant $c_n = \frac{1}{2n!}\left(\frac{n}{2n-1}\right)^{1/2}$ is established in Das [77, Theorem 1]. However, our proof does not yield the better constant as in [77], because there is a difficulty involved in proving the much more general result as given in Theorem 3.6.3, that is, proving (3.6.24), (3.6.25). We also note that the results established in Theorems 3.6.3–3.6.5 can be extended to the case when we replace the conditions

$$D_r^{(i)}u(a) = D_r^{(i)}v(a) = 0 \quad \text{by} \quad D_r^{(i)}u(b) = D_r^{(i)}v(b) = 0, \quad i = 0, 1, \ldots, n-1.$$

The following theorem established by Pachpatte in [312] involves a generalization of Opial's inequality involving a function and its higher-order derivatives.

THEOREM 3.6.6. *Let $u \in C^{(n)}[a,b]$ be a real-valued function such that $u^{(k)}(a) = 0$ for $k = 0, 1, \ldots, n-1$ and $n \geqslant 1$. Let w, v be positive and continuous functions defined on $[a,b]$. Let $p \geqslant 1$ and $q > 0$ be real numbers and $r_k \geqslant 0$, $k = 0, 1, \ldots, n-1$, be real numbers with $\sum_{k=0}^{n-1} r_k = 1$. Then*

$$\int_a^b w(t)\left[\prod_{k=0}^{n-1} \left|u^{(k)}(t)\right|^{r_k}\right]^p \left|u^{(n)}(t)\right|^q \, dt \leqslant C(p,q) \int_a^b v(t)\left|u^{(n)}(t)\right|^{p+q} dt, \tag{3.6.32}$$

where

$$C(p,q) = \left(\frac{q}{p+q}\right)^{q/(p+q)} \left\{\int_a^b w^{(p+q)/p}(t) v^{-q/p}(t) \left\{\sum_{k=0}^{n-1} r_k \left[(n-k-1)!\right]^{-p} \times \left[\int_a^t v^{-1/(p+q-1)}(s) \times (t-s)^{(p+q)(n-k-1)/(p+q-1)} \, ds\right]^{p(p+q-1)/(p+q)}\right\}^{(p+q)/p} dt\right\}^{p/(p+q)} \tag{3.6.33}$$

is finite.

PROOF. From the hypotheses, on u we have

$$u^{(k)}(t) = \frac{1}{(n-k-1)!} \int_a^t (t-s)^{n-k-1} u^{(n)}(s) \, ds \tag{3.6.34}$$

for $k = 0, 1, \ldots, n-1$. From (3.6.34) and using the elementary inequality

$$\prod_{k=0}^{n-1} a_k^{r_k} \leqslant \sum_{k=0}^{n-1} r_k a_k \leqslant \left\{\sum_{k=0}^{n-1} r_k a_k^p\right\}^{1/p},$$

where $a_k \geqslant 0$, $k = 0, 1, \ldots, n-1$, and $p \geqslant 1$ be any real number and r_k as stated in theorem. By using Hölder's inequality with indices $p+q$, $(p+q)/(p+q-1)$,

we observe that

$$\begin{aligned}
&\left[\prod_{k=0}^{n-1}\left|u^{(k)}(t)\right|^{r_k}\right]^{p} \\
&\quad\leqslant \sum_{k=0}^{n-1} r_k\left|u^{(k)}(t)\right|^{p} \\
&\quad\leqslant \sum_{k=0}^{n-1} r_k\left[(n-k-1)!\right]^{-p} \\
&\qquad\quad \times\left\{\int_a^t \left(v^{-1/(p+q)}(s)(t-s)^{n-k-1}\right)\left(v^{1/(p+q)}(s)\left|u^{(n)}(s)\right|\right)\mathrm{d}s\right\}^{p} \\
&\quad\leqslant \sum_{k=0}^{n-1} r_k\left[(n-k-1)!\right]^{-p} \\
&\qquad\quad \times\left[\int_a^t v^{-1/(p+q-1)}(s)(t-s)^{(p+q)(n-k-1)/(p+q-1)}\right]^{p(p+q-1)/(p+q)} \\
&\qquad\quad \times\left[\int_a^t v(s)\left|u^{(n)}(s)\right|^{p+q}\mathrm{d}s\right]^{p/(p+q)}. \qquad (3.6.35)
\end{aligned}$$

Multiplying both sides of (3.6.35) by $w(t)|u^{(n)}(t)|^q$ and integrating the resulting inequality from a to b, rewriting and then using Hölder's inequality with indices $(p+q)/p$, $(p+q)/q$, we observe that

$$\begin{aligned}
&\int_a^b w(t)\left[\prod_{k=0}^{n-1}\left|u^{(k)}(t)\right|^{r_k}\right]^{p}\left|u^{(n)}(t)\right|^{q}\mathrm{d}t \\
&\quad\leqslant \int_a^b\Bigg[w(t)v^{-q/(p+q)}(t)\sum_{k=0}^{n-1} r_k\left[(n-k-1)!\right]^{-p} \\
&\qquad \times\left\{\int_a^t v^{-1/(p+q-1)}(s)(t-s)^{(p+q)(n-k-1)/(p+q-1)}\mathrm{d}s\right\}^{p(p+q-1)/(p+q)}\Bigg] \\
&\qquad \times\left[v^{q/(p+q)}(t)\left|u^{(n)}(t)\right|^{q}\left\{\int_a^t v(s)\left|u^{(n)}(s)\right|^{p+q}\mathrm{d}s\right\}^{p/(p+q)}\right]\mathrm{d}t
\end{aligned}$$

$$
\begin{aligned}
&\leqslant \left\{\int_a^b w^{(p+q)/p}(t)v^{-q/p}(t)\left\{\sum_{k=0}^{n-1} r_k\big[(n-k-1)!\big]^{-p}\right.\right.\\
&\quad\times\left[\int_a^t v^{-1/(p+q-1)}(s)\right.\\
&\quad\left.\left.\left.\times (t-s)^{(p+q)(n-k-1)/(p+q-1)}\,\mathrm{d}s\right]^{p(p+q-1)/(p+q)}\right\}^{(p+q)/p}\mathrm{d}t\right\}^{p/(p+q)}\\
&\quad\times\left\{\int_a^b v(t)\big|u^{(n)}(t)\big|^{p+q}\left[\int_a^t v(s)\big|u^{(n)}(s)\big|^{p+q}\,\mathrm{d}s\right]^{p/q}\mathrm{d}t\right\}^{q/(p+q)}\\
&= C(p,q)\int_a^b v(t)\big|u^{(n)}(t)\big|^{p+q}\,\mathrm{d}t.
\end{aligned}
$$

The proof is complete. □

REMARK 3.6.6. We note that by specializing inequality (3.6.32) we get the various inequalities established earlier by different investigators. We also note that Theorem 3.6.6 can be extended to the case when we replace the conditions $u^{(k)}(a)=0$ by $u^{(k)}(b)=0$ for $k=0,1,\dots,n-1$. For more details, see [312].

The inequalities in the following theorems are established by Pachpatte in [317].

To formulate the results conveniently, we set

$$
\begin{aligned}
M_1 &= \big[(n-i-1)!\big]^{-p_1}\big[(n-j-1)!\big]^{-p_2}\\
&\quad\times\left\{\int_a^b w^{p/(p-p_3)}(t)\right.\\
&\qquad\times\left(\int_a^t (t-s)^{p(n-i-1)/(p-1)}v^{-1/(p-1)}(s)\,\mathrm{d}s\right)^{p_1(p-1)/(p-p_3)}\\
&\qquad\times\left(\int_a^t (t-s)^{p(n-j-1)/(p-1)}v^{-1/(p-1)}(s)\,\mathrm{d}s\right)^{p_2(p-1)/(p-p_3)}\\
&\qquad\left.\times v^{-p_3/(p-p_3)}(t)\,\mathrm{d}t\right\}^{(p-p_3)/p}
\end{aligned}
\tag{3.6.36}
$$

and

$$M_2 = \left[(n-1)!\right]^{-2p}$$
$$\times \left\{ \int_a^b w^{(p+q)/p}(t) \left(\int_a^t \left[\int_\tau^t \frac{(t-s)^{n-1}(s-\tau)^{n-1}}{r(s)} \, \mathrm{d}s \right]^{(p+q)/(p+q-1)} \right.\right.$$
$$\left.\left. \times\, v^{-1/(p+q-1)}(\tau) \, \mathrm{d}\tau \right)^{p+q-1} v^{-q/p}(t) \, \mathrm{d}t \right\}^{p/(p+q)}, \qquad (3.6.37)$$

where $p_1, p_2, p_3, p = p_1 + p_2 + p_3$; p, q, n, i, j are suitable constants and $w(t)$, $v(t)$, $r(t)$ are suitable functions defined on $I = [a, b]$, $a < b$ are real constants, and

$$M_3 = \left\{ \int_a^b w^{p/(p-p_n)}(t) \left(\prod_{i=0}^{n-2} R_i^{p_i}(a,t) \right)^{p/(p-p_n)} \right.$$
$$\left. \times \left(\int_a^t v^{-1/(p-1)}(s) \, \mathrm{d}s \right)^{p-1} v^{-p_n/(p-p_n)}(t) \right\}^{(p-p_n)/p}, \qquad (3.6.38)$$

where $p_0, p_1, \ldots, p_n, p = \sum_{i=0}^n p_i$ are suitable constants, $w(t)$ and $v(t)$ are suitable functions defined on I and R_i is as defined in (3.6.11).

THEOREM 3.6.7. *Let p_1, p_2, p_3 be nonnegative real numbers satisfying $p = p_1 + p_2 + p_3 > p_3 > 0$, $p > 1$, and let $n \geqslant 2$ and $0 \leqslant i \leqslant j \leqslant n-1$ be integers. Let $f(t)$ be of class C^n on I satisfying $f(a) = f'(a) = \cdots = f^{(n-1)}(a) = 0$. Suppose that $w(t)$ and $v(t)$ are positive and continuous functions defined on I. Then*

$$\int_a^b w(t) \left| f^{(i)}(t) \right|^{p_1} \left| f^{(j)}(t) \right|^{p_2} \left| f^{(n)}(t) \right|^{p_3} \mathrm{d}t$$
$$\leqslant \left(\frac{p_3}{p} \right)^{p_3/p} M_1 \int_a^b v(t) \left| f^{(n}(t) \right|^p \mathrm{d}t, \qquad (3.6.39)$$

where M_1 is finite and defined by (3.6.36).

THEOREM 3.6.8. *Let p, q be positive real numbers satisfying $p + q > 1$ and let $n \geqslant 1$ be an integer. Let $r(t) > 0$ be of class C^n on I and $f(t)$ be of class C^{2n} on I satisfying $f^{(i-1)}(a) = 0$, $(r(a) f^{(n)}(a))^{(i-1)} = 0$ for $i = 1, 2, \ldots, n$. Suppose* *that $w(t)$ and $v(t)$ are positive and continuous functions defined on I. Then*

$$\int_a^b w(t)|f(t)|^p \left|\left(r(t)f^{(n)}(t)\right)^{(n)}\right|^q \mathrm{d}t$$

$$\leqslant \left(\frac{q}{p+q}\right)^{q/(p+q)} M_2 \int_a^b v(t)\left|\left(r(t)f^{(n)}(t)\right)^{(n)}\right|^{p+q} \mathrm{d}t, \quad (3.6.40)$$

where M_2 is finite and defined by (3.6.37).

The following corollary to Theorem 3.6.8 given by Pachpatte in [317] is of independent interest.

COROLLARY 3.6.1. *Let $r(t) > 0$ be of class C^n on I and $f(t)$ be of class C^{2n} on I satisfying $f^{(i-1)}(a) = 0$, $(r(a)f^{(n)}(a))^{(i-1)} = 0$ for $i = 1, 2, \ldots, n$. Suppose that $M_2^*(t)$ defined by*

$$M_2^*(t) = \left[(n-1)!\right]^{-2}\left\{\int_a^t \left(\int_a^x \left[\int_\tau^x \frac{(x-s)^{n-1}(s-\tau)^{n-1}}{r(s)} \mathrm{d}s\right]^2 \mathrm{d}\tau\right) \mathrm{d}x\right\}^{1/2}$$

is finite, then

$$\int_a^t |f(s)|\left|\left(r(s)f^{(n)}(s)\right)^{(n)}\right| \mathrm{d}s \leqslant \frac{1}{\sqrt{2}} M_2^*(t) \int_a^t \left|\left(r(t)f^{(n)}(s)\right)^{(n)}\right|^2 \mathrm{d}s$$

for $n \geqslant 1$ and $t \in I$.

This situation is the case of Theorem 3.6.8 in which $p = q = 1$ and $w(t) = v(t) = 1$ and b is replaced by a variable t and M_2 is replaced by $M_2^*(t)$.

THEOREM 3.6.9. *Let $p_0, p_1, \ldots, p_n$ be nonnegative real numbers satisfying $p = \sum_{i=0}^n p_i > p_n$, $p > 1$. Let $r_i(t) > 0$, $i = 1, \ldots, n-1$, $n \geqslant 1$, and $f(t)$ be continuous function defined on I. Let r-derivatives of $f(t)$ exist, be continuous on I and such that $D_r^{(i)} f(a) = 0$, $i = 0, 1, \ldots, n-1$. Suppose that $w(t)$ and $v(t)$ are positive and continuous functions defined on I. Then*

$$\int_a^b w(t) \prod_{i=0}^n \left|\left(D_r^{(i)} f\right)(t)\right|^{p_i} \mathrm{d}t$$

$$\leqslant \left(\frac{p_n}{p}\right)^{p_n/p} M_3 \int_a^b v(t)\left|\left(D_r^{(n)} f\right)(t)\right|^p \mathrm{d}t, \quad (3.6.41)$$

where M_3 is finite and defined by (3.6.38).

PROOFS OF THEOREMS 3.6.7–3.6.9. From the hypotheses of Theorem 3.6.7 and Taylor expansion, we have the representation

$$f^{(i)}(t)=\frac{1}{(n-i-1)!}\int_a^t (t-s)^{n-i-1}f^{(n)}(s)\,\mathrm{d}s \tag{3.6.42}$$

for $0\leqslant i\leqslant n-1$. From (3.6.42) and using Hölder's inequality with indices $p/(p-1)$, p, we observe that, for $t\in I$,

$$\begin{aligned}
&\left|f^{(i)}(t)\right|^{p_1}\\
&\leqslant\left[(n-i-1)!\right]^{-p_1}\left\{\int_a^t\left[(t-s)^{n-i-1}v^{-1/p}(s)\right]\left[v^{1/p}(s)\left|f^{(n)}(s)\right|\right]\mathrm{d}s\right\}^{p_1}\\
&\leqslant\left[(n-i-1)!\right]^{-p_1}\left(\int_a^t(t-s)^{p(n-i-1)/(p-1)}v^{-1/(p-1)}(s)\,\mathrm{d}s\right)^{p_1(p-1)/p}\\
&\quad\times\left(\int_a^t v(s)\left|f^{(n)}(s)\right|^p\mathrm{d}s\right)^{p_1/p}.
\end{aligned}\tag{3.6.43}$$

From (3.6.43) and rewriting (3.6.43) by replacing i by $j, 0\leqslant i\leqslant j\leqslant n-1$, and p_1 by p_2, we observe that

$$\begin{aligned}
&w(t)\left|f^{(i)}(t)\right|^{p_1}\left|f^{(j)}(t)\right|^{p_2}\left|f^{(n)}(t)\right|^{p_3}\\
&\quad\leqslant\left[(n-i-1)!\right]^{-p_1}\left[(n-j-1)!\right]^{-p_2}\\
&\qquad\times\left[w(t)\left(\int_a^t(t-s)^{p(n-i-1)/(p-1)}v^{-1/(p-1)}(s)\,\mathrm{d}s\right)^{p_1(p-1)/p}\right.\\
&\qquad\times\left.\left(\int_a^t(t-s)^{p(n-j-1)/(p-1)}v^{-1/(p-1)}(s)\,\mathrm{d}s\right)^{p_2(p-1)/p}v^{-p_3/p}(t)\right]\\
&\qquad\times\left[v^{p_3/p}(t)\left|f^{(n)}(t)\right|^{p_3}\left(\int_a^t v(s)\left|f^{(n)}(s)\right|^p\mathrm{d}s\right)^{(p-p_3)/p}\right].
\end{aligned}\tag{3.6.44}$$

Now, integrating both sides of (3.6.44) from a to b and using Hölder's inequality with indices $p/(p-p_3)$, p/p_3 on the right-hand side of the resulting inequality, we observe that

$$\begin{aligned}
&\int_a^b w(t)\left|f^{(i)}(t)\right|^{p_1}\left|f^{(j)}(t)\right|^{p_2}\left|f^{(n)}(t)\right|^{p_3}\mathrm{d}t\\
&\quad\leqslant M_1\left\{\int_a^b v(t)\left|f^{(n)}(t)\right|^p\left(\int_a^t v(s)\left|f^{(n)}(s)\right|^p\mathrm{d}s\right)^{(p-p_3)/p_3}\mathrm{d}t\right\}^{p_3/p}
\end{aligned}$$

$$= \left(\frac{p_3}{p}\right)^{p_3/p} M_1 \int_a^b v(t)\left|f^{(n)}(t)\right|^p \mathrm{d}t,$$

where M_1 is defined in (3.6.36). The proof of Theorem 3.6.7 is complete.

From the hypotheses of Theorem 3.6.8 and Taylor expansion, we have the representations

$$f(t) = \frac{1}{(n-1)!}\int_a^t (t-s)^{n-1} f^{(n)}(s)\,\mathrm{d}s \tag{3.6.45}$$

and

$$r(s)f^{(n)}(s) = \frac{1}{(n-1)!}\int_a^s (s-\tau)^{n-1}\big(r(\tau)f^{(n)}(\tau)\big)^{(n)}\,\mathrm{d}\tau. \tag{3.6.46}$$

By substituting (3.6.46) into (3.6.45) and reversing the order of integration on the double integral, we have the representation formula (see [97, p. 315])

$$f(t) = \frac{1}{[(n-1)!]^2} \times \int_a^t \left[\int_\tau^t \frac{(t-s)^{n-1}(s-\tau)^{n-1}}{r(s)}\,\mathrm{d}s\right]\big(r(\tau)f^{(n)}(\tau)\big)^{(n)}\,\mathrm{d}\tau \tag{3.6.47}$$

for $a \leqslant \tau \leqslant s \leqslant t \leqslant b$. From (3.6.47) and using Hölder's inequality with indices $(p+q)/(p+q-1)$, $p+q$, we observe that

$$\begin{aligned}
\left|f(t)\right|^p &\leqslant \big[(n-1)!\big]^{-2p} \\
&\quad\times \left\{\int_a^t \left[\left(\int_\tau^t \frac{(t-s)^{n-1}(s-\tau)^{n-1}}{r(s)}\,\mathrm{d}s\right) v^{-1/(p+q)}(\tau)\right]\right. \\
&\quad\times \left.\left[v^{1/(p+q)}(\tau)\left|\big(r(\tau)f^{(n)}(\tau)\big)^{(n)}\right|\right]\mathrm{d}\tau\right\}^p \\
&\leqslant \big[(n-1)!\big]^{-2p}\left\{\int_a^t \left(\int_\tau^t \frac{(t-s)^{n-1}(s-\tau)^{n-1}}{r(s)}\,\mathrm{d}s\right)^{(p+q)/(p+q-1)}\right. \\
&\quad\times \left. v^{-1/(p+q-1)}(\tau)\,\mathrm{d}\tau\right\}^{p(p+q-1)/(p+q)} \\
&\quad\times \left\{\int_a^t v(\tau)\left|\big(r(\tau)f^{(n)}(\tau)\big)^{(n)}\right|^{p+q}\mathrm{d}\tau\right\}^{p/(p+q)}.
\end{aligned} \tag{3.6.48}$$

From (3.6.48) we observe that

$$\begin{aligned}
&w(t)\big|f(t)\big|^p\big|\big(r(t)f^{(n)}(t)\big)^{(n)}\big|^q \\
&\quad\leqslant \big[(n-1)!\big]^{-2p}\Bigg[w(t)\bigg\{\int_a^t\bigg(\int_\tau^t \frac{(t-s)^{n-1}(s-\tau)^{n-1}}{r(s)}\,\mathrm{d}s\bigg)^{(p+q)/(p+q-1)} \\
&\qquad\qquad \times v^{-1/(p+q-1)}(\tau)\,\mathrm{d}\tau\bigg\}^{p(p+q-1)/(p+q)} v^{-q/(p+q)}(t)\Bigg] \\
&\quad\times\Bigg[v^{q/(p+q)}(t)\big|\big(r(t)f^{(n)}(t)\big)^{(n)}\big|^q \\
&\quad\times\bigg\{\int_a^t v(\tau)\big|\big(r(\tau)f^{(n)}(\tau)\big)^{(n)}\big|^{p+q}\,\mathrm{d}\tau\bigg\}^{p/(p+q)}\Bigg]. \qquad (3.6.49)
\end{aligned}$$

Now, integrating both sides of (3.6.49) from a to b and using Hölder's inequality with indices $(p+q)/p$, $(p+q)/q$ on the right-hand side of the resulting inequality, we observe that

$$\begin{aligned}
&\int_a^b w(t)\big|f(t)\big|^p\big|\big(r(t)f^{(n)}(t)\big)^{(n)}\big|^q\,\mathrm{d}t \\
&\quad\leqslant M_2\bigg\{\int_a^b v(t)\big|\big(r(t)f^{(n)}(t)\big)^{(n)}\big|^{p+q} \\
&\qquad\qquad\times\bigg(\int_a^t v(\tau)\big|\big(r(\tau)f^{(n)}(\tau)\big)^{(n)}\big|^{p+q}\,\mathrm{d}\tau\bigg)^{p/q}\,\mathrm{d}t\bigg\}^{q/(p+q)} \\
&\quad=\bigg(\frac{q}{p+q}\bigg)^{p/(p+q)} M_2\int_a^b v(t)\big|\big(r(t)f^{(n)}(t)\big)^{(n)}\big|^{p+q}\,\mathrm{d}t,
\end{aligned}$$

where M_2 is defined by (3.6.37). This result is the required inequality in (3.6.40) and the proof of Theorem 3.6.8 is complete.

From the hypotheses of the Theorem 3.6.9, for any $t \in I$, we have

$$\big(D_r^{(i)}f\big)(t) = \begin{cases} \int_a^t \big(D_r^{(n)}f\big)(s)\,\mathrm{d}s, & \text{if } i = n-1, \\ \int_a^t \frac{\mathrm{d}s_{i+1}}{r_{i+1}(s_{i+1})}\int_a^{s_{i+1}} \frac{\mathrm{d}s_{i+2}}{r_{i+2}(s_{i+2})}\cdots\int_a^{s_{n-2}} \frac{\mathrm{d}s_{n-1}}{r_{i-1}(s_{n-1})} & \\ \quad\times\int_a^{s_{n-1}}\big(D_r^{(n)}f\big)(s)\,\mathrm{d}s, & \text{if } 0 \leqslant i < n-1. \end{cases} \qquad (3.6.50)$$

From (3.6.50) we observe that

$$\left|\left(D_r^{(i)} f\right)(t)\right| \leqslant \begin{cases} \int_a^t \left|\left(D_r^{(n)} f\right)(s)\right| \mathrm{d}s, & \text{if } i = n-1, \\ R_i(a,t) \int_a^t \left|\left(D_r^{(n)} f\right)(s)\right| \mathrm{d}s, & \text{if } 0 \leqslant i < n-1. \end{cases} \tag{3.6.51}$$

From (3.6.51) we observe that

$$\begin{aligned}
\left|\left(D_r^{(0)} f\right)(t)\right|^{p_0} &\leqslant R_0^{p_0}(a,t)\left(\int_a^t \left|\left(D_r^{(n)} f\right)(s)\right| \mathrm{d}s\right)^{p_0}, \\
\left|\left(D_r^{(1)} f\right)(t)\right|^{p_1} &\leqslant R_1^{p_1}(a,t)\left(\int_a^t \left|\left(D_r^{(n)} f\right)(s)\right| \mathrm{d}s\right)^{p_1}, \\
&\vdots \\
\left|\left(D_r^{(n-2)} f\right)(t)\right|^{p_{n-2}} &\leqslant R_{n-2}^{p_{n-2}}(a,t)\left(\int_a^t \left|\left(D_r^{(n)} f\right)(s)\right| \mathrm{d}s\right)^{p_{n-2}}, \\
\left|\left(D_r^{(n-1)} f\right)(t)\right|^{p_{n-1}} &\leqslant \left(\int_a^t \left|\left(D_r^{(n)} f\right)(s)\right| \mathrm{d}s\right)^{p_{n-1}}.
\end{aligned}$$

From these inequalities and using Hölder's inequality with indices $p/(p-1)$, p, we observe that

$$\begin{aligned}
& w(t) \prod_{i=0}^{n} \left|\left(D_r^{(i)} f\right)(t)\right|^{p_i} \\
&\leqslant w(t)\left|\left(D_r^{(n)} f\right)(t)\right|^{p_n} \\
&\quad \times \prod_{i=0}^{n-2} R_i^{p_i}(a,t)\left(\int_a^t v^{-1/p}(s) v^{1/p}(s)\left|\left(D_r^{(n)} f\right)(s)\right| \mathrm{d}s\right)^{p-p_n} \\
&\leqslant \left[w(t) \prod_{i=0}^{n-2} R_i^{p_i}(a,t)\left(\int_a^t v^{-1/(p-1)}(s)\,\mathrm{d}s\right)^{(p-1)(p-p_n)/p} v^{-p_n/p}(t)\right] \\
&\quad \times \left[v^{p_n/p}(t)\left|\left(D_r^{(n)} f\right)(t)\right|^{p_n}\left(\int_a^t v(s)\left|\left(D_r^{(n)} f\right)(s)\right|^p \mathrm{d}s\right)^{(p-p_n)/p}\right].
\end{aligned} \tag{3.6.52}$$

Now, integrating both sides of (3.6.52) from a to b and using Hölder's inequality with indices $p/(p-p_n)$, p/p_n on the right-hand side of the resulting inequality,

we observe that

$$\int_a^b w(t)\prod_{i=0}^{n}\left|\left(D_r^{(i)}f\right)(t)\right|^{p_i}\,\mathrm{d}t$$

$$\leqslant M_3\left\{\int_a^b v(t)\left|\left(D_r^{(n)}f\right)(t)\right|^p\left(\int_a^t v(s)\left|\left(D_r^{(n)}f\right)(s)\right|^p\,\mathrm{d}s\right)^{(p-p_n)/p_n}\mathrm{d}t\right\}^{p_n/p}$$

$$=\left(\frac{p_n}{p}\right)^{p_n/p}M_3\int_a^b v(t)\left|\left(D_r^{(n)}f\right)(t)\right|^p\,\mathrm{d}t,$$

where M_3 is defined by (3.6.38). The proof of Theorem 3.6.9 is complete. □

3.7 Opial-Type Inequalities in Two and Many Independent Variables

In the past few years, a number of papers have been written dealing with Opial-type inequalities, involving functions of two and many independent variables and their partial derivatives. In this section, we offer basic integral inequalities involving functions of two and many independent variables established by Yang [429] and Pachpatte [233,261,267,284] which claim their origin in Opial's inequality.

First we introduce some of the notations used in our subsequent discussion:

Let $\Delta=[a,b]\times[c,d]$, $\Delta_1=[a,X]\times[c,Y]$, $\Delta_2=[a,X]\times[Y,d]$, $\Delta_3=[X,b]\times[c,Y]$, $\Delta_4=[X,b]\times[Y,d]$ for $a\leqslant X\leqslant b$, $c\leqslant Y\leqslant d$; $a,b,c,d,X,Y\in\mathbb{R}$ ($\mathbb{R}$ the set of real numbers). Further, let $Dp(r)=\frac{\mathrm{d}}{\mathrm{d}r}p(r)$, $D_1h(s,t)=\frac{\partial}{\partial s}h(s,t)$, $D_2h(s,t)=\frac{\partial}{\partial t}h(s,t)$, $D_2D_1h(s,t)=\frac{\partial^2}{\partial s\,\partial t}h(s,t)$, for functions $p(r)$, $h(s,t)$ defined on $\mathbb{R}$ and Δ respectively.

In 1982, Yang [429] has obtained the following analogue of Opial's inequality in two independent variables.

THEOREM 3.7.1. *If $f(s,t)$, $D_1f(s,t)$ and $D_2D_1f(s,t)$ are continuous functions on Δ and $f(a,t)=f(b,t)=D_1f(s,c)=D_1f(s,d)=0$ for $a\leqslant s\leqslant b$, $c\leqslant t\leqslant d$, then*

$$\int_a^b\int_c^d\left|f(s,t)\right|\left|D_2D_1f(s,t)\right|\,\mathrm{d}t\,\mathrm{d}s$$

$$\leqslant\frac{(b-a)(d-c)}{8}\int_a^b\int_c^d\left|D_2D_1f(s,t)\right|^2\,\mathrm{d}t\,\mathrm{d}s. \tag{3.7.1}$$

PROOF. In order to prove (3.7.1), we consider the following four cases.

Case I. Let $(s,t)\in\Delta_1$ and define

$$z(s,t)=\int_a^s\int_c^t\left|D_2D_1f(u,v)\right|\mathrm{d}v\,\mathrm{d}u. \tag{3.7.2}$$

Then

$$\begin{aligned}D_1z(s,t)&=\int_c^t\left|D_2D_1f(s,v)\right|\mathrm{d}v,\\ D_2z(s,t)&=\int_a^s\left|D_2D_1f(u,t)\right|\mathrm{d}u.\end{aligned} \tag{3.7.3}$$

From (3.7.3) we observe that, for each fixed s, $z(s,t)$ is nondecreasing for t on $[c,Y]$. Since $f(a,t)=0$ and $D_1f(s,c)=0$ for $(s,t)\in\Delta_1$, we have

$$\left|f(s,t)\right|\leqslant\int_a^s\left|D_1f(u,t)\right|\mathrm{d}u, \tag{3.7.4}$$

$$\left|D_1f(s,t)\right|\leqslant\int_c^t\left|D_2D_1f(s,v)\right|\mathrm{d}v=D_1z(s,t). \tag{3.7.5}$$

From (3.7.4) and (3.7.5), we observe that, for $(s,t)\in\Delta_1$,

$$\left|f(s,t)\right|\leqslant\int_a^s\left|D_1z(u,t)\right|\mathrm{d}u=z(s,t). \tag{3.7.6}$$

From (3.7.6), (3.7.5), (3.7.2) and applying Schwarz inequality, we have

$$\begin{aligned}&\int_a^X\int_c^Y\left|f(s,t)\right|\left|D_2D_1f(s,t)\right|\mathrm{d}t\,\mathrm{d}s\\ &\quad\leqslant\int_a^X\int_c^Y z(s,t)\left|D_2D_1f(s,t)\right|\mathrm{d}t\,\mathrm{d}s\\ &\quad\leqslant\int_a^X z(s,Y)\int_c^Y\left|D_2D_1f(s,t)\right|\mathrm{d}t\,\mathrm{d}s\\ &\quad=\int_a^X z(s,Y)D_1z(s,Y)\,\mathrm{d}s\\ &\quad=\frac{1}{2}z^2(X,Y)\end{aligned}$$

$$= \frac{1}{2}\left(\int_a^X \int_c^Y \left|D_2 D_1 f(s,t)\right| \mathrm{d}t\, \mathrm{d}s\right)^2$$

$$\leqslant \frac{(X-a)(Y-c)}{2} \int_a^X \int_c^Y \left|D_2 D_1 f(s,t)\right|^2 \mathrm{d}t\, \mathrm{d}s. \tag{3.7.7}$$

Case II. Let $(s,t) \in \Delta_2$ and define

$$z(s,t) = \int_a^s \int_t^d \left|D_2 D_1 f(u,v)\right| \mathrm{d}v\, \mathrm{d}u. \tag{3.7.8}$$

Case III. Let $(s,t) \in \Delta_3$ and define

$$z(s,t) = \int_s^b \int_c^t \left|D_2 D_1 f(u,v)\right| \mathrm{d}v\, \mathrm{d}u. \tag{3.7.9}$$

Case IV. Let $(s,t) \in \Delta_4$ and define

$$z(s,t) = \int_s^b \int_t^d \left|D_2 D_1 f(u,v)\right| \mathrm{d}v\, \mathrm{d}u. \tag{3.7.10}$$

Now, by following similar arguments to those in the proof of Case I, but with suitable modifications, we obtain the following estimates in Cases II–IV:

$$\int_a^X \int_Y^d \left|f(s,t)\right|\left|D_2 D_1 f(s,t)\right| \mathrm{d}t\, \mathrm{d}s$$
$$\leqslant \frac{(X-a)(d-Y)}{2} \int_a^X \int_Y^d \left|D_2 D_1 f(s,t)\right|^2 \mathrm{d}t\, \mathrm{d}s, \tag{3.7.11}$$

$$\int_X^b \int_c^Y \left|f(s,t)\right|\left|D_2 D_1 f(s,t)\right| \mathrm{d}t\, \mathrm{d}s$$
$$\leqslant \frac{(b-X)(Y-c)}{2} \int_X^b \int_c^Y \left|D_2 D_1 f(s,t)\right|^2 \mathrm{d}t\, \mathrm{d}s \tag{3.7.12}$$

and

$$\int_X^b \int_Y^d \left|f(s,t)\right|\left|D_2 D_1 f(s,t)\right| \mathrm{d}t\, \mathrm{d}s$$
$$\leqslant \frac{(b-X)(d-Y)}{2} \int_X^b \int_Y^d \left|D_2 D_1 f(s,t)\right|^2 \mathrm{d}t\, \mathrm{d}s, \tag{3.7.13}$$

respectively.

Let $X=(a+b)/2$, $Y=(c+d)/2$. Then $X-a=b-X=(b-a)/2$, $Y-c=d-Y=(d-c)/2$. It follows from (3.7.7), (3.7.11)–(3.7.13) that

$$\begin{aligned}
&\int_a^b\int_c^d |f(s,t)||D_2D_1f(s,t)|\,dt\,ds\\
&\quad=\int_a^X\int_c^Y |f(s,t)||D_2D_1f(s,t)|\,dt\,ds\\
&\qquad+\int_a^X\int_Y^d |f(s,t)||D_2D_1f(s,t)|\,dt\,ds\\
&\qquad+\int_X^b\int_c^Y |f(s,t)||D_2D_1f(s,t)|\,dt\,ds\\
&\qquad+\int_X^b\int_Y^d |f(s,t)||D_2D_1f(s,t)|\,dt\,ds\\
&\quad\leqslant\frac{(b-a)(d-c)}{8}\int_a^b\int_c^d |D_2D_1f(s,t)|^2\,dt\,ds.
\end{aligned}$$

This result is the desired inequality in (3.7.1) and the proof is complete. □

In 1985, Pachpatte [233] has established some integral inequalities of Opial type in two independent variables. To formulate the results in [233], we list the following hypotheses.

(H_1) Let $f(s,t)$, $D_1f(s,t)$, $D_2D_1f(s,t)$ and $g(s,t)$, $D_1g(s,t)$, $D_2D_1g(s,t)$ be continuous on Δ and $f(a,t)=f(b,t)=D_1f(s,c)=D_1f(s,d)=0$, $g(a,t)=g(b,t)=D_1g(s,c)=D_1g(s,d)=0$ for $(s,t)\in\Delta$.

(H_2) Let $h(s,t)$, $D_1h(s,t)$, $D_2D_1h(s,t)$ be continuous on Δ and $h(a,t)=h(b,t)=D_1h(s,c)=D_1h(s,d)=0$ for $(s,t)\subset\Delta$.

Before stating the results in [233], we introduce the following notation for convenience:

$$\begin{aligned}
&L\bigl[m,f(s,t),g(s,t),D_2D_1f(s,t),D_2D_1g(s,t)\bigr]\\
&\quad=\bigl\{|f(s,t)||g(s,t)|\bigr\}^m\bigl[|f(s,t)||D_2D_1g(s,t)|+|g(s,t)||D_2D_1f(s,t)|\bigr];\\
&D_2D_1L\bigl[m,D_2D_1f(s,t),D_2D_1g(s,t)\bigr]\\
&\quad=|D_2D_1f(s,t)|^{2(m+1)}+|D_2D_1g(s,t)|^{2(m+1)};
\end{aligned}$$

$$\begin{aligned}
&M\big[m, f(s,t), g(s,t), h(s,t), D_2D_1f(s,t), D_2D_1g(s,t), D_2D_1h(s,t)\big]\\
&\quad = \big\{|f(s,t)||g(s,t)|\big\}^m\big[|f(s,t)||D_2D_1g(s,t)| + |g(s,t)||D_2D_1f(s,t)|\big]\\
&\qquad + \big\{|g(s,t)||h(s,t)|\big\}^m\big[|g(s,t)||D_2D_1h(s,t)| + |h(s,t)||D_2D_1g(s,t)|\big]\\
&\qquad + \big\{|h(s,t)||f(s,t)|\big\}^m\big[|h(s,t)||D_2D_1f(s,t)| + |f(s,t)||D_2D_1h(s,t)|\big];\\
&D_2D_1M\big[m, D_2D_1f(s,t), D_2D_1g(s,t), D_2D_1h(s,t)\big]\\
&\quad = |D_2D_1f(s,t)|^{2(m+1)} + |D_2D_1g(s,t)|^{2(m+1)} + |D_2D_1h(s,t)|^{2(m+1)};\\
&N\big[f(s,t), g(s,t), h(s,t), D_2D_1f(s,t), D_2D_1g(s,t), D_2D_1h(s,t)\big]\\
&\quad = |f(s,t)||g(s,t)||h(s,t)|\big[|D_2D_1f(s,t)| + |D_2D_1g(s,t)| + |D_2D_1h(s,t)|\big]\\
&\qquad + \big[|f(s,t)| + |g(s,t)| + |h(s,t)|\big]\\
&\qquad \times \big[|f(s,t)||g(s,t)||D_2D_1h(s,t)|\\
&\qquad + |g(s,t)||h(s,t)||D_2D_1f(s,t)| + |h(s,t)||f(s,t)||D_2D_1g(s,t)|\big];
\end{aligned}$$

where $m \geqslant 0$ is a constant.

The results established in [233] are embodied in the following theorems.

THEOREM 3.7.2. *Assume that* (H_1) *holds. Then*

$$\begin{aligned}
&\int_a^b\int_c^d L\big[m, f(s,t), g(s,t), D_2D_1f(s,t), D_2D_1g(s,t)\big]\,dt\,ds\\
&\quad \leqslant K_m \int_a^b\int_c^d D_2D_1L\big[m, D_2D_1f(s,t), D_2D_1g(s,t)\big]\,dt\,ds, \qquad (3.7.14)
\end{aligned}$$

where

$$K_m = \frac{1}{2(m+1)}\left[\frac{(b-a)(d-c)}{4}\right]^{2m+1}. \qquad (3.7.15)$$

REMARK 3.7.1. In the special case when $g(s,t) = f(s,t)$ and $2m+1 = n$, the inequality established in Theorem 3.7.2 reduces to the following inequality

$$\begin{aligned}
&\int_a^b\int_c^d |f(s,t)|^n|D_2D_1f(s,t)|\,dt\,ds\\
&\quad \leqslant \frac{1}{n+1}\left[\frac{(b-a)(d-c)}{4}\right]^n \int_a^b\int_c^d |D_2D_1f(s,t)|^{n+1}\,dt\,ds, \qquad (3.7.16)
\end{aligned}$$

which in turn contains as a special case Yang's inequality given in (3.7.1) when $n = 1$. We also note that the inequality obtained in (3.7.16) is a two independent variable analogue of Yang's generalization of the Opial inequality [428, Theorem 4] (see also [430]).

As a consequence of Theorem 3.7.2, we have the following corollary.

COROLLARY 3.7.1. *Assume that* (H_1) *and* (H_2) *hold. Then*

$$\int_a^b \int_c^d M\big[m, f(s,t), g(s,t), h(s,t), \\ D_2D_1 f(s,t), D_2D_1 g(s,t), D_2D_1 h(s,t)\big]\,dt\,ds \\ \leqslant 2K_m \int_a^b \int_c^d D_2D_1 M\big[m, D_2D_1 f(s,t), D_2D_1 g(s,t), D_2D_1 h(s,t)\big]\,dt\,ds, \tag{3.7.17}$$

where K_m *is as defined in* (3.7.15).

THEOREM 3.7.3. *Assume that* (H_1) *and* (H_2) *hold. Then*

$$\int_a^b \int_c^d N\big[f(s,t), g(s,t), h(s,t), D_2D_1 f(s,t), D_2D_1 g(s,t), D_2D_1 h(s,t)\big]\,dt\,ds \\ \leqslant 4K_1 \int_a^b \int_c^d D_2D_1 M\big[1, D_2D_1 f(s,t), D_2D_1 g(s,t), D_2D_1 h(s,t)\big]\,dt\,ds, \tag{3.7.18}$$

where K_1 *is obtained by substituting* $m = 1$ *in* (3.7.15).

REMARK 3.7.2. Note that the inequalities established in Theorems 3.7.2 and 3.7.3 and Corollary 3.7.1 are the two independent variable analogues of the inequalities established earlier by Pachpatte in [239,240,256].

PROOFS OF THEOREMS 3.7.2 AND 3.7.3. In order to prove Theorem 3.7.2, we consider the following four cases.

Case I. Let $(s,t) \in \Delta_1$ and define

$$z(s,t) = \int_a^s \int_c^t \big|D_2D_1 f(u,v)\big|\,dv\,du, \tag{3.7.19}$$

and

$$w(s,t)=\int_a^s\int_c^t \left|D_2D_1g(u,v)\right|\,\mathrm{d}v\,\mathrm{d}u. \tag{3.7.20}$$

Then

$$D_1z(s,t)=\int_c^t \left|D_2D_1f(s,v)\right|\mathrm{d}v,\qquad D_2z(s,t)=\int_a^s \left|D_2D_1f(u,t)\right|\mathrm{d}u, \tag{3.7.21}$$

and

$$D_1w(s,t)=\int_c^t \left|D_2D_1g(s,v)\right|\mathrm{d}v,\qquad D_2w(s,t)=\int_a^s \left|D_2D_1g(u,t)\right|\mathrm{d}u. \tag{3.7.22}$$

From (3.7.21) and (3.7.22), we observe that for each fixed s, $z(s,t)$ and $w(s,t)$ are nondecreasing for t on $[c,Y]$. Since $f(a,t)=g(a,t)=0$ and $D_1f(s,c)=D_1g(s,c)=0$, for $(s,t)\in\Delta_1$, we have

$$\begin{aligned}\left|f(s,t)\right|&\leqslant\int_a^s \left|D_1f(u,t)\right|\mathrm{d}u,\\ \left|g(s,t)\right|&\leqslant\int_a^s \left|D_1g(u,t)\right|\mathrm{d}u,\end{aligned} \tag{3.7.23}$$

$$\left|D_1f(s,t)\right|\leqslant\int_c^t \left|D_2D_1f(s,v)\right|\mathrm{d}v=D_1z(s,t), \tag{3.7.24}$$

$$\left|D_1g(s,t)\right|\leqslant\int_c^t \left|D_2D_1g(s,v)\right|\mathrm{d}v=D_1w(s,t). \tag{3.7.25}$$

From (3.7.23)–(3.7.25), we observe that

$$\left|f(s,t)\right|\leqslant\int_a^s \left|D_1z(u,t)\right|\mathrm{d}u=z(s,t), \tag{3.7.26}$$

$$\left|g(s,t)\right|\leqslant\int_a^s \left|D_1w(u,t)\right|\mathrm{d}u=w(s,t). \tag{3.7.27}$$

From (3.7.26), (3.7.27), (3.7.24), (3.7.25) and applying the elementary inequality $\alpha\beta\leqslant\frac{1}{2}(\alpha^2+\beta^2)$ (for α,β reals), (3.7.19), (3.7.20) followed by two applications each of the Schwarz inequality and the Hölder inequality with indices $m+1$ and

$(m+1)/m$, we obtain

$$
\begin{aligned}
&\int_a^X\int_c^Y L\big[m, f(s,t), g(s,t), D_2D_1f(s,t), D_2D_1g(s,t)\big]\,\mathrm{d}t\,\mathrm{d}s \\
&\quad\leqslant \int_a^X\int_c^Y \{z(s,t)w(s,t)\}^m \\
&\qquad\qquad \times \big[z(s,t)\big|D_2D_1g(s,t)\big| + w(s,t)\big|D_2D_1f(s,t)\big|\big]\,\mathrm{d}t\,\mathrm{d}s \\
&\quad\leqslant \int_a^X z^{m+1}(s,Y)w^m(s,Y)\bigg(\int_c^Y \big|D_2D_1g(s,t)\big|\,\mathrm{d}t\bigg)\,\mathrm{d}s \\
&\qquad + \int_a^X z^m(s,Y)w^{m+1}(s,Y)\bigg(\int_c^Y \big|D_2D_1f(s,t)\big|\,\mathrm{d}t\bigg)\,\mathrm{d}s \\
&\quad= \int_a^X \big[z^{m+1}(s,Y)w^m(s,Y)D_1w(s,Y) + z^m(s,Y)w^{m+1}(s,Y)D_1z(s,Y)\big]\,\mathrm{d}s \\
&\quad= \int_a^X \frac{\partial}{\partial s}\bigg[\frac{1}{m+1}z^{m+1}(s,Y)w^{m+1}(s,Y)\bigg]\,\mathrm{d}s \\
&\quad= \frac{1}{m+1}z^{m+1}(X,Y)w^{m+1}(X,Y) \\
&\quad\leqslant \frac{1}{2(m+1)}\big[\big(z^{m+1}(X,Y)\big)^2 + \big(w^{m+1}(X,Y)\big)^2\big] \\
&\quad= \frac{1}{2(m+1)}\bigg[\bigg\{\bigg(\int_a^X\int_c^Y \big|D_2D_1f(s,t)\big|\,\mathrm{d}t\,\mathrm{d}s\bigg)^2\bigg\}^{m+1} \\
&\qquad\qquad + \bigg\{\bigg(\int_a^X\int_c^Y \big|D_2D_1g(s,t)\big|\,\mathrm{d}t\,\mathrm{d}s\bigg)^2\bigg\}^{m+1}\bigg] \\
&\quad\leqslant \frac{\{(X-a)(Y-c)\}^{m+1}}{2(m+1)}\bigg[\bigg\{\int_a^X\int_c^Y \big|D_2D_1f(s,t)\big|^2\,\mathrm{d}t\,\mathrm{d}s\bigg\}^{m+1} \\
&\qquad\qquad + \bigg\{\int_a^X\int_c^Y \big|D_2D_1g(s,t)\big|^2\,\mathrm{d}t\,\mathrm{d}s\bigg\}^{m+1}\bigg] \\
&\quad\leqslant \frac{\{(X-a)(Y-c)\}^{2m+1}}{2(m+1)} \\
&\qquad \times \int_a^X\int_c^Y D_2D_1L\big[m, D_2D_1f(s,t), D_2D_1g(s,t)\big]\,\mathrm{d}t\,\mathrm{d}s. \qquad (3.7.28)
\end{aligned}
$$

Case II. Let $(s,t) \in \Delta_2$ and define

$$z(s,t) = \int_a^s \int_t^d \left|D_2 D_1 f(u,v)\right| \mathrm{d}v\, \mathrm{d}u, \tag{3.7.29}$$

$$w(s,t) = \int_a^s \int_t^d \left|D_2 D_1 g(u,v)\right| \mathrm{d}v\, \mathrm{d}u. \tag{3.7.30}$$

Case III. Let $(s,t) \in \Delta_3$ and define

$$z(s,t) = \int_s^b \int_c^t \left|D_2 D_1 f(u,v)\right| \mathrm{d}v\, \mathrm{d}u, \tag{3.7.31}$$

$$w(s,t) = \int_s^b \int_c^t \left|D_2 D_1 g(u,v)\right| \mathrm{d}v\, \mathrm{d}u. \tag{3.7.32}$$

Case IV. Let $(s,t) \in \Delta_4$ and define

$$z(s,t) = \int_s^b \int_t^d \left|D_2 D_1 f(u,v)\right| \mathrm{d}v\, \mathrm{d}u, \tag{3.7.33}$$

$$w(s,t) = \int_s^b \int_t^d \left|D_2 D_1 g(u,v)\right| \mathrm{d}v\, \mathrm{d}u. \tag{3.7.34}$$

Now, by following similar arguments to those in the proof of Case I, but with suitable modifications, we obtain the following estimates in Cases II–IV:

$$\begin{aligned}
&\int_a^X \int_Y^d L\big[m, f(s,t), g(s,t), D_2 D_1 f(s,t), D_2 D_1 g(s,t)\big]\,\mathrm{d}t\,\mathrm{d}s \\
&\quad\leqslant \frac{\{(X-a)(d-Y)\}^{2m+1}}{2(m+1)} \\
&\qquad\times \int_a^X \int_Y^d D_2 D_1 L\big[m, D_2 D_1 f(s,t), D_2 D_1 g(s,t)\big]\,\mathrm{d}t\,\mathrm{d}s,
\end{aligned} \tag{3.7.35}$$

$$\begin{aligned}
&\int_X^b \int_c^Y L\big[m, f(s,t), g(s,t), D_2 D_1 f(s,t), D_2 D_1 g(s,t)\big]\,\mathrm{d}t\,\mathrm{d}s \\
&\quad\leqslant \frac{\{(b-X)(Y-c)\}^{2m+1}}{2(m+1)} \\
&\qquad\times \int_X^b \int_c^Y D_2 D_1 L\big[m, D_2 D_1 f(s,t), D_2 D_1 g(s,t)\big]\,\mathrm{d}t\,\mathrm{d}s
\end{aligned} \tag{3.7.36}$$

and

$$\int_X^b \int_Y^d L\big[m, f(s,t), g(s,t), D_2D_1f(s,t), D_2D_1g(s,t)\big]\,dt\,ds$$
$$\leqslant \frac{\{(b-X)(d-Y)\}^{2m+1}}{2(m+1)}$$
$$\times \int_X^b \int_Y^d D_2D_1L\big[m, D_2D_1f(s,t), D_2D_1g(s,t)\big]\,dt\,ds, \qquad (3.7.37)$$

respectively.

Let $X = (a+b)/2$, $Y = (c+d)/2$. Then $X - a = b - X = (b-a)/2$, $Y - c = d - Y = (d-c)/2$. It follows from (3.7.28) and (3.7.35)–(3.7.37) that

$$\int_a^b \int_c^d L\big[m, f(s,t), g(s,t), D_2D_1f(s,t), D_2D_1g(s,t)\big]\,dt\,ds$$
$$= \int_a^X \int_c^Y L\big[m, f(s,t), g(s,t), D_2D_1f(s,t), D_2D_1g(s,t)\big]\,dt\,ds$$
$$+ \int_a^X \int_Y^d L\big[m, f(s,t), g(s,t), D_2D_1f(s,t), D_2D_1g(s,t)\big]\,dt\,ds$$
$$+ \int_X^b \int_c^Y L\big[m, f(s,t), g(s,t), D_2D_1f(s,t), D_2D_1g(s,t)\big]\,dt\,ds$$
$$+ \int_X^b \int_Y^d L\big[m, f(s,t), g(s,t), D_2D_1f(s,t), D_2D_1g(s,t)\big]\,dt\,ds$$
$$\leqslant K_m \int_a^b \int_c^d D_2D_1L\big[m, D_2D_1f(s,t), D_2D_1g(s,t)\big]\,dt\,ds.$$

The proof of Theorem 3.7.2 is complete.

In order to prove Theorem 3.7.3, we make the following definitions in corresponding Cases I–IV considered in the proof of Theorem 3.7.2:

$$r(s,t) = \int_a^s \int_c^t \big|D_2D_1h(u,v)\big|\,dv\,du \quad \text{for } (s,t) \in \Delta_1, \qquad (3.7.38)$$

$$r(s,t) = \int_a^s \int_t^d \big|D_2D_1h(u,v)\big|\,dv\,du \quad \text{for } (s,t) \in \Delta_2, \qquad (3.7.39)$$

$$r(s,t) = \int_s^b \int_c^t \big|D_2D_1h(u,v)\big|\,dv\,du \quad \text{for } (s,t) \in \Delta_3, \qquad (3.7.40)$$

$$r(s,t)=\int_s^b\int_t^d \left|D_2D_1h(u,v)\right|\mathrm{d}v\,\mathrm{d}u \quad \text{for } (s,t)\in\Delta_4. \tag{3.7.41}$$

Now we consider the details of the proof of the following case corresponding to Case I in the proof of Theorem 3.7.2.

Case I^0. Let $(s,t)\in\Delta_1$ as in Case I. It is easy to observe that for each fixed $s, r(s,t)$, $(s,t)\in\Delta_1$ as in Case I, we have

$$\left|h(s,t)\right|\leqslant r(s,t) \tag{3.7.42}$$

and

$$\begin{aligned}\left|D_1h(s,t)\right| &\leqslant \int_c^t \left|D_2D_1h(s,v)\right|\mathrm{d}v\\ &= D_1r(s,t).\end{aligned} \tag{3.7.43}$$

From (3.7.26), (3.7.27), (3.7.42), (3.7.24), (3.7.25), (3.7.43) and using the elementary inequalities $\alpha_1\alpha_2\alpha_3(\alpha_1+\alpha_2+\alpha_3)\leqslant\frac{1}{3}(\alpha_1\alpha_2+\alpha_2\alpha_3+\alpha_3\alpha_1)^2$, $\alpha_1\alpha_2+\alpha_2\alpha_3+\alpha_3\alpha_1\leqslant\alpha_1^2+\alpha_2^2+\alpha_3^2$, $(\alpha_1+\alpha_2+\alpha_3)^2\leqslant 3(\alpha_1^2+\alpha_2^2+\alpha_3^2)$ (for $\alpha_1,\alpha_2,\alpha_3$ reals), (3.7.19), (3.7.20), (3.7.38) and repeated application of Schwarz inequality, we obtain

$$\begin{aligned}
&\int_a^X\int_c^Y N\big[f(s,t),g(s,t),h(s,t),\\
&\qquad D_2D_1f(s,t),D_2D_1g(s,t),D_2D_1h(s,t)\big]\,\mathrm{d}t\,\mathrm{d}s\\
&\quad\leqslant\int_a^X\Big[z(s,Y)w(s,Y)r(s,Y)\big[D_1z(s,Y)+D_1w(s,Y)+D_1r(s,Y)\big]\\
&\qquad+\big[z(s,Y)+w(s,Y)+r(s,Y)\big]\\
&\qquad\times\big[z(s,Y)w(s,Y)D_1r(s,Y)\\
&\qquad+w(s,Y)r(s,Y)D_1z(s,Y)+r(s,Y)z(s,Y)D_1w(s,Y)\big]\Big]\,\mathrm{d}s\\
&\quad=\int_a^X\frac{\partial}{\partial s}\Big[z(s,Y)w(s,Y)r(s,Y)\big[z(s,Y)+w(s,Y)+r(s,Y)\big]\Big]\,\mathrm{d}s\\
&\quad=z(X,Y)w(X,Y)r(X,Y)\big[z(X,Y)+w(X,Y)+r(X,Y)\big]\\
&\quad\leqslant\frac{1}{3}\big[z(X,Y)w(X,Y)+w(X,Y)r(X,Y)+r(X,Y)z(X,Y)\big]^2
\end{aligned}$$

$$
\begin{aligned}
&\leqslant \frac{1}{3}\left[z^2(X,Y)+w^2(X,Y)+r^2(X,Y)\right]^2 \\
&\leqslant \left[z^2(X,Y)\right]^2+\left[w^2(X,Y)\right]^2+\left[r^2(X,Y)\right]^2 \\
&= \left\{\left[\int_a^X\int_c^Y \left|D_2D_1f(s,t)\right| \mathrm{d}t\,\mathrm{d}s\right]^2\right\}^2 \\
&\quad + \left\{\left[\int_a^X\int_c^Y \left|D_2D_1g(s,t)\right| \mathrm{d}t\,\mathrm{d}s\right]^2\right\}^2 \\
&\quad + \left\{\left[\int_a^X\int_c^Y \left|D_2D_1h(s,t)\right| \mathrm{d}t\,\mathrm{d}s\right]^2\right\}^2 \\
&\leqslant \left\{(X-a)(Y-c)\right\}^3 \\
&\quad \times \int_a^X\int_c^Y D_2D_1M\left[1, D_2D_1f(s,t), D_2D_1g(s,t), D_2D_1h(s,t)\right] \mathrm{d}t\,\mathrm{d}s.
\end{aligned}
$$

The proofs of Cases II^0–IV^0 corresponding to Cases II–IV follow by the same arguments as those given in the proof of Theorem 3.7.2 in view of the above proof of Case I^0 with suitable modifications. We omit the remaining details of the proof of Theorem 3.7.3. □

The inequalities in the following two theorems similar to those of Wirtinger- and Opial-type inequalities are established by Pachpatte [267].

THEOREM 3.7.4. *Let $p(x,y)$ be a real-valued nonnegative continuous function defined on Δ. Let $f_r(x,y)$, $D_1f_r(x,y)$, $D_2D_1f_r(x,y)$ be real-valued continuous functions defined on Δ for $r=1,\dots,n$ with $f_r(a,y)=f_r(b,y)=D_1f_r(x,c)=D_1f_r(x,d)=0$ for $a\leqslant x\leqslant b$, $c\leqslant y\leqslant d$. Then*

$$
\begin{aligned}
&\int_a^b\int_c^d p(x,y)\left(\prod_{r=1}^n \left|f_r(x,y)\right|^{m_r}\right)^{2/n} \mathrm{d}y\,\mathrm{d}x \\
&\quad\leqslant \frac{1}{n}K(a,b,c,d,n,m_1,\dots,m_n)\left(\int_a^b\int_c^d p(x,y)\,\mathrm{d}y\,\mathrm{d}x\right) \\
&\quad\quad\times\left(\int_a^b\int_c^d \sum_{r=1}^n \left|D_2D_1f_r(x,y)\right|^{2m_r} \mathrm{d}y\,\mathrm{d}x\right),
\end{aligned}
\tag{3.7.44}
$$

where $m_r \geqslant 1$, $r = 1, \ldots, n$, are constants and

$$K(a, b, c, d, n, m_1, \ldots, m_n) = \left\{\frac{1}{4}\right\}^{\frac{2}{n}\sum_{r=1}^{n} m_r} \left\{(b-a)(d-c)\right\}^{1+\frac{2}{n}\sum_{r=1}^{n}(m_r - 1)} \tag{3.7.45}$$

is a constant depending on $a, b, c, d, n, m_1, \ldots, m_n$.

REMARK 3.7.3. In the special cases when (i) $m_r = 1$ for $r = 1, \ldots, n$, (ii) $n = 2$, (iii) $n = 1$, (iv) $n = 2$ and $m_1 = m_2 = 1$, and (v) $n = 1$ and $m_1 = 1$, the inequality established in (3.7.44) reduces to some interesting inequalities of the Wirtinger type which are similar to the two independent variable analogues of the inequalities given by Pachpatte in [243] and Traple in [419].

THEOREM 3.7.5. *Let the functions $p(x, y)$, $f_r(x, y)$, $D_1 f_r(x, y)$, $D_2 D_1 f_r(x, y)$ be as in Theorem* 3.7.4. *Then*

$$\begin{aligned}
&\int_a^b \int_c^d p(x, y)\left(\prod_{r=1}^{n} |f_r(x, y)|^{m_r}\right)^{1/n} \left(\sum_{r=1}^{n} |D_2 D_1 f_r(x, y)|^{m_r}\right) \mathrm{d}y\,\mathrm{d}x \\
&\quad \leqslant \left(K(a, b, c, d, n, m_1, \ldots, m_n) \int_a^b \int_c^d p^2(x, y)\,\mathrm{d}y\,\mathrm{d}x\right)^{1/2} \\
&\qquad \times \left(\int_a^b \int_c^d \left(\sum_{r=1}^{n} |D_2 D_1 f_r(x, y)|^{2m_r}\right) \mathrm{d}y\,\mathrm{d}x\right),
\end{aligned} \tag{3.7.46}$$

where $m_r \geqslant 1$ (for $r = 1, \ldots, n$) are constants and $K(a, b, c, d, n, m_1, \ldots, m_n)$ is as defined in (3.7.45).

REMARK 3.7.4. If we take (i) $m_r = 1$ for $r = 1, \ldots, n$, (ii) $n = 1$, (iii) $n = 1$ and $m_1 = 1$ in (3.7.46), then we get Opial-type inequalities similar to that of given by Traple in [419]. Further, in the special case when $p(x, y)$ is constant, $n = 1$ and $m_1 = 1$, the inequality in (3.7.46) reduces to the following inequality

$$\begin{aligned}
&\int_a^b \int_c^d |f_1(x, y)|\,|D_2 D_1 f_1(x, y)|\,\mathrm{d}y\,\mathrm{d}x \\
&\quad \leqslant \frac{(b-a)(d-c)}{4} \int_a^b \int_c^d |D_2 D_1 f_1(x, y)|^2\,\mathrm{d}y\,\mathrm{d}x.
\end{aligned} \tag{3.7.47}$$

Here we note that the constant $(b-a)(d-c)/4$ involved in (3.7.47) is not the best possible constant. Inequality (3.7.47) with better constant $(b-a)(d-c)/(8\sqrt{2})$ is established by Pachpatte in [260]. However, here our proof of inequality (3.7.46) depends on the inequality established in Theorem 3.7.4 which in turn does not yield the better constant obtained in [260].

PROOFS OF THEOREMS 3.7.4 AND 3.7.5. From the hypothesis of Theorem 3.7.4, it is easy to observe that the following identities hold:

$$f_r(x,y) = \int_a^x \int_c^y D_2 D_1 f_r(s,t)\,dt\,ds, \tag{3.7.48}$$

$$f_r(x,y) = -\int_a^x \int_y^d D_2 D_1 f_r(s,t)\,dt\,ds, \tag{3.7.49}$$

$$f_r(x,y) = -\int_x^b \int_c^y D_2 D_1 f_r(s,t)\,dt\,ds, \tag{3.7.50}$$

$$f_r(x,y) = \int_x^b \int_y^d D_2 D_1 f_r(s,t)\,dt\,ds, \tag{3.7.51}$$

for $r = 1, \ldots, n$. From (3.7.48)–(3.7.51), we observe that

$$\left|f_r(x,y)\right| \leqslant \frac{1}{4}\int_a^b \int_c^d \left|D_2 D_1 f_r(s,t)\right| dt\,ds. \tag{3.7.52}$$

From (3.7) and using Hölder's inequality with indices m_r, $m_r/(m_r-1)$ for $r = 1, \ldots, n$, we obtain

$$\left|f_r(x,y)\right|^{m_r} \leqslant \left(\frac{1}{4}\right)^{m_r} \left\{(b-a)(d-c)\right\}^{m_r-1} \int_a^b \int_c^d \left|D_2 D_1 f_r(s,t)\right|^{m_r} dt\,ds. \tag{3.7.53}$$

From (3.7.53) and using the elementary inequalities

$$\left(\prod_{i=1}^n b_i\right)^{1/n} \leqslant \frac{1}{n}\sum_{i=1}^n b_i \tag{3.7.54}$$

(for $b_1, \ldots, b_n \geqslant 0$ reals and $n \geqslant 1$) and

$$\left(\sum_{i=1}^n b_i\right)^2 \leqslant n\sum_{i=1}^n b_i^2 \tag{3.7.55}$$

(for $b_1, \dots, b_n$ reals), and Schwarz inequality, we obtain

$$
\begin{aligned}
&\left(\prod_{r=1}^{n} \left|f_r(x,y)\right|^{m_r}\right)^{2/n} \\
&\leqslant \left(\frac{1}{4}\right)^{\frac{2}{n}\sum_{r=1}^{n} m_r} \left\{(b-a)(d-c)\right\}^{\frac{2}{n}\sum_{r=1}^{n}(m_r-1)} \\
&\quad \times \left(\left\{\prod_{r=1}^{n}\left(\int_a^b\int_c^d \left|D_2D_1 f_r(s,t)\right|^{m_r}\,\mathrm{d}t\,\mathrm{d}s\right)\right\}^{1/n}\right)^2 \\
&\leqslant \left(\frac{1}{4}\right)^{\frac{2}{n}\sum_{r=1}^{n} m_r} \left\{(b-a)(d-c)\right\}^{\frac{2}{n}\sum_{r=1}^{n}(m_r-1)} \\
&\quad \times \left(\frac{1}{n}\sum_{r=1}^{n}\left(\int_a^b\int_c^d \left|D_2D_1 f_r(s,t)\right|^{m_r}\,\mathrm{d}t\,\mathrm{d}s\right)\right)^2 \\
&\leqslant \frac{1}{n}\left(\frac{1}{4}\right)^{\frac{2}{n}\sum_{r=1}^{n} m_r} \left\{(b-a)(d-c)\right\}^{\frac{2}{n}\sum_{r=1}^{n}(m_r-1)} \\
&\quad \times \left(\sum_{r=1}^{n}\left(\int_a^b\int_c^d \left|D_2D_1 f_r(s,t)\right|^{m_r}\,\mathrm{d}t\,\mathrm{d}s\right)^2\right) \\
&\leqslant \frac{1}{n}K(a,b,c,d,n,m_1,\dots,m_n) \\
&\quad \times \int_a^b\int_c^d\left(\sum_{r=1}^{n}\left|D_2D_1 f_r(s,t)\right|^{2m_r}\right)\mathrm{d}t\,\mathrm{d}s. \qquad (3.7.56)
\end{aligned}
$$

Multiplying both sides of (3.7.56) by $p(x, y)$ and integrating the resulting inequality on Δ we have

$$
\begin{aligned}
&\int_a^b\int_c^d p(x,y)\left(\prod_{r=1}^{n}\left|f_r(x,y)\right|^{m_r}\right)^{2/n}\mathrm{d}y\,\mathrm{d}x \\
&\leqslant \frac{1}{n}K(a,b,c,d,n,m_1,\dots,m_n)\left(\int_a^b\int_c^d p(x,y)\,\mathrm{d}y\,\mathrm{d}x\right) \\
&\quad \times \left(\int_a^b\int_c^d\left(\sum_{r=1}^{n}\left|D_2D_1 f_r(x,y)\right|^{2m_r}\right)\mathrm{d}y\,\mathrm{d}x\right).
\end{aligned}
$$

This result is the desired inequality in (3.7.44) and the proof of Theorem 3.7.4 is complete.

From the hypotheses of Theorem 3.7.5 we have inequality (3.7.44). By using Schwarz inequality and inequalities (3.7.44) and (3.7.55), we observe that

$$\begin{aligned}
&\int_a^b\int_c^d p(x,y)\left(\prod_{r=1}^n |f_r(x,y)|^{m_r}\right)^{1/n}\left(\sum_{r=1}^n |D_2D_1f_r(x,y)|^{m_r}\right)\mathrm{d}y\,\mathrm{d}x\\
&\leqslant\left(\int_a^b\int_c^d p^2(x,y)\left(\prod_{r=1}^n |f_r(x,y)|^{m_r}\right)^{2/n}\mathrm{d}y\,\mathrm{d}x\right)^{1/2}\\
&\quad\times\left(\int_a^b\int_c^d\left(\sum_{r=1}^n |D_2D_1f_r(x,y)|^{m_r}\right)^2\mathrm{d}y\,\mathrm{d}x\right)^{1/2}\\
&\leqslant\left\{\frac{1}{n}K(a,b,c,d,n,m_1,\dots,m_n)\left(\int_a^b\int_c^d p^2(x,y)\,\mathrm{d}y\,\mathrm{d}x\right)\right.\\
&\quad\left.\times\left(\int_a^b\int_c^d\left(\sum_{r=1}^n |D_2D_1f_r(x,y)|^{2m_r}\right)\mathrm{d}y\,\mathrm{d}x\right)\right\}^{1/2}\\
&\quad\times\left\{\int_a^b\int_c^d n\left(\sum_{r=1}^n |D_2D_1f_r(x,y)|^{2m_r}\right)\mathrm{d}y\,\mathrm{d}x\right\}^{1/2}\\
&=\left\{K(a,b,c,d,n,m_1,\dots,m_n)\left(\int_a^b\int_c^d p^2(x,y)\,\mathrm{d}y\,\mathrm{d}x\right)\right\}^{1/2}\\
&\quad\times\left\{\int_a^b\int_c^d\left(\sum_{r=1}^n |D_2D_1f_r(x,y)|^{2m_r}\right)\mathrm{d}y\,\mathrm{d}x\right\},
\end{aligned}$$

which is the desired inequality in (3.7.46) and the proof of Theorem 3.7.5 is complete. □

In the following theorems, we shall deal with some Opial-type inequalities involving functions of several independent variables established by Pachpatte in [261,284].

First we will introduce some notations which we will use in our discussion. Let $\mathbb{R}$ denote the set of real numbers and $\mathbb{R}^n$ the n-dimensional Euclidean space. Let B be a bounded domain in $\mathbb{R}^n$ defined by $B = \prod_{i=1}^n [a_i, b_i]$.

Let $x = (x_1, \ldots, x_n)$ denote a variable point in B, $B_x = \prod_{i=1}^{n}[a_i, x_i]$ and $\mathrm{d}x = \mathrm{d}x_1 \cdots \mathrm{d}x_n$. Let $Dh(u) = \frac{\mathrm{d}}{\mathrm{d}u}h(u)$, $D_k h(x_1, \ldots, x_n) = \frac{\partial}{\partial x_k}h(x_1, \ldots, x_n)$, $1 \leqslant k \leqslant n$, and $D^k h(x_1, \ldots, x_n) = \frac{\partial^k}{\partial x_1 \cdots \partial x_n}h(x_1, \ldots, x_n) = D_1 \cdots D_k h(x_1, \ldots, x_n)$, $1 \leqslant k \leqslant n$. By using the above notation we have $D_1 h = D^1 h$. For any real-valued function $u(x)$ defined on B, we denote by $\int_B u(x)\,\mathrm{d}x$ the n-fold integral $\int_{a_n}^{b_n} \cdots \int_{a_1}^{b_1} u(x_1, \ldots, x_n)\,\mathrm{d}x_1 \cdots \mathrm{d}x_n$, and for $x \in B$ we denote by $\int_{B_x} u(y)\,\mathrm{d}y$ the n-fold integral $\int_{a_n}^{x_n} \cdots \int_{a_1}^{x_1} u(y_1, \ldots, y_n)\,\mathrm{d}y_1 \cdots \mathrm{d}y_n$ and $|\operatorname{grad} u(x)| = (\sum_{i=1}^{n} |\frac{\partial}{\partial x_i}|^2)^{1/2}$. We denote by $F(B)$ the class of continuous functions $u(x)\colon B \to \mathbb{R}$ for which $D^n u(x) = D_1 \cdots D_n u(x)$ ($D_i = \frac{\partial}{\partial x_i}$) exists and that, for each i, $1 \leqslant i \leqslant n$, $u(x)|_{x_i = a_i} = 0$.

In [261] Pachpatte has established the following Opial-type integral inequality.

THEOREM 3.7.6. *Let $p \geqslant 1$, $q \geqslant 1$ be constants. Let u be a real-valued function belonging to $C^1(B)$ which vanishes on the boundary ∂B of B. Then the following inequality holds*

$$\int_B |u(x)|^p |\operatorname{grad} u(x)|^q \,\mathrm{d}x \leqslant M \int_B |\operatorname{grad} u(x)|^{p+q}\,\mathrm{d}x, \tag{3.7.57}$$

where

$$M = \frac{1}{n}\left(\frac{1}{2}\right)^p \left[\sum_{j=1}^{n} (b_j - a_j)^{p(p+q)/q}\right]^{q/(p+q)}. \tag{3.7.58}$$

PROOF. From the hypotheses, we have the following identities

$$nu(x) = \sum_{j=1}^{n} \int_{a_j}^{x_j} \frac{\partial}{\partial t_j} u(x_1, \ldots, t_j, \ldots, x_n)\,\mathrm{d}t_j, \tag{3.7.59}$$

$$nu(x) = -\sum_{j=1}^{n} \int_{x_j}^{b_j} \frac{\partial}{\partial t_j} u(x_1, \ldots, t_j, \ldots, x_n)\,\mathrm{d}t_j. \tag{3.7.60}$$

From (3.7.59) and (3.7.60), we observe that

$$|u(x)| \leqslant \frac{1}{2n} \sum_{j=1}^{n} \int_{a_j}^{b_j} \left|\frac{\partial}{\partial t_j} u(x_1, \ldots, t_j, \ldots, x_n)\right| \mathrm{d}t_j. \tag{3.7.61}$$

Using Hölder's inequality with indices $(p+q)/(p+q-1)$ and $p+q$ to the integral on the right-hand side of (3.7.61), we obtain

$$|u(x)| \leqslant \frac{1}{2n}\sum_{j=1}^{n}\left[(b_j-a_j)^{(p+q-1)/(p+q)} \times \left(\int_{a_j}^{b_j}\left|\frac{\partial}{\partial t_j}u(x_1,\ldots,t_j,\ldots,x_n)\right|^{p+q}\mathrm{d}t_j\right)^{1/(p+q)}\right]. \tag{3.7.62}$$

Taking pth power on both sides of (3.8.62) and using the elementary inequality

$$\left(\sum_{i=1}^{n}c_i\right)^k \leqslant M_{k,n}\sum_{i=1}^{n}c_i^k, \tag{3.7.63}$$

where $c_1,\ldots,c_n \geqslant 0$ reals and $M_{k,n}=n^{k-1}$, $k>1$, and $M_{k,n}=1$, $0\leqslant k\leqslant 1$, we obtain

$$|u(x)|^p \leqslant \left(\frac{1}{2n}\right)^p n^{p-1}\sum_{j=1}^{n}\left[(b_j-a_j)^{p(p+q-1)/(p+q)} \times \left(\int_{a_j}^{b_j}\left|\frac{\partial}{\partial t_j}u(x_1,\ldots,t_j,\ldots,x_n)\right|^{p+q}\mathrm{d}t_j\right)^{p/(p+q)}\right]. \tag{3.7.64}$$

Multiplying both sides of (3.7.64) by $|\operatorname{grad} u(x)|^q$ we have

$$|u(x)|^p|\operatorname{grad} u(x)|^q \leqslant \frac{1}{n}\left(\frac{1}{2}\right)^p\sum_{j=1}^{n}\left[(b_j-a_j)^{p(p+q-1)/(p+q)}|\operatorname{grad} u(x)|^q \times \left(\int_{a_j}^{b_j}\left|\frac{\partial}{\partial t_j}u(x_1,\ldots,t_j,\ldots,x_n)\right|^{p+q}\mathrm{d}t_j\right)^{p/(p+q)}\right]. \tag{3.7.65}$$

Integrating both sides of (3.7.65) with respect to $x_1,\ldots,x_n$ on B and using Hölder's inequality for integrals with indices $(p+q)/p$ and $(p+q)/q$ on the right-hand side we get

$$\int_B |u(x)|^p |\operatorname{grad} u(x)|^q \,\mathrm{d}x$$
$$\leqslant \frac{1}{n}\left(\frac{1}{2}\right)^p \sum_{j=1}^n \Bigg[(b_j-a_j)^{p(p+q-1)/(p+q)}\left\{\int_B |\operatorname{grad} u(x)|^{p+q}\,\mathrm{d}x\right\}^{q/(p+q)}$$
$$\times \left\{\int_B \left(\int_{a_j}^{b_j} \left|\frac{\partial}{\partial t_j} u(x_1,\ldots,t_j,\ldots,x_n)\right|^{p+q} \mathrm{d}t_j\right)\mathrm{d}x\right\}^{p/(p+q)}\Bigg]$$
$$= \frac{1}{n}\left(\frac{1}{2}\right)^p \sum_{j=1}^n \Bigg[(b_j-a_j)^{p(p+q-1)/(p+q)}(b_j-a_j)^{p/(p+q)}$$
$$\times \left\{\int_B |\operatorname{grad} u(x)|^{p+q}\,\mathrm{d}x\right\}^{q/(p+q)}$$
$$\times \left\{\int_B \left|\frac{\partial}{\partial x_j} u(x)\right|^{p+q} \mathrm{d}x\right\}^{p/(p+q)}\Bigg]. \tag{3.7.66}$$

Now, using Hölder's inequality for sum with indices $(p+q)/p$ and $(p+q)/q$ on the right-hand side of (3.7.66) and an application of a suitable version of inequality (3.7.63), we obtain

$$\int_B |u(x)|^p |\operatorname{grad} u(x)|^q \,\mathrm{d}x$$
$$\leqslant \frac{1}{n}\left(\frac{1}{2}\right)^p \left\{\sum_{j=1}^n (b_j-a_j)^{p(p+q)/q} \int_B |\operatorname{grad} u(x)|^{p+q}\,\mathrm{d}x\right\}^{q/(p+q)}$$
$$\times \left\{\sum_{j=1}^n \int_B \left|\frac{\partial}{\partial x_j} u(x)\right|^{p+q} \mathrm{d}x\right\}^{p/(p+q)}$$
$$= M\left\{\int_B |\operatorname{grad} u(x)|^{p+q}\,\mathrm{d}x\right\}^{q/(p+q)}$$
$$\times \left\{\int_B \left(\left(\sum_{j=1}^n \left|\frac{\partial}{\partial x_j} u(x)\right|^{p+q}\right)^{2/(p+q)}\right)^{(p+q)/2} \mathrm{d}x\right\}^{p/(p+q)}$$

$$\leqslant M\left\{\int_B |\operatorname{grad} u(x)|^{p+q}\,\mathrm{d}x\right\}^{q/(p+q)}\left\{\int_B |\operatorname{grad} u(x)|^{p+q}\,\mathrm{d}x\right\}^{q/(p+q)}$$

$$= M\int_B |\operatorname{grad} u(x)|^{p+q}\,\mathrm{d}x.$$

This inequality is required in (3.7.57) and the proof is complete. □

A slightly different version of Theorem 3.7.6 also given by Pachpatte in [261] is embodied in the following theorem.

THEOREM 3.7.7. *Let $p_r \geqslant 1$, $q_r \geqslant 1$, $r = 1,\dots,m$, be constants. Let u_r, $r = 1,\dots,m$, be real-valued functions belonging to $C^1(B)$ which vanish on the boundary ∂B of B. Then, the following inequality holds*

$$\int_B \prod_{r=1}^{m} |u_r(x)|^{p_r} |\operatorname{grad} u_r(x)|^{q_r}\,\mathrm{d}x \leqslant \frac{1}{m}\sum_{r=1}^{m} M_r \int_B |\operatorname{grad} u_r(x)|^{m(p_r+q_r)}\,\mathrm{d}x, \tag{3.7.67}$$

where

$$M_r = \frac{1}{n}\left(\frac{1}{2}\right)^{mp_r}\left[\sum_{j=1}^{n}(b_j - a_j)^{mp_r(p_r+q_r)/q_r}\right]^{q_r/(p_r+q_r)} \tag{3.7.68}$$

for $r = 1,\dots,m$.

PROOF. Using the elementary inequality

$$\left(\prod_{i=1}^{m} c_i\right)^{1/m} \leqslant \frac{1}{m}\sum_{i=1}^{m} c_i$$

(for $c_1,\dots,c_n \geqslant 0$ reals), inequality (3.7.63) and the repeated application of inequality (3.7.57) we observe that

$$\int_B \prod_{r=1}^{m} |u_r(x)|^{p_r} |\operatorname{grad} u_r(x)|^{q_r}\,\mathrm{d}x$$

$$= \int_B \left[\left\{\prod_{r=1}^{m} |u_r(x)|^{p_r} |\operatorname{grad} u_r(x)|^{q_r}\right\}^{1/m}\right]^{m}\mathrm{d}x$$

$$\leqslant \int_B \left[\frac{1}{m} \sum_{r=1}^{m} |u_r(x)|^{p_r} \, |\operatorname{grad} u_r(x)|^{q_r} \right]^m \mathrm{d}x$$

$$\leqslant \frac{1}{m} \sum_{r=1}^{m} M_r \int_B |\operatorname{grad} u_r(x)|^{m(p_r+q_r)} \, \mathrm{d}x.$$

This inequality is desired in (3.7.67) and the proof is complete. □

In [284] Pachpatte has established the following Opial-type inequality involving functions of several variables.

THEOREM 3.7.8. *Let $u \in F(B)$. Then the following inequality holds*

$$\int_B |u(x)| |D_1 \cdots D_n u(x)| \, \mathrm{d}x$$
$$\leqslant \left(\int_B \left[\left(\prod_{i=1}^{n} (x_i - a_i) \right) \int_{B_x} |D_1 \cdots D_n u(y)|^2 \, \mathrm{d}y \right] \mathrm{d}x \right)^{1/2}$$
$$\times \left(\int_B |D_1 \cdots D_n u(x)|^2 \, \mathrm{d}x \right)^{1/2}. \tag{3.7.69}$$

PROOF. For any $u \in F(B)$ we have the following identity

$$u(x) = \int_{B_x} D_1 \cdots D_n u(y) \, \mathrm{d}y. \tag{3.7.70}$$

From (3.7.70) and using Schwarz inequality in the integral form, we observe that

$$|u(x)| \leqslant \int_{B_x} |D_1 \cdots D_n u(y)| \, \mathrm{d}y$$
$$\leqslant \left(\prod_{i=1}^{n} (x_i - a_i) \right)^{1/2} \left(\int_{B_x} |D_1 \cdots D_n u(y)|^2 \, \mathrm{d}y \right)^{1/2}. \tag{3.7.71}$$

Now, by using Schwarz inequality in the integral form, we have

$$\int_B |u(x)| |D_1 \cdots D_n u(x)| \, \mathrm{d}x$$
$$\leqslant \left(\int_B |u(x)|^2 \, \mathrm{d}x \right)^{1/2} \left(\int_B |D_1 \cdots D_n u(x)|^2 \, \mathrm{d}x \right)^{1/2}. \tag{3.7.72}$$

Using (3.7.71) on the right-hand side of (3.7.72) we get the required inequality in (3.7.69) and the proof is complete. □

REMARK 3.7.5. It is easy to observe that

$$\begin{aligned}
&\int_B \left[\left(\prod_{i=1}^{n} (x_i - a_i) \right) \int_{B_x} \left| D_1 \cdots D_n u(y) \right|^2 \mathrm{d}y \right] \mathrm{d}x \\
&\leqslant \left(\int_B \left(\prod_{i=1}^{n} (x_i - a_i) \right) \mathrm{d}x \right) \left(\int_B \left| D_1 \cdots D_n u(y) \right|^2 \mathrm{d}y \right) \\
&= \frac{1}{2^n} \prod_{i=1}^{n} (b_i - a_i)^2 \int_B \left| D_1 \cdots D_n u(x) \right|^2 \mathrm{d}x. \qquad (3.7.73)
\end{aligned}$$

Now, using (3.7.73) on the right-hand side of (3.7.69), we have the following inequality

$$\begin{aligned}
&\int_B \left| u(x) \right| \left| D_1 \cdots D_n u(x) \right| \mathrm{d}x \\
&\qquad \leqslant \frac{1}{(\sqrt{2})^n} \prod_{i=1}^{n} (b_i - a_i) \int_B \left| D_1 \cdots D_n u(x) \right|^2 \mathrm{d}x. \qquad (3.7.74)
\end{aligned}$$

If we take $n = 1$ in (3.7.74) and denote by $a_1 = a$, $b_1 = b$, $D_1 u = u'$, $x_1 = x$, then inequality (3.7.74) reduces to the following inequality

$$\int_a^b \left| u(x) \right| \left| u'(x) \right| \mathrm{d}x \leqslant \frac{b-a}{\sqrt{2}} \int_a^b \left| u'(x) \right|^2 \mathrm{d}x. \qquad (3.7.75)$$

Here we note that the constant appearing in (3.7.75) is greater than the constant obtained in Opial's inequality given in Theorem 2′ in [211, p. 154]. The main reason for increase is the difficulty involved in proving the much more general inequality given in (3.7.74).

3.8 Discrete Opial-Type Inequalities

In 1967, Wong [426] has established the following discrete inequality

$$\sum_{i=1}^{n} u_i^p (u_i - u_{i-1}) \leqslant \frac{(n+1)^p}{p+1} \sum_{i=1}^{n} (u_i - u_{i-1})^{p+1}, \qquad (3.8.1)$$

valid for a nondecreasing sequence $\{u_i\}$ of nonnegative real numbers with $u_0 = 0$ and $p \geqslant 1$. Inequality (3.8.1) is a discrete analogue of the variant of Opial's inequality given by Hua in [158]. In the past few years many results have appeared in the literature concerning various extensions and variants of inequality (3.8.1), see [4] and the references therein. In this section we shall deal with some discrete inequalities investigated by Pachpatte in [235,262,280,287,318,347] which claim their origin to the discrete analogue of Opial's inequality.

Before giving the results we first introduce some of the notations and definitions used in our discussion. Let $\mathbb{N} = \{1, 2, \dots\}$, $\mathbb{N}_0 = \{0, 1, 2, \dots\}$, $\mathbb{N}_{0,n} = \{0, 1, 2, \dots, n\}$, $\mathbb{N}_{n+1} = \{1, 2, \dots, n+1\}$, $n \in \mathbb{N}$; Δ and ∇ are the forward and backward difference operators defined by $\Delta u_k = u_{k+1} - u_k$, $k \in \mathbb{N}_0$, $\nabla u_k = u_k - u_{k-1}$, $k \in \mathbb{N}$. The symbol $\Delta^i u_k = \Delta(\Delta^{i-1} u_k) = \Delta^{i-1}(\Delta u_k)$, where $\Delta^0 u_k = u_k$. Throughout, we shall use the convention that the empty sums and products are taken to be 0 and 1, respectively.

An interesting Opial-type discrete inequality involving two sequences and their forward differences, established by Pachpatte in [262], is given in the following theorem.

THEOREM 3.8.1. *Let $\{u_k\}$ and $\{v_k\}$, $k \in \mathbb{N}_0$, be nondecreasing sequences of nonnegative real numbers with $u_0 = v_0 = 0$. Then the following inequality holds*

$$\sum_{k=0}^{n-1} [u_k \Delta v_k + v_{k+1} \Delta u_k] \leqslant \frac{n}{2} \sum_{k=0}^{n-1} [(\Delta u_k)^2 + (\Delta v_k)^2] \tag{3.8.2}$$

for all $n \in \mathbb{N}_0$.

PROOF. It is easy to observe that the following identity holds

$$\Delta(u_k v_k) = u_k \Delta v_k + v_{k+1} \Delta u_k \tag{3.8.3}$$

for $k \in \mathbb{N}_0$. From (3.8.3) we obtain

$$\sum_{k=0}^{n-1} [u_k \Delta v_k + v_{k+1} \Delta u_k] = u_n v_n \tag{3.8.4}$$

for $n \in \mathbb{N}_0$. Using the elementary inequality $\alpha\beta \leqslant \frac{1}{2}(\alpha^2 + \beta^2)$ (for α, β reals) and the facts that $u_n = \sum_{k=0}^{n-1} \Delta u_k$, $v_n = \sum_{k=0}^{n-1} \Delta v_k$, and Schwarz inequality, we observe that

$$u_n v_n \leqslant \frac{1}{2} \left[\left(\sum_{k=0}^{n-1} \Delta u_k \right)^2 + \left(\sum_{k=0}^{n-1} \Delta v_k \right)^2 \right]$$

$$\leqslant \frac{n}{2} \sum_{k=0}^{n-1} [(\Delta u_k)^2 + (\Delta v_k)^2]. \tag{3.8.5}$$

The desired inequality in (3.8.2) follows from (3.8.4) and (3.8.5). The proof is complete. □

As an immediate consequence of Theorem 3.8.1 established in [262] is embodied in the following theorem.

THEOREM 3.8.2. *Let* $\{u_k\}$, $k \in \mathbb{N}_0$, *be a nondecreasing sequence of nonnegative real numbers with* $u_0 = 0$. *Then the following inequality holds*

$$\sum_{k=0}^{n-1} u_{k+1} \Delta u_k \leqslant \frac{n+1}{2} \sum_{k=0}^{n-1} (\Delta u_k)^2 \tag{3.8.6}$$

for all $n \in \mathbb{N}_0$.

PROOF. Setting $v_k = u_k$, $k \in \mathbb{N}_0$, in (3.8.2), we have

$$\sum_{k=0}^{n-1} [u_k + u_{k+1}] \Delta u_k \leqslant n \sum_{k=0}^{n-1} (\Delta u_k)^2. \tag{3.8.7}$$

We observe that

$$\begin{aligned}\sum_{k=0}^{n-1} [u_k + u_{k+1}] \Delta u_k &= \sum_{k=0}^{n-1} [-\Delta u_k + 2u_{k+1}] \Delta u_k \\ &= -\sum_{k=0}^{n-1} (\Delta u_k)^2 + 2 \sum_{k=0}^{n-1} u_{k+1} \Delta u_k. \end{aligned} \tag{3.8.8}$$

From (3.8.7) and (3.8.8) we obtain the desired inequality in (3.8.6) and the proof is complete. □

The following Wirtinger-type discrete inequality established in [347] is useful in the proof of the next result.

THEOREM 3.8.3. *Let* $p \geqslant 1$ *be a given real number and* $\{a_k\}$, $k \in \mathbb{N}_{0,n}$, *be a sequence of nonnegative real numbers. Let* $\{u_k\}$, $k \in \mathbb{N}_{0,n}$, *be a sequence of real numbers with* $u_0 = u_n = 0$. *Then the following inequality holds*

$$\sum_{k=0}^{n-1} a_k |u_k|^p \leqslant \left(\sum_{k=0}^{n-1}\left[k^{1-p} + (n-k)^{1-p}\right]^{-1} a_k\right)\left(\sum_{k=0}^{n-1} |\Delta u_k|^p\right). \tag{3.8.9}$$

PROOF. From the hypotheses, we have

$$u_k = \sum_{\sigma=0}^{k-1} \Delta u_\sigma, \tag{3.8.10}$$

$$u_k = -\sum_{\sigma=k}^{n-1} \Delta u_\sigma, \tag{3.8.11}$$

for $k \in \mathbb{N}_{0,n}$. From (3.8.10), (3.8.11) and using Hölder's inequality with indices $p, p/(p-1)$, we obtain

$$|u_k|^p \leqslant k^{p-1} \sum_{\sigma=0}^{k-1} |\Delta u_\sigma|^p, \tag{3.8.12}$$

$$|u_k|^p \leqslant (n-k)^{p-1} \sum_{\sigma=k}^{n-1} |\Delta u_\sigma|^p, \tag{3.8.13}$$

for $k \in \mathbb{N}_{0,n}$. Multiplying (3.8.12) by k^{1-p} and (3.8.13) by $(n-k)^{1-p}$ and adding the resulting inequalities we obtain

$$\left[k^{1-p} + (n-k)^{1-p}\right] |u_k|^p \leqslant \sum_{\sigma=0}^{n-1} |\Delta u_\sigma|^p \tag{3.8.14}$$

for $k \in \mathbb{N}_{0,n}$. From (3.8.14) we observe that

$$a_k |u_k|^p \leqslant \left[k^{1-p} + (n-k)^{1-p}\right]^{-1} a_k \sum_{\sigma=0}^{n-1} |\Delta u_\sigma|^p \tag{3.8.15}$$

for $k \in \mathbb{N}_{0,n}$. Now summing both sides of (3.8.15) from $k = 0$ to $n-1$ we get inequality (3.8.9). The proof is complete. □

The following result established in [347] deals with the discrete Opial-type inequality.

THEOREM 3.8.4. *Let $q > 0$, $r \geqslant 0$, $\alpha > 0$, $\beta > 0$ be real numbers with $\frac{1}{\alpha} + \frac{1}{\beta} = 1$ and $q\alpha \geqslant 1$, and $\{b_k\}$, $k \in \mathbb{N}_{0,n}$, be a sequence of nonnegative real numbers. Let $\{x_k\}$ and $\{y_k\}$, $k \in \mathbb{N}_{0,n}$, be sequences of real numbers with $x_0 = y_0 = x_n = y_n = 0$. Then the following inequality holds*

$$\begin{aligned}
&\sum_{k=0}^{n-1} b_k\big[|x_k|^q|\Delta y_k|^r + |y_k|^q|\Delta x_k|^r\big] \\
&\quad\leqslant \left(\sum_{k=0}^{n-1}\big[k^{1-q\alpha} + (n-k)^{1-q\alpha}\big]^{-1} b_k^{\alpha}\right)^{1/\alpha} \\
&\qquad\times \left(\sum_{k=0}^{n-1}\left[\frac{1}{\alpha}\big(|\Delta x_k|^{q\alpha} + |\Delta y_k|^{q\alpha}\big) + \frac{1}{\beta}\big(|\Delta x_k|^{r\beta} + |\Delta y_k|^{r\beta}\big)\right]\right).
\end{aligned} \tag{3.8.16}$$

PROOF. From Hölder's inequality with indices α and β, we have

$$\sum_{k=0}^{n-1} b_k|x_k|^q|\Delta y_k|^r \leqslant \left(\sum_{k=0}^{n-1} b_k^{\alpha}|x_k|^{q\alpha}\right)^{1/\alpha}\left(\sum_{k=0}^{n-1}|\Delta y_k|^{r\beta}\right)^{1/\beta}. \tag{3.8.17}$$

From (3.8.17) and (3.8.9) and Young's inequality, we observe that

$$\begin{aligned}
&\sum_{k=0}^{n-1} b_k|x_k|^q|\Delta y_k|^r \\
&\quad\leqslant \left\{\left(\sum_{k=0}^{n-1}\big[k^{1-q\alpha} + (n-k)^{1-q\alpha}\big]^{-1} b_k^{\alpha}\right)\left(\sum_{k=0}^{n-1}|\Delta x_k|^{q\alpha}\right)\right\}^{1/\alpha} \\
&\qquad\times \left\{\sum_{k=0}^{n-1}|\Delta y_k|^{r\beta}\right\}^{1/\beta} \\
&\quad\leqslant \left(\sum_{k=0}^{n-1}\big[k^{1-q\alpha} + (n-k)^{1-q\alpha}\big]^{-1} b_k^{\alpha}\right)^{1/\alpha} \\
&\qquad\times \left(\sum_{k=0}^{n-1}\left[\frac{1}{\alpha}|\Delta x_k|^{q\alpha} + \frac{1}{\beta}|\Delta y_k|^{r\beta}\right]\right).
\end{aligned} \tag{3.8.18}$$

Similarly, we obtain

$$\sum_{k=0}^{n-1} b_k |y_k|^q |\Delta x_k|^r \leqslant \left(\sum_{k=0}^{n-1} \left[k^{1-q\alpha} + (n-k)^{1-q\alpha} \right]^{-1} b_k^{\alpha} \right)^{1/\alpha} \times \left(\sum_{k=0}^{n-1} \left[\frac{1}{\alpha} |\Delta y_k|^{q\alpha} + \frac{1}{\beta} |\Delta x_k|^{r\beta} \right] \right). \tag{3.8.19}$$

Adding (3.8.18) and (3.8.19) gives inequality (3.8.16). The proof is complete. □

The inequality in the following theorem is established in [280].

THEOREM 3.8.5. *Let $\{u_i\}$, $i \in \mathbb{N}_0$, be a sequence of real numbers with $u_0 = 0$. Let $f(t)$ be defined for all $t = u_i$ and for all t of the form $t(j) = \sum_{k=1}^{j} \nabla u_k$, $|f(t)| \leqslant f(|t|)$ for all t and that $f(t)$ is nondecreasing for $t \geqslant 0$, where $\nabla u_k = u_k - u_{k-1}$. Then the following inequality holds*

$$\sum_{i=1}^{n} \left| f(u_i) \nabla u_i \right| \leqslant F\left(\sum_{i=1}^{n} |\nabla u_i| \right) + \sum_{i=1}^{n} \left[f\left(\sum_{k=1}^{i} |\nabla u_k| \right) - f\left(\sum_{k=1}^{i-1} |\nabla u_k| \right) \right] |\nabla u_i|, \tag{3.8.20}$$

where $F(s) = \int_0^s f(\sigma)\,\mathrm{d}\sigma$, $s \geqslant 0$.

REMARK 3.8.1. If we take $f(t) = t$, then we get $F(s) = s^2/2$, and inequality (3.8.20) reduces to the following inequality

$$\sum_{i=1}^{n} |u_i \nabla u_i| \leqslant \left(\frac{n}{2} + 1 \right) \sum_{i=1}^{n} |\nabla u_i|^2. \tag{3.8.21}$$

PROOF OF THEOREM 3.8.5. Since $u_i = \sum_{k=1}^{i} \nabla u_k$, it follows that

$$\sum_{i=1}^{n} \left| f(u_i) \nabla u_i \right| = \sum_{i=1}^{n} \left| f\left(\sum_{k=1}^{i} \nabla u_k \right) \nabla u_i \right| \leqslant \sum_{i=1}^{n} f\left(\sum_{k=1}^{i} |\nabla u_k| \right) |\nabla u_i|. \tag{3.8.22}$$

Now, from the definition of F and the properties of the function f, we obtain

$$\begin{aligned}
F&\left(\sum_{k=1}^{i}|\nabla u_k|\right) - F\left(\sum_{k=1}^{i-1}|\nabla u_k|\right)\\
&= \int_{\sum_{k=1}^{i-1}|\nabla u_k|}^{\sum_{k=1}^{i}|\nabla u_k|} f(\sigma)\,\mathrm{d}\sigma\\
&\geqslant f\left(\sum_{k=1}^{i-1}|\nabla u_k|\right)|\nabla u_i|\\
&= f\left(\sum_{k=1}^{i}|\nabla u_k|\right)|\nabla u_i| + \left[f\left(\sum_{k=1}^{i-1}|\nabla u_k|\right) - f\left(\sum_{k=1}^{i}|\nabla u_k|\right)\right]|\nabla u_i|.
\end{aligned} \tag{3.8.23}$$

From (3.8.23) we observe that

$$\begin{aligned}
f\left(\sum_{k=1}^{i}|\nabla u_k|\right)|\nabla u_i| &\leqslant F\left(\sum_{k=1}^{i}|\nabla u_k|\right) - F\left(\sum_{k=1}^{i-1}|\nabla u_k|\right)\\
&\quad + \left[f\left(\sum_{k=1}^{i}|\nabla u_k|\right) - f\left(\sum_{k=1}^{i-1}|\nabla u_k|\right)\right]|\nabla u_i|.
\end{aligned} \tag{3.8.24}$$

Now, substituting $i = 1, \ldots, n$ on both sides of inequality (3.8.24) and summing up, we obtain

$$\begin{aligned}
\sum_{i=1}^{n} f\left(\sum_{k=1}^{i}|\nabla u_k|\right)|\nabla u_i| &\leqslant F\left(\sum_{k=1}^{n}|\nabla u_k|\right)\\
&\quad + \sum_{i=1}^{n}\left[f\left(\sum_{k=1}^{i}|\nabla u_k|\right) - f\left(\sum_{k=1}^{i-1}|\nabla u_k|\right)\right]|\nabla u_i|.
\end{aligned} \tag{3.8.25}$$

Replacing k by i in the first term on the right-hand side in (3.8.25) and using this bound in (3.8.22) we obtain the desired inequality in (3.8.20). The proof is complete. □

The following theorem deals with the discrete Opial-type inequalities established in [318].

THEOREM 3.8.6. *Let $p \geqslant 1$, $q \geqslant 1$ be constants. Let $\lambda_n > 0$, $a_n \geqslant 0$ and $\Lambda_n = \lambda_1 + \cdots + \lambda_n$, $A_n = \lambda_1 a_1 + \cdots + \lambda_n a_n$ for $n \in \mathbb{N}$. Then the following inequalities hold*

$$\sum_{n=1}^{m} \frac{A_n^p (A_n^q - A_{n-1}^q)}{\Lambda_n^{p+q-1}} \leqslant q \left(\frac{p+q}{p+q-1} \right)^{p+q-1} \sum_{n=1}^{m} \lambda_n a_n^{p+q}, \tag{3.8.26}$$

$$\sum_{n=1}^{m} A_n^p (A_n^q - A_{n-1}^q) \leqslant \frac{q(m+1)^{p+q-1}}{p+q} \sum_{n=1}^{m} (\lambda_n a_n)^{p+q}, \tag{3.8.27}$$

where any number with suffix zero is equal to 0.

PROOF. Since $A_{n-1} \leqslant A_n$, we have

$$\begin{aligned} A_n^p (A_n^q - A_{n-1}^p) &= A_n^p \left(\sum_{k=0}^{q-1} A_n^{q-1-k} A_{n-1}^k \right) (A_n - A_{n-1}) \\ &\leqslant q A_n^{p+q-1} (A_n - A_{n-1}), \end{aligned} \tag{3.8.28}$$

which implies, by using Hölder's inequality with indices $p+q$, $(p+q)/(p+q-1)$ and a suitable version of Theorem 2.2.2, the inequality

$$\begin{aligned} &\sum_{n=1}^{m} \frac{A_n^p (A_n^q - A_{n-1}^p)}{\Lambda_n^{p+q-1}} \\ &\quad \leqslant q \sum_{n=1}^{m} \left(\frac{A_n}{\Lambda_n} \right)^{p+q-1} (\lambda_n a_n) \\ &\quad = q \sum_{n=1}^{m} \lambda_n^{1/(p+q)} a_n \lambda_n^{(p+q-1)/(p+q)} \left(\frac{A_n}{\Lambda_n} \right)^{p+q-1} \\ &\quad \leqslant q \left[\sum_{n=1}^{m} \lambda_n a_n^{p+q} \right]^{1/(p+q)} \left[\sum_{n=1}^{m} \lambda_n \left(\frac{A_n}{\Lambda_n} \right)^{p+q} \right]^{(p+q-1)/(p+q)} \\ &\quad \leqslant q \left[\sum_{n=1}^{m} \lambda_n a_n^{p+q} \right]^{1/(p+q)} \left[\left(\frac{p+q}{p+q-1} \right)^{p+q} \sum_{n=1}^{m} \lambda_n a_n^{p+q} \right]^{(p+q-1)/(p+q)} \\ &\quad = q \left(\frac{p+q}{p+q-1} \right)^{p+q-1} \sum_{n=1}^{m} \lambda_n a_n^{p+q}, \end{aligned}$$

proving (3.8.26).

From (3.8.28), using inequality (3.8.1), we have

$$\begin{aligned}\sum_{n=1}^{m} A_n^p\left(A_n^q - A_{n-1}^q\right) &\leqslant q\sum_{n=1}^{m} A_n^{p+q-1}(A_n - A_{n-1})\\ &\leqslant \frac{q(m+1)^{p+q-1}}{p+q}\sum_{n=1}^{m}(A_n - A_{n-1})^{p+q}\\ &= \frac{q(m+1)^{p+q-1}}{p+q}\sum_{n=1}^{m}(\lambda_n a_n)^{p+q},\end{aligned}$$

which proves (3.8.27). The proof is complete. □

Before giving the next theorem, we introduce the basic notations and definitions needed in our discussion. Let $\mathbb{N} = \{1, 2, \ldots\}$ and $\mathbb{N}_{0,n} = \{0, 1, 2, \ldots, n\}$, $\mathbb{M}_{0,m} = \{0, 1, 2, \ldots, m\}$ for $m, n \in \mathbb{N}$ and $Q = \mathbb{N}_{0,n} \times \mathbb{M}_{0,m}$. We shall use the usual convention of writing $z(x, y) = 0$ if $x \notin \mathbb{N}_{0,n}$ or $y \notin \mathbb{M}_{0,m}$, or both $x \notin \mathbb{N}_{0,n}$ and $y \notin \mathbb{M}_{0,m}$, where $z(x, y)$ is a function defined on Q. We define the operators: $\nabla_1 z(x, y) = z(x, y) - z(x-1, y)$, $\nabla_2 z(x, y) = z(x, y) - z(x, y-1)$, $\nabla_2\nabla_1 z(x, y) = \nabla_1 z(x, y) - \nabla_1 z(x, y-1)$ for $(x, y) \in Q$.

The inequalities in the following theorem are established in [235].

THEOREM 3.8.7. *Let $f(x, y)$ and $g(x, y)$ be real-valued functions defined for $(x, y) \in Q$ such that $f(0, y) = g(0, y) = 0$, $f(n, y) = g(n, y) = 0$, $f(x, 0) = g(x, 0) = 0$, $f(x, m) = g(x, m) = 0$. Then the following inequalities hold*

$$\sum_{x=1}^{n}\sum_{y=1}^{m}\left|f(x, y)\right|\left|g(x, y)\right| \leqslant \frac{1}{2}\left(\frac{nm}{4}\right)^2\sum_{x=1}^{n}\sum_{y=1}^{m}\left[\left|\nabla_2\nabla_1 f(x, y)\right|^2 + \left|\nabla_2\nabla_1 g(x, y)\right|^2\right], \tag{3.8.29}$$

$$\sum_{x=1}^{n}\sum_{y=1}^{m}\left[\left|f(x, y)\right|\left|\nabla_2\nabla_1 g(x, y)\right| + \left|g(x, y)\right|\left|\nabla_2\nabla_1 f(x, y)\right|\right] \leqslant \left(\frac{nm}{4}\right)\sum_{x=1}^{n}\sum_{y=1}^{m}\left[\left|\nabla_2\nabla_1 f(x, y)\right|^2 + \left|\nabla_2\nabla_1 g(x, y)\right|^2\right]. \tag{3.8.30}$$

REMARK 3.8.2. If we take $g(x, y) = f(x, y)$ in (3.8.29) and (3.8.30), then we get respectively the following inequalities

$$\sum_{x=1}^{n}\sum_{y=1}^{m}|f(x, y)|^2 \leqslant \left(\frac{nm}{4}\right)^2 \sum_{x=1}^{n}\sum_{y=1}^{m}|\nabla_2\nabla_1 f(x, y)|^2, \tag{3.8.31}$$

$$\sum_{x=1}^{n}\sum_{y=1}^{m}|f(x, y)||\nabla_2\nabla_1 f(x, y)| \leqslant \left(\frac{nm}{4}\right) \sum_{x=1}^{n}\sum_{y=1}^{m}|\nabla_2\nabla_1 f(x, y)|^2. \tag{3.8.32}$$

We note that the inequalities obtained in (3.8.31) and (3.8.32) are respectively the discrete Wirtinger- and Opial-type inequalities in two independent variables.

PROOF OF THEOREM 3.8.7. From the hypotheses, it is easy to observe that the following identities hold

$$f(x, y) = \sum_{s=1}^{x}\sum_{t=1}^{y}\nabla_2\nabla_1 f(s, t), \tag{3.8.33}$$

$$f(x, y) = -\sum_{s=1}^{x}\sum_{t=y+1}^{m}\nabla_2\nabla_1 f(s, t), \tag{3.8.34}$$

$$f(x, y) = -\sum_{s=x+1}^{n}\sum_{t=1}^{y}\nabla_2\nabla_1 f(s, t), \tag{3.8.35}$$

$$f(x, y) = \sum_{s=x+1}^{n}\sum_{t=y+1}^{m}\nabla_2\nabla_1 f(s, t). \tag{3.8.36}$$

From (3.8.33)–(3.8.36), we obtain

$$4|f(x, y)| \leqslant \sum_{s=1}^{n}\sum_{t=1}^{m}|\nabla_2\nabla_1 f(s, t)|. \tag{3.8.37}$$

Similarly, we obtain

$$4|g(x, y)| \leqslant \sum_{s=1}^{n}\sum_{t=1}^{m}|\nabla_2\nabla_1 g(s, t)|. \tag{3.8.38}$$

From (3.8.37), (3.8.38), and using the elementary inequality $\alpha\beta \leqslant \frac{1}{2}(\alpha^2+\beta^2)$ (for α, β reals) and Schwarz inequality, we obtain

$$\begin{aligned}
&\sum_{x=1}^{n}\sum_{y=1}^{m}|f(x,y)||g(x,y)| \\
&\quad\leqslant \frac{1}{16}\sum_{x=1}^{n}\sum_{y=1}^{m}\left(\sum_{s=1}^{n}\sum_{t=1}^{m}|\nabla_2\nabla_1 f(s,t)|\right)\left(\sum_{s=1}^{n}\sum_{t=1}^{m}|\nabla_2\nabla_1 g(s,t)|\right) \\
&\quad= \left(\frac{nm}{16}\right)\left(\sum_{s=1}^{n}\sum_{t=1}^{m}|\nabla_2\nabla_1 f(s,t)|\right)\left(\sum_{s=1}^{n}\sum_{t=1}^{m}|\nabla_2\nabla_1 g(s,t)|\right) \\
&\quad\leqslant \frac{1}{2}\left(\frac{nm}{16}\right)\left[\left(\sum_{s=1}^{n}\sum_{t=1}^{m}|\nabla_2\nabla_1 f(s,t)|\right)^2 + \left(\sum_{s=1}^{n}\sum_{t=1}^{m}|\nabla_2\nabla_1 g(s,t)|\right)^2\right] \\
&\quad\leqslant \frac{1}{2}\left(\frac{nm}{4}\right)^2\sum_{x=1}^{n}\sum_{y=1}^{m}\left[|\nabla_2\nabla_1 f(x,y)|^2 + |\nabla_2\nabla_1 g(x,y)|^2\right].
\end{aligned}$$

This result completes the proof of inequality (3.8.29).

By using Schwarz inequality, inequality (3.8.31) and the elementary inequality $\alpha^{1/2}\beta^{1/2} \leqslant \frac{1}{2}(\alpha+\beta)$ (for $\alpha, \beta \geqslant 0$ reals), we observe that

$$\begin{aligned}
&\sum_{x=1}^{n}\sum_{y=1}^{m}\left[|f(x,y)||\nabla_2\nabla_1 g(x,y)| + |g(x,y)||\nabla_2\nabla_1 f(x,y)|\right] \\
&\quad\leqslant \left\{\sum_{x=1}^{n}\sum_{y=1}^{m}|f(x,y)|^2\right\}^{1/2}\left\{\sum_{x=1}^{n}\sum_{y=1}^{m}|\nabla_2\nabla_1 g(x,y)|^2\right\}^{1/2} \\
&\qquad + \left\{\sum_{x=1}^{n}\sum_{y=1}^{m}|g(x,y)|^2\right\}^{1/2}\left\{\sum_{x=1}^{n}\sum_{y=1}^{m}|\nabla_2\nabla_1 f(x,y)|^2\right\}^{1/2} \\
&\quad\leqslant \left\{\left(\frac{nm}{4}\right)^2\sum_{x=1}^{n}\sum_{y=1}^{m}|\nabla_2\nabla_1 f(x,y)|^2\right\}^{1/2}\left\{\sum_{x=1}^{n}\sum_{y=1}^{m}|\nabla_2\nabla_1 g(x,y)|^2\right\}^{1/2}
\end{aligned}$$

$$+\left\{\left(\frac{nm}{4}\right)^2\sum_{x=1}^{n}\sum_{y=1}^{m}|\nabla_2\nabla_1 g(x,y)|^2\right\}^{1/2}\left\{\sum_{x=1}^{n}\sum_{y=1}^{m}|\nabla_2\nabla_1 f(x,y)|^2\right\}^{1/2}$$

$$=2\left(\frac{nm}{4}\right)\left\{\sum_{x=1}^{n}\sum_{y=1}^{m}|\nabla_2\nabla_1 f(x,y)|^2\right\}^{1/2}\left\{\sum_{x=1}^{n}\sum_{y=1}^{m}|\nabla_2\nabla_1 g(x,y)|^2\right\}^{1/2}$$

$$\leqslant\left(\frac{nm}{4}\right)\sum_{x=1}^{n}\sum_{y=1}^{m}\left[|\nabla_2\nabla_1 f(x,y)|^2+|\nabla_2\nabla_1 g(x,y)|^2\right],$$

which is the desired inequality in (3.8.30), and the proof is complete. □

In [287] Pachpatte has established the inequalities in the following theorem.

THEOREM 3.8.8. *Let $u(x,y)$ be a real-valued function defined for $(x,y)\in Q$ such that $u(0,y)=u(n,y)=0$, $u(x,0)=u(x,m)=0$ and $1\leqslant p_i<\infty$ for $i=1,2,3,4$ be constants. Then the following inequalities hold*

$$\sum_{x=1}^{n}\sum_{y=1}^{m}|u(x,y)|^{p_1}|\nabla_1 u(x,y)|^{p_2}|\nabla_2 u(x,y)|^{p_3}|\nabla_2\nabla_1 u(x,y)|^{p_4}$$

$$\leqslant K\prod_{i=1}^{4}\left\{\sum_{x=1}^{n}\sum_{y=1}^{m}|\nabla_2\nabla_1 u(x,y)|^{2p_i}\right\}^{1/2}, \tag{3.8.39}$$

$$\sum_{x=1}^{n}\sum_{y=1}^{m}|u(x,y)|^{p_1}|\nabla_1 u(x,y)|^{p_2}|\nabla_2 u(x,y)|^{p_3}$$

$$\leqslant L\prod_{i=1}^{3}\left\{\sum_{x=1}^{n}\sum_{y=1}^{m}|\nabla_2\nabla_1 u(x,y)|^{p_i}\right\}, \tag{3.8.40}$$

where

$$K=\left(\frac{1}{2}\right)^{2p_1+p_2+p_3}n^{p_1+p_3-1}m^{p_1+p_2-1}, \tag{3.8.41}$$

$$L=\left(\frac{1}{2}\right)^{2p_1+p_2+p_3}n^{p_1+p_3-2}m^{p_1+p_2-2}. \tag{3.8.42}$$

PROOF. From the hypotheses, it is easy to observe that the following identities hold:

$$u(x,y)=\sum_{s=1}^{x}\sum_{t=1}^{y}\nabla_2\nabla_1 u(s,t), \tag{3.8.43}$$

$$u(x,y)=-\sum_{s=1}^{x}\sum_{t=y+1}^{m}\nabla_2\nabla_1 u(s,t), \tag{3.8.44}$$

$$u(x,y)=-\sum_{s=x+1}^{n}\sum_{t=1}^{y}\nabla_2\nabla_1 u(s,t), \tag{3.8.45}$$

$$u(x,y)=\sum_{s=x+1}^{n}\sum_{t=y+1}^{m}\nabla_2\nabla_1 u(s,t), \tag{3.8.46}$$

$$\nabla_1 u(x,y)=\sum_{t=1}^{y}\nabla_2\nabla_1 u(x,t), \tag{3.8.47}$$

$$\nabla_1 u(x,y)=-\sum_{t=y+1}^{m}\nabla_2\nabla_1 u(x,t), \tag{3.8.48}$$

$$\nabla_2 u(x,y)=\sum_{s=1}^{x}\nabla_2\nabla_1 u(s,y), \tag{3.8.49}$$

$$\nabla_2 u(x,y)=-\sum_{s=x+1}^{n}\nabla_2\nabla_1 u(s,y), \tag{3.8.50}$$

for $(x,y)\in Q$. From (3.8.43)–(3.8.46), (3.8.47), (3.8.48) and (3.8.49), (3.8.50), we observe that

$$\left|u(x,y)\right|\leqslant\left(\frac{1}{2}\right)^2\sum_{s=1}^{n}\sum_{t=1}^{m}\left|\nabla_2\nabla_1 u(s,t)\right|, \tag{3.8.51}$$

$$\left|\nabla_1 u(x,y)\right|\leqslant\frac{1}{2}\sum_{t=1}^{m}\left|\nabla_2\nabla_1 u(x,t)\right|, \tag{3.8.52}$$

$$\left|\nabla_2 u(x,y)\right|\leqslant\frac{1}{2}\sum_{s=1}^{n}\left|\nabla_2\nabla_1 u(s,y)\right|, \tag{3.8.53}$$

respectively, for $(x, y) \in Q$. Taking on both sides of (3.8.51), (3.8.52) and (3.8.53) the powers p_1, p_2 and p_3, respectively, and using Hölder's inequality with indices $p_1, p_1/(p_1-1)$; $p_2, p_2/(p_2-1)$ and $p_3, p_3/(p_3-1)$, respectively, on the right-hand sides we get

$$\left|u(x,y)\right|^{p_1} \leqslant \left(\frac{1}{2}\right)^{2p_1} (nm)^{p_1-1} \sum_{s=1}^{n}\sum_{t=1}^{m} \left|\nabla_2\nabla_1 u(s,t)\right|^{p_1}, \tag{3.8.54}$$

$$\left|\nabla_1 u(x,y)\right|^{p_2} \leqslant \left(\frac{1}{2}\right)^{p_2} m^{p_2-1} \sum_{t=1}^{m} \left|\nabla_2\nabla_1 u(x,t)\right|^{p_2}, \tag{3.8.55}$$

$$\left|\nabla_2 u(x,y)\right|^{p_3} \leqslant \left(\frac{1}{2}\right)^{p_3} n^{p_3-1} \sum_{s=1}^{n} \left|\nabla_2\nabla_1 u(s,y)\right|^{p_3}, \tag{3.8.56}$$

respectively. From (3.8.54)–(3.8.56), we observe that

$$\begin{aligned}
&\left|u(x,y)\right|^{p_1}\left|\nabla_1 u(x,y)\right|^{p_2}\left|\nabla_2 u(x,y)\right|^{p_3}\left|\nabla_2\nabla_1 u(x,y)\right|^{p_4} \\
&\quad\leqslant L\left\{\sum_{s=1}^{n}\sum_{t=1}^{m}\left|\nabla_2\nabla_1 u(s,t)\right|^{p_1}\right\}\left\{\sum_{t=1}^{m}\left|\nabla_2\nabla_1 u(x,t)\right|^{p_2}\right\} \\
&\quad\quad\times\left\{\sum_{s=1}^{n}\left|\nabla_2\nabla_1 u(s,y)\right|^{p_3}\right\}\left|\nabla_2\nabla_1 u(x,y)\right|^{p_4},
\end{aligned} \tag{3.8.57}$$

where L is defined by (3.8.42). Now, taking the sum on both sides of (3.8.57), first from $y = 1$ to m and then from $x = 1$ to n, and using Schwarz inequality and rewriting the sums, we observe that

$$\begin{aligned}
&\sum_{x=1}^{n}\sum_{y=1}^{m}\left|u(x,y)\right|^{p_1}\left|\nabla_1 u(x,y)\right|^{p_2}\left|\nabla_2 u(x,y)\right|^{p_3}\left|\nabla_2\nabla_1 u(x,y)\right|^{p_4} \\
&\quad\leqslant L\left\{\sum_{s=1}^{n}\sum_{t=1}^{m}\left|\nabla_2\nabla_1 u(s,t)\right|^{p_1}\right\} \\
&\quad\quad\times\sum_{x=1}^{n}\sum_{y=1}^{m}\left\{\sum_{t-1}^{m}\left|\nabla_2\nabla_1 u(x,t)\right|^{p_2}\right\}\left\{\sum_{s=1}^{n}\left|\nabla_2\nabla_1 u(s,y)\right|^{p_3}\right\}\left|\nabla_2\nabla_1 u(x,y)\right|^{p_4} \\
&\quad\leqslant L\left\{\sum_{s=1}^{n}\sum_{t=1}^{m}\left|\nabla_2\nabla_1 u(s,t)\right|^{p_1}\right\}
\end{aligned}$$

$$\times \left\{ \sum_{x=1}^{n} \sum_{y=1}^{m} \left(\sum_{t=1}^{m} \left| \nabla_2 \nabla_1 u(x,t) \right|^{p_2} \right)^2 \left(\sum_{s=1}^{n} \left| \nabla_2 \nabla_1 u(s,y) \right|^{p_3} \right)^2 \right\}^{1/2}$$

$$\times \left\{ \sum_{x=1}^{n} \sum_{y=1}^{m} \left| \nabla_2 \nabla_1 u(x,y) \right|^{2p_4} \right\}^{1/2}$$

$$\leqslant L(nm)^{1/2} \left\{ \sum_{x=1}^{n} \sum_{y=1}^{m} \left| \nabla_2 \nabla_1 u(x,y) \right|^{2p_1} \right\}^{1/2} (nm)^{1/2}$$

$$\times \left\{ \sum_{x=1}^{n} \sum_{y=1}^{m} \left(\sum_{t=1}^{m} \left| \nabla_2 \nabla_1 u(x,t) \right|^{2p_2} \right) \left(\sum_{s=1}^{n} \left| \nabla_2 \nabla_1 u(s,y) \right|^{2p_3} \right) \right\}^{1/2}$$

$$\times \left\{ \sum_{x=1}^{n} \sum_{y=1}^{m} \left| \nabla_2 \nabla_1 u(x,y) \right|^{2p_4} \right\}^{1/2}$$

$$= K \prod_{i=1}^{4} \left\{ \sum_{x=1}^{n} \sum_{y=1}^{m} \left| \nabla_2 \nabla_1 u(x,y) \right|^{2p_i} \right\}^{1/2}.$$

This result completes the proof of inequality (3.8.39). The proof of inequality (3.8.40) follows by closely looking at the proof of inequality (3.8.39) and here we omit the details. □

For various other discrete Opial-type inequalities, see [4,270,284,360] and some of the references given therein.

3.9 Miscellaneous Inequalities

3.9.1 Agarwal and Pang [5]

Let $\alpha \geqslant 1$ be a given real number and let p be a nonnegative and continuous function on $[0, h]$. Further, let u be an absolutely continuous function on $[0, h]$, with $u(0) = u(h) = 0$. Then the following inequality holds

$$\int_0^h p(t) \left| u(t) \right|^\alpha \mathrm{d}t \leqslant \frac{1}{2} \left(\int_0^h \left(t(h-t) \right)^{(\alpha-1)/2} p(t) \, \mathrm{d}t \right) \int_0^h \left| u'(t) \right|^\alpha \mathrm{d}t.$$

For $p(t) = \text{constant}$, the above inequality reduces to

$$\int_0^h |u(t)|^\alpha \, dt \leqslant \frac{h^\alpha}{2} B\left(\frac{\alpha+1}{2}, \frac{\alpha+1}{2}\right) \int_0^h |u'(t)|^2 \, dt,$$

where B is the beta function.

3.9.2 Agarwal and Pang [5]

Let $r > 0$, $s \geqslant 0$, $\alpha > 0$ and $\beta > 0$ be real numbers with $\frac{1}{\alpha} + \frac{1}{\beta} = 1$ and $r\alpha \geqslant 1$, and let $p \in C^0[0, h]$ be nonnegative. Furthermore, let f and g be absolutely continuous on $[0, h]$ with $f(0) = g(0) = f(h) = g(h) = 0$. Then the following inequality holds

$$\int_0^h p(x)\big[|f(x)|^r |g'(x)|^s + |f'(x)|^s |g(x)|^r\big] \, dx$$
$$\leqslant I_1 \left\{ \frac{1}{\alpha} \int_0^h \big[|f'(x)|^{r\alpha} + |g'(x)|^{r\alpha}\big] \, dx + \frac{1}{\beta} \int_0^h \big[|f'(x)|^{s\beta} + |g'(x)|^{s\beta}\big] \, dx \right\},$$

where

$$I_1 = \left(\frac{1}{2} \int_0^h \big[x(h-x)\big]^{(r\alpha-1)/2} p^\alpha(x) \, dx \right)^{1/\alpha}.$$

3.9.3 Alzer [11]

Let $r > 0$, $s \geqslant 0$, $\alpha > 0$ and $\beta > 0$ be real numbers with $\frac{1}{\alpha} + \frac{1}{\beta} = 1$ and $r\alpha \geqslant 1$, and let $p \in C^0[0, h]$ be nonnegative. Furthermore, let f and g be absolutely continuous on $[0, h]$ with $f(0) = g(0) = f(h) = g(h) = 0$. Then the following inequality holds

$$\int_0^h p(x)\big[|f(x)|^r |g'(x)|^s + |f'(x)|^s |g(x)|^r\big] \, dx$$
$$\leqslant I_2 \left\{ \frac{1}{\alpha} \int_0^h \big[|f'(x)|^{r\alpha} + |g'(x)|^{r\alpha}\big] \, dx + \frac{1}{\beta} \int_0^h \big[|f'(x)|^{s\beta} + |g'(x)|^{s\beta}\big] \, dx \right\},$$

where

$$I_2 = \left(\int_0^h \big[x^{1-r\alpha} + (h-x)^{1-r\alpha}\big]^{-1} p^\alpha(x) \, dx \right)^{1/\alpha}.$$

3.9.4 Adams [2]

Let u be a continuously differentiable function on an open interval $(0, h)$ for fixed $h > 0$. If $\beta \geqslant 1$ and $\alpha > 0$, then

$$\int_0^h |u(t)|^\beta t^{\alpha-1}\,\mathrm{d}t \leqslant \frac{\alpha+1}{\alpha h}\int_0^h |u(t)|^\beta t^\alpha\,\mathrm{d}t + \frac{2\beta}{\alpha}\int_0^h |u(t)|^{\beta-1}|u'(t)|t^\alpha\,\mathrm{d}t.$$

If $p \geqslant 1$ and $\alpha > p-1$, then

$$\int_0^h |u(t)|^p t^{\alpha-p}\,\mathrm{d}t \leqslant 2^{p-1}\left(\frac{\alpha+2+p}{\alpha+1-p}\right)^p \int_0^h \left(\frac{|u(t)|^p}{h^p} + |u'(t)|^p\right)t^\alpha\,\mathrm{d}t.$$

3.9.5 Pachpatte [283]

Let f_r for $r = 1, \ldots, m$ be absolutely continuous functions on $[a, b]$ with $f_r(a) = f_r(b) = 0$. If $h_r(u)$ for $r = 1, \ldots, m$ be nonnegative convex and increasing functions on $[0, \infty)$ and $h_r(0) = 0$, then, for every $c \in (a, b)$, the following inequality holds

$$\int_a^b \left\{\prod_{r=1}^m h_r'(|f_r(x)|)|f_r'(x)|\right\}^{1/m}\mathrm{d}x$$
$$\leqslant \frac{1}{m}\sum_{i=1}^m \left[h_r\left(\int_a^c |f_r'(x)|\,\mathrm{d}x\right) + h_r\left(\int_c^b |f_r'(x)|\,\mathrm{d}x\right)\right].$$

Furthermore, let p_r for $r = 1, \ldots, m$ be positive functions on $[a, b]$ and $\int_a^b p_r(x)\,\mathrm{d}x < \infty$. If $\psi_r(u)$ for $r = 1, \ldots, m$ are nonnegative, convex and increasing functions on $[0, \infty)$ and $\psi_r(0) = 0$, then, for every $c \in (a, b)$, the following inequality holds

$$\int_a^b \left\{\prod_{r=1}^m h_r'(|f_r(x)|)|f_r'(x)|\right\}^{1/m}\mathrm{d}x$$
$$\leqslant \frac{1}{m}\sum_{r=1}^m \Bigg[h_r\Bigg(\left(\int_a^c p_r(x)\,\mathrm{d}x\right)\psi_r^{-1}$$
$$\times \left\{\int_a^c p_r(x)\psi_r\left(\frac{|f_r'(x)|}{p_r(x)}\right)\mathrm{d}x \Big/ \int_a^c p_r(x)\,\mathrm{d}x\right\}\Bigg)$$
$$+ h_r\left(\left(\int_c^b p_r(x)\,\mathrm{d}x\right)\psi_r^{-1}\left\{\int_c^b p_r(x)\psi_r\left(\frac{|f_r'(x)|}{p_r(x)}\right)\mathrm{d}x \Big/ \int_c^b p_r(x)\,\mathrm{d}x\right\}\right)\Bigg].$$

3.9.6 Pachpatte [256]

If f, g, h are absolutely continuous functions on $[a, b]$ and $f(a) = f(b) = 0$, $g(a) = g(b) = 0$, $h(a) = h(b) = 0$, then

$$\begin{aligned}\int_a^b &\big[\big|f(t)g(t)h(t)\big|\big(\big|f'(t)\big| + \big|g'(t)\big| + \big|h'(t)\big|\big)\\ &+ \big(\big|f(t)\big| + \big|g(t)\big| + \big|h(t)\big|\big)\\ &\times \big(\big|f(t)g(t)h'(t)\big| + \big|g(t)h(t)f'(t)\big| + \big|h(t)f(t)g'(t)\big|\big)\big]\,\mathrm{d}t\\ &\leqslant \left(\frac{b-a}{2}\right)^3 \int_a^b \big[\big|f'(t)\big|^4 + \big|g'(t)\big|^4 + \big|h'(t)\big|^4\big]\,\mathrm{d}t. \qquad (3.9.1)\end{aligned}$$

Equality holds in (3.9.1) if and only if

$$f(t) = g(t) = h(t) = \begin{cases} M(t-a), & a \leqslant t \leqslant \frac{a+b}{2},\\ M(b-t), & \frac{a+b}{2} \leqslant t \leqslant b,\end{cases}$$

where M is a constant.

3.9.7 Love [197]

Let $p > 0, q > 0$, $p + q = r \geqslant 1$, $0 \leqslant a < b \leqslant \infty$, $\gamma < r$, $w(x)$ be decreasing and positive in (a, b). Let m and n be integers, $m \geqslant n \geqslant 0$, F be complex-valued and has $(m-1)$th derivative locally absolutely continuous in $[a, b)$, and $F(a) = F'(a) = \cdots = F^{(m-1)}(a) = 0$. Then the following inequality holds

$$\begin{aligned}\int_a^b &\big|F(x)\big|^p \big|F^{(m-n)}(x)\big|^q x^{\gamma - mp - nq - 1} w(x)\,\mathrm{d}x\\ &\leqslant \frac{1}{\{(1-\gamma/r)_m\}^p\{(1-\gamma/r)_n\}^q} \int_a^b \big|F^{(m)}(x)\big|^r x^{\gamma-1} w(x)\,\mathrm{d}x,\end{aligned}$$

where $(k)_m = k(k+1)(k+2)\cdots(k+m-1)$ and $(k)_0 = 1$.

3.9.8 Pachpatte [245]

Let $u, v \in C^{(n-1)}[a, b]$ be such that $u^{(k)}(a) = v^{(k)}(a) = 0$ for $k = 0, 1, \ldots, n-1$, where $n \geqslant 1$, $u^{(n-1)}$, $v^{(n-1)}$ be absolutely continuous and $\int_a^b |u^{(n)}(t)|^2\,\mathrm{d}t < \infty$,

$\int_a^b |v^{(n)}(t)|^2 \, \mathrm{d}t < \infty$. Then the following inequality holds

$$\int_a^b \left[\left|u^{(k)}(t)v^{(n)}(t)\right| + \left|v^{(k)}(t)u^{(n)}(t)\right|\right] \mathrm{d}t$$

$$\leqslant \frac{1}{2} M_k (b-a)^{n-k} \int_a^b \left[\left|u^{(n)}(t)\right|^2 + \left|v^{(n)}(t)\right|^2\right] \mathrm{d}t, \tag{3.9.2}$$

where

$$M_k = \frac{1}{(n-k)!}\left(\frac{n-k}{2n-2k-1}\right)^{1/2} \tag{3.9.3}$$

for $0 \leqslant k \leqslant n-1$. Equality holds in (3.9.2) if and only if $k = 0$, $n = 1$ and $u^{(n)}(t) = v^{(n)}(t) = c$, where c is a constant.

3.9.9 Pachpatte [245]

Assume that the hypotheses of Section 3.9.8 hold. Then the following inequality holds

$$\int_a^b \left[\left|u^{(k)}(t)u^{(n)}(t)\right| + \left|v^{(k)}(t)v^{(n)}(t)\right| + \left|u^{(k)}(t)v^{(n)}(t)\right| + \left|v^{(k)}(t)u^{(n)}(t)\right|\right] \mathrm{d}t$$

$$\leqslant M_k (b-a)^{n-k} \int_a^b \left[\left|u^{(n)}(t)\right|^2 + \left|v^{(n)}(t)\right|^2\right] \mathrm{d}t, \tag{3.9.4}$$

where M_k is as given in (3.9.3) for $0 \leqslant k \leqslant n-1$. Equality holds in (3.9.4) if and only if $k = 0$, $n = 1$ and $u^{(n)}(t) = v^{(n)}(t) = c$, where c is a constant.

3.9.10 Pachpatte [245]

Let $u \in C^{(n-1)}[a,b]$ be such that $u^{(k)}(a) = 0$ for $k = 0, 1, \ldots, n-1$, where $n \geqslant 1$, $u^{(n-1)}$ be absolutely continuous and $\int_a^b |u^{(n)}(t)|^2 \, \mathrm{d}t < \infty$. Then the following inequality holds

$$\int_a^b \left|\prod_{k=0}^{n} u^{(k)}(t)\right| \mathrm{d}t \leqslant N_k (b-a)^{(n^2+1)/2} \left(\int_a^b \left|u^{(n)}(t)\right|^2 \mathrm{d}t\right)^{(n+1)/2}, \tag{3.9.5}$$

where

$$N_k = \frac{1}{(n^2+1)(n+1)\prod_{k=0}^{n-1}(n-k-1)!}\left(\frac{(n^2+1)(n+1)}{\prod_{k=0}^{n-1}(2n-2k-1)}\right)^{1/2}$$

for $0 \leqslant k \leqslant n-1$. Equality holds in (3.9.5) if and only if $n=1$ and $u^{(n)}(t)$ is a constant.

3.9.11 Pachpatte [323]

Let $u \in C^{(n-1)}[a,b]$ be such that $u^{(i)}(a)=0$ for $i=0,1,\ldots,n-1$ where $n \geqslant 1$. Let $u^{(n-1)}$ be absolutely continuous and $\int_a^b |u^{(n)}(t)|^2\,dt < \infty$. Then

$$\int_a^b \sum_{k=1}^{n} \left|u^{(n-k)}(t)\right|\left|u^{(n)}(t)\right| dt \leqslant M \int_a^b \left|u^{(n)}(t)\right|^2 dt,$$

where

$$M = \frac{\sqrt{n}}{2}\left[\sum_{k=1}^{n} \frac{(b-a)^{2k}}{(k-1)!k!(2k-1)}\right]^{1/2}.$$

3.9.12 Pachpatte [323]

Let u be as in Section 3.9.11 and $\int_a^b |u^{(n)}(t)|^4\,dt < \infty$. Then, for $0 \leqslant i \leqslant j \leqslant n-1$,

$$\int_a^b \left|u^{(i)}(t)u^{(j)}(t)\right| dt \leqslant \frac{1}{\sqrt{3}} C(n,i,j)(b-a)^{2n-i-j+1/2}\left(\int_a^b \left|u^{(n)}(t)\right|^4 dt\right)^{1/2},$$

where

$$C(n,i,j) = \frac{1}{(n-i-1)!(n-j-1)!\sqrt{2(n-i)-1}\sqrt{2(n-j)-1}}.$$

3.9.13 Pachpatte [324]

Let r_j, $j=1,\ldots,n-1$, u be real-valued, continuous functions defined on $I=[a,b]$ and r-derivatives of u exist, be continuous on I and such that $D_r^{(i)}u(a)=0$, $i=0,1,\ldots,n-1$, for $n \geqslant 1$ and $a \in I$. Then

$$\int_a^b \sum_{k=1}^{n-1} \left|(D_r^{(k)}u)(t)\right|\left|(D_r^{(n)}u)(t)\right| dt \leqslant \frac{M}{\sqrt{2}} \int_a^b \left|(D_r^{(n)}u)(t)\right|^2 dt,$$

where

$$M = \left[\int_a^b \left(\sum_{k=0}^{n-1} R_k(a,t)\right)^2 (t-a)\,dt\right]^{1/2}$$

and $R_k(a,t)$ is defined by (3.6.11).

3.9.14 Pachpatte [324]

Let r_j, u be as in Section 3.9.13. Then, for $0 \leqslant i \leqslant j \leqslant n-1$,

$$\int_a^b \left|\left(D_r^{(i)}u\right)(t)\right|\left|\left(D_r^{(j)}u\right)(t)\right|\left|\left(D_r^{(n)}u\right)(t)\right| \mathrm{d}t$$

$$\leqslant \begin{cases} N_1\left(\int_a^b \left|\left(D_r^{(n)}u\right)(t)\right|^2 \mathrm{d}t\right)^{3/2}, & \text{if } i=j=n-1, \\ N_2\left(\int_a^b \left|\left(D_r^{(n)}u\right)(t)\right|^2 \mathrm{d}t\right)^{3/2}, & \text{if } 0 \leqslant i \leqslant j < n-1, \end{cases}$$

where

$$N_1 = \frac{1}{3}(b-a)^{3/2}, \qquad N_2 = \frac{1}{\sqrt{3}}\left(\int_a^b \left[R_i(a,t)R_j(a,t)(t-a)\right]^2 \mathrm{d}t\right)^{1/2}$$

and $R_k(a,t)$ is defined by (3.6.11).

3.9.15 Pachpatte [316]

Let p, q be positive real numbers satisfying $p+q>1$ and let $n \geqslant 1$ be an integer. Let $u \in C^n[a,b]$ be a real-valued function such that $u^{(i)}(a)=0$ for $i=0,1,\dots,n-1$. Let $w(t)$ and $v(t)$ be positive and continuous functions defined on $[a,b]$. Then the following inequality holds

$$\int_a^b w(t)\sum_{k=1}^{n} \left|u^{(n-k)}(t)\right|^p \left|u^{(n)}(t)\right|^q \mathrm{d}t$$

$$\leqslant \left(\frac{q}{p+q}\right)^{q/(p+q)} M(p,q)\int_a^b v(t)\left|u^{(n)}(t)\right|^{p+q} \mathrm{d}t,$$

where

$$M(p,q) = \left[\int_a^b w^{(p+q)/p}(t)v^{-q/p}(t)\left[\left((k-1)!\right)^{-p}\right.\right.$$

$$\times\left\{\int_a^t v^{-1/(p+q-1)}(s)\right.$$

$$\left.\left.\left.\times (t-s)^{(k-1)(p+q)/(p+q-1)}\,\mathrm{d}s\right\}^{p(p+q-1)/(p+q)}\right]^{(p+q)/p}\right]^{p/(p+q)} \mathrm{d}t$$

is finite.

3.9.16 Cheung [60]

Let $p \geqslant 1$ and $q > 0$ be any real numbers and $n \geqslant 1$ be a fixed integer. Let $r_k \geqslant 0$, $k = 0, 1, \ldots, n-1$, be real numbers with $\sum_{k=0}^{n-1} r_k = 1$. Let $f \in C^n$ be a real-valued function and w be a positive continuous weight function on $[a, b]$. If $f^{(k)}(a) = 0$ for all $k = 0, 1, \ldots, n-1$, and if w is nonincreasing, then the following inequality holds

$$\int_a^b w(x)\left(\prod_{k=0}^{n-1} \left|f^{(k)}(x)\right|^{r_k}\right)^p \left|f^{(n)}(x)\right|^q \mathrm{d}x \\ \leqslant \sum_{k=0}^{n-1} M_k r_k (b-a)^{(n-k)p} \int_a^b w(x)\left|f^{(n)}(x)\right|^{p+q} \mathrm{d}x,$$

where

$$M_k = \alpha q^{\alpha q}\left[\frac{(n-k)(1-\alpha)}{n-k-\alpha}\right]^{(1-\alpha)p} \left[(n-k)!\right]^{-p}$$

and

$$\alpha = \frac{1}{p+q}.$$

3.9.17 Fink [117]

Let $0 \leqslant k < r < n$, $n \geqslant 2$, but fixed, and let $x(t) \in C^{(n-1)}[0, a]$, $x^{(i)}(0) = 0$, $0 \leqslant i \leqslant n-1$, $x^{(n-1)}(t)$ absolutely continuous and $\int_0^a |x^{(n)}(t)|^\mu \, \mathrm{d}t < \infty$, where $\mu \geqslant 1$ and $\frac{1}{\mu} + \frac{1}{\nu} = 1$. Then the following inequality holds

$$\int_0^a \left|x^{(k)}(t)x^{(r)}(t)\right| \mathrm{d}t \leqslant C(n, k, r, \mu) a^{2n-k-r+1-2/\mu}\left(\int_0^a \left|x^{(n)}(t)\right|^\mu \mathrm{d}t\right)^{2/\mu},$$

where

$$C(n, k, k+1, \mu) = \frac{1}{2((n-k-1)!)^2[(n-k-1)\nu+1]^{2/\nu}}$$

and

$$C(n, k, r, \mu) \\ \leqslant \frac{1}{(n-k-1)!(n-r)![(n-r)\nu+1]^{1/\nu}[(2n-k-r-1)\nu+2]^{1/\nu}}.$$

3.9.18 Pachpatte [359]

Let $n \geqslant 2$, $0 \leqslant m \leqslant n-2$ be integers. Let $r(t) > 0$ be of class C^1 on $I = [a, b]$, $a < b$ and $y(t)$ be of class C^n on I satisfying $y(a) = 0$, $y^{(i-1)}(a) = 0$, $i = 2, 3, \ldots, n$, and $\int_a^b |(r(t)y^{(n-1)}(t))'|^2 \, dt < \infty$. Then the following inequality holds

$$\int_a^b |y^{(m)}(t)||r(t)y^{(n-1)}(t)||(r(t)y^{(n-1)}(t))'| \, dt \leqslant M_1 \left(\int_a^b |(r(t)y^{(n-1)}(t))'|^2 \, dt \right)^{3/2},$$

where

$$M_1 = \frac{1}{\sqrt{3}(n-m-2)!} \left\{ \int_a^b (t-a) \left(\int_a^t (t-s)^{n-m-2}(s-a)^{1/2} \frac{1}{r(s)} \, ds \right)^2 dt \right\}^{1/2}$$

is finite.

3.9.19 Pachpatte [359]

Let $n \geqslant 2$, $0 \leqslant m \leqslant n-2$ be integers. Let $r(t) > 0$ be of class C^{n-1} on $I = [a, b]$, $a < b$ and $y(t)$ be of class C^{n-1} on I satisfying $y(a) = 0$, $(r(a)y'(a))^{(i-2)} = 0$, $i = 2, 3, \ldots, n$, and $\int_a^b |(r(t)y'(t))^{(n-1)}|^2 \, dt < \infty$. Then the following inequality holds

$$\int_a^b |y(t)||(r(t)y'(t))^{(m)}||(r(t)y'(t))^{(n-1)}| \, dt \leqslant M_2 \left(\int_a^b |(r(t)y'(t))^{(n-1)}|^2 \, dt \right)^{3/2},$$

where

$$M_2 = \frac{1}{\sqrt{3}\sqrt{2(n-2)+1}\sqrt{2(n-m-2)+1}(n-2)!(n-m-2)!} \times \left\{ \int_a^b (t-a)^{2(n-m-2)+1} \left(\int_a^s \frac{1}{r(s)} (s-a)^{[2(n-2)+1]/2} \, ds \right)^2 dt \right\}^{1/2}$$

is finite.

3.9.20 Pachpatte [359]

Let $n \geqslant 1$, $0 \leqslant k \leqslant n-1$, $0 \leqslant m \leqslant n-1$ be integers. Let $r(t) > 0$ be of class C^n on $I = [a, b]$, $a < b$, and $y(t)$ be of class C^{2n} on I satisfying $y^{(i-1)}(a) = 0$, $(r(a)y^{(n)}(a))^{(i-1)} = 0$, for $i = 1, 2, \ldots, n$, and $\int_a^b |(r(t)y^{(n)}(t))^{(n)}|^2 \, dt < \infty$. Then the following inequality holds

$$\int_a^b |y^{(k)}(t)| \left|\left(r(t)y^{(n)}(t)\right)^{(m)}\right| \left|\left(r(t)y^{(n)}(t)\right)^{(n)}\right| dt \leqslant M_3 \left(\int_a^b \left|\left(r(t)y^{(n)}(t)\right)^{(n)}\right|^2 dt\right)^{3/2},$$

where

$$M_3 = \frac{1}{\sqrt{3}\sqrt{2n-1}\sqrt{2n-2m-1}(n-1)!(n-k-1)!(n-m-1)!} \times \left\{\int_a^b (t-a)^{2n-2m-1} \left(\int_a^t (t-s)^{n-k-1} (s-a)^{(2n-1)/2} \frac{1}{r(s)} \, ds\right)^2 dt\right\}^{1/2}$$

is finite.

3.9.21 Pachpatte [260]

Suppose p, q are positive and continuous functions on $\Delta_1 = [a, X] \times [c, Y]$. Let f, $D_1 f$, $D_2 D_1 f$ be continuous functions on Δ_1 with $f(a, t) = D_1 f(s, c) = 0$ for $a \leqslant s \leqslant X$, $c \leqslant t \leqslant Y$. Then, if $m, n > 0$, $m + n > 1$, we have

$$\int_a^X \int_c^Y p|f|^m |D_2 D_1 f|^n \, dt \, ds \leqslant K_1(X, Y, m, n) \int_a^X \int_c^Y q |D_2 D_1 f|^{m+n} \, dt \, ds, \tag{3.9.6}$$

where

$$K_1(X, Y, m, n) = \left(\frac{n}{m+n}\right)^{n/(m+n)} \left\{\int_a^X \int_c^Y p^{(m+n)/m} q^{-n/m} \times \left(\int_a^s \int_c^t q^{-1/(m+n-1)} \, dv \, du\right)^{m+n-1} dt \, ds\right\}^{m/(m+n)} \tag{3.9.7}$$

is finite. In case $m < 0$, $m + n > 1$, inequality (3.9.6) holds with "$\leqslant$" replaced by "$\geqslant$".

3.9.22 Pachpatte [260]

Suppose p, q are positive and continuous functions on $\Delta_2 = [a, X] \times [Y, d]$. Let f, $D_1 f$, $D_2 D_1 f$ be continuous functions on Δ_2 with $f(a, t) = D_1 f(s, d) = 0$ for $a \leqslant s \leqslant X$, $Y \leqslant t \leqslant d$. Then, if $m, n > 0$, $m + n > 1$, we have

$$\begin{aligned}\int_a^X \int_Y^d & p|f|^m |D_2 D_1 f|^n \, \mathrm{d}t \, \mathrm{d}s \\ & \leqslant K_2(X, Y, m, n) \int_a^X \int_Y^d q|D_2 D_1 f|^{m+n} \, \mathrm{d}t \, \mathrm{d}s, \qquad (3.9.8)\end{aligned}$$

where

$$\begin{aligned}K_2&(X, Y, m, n) \\ &= \left(\frac{n}{m+n}\right)^{n/(m+n)} \left\{ \int_a^X \int_Y^d p^{(m+n)/m} q^{-n/m} \right. \\ &\qquad \left. \times \left(\int_a^s \int_t^d q^{-1/(m+n-1)} \, \mathrm{d}v \, \mathrm{d}u \right)^{m+n-1} \mathrm{d}t \, \mathrm{d}s \right\}^{m/(m+n)} \qquad (3.9.9)\end{aligned}$$

is finite. In case $m < 0$, $m + n > 1$, inequality (3.9.8) holds with "$\leqslant$" replaced by "$\geqslant$".

3.9.23 Pachpatte [260]

Suppose p, q are positive and continuous functions on $\Delta_3 = [X, b] \times [c, Y]$. Let f, $D_1 f$, $D_2 D_1 f$ be continuous functions on Δ_3 with $f(b, t) = D_1 f(s, c) = 0$ for $X \leqslant s \leqslant b$, $c \leqslant t \leqslant Y$. Then, if $m, n > 0$, $m + n > 1$, we have

$$\begin{aligned}\int_X^b \int_c^Y & p|f|^m |D_2 D_1 f|^n \, \mathrm{d}t \, \mathrm{d}s \\ & \leqslant K_3(X, Y, m, n) \int_X^b \int_c^Y q|D_2 D_1 f|^{m+n} \, \mathrm{d}t \, \mathrm{d}s, \qquad (3.9.10)\end{aligned}$$

where

$$K_3(X,Y,m,n)$$
$$= \left(\frac{n}{m+n}\right)^{n/(m+n)} \left\{ \int_X^b \int_c^Y p^{(m+n)/m} q^{-n/m} \times \left(\int_s^b \int_c^t q^{-1/(m+n-1)} \,\mathrm{d}v\,\mathrm{d}u \right)^{m+n-1} \mathrm{d}t\,\mathrm{d}s \right\}^{m/(m+n)} \tag{3.9.11}$$

is finite. In case $m < 0$, $m + n > 1$, inequality (3.9.10) holds with "$\leqslant$" replaced by "$\geqslant$".

3.9.24 Pachpatte [260]

Suppose p, q are positive and continuous functions on $\Delta_4 = [X, b] \times [Y, d]$. Let f, $D_1 f$, $D_2 D_1 f$ be continuous functions on Δ_4 with $f(b,t) = D_1 f(s,d) = 0$ for $X \leqslant s \leqslant b$, $Y \leqslant t \leqslant d$. Then, if $m, n > 0$, $m + n > 1$, we have

$$\int_X^b \int_Y^d p|f|^m |D_2 D_1 f|^n \,\mathrm{d}t\,\mathrm{d}s \leqslant K_4(X,Y,m,n) \int_X^b \int_Y^d q|D_2 D_1 f|^{m+n} \,\mathrm{d}t\,\mathrm{d}s, \tag{3.9.12}$$

where

$$K_4(X,Y,m,n)$$
$$= \left(\frac{n}{m+n}\right)^{n/(m+n)} \left\{ \int_X^b \int_Y^d p^{(m+n)/m} q^{-n/m} \times \left(\int_s^b \int_t^d q^{-1/(m+n-1)} \,\mathrm{d}v\,\mathrm{d}u \right)^{m+n-1} \mathrm{d}t\,\mathrm{d}s \right\}^{m/(m+n)} \tag{3.9.13}$$

is finite. In case $m < 0$, $m + n > 1$, inequality (3.9.12) holds with "$\leqslant$" replaced by "$\geqslant$".

3.9.25 Pachpatte [260]

Suppose p, q are positive and continuous functions on $\Delta = [a, b] \times [c, d]$. Let f, $D_1 f$, $D_2 D_1 f$ be continuous functions on Δ with $f(a,t) = f(b,t) =$

$D_1 f(s,c) = D_1 f(s,d) = 0$ for $a \leqslant s \leqslant b$, $c \leqslant t \leqslant d$. Suppose there exist $X_0 \in (a,b)$ and $Y_1, Y_2 \in (c,d)$ such that

$$\begin{aligned} K_1(X_0, Y_1, m, n) &= K_2(X_0, Y_1, m, n) \\ &= K_3(X_0, Y_2, m, n) \\ &= K_4(X_0, Y_2, m, n), \end{aligned} \tag{3.9.14}$$

where K_1, K_2, K_3 and K_4 are defined by (3.9.7), (3.9.9), (3.9.11) and (3.9.13), respectively. Then, if $m, n > 0$, $m + n > 1$, we have

$$\int_a^b \int_c^d p|f|^m |D_2 D_1 f|^n \,\mathrm{d}t\,\mathrm{d}s \leqslant K(m,n) \int_a^b \int_c^d q|D_2 D_1 f|^{m+n} \,\mathrm{d}t\,\mathrm{d}s, \tag{3.9.15}$$

where $K(m,n)$ denotes any common value of the four constants given in (3.9.14). In case $m < 0$, $m + n > 1$, inequality (3.9.15) holds with "$\leqslant$" replaced by "$\geqslant$".

3.9.26 Pachpatte [257]

Let $f, D_1 f, D_2 D_1 f$ be real-valued continuous functions on $\Delta = [a,b] \times [c,d]$ and $f(a,t) = f(b,t) = D_1 f(s,c) = D_1 f(s,d) = 0$ for $a \leqslant s \leqslant b$, $c \leqslant t \leqslant d$. Let $H(r)$ be a real-valued continuous function defined for all $r(s,t)$ of the form $\int_a^s \int_c^t |D_2 D_1 f(m,n)| \,\mathrm{d}n\,\mathrm{d}m$, $(s,t) \in \Delta_1 = [a,X] \times [c,Y]$, and similar integrals for (t,s) on $\Delta_2 = [a,X] \times [Y,d]$, $\Delta_3 = [X,b] \times [c,Y]$ and $\Delta_4 = [X,b] \times [Y,d]$, and $|H(r)| \leqslant H(|r|)$ for all r and that $H(r_1) \leqslant H(r_2)$ for $0 \leqslant r_1 \leqslant r_2$. Then the following inequality holds

$$\begin{aligned} &\int_a^b \int_c^d \left|H\big(f(s,t)\big) D_2 D_1 f(s,t)\right| \mathrm{d}t\,\mathrm{d}s \\ &\quad\leqslant F\left(\int_a^X \int_c^Y \left|D_2 D_1 f(s,t)\right| \mathrm{d}t\,\mathrm{d}s\right) + F\left(\int_a^X \int_Y^d \left|D_2 D_1 f(s,t)\right| \mathrm{d}t\,\mathrm{d}s\right) \\ &\qquad + F\left(\int_X^b \int_c^Y \left|D_2 D_1 f(s,t)\right| \mathrm{d}t\,\mathrm{d}s\right) + F\left(\int_X^b \int_Y^d \left|D_2 D_1 f(s,t)\right| \mathrm{d}t\,\mathrm{d}s\right), \end{aligned}$$

where $F(r) = \int_0^r H(\sigma)\,\mathrm{d}\sigma$, $r > 0$.

3.9.27 Pachpatte [266]

Let $f, D_1 f, D_2 D_1 f$ be real-valued continuous functions on $\Delta = [a,b] \times [c,d]$ and $f(a,t) = f(b,t) = D_1 f(s,c) = D_1 f(s,d) = 0$ for $a \leqslant s \leqslant b$, $c \leqslant t \leqslant d$.

If $H(r)$ is a convex increasing function on $[0,\infty)$ with $H(0)=0$, then, for $(X,Y)\in\Delta$, the following inequality holds

$$\begin{aligned}&\int_a^b\int_c^d H'(|f(s,t)|)|D_2D_1f(s,t)|\,\mathrm{d}t\,\mathrm{d}s\\&\quad\leqslant H\left(\int_a^X\int_c^Y|D_2D_1f(s,t)|\,\mathrm{d}t\,\mathrm{d}s\right)+H\left(\int_a^X\int_Y^d|D_2D_1f(s,t)|\,\mathrm{d}t\,\mathrm{d}s\right)\\&\qquad+H\left(\int_X^b\int_c^Y|D_2D_1f(s,t)|\,\mathrm{d}t\,\mathrm{d}s\right)+H\left(\int_X^b\int_Y^d|D_2D_1f(s,t)|\,\mathrm{d}t\,\mathrm{d}s\right).\end{aligned}$$

3.9.28 Pachpatte [271]

Let $f(s,t)$, $D_1f(s,t)$, $D_2D_1f(s,t)$ be real-valued continuous functions on $\Delta=[a,b]\times[c,d]$ and $f(a,t)=f(b,t)=D_1f(s,c)=D_1f(s,d)=0$ for $a\leqslant s\leqslant b$, $c\leqslant t\leqslant d$. Let $H(r)$ be a real-valued continuously differentiable function on $[0,\infty)$ with $H(0)=0$, $H'(r)\geqslant 0$ and $H'(r)$ nondecreasing on $[0,\infty]$. Then, for $(X,Y)\in\Delta$, the following inequality holds

$$\begin{aligned}&\int_a^b\int_c^d\{H(|f(s,t)|)\}^{n-1}H'(|f(s,t)|)|D_2D_1f(s,t)|\,\mathrm{d}t\,\mathrm{d}s\\&\quad\leqslant\frac{1}{n}\sum_{j=1}^{4}\{H(I_jD_2D_1f)\}^n,\end{aligned}$$

where $n\geqslant 2$ and

$$\begin{aligned}I_1D_2D_1f&=\int_a^X\int_c^Y|D_2D_1f(s,t)|\,\mathrm{d}t\,\mathrm{d}s,\\I_2D_2D_1f&=\int_a^X\int_Y^d|D_2D_1f(s,t)|\,\mathrm{d}t\,\mathrm{d}s,\\I_3D_2D_1f&=\int_X^b\int_c^Y|D_2D_1f(s,t)|\,\mathrm{d}t\,\mathrm{d}s,\\I_4D_2D_1f&=\int_X^b\int_Y^d|D_2D_1f(s,t)|\,\mathrm{d}t\,\mathrm{d}s.\end{aligned}$$

3.9.29 Pachpatte [294]

Let $u(x, y)$, $D_1u(x, y)$, $D_2u(x, y)$, $D_2D_1u(x, y)$ be real-valued continuous functions defined on $\Delta = [a, b] \times [c, d]$ with $u(a, y) = u(b, y) = D_1u(x, c) = D_1u(x, d) = 0$ for $a \leqslant x \leqslant b$, $c \leqslant y \leqslant d$.

(i) Let $1 \leqslant p_i < \infty$ for $i = 1, 2, 3$ be constants. Then

$$\int_a^b \int_c^d |u(x, y)|^{p_1} |D_1u(x, y)|^{p_2} |D_2u(x, y)|^{p_3} \, \mathrm{d}y \, \mathrm{d}x \leqslant M_1 \prod_{i=1}^{3} \left(\int_a^b \int_c^d |D_2D_1u(x, y)|^{p_i} \, \mathrm{d}y \, \mathrm{d}x \right),$$

where

$$M_1 = \left(\frac{1}{2}\right)^{2p_1+p_2+p_3} (b-a)^{p_1+p_3-2}(d-c)^{p_1+p_2-2}.$$

(ii) Let $1 \leqslant p_i < \infty$ for $i = 1, 2, 3, 4$ be constants. Then

$$\int_a^b \int_c^d |u(x, y)|^{p_1} |D_1u(x, y)|^{p_2} |D_2u(x, y)|^{p_3} |D_2D_1u(x, y)|^{p_4} \, \mathrm{d}y \, \mathrm{d}x \leqslant M_2 \prod_{i=1}^{4} \left(\int_a^b \int_c^d |D_2D_1u(x, y)|^{2p_i} \, \mathrm{d}y \, \mathrm{d}x \right)^{1/2},$$

where

$$M_2 = \left(\frac{1}{2}\right)^{2p_1+p_2+p_3} (b-a)^{p_1+p_3-1}(d-c)^{p_1+p_2-1}.$$

3.9.30 Pachpatte [244]

Let $u_i(x, y)$ (for $i = 1, 2$) and their partial derivatives $D_1u_i(x, y)$, $D_2u_i(x, y)$ be absolutely continuous real-valued functions defined on $\Delta = [a, b] \times [c, d]$ and $u_i(a, y) = u_i(b, y) = 0$, $u_i(x, c) = u_i(x, d) = 0$. Then the following inequality holds

$$\int_a^b \int_c^d \big[|u_1(x, y)D_2D_1u_2(x, y)| + |u_2(x, y)D_2D_1u_1(x, y)| + |D_1u_1(x, y)D_2u_2(x, y)| + |D_1u_2(x, y)D_2u_1(x, y)| \big] \, \mathrm{d}y \, \mathrm{d}x$$

$$\leqslant \frac{(b-a)(d-c)}{8} \int_a^b \int_c^d \left[\left| D_2 D_1 u_1(x,y) \right|^2 + \left| D_2 D_1 u_2(x,y) \right|^2 \right] \mathrm{d}y \, \mathrm{d}x.$$

3.9.31 Pachpatte [313]

Let $m, n \geqslant 1$ be integers and $\Delta = [0,a] \times [0,b]$, $a > 0$, $b > 0$. Let $u(x,y) \in F(\Delta)$, where $F(\Delta)$ denote the class of continuous functions $u : \Delta \to \mathbb{R}$ for which $D_2^m D_1^n u(x,y)$ exists and continuous on Δ and such that $D_2^j u(x,0) = 0$, $0 \leqslant j \leqslant m-1$, $D_1^i u(0,y) = 0$, $0 \leqslant i \leqslant n-1$, where

$$D_1^n u(x,y) = \frac{\partial^n u(x,y)}{\partial x^n}, \qquad D_2^m u(x,y) = \frac{\partial^m u(x,y)}{\partial x^m}$$

and

$$D_2^m D_1^n u(x,y) = \frac{\partial^{n+m} u(x,y)}{\partial x^m \, \partial y^n}.$$

Then the following inequalities hold

$$\int_0^a \int_0^b \left| u(x,y) \right| \left| D_2^m D_1^n u(x,y) \right| \mathrm{d}y \, \mathrm{d}x \leqslant \sqrt{L} \int_0^a \int_0^b \left| D_2^m D_1^n u(x,y) \right|^2 \mathrm{d}y \, \mathrm{d}x,$$

$$\int_0^a \int_0^b \left| D_1^n u(x,y) \right| \left| D_2^m D_1^n u(x,y) \right| \mathrm{d}y \, \mathrm{d}x \leqslant \sqrt{M} \int_0^a \int_0^b \left| D_2^m D_1^n u(x,y) \right|^2 \mathrm{d}y \, \mathrm{d}x,$$

$$\int_0^a \int_0^b \left| D_2^m u(x,y) \right| \left| D_2^m D_1^n u(x,y) \right| \mathrm{d}y \, \mathrm{d}x \leqslant \sqrt{N} \int_0^a \int_0^b \left| D_2^m D_1^n u(x,y) \right|^2 \mathrm{d}y \, \mathrm{d}x,$$

where

$$L = \frac{a^{2n} b^{2m}}{[(n-1)!(m-1)!]^2 2n(2n-1) 2m(2m-1)},$$

$$M = \frac{b^{2m}}{[(m-1)!]^2 2m(2m-1)}, \qquad N = \frac{a^{2n}}{[(n-1)!]^2 2n(2n-1)}.$$

3.9.32 Pachpatte [280]

Let $\{u_i\}$ and $\{v_i\}$ for $i \in \mathbb{N}$ be sequences of real numbers with $u_0 = v_0 = 0$. Then the following inequality holds

$$\sum_{i=1}^{n}[|u_i||\nabla v_i| + |v_i||\nabla u_i|] \leqslant \frac{n+1}{2}\sum_{i=1}^{n}[|\nabla u_i|^2 + |\nabla v_i|^2]$$

for all $n \in \mathbb{N}$, where $\mathbb{N}$ and ∇ are as defined in Section 3.8.

3.9.33 Pachpatte [234]

Let $\{p_k\}$, $k = 1, \ldots, n$, be a sequence of nonnegative real numbers. If $\{f_k\}$, $\{g_k\}$, $k \in \mathbb{N}_{n+1}$, are the sequences of real numbers such that $f_1 = f_{n+1} = 0$, $g_1 = g_{n+1} = 0$, then

$$\sum_{k=1}^{n} p_k|f_k||g_k| \leqslant \frac{n}{8}\left(\sum_{k=1}^{n} p_k\right)\left(\sum_{k=1}^{n}[|\Delta f_k|^2 + |\Delta g_k|^2]\right)$$

for all $n \in \mathbb{N}$. Further, let $q > 1$, $r > 1$ be constants, then

$$\sum_{k=1}^{n} p_k|f_k|^q|g_k|^r \leqslant \left(\frac{1}{2}\right)^{q+r+1} n^{q+r-1}\left(\sum_{k=1}^{n} p_k\right)\left(\sum_{k=1}^{n}[|\Delta f_k|^{2q} + |\Delta g_k|^{2r}]\right)$$

for all $n \in \mathbb{N}$, where $\mathbb{N}_{n+1}$ and $\mathbb{N}$ are as defined in Section 3.8.

3.9.34 Pachpatte [234]

Let $\{p_k\}$, $\{f_k\}$, $\{g_k\}$ be the sequences as defined in Section 3.9.33. Then the following inequality holds

$$\sum_{k=1}^{n} p_k[|f_k||\Delta g_k| + |g_k||\Delta f_k|] \leqslant \left(\frac{n}{4}\sum_{k=1}^{n} p_k\right)^{1/2}\left(\sum_{k=1}^{n}[|\Delta f_k|^2 + |\Delta g_k|^2]\right).$$

3.10 Notes

Inequality (3.2.1) was first proved by the Polish mathematician Z. Opial [231] in 1960. An interesting feature of Opial's result is that it yields the best possible constant. The original proof of Opial's inequality can also be found in [4].

Theorem 3.2.1 covers a weaker form of Opial's inequality due to Olech [230]. Moreover, Olech's proof is simpler than that of Opial. Theorem 3.2.2 which deals with Wirtinger- and Opial-type inequalities is due to Traple [419]. Theorems 3.2.3 and 3.2.4 are due to Pachpatte [348] which provide new estimates on Opial-type inequalities. The inequalities in Theorems 3.2.5–3.5.7 are due to Pachpatte [304] which claim their origin to the well-known Weyl inequality [141, p. 165] and in the special cases contains the inequalities of Weyl, Opial and Hardy type.

Theorems 3.3.1–3.3.4 deal with Wirtinger- and Opial-type inequalities and are taken from Pachpatte [238,241]. Theorems 3.3.5 and 3.3.6 are due to Calvert [51], and Theorem 3.3.7 is taken from Pachpatte [283]. The inequalities in Theorems 3.4.1–3.4.8 are taken from Pachpatte [239,303] which are motivated by the various generalizations and extensions of Opial's inequality. The inequalities in Theorems 3.5.1 and 3.5.2 are due to Godunov and Levin [130]. Theorem 3.5.3 is a generalization of Theorem 3.5.1 given by Rozanova in [399]. Theorems 3.5.4–3.5.7 deal with generalized Opial-type integral inequalities established by Pachpatte in [346]. The results given in Section 3.6 cover basic Opial-type inequalities involving functions and their higher-order derivatives established by Pachpatte. Theorems 3.6.1 and 3.6.2 are taken from [239], Theorems 3.6.3–3.6.5 are taken from [296], Theorem 3.6.6 is taken from [312] and Theorems 3.6.7–3.6.9 are taken from [317].

In 1982, Yang [429] obtained an analogue of Opial's inequality involving functions of two independent variables. Theorem 3.7.1 is due to Yang [429]. Theorems 3.7.2–3.7.5 are taken from Pachpatte [233,267]. Theorems 3.7.6 and 3.7.7 cover Opial-type inequalities in several independent variables and established by Pachpatte [261]. Theorem 3.7.8 is about another version of Opial-type inequality involving functions of many independent variables and is due to Pachpatte [284]. Discrete analogues of Opial's inequality and its generalizations are established by Wong [426], Lee [183] and others, see [4]. Inequality (3.8.1) is due to Wong [426] and is a discrete variant of Opial's inequality given by Hua in [158]. All the results given in Section 3.8 are due to Pachpatte [235,262,280,287,318,347] which claim their origin to the discrete analogue of Opial's inequality, see [4]. Section 3.9.9 is about some useful miscellaneous inequalities related to Opial's inequality investigated by various investigators.

Chapter 4

Poincaré- and Sobolev-Type Inequalities

4.1 Introduction

In the development of the theory of partial differential equations and in establishing the foundations of the finite element analysis, the fundamental role played by certain inequalities and variational principles involving functions and their partial derivatives is well known. In particular, the integral inequalities originally due to Poincaré and Sobolev and their various generalizations and variants have been extensively used in the study of problems in the theory of partial differential equations and finite element analysis. Because of the dominance of such inequalities in the qualitative analysis of partial differential equations and in finite element analysis, numerous studies have been made of various types of new inequalities related to Poincaré- and Sobolev-type inequalities. These investigations have achieved a diversity of desired goals. Over the years a number of papers have appeared in the literature which deals with the far-reaching generalizations, extensions and variants of Poincaré and Sobolev inequalities and their various applications. This chapter deals with a number of new inequalities recently discovered in the literature which claim their origin to the inequalities of Poincaré and Sobolev.

Let $\mathbb{R}$ be the set of real numbers and B be a bounded domain in $\mathbb{R}^n$, the n-dimensional Euclidean space, defined by $B = \prod_{i=1}^{n}[a_i, b_i]$. For $x_i \in \mathbb{R}$, $x = (x_1, \dots, x_n)$ is a variable point in B and $\mathrm{d}x = \mathrm{d}x_1 \cdots \mathrm{d}x_n$. For any continuous real-valued function $u(x)$ defined on B, we denote by $\int_B u(x)\,\mathrm{d}x$ the n-fold integral $\int_{a_n}^{b_n} \cdots \int_{a_1}^{b_1} u(x_1, \dots, x_n)\,\mathrm{d}x_1 \cdots \mathrm{d}x_n$. The notation $\int_{a_i}^{b_i} u(x_1, \dots, t_i, \dots, x_n)\,\mathrm{d}t_i$ for $i = 1, \dots, n$ we mean, for $i = 1$, it is $\int_{a_1}^{b_1} u(t_1, x_2, \dots, x_n)\,\mathrm{d}t_1$ and so on, and for $i = n$, it is $\int_{a_n}^{b_n} u(x_1, \dots, x_{n-1}, t_n)\,\mathrm{d}t_n$. For any continuous real-valued function $u(x)$ defined on $\mathbb{R}^n$, we denote by $\int_i u(x_1, \dots, t_i, \dots, x_n)\,\mathrm{d}t_i$ the integral $\int_{-\infty}^{\infty} u(x_1, \dots, t_i, \dots, x_n)\,\mathrm{d}t_i$, $i = 1, \dots, n$, taken along the whole line through $x = (x_1, \ldots, x_i, \ldots, x_n)$ parallel to the x_i-axis, and denote by $\int_{\mathbb{R}^n} u(x)\,\mathrm{d}x$ the n-fold integral $\int_{-\infty}^{\infty} \cdots \int_{-\infty}^{\infty} u(x_1, \ldots, x_n)\,\mathrm{d}x_1 \cdots \mathrm{d}x_n$. For any function $u(x)$ defined on B or $\mathbb{R}^n$, we define $|\operatorname{grad} u(x)| = (\sum_{i=1}^n |\frac{\partial u(x)}{\partial x_i}|^2)^{1/2}$. We say that a function is of compact support in S if it is nonzero only on a bounded subdomain S' of the domain S, where S' lies at a positive distance ∂S, the boundary of S. We assume without further mention that all the integrals exist on the respective domains of their definitions.

4.2 Inequalities of Poincaré, Sobolev and Others

There exists a vast literature on the various generalizations, extensions and variants of Poincaré's inequality (10), see Introduction. We start with the following useful version of Poincaré's inequality given in Friedman [120, p. 284].

THEOREM 4.2.1. *Let* $Q = \{x = (x_1, \ldots, x_n) \in \mathbb{R}^n\colon\ 0 \leqslant x_i \leqslant \sigma,\ i = 1, \ldots, n\}$ *and let* u *be a real-valued function belonging to* $C^1(Q)$. *Then*

$$\int_Q u^2(x)\,\mathrm{d}x \leqslant \frac{1}{\sigma^n}\left(\int_Q u(x)\,\mathrm{d}x\right)^2 + \frac{n}{2}\sigma^2 \int_Q \left|\operatorname{grad} u(x)\right|^2 \mathrm{d}x. \tag{4.2.1}$$

PROOF. For any $x = (x_1, \ldots, x_n)$, $y = (y_1, \ldots, y_n) \in Q$, the following identity holds

$$u(x) - u(y) = \sum_{i=1}^{n} \int_{y_i}^{x_i} \frac{\partial}{\partial t_i} u(y_1, \ldots, y_{i-1}, t_i, x_{i+1}, \ldots, x_n)\,\mathrm{d}t_i. \tag{4.2.2}$$

Taking square on both sides of (4.2.2) and using the elementary inequality $(\sum_{i=1}^n a_i)^2 \leqslant n \sum_{i=1}^n a_i^2$, where a_i are reals and Schwarz inequality, we have

$$\begin{aligned} &u^2(x) + u^2(y) - 2u(x)u(y) \\ &\quad \leqslant n\sigma \sum_{i=1}^{n} \int_0^{\sigma} \left(\frac{\partial}{\partial t_i} u(y_1, \ldots, y_{i-1}, t_i, x_{i+1}, \ldots, x_n)\right)^2 \mathrm{d}t_i. \end{aligned} \tag{4.2.3}$$

Integrating (4.2.3) with respect to $x_1, \ldots, x_n, y_1, \ldots, y_n$, we get

$$2\sigma^n \int_Q u^2(x)\,\mathrm{d}x - 2\left(\int_Q u(x)\,\mathrm{d}x\right)^2 \leqslant n\sigma^{n+2} \sum_{i=1}^{n} \int_Q \left(\frac{\partial}{\partial x_i} u(x)\right)^2 \mathrm{d}x,$$

from which (4.2.1) follows. □

In [247] Pachpatte has given the following variant of Theorem 4.2.1.

THEOREM 4.2.2. *Let Q be as defined in Theorem 4.2.1 and f, g be real-valued functions belonging to $C^1(Q)$. Then*

$$\begin{aligned}\int_Q f(x)g(x)\,\mathrm{d}x &\leqslant \frac{1}{\sigma^n}\left(\int_Q f(x)\,\mathrm{d}x\right)\left(\int_Q g(x)\,\mathrm{d}x\right)\\ &\quad+\frac{n}{4}\sigma^2\int_Q\left[|\operatorname{grad} f(x)|^2+|\operatorname{grad} g(x)|^2\right]\mathrm{d}x. \qquad (4.2.4)\end{aligned}$$

PROOF. For any $x, y \in Q$ and $h \in C^1(Q)$, the following identity holds

$$h(x)-h(y)=\sum_{i=1}^n\int_{y_i}^{x_i}\frac{\partial}{\partial t_i}h(y_1,\ldots,y_{i-1},t_i,x_{i+1},\ldots,x_n)\,\mathrm{d}t_i. \qquad (4.2.5)$$

Writing (4.2.5) for the functions f and g, and then by multiplying the results and using the elementary inequalities $ab \leqslant \frac{1}{2}(a^2+b^2)$, $(\sum_{i=1}^n a_i)^2 \leqslant n\sum_{i=1}^n a_i^2$ (a, b, a_i are reals) and Schwarz inequality, we obtain

$$\begin{aligned}&f(x)g(x)+f(y)g(y)-f(x)g(y)-f(y)g(x)\\ &\leqslant \frac{n}{2}\sigma\sum_{i=1}^n\int_0^\sigma\left[\left\{\frac{\partial}{\partial t_i}f(y_1,\ldots,y_{i-1},t_i,x_{i+1},\ldots,x_n)\right\}^2\right.\\ &\qquad\left.+\left\{\frac{\partial}{\partial t_i}g(y_1,\ldots,y_{i-1},t_i,x_{i+1},\ldots,x_n)\right\}^2\right]\mathrm{d}t_i. \qquad (4.2.6)\end{aligned}$$

Integrating both sides of (4.2.6) with respect to $x_1,\ldots,x_n, y_1,\ldots,y_n$, we get

$$\begin{aligned}&2\sigma^n\int_Q f(x)g(x)\,\mathrm{d}x-2\left(\int_Q f(x)\,\mathrm{d}x\right)\left(\int_Q g(x)\,\mathrm{d}x\right)\\ &\quad\leqslant\frac{n}{2}\sigma^{n+2}\int_Q\left[|\operatorname{grad} f(x)|^2+|\operatorname{grad} g(x)|^2\right]\mathrm{d}x. \qquad (4.2.7)\end{aligned}$$

The desired inequality (4.2.4) follows from inequality (4.2.7). □

REMARK 4.2.1. We note that in the special case when $g(x) = f(x)$, the inequality established in Theorem 4.2.2 reduces to the inequality given in Theorem 4.2.1.

In [236] Pachpatte has established the following Poincaré-type inequality.

THEOREM 4.2.3. *Let Q be as defined in Theorem 4.2.1 and f, g be real-valued functions belonging to $C^1(Q)$ which vanish on the boundary ∂Q of Q. Then*

$$\int_Q |f(x)||g(x)|\,\mathrm{d}x \leqslant \frac{\sigma^2}{8n}\int_Q \left[|\mathrm{grad}\, f(x)|^2 + |\mathrm{grad}\, g(x)|^2\right]\mathrm{d}x. \tag{4.2.8}$$

PROOF. If $x \in Q$, then we have the following identities

$$nf(x) = \sum_{i=1}^{n}\int_0^{x_i}\frac{\partial}{\partial t_i} f(x_1,\ldots,t_i,\ldots,x_n)\,\mathrm{d}t_i, \tag{4.2.9}$$

$$nf(x) = -\sum_{i=1}^{n}\int_{x_i}^{\sigma}\frac{\partial}{\partial t_i} f(x_1,\ldots,t_i,\ldots,x_n)\,\mathrm{d}t_i. \tag{4.2.10}$$

From (4.2.9) and (4.2.10), we obtain

$$2n|f(x)| \leqslant \sum_{i=1}^{n}\int_0^{\sigma}\left|\frac{\partial}{\partial t_i} f(x_1,\ldots,t_i,\ldots,x_n)\right|\mathrm{d}t_i. \tag{4.2.11}$$

Similarly, we obtain

$$2n|g(x)| \leqslant \sum_{i=1}^{n}\int_0^{\sigma}\left|\frac{\partial}{\partial t_i} g(x_1,\ldots,t_i,\ldots,x_n)\right|\mathrm{d}t_i. \tag{4.2.12}$$

From (4.2.11), (4.2.12) and using the elementary inequalities $ab \leqslant \frac{1}{2}(a^2+b^2)$, $(\sum_{i=1}^{n} a_i)^2 \leqslant n\sum_{i=1}^{n} a_i^2$ (for a, b, a_i reals) and Schwarz inequality, we obtain

$$\begin{aligned}
&|f(x)||g(x)| \\
&\quad\leqslant \frac{1}{8n}\sigma\left[\sum_{i=1}^{n}\int_0^{\sigma}\left|\frac{\partial}{\partial t_i} f(x_1,\ldots,t_i,\ldots,x_n)\right|^2\mathrm{d}t_i\right. \\
&\qquad\left.+\sum_{i=1}^{n}\int_0^{\sigma}\left|\frac{\partial}{\partial t_i} g(x_1,\ldots,t_i,\ldots,x_n)\right|^2\mathrm{d}t_i\right].
\end{aligned} \tag{4.2.13}$$

Integrating both sides of (4.2.13) with respect to $x_1,\ldots,x_n$ we get

$$\int_Q |f(x)||g(x)|\,\mathrm{d}x \leqslant \frac{\sigma^2}{8n}\int_Q \left[|\mathrm{grad}\, f(x)|^2 + |\mathrm{grad}\, g(x)|^2\right]\mathrm{d}x.$$

The proof is complete. □

REMARK 4.2.2. In the special case when $g(x) = f(x)$, the inequality established in Theorem 4.2.3 reduces to the following Poincaré-type integral inequality

$$\int_Q |f(x)|^2 \,\mathrm{d}x \leqslant \frac{\sigma^2}{4n} \int_Q |\operatorname{grad} f(x)|^2 \,\mathrm{d}x. \tag{4.2.14}$$

One of the many mathematical discoveries of S.L. Sobolev is the following integral inequality (see [157, p. 101])

$$\int_{-\infty}^{\infty}\int_{-\infty}^{\infty} u^4 \,\mathrm{d}x\,\mathrm{d}y \leqslant \frac{\alpha}{2}\left(\int_{-\infty}^{\infty}\int_{-\infty}^{\infty} u^2 \,\mathrm{d}x\,\mathrm{d}y\right)\left(\int_{-\infty}^{\infty}\int_{-\infty}^{\infty} |\operatorname{grad} u|^2 \,\mathrm{d}x\,\mathrm{d}y\right), \tag{4.2.15}$$

where $u(x, y)$ is any smooth function of compact support in two-dimensional Euclidean space and α is a dimensionless constant.

Inequality (4.2.15) is known as Sobolev's inequality, although the same name is attached to the above inequality in n-dimensional Euclidean space. Inequalities of the form (4.2.15) or its variants have been applied with considerable success to the study of many problems in the theory of partial differential equations and in establishing the foundations of the finite element analysis. There is a vast literature which deals with various generalizations, extensions and variants of inequality (4.2.15).

In 1964, Payne [362] has given the following version of inequality (4.2.15).

THEOREM 4.2.4. *Let $u(x, y)$ be any smooth function of compact support in two-dimensional Euclidean space E_2. Then*

$$\int_{-\infty}^{\infty}\int_{-\infty}^{\infty} u^4 \,\mathrm{d}x\,\mathrm{d}y \leqslant \frac{1}{2}\left(\int_{-\infty}^{\infty}\int_{-\infty}^{\infty} u^2 \,\mathrm{d}x\,\mathrm{d}y\right)\left(\int_{-\infty}^{\infty}\int_{-\infty}^{\infty} |\operatorname{grad} u|^2 \,\mathrm{d}x\,\mathrm{d}y\right). \tag{4.2.16}$$

PROOF. From the hypotheses, we have the following identities

$$u^2(x, y) = 2\int_{-\infty}^{x} u(s, y)\frac{\partial}{\partial s}u(s, y)\,\mathrm{d}s = -2\int_{x}^{\infty} u(s, y)\frac{\partial}{\partial s}u(s, y)\,\mathrm{d}s, \tag{4.2.17}$$

$$u^2(x, y) = 2\int_{-\infty}^{y} u(x, t)\frac{\partial}{\partial t}u(x, t)\,\mathrm{d}t = -2\int_{y}^{\infty} u(x, t)\frac{\partial}{\partial t}u(x, t)\,\mathrm{d}t. \tag{4.2.18}$$

From (4.2.17) and (4.2.18), we obtain

$$u^2(x, y) \leqslant \int_{-\infty}^{\infty} |u(s, y)| \left| \frac{\partial}{\partial s} u(s, y) \right| \mathrm{d}s \tag{4.2.19}$$

and

$$u^2(x, y) \leqslant \int_{-\infty}^{\infty} |u(x, t)| \left| \frac{\partial}{\partial t} u(x, t) \right| \mathrm{d}t. \tag{4.2.20}$$

From (4.2.19) and (4.2.20), we observe that

$$\begin{aligned} &\int_{-\infty}^{\infty}\int_{-\infty}^{\infty} u^4(x, y)\,\mathrm{d}x\,\mathrm{d}y \\ &\quad \leqslant \int_{-\infty}^{\infty}\int_{-\infty}^{\infty} \left\{ \left(\int_{-\infty}^{\infty} |u(s, y)| \left| \frac{\partial}{\partial s} u(s, y) \right| \mathrm{d}s \right) \right. \\ &\qquad \left. \times \left(\int_{-\infty}^{\infty} |u(x, t)| \left| \frac{\partial}{\partial t} u(x, t) \right| \mathrm{d}t \right) \right\} \mathrm{d}x\,\mathrm{d}y. \end{aligned} \tag{4.2.21}$$

By using the Schwarz inequality on the right-hand side of (4.2.21), we get

$$\begin{aligned} &\int_{-\infty}^{\infty}\int_{-\infty}^{\infty} u^4(x, y)\,\mathrm{d}x\,\mathrm{d}y \\ &\quad \leqslant \left\{ \int_{-\infty}^{\infty}\int_{-\infty}^{\infty} u^2(x, y)\,\mathrm{d}x\,\mathrm{d}y \right\} \\ &\qquad \times \left\{ \left(\int_{-\infty}^{\infty}\int_{-\infty}^{\infty} \left(\frac{\partial u}{\partial x} \right)^2 \mathrm{d}x\,\mathrm{d}y \right) \left(\int_{-\infty}^{\infty}\int_{-\infty}^{\infty} \left(\frac{\partial u}{\partial y} \right)^2 \mathrm{d}x\,\mathrm{d}y \right) \right\}^{1/2}. \end{aligned} \tag{4.2.22}$$

Now an application of the arithmetic mean and geometric mean inequality on the last term on the right-hand side of (4.2.22) leads to the desired inequality in (4.2.16). □

In 1963, Serrin [405] proved the following useful multidimensional integral inequality.

THEOREM 4.2.5. *Let E be a bounded domain in $\mathbb{R}^n$, $n \geqslant 2$, and u be a real-valued function such that $u \in C^1(E)$ and $u = 0$ on ∂E, the boundary of E, then*

$$\left(\int_E |u|^{n/(n-1)}(x)\,\mathrm{d}x \right)^{(n-1)/n} \leqslant \left(\frac{1}{4n} \right)^{1/2} \int_E \left| \operatorname{grad} u(x) \right| \mathrm{d}x. \tag{4.2.23}$$

PROOF. From the hypotheses, we have the following identities

$$u(x) = \int_{-\infty}^{x_1} \frac{\partial}{\partial t_1} u(t_1, x_2, \ldots, x_n)\, \mathrm{d}t_1, \tag{4.2.24}$$

$$u(x) = -\int_{x_1}^{\infty} \frac{\partial}{\partial t_1} u(t_1, x_2, \ldots, x_n)\, \mathrm{d}t_1. \tag{4.2.25}$$

From (4.2.24) and (4.2.25), we obtain

$$|u(x)| \leqslant \frac{1}{2} \int_1 \left| \frac{\partial}{\partial t_1} u(t_1, x_2, \ldots, x_n) \right| \mathrm{d}t_1. \tag{4.2.26}$$

Similarly, we obtain

$$|u(x)| \leqslant \frac{1}{2} \int_i \left| \frac{\partial}{\partial t_i} u(x_1, \ldots, t_i, \ldots, x_n) \right| \mathrm{d}t_i \tag{4.2.27}$$

for $i = 2, 3, \ldots, n$. From (4.2.26) and (4.2.27), we observe that

$$\begin{aligned} |u(x)|^{n/(n-1)} \leqslant \left(\frac{1}{2}\right)^{n/(n-1)} & \left\{ \int_1 \left| \frac{\partial}{\partial t_1} u(t_1, x_2, \ldots, x_n) \right| \mathrm{d}t_1 \right\}^{1/(n-1)} \\ & \times \cdots \times \left\{ \int_n \left| \frac{\partial}{\partial t_n} u(x_1, \ldots, x_{n-1}, t_n) \right| \mathrm{d}t_n \right\}^{1/(n-1)}. \end{aligned} \tag{4.2.28}$$

We integrate both sides of (4.2.28) with respect to x_1 and use on the right-hand side the general version of Hölder's inequality (see [179, p. 40])

$$\int_i |f_1 \cdots f_k|\, \mathrm{d}k \leqslant \left\{ \int_i |f_1|^k\, \mathrm{d}t \right\}^{1/k} \cdots \left\{ \int_i |f_k|^k\, \mathrm{d}t \right\}^{1/k}, \tag{4.2.29}$$

where $k = n - 1$. We then integrate the resulting inequality with respect to x_2 and use inequality (4.2.29) on the right-hand side. We repeat this procedure, integrating with respect to $x_3, \ldots, x_n$, and obtain (see [121, Chapter 1, Theorem 9.3])

$$\begin{aligned} & \int_E |u(x)|^{n/(n-1)}\, \mathrm{d}x \\ & \leqslant \left(\frac{1}{2}\right)^{n/(n-1)} \left\{ \int_E \left| \frac{\partial}{\partial x_1} u(x) \right| \mathrm{d}x \right\}^{1/(n-1)} \cdots \left\{ \int_E \left| \frac{\partial}{\partial x_n} u(x) \right| \mathrm{d}x \right\}^{1/(n-1)}. \end{aligned} \tag{4.2.30}$$

From (4.2.30) and using the elementary inequalities

$$\left\{\prod_{i=1}^{n} c_i\right\}^{1/n} \leqslant \frac{1}{n}\sum_{i=1}^{n} c_i$$

for nonnegative reals $c_1, \ldots, c_n$ and $n \geqslant 1$ and

$$\left(\sum_{i=1}^{n} c_i\right)^2 \leqslant n\sum_{i=1}^{n} c_i^2$$

for $c_1, \ldots, c_n$ reals, we obtain

$$\begin{aligned}
&\left(\int_E |u(x)|^{n/(n-1)}\,\mathrm{d}x\right)^{(n-1)/n} \\
&\leqslant \frac{1}{2}\left\{\int_E \left|\frac{\partial}{\partial x_1}u(x)\right|\mathrm{d}x\right\}^{1/n} \cdots \left\{\int_E \left|\frac{\partial}{\partial x_n}u(x)\right|\mathrm{d}x\right\}^{1/n} \\
&\leqslant \frac{1}{2n}\sum_{i=1}^{n}\int_E \left|\frac{\partial}{\partial x_i}u(x)\right|\mathrm{d}x \\
&= \frac{1}{2n}\int_E \left[\left\{\sum_{i=1}^{n}\left|\frac{\partial}{\partial x_i}u(x)\right|\right\}^2\right]^{1/2}\mathrm{d}x \\
&\leqslant \frac{1}{2n}\int_E \left[n\sum_{i=1}^{n}\left|\frac{\partial}{\partial x_i}u(x)\right|^2\right]^{1/2}\mathrm{d}x \\
&= \left(\frac{1}{4n}\right)^{1/2}\int_E |\operatorname{grad} u(x)|\,\mathrm{d}x.
\end{aligned}$$

The proof is complete. □

REMARK 4.2.3. We note that on employing Schwarz inequality on the right-hand side of (4.2.23) we get the following inequality

$$\left(\int_E |u(x)|^{n/(n-1)}\,\mathrm{d}x\right)^{(n-1)/n} \leqslant \left(\frac{V(D)}{4n}\right)^{1/2}\left(\int_E |\operatorname{grad} u(x)|^2\,\mathrm{d}x\right)^{1/2}, \tag{4.2.31}$$

where $V(D)$ is the n-dimensional measure of E. By taking $n = 3$ and $u = \phi^2$ in (4.2.23) and using the Schwarz inequality, we obtain

$$\left(\int_E |\phi(x)|^3 \,\mathrm{d}x\right)^{2/3} \leqslant 3^{-1/2}\left(\int_E \phi^2(x)\,\mathrm{d}x\right)^{1/2}\left(\int_E |\operatorname{grad}\phi(x)|^2\,\mathrm{d}x\right)^{1/2}$$

and so

$$\left(\int_E |\phi(x)|^3 \,\mathrm{d}x\right) \leqslant 3^{-3/4}\left(\int_E \phi^2(x)\,\mathrm{d}x\right)^{3/4}\left(\int_E |\operatorname{grad}\phi(x)|^2\,\mathrm{d}x\right)^{3/4}. \tag{4.2.32}$$

In 1991, Pachpatte [290] has established the following inequality.

THEOREM 4.2.6. *Let u be a real-valued sufficiently smooth function of compact support in E, the n-dimensional Euclidean space with $n \geqslant 2$, and $p \geqslant 0$, $q \geqslant 1$ and $q < n$. Then*

$$\left(\int_E |u(x)|^{(p+q)n/(n-q)}\,\mathrm{d}x\right)^{(n-q)/n} \leqslant M\sum_{i=1}^{n}\int_E |u(x)|^p\left|\frac{\partial}{\partial x_i}u(x)\right|^q \mathrm{d}x, \tag{4.2.33}$$

where

$$M = \frac{1}{n}\left[\frac{(p+q)(n-1)}{2(n-q)}\right]^q.$$

PROOF. First we establish inequality (4.2.33) for $p = 0$, $q = 1$ and by taking $u(x) = v(x)$. Since $v(x)$ is a smooth function of compact support in E, we have the following identities

$$v(x) = \int_{-\infty}^{x_1} \frac{\partial}{\partial t_1} v(t_1, x_2, \ldots, x_n)\,\mathrm{d}t_1, \tag{4.2.34}$$

$$v(x) = -\int_{x_1}^{\infty} \frac{\partial}{\partial t_1} v(t_1, x_2, \ldots, x_n)\,\mathrm{d}t_1. \tag{4.2.35}$$

From (4.2.34) and (4.2.35), we obtain

$$|v(x)| \leqslant \frac{1}{2}\int_1 \left|\frac{\partial}{\partial t_1} v(t_1, x_2, \ldots, x_n)\right| \mathrm{d}t_1. \tag{4.2.36}$$

Similarly, we obtain

$$\left|v(x)\right| \leqslant \frac{1}{2}\int_i \left|\frac{\partial}{\partial t_i} v(x_1,\ldots,t_i,\ldots,x_n)\right| \mathrm{d}t_i \tag{4.2.37}$$

for $i = 2, \ldots, n$. Now, by following exactly the same steps as in the proof of Theorem 4.2.5 below inequality (4.2.27), we obtain

$$\left(\int_E \left|v(x)\right|^{n/(n-1)} \mathrm{d}x\right)^{(n-1)/n} \leqslant \frac{1}{2n}\sum_{i=1}^n \int_E \left|\frac{\partial}{\partial x_i} v(x)\right| \mathrm{d}x. \tag{4.2.38}$$

This result proves inequality (4.2.33) for $p = 0$, $q = 1$ and $u(x) = v(x)$. To prove (4.2.33), we take

$$v(x) = \left\{u(x)\right\}^{(p+q)(n-1)/(n-q)}$$

and hence

$$\frac{\partial}{\partial x_i} v(x) = \frac{(p+q)(n-1)}{n-q}\left\{u(x)\right\}^{(p+q)(n-1)/(n-q)-1}\frac{\partial}{\partial x_i} u(x)$$

in inequality (4.2.38), and rewriting the resulting inequality we have

$$\begin{aligned}&\left(\int_E \left|u(x)\right|^{(p+q)n/(n-q)} \mathrm{d}x\right)^{(n-1)/n}\\ &\quad\leqslant \frac{(p+q)(n-1)}{2n(n-q)}\sum_{i=1}^n \int_E \left|u(x)\right|^{p/q}\left|\frac{\partial}{\partial x_i} u(x)\right|\left|u(x)\right|^{(p+q)(n-1)/(n-q)-1-p/q} \mathrm{d}x.\end{aligned} \tag{4.2.39}$$

Using Hölder's inequality with indices q, $q/(q-1)$ on the right-hand side of (4.2.39) we obtain

$$\begin{aligned}&\left(\int_E \left|u(x)\right|^{(p+q)n/(n-q)} \mathrm{d}x\right)^{(n-1)/n}\\ &\quad\leqslant \frac{(p+q)(n-1)}{2n(n-q)}\sum_{i=1}^n \left\{\int_E \left|u(x)\right|^p \left|\frac{\partial}{\partial x_i} u(x)\right|^q \mathrm{d}x\right\}^{1/q}\\ &\qquad\times\left\{\int_E \left|u(x)\right|^{(p+q)n/(n-q)} \mathrm{d}x\right\}^{(q-1)/q}.\end{aligned} \tag{4.2.40}$$

If $\int_E |u(x)|^{(p+q)n/(n-q)}\,\mathrm{d}x = 0$ then (4.2.33) is trivially true; otherwise, we divide both sides of (4.2.40) by $\{\int_E |u(x)|^{(p+q)n/(n-q)}\,\mathrm{d}x\}^{(q-1)/q}$ and then raise both sides to the power q and use the elementary inequality

$$\left\{\sum_{i=1}^{n} c_i\right\}^q \leqslant d_{q,n} = \sum_{i=1}^{n} c_i^q,$$

where c_i are nonnegative reals, $d_{q,n} = n^{q-1}$, $q > 1$, and $d_{q,n} = 1$, $0 \leqslant q \leqslant 1$, on the right-hand side to get (4.2.33). The proof is complete. □

REMARK 4.2.4. By taking $p = 0$, $q = 2$ and $n \geqslant 3$ in (4.2.33) and then raising the power $1/2$ on both sides of the resulting inequality, we get

$$\left(\int_E |u(x)|^{2n/(n-2)}\,\mathrm{d}x\right)^{(n-2)/(2n)} \leqslant \frac{(n-1)}{\sqrt{n}(n-2)}\left(\int_E |\operatorname{grad} u(x)|^2\,\mathrm{d}x\right)^{1/2}. \tag{4.2.41}$$

Further, by taking $p = 1$, $q = 1$ in (4.2.33) and raising the power $1/2$ on both sides of the resulting inequality, we get

$$\left(\int_E |u(x)|^{2n/(n-1)}\,\mathrm{d}x\right)^{(n-1)/(2n)} \leqslant \frac{1}{\sqrt{n}}\left(\int_E |u(x)|\sum_{i=1}^{n}\left|\frac{\partial}{\partial x_i}u(x)\right|\mathrm{d}x\right)^{1/2}. \tag{4.2.42}$$

We note that inequality (4.2.41) is established by Nirenberg in [229] and inequality (4.2.42) provides a new estimate on the Nirenberg-type inequality.

4.3 Poincaré- and Sobolev-Type Inequalities I

The importance of the Poincaré and Sobolev inequalities in the theory of partial differential equations is well known, and over the years much effort has been devoted to the study of these inequalities. In this section we present some Poincaré- and Sobolev-type inequalities established by Pachpatte in [249,265,290].

In 1987, Pachpatte [265] established the following Poincaré-type inequality.

THEOREM 4.3.1. *Let $p \geqslant 2$ be a constant and $B = \prod_{i=1}^{n}[0, a_i]$ be a bounded domain in $\mathbb{R}^n$. Let u be a real-valued function belonging to $C^1(B)$ which vanishes on the boundary ∂B of B. Then*

$$\int_B |u(x)|^p\,\mathrm{d}x \leqslant \frac{1}{n}\left(\frac{\alpha}{2}\right)^p \int_B |\operatorname{grad} u(x)|^p\,\mathrm{d}x, \tag{4.3.1}$$

where $\alpha = \max\{a_1, \ldots, a_n\}$.

PROOF. From the hypotheses, we have the following identities

$$nu(x) = \sum_{i=1}^{n} \int_{0}^{x_i} \frac{\partial}{\partial t_i} u(x_1, \ldots, t_i, \ldots, x_n)\, \mathrm{d}t_i, \tag{4.3.2}$$

$$nu(x) = -\sum_{i=1}^{n} \int_{x_i}^{a_i} \frac{\partial}{\partial t_i} u(x_1, \ldots, t_i, \ldots, x_n)\, \mathrm{d}t_i. \tag{4.3.3}$$

From (4.3.2) and (4.3.3), we observe that

$$|u(x)| \leqslant \frac{1}{2n} \sum_{i=1}^{n} \int_{0}^{a_i} \left| \frac{\partial}{\partial t_i} u(x_1, \ldots, t_i, \ldots, x_n) \right| \mathrm{d}t_i. \tag{4.3.4}$$

From (4.3.4) and using the elementary inequality (see [79,211])

$$\left(\sum_{i=1}^{n} c_i \right)^{k} \leqslant C_{k,n} \sum_{i=1}^{n} c_i^{k}, \tag{4.3.5}$$

where c_i are nonnegative reals and $C_{k,n} = n^{k-1}$, $k > 1$, and $C_{k,n} = 1$, $0 \leqslant k \leqslant 1$, Hölder's inequality with indices p, $p/(p-1)$ and using the definition of α, we obtain

$$\begin{aligned}
|u(x)|^{p} &\leqslant \left(\frac{1}{2n}\right)^{p} n^{p-1} \left[\sum_{i=1}^{n} \left(\int_{0}^{a_i} \left| \frac{\partial}{\partial t_i} u(x_1, \ldots, t_i, \ldots, x_n) \right| \mathrm{d}t_i \right)^{p} \right] \\
&\leqslant \left(\frac{1}{2n}\right)^{p} n^{p-1} \left[\sum_{i=1}^{n} (a_i)^{p-1} \left(\int_{0}^{a_i} \left| \frac{\partial}{\partial t_i} u(x_1, \ldots, t_i, \ldots, x_n) \right|^{p} \mathrm{d}t_i \right) \right] \\
&\leqslant \frac{1}{n} \left(\frac{1}{2}\right)^{p} \alpha^{p-1} \left[\sum_{i=1}^{n} \int_{0}^{a_i} \left| \frac{\partial}{\partial t_i} u(x_1, \ldots, t_i, \ldots, x_n) \right|^{p} \mathrm{d}t_i \right].
\end{aligned} \tag{4.3.6}$$

Integrating both sides of (4.3.6) with respect to $x_1, \ldots, x_n$ on B and using the definition of α and inequality (4.3.5) we have

$$\begin{aligned}
\int_{B} |u(x)|^{p}\, \mathrm{d}x &\leqslant \frac{1}{n} \left(\frac{1}{2}\right)^{p} \alpha^{p-1} \left[\alpha \sum_{i-1}^{n} \int_{B} \left| \frac{\partial}{\partial x_i} u(x) \right|^{p} \mathrm{d}x \right] \\
&= \frac{1}{n} \left(\frac{1}{2}\right)^{p} \alpha^{p} \left[\sum_{i=1}^{n} \int_{B} \left| \frac{\partial}{\partial x_i} u(x) \right|^{p} \mathrm{d}x \right]
\end{aligned}$$

$$= \frac{1}{n}\left(\frac{\alpha}{2}\right)^p \int_B \left[\left(\sum_{i=1}^n \left|\frac{\partial}{\partial x_i} u(x)\right|^p\right)^{2/p}\right]^{p/2} \mathrm{d}x$$

$$\leqslant \frac{1}{n}\left(\frac{\alpha}{2}\right)^p \int_B \left(\sum_{i=1}^n \left|\frac{\partial}{\partial x_i} u(x)\right|^2\right)^{p/2} \mathrm{d}x$$

$$= \frac{1}{n}\left(\frac{\alpha}{2}\right)^p \int_B |\operatorname{grad} u(x)|^p \,\mathrm{d}x.$$

This result is the desired inequality in (4.3.1) and the proof is complete. □

The following theorem deals with the Poincaré-type inequality which is an integral analogue of the discrete inequality given by Pachpatte in [242, Theorem 1].

THEOREM 4.3.2. *Let u, p, B, α be as in Theorem* 4.3.1. *Then*

$$\left(\int_B |u(x)|^{p/(p-1)} \,\mathrm{d}x\right)^{(p-1)/p} \leqslant \frac{1}{2}\left(\frac{1}{n}\right)^{1/2} \alpha^{(2(p-n)+np)/(2p)} \left(\int_B |\operatorname{grad} u(x)|^2 \,\mathrm{d}x\right)^{1/2}. \tag{4.3.7}$$

PROOF. From the hypotheses and by following the proof of Theorem 4.3.1, we have (4.3.4). From (4.3.4) and using inequality (4.3.5) and Hölder's inequality with indices p, $p/(p-1)$, we have

$$\begin{aligned}
&|u(x)|^{p/(p-1)} \\
&\quad\leqslant \left(\frac{1}{2n}\right)^{p/(p-1)} \left(\sum_{i=1}^n \int_0^{a_i} \left|\frac{\partial}{\partial t_i} u(x_1, \ldots, t_i, \ldots, x_n)\right| \mathrm{d}t_i\right)^{p/(p-1)} \\
&\quad\leqslant \left(\frac{1}{2n}\right)^{p/(p-1)} n^{p/(p-1)-1} \sum_{i=1}^n \left(\int_0^{a_i} \left|\frac{\partial}{\partial t_i} u(x_1, \ldots, t_i, \ldots, x_n)\right| \mathrm{d}t_i\right)^{p/(p-1)} \\
&\quad\leqslant \left(\frac{1}{2n}\right)^{p/(p-1)} n^{1/(p-1)} \sum_{i=1}^n \left(\left\{\int_0^{a_i} \mathrm{d}t_i\right\}^{1/p}\right. \\
&\qquad\qquad \times \left.\left\{\int_0^{a_i} \left|\frac{\partial}{\partial t_i} u(x_1, \ldots, t_i \ldots, x_n)\right|^{p/(p-1)} \mathrm{d}t_i\right\}^{(p-1)/p}\right)^{p/(p-1)} \\
&\quad\leqslant \left(\frac{1}{2n}\right)^{p/(p-1)} (\alpha n)^{1/(p-1)} \sum_{i=1}^n \int_0^{a_i} \left|\frac{\partial}{\partial t_i} u(x_1, \ldots, t_i, \ldots, x_n)\right|^{p/(p-1)} \mathrm{d}t_i.
\end{aligned} \tag{4.3.8}$$

Integrating both sides of (4.3.8) with respect to $x_1, \ldots, x_n$ on B, using the definition of α and inequality (4.3.5) we observe that

$$
\begin{aligned}
&\int_B |u(x)|^{p/(p-1)}\,\mathrm{d}x \\
&\leqslant \left(\frac{1}{2n}\right)^{p/(p-1)} (\alpha n)^{1/(p-1)}\alpha \sum_{i=1}^{n} \int_B \left|\frac{\partial}{\partial x_i} u(x)\right|^{p/(p-1)} \mathrm{d}x \\
&= \left(\frac{1}{2n}\right)^{p/(p-1)} (\alpha n)^{1/(p-1)}\alpha \\
&\quad \times \int_B \left\{\left[\sum_{i=1}^{n} \left|\frac{\partial}{\partial x_i} u(x)\right|^{p/(p-1)}\right]^{2(p-1)/p}\right\}^{p/(2(p-1))} \mathrm{d}x \\
&\leqslant \left(\frac{1}{2n}\right)^{p/(p-1)} (\alpha n)^{1/(p-1)}\alpha \int_B \left\{n^{2(p-1)/p-1} \sum_{i=1}^{n} \left|\frac{\partial}{\partial x_i} u(x)\right|^{2}\right\}^{p/(2(p-1))} \mathrm{d}x \\
&= \left(\frac{\alpha}{2}\right)^{p/(p-1)} n^{-p/(2(p-1))} \int_B |\operatorname{grad} u(x)|^{p/(p-1)}\,\mathrm{d}x. \qquad (4.3.9)
\end{aligned}
$$

From (4.3.9) and using Hölder's inequality with indices $2(p-1)/p$, $2(p-1)/(p-2)$ and the definition of α we observe that

$$
\begin{aligned}
&\left(\int_B |u(x)|^{p/(p-1)}\,\mathrm{d}x\right)^{(p-1)/p} \\
&\leqslant \left(\frac{\alpha}{2}\right) n^{-1/2} \left(\int_B |\operatorname{grad} u(x)|^{p/(p-1)}\,\mathrm{d}x\right)^{(p-1)/p} \\
&\leqslant \left(\frac{\alpha}{2}\right) n^{-1/2} \left(\left\{\int_B \mathrm{d}x\right\}^{(p-2)/(2(p-1))} \left\{\int_B |\operatorname{grad} u(x)|^2\,\mathrm{d}x\right\}^{p/(2(p-1))}\right)^{(p-1)/p} \\
&\leqslant \left(\frac{\alpha}{2}\right) n^{-1/2} (\alpha^n)^{(p-2)/(2p)} \left(\int_B |\operatorname{grad} u(x)|^2\,\mathrm{d}x\right)^{1/2} \\
&= \frac{1}{2}\left(\frac{1}{n}\right)^{1/2} \alpha^{(2(p-n)+pn)/(2p)} \left(\int_B |\operatorname{grad} u(x)|^2\,\mathrm{d}x\right)^{1/2}.
\end{aligned}
$$

This result is the required inequality in (4.3.7) and the proof is complete. □

The following theorem established in [265] deals with the Sobolev-type inequality.

THEOREM 4.3.3. *Let $p \geqslant 1$ be a constant and u, B, α be as in Theorem* 4.3.1. *Then*

$$\int_B |u(x)|^p \, dx \leqslant \frac{p\alpha}{2\sqrt{n}} \left(\int_B |u(x)|^{2(p-1)} \, dx \right)^{1/2} \left(\int_B |\operatorname{grad} u(x)|^2 \, dx \right)^{1/2}. \tag{4.3.10}$$

PROOF. From the hypotheses, we have the following identities

$$nu^p(x) = p \sum_{i=1}^{n} \int_0^{x_i} u^{p-1}(x_1, \ldots, t_i, \ldots, x_n) \frac{\partial}{\partial t_i} u(x_1, \ldots, t_i, \ldots, x_n) \, dt_i, \tag{4.3.11}$$

$$nu^p(x) = -p \sum_{i=1}^{n} \int_{x_i}^{a_i} u^{p-1}(x_1, \ldots, t_i, \ldots, x_n) \frac{\partial}{\partial t_i} u(x_1, \ldots, t_i, \ldots, x_n) \, dt_i. \tag{4.3.12}$$

From (4.3.11) and (4.3.12), we observe that

$$|u(x)|^p \leqslant \frac{p}{2n} \sum_{i=1}^{n} \int_0^{a_i} |u(x_1, \ldots, t_i, \ldots, x_n)|^{p-1} \times \left| \frac{\partial}{\partial t_i} u(x_1, \ldots, t_i, \ldots, x_n) \right| dt_i. \tag{4.3.13}$$

Integrating both sides of (4.3.13) with respect to $x_1, \ldots, x_n$ on B and using the definition of α we have

$$\int_B |u(x)|^p \, dx \leqslant \frac{p\alpha}{2n} \sum_{i=1}^{n} \int_B |u(x)|^{p-1} \left| \frac{\partial}{\partial x_i} u(x) \right| dx. \tag{4.3.14}$$

From (4.3.14) and using Schwarz inequality and the elementary inequality $(\sum_{i=1}^n b_i)^2 \leqslant n \sum_{i=1}^n b_i^2$ (for $b_1, \ldots, b_n$ reals), we obtain

$$\int_B |u(x)|^p \, dx \leqslant \frac{p\alpha}{2n} \left(\int_B |u(x)|^{2(p-1)} \, dx \right)^{1/2} \left(\int_B \left(\sum_{i=1}^{n} \left| \frac{\partial}{\partial x_i} u(x) \right| \right)^2 dx \right)^{1/2}$$

$$\leqslant \frac{p\alpha}{2n} n^{1/2} \left(\int_B |u(x)|^{2(p-1)} \, dx \right)^{1/2} \left(\int_B \sum_{i=1}^n \left| \frac{\partial}{\partial x_i} u(x) \right|^2 dx \right)^{1/2}$$

$$= \frac{p\alpha}{2\sqrt{n}} \left(\int_B |u(x)|^{2(p-1)} \, dx \right)^{1/2} \left(\int_B |\operatorname{grad} u(x)|^2 \, dx \right)^{1/2}.$$

This result is the desired inequality in (4.3.10) and the proof is complete. □

The following variant of Sobolev's inequality is established in [290].

THEOREM 4.3.4. *Let $p \geqslant 0$, $q \geqslant 1$ be constants and u, B, α be as in Theorem* 4.3.1. *Then*

$$\int_B |u(x)|^{p+q} \, dx \leqslant \frac{1}{n} \left\{ \left(\frac{p+q}{2} \right) \alpha \right\}^q \sum_{i=1}^n \int_B |u(x)|^p \left| \frac{\partial}{\partial x_i} u(x) \right|^q dx. \tag{4.3.15}$$

PROOF. From the hypotheses, we have the following identities

$$n u^{p+q}(x) = (p+q) \sum_{i=1}^n \int_0^{x_i} u^{p+q-1}(x_1, \ldots, t_i, \ldots, x_n) \times \frac{\partial}{\partial t_i} u(x_1, \ldots, t_i, \ldots, x_n) \, dt_i \tag{4.3.16}$$

and

$$n u^{p+q}(x) = -(p+q) \sum_{i=1}^n \int_{x_i}^{a_i} u^{p+q-1}(x_1, \ldots, t_i, \ldots, x_n) \times \frac{\partial}{\partial t_i} u(x_1, \ldots, t_i, \ldots, x_n) \, dt_i. \tag{4.3.17}$$

From (4.3.16) and (4.3.17), we observe that

$$|u(x)|^{p+q} \leqslant \frac{p+q}{2n} \sum_{i=1}^n \int_0^{a_i} |u(x_1, \ldots, t_i, \ldots, x_n)|^{p+q-1} \times \left| \frac{\partial}{\partial t_i} u(x_1, \ldots, t_i, \ldots, x_n) \right| dt_i. \tag{4.3.18}$$

Integrating both sides of (4.3.18) over B and using the definition of α and rewriting the resulting inequality we have

$$\int_B |u(x)|^{p+q}\,\mathrm{d}x \leqslant \left(\frac{p+q}{2n}\right)\alpha \sum_{i=1}^{n} \int_B |u(x)|^{p/q} \left|\frac{\partial}{\partial x_i} u(x)\right| |u(x)|^{p+q-p/q-1}\,\mathrm{d}x. \tag{4.3.19}$$

By using Hölder's inequality on the right-hand side of (4.3.19) with indices q, $q/(q-1)$, we have

$$\int_B |u(x)|^{p+q}\,\mathrm{d}x \leqslant \left(\frac{p+q}{2n}\right)\alpha \sum_{i=1}^{n} \left\{\int_B |u(x)|^{p} \left|\frac{\partial}{\partial x_i} u(x)\right|^q \mathrm{d}x\right\}^{1/q} \left\{\int_B |u(x)|^{p+q}\,\mathrm{d}x\right\}^{(q-1)/q}. \tag{4.3.20}$$

If $\int_B |u(x)|^{p+q}\,\mathrm{d}x = 0$ then (4.3.15) is trivially true; otherwise, we divide both sides of (4.3.20) by $\{\int_B |u(x)|^{p+q}\,\mathrm{d}x\}^{(q-1)/q}$ and then raise both sides to the power q and use the elementary inequality $(\sum_{i=1}^{n} c_i)^k \leqslant n^{k-1}\sum_{i=1}^{n} c_i^k$ (for $c_i \geqslant 0$ reals and $k \geqslant 1$) to get (4.3.15). The proof is complete. □

REMARK 4.3.1. We note that, in the special cases when (i) $p = 2$, $q = 2$ and (ii) $p = 0$, inequality (4.3.15) reduces, respectively, to the following inequalities

$$\int_B |u(x)|^4\,\mathrm{d}x \leqslant \frac{1}{n}(2\alpha)^2 \sum_{i=1}^{n} \int_B |u(x)|^2 \left|\frac{\partial}{\partial x_i} u(x)\right|^2 \mathrm{d}x \tag{4.3.21}$$

and

$$\int_B |u(x)|^q\,\mathrm{d}x \leqslant \frac{1}{n}\left(\frac{q\alpha}{2}\right)^q \int_B \sum_{i=1}^{n} \left|\frac{\partial}{\partial x_i} u(x)\right|^q \mathrm{d}x. \tag{4.3.22}$$

Inequality (4.3.21) is a Sobolev-type inequality, while inequality (4.3.22) is a variant of Poincaré-type inequality given in Theorem 4.3.1.

The following Poincaré- and Sobolev-type inequalities involving two functions are established in [249].

THEOREM 4.3.5. *Let $p, q \geqslant 1$ be constants and $B = \prod_{i=1}^{n}[a_i, b_i]$ be a bounded domain in $\mathbb{R}^n$. Let u, v be sufficiently smooth functions defined on B which vanish on the boundary ∂B of B. Then*

$$\int_B |u(x)|^p |v(x)|^q \,\mathrm{d}x \leqslant \frac{1}{n}\left(\frac{1}{2}\right)^{p+q+1} \alpha^{p+q} \int_B \left[|\operatorname{grad} u(x)|^{2p} + |\operatorname{grad} v(x)|^{2q}\right] \mathrm{d}x, \tag{4.3.23}$$

where $\alpha = \max\{b_1 - a_1, \ldots, b_n - a_n\}$.

THEOREM 4.3.6. *Let p, q, B, u, v, α be as in Theorem 4.3.5. Then*

$$\begin{aligned}\int_B |u(x)|^p |v(x)|^q \,\mathrm{d}x &\leqslant \frac{pq\alpha^2}{8n}\left[\left(\int_B |u(x)|^{2p}\,\mathrm{d}x\right)^{(p-1)/p}\left(\int_B |\operatorname{grad} u(x)|^{2p}\,\mathrm{d}x\right)^{1/p}\right.\\ &\quad\left.+\left(\int_B |v(x)|^{2q}\,\mathrm{d}x\right)^{(q-1)/q}\left(\int_B |\operatorname{grad} v(x)|^{2q}\,\mathrm{d}x\right)^{1/q}\right].\end{aligned} \tag{4.3.24}$$

REMARK 4.3.2. In the special cases when $p = q = 1$ and $a_i = 0$, inequalities (4.3.23) and (4.3.24) reduce to the Poincaré-type inequality given in Theorem 4.2.3.

PROOFS OF THEOREMS 4.3.5 AND 4.3.6. From the hypotheses of Theorem 4.3.5, we have the following identities

$$nu(x) = \sum_{i=1}^{n} \int_{a_i}^{x_i} \frac{\partial}{\partial t_i} u(x_1, \ldots, t_i, \ldots, x_n)\,\mathrm{d}t_i, \tag{4.3.25}$$

$$nu(x) = -\sum_{i=1}^{n} \int_{x_i}^{b_i} \frac{\partial}{\partial t_i} u(x_1, \ldots, t_i, \ldots, x_n)\,\mathrm{d}t_i. \tag{4.3.26}$$

From (4.3.25) and (4.3.26), we observe that

$$|u(x)| \leqslant \frac{1}{2n} \sum_{i=1}^{n} \int_{a_i}^{b_i} \left|\frac{\partial}{\partial t_i} u(x_1, \ldots, t_i, \ldots, x_n)\right| \mathrm{d}t_i. \tag{4.3.27}$$

From (4.3.27) and using inequality (4.3.5), Hölder's inequality with indices p, $p/(p-1)$ (see [74, p. 126]) and the definition of α, we obtain

$$\begin{aligned}|u(x)|^p &\leqslant \left(\frac{1}{2n}\right)^p n^{p-1}\left[\sum_{i=1}^{n}\left\{\int_{a_i}^{b_i}\left|\frac{\partial}{\partial t_i}u(x_1,\ldots,t_i,\ldots,x_n)\right|\mathrm{d}t_i\right\}^p\right]\\ &\leqslant \left(\frac{1}{2}\right)^p\frac{1}{n}\alpha^{p-1}\left[\sum_{i=1}^{n}\int_{a_i}^{b_i}\left|\frac{\partial}{\partial t_i}u(x_1,\ldots,t_i,\ldots,x_n)\right|^p\mathrm{d}t_i\right].\end{aligned}\tag{4.3.28}$$

Similarly, we obtain

$$|v(x)|^q \leqslant \left(\frac{1}{2}\right)^q\frac{1}{n}\alpha^{q-1}\left[\sum_{i=1}^{n}\int_{a_i}^{b_i}\left|\frac{\partial}{\partial t_i}v(x_1,\ldots,t_i,\ldots,x_n)\right|^q\mathrm{d}t_i\right].\tag{4.3.29}$$

From (4.3.28), (4.3.29) and using the elementary inequality $cd \leqslant \frac{1}{2}(c^2+d^2)$ (for c, d reals), inequality (4.3.5) with $k=2$, the Schwarz inequality and the definition of α, we obtain

$$\begin{aligned}&|u(x)|^p|v(x)|^q\\ &\leqslant \frac{1}{2}\left(\frac{1}{2}\right)^{p+q}\left(\frac{1}{n}\right)^2\alpha^{p+q-2}\left\{\left[\sum_{i=1}^{n}\int_{a_i}^{b_i}\left|\frac{\partial}{\partial t_i}u(x_1,\ldots,t_i,\ldots,x_n)\right|^p\mathrm{d}t_i\right]^2\right.\\ &\qquad\left.+\left[\sum_{i=1}^{n}\int_{a_i}^{b_i}\left|\frac{\partial}{\partial t_i}v(x_1,\ldots,t_i,\ldots,x_n)\right|^q\mathrm{d}t_i\right]^2\right\}\\ &\leqslant \left(\frac{1}{2}\right)^{p+q+1}\left(\frac{1}{n}\right)^2\alpha^{p+q-2}n\left\{\sum_{i=1}^{n}\left\{\int_{a_i}^{b_i}\left|\frac{\partial}{\partial t_i}u(x_1,\ldots,t_i,\ldots,x_n)\right|^p\mathrm{d}t_i\right\}^2\right.\\ &\qquad\left.+\sum_{i=1}^{n}\left\{\int_{a_i}^{b_i}\left|\frac{\partial}{\partial t_i}v(x_1,\ldots,t_i,\ldots,x_n)\right|^q\mathrm{d}t_i\right\}^2\right\}\\ &\leqslant \frac{1}{n}\left(\frac{1}{2}\right)^{p+q+1}\alpha^{p+q-2}\alpha\left\{\sum_{i=1}^{n}\int_{a_i}^{b_i}\left|\frac{\partial}{\partial t_i}u(x_1,\ldots,t_i,\ldots,x_n)\right|^{2p}\mathrm{d}t_i\right.\\ &\qquad\left.+\sum_{i=1}^{n}\int_{a_i}^{b_i}\left|\frac{\partial}{\partial t_i}v(x_1,\ldots,t_i,\ldots,x_n)\right|^{2q}\mathrm{d}t_i\right\}.\end{aligned}\tag{4.3.30}$$

Integrating both sides of (4.3.30) with respect to $x_1, \ldots, x_n$ on B, using the definition of α and a suitable version of inequality (4.3.5) we get

$$\int_B |u(x)|^p |v(x)|^q \,\mathrm{d}x$$

$$\leqslant \frac{1}{n}\left(\frac{1}{2}\right)^{p+q+1} \alpha^{p+q-1}\alpha$$

$$\times \left\{\int_B \left\{\left[\sum_{i=1}^n \left|\frac{\partial}{\partial x_i}u(x)\right|^{2p}\right]^{1/p}\right\}^p \mathrm{d}x + \int_B \left\{\left[\sum_{i=1}^n \left|\frac{\partial}{\partial x_i}v(x)\right|^{2q}\right]^{1/q}\right\}^q \mathrm{d}x\right\}$$

$$\leqslant \frac{1}{n}\left(\frac{1}{2}\right)^{p+q+1} \alpha^{p+q} \int_B \left[|\operatorname{grad} u(x)|^{2p} + |\operatorname{grad} v(x)|^{2q}\right] \mathrm{d}x.$$

The proof of Theorem 4.3.5 is complete.

From the hypotheses of Theorem 4.3.6, for any x in B, we have the following identities

$$nu^p(x) = p \sum_{i=1}^n \int_{a_i}^{x_i} u^{p-1}(x_1, \ldots, t_i, \ldots, x_n) \frac{\partial}{\partial t_i} u(x_1, \ldots, t_i, \ldots, x_n)\,\mathrm{d}t_i, \tag{4.3.31}$$

$$nu^p(x) = -p \sum_{i=1}^n \int_{x_i}^{b_i} u^{p-1}(x_1, \ldots, t_i, \ldots, x_n) \frac{\partial}{\partial t_i} u(x_1, \ldots, t_i, \ldots, x_n)\,\mathrm{d}t_i. \tag{4.3.32}$$

From (4.3.31) and (4.3.32), we observe that

$$|u(x)|^p \leqslant \frac{p}{2n} \sum_{i=1}^n \int_{a_i}^{b_i} |u(x_1, \ldots, t_i, \ldots, x_n)|^{p-1} \times \left|\frac{\partial}{\partial t_i} u(x_1, \ldots, t_i, \ldots, x_n)\right| \mathrm{d}t_i. \tag{4.3.33}$$

Similarly, for any x in B, we obtain the following inequality

$$|v(x)|^q \leqslant \frac{q}{2n} \sum_{i=1}^n \int_{a_i}^{b_i} |v(x_1, \ldots, t_i, \ldots, x_n)|^{q-1} \times \left|\frac{\partial}{\partial t_i} v(x_1, \ldots, t_i, \ldots, x_n)\right| \mathrm{d}t_i. \tag{4.3.34}$$

From (4.3.33), (4.3.34) and using the elementary inequality $cd \leqslant \frac{1}{2}(c^2 + d^2)$ (for c, d reals), inequality (4.3.5) with $k = 2$, the Schwarz inequality and the definition of α, we obtain

$$
\begin{aligned}
|u(x)|^p |v(x)|^q \leqslant \frac{pq}{8n^2} \Bigg\{ & \Bigg[\sum_{i=1}^{n} \int_{a_i}^{b_i} |u(x_1, \dots, t_i, \dots, x_n)|^{p-1} \\
& \times \left| \frac{\partial}{\partial t_i} u(x_1, \dots, t_i, \dots, x_n) \right| \mathrm{d}t_i \Bigg]^2 \\
& + \Bigg[\sum_{i=1}^{n} \int_{a_i}^{b_i} |v(x_1, \dots, t_i, \dots, x_n)|^{q-1} \\
& \times \left| \frac{\partial}{\partial t_i} v(x_1, \dots, t_i, \dots, x_n) \right| \mathrm{d}t_i \Bigg]^2 \Bigg\} \\
\leqslant \frac{npq}{8n^2} \Bigg\{ & \sum_{i=1}^{n} \Bigg\{ \int_{a_i}^{b_i} |u(x_1, \dots, t_i, \dots, x_n)|^{p-1} \\
& \times \left| \frac{\partial}{\partial t_i} u(x_1, \dots, t_i, \dots, x_n) \right| \mathrm{d}t_i \Bigg\}^2 \\
& + \sum_{i=1}^{n} \Bigg\{ \int_{a_i}^{b_i} |v(x_1, \dots, t_i, \dots, x_n)|^{q-1} \\
& \times \left| \frac{\partial}{\partial t_i} v(x_1, \dots, t_i, \dots, x_n) \right| \mathrm{d}t_i \Bigg\}^2 \Bigg\} \\
\leqslant \frac{pq}{8n} \alpha \Bigg\{ & \sum_{i=1}^{n} \int_{a_i}^{b_i} |u(x_1, \dots, t_i, \dots, x_n)|^{2(p-1)} \\
& \times \left| \frac{\partial}{\partial t_i} u(x_1, \dots, t_i, \dots, x_n) \right|^2 \mathrm{d}t_i \\
& + \sum_{i=1}^{n} \int_{a_i}^{b_i} |v(x_1, \dots, t_i, \dots, x_n)|^{2(q-1)} \\
& \times \left| \frac{\partial}{\partial t_i} v(x_1, \dots, t_i, \dots, x_n) \right|^2 \mathrm{d}t_i \Bigg\}. \qquad (4.3.35)
\end{aligned}
$$

Integrating both sides of (4.3.35) with respect to $x_1, \ldots, x_n$ on B, using the definition of α and Hölder's inequality on the right-hand side with indices p, $p/(p-1)$ and q, $q/(q-1)$ (see [74, p. 126]) we obtain

$$\begin{aligned}
&\int_B |u(x)|^p |v(x)|^q \,\mathrm{d}x \\
&\quad\leqslant \frac{pq\alpha^2}{8n}\left\{\int_B |u(x)|^{2(p-1)} |\operatorname{grad} u(x)|^2 \,\mathrm{d}x + \int_B |v(x)|^{2(q-1)} |\operatorname{grad} v(x)|^2 \,\mathrm{d}x\right\} \\
&\quad\leqslant \frac{pq\alpha^2}{8n}\left[\left(\int_B |u(x)|^{2p} \,\mathrm{d}x\right)^{(p-1)/p} \left(\int_B |\operatorname{grad} u(x)|^{2p} \,\mathrm{d}x\right)^{1/p}\right. \\
&\qquad\left. + \left(\int_B |v(x)|^{2q} \,\mathrm{d}x\right)^{(q-1)/q} \left(\int_B |\operatorname{grad} v(x)|^{2q} \,\mathrm{d}x\right)^{1/q}\right].
\end{aligned}$$

This inequality is the desired inequality in (4.3.24) and the proof of Theorem 4.3.6 is complete. □

4.4 Poincaré- and Sobolev-Type Inequalities II

In the recent past, several authors have presented numerous integral inequalities of Poincaré and Sobolev type. In this section we present some Poincaré- and Sobolev-type inequalities investigated by Pachpatte in [237,246].

In 1986, Pachpatte [237] has established the following inequalities of the Poincaré and Sobolev type, involving functions of several independent variables.

THEOREM 4.4.1. *Let $B = \prod_{i=1}^n [0, a_i]$ be a bounded domain in $\mathbb{R}^n$, $n \geqslant 3$. Let $1 \leqslant p < Q$ and u be a real-valued function belonging to $C^1(B)$ which vanishes on the boundary ∂B of B. Then*

$$\begin{aligned}
&\left(\int_B |u(x)|^p \,\mathrm{d}x\right)^{1/p} \\
&\quad\leqslant \frac{(\prod_{i=1}^n a_i)^{(Q-p)/(pQ)}}{2n^{1/p}} \\
&\qquad\times \left(\sum_{i=1}^n a_i^{pQ/(Q-p)}\right)^{(Q-p)/(pQ)} \left(\int_D (\|\nabla u(x)\|_Q)^Q \,\mathrm{d}x\right)^{1/Q},
\end{aligned} \tag{4.4.1}$$

where $(\|\nabla u(x)\|_Q)^Q = \sum_{i=1}^n |\frac{\partial}{\partial t_i} u(x)|^Q$.

REMARK 4.4.1. In the special case when $Q = 2$ and $p = n/(n-1)$ (for $n \geqslant 3$), we see that $1 < p < Q$ holds and inequality (4.4.1) reduces to

$$\left(\int_B |u(x)|^{n/(n-1)}\,\mathrm{d}x\right)^{(n-1)/n} \leqslant \frac{(\prod_{i=1}^n a_i)^{(n-2)/(2n)}}{2n^{(n-1)/n}} \times \left(\sum_{i=1}^n a_i^{2n/(n-2)}\right)^{(n-2)/(2n)} \left(\int_B |\operatorname{grad} u(x)|^2\,\mathrm{d}x\right)^{1/2}. \tag{4.4.2}$$

THEOREM 4.4.2. *Let $p \geqslant 1$, P, $Q > 1$, $P^{-1} + Q^{-1} = 1$, B be as in Theorem 4.4.1 and u be a real-valued function belonging to $C^1(B)$ which vanishes on the boundary ∂B of B. Then*

$$\int_B |u(x)|^p\,\mathrm{d}x \leqslant \frac{p}{2n}\left(\sum_{j=1}^n a_j^P\right)^{1/P} \left(\int_B |u(x)|^{P(p-1)}\,\mathrm{d}x\right)^{1/P} \times \left(\int_B (\|\nabla u(x)\|_Q)^Q\,\mathrm{d}x\right)^{1/Q}, \tag{4.4.3}$$

where $(\|\nabla u(x)\|_Q)^Q$ is as defined in Theorem 4.4.1.

REMARK 4.4.2. By taking $p = n/(n-1)$ (for $n \geqslant 3$), $P = Q = 2$ in (4.4.3) and then squaring on both sides of the resulting inequality we have the following inequality

$$\left(\int_B |u(x)|^{n/(n-1)}\,\mathrm{d}x\right)^2 \leqslant \frac{\sum_{i=1}^n a_i^2}{4(n-1)^2}\left(\int_B |u(x)|^{2/(n-1)}\,\mathrm{d}x\right)\left(\int_B |\operatorname{grad} u(x)|^2\,\mathrm{d}x\right). \tag{4.4.4}$$

Further, for $n = 3$, inequality (4.4.4) reduces to

$$\left(\int_B |u(x)|^{3/2}\,\mathrm{d}x\right)^2 \leqslant \frac{\sum_{i=1}^3 a_i^2}{16}\left(\int_B |u(x)|\,\mathrm{d}x\right)\left(\int_B |\operatorname{grad} u(x)|^2\,\mathrm{d}x\right). \tag{4.4.5}$$

For the inequalities of the forms (4.4.4) and (4.4.5), see [152,155].

PROOFS OF THEOREMS 4.4.1 AND 4.4.2. If $u \in C^1(B)$, then we have the following identities

$$nu(x) = \sum_{i=1}^{n} \int_0^{x_i} \frac{\partial}{\partial t_i} u(x_1, \ldots, t_i, \ldots, x_n)\, \mathrm{d}t_i, \tag{4.4.6}$$

$$nu(x) = -\sum_{i=1}^{n} \int_{x_i}^{a_i} \frac{\partial}{\partial t_i} u(x_1, \ldots, t_i, \ldots, x_n)\, \mathrm{d}t_i. \tag{4.4.7}$$

From (4.4.6) and (4.4.7), we obtain

$$|u(x)| \leqslant \frac{1}{2n} \sum_{i=1}^{n} \int_0^{a_i} \left| \frac{\partial}{\partial t_i} u(x_1, \ldots, t_i, \ldots, x_n) \right| \mathrm{d}t_i. \tag{4.4.8}$$

From (4.4.8) and using the elementary inequality (see [3, p. 338])

$$\left(\sum_{i=1}^{n} c_i \right)^k \leqslant n^{k-1} \sum_{i=1}^{n} c_i^k \tag{4.4.9}$$

(for $c_i \geqslant 0$ reals and $k \geqslant 1$), we obtain

$$|u(x)|^p \leqslant \left(\frac{1}{2n}\right)^p n^{p-1} \sum_{i=1}^{n} \left(\int_0^{a_i} \left| \frac{\partial}{\partial t_i} u(x_1, \ldots, t_i, \ldots, x_n)\, \mathrm{d}t_i \right| \right)^p \tag{4.4.10}$$

for any $p \geqslant 1$. Applying Hölder's inequality with indices P, $Q > 1$ $(P^{-1} + Q^{-1} = 1)$ to each integral in (4.4.10) we get

$$\begin{aligned} |u(x)|^p &\leqslant \frac{1}{2^p n} \sum_{i=1}^{n} \left(\left(\int_0^{a_i} 1\, \mathrm{d}t_i \right)^{1/P} \left(\int_0^{a_i} \left| \frac{\partial}{\partial t_i} u(x_1, \ldots, t_i, \ldots, x_n) \right|^Q \mathrm{d}t_i \right)^{1/Q} \right)^p \\ &= \frac{1}{2^p n} \sum_{i=1}^{n} a_i^{p/P} \left(\int_0^{a_i} \left| \frac{\partial}{\partial t_i} u(x_1, \ldots, t_i, \ldots, x_n) \right|^Q \mathrm{d}t_i \right)^{p/Q}. \end{aligned} \tag{4.4.11}$$

Integrating both sides of (4.4.11) over B we get

$$\begin{aligned} &\int_B |u(x)|^p\, \mathrm{d}x \\ &\quad \leqslant \frac{1}{2^p n} \sum_{i=1}^{n} a_i^{p/P} \int_B 1 \cdot \left(\int_0^{a_i} \left| \frac{\partial}{\partial t_i} u(x_1, \ldots, t_i, \ldots, x_n) \right|^Q \mathrm{d}t_i \right)^{p/Q} \mathrm{d}x. \end{aligned} \tag{4.4.12}$$

Now, applying Hölder's inequality with indices $p_1 = Q/(Q-p)$, $q_1 = Q/p$ to each integral on the right-hand side in (4.4.12), we get

$$\begin{aligned}
&\int_B |u(x)|^p \,\mathrm{d}x \\
&\quad\leqslant \frac{(\prod_{i=1}^n a_i)^{(Q-p)/Q}}{2^p n} \sum_{i=1}^n a_i^{p/P} a_i^{p/Q} \left(\int_B \left|\frac{\partial}{\partial x_i} u(x)\right|^Q \mathrm{d}x\right)^{p/Q} \\
&\quad= \frac{(\prod_{i=1}^n a_i)^{(Q-p)/Q}}{2^p n} \sum_{i=1}^n a_i^p \left(\int_B \left|\frac{\partial}{\partial x_i} u(x)\right|^Q \mathrm{d}x\right)^{p/Q}. \qquad (4.4.13)
\end{aligned}$$

Now, applying Hölder's inequality to the sum on the right-hand side in (4.4.13) with indices p_1 and q_1 again, we obtain

$$\begin{aligned}
&\int_B |u(x)|^p \,\mathrm{d}x \\
&\quad\leqslant \frac{(\prod_{i=1}^n a_i)^{(Q-p)/Q}}{2^p n} \\
&\qquad\times \left(\sum_{i=1}^n a_i^{pQ/(Q-p)}\right)^{(Q-p)/Q} \left(\sum_{i=1}^n \left(\int_B \left|\frac{\partial}{\partial x_i} u(x)\right|^Q \mathrm{d}x\right)\right)^{p/Q}. \qquad (4.4.14)
\end{aligned}$$

From (4.4.14) we have

$$\begin{aligned}
&\left(\int_B |u(x)|^p \,\mathrm{d}x\right)^{1/p} \\
&\quad\leqslant \frac{(\prod_{i=1}^n a_i)^{(Q-p)/(pQ)}}{2n^{1/p}} \\
&\qquad\times \left(\sum_{i=1}^n a_i^{pQ/(Q-p)}\right)^{(Q-p)/(pQ)} \left(\int_B (\|\nabla u(x)\|_Q)^Q \,\mathrm{d}x\right)^{1/Q}.
\end{aligned}$$

The proof of Theorem 4.4.1 is complete.

From the hypotheses of Theorem 4.4.2, if $u \in C^1(B)$ then we have the following identities

$$\begin{aligned}
nu^p(x) = p \sum_{i=1}^n \int_0^{x_i} &\{u(x_1, \ldots, t_i, \ldots, x_n)\}^{p-1} \\
&\times \frac{\partial}{\partial t_i} u(x_1, \ldots, t_i, \ldots, x_n) \,\mathrm{d}t_i, \qquad (4.4.15)
\end{aligned}$$

$$nu^p(x) = -p\sum_{i=1}^{n}\int_{x_i}^{a_i}\{u(x_1,\ldots,t_i,\ldots,x_n)\}^{p-1} \times \frac{\partial}{\partial t_i}u(x_1,\ldots,t_i,\ldots,x_n)\,\mathrm{d}t_i. \tag{4.4.16}$$

From (4.4.15) and (4.4.16), we obtain

$$|u(x)|^p \leqslant \frac{p}{2n}\sum_{i=1}^{n}\int_0^{a_i}|u(x_1,\ldots,t_i,\ldots,x_n)|^{p-1}\left|\frac{\partial}{\partial t_i}u(x_1,\ldots,t_i,\ldots,x_n)\right|\mathrm{d}t_i. \tag{4.4.17}$$

Integrating both sides of (4.4.17) over B we get

$$\int_B|u(x)|^p\,\mathrm{d}x \leqslant \frac{p}{2n}\sum_{i=1}^{n}a_i\int_B|u(x)|^{p-1}\left|\frac{\partial}{\partial x_i}u(x)\right|\mathrm{d}x. \tag{4.4.18}$$

Applying Hölder's inequality with indices P, $Q>1$ ($P^{-1}+Q^{-1}=1$) to each integral on the right-hand side in (4.4.18) we get

$$\begin{aligned}\int_B|u(x)|^p\,\mathrm{d}x &\leqslant \frac{p}{2n}\sum_{i=1}^{n}a_i\left(\int_B|u(x)|^{P(p-1)}\,\mathrm{d}x\right)^{1/P}\left(\int_B\left|\frac{\partial}{\partial x_i}u(x)\right|^Q\mathrm{d}x\right)^{1/Q}\\ &= \frac{p}{2n}\left(\int_B|u(x)|^{P(p-1)}\,\mathrm{d}x\right)^{1/P}\sum_{i=1}^{n}a_i\left(\int_B\left|\frac{\partial}{\partial x_i}u(x)\right|^Q\mathrm{d}x\right)^{1/Q}.\end{aligned}$$

Now, applying Hölder's inequality to the sum on the right-hand side in the above inequality with the same indices P, Q, we obtain

$$\int_B|u(x)|^p\,\mathrm{d}x \leqslant \frac{p}{2n}\left(\sum_{i=1}^{n}a_i^P\right)^{1/P}\left(\int_B|u(x)|^{P(p-1)}\,\mathrm{d}x\right)^{1/P}\left(\int_B(\|\nabla u(x)\|_Q)^Q\,\mathrm{d}x\right)^{1/Q}.$$

This inequality is the desired inequality in (4.4.3) and the proof of Theorem 4.4.2 is complete. □

The following Poincaré- and Sobolev-type inequalities in which the constants appearing do not depend on the size of the domain of definitions of the functions are established in [246].

THEOREM 4.4.3. *Let u_r, $r = 1, \ldots, m$, be sufficiently smooth functions of compact support in E, the n-dimensional Euclidean space with $n \geqslant 2$. Then*

$$\left\{\int_E \left(\prod_{i=1}^{m} |u_r(x)|^{n/(n-1)}\right)^{1/m} \mathrm{d}x\right\}^{(n-1)/n} \leqslant \frac{1}{\sqrt{4n}}\left(\frac{1}{m}\right)^{(n-1)/n} \sum_{r=1}^{m} \int_E |\operatorname{grad} u_r(x)|\, \mathrm{d}x. \tag{4.4.19}$$

REMARK 4.4.3. We note that in the special case when $m = 1$, $u_1 = u$, inequality (4.4.19) reduces to the following inequality

$$\left\{\int_E |u(x)|^{n/(n-1)}\, \mathrm{d}x\right\}^{(n-1)/n} \leqslant \mu \int_E |\operatorname{grad} u(x)|\, \mathrm{d}x,$$

where $\mu = 1/\sqrt{4n}$, which is given in Theorem 4.2.5.

THEOREM 4.4.4. *Let u_r, $r = 1, \ldots, m$, be sufficiently smooth functions of compact support in E, the n-dimensional Euclidean space with $n \geqslant 2$, and let $p_r \geqslant 1$ be constants. Then*

$$\begin{aligned} &\left\{\int_E \left(\prod_{r=1}^{m} |u_r(x)|^{((p_r+2)/2)(n/(n-1))}\right)^{1/m} \mathrm{d}x\right\}^{(n-1)/n} \\ &\quad\leqslant \frac{1}{\sqrt{n}}\left(\frac{1}{m}\right)^{(n-1)/n}\left(\prod_{r=1}^{m}\left(\frac{p_r+2}{4}\right)\right)^{1/m} \\ &\quad\quad\times \sum_{r=1}^{m}\left\{\int_E |u_r(x)|^{p_r}\, \mathrm{d}x\right\}^{1/2}\left\{\int_E |\operatorname{grad} u_r(x)|^2\, \mathrm{d}x\right\}^{1/2}. \end{aligned} \tag{4.4.20}$$

REMARK 4.4.4. We note that in the special case where $m = 1$, inequality (4.4.20) reduces to

$$\begin{aligned} &\left\{\int_E |u_1(x)|^{((p_1+2)/2)(n/(n-1))}\, \mathrm{d}x\right\}^{(n-1)/n} \\ &\quad\leqslant \frac{1}{\sqrt{n}}\left(\frac{p_1+2}{4}\right)\left\{\int_E |u_1(x)|^{p_1}\, \mathrm{d}x\right\}^{1/2}\left\{\int_E |\operatorname{grad} u_1(x)|\, \mathrm{d}x\right\}^{1/2}. \end{aligned} \tag{4.4.21}$$

On taking $n = 2$ and $p_1 = 2$ in (4.4.21) and squaring both sides of the resulting inequality, we obtain the sharpened version of Sobolev's inequality established by Payne in [362].

The next two theorems established in [246] deal with the Poincaré- and Sobolev-type inequalities in which the constants appearing depend on the size of the domain of definitions of the function.

THEOREM 4.4.5. *Let $B = \prod_{i=1}^{n}[a_i, b_i]$ be a bounded domain in $\mathbb{R}^n$ with $n \geqslant 2$, u_r, $r = 1, \ldots, m$, be sufficiently smooth functions defined on B which vanish on the boundary ∂B of B and $p_r \geqslant 2$ be constants. Then*

$$\int_B \left(\prod_{r=1}^{m} |u_r(x)|^{p_r} \right)^{1/m} \mathrm{d}x \leqslant \frac{1}{mn} \left(\frac{\alpha}{2} \right)^{1/m(\sum_{r=1}^{m} p_r)} \sum_{r=1}^{m} \int_B |\operatorname{grad} u_r(x)|^{p_r} \,\mathrm{d}x, \tag{4.4.22}$$

where $\alpha = \max\{b_1 - a_1, \ldots, b_n - a_n\}$.

THEOREM 4.4.6. *Let B, u_r, $r = 1, \ldots, m$, be as in Theorem 4.4.5 and $p_r \geqslant 1$ be constants. Then*

$$\begin{aligned} &\int_B \left(\prod_{r=1}^{m} |u_r(x)|^{(p_r+2)/2} \right)^{1/m} \mathrm{d}x \\ &\quad \leqslant \frac{\alpha}{2m\sqrt{n}} \left(\prod_{r=1}^{m} \left(\frac{p_r + 2}{2} \right) \right)^{1/m} \\ &\qquad \times \sum_{r=1}^{m} \left\{ \int_B |u_r(x)|^{p_r} \,\mathrm{d}x \right\}^{1/2} \left\{ \int_B |\operatorname{grad} u_r(x)|^2 \,\mathrm{d}x \right\}^{1/2}, \end{aligned} \tag{4.4.23}$$

where α is as defined in Theorem 4.4.5.

REMARK 4.4.5. In the special case when $m = 1$, inequalities (4.4.22) and (4.4.23) reduce respectively to the following Poincaré- and Sobolev-type inequalities

$$\int_B |u_1(x)|^{p_1} \,\mathrm{d}x \leqslant \frac{1}{n} \left(\frac{\alpha}{2} \right)^{p_1} \int_B |\operatorname{grad} u_1(x)|^{p_1} \,\mathrm{d}x, \quad p_1 \geqslant 2, \tag{4.4.24}$$

and

$$\int_B |u_1(x)|^{(p_r+2)/2}\,\mathrm{d}x$$
$$\leqslant \frac{\alpha}{2\sqrt{n}}\left(\frac{p_1+2}{2}\right)\left\{\int_B |u_1(x)|^{p_1}\,\mathrm{d}x\right\}^{1/2}\left\{\int_B |\operatorname{grad} u_1(x)|^2\,\mathrm{d}x\right\}^{1/2}, \quad p_1 \geqslant 1. \tag{4.4.25}$$

For similar inequalities, see [73,120,121,152–157,178,179,418].

PROOFS OF THEOREMS 4.4.3 AND 4.4.4. From the hypotheses of Theorem 4.4.3, we have the following identities

$$u_r(x) = \int_{-\infty}^{x_1} \frac{\partial}{\partial t_1} u_r(t_1, x_2, \ldots, x_n)\,\mathrm{d}t_1, \tag{4.4.26}$$

$$u_r(x) = -\int_{x_1}^{\infty} \frac{\partial}{\partial t_1} u_r(t_1, x_2, \ldots, x_n)\,\mathrm{d}t_1, \tag{4.4.27}$$

for $r = 1, \ldots, m$. From (4.4.26) and (4.4.27), we obtain

$$|u_r(x)| \leqslant \frac{1}{2}\int_1 \left|\frac{\partial}{\partial t_1} u_r(t_1, x_2, \ldots, x_n)\right| \mathrm{d}t_1. \tag{4.4.28}$$

Similarly, we obtain

$$|u_r(x)| \leqslant \frac{1}{2}\int_i \left|\frac{\partial}{\partial t_i} u_r(x_1, \ldots, t_i, \ldots, x_n)\right| \mathrm{d}t_i \tag{4.4.29}$$

for $i = 2, \ldots, n$. From (4.4.28) and (4.4.29), we observe that

$$|u_r(x)|^{n/(n-1)}$$
$$\leqslant \left(\frac{1}{2}\right)^{n/(n-1)}\left\{\int_1 \left|\frac{\partial}{\partial t_1} u_r(t_1, x_2, \ldots, x_n)\right| \mathrm{d}t_1\right\}^{1/(n-1)}$$
$$\times \cdots \times \left\{\int_n \left|\frac{\partial}{\partial t_n} u_r(x_1, \ldots, x_{n-1}, t_n)\right| \mathrm{d}t_n\right\}^{1/(n-1)} \tag{4.4.30}$$

for $r = 1, \ldots, m$. From (4.4.30) and using the inequality

$$\left(\prod_{i=1}^{k} c_i\right)^{1/k} \leqslant \frac{1}{k}\sum_{i=1}^{k} c_i \tag{4.4.31}$$

(for c_i nonnegative reals and $k \geqslant 1$), we obtain

$$
\begin{aligned}
&\left\{\prod_{r=1}^{m} |u_r(x)|^{n/(n-1)}\right\}^{1/m} \\
&\quad \leqslant \frac{1}{m}\left(\frac{1}{2}\right)^{n/(n-1)} \left\{\left\{\int_{1}\left|\frac{\partial}{\partial t_1} u_1(t_1, x_2, \ldots, x_n)\right| \mathrm{d}t_1\right\}^{1/(n-1)}\right. \\
&\qquad\qquad \times \cdots \times \left\{\int_{n}\left|\frac{\partial}{\partial t_n} u_1(x_1, \ldots, x_{n-1}, t_n)\right| \mathrm{d}t_n\right\}^{1/(n-1)} \\
&\qquad\qquad \vdots \\
&\qquad\qquad + \left\{\int_{1}\left|\frac{\partial}{\partial t_1} u_m(t_1, x_2, \ldots, x_n)\right| \mathrm{d}t_1\right\}^{1/(n-1)} \\
&\qquad\qquad \left. \times \cdots \times \left\{\int_{n}\left|\frac{\partial}{\partial t_n} u_m(x_1, \ldots, x_{n-1}, t_n)\right| \mathrm{d}t_n\right\}^{1/(n-1)}\right\}.
\end{aligned}
\tag{4.4.32}
$$

We integrate both sides of (4.4.32) with respect to x_1 and use on the right-hand side the general version of Hölder's inequality (4.2.29) (see [179, p. 40]) with $k = n - 1$. We then integrate the resulting inequality with respect to x_2 and use inequality (4.4.31) on the right-hand side. We repeat this procedure, integrating with respect to $x_3, \ldots, x_n$, and obtain (see [121, Chapter 1, Theorem 9.3])

$$
\begin{aligned}
&\int_E \left(\prod_{r=1}^{m} |u_r(x)|^{n/(n-1)}\right)^{1/m} \mathrm{d}x \\
&\quad \leqslant \frac{1}{m}\left(\frac{1}{2}\right)^{n/(n-1)} \\
&\qquad \times \left\{\left\{\int_E\left|\frac{\partial}{\partial x_1} u_1(x)\right| \mathrm{d}x\right\}^{1/(n-1)} \cdots \left\{\int_E\left|\frac{\partial}{\partial x_n} u_1(x)\right| \mathrm{d}x\right\}^{1/(n-1)}\right. \\
&\qquad\quad \vdots \\
&\qquad\quad \left. + \left\{\int_E\left|\frac{\partial}{\partial x_1} u_m(x)\right| \mathrm{d}x\right\}^{1/(n-1)} \cdots \left\{\int_E\left|\frac{\partial}{\partial x_n} u_m(x)\right| \mathrm{d}x\right\}^{1/(n-1)}\right\}.
\end{aligned}
\tag{4.4.33}
$$

From (4.4.33) and using inequalities (4.3.5), (4.4.31) and the inequality

$$\left(\sum_{i=1}^{n} c_i\right)^2 \leqslant n \sum_{i=1}^{n} c_i^2, \tag{4.4.34}$$

where $c_1, \ldots, c_n$ are reals, we obtain

$$\begin{aligned}
&\left\{\int_E \left(\prod_{r=1}^{m} |u_r(x)|^{n/(n-1)}\right)^{1/m} \mathrm{d}x\right\}^{(n-1)/n} \\
&\leqslant \frac{1}{2}\left(\frac{1}{m}\right)^{(n-1)/n} \left\{\left\{\int_E \left|\frac{\partial}{\partial x_1} u_1(x)\right| \mathrm{d}x\right\}^{1/n} \cdots \left\{\int_E \left|\frac{\partial}{\partial x_n} u_1(x)\right| \mathrm{d}x\right\}^{1/n}\right. \\
&\qquad \vdots \\
&\qquad \left. + \left\{\int_E \left|\frac{\partial}{\partial x_1} u_m(x)\right| \mathrm{d}x\right\}^{1/n} \cdots \left\{\int_E \left|\frac{\partial}{\partial x_n} u_m(x)\right| \mathrm{d}x\right\}^{1/n}\right\} \\
&\leqslant \frac{1}{2n}\left(\frac{1}{m}\right)^{(n-1)/n} \left\{\int_E \left[\left|\frac{\partial}{\partial x_1} u_1(x)\right| + \cdots + \left|\frac{\partial}{\partial x_n} u_1(x)\right|\right] \mathrm{d}x\right. \\
&\qquad \vdots \\
&\qquad \left. + \int_E \left[\left|\frac{\partial}{\partial x_1} u_m(x)\right| + \cdots + \left|\frac{\partial}{\partial x_n} u_m(x)\right|\right] \mathrm{d}x\right\} \\
&= \frac{1}{2n}\left(\frac{1}{m}\right)^{(n-1)/n} \left\{\int_E \left\{\left[\left|\frac{\partial}{\partial x_1} u_1(x)\right| + \cdots + \left|\frac{\partial}{\partial x_n} u_1(x)\right|\right]^2\right\}^{1/2} \mathrm{d}x\right. \\
&\qquad \vdots \\
&\qquad \left. + \int_E \left\{\left[\left|\frac{\partial}{\partial x_1} u_m(x)\right| + \cdots + \left|\frac{\partial}{\partial x_n} u_m(x)\right|\right]^2\right\}^{1/2} \mathrm{d}x\right\} \\
&\leqslant \frac{1}{\sqrt{4n}}\left(\frac{1}{m}\right)^{(n-1)/n} \sum_{r=1}^{m} \int_E |\operatorname{grad} u_r(x)|\, \mathrm{d}x.
\end{aligned}$$

The proof of Theorem 4.4.3 is complete.

From the assumptions of Theorem 4.4.4, we have the following identities

$$\{u_r(x)\}^{(p_r+2)/2} = \left(\frac{p_r+2}{2}\right)\int_{-\infty}^{x_1} u_r^{p_r/2}(t_1, x_2, \ldots, x_n) \times \frac{\partial}{\partial t_1} u_r(t_1, x_2, \ldots, x_n)\,\mathrm{d}t_1, \tag{4.4.35}$$

$$\{u_r(x)\}^{(p_r+2)/2} = -\left(\frac{p_r+2}{2}\right)\int_{x_1}^{\infty} u_r^{p_r/2}(t_1, x_2, \ldots, x_n) \times \frac{\partial}{\partial t_1} u_r(t_1, x_2, \ldots, x_n)\,\mathrm{d}t_1, \tag{4.4.36}$$

for $r = 1, \ldots, m$. From (4.4.35) and (4.4.36), we obtain

$$|u_r(x)|^{(p_r+2)/2} \leqslant \left(\frac{p_r+2}{4}\right)\int_1 |u_r(t_1, x_2, \ldots, x_n)|^{p_r/2}\left|\frac{\partial}{\partial t_1} u_r(t_1, x_2, \ldots, x_n)\right|\mathrm{d}t_1. \tag{4.4.37}$$

Similarly, we obtain

$$|u_r(x)|^{(p_r+2)/2} \leqslant \left(\frac{p_r+2}{4}\right)\int_i |u_r(x_1, \ldots, t_i, \ldots, x_n)|^{p_r/2}\left|\frac{\partial}{\partial t_i} u_r(x_1, \ldots, t_i, \ldots, x_n)\right|\mathrm{d}t_i \tag{4.4.38}$$

for $i = 2, \ldots, n$. From (4.4.37) and (4.4.38), we observe that

$$\begin{aligned} |u_r(x)|^{((p_r+2)/2)(n/(n-1))} &\leqslant \left(\frac{p_r+2}{4}\right)^{n/(n-1)} \\ &\quad \times \left\{\int_1 |u_r(t_1, x_2, \ldots, x_n)|^{p_r/2}\left|\frac{\partial}{\partial t_1} u_r(t_1, x_2, \ldots, x_n)\right|\mathrm{d}t_1\right\}^{1/(n-1)} \\ &\quad \times \cdots \times \left\{\int_n |u_r(x_1, \ldots, x_{n-1}, t_n)|^{p_r/2}\left|\frac{\partial}{\partial t_n} u_r(x_1, \ldots, x_{n-1}, t_n)\right|\mathrm{d}t_n\right\}^{1/(n-1)} \end{aligned} \tag{4.4.39}$$

for $r = 1, \ldots, m$. From (4.4.39) and inequality (4.4.31), we obtain

$$
\begin{aligned}
&\left\{\prod_{r=1}^{m} |u_r(x)|^{((p_r+2)/2)(n/(n-1))}\right\}^{1/m} \\
&\quad \leqslant \frac{1}{m}\left\{\prod_{r=1}^{m}\left(\frac{p_r+2}{4}\right)^{n/(n-1)}\right\}^{1/m} \\
&\qquad \times \left[\left\{\int_{1} |u_1(t_1, x_2, \ldots, x_n)|^{p_1/2}\left|\frac{\partial}{\partial t_1}u_1(t_1, x_2, \ldots, x_n)\right| \mathrm{d}t_1\right\}^{1/(n-1)}\right. \\
&\qquad\qquad \times \cdots \times \left\{\int_{n} |u_1(x_1, \ldots, x_{n-1}, t_n)|^{p_1/2}\right. \\
&\qquad\qquad\qquad \left.\times \left|\frac{\partial}{\partial t_n}u_1(x_1, \ldots, x_{n-1}, t_n)\right| \mathrm{d}t_n\right\}^{1/(n-1)} \\
&\qquad\qquad + \cdots + \left\{\int_{1} |u_m(t_1, x_2, \ldots, x_n)|^{p_m/2}\left|\frac{\partial}{\partial t_1}u_m(t_1, x_2, \ldots, x_n)\right| \mathrm{d}t_1\right\}^{1/(n-1)} \\
&\qquad\qquad \times \cdots \times \left\{\int_{n} |u_m(x_1, \ldots, x_{n-1}, t_n)|^{p_m/2}\right. \\
&\qquad\qquad\qquad \left.\left.\times \left|\frac{\partial}{\partial t_n}u_m(x_1, \ldots, x_{n-1}, t_n)\right| \mathrm{d}t_n\right\}^{1/(n-1)}\right].
\end{aligned}
\tag{4.4.40}
$$

We integrate both sides of (4.4.40) with respect to x_1, use Hölder's inequality (4.2.29) on the right-hand side, then integrate the resulting inequality with respect to x_2, and finally use inequality (4.2.29) on the right-hand side once more. We repeat this process, integrating with respect to $x_3, \ldots, x_n$, we obtain

$$
\begin{aligned}
&\int_E \left\{\prod_{r=1}^{m} |u_r(x)|^{((p_r+2)/2)(n/(n-1))}\right\}^{1/m} \mathrm{d}x \\
&\quad \leqslant \frac{1}{m}\left\{\prod_{r=1}^{m}\left(\frac{p_r+2}{4}\right)^{n/(n-1)}\right\}^{1/m} \\
&\qquad \times \left[\left\{\int_E |u_1(x)|^{p_1/2}\left|\frac{\partial}{\partial x_1}u_1(x)\right| \mathrm{d}x\right\}^{1/(n-1)}\right. \\
&\qquad\qquad \times \cdots \times \left\{\int_E |u_1(x)|^{p_1/2}\left|\frac{\partial}{\partial x_n}u_1(x)\right| \mathrm{d}x\right\}^{1/(n-1)}
\end{aligned}
$$

$$+\cdots+\left\{\int_E |u_m(x)|^{p_m/2}\left|\frac{\partial}{\partial x_1}u_m(x)\right| \mathrm{d}x\right\}^{1/(n-1)}$$

$$\left.\times\cdots\times\left\{\int_E |u_m(x)|^{p_m/2}\left|\frac{\partial}{\partial x_n}u_m(x)\right| \mathrm{d}x\right\}^{1/(n-1)}\right]. \tag{4.4.41}$$

From (4.4.41), (4.3.5), (4.4.31), the Schwarz inequality and (4.4.34), we obtain

$$\left\{\int_E \left\{\prod_{r=1}^{m} |u_r(x)|^{((p_r+2)/2)(n/(n-1))}\right\}^{1/m} \mathrm{d}x\right\}^{(n-1)/n}$$

$$\leqslant \left(\frac{1}{m}\right)^{(n-1)/n} \left\{\prod_{r=1}^{m}\left(\frac{p_r+2}{4}\right)\right\}^{1/m}$$

$$\times\left[\left\{\int_E |u_1(x)|^{p_1/2}\left|\frac{\partial}{\partial x_1}u_1(x)\right| \mathrm{d}x\right\}^{1/n}\right.$$

$$\times\cdots\times\left\{\int_E |u_1(x)|^{p_1/2}\left|\frac{\partial}{\partial x_n}u_1(x)\right| \mathrm{d}x\right\}^{1/n}$$

$$\vdots$$

$$+\left\{\int_E |u_m(x)|^{p_m/2}\left|\frac{\partial}{\partial x_1}u_m(x)\right| \mathrm{d}x\right\}^{1/n}$$

$$\left.\times\cdots\times\left\{\int_E |u_m(x)|^{p_m/2}\left|\frac{\partial}{\partial x_n}u_m(x)\right| \mathrm{d}x\right\}^{1/n}\right]$$

$$\leqslant \frac{1}{n}\left(\frac{1}{m}\right)^{(n-1)/n} \left\{\prod_{r=1}^{m}\left(\frac{p_r+2}{4}\right)\right\}^{1/m}$$

$$\times\left\{\int_E |u_1(x)|^{p_1/2}\left[\left|\frac{\partial}{\partial x_1}u_1(x)\right|+\cdots+\left|\frac{\partial}{\partial x_n}u_1(x)\right|\right] \mathrm{d}x\right.$$

$$\left.+\cdots+\int_E |u_m(x)|^{p_m/2}\left[\left|\frac{\partial}{\partial x_1}u_m(x)\right|+\cdots+\left|\frac{\partial}{\partial x_n}u_m(x)\right|\right] \mathrm{d}x\right\}$$

$$
\begin{aligned}
&\leqslant \frac{1}{n}\left(\frac{1}{m}\right)^{(n-1)/n}\left\{\prod_{r=1}^{m}\left(\frac{p_r+2}{4}\right)\right\}^{1/m}\\
&\quad\times\left\{\int_E |u_1(x)|^{p_1}\,\mathrm{d}x\right\}^{1/2}\\
&\quad\times\left\{\int_E\left[\left|\frac{\partial}{\partial x_1}u_1(x)\right|+\cdots+\left|\frac{\partial}{\partial x_n}u_1(x)\right|\right]^2\mathrm{d}x\right\}^{1/2}\\
&\quad\vdots\\
&\quad+\left\{\int_E |u_m(x)|^{p_m}\,\mathrm{d}x\right\}^{1/2}\\
&\quad\times\left\{\int_E\left[\left|\frac{\partial}{\partial x_1}u_m(x)\right|+\cdots+\left|\frac{\partial}{\partial x_n}u_m(x)\right|\right]^2\mathrm{d}x\right\}^{1/2}\\
&\leqslant \frac{1}{\sqrt{n}}\left(\frac{1}{m}\right)^{(n-1)/n}\left\{\prod_{r=1}^{m}\left(\frac{p_r+2}{4}\right)\right\}^{1/m}\\
&\quad\times\sum_{r=1}^{m}\left\{\int_E |u_r(x)|^{p_r}\,\mathrm{d}x\right\}^{1/2}\left\{\int_E |\operatorname{grad} u_r(x)|^2\,\mathrm{d}x\right\}^{1/2}.
\end{aligned}
$$

This inequality is the desired inequality (4.4.20) and the proof of Theorem 4.4.4 is complete. □

PROOFS OF THEOREMS 4.4.5 AND 4.4.6. From the hypotheses of Theorem 4.4.5, since $u_r(x)$ are smooth functions defined on B which vanish on the boundary ∂B of B, we have the following identities

$$
nu_r(x)=\sum_{i=1}^{n}\int_{a_i}^{x_i}\frac{\partial}{\partial t_i}u_r(x_1,\ldots,t_i,\ldots,x_n)\,\mathrm{d}t_i \tag{4.4.42}
$$

and

$$
nu_r(x)=-\sum_{i=1}^{n}\int_{x_i}^{b_i}\frac{\partial}{\partial t_i}u_r(x_1,\ldots,t_i,\ldots,x_n)\,\mathrm{d}t_i, \tag{4.4.43}
$$

for $r = 1, \ldots, m$. From (4.4.42) and (4.4.43), we observe that

$$2n|u_r(x)| \leqslant \sum_{i=1}^{n} \int_{a_i}^{b_i} \left| \frac{\partial}{\partial t_i} u_r(x_1, \ldots, t_i, \ldots, x_n) \right| \mathrm{d}t_i. \qquad (4.4.44)$$

From (4.4.44) and on using inequality (4.3.5), Hölder's inequality with indices p_r, $p_r/(p_r - 1)$ and the definition of α, we obtain

$$\begin{aligned} |u_r(x)|^{p_r} &\leqslant \left(\frac{1}{2n}\right)^{p_r} n^{p_r - 1} \left[\left\{ \sum_{i=1}^{n} \int_{a_i}^{b_i} \left| \frac{\partial}{\partial t_i} u_r(x_1, \ldots, t_i, \ldots, x_n) \right| \mathrm{d}t_i \right\}^{p_r} \right] \\ &\leqslant \frac{1}{n} \left(\frac{1}{2}\right)^{p_r} \alpha^{p_r - 1} \left[\sum_{i=1}^{n} \int_{a_i}^{b_i} \left| \frac{\partial}{\partial t_i} u_r(x_1, \ldots, t_i, \ldots, x_n) \right|^{p_r} \mathrm{d}t_i \right]. \end{aligned} \qquad (4.4.45)$$

From (4.4.45) and inequality (4.4.31), we obtain

$$\begin{aligned} &\left(\prod_{r=1}^{m} |u_r(x)|^{p_r} \right)^{1/m} \\ &\leqslant \frac{1}{nm} \left(\frac{1}{2}\right)^{\frac{1}{m} \sum_{r=1}^{m} p_r} \alpha^{\frac{1}{m}(\sum_{r=1}^{m} p_r) - 1} \\ &\quad \times \left\{ \left[\sum_{i=1}^{n} \int_{a_i}^{b_i} \left| \frac{\partial}{\partial t_i} u_1(x_1, \ldots, t_i, \ldots, x_n) \right|^{p_1} \mathrm{d}t_i \right] \right. \\ &\qquad \vdots \\ &\quad \left. + \left[\sum_{i=1}^{n} \int_{a_i}^{b_i} \left| \frac{\partial}{\partial t_i} u_m(x_1, \ldots, t_i, \ldots, x_n) \right|^{p_m} \mathrm{d}t_i \right] \right\}. \end{aligned} \qquad (4.4.46)$$

By integrating both sides of (4.4.46) over B, using the definition of α and inequality (4.3.5) with $k = 2/p_r \leqslant 1$, we have

$$\begin{aligned} &\int_B \left(\prod_{r=1}^{m} |u_r(x)|^{p_r} \right)^{1/m} \mathrm{d}x \\ &\leqslant \frac{1}{nm} \left(\frac{1}{2}\right)^{\frac{1}{m} \sum_{r=1}^{m} p_r} \alpha^{\frac{1}{m}(\sum_{r=1}^{m} p_r) - 1} \end{aligned}$$

$$
\begin{aligned}
&\times \alpha\left\{\int_B\left\{\left[\left|\frac{\partial}{\partial x_1}u_1(x)\right|^{p_1}+\cdots+\left|\frac{\partial}{\partial x_n}u_1(x)\right|^{p_1}\right]^{2/p_1}\right\}^{p_1/2}\mathrm{d}x\right.\\
&\qquad\vdots\\
&\qquad\left.+\int_B\left\{\left[\left|\frac{\partial}{\partial x_1}u_m(x)\right|^{p_m}+\cdots+\left|\frac{\partial}{\partial x_n}u_m(x)\right|^{p_m}\right]^{2/p_m}\right\}^{p_m/2}\mathrm{d}x\right\}\\
&\leqslant \frac{1}{nm}\left(\frac{\alpha}{2}\right)^{\frac{1}{m}\sum_{r=1}^{m}p_r}\sum_{r=1}^{m}\int_B\left|\operatorname{grad}u_r(x)\right|^{p_r}\mathrm{d}x.
\end{aligned}
$$

The proof of Theorem 4.4.5 is complete.

From the assumptions on the functions $u_r(x)$ in Theorem 4.4.6, we have the following identities

$$
n\big(u_r(x)\big)^{(p_r+2)/2}=\left(\frac{p_r+2}{2}\right)\left[\sum_{i=1}^{n}\int_{a_i}^{x_i}u_r^{p_r/2}(x_1,\ldots,t_i,\ldots,x_n)\times\frac{\partial}{\partial t_i}u_r(x_1,\ldots,t_i,\ldots,x_n)\,\mathrm{d}t_i\right], \tag{4.4.47}
$$

$$
n\big(u_r(x)\big)^{(p_r+2)/2}=-\left(\frac{p_r+2}{2}\right)\left[\sum_{i=1}^{n}\int_{x_i}^{b_i}u_r^{p_r/2}(x_1,\ldots,t_i,\ldots,x_n)\times\frac{\partial}{\partial t_i}u_r(x_1,\ldots,t_i,\ldots,x_n)\,\mathrm{d}t_i\right], \tag{4.4.48}
$$

for $r=1,\ldots,m$. From (4.4.47) and (4.4.48), we observe that

$$
\begin{aligned}
&2n\left|u_r(x)\right|^{(p_r+2)/2}\\
&\leqslant\left(\frac{p_r+2}{2}\right)\left[\sum_{i=1}^{n}\int_{a_i}^{b_i}\left|u_r(x_1,\ldots,t_i,\ldots,x_n)\right|^{p_r/2}\times\left|\frac{\partial}{\partial t_i}u_r(x_1,\ldots,t_i,\ldots,x_n)\right|\mathrm{d}t_i\right].
\end{aligned} \tag{4.4.49}
$$

From (4.4.49) and inequality (4.4.31), we obtain

$$
\begin{aligned}
&\left(\prod_{r=1}^{m}\left|u_r(x)\right|^{(p_r+2)/2}\right)^{1/m} \\
&\quad \leqslant \frac{1}{2nm}\left(\prod_{r=1}^{m}\left(\frac{p_r+2}{2}\right)\right)^{1/m} \\
&\quad \times \left\{\left[\sum_{i=1}^{n}\int_{a_i}^{b_i}\left|u_1(x_1,\ldots,t_i,\ldots,x_n)\right|^{p_1/2}\left|\frac{\partial}{\partial t_i}u_1(x_1,\ldots,t_i,\ldots,x_n)\right|\,\mathrm{d}t_i\right]\right. \\
&\qquad \vdots \\
&\qquad + \left[\sum_{i=1}^{n}\int_{a_i}^{b_i}\left|u_m(x_1,\ldots,t_i,\ldots,x_n)\right|^{p_m/2}\right. \\
&\qquad\qquad \left.\left.\times\left|\frac{\partial}{\partial t_i}u_m(x_1,\ldots,t_i,\ldots,x_n)\right|\,\mathrm{d}t_i\right]\right\}. \qquad (4.4.50)
\end{aligned}
$$

Integrating both sides of (4.4.50) over B, using the definition of α, the Schwarz inequality and inequality (4.4.34) we have

$$
\begin{aligned}
&\int_B\left(\prod_{r=1}^{m}\left|u_r(x)\right|^{(p_r+2)/2}\right)^{1/m}\mathrm{d}x \\
&\quad \leqslant \frac{1}{2nm}\left(\prod_{r=1}^{m}\left(\frac{p_r+2}{2}\right)\right)^{1/m} \\
&\quad \times \alpha\left\{\int_B\left|u_1(x)\right|^{p_1/2}\left[\left|\frac{\partial}{\partial x_1}u_1(x)\right|+\cdots+\left|\frac{\partial}{\partial x_n}u_1(x)\right|\right]\mathrm{d}x\right. \\
&\qquad \vdots \\
&\qquad \left.+\int_D\left|u_m(x)\right|^{p_m/2}\left[\left|\frac{\partial}{\partial x_1}u_m(x)\right|+\cdots+\left|\frac{\partial}{\partial x_n}u_m(x)\right|\right]\mathrm{d}x\right\} \\
&\quad \leqslant \frac{\alpha}{2nm}\left(\prod_{r=1}^{m}\left(\frac{p_r+2}{2}\right)\right)^{1/m}
\end{aligned}
$$

$$\times \left[\left\{ \int_B |u_1(x)|^{p_1} \, \mathrm{d}x \right\}^{1/2} \left\{ \int_B \left[\left| \frac{\partial}{\partial x_1} u_1(x) \right| + \cdots + \left| \frac{\partial}{\partial x_n} u_1(x) \right| \right]^2 \mathrm{d}x \right\}^{1/2} \right.$$

$$\vdots$$

$$+ \left\{ \int_B |u_m(x)|^{p_m} \, \mathrm{d}x \right\}^{1/2}$$

$$\left. \times \left\{ \int_B \left[\left| \frac{\partial}{\partial x_1} u_m(x) \right| + \cdots + \left| \frac{\partial}{\partial x_n} u_m(x) \right| \right]^2 \mathrm{d}x \right\}^{1/2} \right]$$

$$\leqslant \frac{1}{2m\sqrt{n}} \left(\prod_{r=1}^{m} \left(\frac{p_r + 2}{2} \right) \right)^{1/m}$$

$$\times \sum_{r=1}^{m} \left\{ \int_B |u_r(x)|^{p_r} \, \mathrm{d}x \right\}^{1/2} \left\{ \int_B |\operatorname{grad} u_r(x)|^2 \, \mathrm{d}x \right\}^{1/2}.$$

This inequality is the required inequality in (4.4.23) and the proof of Theorem 4.4.6 is complete. □

4.5 Inequalities of Dubinskii and Others

Integral inequalities of Poincaré and Sobolev type play a fundamental role in the theory and applications of partial differential equations. A large number of inequalities related to these inequalities are established by several authors in the literature. In this section we deal with certain inequalities established by Dubinskii [95], Alzer [10] and Pachpatte [345].

In what follows, we let $x = (x_1, \ldots, x_n)$ be a variable point in $\mathbb{R}^n$, an n-dimensional Euclidean space, G be a bounded region in $\mathbb{R}^n$ with boundary ∂G satisfying the cone condition (see [95]), $C^m(G)$ is the space of functions $u(x)$ with bounded derivatives in $\overline{G}$ (the closure of G) up to order m inclusive, $\mathrm{d}x = \mathrm{d}x_1 \cdots \mathrm{d}x_n$ is the volume element, and $\mathrm{d}s$ is the surface element corresponding to ∂G. Constant quantities, not depending on $u(x)$, will be denoted by the symbol K. In different inequalities their meaning will be different.

The inequalities in the following theorems are established by Dubinskii [95].

THEOREM 4.5.1. *Let* $-\infty < \alpha_0 < +\infty$, $\alpha_1 \geqslant 1$, $u(x), |u(x)|^{\alpha_0+\alpha_1} \in C^1(G)$. *Then the following inequality is valid*

$$\int_G |u|^{\alpha_0+\alpha_1} \, \mathrm{d}x \leqslant K \left[\int_G |u|^{\alpha_0} \left| \frac{\partial u}{\partial x_i} \right|^{\alpha_1} \mathrm{d}x + \int_{\partial G} |u|^{\alpha_0+\alpha_1} \, \mathrm{d}s \right] \tag{4.5.1}$$

for $i = 1, \dots, n$. *The constant* K *depends on* α_0, α_1 *and* G.

PROOF. From the divergence theorem we have

$$\int_G \frac{\partial}{\partial x_i}\left(x_i |u|^{\alpha_0+\alpha_1}\right) \mathrm{d}x = \int_{\partial G} x_i |u|^{\alpha_0+\alpha_1}\,\mathrm{d}s. \tag{4.5.2}$$

From (4.5.2) it is easy to observe that

$$\int_G |u|^{\alpha_0+\alpha_1}\,\mathrm{d}x \leqslant K\left[\int_G |u|^{\alpha_0+\alpha_1-1}\left|\frac{\partial u}{\partial x_i}\right| \mathrm{d}x + \int_{\partial G} |u|^{\alpha_0+\alpha_1}\,\mathrm{d}s\right], \tag{4.5.3}$$

from which, for $\alpha_1 = 1$, we obtain inequality (4.5.1). If $\alpha_1 > 1$, then

$$\begin{aligned}
&\int_G |u|^{\alpha_0+\alpha_1-1}\left|\frac{\partial u}{\partial x_i}\right| \mathrm{d}x \\
&= \int_G |u|^{\alpha_0/\alpha_1}\left|\frac{\partial u}{\partial x_i}\right| |u|^{\alpha_0+\alpha_1-1-\alpha_0/\alpha_1}\,\mathrm{d}x \\
&\leqslant \frac{\varepsilon^{\alpha_1}}{\alpha_1}\int_G |u|^{\alpha_0}\left|\frac{\partial u}{\partial x_i}\right|^{\alpha_1} \mathrm{d}x + \left(\frac{\alpha_1-1}{\alpha_1}\right)\varepsilon^{-\alpha_1/(\alpha_1-1)}\int_G |u|^{\alpha_0+\alpha_1}\,\mathrm{d}x.
\end{aligned} \tag{4.5.4}$$

Here we have used Young's inequality

$$ab \leqslant \frac{\varepsilon^p}{p}a^p + \frac{1}{q}\varepsilon^{-q}b^q, \quad a, b \geqslant 0,\ \frac{1}{p} + \frac{1}{q} = 1,\ \varepsilon > 0,$$

for $p = \alpha_1$. Choosing $\varepsilon > 0$ sufficiently large, from (4.5.3) and (4.5.4), we obtain inequality (4.5.1). □

REMARK 4.5.1. We note that, for the case when α_0, α_1 are even and $u|_{\partial G} = 0$, inequality (4.5.1) was obtained earlier by Visik [421].

THEOREM 4.5.2. *Let* $-\infty < \alpha_0 < +\infty$, $\alpha_1 \geqslant 0$, $\alpha_2 \geqslant 0$, $\alpha_1 + \alpha_2 \geqslant 1$, $u(x)$, $|u(x)|^{\alpha_0+\alpha_1+\alpha_2} \in C^1(G)$. *Then the following inequality is valid*

$$\int_G |u|^{\alpha_0+\alpha_1}\left|\frac{\partial u}{\partial x_i}\right|^{\alpha_2} \leqslant K\left[\int_G |u|^{\alpha_0}\left|\frac{\partial u}{\partial x_i}\right|^{\alpha_1+\alpha_2} \mathrm{d}x + \int_{\partial G} |u|^{\alpha_0+\alpha_1+\alpha_2}\,\mathrm{d}s\right] \tag{4.5.5}$$

for $i = 1, \dots, n$.

The proof follows by estimating the integral on the left-hand side of (4.5.5), by using Young's inequality with index $p = (\alpha_1 + \alpha_2)\alpha_2^{-1}$ and Theorem 4.5.1.

THEOREM 4.5.3. *Let α_0, α_1 and α_2 be nonnegative numbers, $\alpha_3 \geqslant 1$, $\alpha_3 \geqslant \alpha_1$, $\alpha_0 + \alpha_2 + \alpha_3 - (\alpha_3 - \alpha_1)(\alpha_3 - 1)^{-1} \geqslant 0$, $u(x) \in C^2(G)$. Then the following inequality is valid*

$$\int_G |u|^{\alpha_0} \left| \frac{\partial u}{\partial x_i} \right|^{\alpha_1+\alpha_2+\alpha_3} \mathrm{d}x$$
$$\leqslant K \left[\int_G |u|^{\alpha_0+\alpha_1} \left| \frac{\partial u}{\partial x_i} \right|^{\alpha_2} \left| \frac{\partial^2 u}{\partial x_i^2} \right|^{\alpha_3} \mathrm{d}x + \int_{\partial G} |u|^{\alpha_0+\alpha_1+\alpha_2+\alpha_3} \mathrm{d}s + \int_{\partial G} \left| \frac{\partial u}{\partial x_i} \right|^{\alpha_0+\alpha_1+\alpha_2+\alpha_3} \mathrm{d}s \right] \quad (4.5.6)$$

for $i = 1, \ldots, n$.

PROOF. We have the obvious equalities

$$|u|^{\alpha_0} \left| \frac{\partial u}{\partial x_i} \right|^{\alpha_1+\alpha_2+\alpha_3} = |u|^{\alpha_0} \frac{\partial u}{\partial x_i} \left| \frac{\partial u}{\partial x_i} \right|^{\alpha_1+\alpha_2+\alpha_3-1} \operatorname{sign} \frac{\partial u}{\partial x_i}$$
$$= \frac{1}{\alpha_0+1} \frac{\partial}{\partial x_i} \left[|u|^{\alpha_0+1} \operatorname{sign} u \right] \left| \frac{\partial u}{\partial x_i} \right|^{\alpha_1+\alpha_2+\alpha_3-1} \operatorname{sign} \frac{\partial u}{\partial x_i}.$$

Integrating this equation over G we obtain

$$\int_G |u|^{\alpha_0} \left| \frac{\partial u}{\partial x_i} \right|^{\alpha_1+\alpha_2+\alpha_3} \mathrm{d}x$$
$$= \frac{1}{\alpha_0+1} \int_G \frac{\partial}{\partial x_i} \left[|u|^{\alpha_0+1} \operatorname{sign} u \right] \left| \frac{\partial u}{\partial x_i} \right|^{\alpha_1+\alpha_2+\alpha_3-1} \operatorname{sign} \frac{\partial u}{\partial x_i} \, \mathrm{d}x.$$

From this equation, integrating by parts, we have

$$\int_G |u|^{\alpha_0} \left| \frac{\partial u}{\partial x_i} \right|^{\alpha_1+\alpha_2+\alpha_3} \mathrm{d}x$$
$$\leqslant K \left[\int_G |u|^{\alpha_0+1} \left| \frac{\partial u}{\partial x_i} \right|^{\alpha_1+\alpha_2+\alpha_3-2} \left| \frac{\partial^2 u}{\partial x_i^2} \right| \mathrm{d}x + \int_{\partial G} |u|^{\alpha_0+1} \left| \frac{\partial u}{\partial x_i} \right|^{\alpha_1+\alpha_2+\alpha_3-1} \mathrm{d}s \right]$$
$$\leqslant K \left[\int_G |u|^{\alpha_0+1} \left| \frac{\partial u}{\partial x_i} \right|^{\alpha_1+\alpha_2+\alpha_3-2} \left| \frac{\partial^2 u}{\partial x_i^2} \right| \mathrm{d}x + \int_{\partial G} |u|^{\alpha_0+\alpha_1+\alpha_2+\alpha_3} \mathrm{d}s + \int_{\partial G} \left| \frac{\partial u}{\partial x_i} \right|^{\alpha_0+\alpha_1+\alpha_2+\alpha_3} \mathrm{d}s \right]. \quad (4.5.7)$$

Here we use Young's inequality with index $p = (\alpha_0 + \alpha_1 + \alpha_2 + \alpha_3)(\alpha_0 + 1)^{-1}$ to the integral over ∂G. Inequality (4.5.7) yields inequality (4.5.6) for the case

$\alpha_3 = \alpha_1 = 1$. If $\alpha_3 > 1$, then applying Young's inequality and Theorem 4.5.2, we obtain

$$
\begin{aligned}
&\int_G |u|^{\alpha_0+1}\left|\frac{\partial u}{\partial x_i}\right|^{\alpha_1+\alpha_2+\alpha_3-2}\left|\frac{\partial^2 u}{\partial x_i^2}\right| \mathrm{d}x \\
&= \int_G \left[|u|^{(\alpha_0+\alpha_1)/\alpha_3}\left|\frac{\partial u}{\partial x_i}\right|^{\alpha_2/\alpha_3}\left|\frac{\partial^2 u}{\partial x_i^2}\right|\right] \\
&\qquad \times \left[|u|^{\alpha_0+1-(\alpha_0+\alpha_1)/\alpha_3}\left|\frac{\partial u}{\partial x_i}\right|^{\alpha_1+\alpha_2+\alpha_3-2-\alpha_2/\alpha_3}\right] \mathrm{d}x \\
&\leqslant \frac{\varepsilon^{\alpha_3}}{\alpha_3}\int_G |u|^{\alpha_0+\alpha_1}\left|\frac{\partial u}{\partial x_i}\right|^{\alpha_2}\left|\frac{\partial^2 u}{\partial x_i^2}\right|^{\alpha_3} \mathrm{d}x \\
&\quad + \left(\frac{\alpha_3-1}{\alpha_3}\right)\varepsilon^{-\alpha_3/(\alpha_3-1)} \\
&\qquad \times \int_G |u|^{\alpha_0+(\alpha_3-\alpha_1)/(\alpha_3-1)}\left|\frac{\partial u}{\partial x_i}\right|^{\alpha_1+\alpha_2+\alpha_3-(\alpha_3-\alpha_1)/(\alpha_3-1)} \mathrm{d}x \\
&\leqslant \frac{\varepsilon^{\alpha_3}}{\alpha_3}\int_G |u|^{\alpha_0+\alpha_1}\left|\frac{\partial u}{\partial x_i}\right|^{\alpha_2}\left|\frac{\partial^2 u}{\partial x_i^2}\right|^{\alpha_3} \mathrm{d}x \\
&\quad + K\varepsilon^{-\alpha_3/(\alpha_3-1)}\int_G |u|^{\alpha_0}\left|\frac{\partial u}{\partial x_i}\right|^{\alpha_1+\alpha_2+\alpha_3} \mathrm{d}x \\
&\quad + K\varepsilon^{-\alpha_3/(\alpha_3-1)}\int_{\partial G} |u|^{\alpha_0+\alpha_1+\alpha_2+\alpha_3}\,\mathrm{d}s. \qquad (4.5.8)
\end{aligned}
$$

Inequality (4.5.6) follows from (4.5.7) and (4.5.8) if $\varepsilon > 0$ is taken sufficiently large. □

THEOREM 4.5.4. *Let* $\alpha_0, \alpha_1, \alpha_2, \alpha_3$ *be nonnegative numbers,* $\alpha_4 \geqslant 1$, $\alpha_3 \geqslant \alpha_1$, $\alpha_0\alpha_3 - \alpha_1\alpha_4 \geqslant 0$, $\sum_{i=0}^4 \alpha_i - \alpha_4(\alpha_3-\alpha_1)(\alpha_4-1)^{-1} \geqslant 0$, *and* $u(x) \in C^2(G)$. *Then the following inequality is valid*

$$
\begin{aligned}
&\int_G |u|^{\alpha_0}\left|\frac{\partial u}{\partial x_i}\right|^{\alpha_1+\alpha_2+\alpha_3}\left|\frac{\partial^2 u}{\partial x_i^2}\right|^{\alpha_4} \mathrm{d}x \\
&\leqslant K\left[\int_G |u|^{\alpha_0+\alpha_1}\left|\frac{\partial u}{\partial x_i}\right|^{\alpha_2}\left|\frac{\partial^2 u}{\partial x_i^2}\right|^{\alpha_3+\alpha_4} \mathrm{d}x\right. \\
&\qquad \left. + \int_{\partial G} |u|^{\sum_{i=1}^4 \alpha_i}\,\mathrm{d}s + \int_{\partial G}\left|\frac{\partial u}{\partial x_i}\right|^{\sum_{i=1}^4 \alpha_i} \mathrm{d}s\right] \qquad (4.5.9)
\end{aligned}
$$

for $i = 1, \ldots, n$.

The proof of inequality (4.5.9) follows from Young's inequality with index $p = (\alpha_3 + \alpha_4)\alpha_4^{-1}$ and Theorem 4.5.3.

In the following theorems, we let

$$\|\operatorname{grad} u(x)\|_\mu = \left(\sum_{i=1}^n \left| \frac{\partial}{\partial x_i} u(x) \right|^\mu \right)^{1/\mu}, \quad \mu > 0,$$

S be the set of all real-valued functions $u(x)$ which are continuous on $B = \prod_{i=1}^n [a_i, b_i]$ (the bounded domain in $\mathbb{R}^n$) which satisfy $u(x)|_{x_i = a_i} = u(x)|_{x_i = b_i} = 0$ for each $i \in \{1, \ldots, n\}$, and for which the partial derivatives $\frac{\partial}{\partial x_i} u(x)$ exist.

The following theorem is given by Alzer in [10].

THEOREM 4.5.5. *Let* $\lambda \geqslant 1$ *and* $\mu > 0$ *be real numbers. Then we have for all* $u \in S$

$$\int_B |u(x)|^\lambda \, dx \leqslant K_n(\lambda, \mu; a, b) \int_B \|\operatorname{grad} u(x)\|_\mu^\lambda \, dx, \tag{4.5.10}$$

where

$$K_n(\lambda, \mu; a, b) = I(\lambda) n^{-\min(1, \lambda/\mu)} \prod_{i=1}^n (b_i - a_i)^{\lambda/n},$$

in which

$$I(\lambda) = \int_0^1 \left[t^{1-\lambda} + (1-t)^{1-\lambda} \right]^{-1} dt.$$

PROOF. From the hypotheses, we have the following identities

$$u(x) = \int_{a_i}^{x_i} \frac{\partial}{\partial t_i} u(x_1, \ldots, t_i, \ldots, x_n) \, dt_i \tag{4.5.11}$$

and

$$u(x) = -\int_{x_i}^{b_i} \frac{\partial}{\partial t_i} u(x_1, \ldots, t_i, \ldots, x_n) \, dt_i. \tag{4.5.12}$$

From (4.5.11), (4.5.12) and using Hölder's inequality with indices λ, $\lambda/(\lambda-1)$, we get

$$\begin{aligned}\left|u(x)\right|^{\lambda} &\leqslant \left(\int_{a_i}^{x_i}\left|\frac{\partial}{\partial t_i}u(x_1,\ldots,t_i,\ldots,x_n)\right|\mathrm{d}t_i\right)^{\lambda}\\ &\leqslant (x_i-a_i)^{\lambda-1}\int_{a_i}^{x_i}\left|\frac{\partial}{\partial t_i}u(x_1,\ldots,t_i,\ldots,x_n)\right|^{\lambda}\mathrm{d}t_i, \qquad (4.5.13)\end{aligned}$$

$$\left|u(x)\right|^{\lambda} \leqslant (b_i-x_i)^{\lambda-1}\int_{x_i}^{b_i}\left|\frac{\partial}{\partial t_i}u(x_1,\ldots,t_i,\ldots,x_n)\right|^{\lambda}\mathrm{d}t_i. \qquad (4.5.14)$$

From (4.5.13) and (4.5.14), we obtain for $x_i \in (a_i, b_i)$

$$\left|u(x)\right|^{\lambda}\left[(x_i-a_i)^{1-\lambda}+(b_i-x_i)^{1-\lambda}\right] \leqslant \int_{a_i}^{b_i}\left|\frac{\partial}{\partial t_i}u(x_1,\ldots,t_i,\ldots,x_n)\right|^{\lambda}\mathrm{d}t_i. \qquad (4.5.15)$$

Next, we multiply both sides of (4.5.15) by $[(x_i-a_i)^{1-\lambda}+(b_i-x_i)^{1-\lambda}]^{-1}$ and integrate over B. Then we have

$$\int_B\left|u(x)\right|^{\lambda}\mathrm{d}x \leqslant \int_{a_i}^{b_i}\left[(x_i-a_i)^{1-\lambda}+(b_i-x_i)^{1-\lambda}\right]^{-1}\mathrm{d}x_i\int_B\left|\frac{\partial}{\partial x_i}u(x)\right|^{\lambda}\mathrm{d}x.$$

Now, by taking $i=1,\ldots,n$ in the above inequality and multiplying the resulting inequalities and applying arithmetic mean–geometric mean inequality, we obtain

$$\begin{aligned}&\int_B\left|u(x)\right|^{\lambda}\mathrm{d}x\\ &\quad\leqslant \prod_{i=1}^{n}\left(\int_{a_i}^{b_i}\left[(x_i-a_i)^{1-\lambda}+(b_i-x_i)^{1-\lambda}\right]^{-1}\mathrm{d}x_i\right)^{1/n}\\ &\quad\quad\times\prod_{i=1}^{n}\left(\int_B\left|\frac{\partial}{\partial x_i}u(x)\right|^{\lambda}\mathrm{d}x\right)^{1/n}\\ &\quad\leqslant \frac{1}{n}\prod_{i=1}^{n}\left(\int_{a_i}^{b_i}\left[(x_i-a_i)^{1-\lambda}+(b_i-x_i)^{1-\lambda}\right]^{-1}\mathrm{d}x_i\right)^{1/n}\int_B\sum_{i=1}^{n}\left|\frac{\partial}{\partial x_i}u(x)\right|^{\lambda}\mathrm{d}x\\ &\quad= \frac{1}{n}\left(\prod_{i=1}^{n}(b_i-a_i)^{\lambda/n}\right)\int_0^1\left[t^{1-\lambda}+(1-t)^{1-\lambda}\right]^{-1}\mathrm{d}t\int_B\left\|\operatorname{grad}u(x)\right\|_{\lambda}^{\lambda}\mathrm{d}x.\end{aligned} \qquad (4.5.16)$$

Finally, we use the inequality (see [48, pp. 143 and 159])

$$\sum_{i=1}^{n} a_i^{\alpha} \leqslant n^{1-\min(1,\alpha)} \left(\sum_{i=1}^{n} a_i \right)^{\alpha}, \quad a_i \geqslant 0,\ i = 1, \ldots, n,\ \alpha > 0, \tag{4.5.17}$$

to obtain

$$\begin{aligned} \|\operatorname{grad} u(x)\|_{\lambda}^{\lambda} &= \sum_{i=1}^{n} \left| \frac{\partial}{\partial x_i} u(x) \right|^{\lambda} = \sum_{i=1}^{n} \left| \frac{\partial}{\partial x_i} u(x) \right|^{(\lambda/\mu)\mu} \\ &\leqslant n^{1-\min(1,\lambda/\mu)} \left(\sum_{i=1}^{n} \left| \frac{\partial}{\partial x_i} u(x) \right|^{\mu} \right)^{\lambda/\mu} \\ &= n^{1-\min(1,\lambda/\mu)} \|\operatorname{grad} u(x)\|_{\mu}^{\lambda}, \end{aligned} \tag{4.5.18}$$

so that (4.5.16) and (4.5.18) imply

$$\begin{aligned} \int_B |u(x)|^{\lambda} \, dx \leqslant \left(\prod_{i=1}^{n} (b_i - a_i)^{\lambda/n} \right) \int_0^1 \left[t^{1-\lambda} + (1-t)^{1-\lambda} \right]^{-1} dt \\ \times n^{-\min(1,\lambda/\mu)} \int_B \|\operatorname{grad} u(x)\|_{\mu}^{\lambda} \, dx. \end{aligned}$$

The proof is complete. □

REMARK 4.5.2. We note that inequality (4.5.10) sharpens the inequality given by Agarwal and Sheng in [6]. For further results, see [7,10] and the references given therein.

In [345] Pachpatte has established the inequalities in the following theorems.

THEOREM 4.5.6. *Let $p \geqslant 0$, $q \geqslant 1$, $r \geqslant 1$, $\mu > 0$ be constants and $u \in S$. Then*

$$\int_B |u(x)|^{r(p+q)} \, dx \leqslant M^q \int_B |u(x)|^{rp} \|\operatorname{grad} u(x)\|_{\mu}^{rq} \, dx, \tag{4.5.19}$$

$$\int_B |u(x)|^{r(p+q)} \, dx \leqslant M^{p+q} \int_B \|\operatorname{grad} u(x)\|_{\mu}^{r(p+q)} \, dx, \tag{4.5.20}$$

where

$$M = \left[(p+q)^r n^{-\min(1,r/\mu)} \left(\prod_{i=1}^{n} (b_i - a_i)^{r/n} \right) I(r) \right], \tag{4.5.21}$$

in which

$$I(r) = \int_0^1 \left[t^{1-r} + (1-t)^{1-r} \right]^{-1} \mathrm{d}t. \tag{4.5.22}$$

PROOF. From the hypotheses, we have the following identities

$$u^{p+q}(x) = (p+q) \int_{a_i}^{x_i} u^{p+q-1}(x_1, \ldots, t_i, \ldots, x_n) \times \frac{\partial}{\partial t_i} u(x_1, \ldots, t_i, \ldots, x_n)\, \mathrm{d}t_i, \tag{4.5.23}$$

$$u^{p+q}(x) = -(p+q) \int_{x_i}^{b_i} u^{p+q-1}(x_1, \ldots, t_i, \ldots, x_n) \times \frac{\partial}{\partial t_i} u(x_1, \ldots, t_i, \ldots, x_n)\, \mathrm{d}t_i, \tag{4.5.24}$$

for $i = 1, \ldots, n$. From (4.5.23), (4.5.24) and using Hölder's inequality with indices r, $r/(r-1)$, we observe that

$$\left|u(x)\right|^{r(p+q)} \leqslant (p+q)^r (x_i - a_i)^{r-1} \int_{a_i}^{x_i} \left|u(x_1, \ldots, t_i, \ldots, x_n)\right|^{r(p+q-1)} \times \left| \frac{\partial}{\partial t_i} u(x_1, \ldots, t_i, \ldots, x_n) \right|^r \mathrm{d}t_i, \tag{4.5.25}$$

$$\left|u(x)\right|^{r(p+q)} \leqslant (p+q)^r (b_i - x_i)^{r-1} \int_{x_i}^{b_i} \left|u(x_1, \ldots, t_i, \ldots, x_n)\right|^{r(p+q-1)} \times \left| \frac{\partial}{\partial t_i} u(x_1, \ldots, t_i, \ldots, x_n) \right|^r \mathrm{d}t_i, \tag{4.5.26}$$

for $i = 1, \ldots, n$. From (4.5.25) and (4.5.26), we obtain for $x_i \in (a_i, b_i)$

$$\left[(x_i - a_i)^{1-r} + (b_i - x_i)^{1-r} \right] \left|u(x)\right|^{p+q}$$

$$\leqslant (p+q)^r \int_{a_i}^{b_i} \left|u(x_1, \ldots, t_i, \ldots, x_n)\right|^{r(p+q-1)} \times \left|\frac{\partial}{\partial t_i} u(x_1, \ldots, t_i, \ldots, x_n)\right|^r \mathrm{d}t_i. \tag{4.5.27}$$

Next, we multiply both sides of (4.5.27) by $[(x_i - a_i)^{1-r} + (b_i - x_i)^{1-r}]^{-1}$ and integrate over B. Then we have

$$\int_B |u(x)|^{r(p+q)} \,\mathrm{d}x \leqslant (p+q)^r \int_{a_i}^{b_i} \left[(x_i - a_i)^{1-r} + (b_i - x_i)^{1-r}\right]^{-1} \mathrm{d}x_i \times \int_B |u(x)|^{r(p+q-1)} \left|\frac{\partial}{\partial x_i} u(x)\right|^r \mathrm{d}x. \tag{4.5.28}$$

Now, by taking $i = 1, \ldots, n$ in (4.5.28) and multiplying the resulting inequalities and applying the arithmetic mean–geometric mean inequality, we obtain

$$\begin{aligned}
&\int_B |u(x)|^{r(p+q)} \,\mathrm{d}x \\
&\quad \leqslant (p+q)^r \prod_{i=1}^n \left(\int_{a_i}^{b_i} \left[(x_i - a_i)^{1-r} + (b_i - x_i)^{1-r}\right]^{-1} \mathrm{d}x_i \right)^{1/n} \\
&\qquad \times \prod_{i=1}^n \left(\int_B |u(x)|^{r(p+q-1)} \left|\frac{\partial}{\partial x_i} u(x)\right|^r \mathrm{d}x \right)^{1/n} \\
&\quad \leqslant \frac{1}{n}(p+q)^r \prod_{i=1}^n \left(\int_{a_i}^{b_i} \left[(x_i - a_i)^{1-r} + (b_i - x_i)^{1-r}\right]^{-1} \mathrm{d}x_i \right)^{1/n} \\
&\qquad \times \int_B \sum_{i=1}^n |u(x)|^{r(p+q-1)} \left|\frac{\partial}{\partial x_i} u(x)\right|^r \mathrm{d}x \\
&\quad = \frac{1}{n}(p+q)^r \left(\prod_{i=1}^n (b_i - a_i)^{r/n} \right) \int_0^1 \left[t^{1-r} + (1-t)^{1-r}\right]^{-1} \mathrm{d}t \\
&\qquad \times \int_B |u(x)|^{r(p+q-1)} \left\|\operatorname{grad} u(x)\right\|_r^r \mathrm{d}x.
\end{aligned} \tag{4.5.29}$$

By using inequality (4.5.17), as in the proof of Theorem 4.5.5, we obtain

$$\left\|\operatorname{grad} u(x)\right\|_r^r \leqslant n^{1-\min(1, r/\mu)} \left\|\operatorname{grad} u(x)\right\|_\mu^r. \tag{4.5.30}$$

From (4.5.29) and (4.5.30), we obtain

$$\begin{aligned}
&\int_B |u(x)|^{r(p+q)}\,\mathrm{d}x \\
&\quad \leqslant (p+q)^r n^{-\min(1,r/\mu)} \\
&\qquad \times \left(\prod_{i=1}^{n}(b_i-a_i)^{r/n}\right) I(r) \int_B |u(x)|^{r(p+q-1)} \|\operatorname{grad} u(x)\|_\mu^r\,\mathrm{d}x \\
&\quad = M \int_B \big[|u(x)|^{rp/q} \|\operatorname{grad} u(x)\|_\mu^r\big]\big[|u(x)|^{r(p+q-1)-rp/q}\big]\,\mathrm{d}x. \qquad (4.5.31)
\end{aligned}$$

Using Hölder's inequality with indices q, $q/(q-1)$ on the right-hand side of (4.5.31) we have

$$\begin{aligned}
&\int_B |u(x)|^{r(p+q)}\,\mathrm{d}x \\
&\quad \leqslant M\left[\int_B |u(x)|^{rp} \|\operatorname{grad} u(x)\|_\mu^{rq}\right]^{1/q}\left[\int_B |u(x)|^{r(p+q)}\,\mathrm{d}x\right]^{(q-1)/q}. \qquad (4.5.32)
\end{aligned}$$

If $\int_B |u(x)|^{r(p+q)}\,\mathrm{d}x = 0$ then (4.5.19) is trivially true; otherwise, we divide both sides of (4.5.32) by $[\int_B |u(x)|^{r(p+q)}\,\mathrm{d}x]^{(q-1)/q}$ and then raise both sides to the power q to get the required inequality in (4.5.19).

By using Hölder's inequality with indices $(p+q)/p$, $(p+q)/q$ to the right-hand side of (4.5.19), we get

$$\begin{aligned}
&\int_B |u(x)|^{r(p+q)}\,\mathrm{d}x \\
&\quad \leqslant M^q\left[\int_B |u(x)|^{r(p+q)}\,\mathrm{d}x\right]^{p/(p+q)}\left[\int_B \|\operatorname{grad} u(x)\|_\mu^{r(p+q)}\,\mathrm{d}x\right]^{q/(p+q)}.
\end{aligned} \qquad (4.5.33)$$

If $\int_B |u(x)|^{r(p+q)}\,\mathrm{d}x = 0$ then (4.5.20) is trivially true; otherwise, we divide both sides of (4.5.33) by $[\int_B |u(x)|^{r(p+q)}\,\mathrm{d}x]^{p/(p+q)}$ and then raise both sides to the power $(p+q)/q$ to get the required inequality in (4.5.20). The proof is complete. □

THEOREM 4.5.7. *Let $p \geqslant 0$, $q \geqslant 1$, $m \geqslant 0$, $r \geqslant 1$, $\mu > 0$ be constants and $u \in S$. Then*

$$\int_B |u(x)|^{r(p+q)} \|\operatorname{grad} u(x)\|_\mu^{rm}\,\mathrm{d}x \leqslant L^q \int_B |u(x)|^{rp} \|\operatorname{grad} u(x)\|_\mu^{r(q+m)}\,\mathrm{d}x, \qquad (4.5.34)$$

and

$$\int_B |u(x)|^{r(p+q)} \|\operatorname{grad} u(x)\|_\mu^{rm} \,\mathrm{d}x \leqslant L^{p+q} \int_B \|\operatorname{grad} u(x)\|_\mu^{r(p+q+m)} \,\mathrm{d}x, \tag{4.5.35}$$

where

$$L = \left[(p+q+m)^r n^{-\min(1,r/\mu)} \left(\prod_{i=1}^{n} (b_i - a_i)^{r/n} \right) I(r) \right], \tag{4.5.36}$$

in which $I(r)$ is defined by (4.5.22).

PROOF. By rewriting the integral on the left-hand side of (4.5.34) and using Hölder's inequality with indices $(q+m)/m$, $(q+m)/q$ and inequality (4.5.19), we observe that

$$\begin{aligned}
&\int_B |u(x)|^{r(p+q)} \|\operatorname{grad} u(x)\|_\mu^{rm} \,\mathrm{d}x \\
&= \int_B \left[|u(x)|^{r(pm/(q+m))} \|\operatorname{grad} u(x)\|_\mu^{rm}\right]\left[|u(x)|^{r(p+q)-r(pm/(q+m))}\right] \mathrm{d}x \\
&\leqslant \left[\int_B |u(x)|^{rp} \|\operatorname{grad} u(x)\|_\mu^{r(q+m)} \,\mathrm{d}x \right]^{m/(q+m)} \\
&\quad \times \left[\int_B |u(x)|^{r(p+q+m)} \,\mathrm{d}x \right]^{q/(q+m)} \\
&\leqslant \left[\int_B |u(x)|^{rp} \|\operatorname{grad} u(x)\|_\mu^{r(q+m)} \,\mathrm{d}x \right]^{m/(q+m)} \\
&\quad \times \Bigg[\left[(p+q+m)^r n^{-\min(1,r/\mu)} \left(\prod_{i=1}^{n} (b_i - a_i)^{r/n} \right) I(r) \right]^{q+m} \\
&\qquad \times \int_B |u(x)|^{rp} \|\operatorname{grad} u(x)\|_\mu^{r(q+m)} \,\mathrm{d}x \Bigg]^{q/(q+m)} \\
&= L^q \int_B |u(x)|^{rp} \|\operatorname{grad} u(x)\|_\mu^{r(q+m)} \,\mathrm{d}x.
\end{aligned}$$

The proof of inequality (4.5.34) is complete.

By rewriting inequality (4.5.34) and using Hölder's inequality with indices $(p+q)/p$, $(p+q)/q$, we observe that

$$\begin{aligned}
&\int_B |u(x)|^{r(p+q)} \|\operatorname{grad} u(x)\|_\mu^{rm}\,\mathrm{d}x \\
&\leqslant L^q \int_B \big[|u(x)|^{rp} \|\operatorname{grad} u(x)\|_\mu^{r(mp/(p+q))}\big]\big[\|\operatorname{grad} u(x)\|_\mu^{r(q+m)-r(mp/(p+q))}\big]\,\mathrm{d}x \\
&\leqslant L^q \left[\int_B |u(x)|^{r(p+q)} \|\operatorname{grad} u(x)\|_\mu^{rm}\,\mathrm{d}x\right]^{p/(p+q)} \\
&\quad\times \left[\int_B \|\operatorname{grad} u(x)\|_\mu^{r(p+q+m)}\,\mathrm{d}x\right]^{q/(p+q)}.
\end{aligned}$$

Now, by following the arguments as in the last part of the proof of inequality (4.5.20) with suitable modifications, we get the required inequality in (4.5.35). The proof is complete. □

4.6 Poincaré- and Sobolev-Like Inequalities

In the present section we shall deal with the Poincaré- and Sobolev-like inequalities established by Pachpatte in [276,289].

The following inequalities are established in [276].

THEOREM 4.6.1. *Let u_r, $r = 1, \ldots, N$, be sufficiently smooth functions defined on $B = \prod_{i=1}^n [a_i, b_i]$, the bounded domain in $\mathbb{R}^n$, which vanish on the boundary ∂B of B and let $m \geqslant 1$, $p \geqslant 2$ be real constants. Then*

$$\left[\int_B \left\{\sum_{r=1}^N |u_r(x)|^2\right\}^{p/(p-1)} \mathrm{d}x\right]^{2m(p-1)/p} \leqslant k_1 \sum_{r=1}^N \int_B \sum_{i=1}^n \left|\frac{\partial}{\partial x_i} u_r(x)\right|^{4m} \mathrm{d}x, \tag{4.6.1}$$

where

$$k_1 = \frac{1}{nN}\left(\frac{N}{4}\right)^{2m} \alpha^{(p(4m+2nm-n)-2nm)/p},$$

in which $\alpha = \max\{b_1 - a_1, \ldots, b_n - a_n\}$.

REMARK 4.6.1. In the special case when $m = 1$, inequality (4.6.1) reduces to the following inequality

$$\left[\int_B \left\{\sum_{r=1}^N |u_r(x)|^2\right\}^{p/(p-1)} \mathrm{d}x\right]^{2(p-1)/p} \leqslant k_1^0 \sum_{r=1}^N \int_B \sum_{i=1}^n \left|\frac{\partial}{\partial x_i} u_r(x)\right|^4 \mathrm{d}x, \tag{4.6.2}$$

where

$$k_1^0 = \frac{N}{16n}\alpha^{(n+4)-2n/p}.$$

THEOREM 4.6.2. *Let u_r, m, p be as in Theorem* 4.6.1. *Then*

$$\left[\int_B \left\{\sum_{r=1}^N |u_r(x)|^2\right\}^{p/(p-1)} \mathrm{d}x\right]^{2m(p-1)/p} \leqslant k_2 \sum_{r=1}^N \int_B \sum_{i=1}^n |u_r(x)|^{2m} \left|\frac{\partial}{\partial x_i} u_r(x)\right|^{2m} \mathrm{d}x, \tag{4.6.3}$$

where

$$k_2 = \frac{1}{n} N^{2m-1} \alpha^{(p(2m+2nm-n)-2nm)/p},$$

in which $\alpha = \max\{b_1 - a_1, \ldots, b_n - a_n\}$.

REMARK 4.6.2. We note that in the special case when $m = 1$, inequality (4.6.3) reduces to

$$\left[\int_B \left\{\sum_{r=1}^N |u_r(x)|^2\right\}^{p/(p-1)} \mathrm{d}x\right]^{2(p-1)/p} \leqslant k_2^0 \sum_{r=1}^N \int_B \sum_{i=1}^n |u_r(x)|^2 \left|\frac{\partial}{\partial x_i} u_r(x)\right|^2 \mathrm{d}x, \tag{4.6.4}$$

where

$$k_2^0 = \frac{N}{n}\alpha^{(n+2)-2n/p}.$$

PROOFS OF THEOREMS 4.6.1 AND 4.6.2. From the hypotheses of Theorem 4.6.1, we have the following identities

$$nu_r(x) = \sum_{i=1}^{n} \int_{a_i}^{x_i} \frac{\partial}{\partial t_i} u_r(x_1, \ldots, t_i, \ldots, x_n)\, \mathrm{d}t_i \tag{4.6.5}$$

and

$$nu_r(x) = -\sum_{i=1}^{n} \int_{x_i}^{b_i} \frac{\partial}{\partial t_i} u_r(x_1, \ldots, t_i, \ldots, x_n)\, \mathrm{d}t_i, \tag{4.6.6}$$

for $r = 1, \ldots, N$. From (4.6.5) and (4.6.6), we observe that

$$2n\left|u_r(x)\right| \leqslant \sum_{i=1}^{n} \int_{a_i}^{b_i} \left|\frac{\partial}{\partial t_i} u_r(x_1, \ldots, t_i, \ldots, x_n)\right| \mathrm{d}t_i \tag{4.6.7}$$

for $r = 1, \ldots, N$. From (4.6.7), using inequality (4.3.5), the Schwarz inequality and the definition of α, we obtain

$$\begin{aligned}
\left|u_r(x)\right|^2 &\leqslant \left(\frac{1}{2n}\right)^2 \left[\sum_{i=1}^{n} \int_{a_i}^{b_i} \left|\frac{\partial}{\partial t_i} u_r(x_1, \ldots, t_i, \ldots, x_n)\right| \mathrm{d}t_i\right]^2 \\
&\leqslant \left(\frac{1}{2n}\right)^2 n \sum_{i=1}^{n} \left\{\int_{a_i}^{b_i} \left|\frac{\partial}{\partial t_i} u_r(x_1, \ldots, t_i, \ldots, x_n)\right| \mathrm{d}t_i\right\}^2 \\
&\leqslant \left(\frac{\alpha}{4n}\right) \sum_{i=1}^{n} \int_{a_i}^{b_i} \left|\frac{\partial}{\partial t_i} u_r(x_1, \ldots, t_i, \ldots, x_n)\right|^2 \mathrm{d}t_i.
\end{aligned} \tag{4.6.8}$$

From (4.6.8), using inequality (4.3.5) repeatedly, Hölder's inequality with indices p, $p/(p-1)$ and the definition of α, we obtain

$$\begin{aligned}
&\left\{\sum_{r=1}^{N} \left|u_r(x)\right|^2\right\}^{p/(p-1)} \\
&\quad\leqslant \left(\frac{\alpha}{4n}\right)^{p/(p-1)} (Nn)^{p/(p-1)-1} \\
&\qquad\times \sum_{r=1}^{N} \left\{\sum_{i=1}^{n} \left\{\int_{a_i}^{b_i} \left|\frac{\partial}{\partial t_i} u_r(x_1, \ldots, t_i, \ldots, x_n)\right|^2 \mathrm{d}t_i\right\}^{p/(p-1)}\right\}
\end{aligned}$$

$$\leqslant \left(\frac{\alpha}{4n}\right)^{p/(p-1)} (Nn)^{1/(p-1)} \alpha^{1/(p-1)}$$
$$\times \sum_{r=1}^{N} \left\{ \sum_{i=1}^{n} \left\{ \int_{a_i}^{b_i} \left| \frac{\partial}{\partial t_i} u_r(x_1, \ldots, t_i, \ldots, x_n) \right|^{2p/(p-1)} \mathrm{d}t_i \right\}\right\}. \tag{4.6.9}$$

Integrating both sides of (4.6.9) over B and using the definition of α we have

$$\int_B \left\{ \sum_{r=1}^{N} |u_r(x)|^2 \right\}^{p/(p-1)} \mathrm{d}x$$
$$\leqslant \left(\frac{\alpha}{4n}\right)^{p/(p-1)} (Nn\alpha)^{1/(p-1)} \alpha \sum_{r=1}^{N} \left\{ \sum_{i=1}^{n} \left\{ \int_B \left| \frac{\partial}{\partial x_i} u_r(x) \right|^{2p/(p-1)} \mathrm{d}x \right\}\right\}. \tag{4.6.10}$$

From (4.6.10), using inequality (4.3.5) repeatedly, Hölder's inequality with indices $2m(p-1)/p$, $2m(p-1)/(2m(p-1)-p)$ and the definition of α, we obtain

$$\left[\int_B \left\{ \sum_{r=1}^{N} |u_r(x)|^2 \right\}^{p/(p-1)} \mathrm{d}x \right]^{2m(p-1)/p}$$
$$\leqslant \left\{ \left(\frac{\alpha}{4n}\right)^{p/(p-1)} (Nn\alpha)^{1/(p-1)} \alpha \right\}^{2m(p-1)/p}$$
$$\times \left[\sum_{r=1}^{N} \left\{ \sum_{i=1}^{n} \left\{ \int_B \left| \frac{\partial}{\partial x_i} u_r(x) \right|^{2p/(p-1)} \mathrm{d}x \right\}\right\}\right]^{2m(p-1)/p}$$
$$\leqslant \left\{ \left(\frac{\alpha}{4n}\right)^{p/(p-1)} (Nn\alpha)^{1/(p-1)} \alpha \right\}^{2m(p-1)/p} (Nn)^{2m(p-1)/p-1}$$
$$\times \sum_{r=1}^{N} \left\{ \sum_{i=1}^{n} \left\{ \int_B \left| \frac{\partial}{\partial x_i} u_r(x) \right|^{2p/(p-1)} \mathrm{d}x \right\}^{2m(p-1)/p} \right\}$$
$$\leqslant \left\{ \left(\frac{\alpha}{4n}\right)^{p/(p-1)} (Nn\alpha)^{1/(p-1)} \alpha \right\}^{2m(p-1)/p} (Nn)^{2m(p-1)/p-1}$$
$$\times (\alpha^n)^{(2m(p-1)-p)/p} \sum_{r=1}^{N} \left\{ \sum_{i=1}^{n} \int_B \left| \frac{\partial}{\partial x_i} u_r(x) \right|^{4m} \mathrm{d}x \right\}$$
$$= k_1 \sum_{r=1}^{N} \int_B \sum_{i=1}^{n} \left| \frac{\partial}{\partial x_i} u_r(x) \right|^{4m} \mathrm{d}x. \tag{4.6.11}$$

The proof of Theorem 4.6.1 is complete.

From the assumptions on the functions $u_r(x)$ in Theorem 4.6.2, we have the following identities

$$nu_r^2(x) = 2\sum_{i=1}^{n}\int_{a_i}^{x_i} u_r(x_1,\ldots,t_i,\ldots,x_n)\frac{\partial}{\partial t_i}u_r(x_1,\ldots,t_i,\ldots,x_n)\,\mathrm{d}t_i, \tag{4.6.12}$$

$$nu_r^2(x) = -2\sum_{i=1}^{n}\int_{x_i}^{b_i} u_r(x_1,\ldots,t_i,\ldots,x_n)\frac{\partial}{\partial t_i}u_r(x_1,\ldots,t_i,\ldots,x_n)\,\mathrm{d}t_i, \tag{4.6.13}$$

for $r = 1,\ldots,N$. From (4.6.12) and (4.6.13), we observe that

$$n\left|u_r(x)\right|^2 \leqslant \sum_{i=1}^{n}\int_{a_i}^{b_i} \left|u_r(x_1,\ldots,t_i,\ldots,x_n)\right|\left|\frac{\partial}{\partial t_i}u_r(x_1,\ldots,t_i,\ldots,x_n)\right|\mathrm{d}t_i. \tag{4.6.14}$$

From (4.6.14), using inequality (4.3.5) repeatedly, Hölder's inequality with indices p, $p/(p-1)$ and the definition of α, we obtain as in (4.6.9)

$$\begin{aligned}
&\left\{\sum_{r=1}^{N}\left|u_r(x)\right|^2\right\}^{p/(p-1)} \\
&\quad\leqslant \left(\frac{1}{n}\right)^{p/(p-1)}(Nn\alpha)^{1/(p-1)} \\
&\qquad\times\sum_{r=1}^{N}\left(\sum_{i=1}^{n}\left\{\int_{a_i}^{b_i}\left[\left|u_r(x_1,\ldots,t_i,\ldots,x_n)\right|\right.\right.\right. \\
&\qquad\qquad\left.\left.\left.\times\left|\frac{\partial}{\partial t_i}u_r(x_1,\ldots,t_i,\ldots,x_n)\right|\right]^{p/(p-1)}\mathrm{d}t_i\right\}\right).
\end{aligned} \tag{4.6.15}$$

Integrating both sides of (4.6.15) over B and using the definition of α, we have

$$\begin{aligned}
&\int_B\left\{\sum_{r=1}^{N}\left|u_r(x)\right|^2\right\}^{p/(p-1)}\mathrm{d}x \\
&\quad\leqslant \left(\frac{1}{n}\right)^{p/(p-1)}(Nn\alpha)^{1/(p-1)}\alpha
\end{aligned}$$

$$\times \sum_{r=1}^{N} \left\{ \sum_{i=1}^{n} \left\{ \int_B \left[|u_r(x)| \left| \frac{\partial}{\partial x_i} u_r(x) \right| \right]^{p/(p-1)} \mathrm{d}x \right\} \right\}. \qquad (4.6.16)$$

From (4.6.16), using inequality (4.3.5) repeatedly, Hölder's inequality with indices $2m(p-1)/p$, $2m(p-1)/(2m(p-1)-p)$ and the definition of α and following the same steps as in the proof of inequality (4.6.11) we get the required inequality in (4.6.3). The proof of Theorem 4.6.2 is complete. □

In our further discussion, we make use of the following fundamental result. In what follows, an open, simply connected, bounded set B of points in $\mathbb{R}^n$ is said to be a normal domain if B admits the application of the following Gauss integral theorem (see [149, p. 49]).

On the set of boundary points $x \in \partial B$ with $\overline{B} = B + \partial B$ (union of B and ∂B) there is a real-valued vector field

$$z(x) = \big(z_1(x), \ldots, z_n(x)\big) \quad \text{with } |z| = \left\{ \sum_{i=1}^{n} z_i^2(x) \right\}^{1/2} = 1$$

such that, for all complex-valued $w(x) = w(x_1, \ldots, x_n) \in C^1(\overline{B})$,

$$\int_B \frac{\partial}{\partial x_i} w(x)\, \mathrm{d}x = \int_{\partial B} w(x) z_i(x)\, \mathrm{d}s, \quad i = 1, \ldots, n, \qquad (4.6.17)$$

where $\mathrm{d}x = \mathrm{d}x_1 \cdots \mathrm{d}x_n$ is the volume element and $\mathrm{d}s$ the surface element corresponding to ∂B.

The following inequalities are also established in [276].

THEOREM 4.6.3. *Let B be a normal domain in $\mathbb{R}^n$ with boundary ∂B and $\overline{B} = B + \partial B$. Let $m \geqslant 1$, $p \geqslant 2$ be real constants and u_r, $r = 1, \ldots, N$, be real-valued functions such that u_r, $\{\sum_{r=1}^N |u_r|^2\}^{p/(p-1)} \in C^1(\overline{B})$. Then*

$$\left[\int_B \left\{ \sum_{r=1}^{N} |u_r(x)|^2 \right\}^{p/(p-1)} \mathrm{d}x \right]^{2m(p-1)/p}$$
$$\leqslant k_3 \left\{ \left[\int_{\partial B} \left\{ \sum_{r=1}^{N} |u_r(x)|^2 \right\}^{p/(p-1)} \mathrm{d}s \right]^{2m(p-1)/p} \right.$$
$$\left. + \sum_{r=1}^{N} \int_B \sum_{i=1}^{n} |u_r(x)|^{2m} \left| \frac{\partial}{\partial x_i} u_r(x) \right|^{2m} \mathrm{d}x \right\}, \qquad (4.6.18)$$

where

$$k_3 = \max\Bigg\{2^{2m(p-1)/p-1}(2\delta)^{2m(p-1)/p},$$
$$2^{2m(p-1)/p-1}\left(\frac{4\delta}{n}\right)^{2m(p-1)/p}\left(\frac{n(p-1)}{4\delta}\right)^{-2m/\delta}$$
$$\times\big(D(B)\big)^{(2m(p-1)-p)/p}(Nn)^{2m-1}\Bigg\},$$

in which $\delta = \max\{|x_1|,\ldots,|x_n|\}$ *and* $D(B)$ *is the n-dimensional measure of* B.

THEOREM 4.6.4. *Let* B *be a normal domain in* $\mathbb{R}^n$ *with sufficiently smooth boundary* ∂B *and* $\overline{B} = B + \partial B$. *Let* $\alpha_i(x) \in C^1(\overline{B})$, $i = 1,\ldots,n$, *be auxiliary functions such that* $\alpha_i(x) = z_i(x)$ *for* $x \in \partial B$. *Let* $m \geqslant 1$, $p \geqslant 2$ *be real constants and* u_r, $r = 1,\ldots,N$, *be real-valued functions such that* u_r, $\{\sum_{r=1}^N |u_r(x)|^2\}^{p/(p-1)} \in C^1(B)$. *Then*

$$\left[\int_{\partial B}\left\{\sum_{r=1}^N |u_r(x)|^2\right\}^{p/(p-1)} \mathrm{d}s\right]^{2m(p-1)/p}$$
$$\leqslant k_4\Bigg\{\left[\int_B\left\{\sum_{r=1}^N |u_r(x)|^2\right\}^{p/(p-1)} \mathrm{d}x\right]^{2m(p-1)/p}$$
$$+\sum_{r=1}^N\int_B\sum_{i=1}^n |u_r(x)|^{2m}\left|\frac{\partial}{\partial x_i}u_r(x)\right|^{2m}\mathrm{d}x\Bigg\}, \tag{4.6.19}$$

where

$$k_4 = \max\Bigg\{2^{2m(p-1)/p-1}\left(c_0 + \frac{2c_1c^p}{p-1}\right)^{2m(p-1)/p},$$
$$2^{2m(p-1)/p}\big(2c_1c^{-p/(p-1)}\big)^{2m(p-1)/p}$$
$$\times\big(D(B)\big)^{(2m(p-1)-p)/p}(Nn)^{2m-1}\Bigg\},$$

in which $D(B)$ *is as in Theorem* 4.6.3, $c > 0$ *is an arbitrary constant and*

$$c_0 = \sup_{x\in B}\left|\sum_{i=1}^n \frac{\partial}{\partial x_i}\alpha_i(x)\right|, \qquad c_1 = \sup_{i=1,\ldots,n}\left\{\sup_{x\in B}|\alpha_i(x)|\right\}.$$

PROOFS OF THEOREMS 4.6.3 AND 4.6.4. If we set $w(x) = x_i \times \{\sum_{r=1}^N |u_r(x)|^2\}^{p/(p-1)}$ in Gauss integral formula (4.6.17), then we have

$$\int_B \left\{\sum_{r=1}^N |u_r(x)|^2\right\}^{p/(p-1)} dx = \int_{\partial B} x_i \left\{\sum_{r=1}^N |u_r(x)|^2\right\}^{p/(p-1)} z_i(x)\, ds - \int_B x_i \left(\frac{p}{p-1}\right) \left\{\sum_{r=1}^N |u_r(x)|^2\right\}^{p/(p-1)-1} \times \sum_{r=1}^N 2|u_r(x)| \frac{\partial}{\partial x_i} u_r(x) \operatorname{sign} u_r(x)\, dx \tag{4.6.20}$$

for $i = 1, \dots, n$. From (4.6.20) we observe that

$$\begin{aligned} n \int_B \left\{\sum_{r=1}^N |u_r(x)|^2\right\}^{p/(p-1)} dx &= \int_{\partial B} \left\{\sum_{i=1}^n x_i \left\{\sum_{r=1}^N |u_r(x)|^2\right\}^{p/(p-1)} z_i(x)\right\} ds \\ &\quad - \frac{2p}{p-1} \\ &\quad \times \int_B \left\{\sum_{i=1}^n x_i \left\{\sum_{r=1}^N |u_r(x)|^2\right\}^{1/(p-1)} \sum_{r=1}^N |u_r(x)| \frac{\partial}{\partial x_i} u_r(x) \operatorname{sign} u_r(x)\right\} dx. \end{aligned} \tag{4.6.21}$$

From (4.6.21), the definition of δ and the fact that $|z_i(x)| \leqslant 1$ for $i = 1, \dots, n$, we obtain

$$\begin{aligned} n \int_B \left\{\sum_{r=1}^N |u_r(x)|^2\right\}^{p/(p-1)} dx &\leqslant n\delta \int_{\partial B} \left\{\sum_{r=1}^N |u_r(x)|^2\right\}^{p/(p-1)} ds \\ &\quad + \frac{2p\delta}{p-1} \int_B \left\{\sum_{r=1}^N |u_r(x)|^2\right\}^{1/(p-1)} \sum_{r=1}^N \left\{ |u_r(x)| \sum_{i=1}^n \left| \frac{\partial}{\partial x_i} u_r(x) \right| \right\} dx. \end{aligned} \tag{4.6.22}$$

From (4.5.22) and using the following version of Young's inequality

$$d_1 d_2 \leqslant \frac{1}{p}\varepsilon^p d_1^p + \left(\frac{p-1}{p}\right)\varepsilon^{-p/(p-1)} d_2^{p/(p-1)}, \tag{4.6.23}$$

where $d_1, d_2 \geqslant 0$, $p \geqslant 2$, $\varepsilon > 0$, and setting $\varepsilon = \{n(p-1)/(4\delta)\}^{1/p}$, we observe that

$$\begin{aligned}
n \int_B &\left\{\sum_{r=1}^N |u_r(x)|^2\right\}^{p/(p-1)} \mathrm{d}x \\
&\leqslant n\delta \int_{\partial B} \left\{\sum_{r=1}^N |u_r(x)|^2\right\}^{p/(p-1)} \mathrm{d}s \\
&\quad + \frac{2p\delta}{p-1} \int_B \left[\frac{1}{p}\left(\frac{n(p-1)}{4\delta}\right)\left\{\sum_{r=1}^N |u_r(x)|^2\right\}^{p/(p-1)}\right. \\
&\quad\quad + \left(\frac{p-1}{p}\right)\left(\frac{n(p-1)}{4\delta}\right)^{-1/(p-1)} \\
&\quad\quad \left. \times \left\{\sum_{r=1}^N \left\{|u_r(x)| \sum_{i=1}^n \left|\frac{\partial}{\partial x_i} u_r(x)\right|\right\}\right\}^{p/(p-1)}\right] \mathrm{d}x.
\end{aligned} \tag{4.6.24}$$

From (4.6.24) we observe that

$$\begin{aligned}
\int_B &\left\{\sum_{r=1}^N |u_r(x)|^2\right\}^{p/(p-1)} \mathrm{d}x \\
&\leqslant 2\delta \int_{\partial B} \left\{\sum_{r=1}^N |u_r(x)|^2\right\}^{p/(p-1)} \mathrm{d}s \\
&\quad + \frac{4\delta}{n}\left(\frac{n(p-1)}{4\delta}\right)^{-1/(p-1)} \int_B \left\{\sum_{r=1}^N \left\{|u_r(x)| \sum_{i=1}^n \left|\frac{\partial}{\partial x_i} u_r(x)\right|\right\}\right\}^{p/(p-1)} \mathrm{d}x.
\end{aligned} \tag{4.6.25}$$

From (4.6.25) and using the following inequality repeatedly,

$$\left\{\sum_{i=1}^n a_i\right\}^k \leqslant M_{k,n} \sum_{i=1}^n a_i^k, \tag{4.6.26}$$

where a_i are nonnegative reals and $M_{k,n} = n^{k-1}$, $k \geqslant 1$, $M_{k,n} = 1$, $0 \leqslant k \leqslant 1$, Hölder's inequality with indices $2m(p-1)/p$, $2m(p-1)/(2m(p-1)-p)$, we observe that

$$\begin{aligned}
&\left[\int_B \left\{\sum_{r=1}^N |u_r(x)|^2\right\}^{p/(p-1)} \mathrm{d}x\right]^{2m(p-1)/p} \\
&\quad \leqslant 2^{2m(p-1)/p-1} \\
&\qquad \times \left\{(2\delta)^{2m(p-1)/p}\left[\int_{\partial B}\left\{\sum_{r=1}^N |u_r(x)|^2\right\}^{p/(p-1)} \mathrm{d}s\right]^{2m(p-1)/p}\right. \\
&\qquad\quad + \left\{\frac{4\delta}{n}\left(\frac{n(p-1)}{4\delta}\right)^{-1/(p-1)}\right\}^{2m(p-1)/p} \\
&\qquad\quad \left.\times \left[\int_B \left\{\sum_{r=1}^N \left\{|u_r(x)| \sum_{i=1}^n \left|\frac{\partial}{\partial x_i} u_r(x)\right|\right\}\right\}^{p/(p-1)} \mathrm{d}x\right]^{2m(p-1)/p}\right\} \\
&\quad \leqslant 2^{2m(p-1)/p-1}(2\delta)^{2m(p-1)/p}\left[\int_{\partial B}\left\{\sum_{r=1}^N |u_r(x)|^2\right\}^{p/(p-1)} \mathrm{d}s\right]^{2m(p-1)/p} \\
&\qquad + 2^{2m(p-1)/p-1}\left\{\frac{4\delta}{n}\left(\frac{n(p-1)}{4\delta}\right)^{-1/(p-1)}\right\}^{2m(p-1)/p} \\
&\qquad \times \left[\left\{\int_B 1\,\mathrm{d}x\right\}^{(2m(p-1)-p)/(2m(p-1))}\right. \\
&\qquad \left.\times \left\{\int_B \left\{\sum_{r=1}^N \left\{|u_r(x)| \sum_{i=1}^n \left|\frac{\partial}{\partial x_i} u_r(x)\right|\right\}\right\}^{2m} \mathrm{d}x\right\}^{p/(2m(p-1))}\right]^{2m(p-1)/p} \\
&\quad \leqslant 2^{2m(p-1)/p-1}(2\delta)^{2m(p-1)/p}\left[\int_{\partial B}\left\{\sum_{r=1}^N |u_r(x)|^2\right\}^{p/(p-1)} \mathrm{d}s\right]^{2m(p-1)/p} \\
&\qquad + 2^{2m(p-1)/p-1}\left\{\frac{4\delta}{n}\left(\frac{n(p-1)}{4\delta}\right)^{-1/(p-1)}\right\}^{2m(p-1)/p} \\
&\qquad \times \left(D(B)\right)^{(2m(p-1)-p)/p}(Nn)^{2m-1}\sum_{r=1}^N \int_B \sum_{i=1}^n |u_r(x)|^{2m}\left|\frac{\partial}{\partial x_i} u_r(x)\right|^{2m} \mathrm{d}x.
\end{aligned} \tag{4.6.27}$$

From (4.6.27) and the definition of k_3, the desired inequality in (4.6.18) follows. The proof of Theorem 4.6.3 is complete.

From the hypotheses of Theorem 4.6.4, since B has sufficiently smooth boundary, we have auxiliary functions $\alpha_i(x) = z_i(x)$ for $x \in \partial B$ (see [149, p. 69]). Then, by making use of formula (4.6.17), we have

$$\begin{aligned}
&\int_{\partial B}\left\{\sum_{r=1}^{N}|u_r(x)|^2\right\}^{p/(p-1)} \mathrm{d}s \\
&\quad= \int_{\partial B}\left\{\sum_{i=1}^{n} z_i^2(x)\right\}\left\{\sum_{r=1}^{N}|u_r(x)|^2\right\}^{p/(p-1)} \mathrm{d}s \\
&\quad= \int_{\partial B}\left\{\sum_{i=1}^{n}\alpha_i(x)\left\{\sum_{r=1}^{N}|u_r(x)|^2\right\}^{p/(p-1)} z_i(x)\right\} \mathrm{d}s \\
&\quad= \int_{B}\frac{\partial}{\partial x_i}\left\{\sum_{i=1}^{n}\alpha_i(x)\left\{\sum_{r=1}^{N}|u_r(x)|^2\right\}^{p/(p-1)}\right\} \mathrm{d}x \\
&\quad= \int_{B}\left\{\sum_{i=1}^{n}\frac{\partial}{\partial x_i}\alpha_i(x)\right\}\left\{\sum_{r=1}^{N}|u_r(x)|^2\right\}^{p/(p-1)} \mathrm{d}x \\
&\qquad+ \frac{2p}{p-1}\int_{B}\left\{\sum_{i=1}^{n}\alpha_i(x)\left\{\sum_{r=1}^{N}|u_r(x)|^2\right\}^{1/(p-1)}\right. \\
&\qquad\qquad\left.\times\sum_{r=1}^{n}|u_r(x)|\frac{\partial}{\partial x_i}u_r(x)\,\mathrm{sign}\,u_r(x)\right\} \mathrm{d}x. \qquad (4.6.28)
\end{aligned}$$

From (4.6.28) and using the definitions of c_0, c_1, and Young's inequality (4.6.23) with $\varepsilon = c$, we observe that

$$\begin{aligned}
&\int_{\partial B}\left\{\sum_{r=1}^{N}|u_r(x)|^2\right\}^{p/(p-1)} \mathrm{d}s \\
&\quad\leqslant c_0\int_{B}\left\{\sum_{r=1}^{N}|u_r(x)|^2\right\}^{p/(p-1)} \mathrm{d}x \\
&\qquad+ \frac{2pc_1}{p-1}\int_{B}\left\{\sum_{r=1}^{N}|u_r(x)|^2\right\}^{1/(p-1)}\sum_{r=1}^{N}\left\{|u_r(x)|\sum_{i=1}^{n}\left|\frac{\partial}{\partial x_i}u_r(x)\right|\right\} \mathrm{d}x
\end{aligned}$$

$$\leqslant c_0 \int_B \left\{ \sum_{r=1}^{N} |u_r(x)|^2 \right\}^{p/(p-1)} \mathrm{d}x$$
$$+ \frac{2pc_1}{p-1} \int_B \left[\frac{1}{p} c^p \left\{ \sum_{r=1}^{N} |u_r(x)|^2 \right\}^{p/(p-1)} \right.$$
$$+ \left(\frac{p-1}{p}\right) c^{-p/(p-1)}$$
$$\left. \times \left\{ \sum_{r=1}^{N} \left\{ |u_r(x)| \sum_{i=1}^{n} \left| \frac{\partial}{\partial x_i} u_r(x) \right| \right\} \right\}^{p/(p-1)} \right] \mathrm{d}x$$
$$= \left(c_0 + \frac{2c_1 c^p}{p-1}\right) \int_B \left\{ \sum_{r=1}^{N} |u_r(x)|^2 \right\}^{p/(p-1)} \mathrm{d}x$$
$$+ 2c_1 c^{-p/(p-1)} \int_B \left\{ \sum_{r=1}^{N} \left\{ |u_r(x)| \sum_{i=1}^{n} \left| \frac{\partial}{\partial x_i} u_r(x) \right| \right\} \right\}^{p/(p-1)} \mathrm{d}x. \quad (4.6.29)$$

From (4.6.29), using inequality (4.6.26) repeatedly, Hölder's inequality with indices $2m(p-1)/p$, $2m(p-1)/(2m(p-1)-p)$ and following the same steps as in the proof of inequality (4.6.27) with suitable changes, we get the desired inequality in (4.6.19). The proof of Theorem 4.6.4 is complete. □

The inequalities in the following theorems are established by Pachpatte in [289].

THEOREM 4.6.5. *Let B be a normal domain in $\mathbb{R}^n$ with boundary ∂B and $\overline{B} = B + \partial B$. Let u_r, $r = 1, \ldots, m$, be real-valued functions belonging to $C^1(\overline{B})$. Then*

$$\int_B \left[\prod_{r=1}^{m} u_r^2(x) \right]^{1/m} \mathrm{d}x \leqslant \mu \left[\int_{\partial B} \left(\sum_{r=1}^{m} u_r^2(x) \right) \mathrm{d}s + \int_B \left(\sum_{r=1}^{m} |\operatorname{grad} u_r(x)|^2 \right) \mathrm{d}x \right], \quad (4.6.30)$$

where $\mu = \max\{\frac{4\delta^2}{mn}, \frac{2\delta}{m}\}$, $\delta = \max\{|x_1|, \ldots, |x_n|\}$.

REMARK 4.6.3. In the special case when $m = 1$ and $u_1(x) = u(x)$, inequality (4.6.30) reduces to the following inequality

$$\int_B u^2(x)\, \mathrm{d}x \leqslant \mu_1 \left[\int_{\partial B} u^2(x)\, \mathrm{d}s + \int_B |\operatorname{grad} u(x)|^2\, \mathrm{d}x \right], \quad (4.6.31)$$

where μ_1 is the constant defined by μ when $m = 1$. An inequality closely related to (4.6.31) in which the multiplicative constant on the right-hand side is different was first used by Friedrichs (see [392, p. 242]) to study the problem of boundedness from below of a differential operator and is now known in the literature as Friedrich's second inequality.

THEOREM 4.6.6. *Let B be a normal domain in $\mathbb{R}^n$ with sufficiently smooth boundary ∂B and $\overline{B} = B + \partial B$. Let $\alpha_i(x) \in C^1(\overline{B})$, $i = 1, \dots, n$, be auxiliary functions such that $\alpha_i(x) = z_i(x)$ for $x \in \partial B$. Let u_r, $r = 1, \dots, m$, be real-valued functions belonging to $C^1(\overline{B})$. Then*

$$\int_{\partial B} \left[\prod_{r=1}^{m} u_r^2(x) \right]^{1/m} \mathrm{d}s$$

$$\leqslant \lambda \left[\int_B \left(\sum_{r=1}^{m} u_r^2(x) \right) \mathrm{d}x + \int_B \left(\sum_{r=1}^{m} \left| \operatorname{grad} u_r(x) \right|^2 \right) \mathrm{d}x \right], \tag{4.6.32}$$

where $\lambda = \max\{c_0 c + \frac{c_1}{mc}, \frac{cc_1 n}{m}\}$, $c > 0$ is arbitrary constant and

$$c_0 = \sup_{x \in B} \left| \sum_{i=1}^{n} \frac{\partial}{\partial x_i} \alpha_i(x) \right|, \qquad c_1 = \sup_{i=1,\dots,n} \left\{ \sup_{x \in B} \left| \alpha_i(x) \right| \right\}.$$

REMARK 4.6.4. We note that in the special case when $m = 1$ and $u_1(x) = u(x)$, inequality (4.6.32) reduces to

$$\int_{\partial B} u^2(x)\,\mathrm{d}s \leqslant \lambda_1 \int_B \left[u^2(x) + \left| \operatorname{grad} u(x) \right|^2 \right] \mathrm{d}x, \tag{4.6.33}$$

where λ_1 is the constant defined by λ when $m = 1$. The inequalities of the forms (4.6.33) are established by many authors by using Trace theorem in interpolation spaces (see, e.g., [154]). For different forms, see [149,208,392] and the references given therein.

PROOFS OF THEOREMS 4.6.5 AND 4.6.6. If we set $u(x) = x_i u_r^2(x)$ in Gauss integral formula (4.6.17), then we have

$$\int_B u_r^2(x)\,\mathrm{d}x = \int_{\partial B} x_i u_r^2(x) z_i(x)\,\mathrm{d}s - \int_B 2x_i u_r(x) \frac{\partial}{\partial x_i} u_r(x)\,\mathrm{d}x \tag{4.6.34}$$

for $i = 1, \ldots, n$ and $r = 1, \ldots, m$. From (4.6.34) we observe that

$$
\begin{aligned}
n & \int_B u_r^2(x)\,\mathrm{d}x \\
& = \int_{\partial B} \left(\sum_{i=1}^{n} x_i u_r^2(x) z_i(x) \right) \mathrm{d}s - \int_B \left(\sum_{i=1}^{n} 2x_i u_r(x) \frac{\partial}{\partial x_i} u_r(x) \right) \mathrm{d}x. \qquad (4.6.35)
\end{aligned}
$$

Using the elementary inequality

$$
|2ab| \leqslant ca^2 + \frac{1}{c} b^2, \qquad (4.6.36)
$$

where a, b, c are arbitrary real numbers and $c > 0$. Setting $c = \frac{1}{2\delta}$, we observe that

$$
\begin{aligned}
\left| -2x_i u_r(x) \frac{\partial}{\partial x_i} u_r(x) \right| & \leqslant \delta \left| 2u_r(x) \frac{\partial}{\partial x_i} u_r(x) \right| \\
& \leqslant \delta \left[\frac{1}{2\delta} u_r^2(x) + 2\delta \left\{ \frac{\partial}{\partial x_i} u_r(x) \right\}^2 \right] \\
& = \frac{1}{2} u_r^2(x) + 2\delta^2 \left\{ \frac{\partial}{\partial x_i} u_r(x) \right\}^2. \qquad (4.6.37)
\end{aligned}
$$

From (4.6.35), (4.6.37), the definition of δ and the fact that $|z_i(x)| \leqslant 1$ for $i = 1, \ldots, n$, we obtain

$$
\begin{aligned}
n & \int_B u_r^2(x)\,\mathrm{d}x \\
& \leqslant \delta n \int_{\partial B} u_r^2(x)\,\mathrm{d}s + \frac{1}{2} n \int_B u_r^2(x)\,\mathrm{d}x + 2\delta^2 \int_B \left(\sum_{i=1}^{n} \left\{ \frac{\partial}{\partial x_i} u_r(x) \right\}^2 \right) \mathrm{d}x. \\
& \qquad (4.6.38)
\end{aligned}
$$

From (4.6.38) we obtain the inequality

$$
\int_B u_r^2(x)\,\mathrm{d}x \leqslant 2\delta \int_{\partial B} u_r^2(x)\,\mathrm{d}s + \frac{4\delta^2}{n} \int_B \left| \operatorname{grad} u_r(x) \right|^2 \mathrm{d}x \qquad (4.6.39)
$$

for $r = 1, \ldots, m$. From (4.6.39) and the elementary inequality

$$
\left[\prod_{r=1}^{m} a_r \right]^{1/m} \leqslant \frac{1}{m} \sum_{r=1}^{m} a_r, \quad a_r \geqslant 0, \qquad (4.6.40)
$$

we observe that

$$\int_B \left[\prod_{r=1}^{m} u_r^2(x)\right]^{1/m} \mathrm{d}x \leqslant \frac{1}{m}\int_B \left(\sum_{r=1}^{m} u_r^2(x)\right) \mathrm{d}x$$
$$\leqslant \mu\left[\int_{\partial B}\left(\sum_{r=1}^{m} u_r^2(x)\right)\mathrm{d}s + \int_B \left(\sum_{r=1}^{m} |\operatorname{grad} u_r(x)|^2\right)\mathrm{d}x\right].$$

The proof of Theorem 4.6.5 is complete.

From the hypotheses of Theorem 4.6.6, since B has a sufficiently smooth boundary, we choose auxiliary functions $\alpha_i(x)$ so that $\alpha_i(x) \in C^1(\overline{B})$ and $\alpha_i(x) = z_i(x)$ for $x \in \partial B$ (see [149, p. 69]), we have

$$\begin{aligned}
&\int_{\partial B} u_r^2(x)\,\mathrm{d}s \\
&= \int_{\partial B}\left(\sum_{i=1}^{n} z_i^2(x) u_r^2(x)\right)\mathrm{d}s \\
&= \int_{\partial B}\left(\sum_{i=1}^{n} \alpha_i(x) u_r^2(x) z_i(x)\right)\mathrm{d}s \\
&= \int_B \left\{\sum_{i=1}^{n} \frac{\partial}{\partial x_i}\alpha_i(x)\right\} u_r^2(x)\,\mathrm{d}x + \int_B \left\{\sum_{i=1}^{n} 2\alpha_i(x) u_r(x) \frac{\partial}{\partial x_i} u_r(x)\right\}\mathrm{d}x.
\end{aligned} \tag{4.6.41}$$

Here, a suitable version of the Gauss integral formula (4.6.17) has been used to get the last equality in (4.6.41). Using (4.6.36), Schwarz inequality for sums and the definition of c_1 we observe that

$$\begin{aligned}
\left|\sum_{i=1}^{n} 2\alpha_i(x) u_r(x) \frac{\partial}{\partial x_i} u_r(x)\right| &\leqslant 2c_1 |u_r(x)| \sum_{i=1}^{n}\left|\frac{\partial}{\partial x_i} u_r(x)\right| \\
&\leqslant c_1\left[c\left\{\sum_{i=1}^{n}\left|\frac{\partial}{\partial x_i} u_r(x)\right|\right\}^2 + \frac{1}{c} u_r^2(x)\right] \\
&\leqslant c_1\left[cn\sum_{i=1}^{n}\left\{\frac{\partial}{\partial x_i} u_r(x)\right\}^2 + \frac{1}{c} u_r^2(x)\right].
\end{aligned} \tag{4.6.42}$$

From (4.6.41), (4.6.42) and using the definition of c_0, we obtain

$$\int_{\partial B} u_r^2(x)\,\mathrm{d}s \leqslant \left(\frac{c_0 c + c_1}{c}\right)\int_B u_r^2(x)\,\mathrm{d}x + c c_1 n \int_B \left|\operatorname{grad} u_r(x)\right|^2 \mathrm{d}x. \quad (4.6.43)$$

From (4.6.40) and (4.6.43), we observe that

$$\begin{aligned}\int_{\partial B}\left[\prod_{r=1}^{m} u_r^2(x)\right]^{1/m} \mathrm{d}s &\leqslant \frac{1}{m}\int_{\partial B}\left(\sum_{r=1}^{m} u_r^2(x)\right)\mathrm{d}s \\ &\leqslant \lambda\left[\int_B\left(\sum_{r=1}^{m} u_r^2(x)\right)\mathrm{d}x + \int_B\left(\sum_{r=1}^{m}\left|\operatorname{grad} u_r(x)\right|^2\right)\mathrm{d}x\right].\end{aligned} \quad (4.6.44)$$

This inequality is the desired inequality in (4.6.32) and the proof of Theorem 4.6.6 is complete. □

For various other inequalities similar to that of Poincaré and Sobolev, see [56,98,127,154,193,264] and the references given therein.

4.7 Some Extensions of Rellich's Inequality

In his fundamental work on perturbations theory of eigenvalue problems F. Rellich [396] established the following inequality:

$$\int_{\mathbb{R}^n} |\Delta u|^2\,\mathrm{d}x \geqslant \frac{n^2(n-4)^2}{16}\int_{\mathbb{R}^n} |x|^{-4}|u|^2\,\mathrm{d}x, \quad n \neq 2, \quad (\mathrm{R})$$

where $u(x)$ is a function in $C_0^\infty(\mathbb{R}^n \setminus \{0\})$ which is not identically zero, C_0^∞ denote the vector space of infinitely differentiable functions with compact support (see [3, p. 9]) and $\Delta = \sum_{i=1}^n \frac{\partial^2}{\partial x_i^2}$.

In this section we deal with extensions of inequality (R) established by Pachpatte in [286,288]. In what follows, we assume that H is an open, connected subset of $\mathbb{R}^n$ that is not necessarily bounded and that the boundary ∂H of H is sufficiently smooth in order that the Green formulas applies. Let $\nabla = (\frac{\partial}{\partial x_1}, \ldots, \frac{\partial}{\partial x_n})$ and $\Delta = \sum_{i=1}^n \frac{\partial^2}{\partial x_i^2}$. A point in $\mathbb{R}^n$ is denoted by $x = (x_1, \ldots, x_n)$ and its norm is given by $|x| = (\sum_{i=1}^n |x_i|^2)^{1/2}$. For any nonnegative integer m, we denote by $C^m(H)$ the vector space consisting of all functions ϕ which, together with all their partial derivatives $D^\alpha\phi$ of order $|\alpha| \leqslant m$, are continuous on H and denote by $C_0^\infty(H)$ the vector space of infinitely differentiable functions with compact support (see [3, p. 9]).

We begin with the following useful inequalities established in [286].

THEOREM 4.7.1. *Let $p \geqslant 0$, $q \geqslant 1$ be constants, $g \in C^2(H)$, $\Delta g \neq 0$ in H and $u \in C_0^\infty(H)$ be a real-valued function. Then*

$$\int_H |\Delta g||u|^{p+q}\,\mathrm{d}x \leqslant (p+q)^q \int_H |\Delta g|^{-(q-1)}|\nabla g|^q|u|^p|\nabla u|^q\,\mathrm{d}x. \qquad (4.7.1)$$

THEOREM 4.7.2. *Let p, q, g, u be as in Theorem 4.7.1. Then*

$$\int_H |\Delta g||u|^{p+q}\,\mathrm{d}x$$
$$\leqslant (p+q)^{p+q}\int_H |\Delta g|^{-(p+q-1)}|\nabla g|^{p+q}|\nabla u|^{p+q}\,\mathrm{d}x. \qquad (4.7.2)$$

REMARK 4.7.1. If we take $g = |x|^{\alpha+2}$, $\alpha \geqslant 0$ is a real constant, and hence $|\nabla g|^2 = (\alpha+2)^2|x|^{2\alpha+2}$ and $\Delta g = (\alpha+n)(\alpha+2)|x|^\alpha$ in (4.7.2), then we get the following Hardy-type inequality (see [27, p. 303])

$$\int_H |x|^\alpha|u|^{p+q}\,\mathrm{d}x \leqslant \left(\frac{p+q}{\alpha+n}\right)^{p+q}\int_H |x|^{p+q+\alpha}|\nabla u|^{p+q}\,\mathrm{d}x. \qquad (4.7.3)$$

The Rellich-type inequalities established in [286] are given in the following theorems.

THEOREM 4.7.3. *If p, q, g, u be as in Theorem 4.7.1, then for any constants $\delta \geqslant 0$, $\varepsilon > 0$,*

$$\int_H |\Delta g|^{-(q-1)}|g|^q|u|^p|\Delta u|^q\,\mathrm{d}x$$
$$\geqslant -\varepsilon^{q-1}q(p+q-1)\,\mathrm{sgn}(\Delta g)\int_H g|u|^{p+q-2}|\nabla u|^2\,\mathrm{d}x$$
$$-\,\varepsilon^{q-1}\frac{q\delta}{p+q}\int_H |\Delta g|^{-(q-1)}|\nabla g|^q|u|^p|\nabla u|^q\,\mathrm{d}x$$
$$+\,\varepsilon^{q-1}\left[\frac{q}{p+q}-(q-1)\varepsilon+\frac{q\delta}{(p+q)^{q+1}}\right]\int_H |\Delta g||u|^{p+q}\,\mathrm{d}x. \qquad (4.7.4)$$

THEOREM 4.7.4. *If p, q, g, u, δ, ε be as in Theorem* 4.7.3, *then*

$$\begin{aligned}
&\int_H |\Delta g|^{-(p+q-1)}|g|^{p+q}|\Delta u|^{p+q}\,\mathrm{d}x \\
&\quad\geqslant -\varepsilon^{(p+q-1)}(p+q)(p+q-1)\operatorname{sgn}(\Delta g)\int_H g|u|^{p+q-2}|\nabla u|^2\,\mathrm{d}x \\
&\qquad - \varepsilon^{(p+q-1)}\delta\int_H |\Delta g|^{-(p+q-1)}|\nabla g|^{p+q}|\nabla u|^{p+q}\,\mathrm{d}x \\
&\qquad + \varepsilon^{(p+q-1)}\left[1-(p+q-1)\varepsilon+\frac{\delta}{(p+q)^{p+q}}\right]\int_H |\Delta g||u|^{p+q}\,\mathrm{d}x. \qquad (4.7.5)
\end{aligned}$$

REMARK 4.7.2. We note that in the special cases when $p=0$, $q=2$ and using the definition $\operatorname{sgn}(\Delta g)=\Delta g/|\Delta g|$, inequalities (4.7.4) and (4.7.5) reduce to the following inequality

$$\begin{aligned}
&\int_H |\Delta g|^{-1}|g|^2|\Delta u|^2\,\mathrm{d}x \\
&\quad\geqslant -\varepsilon\int_H \left[2g\Delta g+\delta|\nabla g|^2\right]|\Delta g|^{-1}|\nabla u|^2\,\mathrm{d}x+\varepsilon\left[1-\varepsilon+\frac{\delta}{4}\right]\int_H |\Delta g||u|^2\,\mathrm{d}x, \qquad (4.7.6)
\end{aligned}$$

which is established by Bannett in [21, Theorem 5]. By taking $p=0$ and $q=4$ in (4.7.4) and $p=2$, $q=2$ in (4.7.5), we get the inequalities of the Rellich type. Furthermore, by specializing the conditions on p, q and the function g in (4.7.4) and (4.7.5) we get different inequalities of some interest in their own right.

PROOFS OF THEOREMS 4.7.1 AND 4.7.2. By applying Green's first formula to $\int_H \Delta g|u|^{p+q}\,\mathrm{d}x$, we have

$$\int_H \Delta g|u|^{p+q}\,\mathrm{d}x=-\int_H \nabla g\nabla\left(|u|^{p+q}\right)\mathrm{d}x. \qquad (4.7.7)$$

From (4.7.7) and using the definition, $\operatorname{sgn}(\Delta g)=\Delta g/|\Delta g|$, the fact that $\nabla(|u|^{p+q})=(p+q)|u|^{p+q-1}\nabla u\operatorname{sgn}u$, and applying Hölder's inequality with indices q, $q/(q-1)$, we observe that

$$\begin{aligned}
&\int_H |\Delta g||u|^{p+q}\,\mathrm{d}x \\
&\quad=-\operatorname{sgn}(\Delta g)\int_H \nabla g\nabla\left(|u|^{p+q}\right)\mathrm{d}x
\end{aligned}$$

$$
\begin{aligned}
&= -(p+q)\operatorname{sgn}(\Delta g)\int_H \nabla g|u|^{p+q-1}\nabla u \operatorname{sgn} u \,\mathrm{d}x \\
&\leqslant (p+q)\int_H |\nabla g||u|^{p+q-1}|\nabla u|\,\mathrm{d}x \\
&= (p+q)\int_H \big[|\Delta g|^{-(q-1)/q}|\nabla g||u|^{p/q}|\nabla u|\big]\big[|\Delta g|^{(q-1)/q}|u|^{p+q-1-p/q}\big]\,\mathrm{d}x \\
&\leqslant (p+q)\left\{\int_H |\Delta g|^{-(q-1)}|\nabla g|^q|u|^p|\nabla u|^q\,\mathrm{d}x\right\}^{1/q}\left\{\int_H |\Delta g||u|^{p+q}\,\mathrm{d}x\right\}^{(q-1)/q}.
\end{aligned} \tag{4.7.8}
$$

If $\int_H |\Delta g||u|^{p+q}\,\mathrm{d}x = 0$ then (4.7.1) is trivially true; otherwise, we divide both sides of (4.7.8) by $\{\int_H |\Delta g||u|^{p+q}\,\mathrm{d}x\}^{(q-1)/q}$ and then raise both sides of the resulting inequality to the power q, to get inequality (4.7.1). The proof of Theorem 4.7.1 is complete.

From the hypotheses of Theorem 4.7.2 and by following the proof of Theorem 4.7.1, we have

$$
\begin{aligned}
\int_H |\Delta g||u|^{p+q}\,\mathrm{d}x &\leqslant (p+q)\int_H |\nabla g||u|^{p+q-1}|\nabla u|\,\mathrm{d}x \\
&= (p+q)\int_H \big[|\Delta g|^{-(p+q-1)/(p+q)}|\nabla g||\nabla u|\big] \\
&\qquad \times \big[|\Delta g|^{(p+q-1)/(p+q)}|u|^{p+q-1}\big]\,\mathrm{d}x.
\end{aligned} \tag{4.7.9}
$$

Now, using Hölder's inequality with indices $p+q$, $(p+q)/(p+q-1)$ on the right-hand side of (4.7.9) and following exactly the same arguments as in the last part of the proof of Theorem 4.7.1 given above with suitable changes, we get the desired inequality in (4.7.2). The proof of Theorem 4.7.2 is complete. □

REMARK 4.7.3. If we take g, $|\nabla g|^2$ and Δg as in Remark 4.7.1, in inequality (4.7.9), then we get

$$
\begin{aligned}
&\int_H |x|^\alpha |u|^{p+q}\,\mathrm{d}x \\
&\quad \leqslant \left(\frac{p+q}{\alpha+n}\right)\int_H |x|^{\alpha+1}|u|^{p+q-1}|\nabla u|\,\mathrm{d}x \\
&\quad = \left(\frac{p+q}{\alpha+n}\right)\int_H \big[|x|^{(\alpha+1)/(p+q)}|\nabla u|\big]\big[|x|^{-(\alpha+1)/(p+q)}|x|^{\alpha+1}|u|^{p+q-1}\big]\,\mathrm{d}x.
\end{aligned} \tag{4.7.10}
$$

Using Hölder's inequality with indices $p+q$, $(p+q)/(p+q-1)$ on the right-hand side of (4.7.10) we get the following Weyl-type inequality (see [27, p. 303])

$$\int_H |x|^\alpha |u|^{p+q}\,\mathrm{d}x \leqslant \left(\frac{p+q}{\alpha+n}\right)\left\{\int_H |x|^{\alpha+1}|\nabla u|^{p+q}\,\mathrm{d}x\right\}^{1/(p+q)} \times\left\{\int_H |x|^{\alpha+1}|u|^{p+q}\,\mathrm{d}x\right\}^{(p+q-1)/(p+q)}. \tag{4.7.11}$$

For a version of Weyl's inequality in one independent variable, see [25].

Proofs of Theorems 4.7.3 and 4.7.4. Let A, B, C, D denote integrals (without the exterior constants) in (4.7.4) successively. Applying Green's second formula to $\int_H \Delta g|u|^{p+q}\,\mathrm{d}x$ we have

$$\int_H \Delta g|u|^{p+q}\,\mathrm{d}x = \int_H g\Delta\big(|u|^{p+q}\big)\,\mathrm{d}x. \tag{4.7.12}$$

Using the definition, $\operatorname{sgn}(\Delta g) = \Delta g/|\Delta g|$ in (4.7.12), we observe that

$$D = \operatorname{sgn}(\Delta g)\int_H g\Delta\big(|u|^{p+q}\big)\,\mathrm{d}x. \tag{4.7.13}$$

Using the fact that

$$\Delta\big(|u|^{p+q}\big) = (p+q)|u|^{p+q-1}\Delta u\operatorname{sgn} u + (p+q)(p+q-1)|u|^{p+q-2}|\nabla u|^2 \tag{4.7.14}$$

in (4.7.13) we have

$$\begin{aligned} D &= \operatorname{sgn}(\Delta g)(p+q)\int_H g|u|^{p+q-1}\Delta u\operatorname{sgn} u\,\mathrm{d}x \\ &\quad + \operatorname{sgn}(\Delta g)(p+q)(p+q-1)\int_H g|u|^{p+q-2}|\nabla u|^2\,\mathrm{d}x \\ &\leqslant (p+q)\int_H |g||u|^{p+q-1}|\Delta u|\,\mathrm{d}x + (p+q)(p+q-1)\operatorname{sgn}(\Delta g)B \\ &= (p+q)\int_H \big[|\Delta g|^{-(q-1)/q}|g||u|^{p/q}|\Delta u|\big]\big[|\Delta g|^{-(q-1)/q}|u|^{p+q-1-p/q}\big]\,\mathrm{d}x \\ &\quad + (p+q)(p+q-1)\operatorname{sgn}(\Delta g)B. \end{aligned} \tag{4.7.15}$$

Now, first applying Hölder's inequality with indices $q, q/(q-1)$ on the right-hand side of (4.7.15) and then using Young's inequality with indices q, $q/(q-1)$, we see that

$$
\begin{aligned}
D &\leqslant (p+q)\left\{\int_H |\Delta g|^{-(q-1)}|g|^q|u|^p|\Delta u|^q\,\mathrm{d}x\right\}^{1/q}\left\{\int_H |\Delta g||u|^{p+q}\,\mathrm{d}x\right\}^{(q-1)/q} \\
&\quad + (p+q)(p+q-1)\,\mathrm{sgn}(\Delta g)B \\
&= (p+q)A^{1/q}D^{(q-1)/q} + (p+q)(p+q-1)\,\mathrm{sgn}(\Delta g)B \\
&= (p+q)\big(\varepsilon^{-(q-1)/q}A^{1/q}\big)\big(\varepsilon^{(q-1)/q}D^{(q-1)/q}\big) \\
&\quad + (p+q)(p+q-1)\,\mathrm{sgn}(\Delta g)B \\
&\leqslant \left(\frac{p+q}{q}\right)\varepsilon^{-(q-1)}A + \frac{(p+q)(q-1)}{q}\varepsilon D \\
&\quad + (p+q)(p+q-1)\,\mathrm{sgn}(\Delta g)B \qquad (4.7.16)
\end{aligned}
$$

for $\varepsilon > 0$. Now, for any $\delta \geqslant 0$, from (4.7.1) we observe that

$$
\delta C - \frac{\delta}{(p+q)^q}D \geqslant 0.
$$

Combining this fact with (4.7.16) we have

$$
\begin{aligned}
D \leqslant{}& \left(\frac{p+q}{q}\right)\varepsilon^{-(q-1)}A + \frac{(p+q)(q-1)}{q}\varepsilon D \\
&+ (p+q)(p+q-1)\,\mathrm{sgn}(\Delta g)B + \delta C - \frac{\delta}{(p+q)^q}D \qquad (4.7.17)
\end{aligned}
$$

for all $\varepsilon > 0$ and $\delta \geqslant 0$. Rewriting (4.7.17) we get the desired inequality in (4.7.4). The proof of Theorem 4.7.3 is complete.

In order to prove Theorem 4.7.4, let A, B, C, D denote the integrals (without the exterior constants) in (4.7.5) successively. By following the arguments in the first part of the proof of Theorem 4.7.3, we have

$$
\begin{aligned}
D &\leqslant (p+q)\int_H |g||u|^{p+q-1}|\Delta u|\,\mathrm{d}x + (p+q)(p+q-1)\,\mathrm{sgn}(\Delta g)B \\
&= (p+q)\int_H \big[|\Delta g|^{-(p+q-1)/(p+q)}|g||\Delta u|\big]\big[|\Delta g|^{(p+q-1)/(p+q)}|u|^{p+q-1}\big]\,\mathrm{d}x \\
&\quad + (p+q)(p+q-1)\,\mathrm{sgn}(\Delta g)B. \qquad (4.7.18)
\end{aligned}
$$

Now, first using Hölder's inequality with indices $p+q$, $(p+q)/(p+q-1)$, then

Young's inequality with indices $p+q$, $(p+q)/(p+q-1)$ on the right-hand side in (4.7.18), inequality (4.7.2) and following closely the arguments in the proof of Theorem 4.7.3 with suitable modifications, we get the required inequality in (4.7.5). The proof of Theorem 4.7.4 is complete. □

REMARK 4.7.4. If we specialize inequalities (4.7.4) and (4.7.5) by putting $g = |x|^{\alpha+2}$, $\alpha \geqslant 0$ real, and hence $|\nabla g|^2 = (\alpha+2)^2|x|^{2\alpha+2}$, $\Delta g = (\alpha+n)(\alpha+2)|x|^{\alpha}$, we get some new inequalities similar to that of inequality given by Bennett [21, p. 992].

The following inequality established in [288] is needed in proving the next theorem.

THEOREM 4.7.5. *Let $p \geqslant 2$ be a constant, $g \in C^2(H)$, $\Delta g \neq 0$ in H and $u_r \in C_0^\infty(H)$, $r = 1, \ldots, N$, be real-valued functions. Then*

$$\int_H |\Delta g| \left\{ \sum_{r=1}^{N} |u_r|^2 \right\}^{p/(p-1)} \mathrm{d}x$$

$$\leqslant \left\{ \frac{2p}{p-1} \right\}^{2p/(p-1)}$$

$$\times \int_H |\Delta g|^{-(p+1)/(p-1)} |\nabla g|^{2p/(p-1)} \left\{ \sum_{r=1}^{N} |\nabla u_r|^2 \right\}^{p/(p-1)} \mathrm{d}x. \quad (4.7.19)$$

PROOF. By applying Green's first formula to $\int_H \Delta g \{\sum_{r=1}^N |u_r|^2\}^{p/(p-1)} \mathrm{d}x$ and using the definition, $\operatorname{sgn}(\Delta g) = \Delta g/|\Delta g|$, we observe that

$$\int_H |\Delta g| \left\{ \sum_{r=1}^{N} |u_r|^2 \right\}^{p/(p-1)} \mathrm{d}x$$

$$= -\operatorname{sgn}(\Delta g) \int_H \nabla g \nabla \left\{ \sum_{r=1}^{N} |u_r|^2 \right\}^{p/(p-1)} \mathrm{d}x$$

$$\leqslant \int_H |\nabla g| \left| \nabla \left\{ \sum_{r=1}^{N} |u_r|^2 \right\}^{p/(p-1)} \right| \mathrm{d}x. \quad (4.7.20)$$

By simple calculation, it is easy to see that

$$\left|\nabla\left\{\sum_{r=1}^{N}|u_r|^2\right\}^{p/(p-1)}\right| \leqslant \left(\frac{2p}{p-1}\right)\left\{\sum_{r=1}^{N}|u_r|^2\right\}^{(p+1)/(2(p-1))}\left\{\sum_{r=1}^{N}|\nabla u_r|^2\right\}^{1/2}. \tag{4.7.21}$$

Using (4.7.21) in (4.7.20) and applying Hölder's inequality with indices $2p/(p+1), 2p/(p-1)$ we have

$$\begin{aligned}
&\int_H |\Delta g|\left\{\sum_{r=1}^{N}|u_r|^2\right\}^{p/(p-1)} \mathrm{d}x \\
&\leqslant \left(\frac{2p}{p-1}\right)\int_H \left[|\Delta g|^{(p+1)/(2p)}\left\{\sum_{r=1}^{N}|u_r|^2\right\}^{(p+1)/(2(p-1))}\right] \\
&\quad \times \left[|\Delta g|^{-(p+1)/(2p)}|\nabla g|\left\{\sum_{r=1}^{N}|\nabla u_r|^2\right\}^{1/2}\right]\mathrm{d}x \\
&\leqslant \left(\frac{2p}{p-1}\right)\left\{\int_H |\Delta g|\left\{\sum_{r=1}^{N}|u_r|^2\right\}^{p/(p-1)} \mathrm{d}x\right\}^{(p+1)/(2p)} \\
&\quad \times \left\{\int_H |\Delta g|^{-(p+1)/(p-1)}|\nabla g|^{2p/(p-1)}\left\{\sum_{r=1}^{N}|\nabla u_r|^2\right\}^{p/(p-1)} \mathrm{d}x\right\}^{(p-1)/(2p)}.
\end{aligned} \tag{4.7.22}$$

If $\int_H |\Delta g|\{\sum_{r=1}^{N}|u_r|^2\}^{p/(p-1)}\,\mathrm{d}x = 0$, then (4.7.19) is trivially true; otherwise, we divide both sides of (4.7.22) by $\{\int_H |\Delta g|\{\sum_{r=1}^{N}|u_r|^2\}^{p/(p-1)}\,\mathrm{d}x\}^{(p+1)/(2p)}$ and raise both sides to the power $2p/(p-1)$, to get the inequality (4.7.19). The proof is complete. □

The Rellich-type inequality established in [288] is given in the following theorem.

THEOREM 4.7.6. *Let p, g, u_r be as in Theorem 4.7.5. Then for any constants $\delta \geqslant 0$, $\varepsilon > 0$,*

$$\int_H |\Delta g|^{-(p+1)/(p-1)}|g|^{2p/(p-1)}\left\{\sum_{r=1}^{N}|\Delta u_r|^2\right\}^{p/(p-1)} \mathrm{d}x$$

$$
\begin{aligned}
&\geqslant -\varepsilon\left[2+\frac{4}{p-1}\varepsilon^{-1/(p-1)}\right] \\
&\quad\times \int_H |\Delta g|^{-1/(p-1)}|g|^{p/(p-1)}\left\{\sum_{r=1}^{N}|\nabla u_r|^2\right\}^{p/(p-1)}\mathrm{d}x \\
&\quad-\delta\varepsilon\int_H |\Delta g|^{-(p+1)/(p-1)}|\nabla g|^{2p/(p-1)}\left\{\sum_{r=1}^{N}|\nabla u_r|^2\right\}^{p/(p-1)}\mathrm{d}x \\
&\quad+\varepsilon\left[1-\varepsilon-\frac{4}{p-1}-\frac{4\varepsilon}{(p-1)^2}+\delta\left(\frac{2p}{p-1}\right)^{-2p/(p-1)}\right] \\
&\quad\times\int_H |\Delta g|\left\{\sum_{r=1}^{N}|u_r|^2\right\}^{p/(p-1)}\mathrm{d}x. \qquad (4.7.23)
\end{aligned}
$$

PROOF. Let A, B, C, D denote the integrals (without the exterior constants) in (4.7.23) successively. Applying Green's second formula to $\int_H \Delta g \times \{\sum_{r=1}^N |u_r|^2\}^{p/(p-1)}\,\mathrm{d}x$ and using the definition, $\operatorname{sgn}(\Delta g)=\Delta g/|\Delta g|$, we observe that

$$
D=\operatorname{sgn}(\Delta g)\int_H g\Delta\left\{\sum_{r=1}^{N}|u_r|^2\right\}^{p/(p-1)}\mathrm{d}x. \qquad (4.7.24)
$$

By the simple partial differentiation, we have the following identity

$$
\begin{aligned}
&\Delta\left\{\sum_{r=1}^{N}|u_r|^2\right\}^{p/(p-1)} \\
&=\left(\frac{2p}{p-1}\right)\left\{\sum_{r=1}^{N}|u_r|^2\right\}^{1/(p-1)}\sum_{r=1}^{N}|u_r|\Delta u_r\operatorname{sgn}u_r \\
&\quad+\left(\frac{2p}{p-1}\right)\left\{\sum_{r=1}^{N}|u_r|^2\right\}^{1/(p-1)}\sum_{r=1}^{N}|\nabla u_r|^2\operatorname{sgn}u_r \\
&\quad+\frac{4p}{(p-1)^2}\left\{\sum_{r=1}^{N}|u_r|^2\right\}^{(-p+2)/(p-1)}\sum_{i=1}^{n}\left\{\sum_{r=1}^{N}|u_r|\frac{\partial u_r}{\partial x_i}\operatorname{sgn}u_r\right\}^2. \qquad (4.7.25)
\end{aligned}
$$

Using (4.7.25) in (4.7.24) and applying Schwarz inequality for sum we see that

$$\begin{aligned}
D \leqslant{}& \left(\frac{2p}{p-1}\right)\int_H |g|\left\{\sum_{r=1}^N |u_r|^2\right\}^{1/(p-1)} \sum_{r=1}^N |u_r||\Delta u_r|\,\mathrm{d}x \\
&+ \left(\frac{2p}{p-1}\right)\int_H |g|\left\{\sum_{r=1}^N |u_r|^2\right\}^{1/(p-1)} \sum_{r=1}^N |\nabla u_r|^2\,\mathrm{d}x \\
&+ \frac{4p}{(p-1)^2}\int_H |g|\left\{\sum_{r=1}^N |u_r|^2\right\}^{(-p+2)/(p-1)} \\
&\qquad \times \sum_{i=1}^n \left(\sum_{r=1}^N |u_r|^2\right)\left(\sum_{r=1}^N \left|\frac{\partial u_r}{\partial x_i}\right|^2\right)\mathrm{d}x. \qquad (4.7.26)
\end{aligned}$$

Let I_1, I_2, I_3 denote the integrals (without the exterior constants) on the right-hand side in (4.7.26) successively. From the definition of I_1 and applying Young's inequality with indices p, $p/(p-1)$, Schwarz inequality first for sum and then for integrals, we observe that

$$\begin{aligned}
I_1 ={}& \int_H \left[|\Delta g|^{1/p}\left\{\sum_{r=1}^N |u_r|^2\right\}^{1/(p-1)}\right]\left[|\Delta g|^{-1/p}|g|\sum_{r=1}^N |u_r||\Delta u_r|\right]\mathrm{d}x \\
\leqslant{}& \int_H \Bigg[\frac{1}{p}|\Delta g|\left\{\sum_{r=1}^N |u_r|^2\right\}^{1/(p-1)} \\
&+ \left(\frac{p-1}{p}\right)|\Delta g|^{-1/(p-1)}|g|^{p/(p-1)}\left\{\sum_{r=1}^N |u_r||\Delta u_r|\right\}^{p/(p-1)}\Bigg]\mathrm{d}x \\
\leqslant{}& \frac{1}{p}D + \left(\frac{p-1}{p}\right)\int_H |\Delta g|^{-1/(p-1)}|g|^{p/(p-1)} \\
&\qquad \times \left\{\left\{\sum_{r=1}^N |u_r|^2\right\}^{1/2}\left\{\sum_{r=1}^N |\Delta u_r|^2\right\}^{1/2}\right\}^{p/(p-1)}\mathrm{d}x \\
={}& \frac{1}{p}D + \left(\frac{p-1}{p}\right)
\end{aligned}$$

$$\times \int_H \left[|\Delta g|^{1/2} \left\{ \sum_{r=1}^N |u_r|^2 \right\}^{p/(2(p-1))} \right]$$

$$\times \left[|\Delta g|^{-(p+1)/(2(p-1))} |g|^{p/(p-1)} \left\{ \sum_{r=1}^N |\Delta u_r|^2 \right\}^{p/(2(p-1))} \right] \mathrm{d}x$$

$$\leqslant \frac{1}{p} D + \left(\frac{p-1}{p} \right) \left\{ \int_H |\Delta g| \left\{ \sum_{r=1}^N |u_r|^2 \right\}^{p/(p-1)} \mathrm{d}x \right\}^{1/2}$$

$$\times \left\{ \int_H |\Delta g|^{-(p+1)/(p-1)} |g|^{2p/(p-1)} \left\{ \sum_{r=1}^N |\Delta u_r|^2 \right\}^{p/(p-1)} \mathrm{d}x \right\}^{1/2}$$

$$= \frac{1}{p} D + \left(\frac{p-1}{p} \right) D^{1/2} A^{1/2}. \tag{4.7.27}$$

Rewriting I_2 and applying Young's inequality with indices p, $p/(p-1)$ we have

$$I_2 = \int_H \left[|\Delta g|^{1/p} \left\{ \sum_{r=1}^N |u_r|^2 \right\}^{1/(p-1)} \right] \left[|\Delta g|^{-1/p} |g| \left\{ \sum_{r=1}^N |\nabla u_r|^2 \right\} \right] \mathrm{d}x$$

$$\leqslant \int_H \left[\frac{1}{p} |\Delta g| \left\{ \sum_{r=1}^N |u_r|^2 \right\}^{p/(p-1)} \right.$$

$$\left. + \left(\frac{p-1}{p} \right) |\Delta g|^{-1/(p-1)} |g|^{p/(p-1)} \left\{ \sum_{r=1}^N |\nabla u_r|^2 \right\}^{p/(p-1)} \right] \mathrm{d}x$$

$$= \frac{1}{p} D + \left(\frac{p-1}{p} \right) B. \tag{4.7.28}$$

Rewriting I_3 and applying Hölder's inequality with indices p, $p/(p-1)$ we have

$$I_3 = \int_H \left[|\Delta g|^{1/p} \left\{ \sum_{r=1}^N |u_r|^2 \right\}^{1/(p-1)} \right] \left[|\Delta g|^{-1/p} |g| \left\{ \sum_{r=1}^N |\nabla u_r|^2 \right\} \right] \mathrm{d}x$$

$$\leqslant \left\{ \int_H |\Delta g| \left\{ \sum_{r=1}^N |u_r|^2 \right\}^{p/(p-1)} \mathrm{d}x \right\}^{1/p}$$

$$\times \left\{ \int_H |\Delta g|^{-1/(p-1)} |g|^{p/(p-1)} \left\{ \sum_{r=1}^{N} |\nabla u_r|^2 \right\}^{p/(p-1)} \mathrm{d}x \right\}^{(p-1)/p}$$
$$= D^{1/p} B^{(p-1)/p}. \tag{4.7.29}$$

Now, using (4.7.27)–(4.7.29) in (4.7.26) and applying the elementary inequality $2ab \leqslant a^2 + b^2$ (a, b reals) and Young's inequality with indices p, $p/(p-1)$, we observe that

$$\begin{aligned} D &\leqslant \left(\frac{4}{p-1}\right) D + 2D^{1/2}A^{1/2} + 2B + \frac{4p}{(p-1)^2} D^{1/p} B^{(p-1)/p} \\ &= \left(\frac{4}{p-1}\right) D + 2\big(\varepsilon^{1/2} D^{1/2}\big)\big(\varepsilon^{-1/2} A^{1/2}\big) + 2B \\ &\quad + \frac{4p}{(p-1)^2} \big(\varepsilon^{1/p} D^{1/p}\big)\big(\varepsilon^{-1/p} B^{(p-1)p}\big) \\ &\leqslant \left(\frac{4}{p-1}\right) D + \varepsilon D + \frac{1}{\varepsilon} A + 2B + \frac{4p}{(p-1)^2} \left[\frac{1}{p} \varepsilon D + \left(\frac{p-1}{p}\right) \varepsilon^{-1/(p-1)} B \right] \\ &= \left[\varepsilon + \frac{4}{p-1} + \frac{4\varepsilon}{(p-1)^2} \right] D + \frac{1}{\varepsilon} A + \left[2 + \frac{4}{(p-1)} \varepsilon^{-1/(p-1)} \right] B \end{aligned} \tag{4.7.30}$$

for $\varepsilon > 0$. Now, for any $\delta \geqslant 0$, from (4.7.19) we observe that

$$\delta C - \delta \left(\frac{2p}{p-1}\right)^{-2p/(p-1)} D \geqslant 0. \tag{4.7.31}$$

From (4.7.30) and (4.7.31), we have

$$\begin{aligned} D &\leqslant \left[\varepsilon + \frac{4}{p-1} + \frac{4\varepsilon}{(p-1)^2} \right] D + \frac{1}{\varepsilon} A \\ &\quad + \left[2 + \frac{4}{(p-1)} \varepsilon^{-1/(p-1)} \right] B + \delta C - \delta \left(\frac{2p}{p-1}\right)^{-2p/(p-1)} D \end{aligned} \tag{4.7.32}$$

for all $\varepsilon > 0$, $\delta \geqslant 0$. Rewriting (4.7.32) we get the desired inequality in (4.7.23). The proof is complete. □

REMARK 4.7.5. We note that in the special cases, when (i) $N = 1$, $u_1 = u$, (ii) $N = 1$, $u_1 = u$ and $p = 2$, inequality (4.7.23) reduces to the new inequalities. If we specialize inequality (4.7.23) by taking $N = 1$, $u_1 = u$ and then by putting $g = |x|^{\alpha+2}$, $\alpha \geqslant 0$ real constant, and hence $|\nabla g|^2 = (\alpha + 2)^2 |x|^{2\alpha+2}$,

$\Delta g = (\alpha + n)(\alpha + 2)|x|^{\alpha}$, we get an inequality similar to that of inequality given by Bennett in [21, p. 992]. For other extensions and variants of the Rellich inequality, see [8,191,387,396,403] and some of the references cited therein.

4.8 Poincaré- and Sobolev-Type Discrete Inequalities

Discrete inequalities involving functions of several independent variables and their forward differences have been investigated by many authors in the literature. This section deals with the Poincaré- and Sobolev-type discrete inequalities established by Pachpatte in [269,275,285].

In what follows $\mathbb{R}$ denote the set of real numbers and $\mathbb{N} = \{1, 2, \ldots\}$. For $x = (x_1, \ldots, x_n) \in \mathbb{N}^n$ and $z(x) : \mathbb{N}^n \to \mathbb{R}$, we define the forward difference operators by $\Delta_1 z(x) = z(x_1 + 1, x_2, \ldots, x_n) - z(x), \ldots, \Delta_n z(x) = z(x_1, \ldots, x_{n-1}, x_n + 1) - z(x)$. The notation $\Delta_i z(x_1, \ldots, y_i, \ldots, x_n)$ for $i = 1, \ldots, n$ we mean, for $i = 1$, it is $\Delta_1 z(y_1, x_2, \ldots, x_n) = z(y_1 + 1, x_2, \ldots, x_n) - z(y_1, x_2, \ldots, x_n)$ and so on, for $i = n$, it is $\Delta_n z(x_1, \ldots, x_{n-1}, y_n) = z(x_1, \ldots, x_{n-1}, y_n + 1) - z(x_1, \ldots, x_{n-1}, y_n)$. Let $B = \prod_{i=1}^{n} [1, a_i + 1]$ be a bounded domain in $\mathbb{N}^n$ with $n \geqslant 1$ as an integer.

We denote by $F(B)$ the class of functions $z(x) : B \to \mathbb{R}$ for which

$$z(1, x_2, \ldots, x_n) = z(x_1, 1, x_3, \ldots, x_n) = \cdots = z(x_1, \ldots, x_{n-1}, 1) = 0,$$

$$\begin{aligned} z(a_1 + 1, x_2, \ldots, x_n) &= z(x_1, a_2 + 1, x_3, \ldots, x_n) \\ &= \cdots = z(x_1, \ldots, x_{n-1}, a_n + 1) = 0. \end{aligned}$$

For $y = (y_1, \ldots, y_n)$ and $z(x) : B \to \mathbb{R}$, we use the following notations

$$\sum_B z(y) = \sum_{y_n=1}^{a_n} \cdots \sum_{y_1=1}^{a_1} z(y_1, \ldots, y_n), \qquad \left|\Delta z(x)\right| = \left(\sum_{i=1}^{n} \left|\Delta_i z(x)\right|^2\right)^{1/2}.$$

Throughout, the empty sum and product are taken to be 0 and 1, respectively.

The following Poincaré-type discrete inequalities are established in [269].

THEOREM 4.8.1. *Let $p_m \geqslant 2$ be constants and $u_m \in F(B)$ for $m = 1, \ldots, r$. Then*

$$\sum_B \left(\prod_{m=1}^{r} \left|u_m(y)\right|^{p_m}\right)^{1/r} \leqslant \frac{1}{nr} \left(\frac{\alpha}{2}\right)^{\frac{1}{r}\sum_{m=1}^{r} p_r} \sum_{m=1}^{r} \left(\sum_B \left|\Delta u_m(y)\right|^{p_m}\right), \tag{4.8.1}$$

where $\alpha = \max\{a_1, \ldots, a_n\}$.

REMARK 4.8.1. In the special case when $r = 1$, inequality (4.8.1) reduces to the following Poincaré-type discrete inequality

$$\sum_B |u_1(y)|^{p_1} \leqslant \frac{1}{n}\left(\frac{\alpha}{2}\right)^{p_1} \sum_B |\Delta u_1(y)|^{p_1}, \tag{4.8.2}$$

in n independent variables.

THEOREM 4.8.2. *Let $u_m \in F(B)$ for $m = 1, \ldots, r$. Then*

$$\left(\sum_B \left[\prod_{m=1}^{r} |u_m(y)|\right]^{1/r}\right)^2 \leqslant \left(\frac{\alpha^{n+2}}{4nr}\right) \sum_{m=1}^{r} \left(\sum_B |\Delta u_m(y)|^2\right), \tag{4.8.3}$$

where α is as in Theorem 4.8.1.

REMARK 4.8.2. In the special case when $r = 1$, inequality (4.8.3) reduces to the following Poincaré-type discrete inequality

$$\left(\sum_B |u_1(y)|\right)^2 \leqslant \left(\frac{\alpha^{n+2}}{4n}\right) \sum_B |\Delta u_1(y)|^2, \tag{4.8.4}$$

in n independent variables.

PROOFS OF THEOREMS 4.8.1 AND 4.8.2. Since $u_m \in F(B)$, we have the following identities

$$nu_m(x) = \sum_{i=1}^{n} \left\{ \sum_{y_i=1}^{x_i-1} \Delta_i u_m(x_1, \ldots, y_i, \ldots, x_n) \right\}, \tag{4.8.5}$$

$$nu_m(x) = -\sum_{i=1}^{n} \left\{ \sum_{y_i=x_i}^{a_i} \Delta_i u_m(x_1, \ldots, y_i, \ldots, x_n) \right\}, \tag{4.8.6}$$

for $m = 1, \ldots, r$. From (4.8.5) and (4.8.6), we obtain

$$|u_m(x)| \leqslant \frac{1}{2n} \sum_{i=1}^{n} \left\{ \sum_{y_i=1}^{a_i} |\Delta_i u_m(x_1, \ldots, y_i, \ldots, x_n)| \right\} \tag{4.8.7}$$

for $m = 1, \ldots, r$. From (4.8.7) and using the elementary inequality

$$\left\{ \sum_{i=1}^{k} b_i \right\}^{\gamma} \leqslant d_{\gamma,k} \left(\sum_{i=1}^{k} b_i^{\gamma} \right), \tag{4.8.8}$$

where $b_i \geqslant 0$ reals, $d_{\gamma,k} = k^{\gamma-1}$, $\gamma > 1$, and $d_{\gamma,k} = 1$, $0 \leqslant \gamma \leqslant 1$, Hölder's inequality with indices p_m, $p_m/(p_m - 1)$ and using the definition of α, we obtain

$$\begin{aligned} |u_m(x)|^{p_m} &\leqslant \left(\frac{1}{2n}\right)^{p_m} n^{p_m-1} \sum_{i=1}^{n} \left\{ \sum_{y_i=1}^{a_i} |\Delta_i u_m(x_1, \ldots, y_i, \ldots, x_n)| \right\}^{p_m} \\ &\leqslant \frac{1}{n}\left(\frac{1}{2}\right)^{p_m} \alpha^{p_m-1} \sum_{i=1}^{n} \left\{ \sum_{y_i=1}^{a_i} |\Delta_i u_m(x_1, \ldots, y_i, \ldots, x_n)|^{p_m} \right\}. \end{aligned} \tag{4.8.9}$$

From (4.8.9) and using the elementary inequality

$$\left\{ \prod_{m=1}^{r} b_m \right\}^{1/r} \leqslant \frac{1}{r} \sum_{m=1}^{r} b_m, \tag{4.8.10}$$

where $b_m \geqslant 0$ reals and $r \geqslant 1$, we obtain

$$\begin{aligned} &\left(\prod_{m=1}^{r} |u_m(x)|^{p_m} \right)^{1/r} \\ &\quad \leqslant \frac{1}{nr}\left(\frac{1}{2}\right)^{\frac{1}{r}\sum_{m=1}^{r} p_m} \alpha^{\frac{1}{r}(\sum_{m=1}^{r} p_m)-1} \\ &\quad\quad \times \sum_{m=1}^{r} \left\{ \sum_{i=1}^{n} \left\{ \sum_{y_i=1}^{a_i} |\Delta_i u_m(x_1, \ldots, y_i, \ldots, x_n)|^{p_m} \right\} \right\}. \end{aligned} \tag{4.8.11}$$

Setting $x_i = y_i$, $i = 1, \ldots, n$, in (4.8.11) and taking the sum over both sides of (4.8.11) with respect to $y_1, \ldots, y_n$ on B and using the definition of α and inequality (4.8.8) with $\gamma = 2/p_r \leqslant 1$ we have

$$\begin{aligned} &\sum_{B} \left(\prod_{m=1}^{r} |u_m(y)|^{p_m} \right)^{1/r} \\ &\quad \leqslant \frac{1}{nr}\left(\frac{1}{2}\right)^{\frac{1}{r}\sum_{m=1}^{r} p_m} \alpha^{\frac{1}{r}(\sum_{m=1}^{r} p_m)-1} \\ &\quad\quad \times \alpha \sum_{m=1}^{r} \left\{ \sum_{B} \left\{ \left[\sum_{i=1}^{n} |\Delta_i u_m(y)|^{p_m} \right]^{2/p_m} \right\}^{p_m/2} \right\} \\ &\quad \leqslant \left(\frac{1}{nr}\right)\left(\frac{\alpha}{2}\right)^{\frac{1}{r}\sum_{m=1}^{r} p_m} \sum_{m=1}^{r} \left\{ \sum_{B} |\Delta u_m(y)|^{p_m} \right\}. \end{aligned}$$

The proof of Theorem 4.8.1 is complete.

From the hypotheses of Theorem 4.8.2, we have inequality (4.8.7). From (4.8.7) and using inequality (4.8.10), we obtain

$$\left[\prod_{m=1}^{r}|u_m(x)|\right]^{1/r} \leqslant \left(\frac{1}{2nr}\right)\sum_{m=1}^{r}\left\{\sum_{i=1}^{n}\left\{\sum_{y_i=1}^{a_i}|\Delta_i u_m(x_1,\ldots,y_i,\ldots,x_n)|\right\}\right\}. \tag{4.8.12}$$

Setting $x_i = y_i$, $i = 1,\ldots,n$, in (4.8.12) and taking the sum over both sides of (4.8.12) with respect to $y_1,\ldots,y_n$ on B, using the definition of α, then taking the square on both sides of the resulting inequality, using inequality (4.4.8), first with $k = r$, $\gamma = 2$, and then with $k = n$, $\gamma = 2$, Schwarz inequality and again the definition of α we have

$$\begin{aligned}
\left(\sum_{B}\left[\prod_{m=1}^{r}|u_m(y)|\right]^{1/r}\right)^2 &\leqslant \left(\frac{\alpha}{2nr}\right)^2\left\{\sum_{m=1}^{r}\left\{\sum_{i=1}^{n}\left\{\sum_{B}|\Delta_i u_m(y)|\right\}\right\}\right\}^2 \\
&\leqslant \left(\frac{\alpha}{2nr}\right)^2 r\sum_{m=1}^{r}\left\{\sum_{i=1}^{n}\left\{\sum_{B}|\Delta_i u_m(y)|\right\}\right\}^2 \\
&\leqslant \left(\frac{\alpha}{2nr}\right)^2 rn\sum_{m=1}^{r}\left\{\sum_{i=1}^{n}\left\{\sum_{B}|\Delta_i u_m(y)|\right\}^2\right\} \\
&\leqslant \left(\frac{\alpha^2}{4nr}\right)\alpha^n\sum_{m=1}^{r}\left\{\sum_{B}\left\{\sum_{i=1}^{n}|\Delta_i u_m(y)|^2\right\}\right\} \\
&= \left(\frac{\alpha^{n+2}}{4nr}\right)\sum_{m=1}^{r}\left\{\sum_{B}|\Delta u_m(y)|^2\right\}.
\end{aligned}$$

This inequality is the required inequality in (4.8.3) and the proof of Theorem 4.8.2 is complete. □

The following discrete inequality is established in [275].

THEOREM 4.8.3. *Let $p, q \geqslant 2$ be constants such that $\frac{1}{p} + \frac{1}{q} = 1$ and suppose that $u, v \in F(B)$. Then*

$$\sum_{B}|u(y)||v(y)| \leqslant \frac{\lambda^p}{np}\sum_{B}|\Delta u(y)|^p + \frac{\lambda^q}{nq}\sum_{B}|\Delta v(y)|^q, \tag{4.8.13}$$

where $\lambda = \alpha/2$, *in which* $\alpha = \max\{a_1, \dots, a_n\}$.

REMARK 4.8.3. If we take $u = v = f$ and $p = q = 2$ in (4.8.13), then we get the following Wirtinger-type discrete inequality in n independent variables

$$\sum_B |f(y)|^2 \leqslant \frac{\lambda^2}{n} \sum_B |\Delta f(y)|^2. \tag{4.8.14}$$

PROOF OF THEOREM 4.8.3. From the hypotheses, it is easy to observe that the following identities hold

$$nu(x) = \sum_{i=1}^{n} \left\{ \sum_{y_i=1}^{x_i-1} \Delta_i u(x_1, \dots, y_i, \dots, x_n) \right\}, \tag{4.8.15}$$

$$nu(x) = -\sum_{i=1}^{n} \left\{ \sum_{y_i=x_i}^{a_i} \Delta_i u(x_1, \dots, y_i, \dots, x_n) \right\}. \tag{4.8.16}$$

From (4.8.15) and (4.8.16), we observe that

$$2n|u(x)| \leqslant \sum_{i=1}^{n} \left\{ \sum_{y_i=1}^{a_i} |\Delta_i u(x_1, \dots, y_i, \dots, x_n)| \right\}. \tag{4.8.17}$$

Similarly, we obtain

$$2n|v(x)| \leqslant \sum_{i=1}^{n} \left\{ \sum_{y_i=1}^{a_i} |\Delta_i v(x_1, \dots, y_i, \dots, x_n)| \right\}. \tag{4.8.18}$$

From (4.8.17), (4.8.18) and using the elementary inequality

$$b_1 b_2 \leqslant \frac{1}{p} b_1^p + \frac{1}{q} b_2^q, \tag{4.8.19}$$

where $b_1, b_2 \geqslant 0$, $p > 1$ and $\frac{1}{p} + \frac{1}{q} = 1$, we obtain

$$\begin{aligned} |u(x)||v(x)| \leqslant{}& \frac{1}{p} \left\{ \frac{1}{2n} \left[\sum_{i=1}^{n} \left\{ \sum_{y_i=1}^{a_i} |\Delta_i u(x_1, \dots, y_i, \dots, x_n)| \right\} \right] \right\}^p \\ &+ \frac{1}{q} \left\{ \frac{1}{2n} \left[\sum_{i=1}^{n} \left\{ \sum_{y_i=1}^{a_i} |\Delta_i v(x_1, \dots, y_i, \dots, x_n)| \right\} \right] \right\}^q. \end{aligned} \tag{4.8.20}$$

From (4.8.20) and using the elementary inequality (4.8.8), Hölder's inequality with indices p, $p/(p-1)$ and q, $q/(q-1)$, we obtain

$$\begin{aligned}
|u(x)||v(x)| &\leqslant \frac{1}{p}\left(\frac{1}{2n}\right)^p n^{p-1}\sum_{i=1}^{n}\left\{\sum_{y_i=1}^{a_i}\left|\Delta_i u(x_1,\ldots,y_i,\ldots,x_n)\right|\right\}^p \\
&\quad + \frac{1}{q}\left(\frac{1}{2n}\right)^q n^{q-1}\sum_{i=1}^{n}\left\{\sum_{y_i=1}^{a_i}\left|\Delta_i v(x_1,\ldots,y_i,\ldots,x_n)\right|\right\}^q \\
&\leqslant \frac{1}{p}\left(\frac{1}{2n}\right)^p n^{p-1}\alpha^{p-1}\sum_{i=1}^{n}\left\{\sum_{y_i=1}^{a_i}\left|\Delta_i u(x_1,\ldots,y_i,\ldots,x_n)\right|^p\right\} \\
&\quad + \frac{1}{q}\left(\frac{1}{2n}\right)^q n^{q-1}\alpha^{q-1}\sum_{i=1}^{n}\left\{\sum_{y_i=1}^{a_i}\left|\Delta_i v(x_1,\ldots,y_i,\ldots,x_n)\right|^q\right\}.
\end{aligned} \tag{4.8.21}$$

Setting $x_i = y_i$, $i = 1,\ldots,n$, in (4.8.21), taking the sum over both sides of (4.8.21) with respect to $y_1,\ldots,y_n$ on B, using the definition of α and the suitable applications of (4.8.8) we get

$$\begin{aligned}
&\sum_B |u(y)||v(y)| \\
&\quad\leqslant \frac{1}{np}\left(\frac{\alpha}{2}\right)^p \sum_B\left\{\sum_{i=1}^{n}\left|\Delta_i u(y)\right|^p\right\} + \frac{1}{nq}\left(\frac{\alpha}{2}\right)^q \sum_B\left\{\sum_{i=1}^{n}\left|\Delta_i v(y)\right|^q\right\} \\
&\quad= \frac{1}{np}\left(\frac{\alpha}{2}\right)^p \sum_B\left\{\left\{\sum_{i=1}^{n}\left|\Delta_i u(y)\right|^p\right\}^{2/p}\right\}^{p/2} \\
&\qquad + \frac{1}{nq}\left(\frac{\alpha}{2}\right)^q \sum_B\left\{\left\{\sum_{i=1}^{n}\left|\Delta_i v(y)\right|^q\right\}^{2/q}\right\}^{q/2} \\
&\quad\leqslant \frac{\lambda^p}{np}\sum_B\left|\Delta u(y)\right|^p + \frac{\lambda^q}{nq}\sum_B\left|\Delta v(y)\right|^q.
\end{aligned}$$

This inequality completes the proof of Theorem 4.8.3. □

The discrete inequalities in the following theorems are established in [285].

THEOREM 4.8.4. *Let $u_r \in F(B)$ for $r = 1, \dots, M$ and let $m \geqslant 1$ and $p \geqslant 2$ be real constants. Then*

$$\left[\sum_B \left\{\sum_{r=1}^{M} |u_r(y)|^2\right\}^{p/(p-1)}\right]^{2m(p-1)/p} \leqslant k_1 \sum_{r=1}^{M} \sum_B \sum_{i=1}^{n} |\Delta_i u_r(y)|^{4m}, \tag{4.8.22}$$

where

$$k_1 = \frac{1}{n}\left(\frac{1}{4}\right)^{2m} M^{2m-1} \alpha^{(p(4m+2nm-n)-2nm)/p}.$$

REMARK 4.8.4. If we take $m = 1$ in (4.8.22) then we get the inequality analogous to the discrete version of the inequality of the form given by Lieb and Thirring in [193]. On taking $u_r(x) = u(x)$ for $r = 1, \dots, M$ and $m = 1$, $p = 2$ in (4.8.22), we get the following discrete Poincaré-type inequality

$$\sum_B |u(x)|^4 \leqslant \frac{1}{n}\left(\frac{\alpha}{2}\right)^4 \sum_B \sum_{i=1}^{n} |\Delta_i u(y)|^4. \tag{4.8.23}$$

THEOREM 4.8.5. *Let u_r, m, p be as in Theorem 4.8.4. Then*

$$\left[\sum_B \left\{\sum_{r=1}^{M} |u_r(y)|^2\right\}^{p/(p-1)}\right]^{2m(p-1)/p}$$
$$\leqslant k_2 \sum_{r=1}^{M} \sum_B \sum_{i=1}^{n} |\Delta_i u_r(y)|^{4m} + k_3 \sum_{r=1}^{M} \sum_B \sum_{i=1}^{n} |u_r(y)|^{2m} |\Delta_i u_r(y)|^{2m}, \tag{4.8.24}$$

where

$$k_2 = \frac{1}{2n} M^{2m-1} \alpha^{(p(2m+2nm-n)-2nm)/p}, \qquad k_3 = 2^{2m} k_2.$$

REMARK 4.8.5. In the special case when $m = 1$, $p = 2$ and $u_r(x) = u(x)$ for $r = 1, \dots, M$, inequality (4.8.24) reduces to the following discrete Sobolev-like inequality

$$\sum_B |u(y)|^4 \leqslant \frac{\alpha^2}{2n} \sum_B \sum_{i=1}^{n} |\Delta_i u(y)|^4 + \frac{2\alpha^2}{n} \sum_B \sum_{i=1}^{n} |u(y)|^2 |\Delta_i u(y)|^2. \tag{4.8.25}$$

PROOFS OF THEOREMS 4.8.4 AND 4.8.5. Since $u_r \in F(B)$, we have the following identities

$$nu_r(x) = \sum_{i=1}^{n} \left\{ \sum_{y_i=1}^{x_i-1} \Delta_i u_r(x_1, \ldots, y_i, \ldots, x_n) \right\}, \tag{4.8.26}$$

$$nu_r(x) = -\sum_{i=1}^{n} \left\{ \sum_{y_i=x_i}^{a_i} \Delta_i u_r(x_1, \ldots, y_i, \ldots, x_n) \right\}, \tag{4.8.27}$$

for $r = 1, \ldots, M$. From (4.8.26) and (4.8.27), we obtain

$$\left| u_r(x) \right| \leqslant \frac{1}{2n} \sum_{i=1}^{n} \left\{ \sum_{y_i=1}^{a_i} \left| \Delta_i u_r(x_1, \ldots, y_i, \ldots, x_n) \right| \right\}. \tag{4.8.28}$$

From (4.8.28) and using inequality (4.8.8), Schwarz inequality and the definition of α, we obtain

$$\begin{aligned} \left| u_r(x) \right|^2 &\leqslant \left(\frac{1}{2n} \right)^2 \left[\sum_{i=1}^{n} \left\{ \sum_{y_i=1}^{a_i} \left| \Delta_i u_r(x_1, \ldots, y_i, \ldots, x_n) \right| \right\} \right]^2 \\ &\leqslant \left(\frac{1}{2n} \right)^2 n \sum_{i=1}^{n} \left\{ \sum_{y_i=1}^{a_i} \left| \Delta_i u_r(x_1, \ldots, y_i, \ldots, x_n) \right| \right\}^2 \\ &\leqslant \left(\frac{\alpha}{4n} \right) \sum_{i=1}^{n} \left\{ \sum_{y_i=1}^{a_i} \left| \Delta_i u_r(x_1, \ldots, y_i, \ldots, x_n) \right|^2 \right\}. \end{aligned} \tag{4.8.29}$$

From (4.8.29) and using (4.8.8) repeatedly, Hölder's inequality with indices p, $p/(p-1)$ and the definition of α, we obtain

$$\begin{aligned} &\left\{ \sum_{r=1}^{M} \left| u_r(x) \right|^2 \right\}^{p/(p-1)} \\ &\quad \leqslant \left(\frac{\alpha}{4n} \right)^{p/(p-1)} (Mn)^{p/(p-1)-1} \\ &\qquad \times \sum_{r=1}^{M} \left\{ \sum_{i=1}^{n} \left\{ \sum_{y_i=1}^{a_i} \left| \Delta_i u_r(x_1, \ldots, y_i, \ldots, x_n) \right|^2 \right\}^{p/(p-1)} \right\} \end{aligned}$$

$$\leqslant \left(\frac{\alpha}{4n}\right)^{p/(p-1)} (Mn)^{1/(p-1)} \alpha^{1/(p-1)}$$

$$\times \sum_{r=1}^{M} \left\{ \sum_{i=1}^{n} \left\{ \sum_{y_i=1}^{a_i} \left| \Delta_i u_r(x_1, \ldots, y_i, \ldots, x_n) \right|^{2p/(p-1)} \right\} \right\}. \quad (4.8.30)$$

Setting $x_i = y_i$, $i = 1, \ldots, n$, in (4.8.30) and taking the sum over both sides of (4.8.30) with respect to $y_1, \ldots, y_n$ on B and using the definition of α, we have

$$\sum_{B} \left\{ \sum_{r=1}^{M} |u_r(y)|^2 \right\}^{p/(p-1)}$$

$$\leqslant \left(\frac{\alpha}{4n}\right)^{p/(p-1)} (Mn\alpha)^{1/(p-1)} \alpha \sum_{r=1}^{M} \left\{ \sum_{i=1}^{n} \left\{ \sum_{B} |\Delta_i u_r(y)|^{2p/(p-1)} \right\} \right\}. \quad (4.8.31)$$

From (4.8.31) and using inequality (4.8.8) repeatedly, Hölder's inequality with indices $2m(p-1)/p$, $2m(p-1)/(2m(p-1)-p)$ and the definition of α, we obtain

$$\left[\sum_{B} \left\{ \sum_{r=1}^{M} |u_r(y)|^2 \right\}^{p/(p-1)} \right]^{2m(p-1)/p}$$

$$\leqslant \left\{ \left(\frac{\alpha}{4n}\right)^{p/(p-1)} (Mn\alpha)^{1/(p-1)} \alpha \right\}^{2m(p-1)/p} (Mn)^{2m(p-1)/p-1}$$

$$\times \sum_{r=1}^{M} \left\{ \sum_{i=1}^{n} \left\{ \sum_{B} |\Delta_i u_r(y)|^{2p/(p-1)} \right\}^{2m(p-1)/p} \right\}$$

$$\leqslant \left\{ \left(\frac{\alpha}{4n}\right)^{p/(p-1)} (Mn\alpha)^{1/(p-1)} \alpha \right\}^{2m(p-1)/p} (Mn)^{2m(p-1)/p-1}$$

$$\times \left(\alpha^n\right)^{(2m(p-1)-p)/p} \sum_{r=1}^{M} \left\{ \sum_{i=1}^{n} \left\{ \sum_{B} |\Delta_i u_r(y)|^{4m} \right\} \right\}$$

$$= k_1 \sum_{r=1}^{M} \sum_{B} \sum_{i=1}^{n} |\Delta_i u_r(y)|^{4m}.$$

This inequality completes the proof of Theorem 4.8.4.

From the hypotheses of Theorem 4.8.5, we have the following identities

$$nu_r^2(x) = \sum_{i=1}^{n}\left\{\sum_{y_i=1}^{x_i-1} \Delta_i u_r^2(x_1, \ldots, y_i, \ldots, x_n)\right\}, \tag{4.8.32}$$

$$nu_r^2(x) = -\sum_{i=1}^{n}\left\{\sum_{y_i=x_i}^{a_i} \Delta_i u_r^2(x_1, \ldots, y_i, \ldots, x_n)\right\}, \tag{4.8.33}$$

for $r = 1, \ldots, M$. From (4.8.32) and (4.8.33), we observe that

$$\begin{aligned}
&|u_r(x)|^2 \\
&\leqslant \frac{1}{2n}\sum_{i=1}^{n}\left\{\sum_{y_i=1}^{a_i}\left|\Delta_i u_r^2(x_1, \ldots, y_i, \ldots, x_n)\right|\right\} \\
&= \frac{1}{2n}\sum_{i=1}^{n}\left\{\sum_{y_i=1}^{a_i}\left|u_r^2(x_1, \ldots, y_i+1, \ldots, x_n) - u_r^2(x_1, \ldots, y_i, \ldots, x_n)\right|\right\} \\
&= \frac{1}{2n}\sum_{i=1}^{n}\left\{\sum_{y_i=1}^{a_i}\left|\Delta_i u_r(x_1, \ldots, y_i, \ldots, x_n)\right.\right. \\
&\qquad \left.\left.\times \left\{u_r(x_1, \ldots, y_i+1, \ldots, x_n) + u_r(x_1, \ldots, y_i, \ldots, x_n)\right\}\right|\right\} \\
&= \frac{1}{2n}\sum_{i=1}^{n}\left\{\sum_{y_i=1}^{a_i}\left|\Delta_i u_r(x_1, \ldots, y_i, \ldots, x_n)\right.\right. \\
&\qquad \left.\left.\times \left\{\Delta_i u_r(x_1, \ldots, y_i, \ldots, x_n) + 2u_r(x_1, \ldots, y_i, \ldots, x_n)\right\}\right|\right\} \\
&= \frac{1}{2n}\sum_{i=1}^{n}\left\{\sum_{y_i=1}^{a_i}\left|\{\Delta_i u_r(x_1, \ldots, y_i, \ldots, x_n)\}^2\right.\right. \\
&\qquad \left.\left.+ 2u_r(x_1, \ldots, y_i, \ldots, x_n)\Delta_i u_r(x_1, \ldots, y_i, \ldots, x_n)\right|\right\}.
\end{aligned} \tag{4.8.34}$$

From (4.8.34) and using (4.8.8) repeatedly, Hölder's inequality with indices p, $p/(p-1)$ and the definition of α, we obtain

$$\begin{aligned}
&\left\{\sum_{r=1}^{M}|u_r(x)|^2\right\}^{p/(p-1)} \\
&\quad\leqslant \left(\frac{1}{2n}\right)^{p/(p-1)}(Mn)^{p/(p-1)-1} \\
&\qquad\times\sum_{r=1}^{M}\left\{\sum_{i=1}^{n}\left\{\sum_{y_i=1}^{a_i}\left|\{\Delta_i u_r(x_1,\ldots,y_i,\ldots,x_n)\}^2\right.\right.\right. \\
&\qquad\qquad + 2u_r(x_1,\ldots,y_i,\ldots,x_n) \\
&\qquad\qquad \left.\left.\left.\times\,\Delta_i u_r(x_1,\ldots,y_i,\ldots,x_n)\right|\right\}^{p/(p-1)}\right\} \\
&\quad\leqslant \left(\frac{1}{2n}\right)^{p/(p-1)}(Mn)^{1/(p-1)}\alpha^{1/(p-1)} \\
&\qquad\times\sum_{r=1}^{M}\left\{\sum_{i=1}^{n}\left\{\sum_{y_i=1}^{a_i}\left|\{\Delta_i u_r(x_1,\ldots,y_i,\ldots,x_n)\}^2\right.\right.\right. \\
&\qquad\qquad + 2u_r(x_1,\ldots,y_i,\ldots,x_n) \\
&\qquad\qquad \left.\left.\left.\times\,\Delta_i u_r(x_1,\ldots,y_i,\ldots,x_n)\right|^{p/(p-1)}\right\}\right\}.
\end{aligned} \tag{4.8.35}$$

Setting $x_i = y_i$, $i = 1,\ldots,n$, in (4.8.35) and taking the sum over both sides of (4.8.35) with respect to $y_1,\ldots,y_n$ on B and using the definition of α, we have

$$\begin{aligned}
&\sum_{B}\left\{\sum_{r=1}^{M}|u_r(x)|^2\right\}^{p/(p-1)} \\
&\quad\leqslant \left(\frac{1}{2n}\right)^{p/(p-1)}(Mn\alpha)^{1/(p-1)}\alpha \\
&\qquad\times\sum_{r=1}^{M}\left\{\sum_{i=1}^{n}\left\{\sum_{B}\left|\{\Delta_i u_r(y)\}^2+2u_r(y)\Delta_i u_r(y)\right|^{p/(p-1)}\right\}\right\}.
\end{aligned} \tag{4.8.36}$$

From (4.8.36) and using inequality (4.8.8) repeatedly, Hölder's inequality with indices $2m(p-1)/p$, $2m(p-1)/(2m(p-1)-p)$ and the definition of α, we have

$$\left[\sum_B \left\{\sum_{r=1}^{M} |u_r(x)|^2\right\}^{p/(p-1)}\right]^{2m(p-1)/p}$$

$$\leqslant \left\{\left(\frac{1}{2n}\right)^{p/(p-1)} (Mn\alpha)^{1/(p-1)}\alpha\right\}^{2m(p-1)/p} (Mn)^{2m(p-1)/p-1}$$

$$\times \sum_{r=1}^{M}\left\{\sum_{i=1}^{n}\left\{\sum_B \left|\{\Delta_i u_r(y)\}^2 + 2u_r(y)\Delta_i u_r(y)\right|^{p/(p-1)}\right\}^{2m(p-1)/p}\right\}$$

$$\leqslant \left\{\left(\frac{1}{2n}\right)^{p/(p-1)} (Mn\alpha)^{1/(p-1)}\alpha\right\}^{2m(p-1)/p}$$

$$\times (Mn)^{2m(p-1)/p-1}(\alpha n)^{(2m(p-1)-p)/p}$$

$$\times \sum_{r=1}^{M}\left\{\sum_{i=1}^{n}\left\{\sum_B \left|\{\Delta_i u_r(y)\}^2 + 2u_r(y)\Delta_i u_r(y)\right|^{2m}\right\}\right\}$$

$$\leqslant \left(\frac{1}{2n}\right)^{2m} (Mn\alpha)^{2m/p}\alpha^{2m(p-1)/p}(Mn)^{(2m(p-1)-p)/p}\left(\alpha^n\right)^{(2m(p-1)-p)/p}$$

$$\times \sum_{r=1}^{M}\left\{\sum_{i=1}^{n}\left\{\sum_B 2^{2m-1}\left[|\Delta_i u_r(y)|^{4m} + 2^{2m}|u_r(y)|^{2m}|\Delta_i u_r(y)|^{2m}\right]\right\}\right\}$$

$$= k_2\sum_{r=1}^{M}\sum_B\sum_{i=1}^{n}|\Delta_i u_r(y)|^{4m} + k_3\sum_{r=1}^{M}\sum_B\sum_{i=1}^{n}|u_r(y)|^{2m}|\Delta_i u_r(y)|^{2m}.$$

This result is the required inequality in (4.8.23) and the proof of Theorem 4.8.5 is complete. □

4.9 Miscellaneous Inequalities

4.9.1 Horgan [152]

Let u be sufficiently smooth function defined on an n-dimensional domain B which vanish on the boundary ∂B of B, then

$$\int_B |u|^3\,\mathrm{d}x \leqslant 3^{-3/4}\left(\int_B u^2\,\mathrm{d}x\right)^{3/4}\left(\int_B |\nabla u|^2\,\mathrm{d}x\right)^{3/4},$$

where $|\nabla u| = (\sum_{i=1}^{n} |\frac{\partial u}{\partial x_i}|^2)^{1/2}$.

4.9.2 Horgan and Nachlinger [155]

Let B be a bounded, three-dimensional domain with boundary ∂B. For any sufficiently smooth function u defined on B which vanish on the boundary ∂B, we have

$$\int_B |u|^3 \, \mathrm{d}V \leqslant M\left[\int_B u^2 \, \mathrm{d}V\right]^{1/2}\left[\int_B |\nabla u|^2 \, \mathrm{d}V\right].$$

Here $|\nabla u| = (\sum_{i=1}^{3} |\frac{\partial u}{\partial x_i}|^2)^{1/2}$, $\mathrm{d}V = \mathrm{d}x_1\, \mathrm{d}x_2\, \mathrm{d}x_3$, the number M is such that

$$M \leqslant (4\pi)^{-1/2}\lambda^{-1/4},$$

where λ is the smallest positive eigenvalue of the problem

$$\nabla^2 w + \lambda w = 0 \quad \text{in } B, \qquad u = 0 \quad \text{on } \partial B.$$

4.9.3 Pachpatte [236]

Let Q, f, g be as in Theorem 4.2.3. Then

$$\int_Q \big[|f(x)||\operatorname{grad} g(x)| + |g(x)||\operatorname{grad} f(x)|\big]\, \mathrm{d}x$$
$$\leqslant \frac{\sigma}{2\sqrt{n}} \int_Q \big[|\operatorname{grad} f(x)|^2 + |\operatorname{grad} g(x)|^2\big]\, \mathrm{d}x.$$

4.9.4 Payne [362]

Let u be any smooth function of compact support in three-dimensional Euclidean space E_3. Then

$$\int_{-\infty}^{\infty}\int_{-\infty}^{\infty}\int_{-\infty}^{\infty} u^4 \, \mathrm{d}x_1\, \mathrm{d}x_2\, \mathrm{d}x_3$$
$$\leqslant \frac{\sqrt{3}}{9}\left[\int_{-\infty}^{\infty}\int_{-\infty}^{\infty}\int_{-\infty}^{\infty} u^2 \, \mathrm{d}x_1\, \mathrm{d}x_2\, \mathrm{d}x_3\right]^{1/2}$$
$$\times \left[\int_{-\infty}^{\infty}\int_{-\infty}^{\infty}\int_{-\infty}^{\infty} |\operatorname{grad} u|^2 \, \mathrm{d}x_1\, \mathrm{d}x_2\, \mathrm{d}x_3\right]^{3/2}.$$

4.9.5 Pachpatte [258]

Let $u_r(x, y)$, $r = 1, \ldots, m$, be any smooth functions of compact support in two-dimensional Euclidean space E_2. Then

$$\left(\int_{-\infty}^{\infty}\int_{-\infty}^{\infty}\prod_{r=1}^{m}|u_r(x, y)|^{2/m}\,dx\,dy\right)^{1/2} \leqslant \frac{1}{2\sqrt{2m}}\int_{-\infty}^{\infty}\int_{-\infty}^{\infty}\left(\sum_{i=1}^{m}|\operatorname{grad} u_r(x, y)|\right)dx\,dy.$$

4.9.6 Pachpatte [258]

Let $p \geqslant 1$ be an integer and $u_r(x, y)$, $r = 1, \ldots, m$, be any smooth functions of compact support in two-dimensional Euclidean space E_2. Then

$$\int_{-\infty}^{\infty}\int_{-\infty}^{\infty}\prod_{r=1}^{m}|u_r(x, y)|^{(p+2)/m}\,dx\,dy \leqslant \frac{1}{2m}\left(\frac{p+2}{4}\right)^{2}\sum_{r=1}^{m}\left[\left(\int_{-\infty}^{\infty}\int_{-\infty}^{\infty}|u_r(x, y)|^{p}\,dx\,dy\right) \times \left(\int_{-\infty}^{\infty}\int_{-\infty}^{\infty}|\operatorname{grad} u_r(x, y)|^{2}\,dx\,dy\right)\right].$$

4.9.7 Pachpatte [265]

Let u, p, B be as in Theorem 4.3.1. Then

$$\int_B |u(x)|^{p+1}\,dx \leqslant \frac{1}{2}\alpha\left(\frac{1}{n}\right)^{1/2}\left(\int_B |u(x)|^{2p}\,dx\right)^{1/2}\left(\int_B |\nabla u(x)|^{2}\,dx\right)^{1/2},$$

where $\alpha = \max\{a_1, \ldots, a_n\}$ and $|\nabla u|$ is as in Section 4.9.1.

4.9.8 Pachpatte [259]

Let B be a bounded domain in $\mathbb{R}^n$, $n \geqslant 2$, be as in Theorem 4.3.1. Let u_m, $m = 1, \ldots, r$, be real-valued functions belonging to $C^1(B)$ which vanish on the boundary ∂B of B. Then

$$\left(\prod_{m=1}^{r}\left(\int_B |u_m(x)|\,\mathrm{d}x\right)\right)^{2/r} \leqslant \left(\frac{\alpha^{n+2}}{4nr}\right)\int_B\left(\sum_{m=1}^{r}|\nabla u_m(x)|^2\right)\mathrm{d}x,$$

where α and $|\nabla u_m|$ are as defined in Section 4.9.7.

4.9.9 Pachpatte [259]

Let the functions u_m, $m = 1,\ldots,r$, be as in Section 4.9.8. Then

$$\left(\prod_{m=1}^{r}\left(\int_B |u_m(x)|^2\,\mathrm{d}x\right)\right)^{2/r}$$
$$\leqslant \frac{1}{n}\left(\frac{\alpha}{r}\right)^2\left(\int_B\left(\sum_{m=1}^{r}|u_m(x)|^2\right)\mathrm{d}x\right)\left(\int_B\left(\sum_{m=1}^{r}|\nabla u_m(x)|^2\right)\mathrm{d}x\right),$$

where α and $|\nabla u_m|$ are as in Section 4.9.8.

4.9.10 Pachpatte [259]

Let D be a bounded region in $\mathbb{R}^n$, $n \geqslant 2$. Let u_m, $m = 1,\ldots,r$, be real-valued twice continuously differentiable functions on the closure $\overline{D}$ of D which vanish on the boundary ∂D of D and $\frac{\partial u_m}{\partial x_i}$, $i = 1,\ldots,n$, belong to $L_2(D)$, the set of functions which are square integrable on D. Then

$$\left(\prod_{m=1}^{r}\left(\int_D |\nabla u_m(x)|^2\,\mathrm{d}x\right)\right)^{1/r}$$
$$\leqslant \frac{1}{2r}\left[\int_D\left(\sum_{m=1}^{r}|u_m(x)|^2\right)\mathrm{d}x + \int_D\left(\sum_{m=1}^{r}|\Delta u_m(x)|^2\right)\mathrm{d}x\right],$$

where

$$|\nabla u_m| = \left(\sum_{i=1}^{n}\left(\frac{\partial u_m}{\partial x_i}\right)^2\right)^{1/2} \quad \text{and} \quad \Delta u_m = \sum_{i=1}^{n}\left(\frac{\partial^2 u_m}{\partial x_i^2}\right).$$

4.9.11 Pachpatte [259]

Let the functions u_m, $m = 1, \dots, r$, $|\nabla u_m|$ and Δu_m be as in Section 4.9.10. Then

$$\left(\prod_{m=1}^{r}\left(\int_D |u_m(x)||\nabla u_m(x)|\,\mathrm{d}x\right)\right)^{1/r} \leqslant \frac{1}{r}\left[\frac{3}{4}\int_D\left(\sum_{m=1}^{r}|u_m(x)|^2\right)\mathrm{d}x + \frac{1}{4}\int_D\left(\sum_{m=1}^{r}|\Delta u_m(x)|^2\right)\mathrm{d}x\right].$$

4.9.12 Pachpatte [248]

Suppose that $u, v \in G(B)$, the class of sufficiently smooth functions $z: B \to \mathbb{R}$, which vanish on the boundary ∂B of B and B is defined as in Theorem 4.3.5.

(i) Let $p, q \geqslant 2$ be constants such that $\frac{1}{p} + \frac{1}{q} = 1$. Then

$$\int_B |u(x)||v(x)|\,\mathrm{d}x \leqslant \frac{\mu^p}{np}\int_B |\mathrm{grad}\, u(x)|^p\,\mathrm{d}x + \frac{\mu^q}{nq}\int_B |\mathrm{grad}\, v(x)|^q\,\mathrm{d}x,$$

where $\alpha = \max\{b_1 - a_1, \dots, b_n - a_n\}$ and $\mu = \alpha/2$.

(ii) Let $p, q \geqslant 1$ be constants. Then

$$\int_B |u(x)|^p|v(x)|^q\,\mathrm{d}x \leqslant \frac{\mu^{p+q}}{n}\left[\left(\frac{p}{p+q}\right)\int_B |\mathrm{grad}\, u(x)|^{p+q}\,\mathrm{d}x + \left(\frac{q}{p+q}\right)\int_B |\mathrm{grad}\, v(x)|^{p+q}\,\mathrm{d}x\right],$$

where μ and α are as given in part (i).

4.9.13 Pachpatte [248]

Suppose that $u, v \in G(B)$, where $G(B)$ is as in Section 4.9.12. Let α be as in Section 4.9.12.

(i) Let $p, q > 1$ be constants such that $\frac{1}{p} + \frac{1}{q} = 1$. Then

$$\int_B |u(x)||v(x)|\,\mathrm{d}x \leqslant \frac{\alpha}{2\sqrt{n}}\left[\left(\int_B |u(x)|^{2(p-1)}\,\mathrm{d}x\right)^{1/2}\left(\int_B |\mathrm{grad}\, u(x)|^2\,\mathrm{d}x\right)^{1/2}\right.$$

$$+\left(\int_B |v(x)|^{2(p-1)}\,\mathrm{d}x\right)^{1/2}\left(\int_B |\operatorname{grad} v(x)|^2\,\mathrm{d}x\right)^{1/2}\Bigg].$$

(ii) Let $p, q \geqslant 1$ be constants. Then

$$\int_B |u(x)|^p |v(x)|^q\,\mathrm{d}x$$
$$\leqslant \left(\frac{p\alpha}{2\sqrt{n}}\right)\left(\int_B |u(x)|^{2(p+q-1)}\,\mathrm{d}x\right)^{1/2}\left(\int_B |\operatorname{grad} u(x)|^2\,\mathrm{d}x\right)^{1/2}$$
$$+\left(\frac{q\alpha}{2\sqrt{n}}\right)\left(\int_B |v(x)|^{2(p+q-1)}\,\mathrm{d}x\right)^{1/2}\left(\int_B |\operatorname{grad} v(x)|^2\,\mathrm{d}x\right)^{1/2}.$$

4.9.14 Pachpatte [251]

Let $p, q, r \geqslant 1$ be constants and suppose that $u, v, w \in G(B)$, where $G(B)$ is as in Section 4.9.12. Then

$$\int_B \left[|u(x)|^p |v(x)|^q + |v(x)|^q |w(x)|^r + |w(x)|^r |u(x)|^p\right]\mathrm{d}x$$
$$\leqslant \frac{1}{n}\lambda^{2p}\int_B |\operatorname{grad} u(x)|^{2p}\,\mathrm{d}x + \frac{1}{n}\lambda^{2q}\int_B |\operatorname{grad} v(x)|^{2q}\,\mathrm{d}x$$
$$+\frac{1}{n}\lambda^{2r}\int_B |\operatorname{grad} w(x)|^{2r}\,\mathrm{d}x,$$

$$\int_B |u(x)|^p |v(x)|^q |w(x)|^r \left[|u(x)|^p + |v(x)|^q + |w(x)|^r\right]\mathrm{d}x$$
$$\leqslant \frac{1}{n}\lambda^{4p}\int_B |\operatorname{grad} u(x)|^{4p}\,\mathrm{d}x + \frac{1}{n}\lambda^{4q}\int_B |\operatorname{grad} v(x)|^{4q}\,\mathrm{d}x$$
$$+\frac{1}{n}\lambda^{4r}\int_B |\operatorname{grad} w(x)|^{4r}\,\mathrm{d}x,$$

where $\lambda = \alpha/2$ and $\alpha = \max\{b_1 - a_1, \ldots, b_n - a_n\}$.

4.9.15 Pachpatte [251]

Let $p, q, r \geqslant 2$ be constants and suppose that $u, v, w \in G(B)$, where $G(B)$ is as in Section 4.9.12. Let λ be as in Section 4.9.14. Then

$$\int_B \left[|u(x)|^p|v(x)|^q + |v(x)|^q|w(x)|^r + |w(x)|^r|u(x)|^p\right]\mathrm{d}x$$

$$\leqslant \frac{p^2}{n}\lambda^2\left(\int_B |u(x)|^{2p}\,\mathrm{d}x\right)^{(p-1)/p}\left(\int_B |\operatorname{grad} u(x)|^{2p}\,\mathrm{d}x\right)^{1/p}$$

$$+ \frac{q^2}{n}\lambda^2\left(\int_B |v(x)|^{2q}\,\mathrm{d}x\right)^{(q-1)/q}\left(\int_B |\operatorname{grad} v(x)|^{2q}\,\mathrm{d}x\right)^{1/q}$$

$$+ \frac{r^2}{n}\lambda^2\left(\int_B |w(x)|^{2r}\,\mathrm{d}x\right)^{(r-1)/r}\left(\int_B |\operatorname{grad} w(x)|^{2r}\,\mathrm{d}x\right)^{1/r},$$

$$\int_B |u(x)|^p|v(x)|^q|w(x)|^r\left[|u(x)|^p + |v(x)|^q + |w(x)|^r\right]\mathrm{d}x$$

$$\leqslant \frac{p^4}{n}\lambda^4\left(\int_B |u(x)|^{4p}\,\mathrm{d}x\right)^{(p-1)/p}\left(\int_B |\operatorname{grad} u(x)|^{4p}\,\mathrm{d}x\right)^{1/p}$$

$$+ \frac{q^4}{n}\lambda^4\left(\int_B |v(x)|^{4q}\,\mathrm{d}x\right)^{(q-1)/q}\left(\int_B |\operatorname{grad} v(x)|^{4q}\,\mathrm{d}x\right)^{1/q}$$

$$+ \frac{r^4}{n}\lambda^4\left(\int_B |w(x)|^{4r}\,\mathrm{d}x\right)^{(r-1)/r}\left(\int_B |\operatorname{grad} w(x)|^{4r}\,\mathrm{d}x\right)^{1/r},$$

where λ is as defined in Section 4.9.14.

4.9.16 Dubinskii [95]

Let G be a bounded region in $\mathbb{R}^n$ with boundary Γ. Let $u(x) \in C^1(G)$, $\alpha_0 \geqslant 0$, $\alpha_1 \geqslant 1$. Then the following inequalities are valid

(a_1) If $\alpha_1 < n$, then

$$\left(\int_G |u|^{(\alpha_0+\alpha_1)n/(n-\alpha_1)}\,\mathrm{d}x\right)^{(n-\alpha_1)/n}$$

$$\leqslant K\sum_{i=1}^{n}\left(\int_G |u|^{\alpha_0}\left|\frac{\partial u}{\partial x_i}\right|^{\alpha_1}\mathrm{d}x + \int_\Gamma |u|^{\alpha_0+\alpha_1}\,\mathrm{d}\gamma\right).$$

(a$_2$) If $\alpha_1 = n$, then

$$\left(\int_G |u|^{(\alpha_0+\alpha_1)p}\,dx\right)^{1/p} \leqslant K\sum_{i=1}^{n}\left(\int_G |u|^{\alpha_0}\left|\frac{\partial u}{\partial x_i}\right|^{\alpha_1} dx + \int_\Gamma |u|^{\alpha_0+\alpha_1}\,d\gamma\right),$$

where $p \geqslant 1$ is arbitrary.

(a$_3$) If $\alpha_1 > n$, then

$$\max\left|u(x)\right| \leqslant K\sum_{i=1}^{n}\left(\int_G |u|^{\alpha_0}\left|\frac{\partial u}{\partial x_i}\right|^{\alpha_1} dx + \int_\Gamma |u|^{\alpha_0+\alpha_1}\,d\gamma\right)^{1/(\alpha_0+\alpha_1)},$$

where the constant K is as explained in Section 4.5. Here, from the uniform boundedness of the right-hand sides of these inequalities, it follows the compactness of the sets of $u(x)$ respectively in the spaces L_q, $q < (\alpha_0 + \alpha_1)n/(n - \alpha_1)$, L_p and C.

4.9.17 Pachpatte [263]

Let E be an n-dimensional Euclidean space with $n \geqslant 2$ and B be a bounded domain in E defined by $B = \{x \in E\colon\ a \leqslant x \leqslant b\}$, $a = (a_1, \ldots, a_n)$, $b = (b_1, \ldots, b_n) \in E$ for $-\infty < a_i \leqslant b_i < \infty$. Let u_r, $r = 1, \ldots, m$, be real-valued sufficiently smooth functions defined on B which vanish only on the boundary ∂B of B, and $p_r \geqslant 1$ be constants. Then

$$\int_B\left[\prod_{r=1}^{m}\left|u_r^{p_r}(x)\ln u_r(x) - \frac{1}{p_r}u_r^{p_r}(x)\right|\right]^{1/m} dx$$
$$\leqslant M\sum_{r=1}^{m}\left[\int_B\left|\operatorname{grad} u_r(x)\right|^2 dx + \int_B\left|u_r(x)\right|^{2(p_r-1)}\left|\ln u_r(x)\right|^2 dx\right],$$

where

$$M = \left(\frac{\alpha\sqrt{n}}{2m}\right)\prod_{r=1}^{m}\left(\frac{p_r}{2n}\right)^{1/m}, \qquad \alpha = \max\{b_1 - a_1, \ldots, b_n - a_n\},$$

and

$$\left|\operatorname{grad} u_r(x)\right| = \left(\sum_{i=1}^{n}\left|\frac{\partial u_r(x)}{\partial x_i}\right|^2\right)^{1/2}.$$

4.9.18 Pachpatte [263]

Let E be an n-dimensional Euclidean space with $n \geqslant 2$. Let u_r, $r = 1, \ldots, m$, be real-valued and sufficiently smooth functions of compact support in E and $p_r \geqslant 1$ be constants. Then

$$\left\{\int_E \left[\prod_{r=1}^{m} \left| u_r^{p_r}(x) \ln u_r(x) - \frac{1}{p_r} u_r^{p_r}(x) \right|^{n/(n-1)}\right]^{1/m} \mathrm{d}x \right\}^{(n-1)/n}$$
$$\leqslant N \sum_{r=1}^{m} \left[\int_E \left|\operatorname{grad} u_r(x)\right|^2 \mathrm{d}x + \int_E \left|u_r(x)\right|^{2(p_r-1)} \left|\ln u_r(x)\right|^2 \mathrm{d}x\right],$$

where

$$N = \frac{1}{2\sqrt{n}} \left(\frac{1}{m}\right)^{(n-1)/n} \prod_{r=1}^{m} \left(\frac{p_r}{2}\right)^{1/m}$$

and $|\operatorname{grad} u_r(x)|$ is given as in Section 4.9.17.

4.9.19 Horgan [154]

Let $u(x)$ be a sufficiently regular function defined on a bounded domain B in $\mathbb{R}^n$, $n > 1$, with boundary ∂B, then we have

$$\int_{\partial B} u^2 \,\mathrm{d}S \leqslant C \int_B \left(u^2 + u_{,i} u_{,i}\right) \mathrm{d}x,$$

where the usual Cartesian tensor notation is used, with subscripts preceded by a comma denoting differentiation with respect to the corresponding coordinate.

4.9.20 Pachpatte [264]

Let B be a normal domain in $\mathbb{R}^n$ with boundary ∂B and $\overline{B} = B + \partial B$ (union of B and ∂B). Let $u(x)$ and $v(x)$ be real-valued functions belonging to $C^1(\overline{B})$. Then

$$\int_B u^2(x) v^2(x) \,\mathrm{d}x$$
$$\leqslant \mu \left[\int_{\partial B} u^2(x) v^2(x) \,\mathrm{d}S + \int_B \left[u^2(x) \left|\operatorname{grad} v(x)\right|^2 + v^2(x) \left|\operatorname{grad} u(x)\right|^2\right] \mathrm{d}x\right],$$

where $\mu = \max\{2\delta, \frac{8\delta^2}{n}\}$, $\delta = \max\{|x_1|, \ldots, |x_n|\}$.

4.9.21 Pachpatte [264]

Let B be a normal domain in $\mathbb{R}^n$ with sufficiently smooth boundary ∂B and $\overline{B} = B + \partial B$ (union of B and ∂B). Let $\alpha_i(x) \in C^1(\overline{B})$, $i = 1, \ldots, n$, be auxiliary functions such that $\alpha_i(x) = z_i(x)$ for $x \in \partial B$. Let $u(x)$ and $v(x)$ be real-valued functions belonging to $C^1(\overline{B})$. Then

$$\int_{\partial B} u^2(x)v^2(x)\,\mathrm{d}S$$
$$\leqslant \lambda\left[\int_B u^2(x)v^2(x)\,\mathrm{d}x + \int_B \left[u^2(x)\left|\operatorname{grad} v(x)\right|^2 + v^2(x)\left|\operatorname{grad} u(x)\right|^2\right]\mathrm{d}x\right],$$

where $\lambda = \max\{c_0 + c_1 c, \frac{2nc_1}{c}\}$, $c > 0$ is an arbitrary constant and

$$c_0 = \sup_{x\in B}\left|\sum_{i=1}^{n} \frac{\partial}{\partial x_i}\alpha_i(x)\right|, \qquad c_1 = \sup_{i=1,\ldots,n}\left\{\sup_{x\in B}\left|\alpha_i(x)\right|\right\}.$$

4.9.22 Allegretto [8]

Let $u \in C_0^\infty(\mathbb{R}^n \setminus \{0\})$, $\alpha \in \mathbb{R}^1$, $\alpha \leqslant 0$. Then the following inequality is valid

$$\int_{\mathbb{R}^n} |x|^\alpha (\Delta u)^2\,\mathrm{d}x \geqslant K(\alpha)\int_{\mathbb{R}^n} |x|^{\alpha-4}u^2\,\mathrm{d}x,$$

where

$$\Delta = \sum_{i=1}^{n}\frac{\partial^2}{\partial x_i^2} \quad \text{and} \quad K(\alpha) = \frac{(4-\alpha-n)^2(n-\alpha)^2}{16} + \tau(\alpha),$$

$$\tau(\alpha) = \inf_{k\in\{0,1,2,\ldots\}}\left\{(k)(k+n-2)\left(k^2 + (n-2)k + \frac{n^2-4n+4\alpha-\alpha^2}{2}\right)\right\}.$$

4.9.23 Schmincke [403]

Suppose $u \in C_0^\infty(B)$ and $s \in [-\frac{n(n-4)}{2}, \infty)$, where $B = \mathbb{R}^n \setminus \{0\}$, $n \geqslant 2$. Then

$$\int_{\mathbb{R}^n} |\Delta u|^2\,\mathrm{d}x \geqslant -s\int_B |\nabla u|^2|x|^{-2}\,\mathrm{d}x + \frac{(n-4)^2}{16}\left(n^2+4s\right)\int_B |u|^2|x|^{-4}\,\mathrm{d}x,$$

where $\nabla = (\frac{\partial}{\partial x_1}, \ldots, \frac{\partial}{\partial x_n})$ and $\Delta = \sum_{i=1}^n \frac{\partial^2}{\partial x_i^2}$.

4.9.24 Chang, Wilson and Wolff [53]

Let $\phi(t)$ be a nonnegative, increasing function on $(0,\infty)$ which satisfies $\int_1^\infty \frac{dt}{t\phi(t)} < \infty$. Suppose that $v(x)$ is a nonnegative function on $\mathbb{R}^n$ such that for every cube I,

$$\int_I \phi\big(|I|^{2/n} v(x)\big) v(x)\,dx \leqslant c|I|^{1-2/n},$$

with c independent of I. Then for $f \in C_0^\infty(\mathbb{R}^n)$,

$$\int_{\mathbb{R}^n} |f(x)|^2 v(x)\,dx \leqslant c \int_{\mathbb{R}^n} |\nabla f(x)|^2\,dx,$$

where c is independent of f and ∇ is as in Section 4.9.23.

4.9.25 Chanillo and Wheeden [55]

Let $1 < p < \infty$ and $\phi(t)$ be a nonnegative and increasing function on $(0,\infty)$ which satisfies

$$\int_1^\infty \frac{dt}{t\phi(t)^{p'-1}} < \infty, \qquad \frac{1}{p} + \frac{1}{p'} = 1,$$

and v and w be weight functions (nonnegative measurable functions) on $\mathbb{R}^n$, $n > 1$, with $w \in A_p$, the mean for all cubes I

$$\left[\frac{1}{|I|}\int_I w(x)\,dx\right]\left[\frac{1}{|I|}\int_I w(x)^{-1/(p-1)}\,dx\right]^{p-1} \leqslant c,$$

with c independent of I. Let f be Lipschitz continuous on a cube $I \subset \mathbb{R}^n$ and suppose that for all cubes $Q \subset 2I$,

$$|Q|^{p/n} \int_Q \phi\left(|Q|^{p/n} \frac{v(x)}{w(x)}\right) v(x)\,dx \leqslant c \int_Q w(x)\,dx,$$

where c is independent of Q. Then

$$\int_I |f(x) - f_I|^p v(x)\,dx \leqslant c \int_I |\nabla f(x)|^p w(x)\,dx, \quad f_I = \frac{1}{|I|}\int_I f(x)\,dx,$$

with c independent of f and ∇ is as in Section 4.9.23. If the constant in the hypothesis is independent of I, then so it is the one in the conclusion.

4.9.26 Lieb and Thirring [193]

Let $\phi_1, \dots, \phi_N$ be a finite family of functions in $H^1(\mathbb{R}^n)$ which are orthonormal in $L^2(\mathbb{R}^n)$, that is, $\int_{\mathbb{R}^n} \phi_i \phi_j \, dx = \delta_{i,j}$, $1 \leqslant i, j \leqslant N$, where $L^2(\mathbb{R}^n)$ is the set of classes of real functions measurable in Ω, a bounded open set of $\mathbb{R}^n$ and square integrable on $\mathbb{R}^n$; $H^1(\mathbb{R}^n)$ is the Sobolev space of order one constructed on $L^2(\mathbb{R}^n)$. Let p be a constant satisfying $\max\{1, n/2\} < p < 1+n/2$. Then there exists a constant $k = k(n, p)$ independent of N and of the ϕ_j's such that

$$\left(\int_{\mathbb{R}^n}\left(\sum_{j=1}^N \phi_j(x)^2\right)^{p/(p-1)} dx\right)^{2(p-1)/n} \leqslant k \sum_{j=1}^N \int_{\mathbb{R}^n} \sum_{i=1}^n \left(\frac{\partial \phi_j(x)}{\partial x_i}\right)^2 dx.$$

4.9.27 Ghidaglia, Marion and Temam [127]

Let Ω be a bounded open set of $\mathbb{R}^n$, $L^2(\Omega)$ be the set of classes of real functions which are measurable in Ω and square integrable on Ω, $H^m(\Omega)$ be the Sobolev space of order m constructed on $L^2(\Omega)$. Suppose that there exists a linear prolongation operator Π_m mapping $H^m(\Omega)$ into $H^m(\mathbb{R}^n)$ such that $\Pi_m \in L(H^r(\Omega), H^r(\mathbb{R}^n))$, $r = 0, 1, 2, \dots, m$, and $\Pi_m u(x) = u(x)$ for a.e. $x \in \Omega$. Let $\{\phi_j\}_{j=1}^N$ in $H^m(\Omega)$, $m \geqslant 1$, is suborthonormal in $L^2(\Omega)$, that is,

$$\sum_{i,j=1}^N \xi_i \xi_j \int_\Omega \phi_i \phi_j \, dx \leqslant \sum_{i=1}^N \xi_i^2$$

for all $\xi_1, \dots, \xi_N \in \mathbb{R}$. Let p be a constant satisfying the condition

$$\max\left\{1, \frac{n}{2m}\right\} < p \leqslant 1 + \frac{n}{2m},$$

then there exist two positive constants k_1 and k_2 such that

$$\left(\int_\Omega \left(\sum_{j=1}^N |\phi_j(x)|\right)^{p/(p-1)} dx\right)^{2m(p-1)/n}$$
$$\leqslant k_1 \sum_{j=1}^N \int_\Omega \sum_{[\alpha]=m} \left|D^\alpha \phi_j(x)\right|^2 dx + \frac{k_2}{\delta(\Omega)^{2m}} \int_\Omega \left(\sum_{j=1}^N \left|\phi_j^2(x)\right|^2\right) dx,$$

where $|\cdot|$ denotes the usual Euclidean norm on $\mathbb{R}^k$, $\delta(\Omega)$ is the diameter of Ω, $D^\alpha = D_1^{\alpha_1} \cdots D_n^{\alpha_n} = \frac{\partial^{[\alpha]}}{\partial x_1^{\alpha_1} \cdots \partial x_n^{\alpha_n}}$, $[\alpha] = \alpha_1 + \cdots + \alpha_n$. The constants k_1, k_2 depend on m, n, p and the space Ω.

4.9.28 Agarwal, Pečarić and Brnetić [7]

Let E be a bounded domain in $\mathbb{R}^n$ defined by $E = \prod_{i=1}^{n}[a_i, b_i]$. Let $H_n(\alpha_i)$ stand for the harmonic means of $\alpha_1, \ldots, \alpha_n$. Let $G(E)$ be the class of continuous functions $u(x): E \to \mathbb{R}$ for which $D^n u(x) = D_1 \cdots D_n u(x)$, $D_i = \frac{\partial}{\partial x_i}$, exists and that for each i, $1 \leqslant i \leqslant n$, $u(x)|_{x_i=a_i} = u(x)|_{x_i=b_i} = 0$.

(i) Let $\lambda, \mu \geqslant 1$ and $u \in G(E)$. Then

$$\int_E |u(x)|^\lambda \,\mathrm{d}x \leqslant k_1(\lambda, \mu) \int_E \|\operatorname{grad} u(x)\|_\mu^\lambda \,\mathrm{d}x,$$

where

$$k_1(\lambda, \mu) = \frac{1}{n} I(\lambda) C\left(\frac{\lambda}{\mu}\right) H_n\big((b_i - a_i)^\lambda\big),$$

$$I(\lambda) = \int_0^1 \left[t^{1-\lambda} + (1-t)^{1-\lambda}\right]^{-1} \mathrm{d}t,$$

and $C(\alpha) = 1$ if $\alpha \geqslant 1$ and $C(\alpha) = n^{1-\alpha}$ if $0 \leqslant \alpha \leqslant 1$.

(ii) Let $p, \lambda \geqslant 1$ and $u \in G(E)$. Then

$$\int_E |u(x)|^{\lambda p} \,\mathrm{d}x \leqslant \frac{p^\lambda H_n((b_i - a_i)^\lambda)}{n} I(\lambda) \left(\int_E |u(x)|^{\lambda p} \,\mathrm{d}x\right)^{(p-1)/p} \left(\int_E \|\operatorname{grad} u(x)\|_\lambda^{\lambda p} \,\mathrm{d}x\right)^{1/p},$$

where $I(\lambda)$ is defined in (i).

(iii) Let $l \geqslant 0$, $m \geqslant 1$ and $u \in G(E)$. Then

$$\int_E |u(x)|^{l+m} \,\mathrm{d}x \leqslant \frac{1}{n}\left(\frac{l+m}{m}\right)^m I(m) \sum_{i=1}^n (b_i - a_i)^m \int_E |u(x)|^l \left|\frac{\partial}{\partial x_i} u(x)\right|^m \mathrm{d}x,$$

where $I(m)$ is defined in (i). Moreover, in (i) and (ii),

$$\|\operatorname{grad} u(x)\|_\mu = \left(\sum_{i=1}^n \left|\frac{\partial}{\partial x_i} u(x)\right|^\mu\right)^{1/\mu}.$$

4.9.29 Pachpatte [275]

Let $\mathbb{N} = \{1, 2, \ldots\}$, $\mathbb{N}_{k+1} = \{1, 2, \ldots, k, k+1\}$, $k \in \mathbb{N}$, $\mathbb{N}_{m+1} = \{1, 2, \ldots, m, m+1\}$, $m \in \mathbb{N}$ and $Q = \mathbb{N}_{k+1} \times \mathbb{N}_{m+1}$. Let $F(Q)$ denote the class of functions $c: Q \to \mathbb{R}$ such that $c(1, y) = c(k+1, y) = 0$ for $1 \leqslant y \leqslant m+1$, $m \in \mathbb{N}$, $\Delta_1 c(x, 1) = 0$, $\Delta_1 c(x, m+1) = 0$ for $1 \leqslant x \leqslant k+1$, $k \in \mathbb{N}$, where $\Delta_1 c(x, y) = c(x+1, y) - c(x, y)$, and let $\Delta_2 c(x, y) = c(x, y+1) - c(x, y)$ and $\Delta_1 \Delta_2 c(x, y) = \Delta_1[\Delta_2 c(x, y)]$.

(i) Let $p, q > 1$ be constants such that $\frac{1}{p} + \frac{1}{q} = 1$ and suppose that $u, v \in F(Q)$. Then

$$\sum_{x=1}^{k} \sum_{y=1}^{m} |u(x, y)||v(x, y)|$$

$$\leqslant \frac{\mu^p}{p} \sum_{x=1}^{k} \sum_{y=1}^{m} |\Delta_2 \Delta_1 u(x, y)|^p + \frac{\mu^q}{q} \sum_{x=1}^{k} \sum_{y=1}^{m} |\Delta_2 \Delta_1 v(x, y)|^q,$$

where $\mu = km/4$.

(ii) Let $p, q \geqslant 1$ be constants and suppose that $u, v \in F(Q)$. Then

$$\sum_{x=1}^{k} \sum_{y=1}^{m} |u(x, y)|^p |v(x, y)|^q$$

$$\leqslant \mu^{p+q} \left[\left(\frac{p}{p+q} \right) \sum_{x=1}^{k} \sum_{y=1}^{m} |\Delta_2 \Delta_1 u(x, y)|^p \right.$$

$$\left. + \left(\frac{q}{p+q} \right) \sum_{x=1}^{k} \sum_{y=1}^{m} |\Delta_2 \Delta_1 v(x, y)|^{p+q} \right],$$

where μ is as defined in (i).

4.9.30 Pachpatte [250]

Let Q, $F(Q)$ be as in Section 4.9.29. Let $p, q, r \geqslant 1$ be constants and suppose that $f, g, h \in F(Q)$. Then

$$\sum_{x=1}^{k} \sum_{y=1}^{m} [|f(x, y)|^p |g(x, y)|^q + |g(x, y)|^q |h(x, y)|^r + |h(x, y)|^r |f(x, y)|^p]$$

$$\leqslant \left(\frac{km}{4} \right)^{2p} \sum_{x=1}^{k} \sum_{y=1}^{m} |\Delta_2 \Delta_1 f(x, y)|^{2p} + \left(\frac{km}{4} \right)^{2q} \sum_{x=1}^{k} \sum_{y=1}^{m} |\Delta_2 \Delta_1 g(x, y)|^{2q}$$

$$+\left(\frac{km}{4}\right)^{2r}\sum_{x=1}^{k}\sum_{y=1}^{m}\left|\Delta_2\Delta_1 h(x,y)\right|^{2r},$$

$$\sum_{x=1}^{k}\sum_{y=1}^{m}\left|f(x,y)\right|^{p}\left|g(x,y)\right|^{q}\left|h(x,y)\right|^{r}\left(\left|f(x,y)\right|^{p}+\left|g(x,y)\right|^{q}+\left|h(x,y)\right|^{r}\right)$$

$$\leqslant\left(\frac{km}{4}\right)^{4p}\sum_{x=1}^{k}\sum_{y=1}^{m}\left|\Delta_2\Delta_1 f(x,y)\right|^{4p}+\left(\frac{km}{4}\right)^{4q}\sum_{x=1}^{k}\sum_{y=1}^{m}\left|\Delta_2\Delta_1 g(x,y)\right|^{4q}$$

$$+\left(\frac{km}{4}\right)^{4r}\sum_{x=1}^{k}\sum_{y=1}^{m}\left|\Delta_2\Delta_1 h(x,y)\right|^{4r}.$$

4.9.31 Pachpatte [242]

Let $p \geqslant 2$ be a constant and $F(B)$, α, Δ be as in Theorem 4.8.1. Let $u \in F(B)$. Then

$$\left(\sum_{B}\left|u(y)\right|^{p/(p-1)}\right)^{(p-1)/p}\leqslant\frac{1}{2}\left(\frac{1}{n}\right)^{1/2}\alpha^{(2(p-n)+np)/(2p)}\left(\sum_{B}\left|\Delta u(y)\right|^{2}\right)^{1/2}.$$

4.9.32 Pachpatte [242]

Let p, u, $F(B)$, α, Δ be as in Section 4.9.31. Then

$$\left(\sum_{B}\left|u(y)\right|^{2(p+2)/p}\right)^{p/(p+2)}\leqslant\left(\frac{\alpha}{2}\right)^{2}n^{-p/(p+2)}\left(\sum_{B}\left|\Delta u(y)\right|^{2(p+2)/p}\right)^{p/(p+2)}.$$

4.10 Notes

Theorem 4.2.1 was formulated by Friedman [120] and is a useful tool in the study of partial differential equations. Different versions of this theorem essentially go back to Poincaré [389], and this type of investigation was first initiated by Schwarz [404]. Theorem 4.2.2 is due to Pachpatte [247] and is a variant of Theorem 4.2.1. Theorem 4.2.3 is taken from Pachpatte [236]. The inequality in Theorem 4.2.4 is due to Payne [362] and was given while studying the uniqueness criteria for Navier–Stokes equations, and the inequality in Theorem 4.2.5 is given by Serrin in [405]. Theorem 4.2.6 is due to Pachpatte [290] and contains, in the special case, the known inequality due to Nirenberg given in [229].

The results given in Section 4.3 are due to Pachpatte [249,265,290]. Theorems 4.3.1–4.3.6 deal with different variants of the Poincaré and Sobolev inequalities. Theorems 4.4.1–4.4.6 are established by Pachpatte in [237,246]. The results in Theorems 4.5.1–4.5.4 are taken from Dubinskii [95]. Theorem 4.5.5 is due to Alzer [10]. Theorems 4.5.6 and 4.5.7 are established by Pachpatte in [265] and are motivated by Dubinskii's inequalities given in [95]. Theorems 4.6.1–4.6.4 which relate to Poincaré- and Sobolev-type inequalities are established by Pachpatte in [276], while Theorems 4.6.5 and 4.6.6 are taken from Pachpatte [289].

Theorems 4.7.1 and 4.7.2 are taken from Pachpatte [286]. Theorems 4.7.3 and 4.7.4 give the Rellich-type inequalities and are established by Pachpatte in [286]. Theorem 4.7.5 is taken from [288] and Theorem 4.7.6 is a more general version of the Rellich-type inequality established by Pachpatte in [288]. Theorems 4.8.1–4.8.5 are due to Pachpatte [269,275,285] which relate to the discrete Poincaré- and Sobolev-type inequalities involving functions of several independent variables and their forward differences. Section 4.9.9 covers miscellaneous inequalities established by various investigators.

Chapter 5

Levin- and Lyapunov-Type Inequalities

5.1 Introduction

The importance of basic comparison inequalities has been long recognized in the study of qualitative behavior of solutions of ordinary second-order differential equations. The history of these inequalities for continuous differential systems goes far back starting with the famous paper of Sturm [414] which gives inequalities for the zeros of solutions of linear second-order differential equations. In his fundamental work [201] Lyapunov has given one of the most basic and inspiring inequalities which provides a lower bound for the distance between consecutive zeros of the solutions of the linear second-order differential equation. Lyapunov's inequality has become a versatile tool in the study of qualitative nature of solutions of ordinary second-order differential equations. Over the years there have appeared a number of generalizations, extensions, variants and applications related to the basic Sturmain comparison theorem and the original Lyapunov inequality. This chapter considers basic inequalities developed in the literature related to the Sturmain comparison theorem and to the Lyapunov inequality which occupies a fundamental place in the theory of ordinary differential equations.

5.2 Inequalities of Levin and Others

In this section we give results involving comparison of the solutions of linear and nonlinear second-order differential equations investigated by Levin [187], Kreith [171] and Ladas [177]. Here we consider only solutions which are defined on the whole interval of definition of the independent variable and their existence and uniqueness will be assumed without further mention. An oscillatory solution is (by definition) one which has arbitrary large zeros.

The basic Sturmain comparison theorem deals with functions $u(x)$ and $v(x)$ satisfying

$$u'' + c(x)u = 0, \tag{5.2.1}$$

$$v'' + \gamma(x)v = 0. \tag{5.2.2}$$

If $\gamma(x) \geqslant c(x)$, then solutions of (5.2.2) oscillate more rapidly than solutions of (5.2.1). More precisely, if $u(x)$ is a nontrivial solution of (5.2.1) for which $u(x_1) = u(x_2) = 0$, $x_1 < x_2$, and $\gamma(x) \geqslant c(x)$ for $x_1 \leqslant x \leqslant x_2$, then $v(x)$ has a zero in $(x_1, x_2]$.

In 1960, Levin [187] extended Sturm's theorem in a direction somewhat different from other earlier investigators. The method used by Levin involves the transformation of the differential equations (5.2.1), (5.2.2) into the Riccati equations

$$w' = w^2 + c(x), \tag{5.2.3}$$

$$z' = z^2 + \gamma(x), \tag{5.2.4}$$

by the substitutions $w = -u'/u$, $z = -v'/v$, respectively, and assuming that $c(x)$ and $\gamma(x)$ are continuous on $[\alpha, \beta]$.

In the following theorems we present the main comparison theorems established by Levin in [187].

THEOREM 5.2.1. *Let u and v be nontrivial solutions of* (5.2.1) *and* (5.2.2), *respectively, such that $u(x)$ does not vanish on $[\alpha, \beta]$, $v(\alpha) \neq 0$ and the inequality*

$$-\frac{u'(\alpha)}{u(\alpha)} + \int_\alpha^x c(t)\,dt > \left| -\frac{v'(\alpha)}{v(\alpha)} + \int_\alpha^x \gamma(t)\,dt \right| \tag{5.2.5}$$

holds for all x on $[\alpha, \beta]$. Then $v(x)$ does not vanish on $[\alpha, \beta]$ and

$$-\frac{u'(x)}{u(x)} > \left| \frac{v'(x)}{v(x)} \right|, \quad \alpha \leqslant x \leqslant \beta. \tag{5.2.6}$$

The same theorem holds if the inequality signs in (5.2.5) *and* (5.2.6) *are replaced by "$\geqslant$".*

PROOF. Since $u(x)$ does not vanish, $w = -u'/u$ is continuous on $[\alpha, \beta]$ and satisfies the Riccati equation (5.2.3), which is equivalent to the integral equation

$$w(x) = w(\alpha) + \int_\alpha^x w^2(t)\,dt + \int_\alpha^x c(t)\,dt. \tag{5.2.7}$$

By the hypothesis (5.2.5),

$$w(x) \geqslant -\frac{u'(\alpha)}{u(\alpha)} + \int_\alpha^x c(t)\,\mathrm{d}t > 0. \tag{5.2.8}$$

Since $v(\alpha) \neq 0$, $z = -v'/v$ is continuous on some interval $[\alpha, \delta]$, $\alpha < \delta \leqslant \beta$. On this interval, (5.2.4) is well defined and implies the integral equation

$$z(x) = z(\alpha) + \int_\alpha^x z^2(t)\,\mathrm{d}t + \int_\alpha^x \gamma(t)\,\mathrm{d}t. \tag{5.2.9}$$

From (5.2.9), (5.2.5) and (5.2.8), we observe that

$$\begin{aligned} z(x) &\geqslant z(\alpha) + \int_\alpha^x \gamma(t)\,\mathrm{d}t \\ &> -w(\alpha) - \int_\alpha^x c(t)\,\mathrm{d}t \\ &\geqslant -w(x), \end{aligned}$$

and consequently, $w(x) > -z(x)$. In order to show that

$$\left|z(x)\right| < w(x) \quad \text{on } \alpha \leqslant x \leqslant \delta, \tag{5.2.10}$$

it is sufficient to show that $w(x) > z(x)$ on this interval. Suppose to the contrary that there exists a point x_0 on $[\alpha, \delta]$ such that $z(x_0) \geqslant w(x_0)$. Then, since $|z(\alpha)| < w(\alpha)$ from (5.2.5) (with $x = \alpha$) and since w and z are continuous on $[\alpha, \delta]$, there exists x_1 in $\alpha < x_1 \leqslant x_0$ such that $z(x_1) = w(x_1)$ and $z(x) < w(x)$ for $\alpha \leqslant x < x_1$. Since $w(x) > -z(x)$ was established, it follows that $|z(x)| < w(x)$ for $\alpha \leqslant x < x_1$, and consequently,

$$\int_\alpha^{x_1} z^2(t)\,\mathrm{d}t < \int_\alpha^{x_1} w^2(t)\,\mathrm{d}t.$$

Using (5.2.9), (5.2.5) and (5.2.7) yields

$$\begin{aligned} z(x_1) &= z(\alpha) + \int_\alpha^{x_1} \gamma(t)\,\mathrm{d}t + \int_\alpha^{x_1} z^2(t)\,\mathrm{d}t \\ &< w(\alpha) + \int_\alpha^{x_1} c(t)\,\mathrm{d}t + \int_\alpha^{x_1} w^2(t)\,\mathrm{d}t \\ &= w(x_1), \end{aligned}$$

contradicting $z(x_1) = w(x_1)$.

Thus (5.2.10) holds on any interval $[\alpha, \delta]$ of continuity of z, $\alpha < \delta \leqslant \beta$, but this implies that z is continuous on the entire interval $[\alpha, \beta]$, since $w(x)$ is bounded and $z(x)$ has only poles at its points of discontinuity (if any). Thus (5.2.10) holds on all of the interval $[\alpha, \beta]$. This result proves (5.2.6), and since the left member is bounded on $[\alpha, \beta]$, $v(x)$ cannot have a zero on this interval.

A slight modification of the proof shows that if "$>$" is replaced by "$\geqslant$" in the hypothesis (5.2.5), then the conclusion is still valid provided "$>$" is replaced by "$\geqslant$" in (5.2.6). The proof is complete. □

THEOREM 5.2.2. *Let u and v be nontrivial solutions of* (5.2.1) *and* (5.2.2), *respectively, such that $u(x)$ does not vanish on $[\alpha, \beta]$, $v(\beta) \neq 0$, and the inequality*

$$\frac{u'(\beta)}{u(\beta)} + \int_x^\beta c(t)\,\mathrm{d}t > \left| \frac{v'(\beta)}{v(\beta)} + \int_x^\beta \gamma(t)\,\mathrm{d}t \right| \tag{5.2.11}$$

holds for all x on $[\alpha, \beta]$. Then $v(x)$ does not vanish on $[\alpha, \beta]$ and

$$\frac{u'(x)}{u(x)} > \left| \frac{v'(x)}{v(x)} \right|, \qquad \alpha \leqslant x \leqslant \beta. \tag{5.2.12}$$

The same result holds if "$>$" in (5.2.11) *and* (5.2.12) *is replaced by "$\geqslant$".*

PROOF. Let new functions u_1, v_1, c_1, γ_1 be defined on $\alpha \leqslant x \leqslant \beta$ by the equations

$$u_1(x) = u(\alpha + \beta - x), \qquad v_1(x) = v(\alpha + \beta - x),$$

$$c_1(x) = c(\alpha + \beta - x), \qquad \gamma_1(x) = \gamma(\alpha + \beta - x).$$

Then $u_1(x)$ does not vanish on $[\alpha, \beta]$, $v_1(\alpha) = v(\beta) \neq 0$ and

$$-\frac{u_1'(\alpha)}{u_1(\alpha)} + \int_\alpha^{\alpha+\beta-x} c_1(t)\,\mathrm{d}t = \frac{u'(\beta)}{u(\beta)} + \int_x^\beta c(t)\,\mathrm{d}t,$$

$$-\frac{v_1'(\alpha)}{v_1(\alpha)} + \int_\alpha^{\alpha+\beta-x} \gamma_1(t)\,\mathrm{d}t = \frac{v'(\beta)}{v(\beta)} + \int_x^\beta \gamma(t)\,\mathrm{d}t.$$

Thus the hypothesis (5.2.11) is equivalent to the hypothesis (5.2.5) of Theorem 5.2.1. Since $x \in [\alpha, \beta]$ if and only if $\alpha + \beta - x \in [\alpha, \beta]$, and the conclusion (5.2.12) follows from Theorem 5.2.1. □

In 1972, Kreith [171] has given the Levin-type comparison theorems for the differential equations of the form

$$u'' - 2b(x)u' + c(x)u = 0, \tag{5.2.13}$$

$$v'' - 2e(x)v' + \gamma(x)v = 0, \tag{5.2.14}$$

whose coefficients are assumed to be real and continuous, satisfying the initial conditions

$$u'(x_1) + \sigma u(x_1) = 0, \tag{5.2.15}$$

$$v'(x_1) + \tau v(x_1) = 0, \tag{5.2.16}$$

respectively, where σ and τ are finite constants. By means of the transformation

$$w = -\frac{u'}{u}, \qquad z = -\frac{v'}{v},$$

equations (5.2.13), (5.2.14) are transformed into Riccati equations

$$w' = w^2 + 2bw + c, \tag{5.2.17}$$

$$z' = z^2 + 2ez + \gamma, \tag{5.2.18}$$

and the initial conditions

$$-\frac{u'(x_1)}{u(x_1)} = \sigma, \qquad -\frac{v'(x_1)}{v(x_1)} = \tau, \tag{5.2.19}$$

for (5.2.13), (5.2.14), become initial values

$$w(x_1) - \sigma, \qquad z(x_1) = \tau, \tag{5.2.20}$$

for (5.2.17) and (5.2.18). The differential equations (5.2.17) and (5.2.18) subject to (5.2.20) can be written as equivalent integral equations

$$w(x) = \sigma + \int_{x_1}^{x} w^2 \,\mathrm{d}t + \int_{x_1}^{x} 2bw \,\mathrm{d}t + \int_{x_1}^{x} c \,\mathrm{d}t, \tag{5.2.21}$$

$$z(x) = \tau + \int_{x_1}^{x} z^2 \,\mathrm{d}t + \int_{x_1}^{x} 2ez \,\mathrm{d}t + \int_{x_1}^{x} \gamma \,\mathrm{d}t. \tag{5.2.22}$$

It is obvious from these equations that if $\tau \geqslant \sigma \geqslant 0$, $e(x) \geqslant b(x) \geqslant 0$ and

$$\int_{x_1}^{x} \gamma(t) \,\mathrm{d}t \geqslant \int_{x_1}^{x} c(t) \,\mathrm{d}t \geqslant 0$$

on an interval $[x_1, x_2]$, then $z(x) \geqslant w(x) \geqslant 0$ as long as $z(x)$ can be continued on $[x_1, x_2]$. Since the singularities of $w(x)$ and $z(x)$ correspond to the zeros of $u(x)$ and $v(x)$, respectively, these observations lead to the following comparison theorem for (5.2.13) and (5.2.14).

THEOREM 5.2.3. *Suppose $u(x)$ is a nontrivial solution of* (5.2.13) *satisfying $-u'(x_1)/u(x_1) = \sigma \geqslant 0$, $u(x_2) = 0$. If*

(i) $$e(x) \geqslant b(x) \geqslant 0 \quad \text{for } x_1 \leqslant x \leqslant x_2,$$

(ii) $$\int_{x_1}^{x} \gamma(t)\,\mathrm{d}t \geqslant \int_{x_1}^{x} c(t)\,\mathrm{d}t \geqslant 0 \quad \text{for } x_1 \leqslant x \leqslant x_2,$$

then every solution of (5.2.14) *satisfying $-v'(x_1)/v(x_1) \geqslant \sigma$ has a zero in $(x_1, x_2]$.*

In [171] Kreith has also given the variation of Theorem 5.2.3 which do not require the nonnegativity of $\sigma, \tau, b(x)$ and $\int_{x_1}^{x} c(t)\,\mathrm{d}t$. We note that the integral equations (5.2.21) and (5.2.22) can be written as

$$w(x) = \sigma + \int_{x_1}^{x} (w + b)^2\,\mathrm{d}t + \int_{x_1}^{x} \left(c - b^2\right)\mathrm{d}t, \tag{5.2.23}$$

$$z(x) = \tau + \int_{x_1}^{x} (z + e)^2\,\mathrm{d}t + \int_{x_1}^{x} \left(\gamma - e^2\right)\mathrm{d}t. \tag{5.2.24}$$

This formulation shows that condition (ii) of Theorem 5.2.3 can be replaced by

$$\int_{x_1}^{x} \left(\gamma - e^2\right)\mathrm{d}t \geqslant \int_{x_1}^{x} \left(c - b^2\right)\mathrm{d}t \geqslant 0.$$

In order to obtain the generalization of Levin's Theorem 5.2.1, Kreith [171] has given the following lemmas which are of independent interest.

LEMMA 5.2.1. *Let $w(x)$ and $z(x)$ be solutions of* (5.2.23) *and* (5.2.24), *respectively, for which $\sigma > -\infty$ and*

(i) $$\tau + \int_{x_1}^{x} \left(\gamma - e^2\right)\mathrm{d}t > \left|\sigma + \int_{x_1}^{x} \left(c - b^2\right)\mathrm{d}t\right| \quad \text{for } x_1 \leqslant x \leqslant x_2,$$

(ii) $$e(x) \geqslant \left|b(x)\right| \quad \text{for } x_1 \leqslant x \leqslant x_2.$$

Then $z(x) \geqslant |w(x)|$ as long as $z(x)$ can be continued on $[x_1, x_2]$.

PROOF. From (5.2.24) we have

$$z(x) \geqslant \tau + \int_{x_1}^{x} (\gamma - e^2)\,\mathrm{d}t \quad \text{for } x_1 \leqslant x \leqslant x_2.$$

Using (i) and (5.2.23) this result implies that

$$\begin{aligned} z(x) &> -\sigma - \int_{x_1}^{x} (c - b^2)\,\mathrm{d}t \\ &> -\sigma - \int_{x_1}^{x} (c - b^2)\,\mathrm{d}t - \int_{x_1}^{x} (w + b)^2\,\mathrm{d}t > -w(x) \end{aligned}$$

for $x_1 \leqslant x \leqslant x_2$. It remains to show that $z(x) > w(x)$. We assume to the contrary that there exists $x_0 \in (x_1, x_2]$ such that $z(x_0) \leqslant w(x_0)$. Then there exists an $\bar{x} \in (x_1, x_0]$ such that $z(\bar{x}) = w(\bar{x})$ and $z(x) > |w(x)|$ for $x_1 \leqslant x < \bar{x}$. Using (ii) we have that

$$z(x) + e(x) > |w(x)| + |b(x)| \geqslant |w(x) + b(x)| \quad \text{for } x_1 \leqslant x < \bar{x},$$

and consequently that $\int_{x_1}^{\bar{x}} (z + e)^2\,\mathrm{d}t > \int_{x_1}^{\bar{x}} (w + b)^2\,\mathrm{d}t$. Using (5.2.24), (i) and (5.2.23) yields

$$\begin{aligned} w(\bar{x}) &= \sigma + \int_{x_1}^{\bar{x}} (c - b^2)\,\mathrm{d}t + \int_{x_1}^{\bar{x}} (w + b)^2\,\mathrm{d}t \\ &< \tau + \int_{x_1}^{\bar{x}} (\gamma - e^2)\,\mathrm{d}t + \int_{x_1}^{\bar{x}} (z + e)^2\,\mathrm{d}t = z(\bar{x}), \end{aligned}$$

which is a contradiction and establishes the lemma. □

A continuity argument can be used to establish the following lemma.

LEMMA 5.2.2. *Let $w(x)$ and $z(x)$ be solutions of* (5.2.23) *and* (5.2.24), *respectively, for which $\sigma > -\infty$ and*

(i) $$\tau + \int_{x_1}^{x} (\gamma - e^2)\,\mathrm{d}t \geqslant \left| \sigma + \int_{x_1}^{x} (c - b^2)\,\mathrm{d}t \right| \quad \textit{for } x_1 \leqslant x \leqslant x_2,$$

(ii) $$e(x) \geqslant |b(x)| \quad \textit{for } x_1 \leqslant x \leqslant x_2.$$

Then $z(x) \geqslant w(x)$ as long as $z(x)$ can be continued on $[x_1, x_2]$.

As an immediate consequence of Lemma 5.2.2 we have the following generalization of Theorem 5.2.1.

THEOREM 5.2.4. *Suppose $u(x)$ and $v(x)$ are nontrivial solutions of* (5.2.13) *and* (5.2.14), *respectively, and that $u(x) \neq 0$ for $x_1 \leqslant x < x_2$, $u(x_2) = 0$. If*

(i)
$$-\frac{v'(x_1)}{v(x_1)} + \int_{x_1}^{x} (\gamma - e^2)\,\mathrm{d}t \geqslant \left| -\frac{u'(x_1)}{u(x_1)} + \int_{x_1}^{x} (c - b^2)\,\mathrm{d}t \right| \quad \textit{for } x_1 \leqslant x \leqslant x_2,$$

(ii)
$$e(x) \geqslant |b(x)| \quad \textit{for } x_1 \leqslant x \leqslant x_2,$$

then $v(x)$ has a zero in $(x_1, x_2]$.

In 1969, Ladas [177] has established the following generalizations of Levin's comparison theorems for the pair of nonlinear differential equations

$$x'' + p(t)g(x) = 0, \tag{5.2.25}$$

$$y'' + q(t)g(y) = 0, \tag{5.2.26}$$

for $t \in [a, b]$, under some suitable conditions on the functions involved in (5.2.25) and (5.2.26).

THEOREM 5.2.5. *Let the following conditions be satisfied:*

(i) *$p(t)$ and $q(t)$ are real-valued continuous functions for $t \in [a, b]$;*

(ii) *$g(s)$ is a real-valued continuous function for $s \in \mathbb{R}$ such that $g'(s) \geqslant 0$ for all $s \in [a, b]$, and $g(0) = 0$;*

(iii) *$x(t)$ and $y(t)$ are solutions of* (5.2.25) *and* (5.2.26), *respectively, such that $x(t) \neq 0$ for $t \in [a, b]$, $y(a) \neq 0$ and, for all $t \in [a, b]$,*

$$-\frac{x'(a)}{g(x(a))} + \int_a^t p(s)\,\mathrm{d}s > \left| -\frac{y'(a)}{g(y(a))} + \int_a^t q(s)\,\mathrm{d}s \right| \tag{5.2.27}$$

and

$$g'(s)\big|_{s=x(t)} \geqslant g'(s)\big|_{s=y(t)}. \tag{5.2.28}$$

Then, for all $t \in [a, b]$, $y(t) \neq 0$ and

$$-\frac{x'(t)}{g(x(t))} > \left| \frac{y'(t)}{g(y(t))} \right|.$$

PROOF. Since $x(t) \neq 0$, it follows that $g(x(t)) \neq 0$, $t \in [a, b]$. Setting

$$w(t) = -\frac{x'(t)}{g(x(t))}, \quad t \in [a, b], \tag{5.2.29}$$

it is easily verified that w satisfies the differential equation

$$w' = p(t) + \left[\frac{\mathrm{d}}{\mathrm{d}x} g\big(x(t)\big)\right] w^2, \quad t \in [a, b]. \tag{5.2.30}$$

The differential equation (5.2.30) will play the role of the Riccati equation to which a linear second-order differential equation is transformed by (5.2.29) in case $g(s) \equiv s$.

The rest of the proof, which we present for completeness, is an adaptation of Levin's proof with minor modifications (see [177]). We set for simplicity $\frac{\mathrm{d}}{\mathrm{d}x} g(x(t)) = g'(x(t))$.

Integrating (5.2.30) over $[a, t]$, $t \leqslant b$, and using (ii) and (iii) we obtain

$$\begin{aligned} w(t) &= w(a) + \int_a^t p(s)\,\mathrm{d}s + \int_a^t g'\big(x(s)\big) w^2(s)\,\mathrm{d}s \\ &\geqslant w(a) + \int_a^t p(s)\,\mathrm{d}s > 0. \end{aligned} \tag{5.2.31}$$

Since $y(a) \neq 0$, it follows from the continuity of $y(t)$ that there is a closed interval $[a, c]$, $a < c \leqslant b$ such that $y(t) \neq 0$, $t \in [a, c]$. Then $g(y(t)) \neq 0$, $t \in [a, b]$, and $z(t) = -y'(t)/g(y(t))$ satisfies the equation

$$z' = q(t) + \left[\frac{\mathrm{d}}{\mathrm{d}y} g\big(y(t)\big)\right] z^2, \quad t \in [a, c]. \tag{5.2.32}$$

Integrating (5.2.32) over $[a, t]$, $t \leqslant c$, and using (ii), (iii) and (5.2.31) we obtain

$$\begin{aligned} z(t) &= z(a) + \int_a^t q(s)\,\mathrm{d}s + \int_a^t g'\big(y(s)\big) z^2(s)\,\mathrm{d}s \\ &\geqslant z(a) + \int_a^t q(s)\,\mathrm{d}s > -w(a) - \int_a^t q(s)\,\mathrm{d}s \\ &\geqslant -w(t). \end{aligned} \tag{5.2.33}$$

We claim now that also

$$z(t) < w(t), \quad t \in [a, c], \tag{5.2.34}$$

so that (5.2.33) and (5.2.34) would imply

$$\big|z(t)\big| \leqslant w(t), \quad t \in [a, c]. \tag{5.2.35}$$

Indeed if (5.2.35) were false, there should exist a point $t_1 \in [a, c]$ such that $z(t_1) = w(t_1)$ and $w(t) \geqslant |z(t)|$ for $t \in [a, t_1]$. (We used the fact that (5.2.27) for $t = a$ gives $w(a) > |z(a)|$.) Then using (5.2.33), (iii) and (5.2.31) we obtain a contradiction

$$\begin{aligned} z(t_1) &= z(a) + \int_a^{t_1} q(s)\,ds + \int_a^{t_1} g'\big(y(s)\big)z^2(s)\,ds \\ &< w(a) + \int_a^{t_1} p(s)\,ds + \int_a^{t_1} g'\big(x(s)\big)w^2(s)\,ds \\ &= w(t_1). \end{aligned}$$

Therefore, inequality (5.2.35) is established for every interval $[a, c]$ of continuity of $z(t)$. But $w(t)$ is bounded on $[a, b]$ and $z(t)$ can have only pole discontinuities on $[a, b]$ so (5.2.35) holds throughout $[a, b]$, that is,

$$-\frac{x'(t)}{g(x(t))} > \left|\frac{y'(t)}{g(y(t))}\right|, \qquad t \in [a, b],$$

and $g(y(t)) \neq 0$, that is, $y(t) \neq 0$. (Here we also used the uniqueness of solutions of (5.2.26).) □

THEOREM 5.2.6. *Let in addition to the hypotheses* (i) *and* (ii) *of Theorem* 5.2.5, *the following condition be satisfied.*

(iii′) $x(t)$ *and* $y(t)$ *are solutions of* (5.2.25) *and* (5.2.26), *respectively, such that* $x(t) \neq 0$, $t \in [a, b]$, $y(b) \neq 0$ *and, for all* $t \in [a, b]$,

$$\frac{x'(b)}{g(x(b))} + \int_t^b p(s)\,ds > \left|\frac{y'(b)}{g(y(b))} + \int_t^b q(s)\,ds\right| \tag{5.2.36}$$

and

$$g'(s)\big|_{s=x(t)} \geqslant g'(s)\big|_{s=y(t)}. \tag{5.2.37}$$

Then, for all $t \in [a, b]$, $y(t) \neq 0$ *and*

$$\frac{x'(t)}{g(x(t))} > \left|\frac{y'(t)}{g(y(t))}\right|.$$

PROOF. It follows from Theorem 5.2.5 by setting $-t + a + b$ in place of t. □

In 1970, Bobisud [34] has established Levin-type results involving comparison of the solutions of nonlinear second-order differential equations and inequalities of the forms

$$y'' + a(t)f(y) \gtrless 0 \tag{5.2.38}$$

and

$$y'' + p(t, y)y' + g(t, y)y \gtrless 0 \tag{5.2.39}$$

under some suitable conditions on the functions involved in (5.2.38) and (5.2.39). Here we do not discuss the details.

5.3 Levin-Type Inequalities

In this section we are concerned with Levin-type inequalities established by Lalli and Jahagirdar [180,181] and Pachpatte [327] for certain second-order nonlinear differential equations. In what follows, it will be assumed that the solutions of the equations considered here exist and are unique on the required interval.

In [180,181] Lalli and Jahagirdar have established Levin-type comparison theorems for the pair of nonlinear differential equations

$$u'' + p(t)f(t, u) = 0, \tag{5.3.1}$$

$$v'' + q(t)f(t, v) = 0, \tag{5.3.2}$$

under some suitable conditions on the functions involved in (5.3.1), (5.3.2).

The results established in [180] are given in the following theorems.

THEOREM 5.3.1. *Let the following conditions be satisfied.*

(i) *$p(t)$ and $q(t)$ are real-valued nonnegative and continuous functions for $t \in [\alpha, \beta]$.*

(ii) *$f(t, x)$ is a real-valued nonnegative continuous function on $[\alpha, \beta] \times \mathbb{R}$ such that $\frac{\partial f}{\partial x} \geqslant 0$, $f(t, x) \neq 0$ for $x \neq 0$, $f(t, 0) = 0$ and f is monotone nondecreasing function of t for each fixed x.*

(iii) *$u(t)$ and $v(t)$ are solutions of (5.3.1) and (5.3.2), respectively, such that $u(t) \neq 0$ for $t \in [\alpha, \beta]$, $v(\alpha) \neq 0$ and, for all $t \in [\alpha, \beta]$,*

$$-\frac{u'(\alpha)}{f(\alpha, u(\alpha))} + \int_\alpha^t p(s)\,ds > \left| -\frac{v'(\alpha)}{f(\alpha, v(\alpha))} + \int_\alpha^t q(s)\,ds \right|. \tag{5.3.3}$$

Then, for all $t \in [\alpha, \beta]$, $v(t) \neq 0$ and

$$-\frac{u'(t)}{f(\alpha, u(t))} > \left| \frac{v'(t)}{f(\alpha, v(t))} \right|. \tag{5.3.4}$$

PROOF. Since $u(t) \neq 0$, it follows that $f(t, u(t)) \neq 0$ for $t \in [\alpha, \beta]$. Let

$$w(t) = -\frac{u'(t)}{f(\alpha, u(t))}, \quad t \in [\alpha, \beta]. \tag{5.3.5}$$

It follows that

$$w'(t) = p(t)\left[\frac{f(t, u(t))}{f(\alpha, u(t))}\right] + \left[\frac{\mathrm{d}}{\mathrm{d}u} f\big(\alpha, u(t)\big)\right] w^2. \tag{5.3.6}$$

Integrating (5.3.6) over $[\alpha, t]$, $t \leqslant \beta$, it follows that

$$w(t) \geqslant w(\alpha) + \int_\alpha^t p(s)\,\mathrm{d}s > 0. \tag{5.3.7}$$

Since $v(\alpha) \neq 0$, it follows from the continuity of $v(t)$ that $v(t) \neq 0$ for t in some interval $[\alpha, \gamma]$, $\alpha < \gamma \leqslant \beta$. We put

$$z(t) = -\frac{v'(t)}{f(\alpha, v(t))}, \quad t \in [\alpha, \gamma].$$

Using (5.3.2) we get the inequality

$$z' \geqslant q(t) + \frac{\mathrm{d}}{\mathrm{d}v} f\big(\alpha, v(t)\big) z^2, \quad t \in [\alpha, \gamma]. \tag{5.3.8}$$

Integrating (5.3.8) from α to t, $\alpha \leqslant t \leqslant \gamma$, we obtain

$$z(t) \geqslant z(\alpha) + \int_\alpha^t q(s)\,\mathrm{d}s, \quad t \in [\alpha, \gamma]. \tag{5.3.9}$$

From (5.3.7) and (5.3.9), in view of (5.3.3), it follows that

$$z(t) > -w(t), \quad t \in [\alpha, \gamma]. \tag{5.3.10}$$

We will show now that

$$z(t) < w(t), \quad t \in [\alpha, \gamma]. \tag{5.3.11}$$

Suppose (5.3.11) fails to hold for all $t \in [\alpha, \gamma]$ then there is a t_1, $\alpha < t_1 \leqslant \gamma$, such that

$$z(t_1) = w(t_1)$$

and

$$w(t) > \big|z(t)\big| \quad \text{for } t \in [\alpha, t_1).$$

For $t = t_1$,

$$z(t_1) = z(\alpha) + \int_\alpha^{t_1} q(s)\frac{f(s,u(s))}{f(\alpha,u(s))}\,\mathrm{d}s + \int_\alpha^{t_1} \frac{\mathrm{d}}{\mathrm{d}v} f\big(\alpha, v(s)\big)z^2(s)\,\mathrm{d}s$$

or

$$\begin{aligned} -z(t_1) &\leqslant -z(\alpha) - \int_\alpha^{t_1} q(s)\,\mathrm{d}s - \int_\alpha^{t_1} \frac{\mathrm{d}}{\mathrm{d}v} f\big(\alpha, v(s)\big)z^2(s)\,\mathrm{d}s \\ &\leqslant \frac{v'(\alpha)}{f(\alpha, v(\alpha))} - \int_\alpha^{t_1} q(s)\,\mathrm{d}s \\ &< -\frac{u'(\alpha)}{f(\alpha, u(\alpha))} + \int_\alpha^{t_1} p(s)\,\mathrm{d}s \\ &= w(\alpha) + \int_\alpha^{t_1} p(s)\,\mathrm{d}s \\ &\leqslant w(t_1), \end{aligned}$$

which is a contradiction. Combining (5.3.10) and (5.3.11) we have

$$w(t) > \big|z(t)\big|, \quad t \in [\alpha, \gamma]. \tag{5.3.12}$$

Therefore, (5.3.12) is true for every interval $[\alpha, \gamma]$ of continuity of $z(t)$. Since $w(t)$ is bounded on $[\alpha, \beta]$ and $z(t)$ can have only poles at points of discontinuities on $[\alpha, \beta]$, it follows that (5.3.12) holds throughout $[\alpha, \beta]$. Thus, $f(\alpha, u(t)) \neq 0$, $t \in [\alpha, \beta]$, and consequently, $u(t) \neq 0$ on $[\alpha, \beta]$. □

REMARK 5.3.1. We note that in the special case when $f(t,x) = x$ or $f(t,x) = g(x)$, the condition that p and q be nonnegative is no longer needed in the proof of Theorem 5.3.1.

A slight variant of Theorem 5.3.1 established in [180] is given in the following theorem.

THEOREM 5.3.2. *In the hypotheses of Theorem* 5.3.1, *let* (iii) *be replaced by*

(iii′) $u(t)$ *and* $v(t)$ *are solutions of* (5.3.1) *and* (5.3.2), *respectively, such that* $u(t) \neq 0$ *for* $t \in [\alpha, \beta]$, $v(\alpha) \neq 0$ *and, for any* $\sigma \in [\alpha, \beta]$ *and that* $v(\sigma) \neq 0$, *we have*

$$-\frac{u'(\sigma)}{f(\sigma, u(\sigma))} + \int_\sigma^t p(s)\,\mathrm{d}s > \left| -\frac{v'(\sigma)}{f(\sigma, v(\sigma))} + \int_\sigma^t q(s)\,\mathrm{d}s \right|, \quad t \in [\sigma, \beta]. \tag{5.3.13}$$

Then, for all $t \in [\alpha, \beta]$, $v(t) \neq 0$ and

$$-\frac{u'(t)}{f(t, u(t))} > \left|\frac{v'(t)}{f(t, v(t))}\right|. \tag{5.3.14}$$

PROOF. If (5.3.14) is false, then there is a $\bar{t}$ such that

$$-\frac{u'(\bar{t})}{f(\bar{t}, u(\bar{t}))} \leqslant \left|\frac{v'(\bar{t})}{f(\bar{t}, v(\bar{t}))}\right|.$$

Since $v(\bar{t}) \neq 0$, from the argument used for proving (5.3.4), it follows that

$$-\frac{u'(t)}{f(\bar{t}, u(t))} > \left|\frac{v'(t)}{f(\bar{t}, v(t))}\right| \quad \text{for } t \in [\bar{t}, \beta].$$

In particular, for $t = \bar{t}$, we get a contradiction. □

In [181] Lalli and Jahagirdar established comparison theorems of Levin type given in the following theorems.

THEOREM 5.3.3. *Let the following conditions be satisfied.*

(i) *$p(t)$, $q(t)$ are real-valued, continuous and nonnegative functions for $t \in [\alpha, \beta]$;*

(ii) *$f(t, x)$ is a real-valued continuous function on $[\alpha, \beta] \times \mathbb{R}$ such that*

(a) *$f(t, x) \neq 0$ for $x \neq 0$, $f(t, 0) = 0$,*
(b) *$1 \leqslant K_1 \leqslant f(t, x)/f(\alpha, x) \leqslant K_2$, $t \in [\alpha, \beta]$,*
(c) *$\frac{\partial}{\partial x} f \geqslant 0$;*

(iii) *$u(t)$, $v(t)$ are solutions of (5.3.1) and (5.3.2), respectively, such that $u(t) \neq 0$ for $t \in [\alpha, \beta]$, $v(\alpha) \neq 0$ and, for all $t \in [\alpha, \beta]$,*

$$\begin{aligned} -\frac{u'(\alpha)}{f(\alpha, u(\alpha))} + K_1 \int_\alpha^t p(s)\,\mathrm{d}s &> -\left[-\frac{v'(\alpha)}{f(\alpha, v(\alpha))} + \int_\alpha^t q(s)\,\mathrm{d}s\right], \\ -\frac{u'(\alpha)}{f(\alpha, u(\alpha))} + \int_\alpha^t p(s)\,\mathrm{d}s &> -\frac{v'(\alpha)}{f(\alpha, v(\alpha))} + K_2 \int_\alpha^t q(s)\,\mathrm{d}s; \end{aligned} \tag{5.3.15}$$

(iv)
$$\left.\frac{\partial f}{\partial x}\right|_{x(t)=u(t)} \geqslant \left.\frac{\partial f}{\partial x}\right|_{x(t)=v(t)}.$$

Then, for all $t \in [\alpha, \beta]$, $v(t) \neq 0$ *and*

$$-\frac{u'(t)}{f(\alpha, u(t))} > \left|\frac{v'(t)}{f(\alpha, v(t))}\right|. \tag{5.3.16}$$

PROOF. Since $u(t) \neq 0$, it follows that $f(t, u(t)) \neq 0$ for $t \in [\alpha, \beta]$. Let the function $w(t)$ be defined by

$$w(t) = -\frac{u'(t)}{f(\alpha, u(t))}, \quad t \in [\alpha, \beta]. \tag{5.3.17}$$

We differentiate (5.3.17) with respect to t and obtain

$$w'(t) = p(t)\frac{f(t, u(t))}{f(\alpha, u(t))} + \frac{\mathrm{d}}{\mathrm{d}u}\big(f(\alpha, u)\big)w^2, \quad t \in [\alpha, \beta],$$

or

$$w'(t) \geqslant K_1 p(t) + \frac{\mathrm{d}}{\mathrm{d}u}\big(f(\alpha, u)\big)w^2. \tag{5.3.18}$$

In view of (c), the above inequality reduces to

$$w'(t) \geqslant K_1 p(t)$$

or

$$w(t) \geqslant w(\alpha) + K_1 \int_\alpha^t p(s)\,\mathrm{d}s, \quad t \in [\alpha, \beta]. \tag{5.3.19}$$

Since $v(\alpha) \neq 0$, it follows from the continuity of $v(t)$ that $v(t) \neq 0$ for t in some closed interval $[\alpha, \gamma]$, $\alpha < \gamma \leqslant \beta$. We put

$$z(t) = -\frac{v'(t)}{f(\alpha, v(t))}, \quad t \in [\alpha, \gamma].$$

In view of (5.3.2) and hypothesis (b), we get the inequality

$$z'(t) \geqslant K_1 q(t) + \frac{\mathrm{d}}{\mathrm{d}v}\big(f(\alpha, v)\big)z^2, \quad t \in [\alpha, \gamma].$$

We integrate the above inequality from α to t, $t \leqslant \gamma$, and use monotonicity of f to obtain

$$z(t) \geqslant z(\alpha) + K_1 \int_\alpha^t q(s)\,\mathrm{d}s, \quad t \in [\alpha, \gamma]. \tag{5.3.20}$$

From (5.3.20), (5.3.15) and (5.3.19), for $t \in [\alpha, \gamma]$, we have

$$\begin{aligned} z(t) &\geqslant z(\alpha) + K_1 \int_\alpha^t q(s)\,\mathrm{d}s \\ &\geqslant z(\alpha) + \int_\alpha^t q(s)\,\mathrm{d}s \\ &> -w(\alpha) - K_1 \int_\alpha^t q(s)\,\mathrm{d}s \\ &\geqslant -w(t), \end{aligned}$$

that is,

$$z(t) > -w(t) \quad \text{for } t \in [\alpha, \gamma]. \tag{5.3.21}$$

Now we shall show that

$$z(t) < w(t) \quad \text{for } t \in [\alpha, \gamma]. \tag{5.3.22}$$

Suppose (5.3.22) fails to hold for all $t \in [\alpha, \gamma]$ then there is a t_1, $\alpha < t_1 < \gamma$, such that

$$z(t_1) = w(t_1) \quad \text{and} \quad w(t) > z(t) \quad \text{for } t \in [\alpha, t_1).$$

For $t = t_1$, we have

$$\begin{aligned} z(t_1) &= z(\alpha) + \int_\alpha^{t_1} q(s) \frac{f(s, v(s))}{f(\alpha, v(s))}\,\mathrm{d}s + \int_\alpha^{t_1} \frac{\partial}{\partial v} f(\alpha, v) z^2\,\mathrm{d}s \\ &\leqslant z(\alpha) + K_2 \int_\alpha^{t_1} q(s)\,\mathrm{d}s + \int_\alpha^{t_1} \frac{\partial}{\partial u} f(\alpha, u) w^2\,\mathrm{d}s \\ &< w(\alpha) + \int_\alpha^{t_1} p(s)\,\mathrm{d}s + \int_\alpha^{t_1} \frac{\partial}{\partial u} f(\alpha, u) w^2\,\mathrm{d}s \\ &\leqslant w(\alpha) + \int_\alpha^{t_1} p(s) \frac{f(s, u(s))}{f(\alpha, u(s))}\,\mathrm{d}s + \int_\alpha^{t_1} \frac{\partial}{\partial u} f(\alpha, u) w^2\,\mathrm{d}s \\ &= w(t_1), \end{aligned}$$

which is a contradiction. Hence from (5.3.21) and (5.3.22), we have

$$w(t) > |z(t)|, \quad t \in [\alpha, \gamma]. \tag{5.3.23}$$

Therefore, (5.3.23) is true for every interval $[\alpha, \gamma]$ of continuity of $z(t)$. Since $w(t)$ is bounded on $[\alpha, \beta]$ and $z(t)$ can have only poles at points of discontinuities on $[\alpha, \beta]$, it follows that (5.3.23) holds throughout $[\alpha, \beta]$. Thus $f(\alpha, v(t)) \neq 0$ for $t \in [\alpha, \beta]$ and consequently, $v(t) \neq 0$ on $[\alpha, \beta]$. □

REMARK 5.3.2. In the special case when $f(t, x) = f(x)$ or x, then we can discard the condition that $p(t)$ and $q(t)$ be nonnegative, and inequalities (5.3.15) reduce to the inequality (5.3.3) with $K_1 = K_2 = 1$. In this case Levin's result and the result of Ladas are special cases of Theorem 5.3.3.

THEOREM 5.3.4. *In addition to the conditions* (i) *and* (iv) *in Theorem* 5.3.3 *assume that*

(ii′) $f(t, x)$ *is a real-valued continuous function on* $[\alpha, \beta] \times \mathbb{R}$ *such that*

- (a′) $f(t, x) \neq 0$ *for* $x \neq 0$, $f(t, 0) = 0$,
- (b′) $1 \leqslant K_1 \leqslant f(t, x)/f(\beta, x) \leqslant K_2$, $t \in [\alpha, \beta]$,
- (c′) $\frac{\partial f}{\partial x} \geqslant 0$;

(iii′) $u(t)$ *and* $v(t)$ *are the solutions of* (5.3.1) *and* (5.3.2), *respectively, such that* $u(t) \neq 0$ *for* $t \in [\alpha, \beta]$, $v(\beta) \neq 0$ *and, for* $t \in [\alpha, \beta]$,

$$\frac{u'(\beta)}{f(\beta, u(\beta))} + K_1 \int_t^\beta p(s)\,ds > -\left[\frac{v'(\beta)}{f(\beta, v(\beta))} + \int_t^\beta q(s)\,ds\right],$$

$$\frac{u'(\beta)}{f(\beta, u(\beta))} + \int_t^\beta p(s)\,ds > \frac{v'(\beta)}{f(\beta, v(\beta))} + K_2 \int_t^\beta q(s)\,ds.$$

Then, for all $t \in [\alpha, \beta]$, $v(t) \neq 0$ *and*

$$\frac{u'(t)}{f(\beta, u(t))} > \left|\frac{v'(t)}{f(\beta, v(t))}\right|, \quad t \in [\alpha, \beta]. \tag{5.3.24}$$

PROOF. The result follows from Theorem 5.3.3 by setting $-t + \alpha + \beta$ in place of t. □

In [327] Pachpatte has established the Levin-type comparison theorems for nonlinear differential equations of the forms

$$\big(a(t)h\big(u'(t)\big)\big)' + p(t)f\big(u(t)\big) = 0, \tag{5.3.25}$$

$$\big(b(t)g\big(v'(t)\big)\big)' + q(t)f\big(v(t)\big) = 0, \tag{5.3.26}$$

where

(i) $p, q \in C(I, \mathbb{R}_+)$, $I = [\alpha, \beta]$, $\mathbb{R}_+ = [0, \infty)$;

(ii) $a, b \in C^1(I, (0, \infty))$;

(iii) $f \in C^1(\mathbb{R}, \mathbb{R})$, $f(x) \neq 0$, $f(0) = 0$ and $f'(x) \geqslant 0$ for all $x \in \mathbb{R} = (-\infty, \infty)$;

(iv) $h, g \in C^1(\mathbb{R}, \mathbb{R})$, $h(-x) = -h(x)$, $g(-x) = -g(x)$; $\operatorname{sgn} h(x) = \operatorname{sgn} x$, $\operatorname{sgn} g(x) = \operatorname{sgn} x$; $0 < x/h(x) \leqslant m_1$, $0 < x/g(x) \leqslant m_2$, for some constants m_1, m_2 in $\mathbb{R}$; $\lim_{x\to\infty} x/h(x)$, $\lim_{x\to\infty} x/g(x)$ exist finitely.

The main results in [327] are given in the following theorems.

THEOREM 5.3.5. *Assume that the hypotheses* (i)–(iv) *hold. If $u(t)$ and $v(t)$ are solutions of* (5.3.25) *and* (5.3.26), *respectively, such that $u(t) \neq 0$ for $t \in I$, $v(\alpha) \neq 0$ and, for all $t \in I$,*

$$-\frac{a(\alpha)h(u'(\alpha))}{f(u(\alpha))} + \int_\alpha^t p(s)\,\mathrm{d}s > \left| -\frac{b(\alpha)g(v'(\alpha))}{f(v(\alpha))} + \int_\alpha^t q(s)\,\mathrm{d}s \right|, \tag{5.3.27}$$

then, for all $t \in I$, $v(t) \neq 0$ and

$$-\frac{a(t)h(u'(t))}{f(u(t))} > \left| \frac{b(t)g(v'(t))}{f(v(t))} \right|. \tag{5.3.28}$$

PROOF. Since $u(t) \neq 0$, it follows that $f(u(t)) \neq 0$, $t \in I$. Let

$$w(t) = -\frac{a(t)h(u'(t))}{f(u(t))}, \quad t \in I. \tag{5.3.29}$$

Differentiating (5.3.29) with respect to t and using (5.3.25), it follows that

$$w'(t) = p(t) + \frac{f'(u(t))}{a(t)} \left(\frac{u'(t)}{h(u'(t))} \right) w^2(t), \quad t \in I. \tag{5.3.30}$$

Integrating (5.3.30) over $[\alpha, t]$, $t \leqslant \beta$, we obtain

$$w(t) = w(\alpha) + \int_\alpha^t p(s)\,\mathrm{d}s + \int_\alpha^t \frac{f'(u(s))}{a(s)} \left(\frac{u'(s)}{h(u'(s))} \right) w^2(s)\,\mathrm{d}s. \tag{5.3.31}$$

From (ii)–(iv), (5.3.27) and (5.3.31), we observe that

$$w(t) \geqslant w(\alpha) + \int_\alpha^t p(s)\,\mathrm{d}s > 0. \tag{5.3.32}$$

Since $v(\alpha) \neq 0$, it follows from the continuity of $v(t)$ that $v(t) \neq 0$ for t in some closed interval $[\alpha, c]$, $\alpha < c \leqslant \beta$. Let

$$z(t) = -\frac{b(t)g(v'(t))}{f(v(t))}, \quad t \in [\alpha, c]. \tag{5.3.33}$$

Differentiating (5.3.33) with respect to t and using (5.3.26), it follows that

$$z'(t) = q(t) + \frac{f'(v(t))}{b(t)}\left(\frac{v'(t)}{g(v'(t))}\right)z^2(t), \quad t \in [\alpha, c]. \tag{5.3.34}$$

Integrating (5.3.34) over $[\alpha, t]$, $t \leqslant c$, we obtain

$$z(t) = z(\alpha) + \int_\alpha^t q(s)\,\mathrm{d}s + \int_\alpha^t \frac{f'(v(s))}{b(s)}\left(\frac{v'(s)}{g(v'(s))}\right)z^2(s)\,\mathrm{d}s. \tag{5.3.35}$$

Using (ii), (iii), (iv), (5.3.27) and (5.3.32) in (5.3.35), we observe that

$$\begin{aligned} z(t) &\geqslant z(\alpha) + \int_\alpha^t q(s)\,\mathrm{d}s \\ &> -w(\alpha) - \int_\alpha^t p(s)\,\mathrm{d}s \\ &\geqslant -w(t) \end{aligned} \tag{5.3.36}$$

for $t \in [\alpha, c]$.

Now we shall show that

$$z(t) < w(t), \quad t \in [\alpha, c]. \tag{5.3.37}$$

Suppose (5.3.37) fails to hold for all $t \in [\alpha, c]$ then there is a t_1, $\alpha < t_1 \leqslant c$, such that

$$z(t_1) = w(t_1) \tag{5.3.38}$$

and $w(t) > |z(t)|$ for $t \in [\alpha, t_1)$. By taking $t = t_1$ in (5.3.35) and using (ii)–(iv), (5.3.27) and (5.3.32), we observe that

$$\begin{aligned} -z(t_1) &= -z(\alpha) - \int_\alpha^{t_1} q(s)\,\mathrm{d}s - \int_\alpha^{t_1} \frac{f'(v(s))}{b(s)}\left(\frac{v'(s)}{g(v'(s))}\right)z^2(s)\,\mathrm{d}s \\ &\leqslant -z(\alpha) - \int_\alpha^{t_1} q(s)\,\mathrm{d}s \\ &< w(\alpha) - \int_\alpha^{t_1} p(s)\,\mathrm{d}s \\ &\leqslant w(t_1), \end{aligned}$$

which is a contradiction to (5.3.38). Thus, from (5.3.36) and (5.3.37), we have

$$w(t) > |z(t)|, \quad t \in [\alpha, c]. \tag{5.3.39}$$

Therefore, (5.3.39) is true for every interval $[\alpha, c]$ of continuity of $z(t)$. Since $w(t)$ is bounded on I and $z(t)$ can have only poles discontinuities on I, it follows that (5.3.39) holds throughout I. Thus $f(v(t)) \neq 0$, $t \in I$, and consequently $v(t) \neq 0$ on I. The proof is complete. □

THEOREM 5.3.6. *Assume that the hypotheses* (i)–(iv) *hold. If $u(t)$ and $v(t)$ are solutions of* (5.3.25) *and* (5.3.26), *respectively, such that $u(t) \neq 0$ for $t \in I$, $v(\beta) \neq 0$ and, for all $t \in I$,*

$$\frac{a(\beta)h(u'(\beta))}{f(u(\beta))} + \int_t^\beta p(s)\,\mathrm{d}s > \left|\frac{b(\beta)g(v'(\beta))}{f(v(\beta))} + \int_t^\beta q(s)\,\mathrm{d}s\right|, \tag{5.3.40}$$

then, for all $t \in I$, $v(t) \neq 0$ and

$$\frac{a(t)h(u'(t))}{f(u(t))} > \left|\frac{b(t)g(v'(t))}{f(v(t))}\right|. \tag{5.3.41}$$

The proof follows by the similar arguments as in the proof of Theorem 5.3.5.

In [327] the following Levin-type comparison theorems are also established for the pair of nonlinear differential equations of the forms

$$\big(a(t)h\big(u(t)\big)u'(t)\big)' + p(t)f\big(u(t)\big) = 0, \tag{5.3.42}$$

$$\big(b(t)g\big(v(t)\big)v'(t)\big)' + q(t)f\big(v(t)\big) = 0, \tag{5.3.43}$$

where a, b, p, q, f are as in equations (5.3.25) and (5.3.26) satisfying the hypotheses (i)–(iii) and

(v) $h, g \in C(\mathbb{R}, \mathbb{R}) \cap C^1(\mathbb{R}, \mathbb{R})$ and $h(x) > 0$, $g(x) > 0$ for $x \neq 0$.

THEOREM 5.3.7. *Assume that the hypotheses* (i)–(iii) *and* (v) *hold. If $u(t)$ and $v(t)$ are solutions of* (5.3.42) *and* (5.3.43), *respectively, such that $u(t) \neq 0$ for $t \in I$, $v(\alpha) \neq 0$ and, for all $t \in I$,*

$$-\frac{a(\alpha)h(u(\alpha))u'(\alpha)}{f(u(\alpha))} + \int_\alpha^t p(s)\,\mathrm{d}s > \left|-\frac{b(\alpha)g(v(\alpha))v'(\alpha)}{f(v(\alpha))} + \int_\alpha^t q(s)\,\mathrm{d}s\right|, \tag{5.3.44}$$

then, for all $t \in I$, $v(t) \neq 0$ and

$$-\frac{a(t)h(u(t))u'(t)}{f(u(t))} > \left|\frac{b(t)g(v(t))v'(t)}{f(v(t))}\right|. \tag{5.3.45}$$

THEOREM 5.3.8. *Assume that the hypotheses* (i)–(iii) *and* (v) *hold. If* $u(t)$ *and* $v(t)$ *are solutions of* (5.3.42) *and* (5.3.43), *respectively, such that* $u(t) \neq 0$ *for* $t \in I$, $v(\beta) \neq 0$ *and, for all* $t \in I$,

$$\frac{a(\beta)h(u(\beta))u'(\beta)}{f(u(\beta))} + \int_t^\beta p(s)\,\mathrm{d}s > \left|\frac{b(\beta)g(v(\beta))v'(\beta)}{f(v(\beta))} + \int_t^\beta q(s)\,\mathrm{d}s\right|, \tag{5.3.46}$$

then, for all $t \in I$, $v(t) \neq 0$ *and*

$$\frac{a(t)h(u(t))u'(t)}{f(u(t))} > \left|\frac{b(t)g(v(t))v'(t)}{f(v(t))}\right|. \tag{5.3.47}$$

The proofs of Theorems 5.3.7 and 5.3.8 follow by the similar arguments as in the proof of Theorem 5.3.5 given above with suitable changes. Here we omit the details.

For further extensions of Levin-type comparison theorems to the following nonlinear differential inequality

$$\big(A(t)\psi\big(u(t)\big)u'(t)\big)' + B(t)f\big(u(t)\big) \leqslant 0 \tag{5.3.48}$$

and to the nonlinear differential equation

$$\big(a(t)\psi\big(v(t)\big)v'(t)\big)' + b(t)f\big(v(t)\big) = 0 \tag{5.3.49}$$

under some suitable conditions on the functions involved in (5.3.48) and (5.3.49); see [433].

5.4 Inequalities Related to Lyapunov's Inequality

In 1893, Lyapunov [201] proved the following remarkable inequality.

If y is a nontrivial solution of

$$y'' + q(t)y = 0 \tag{5.4.1}$$

on an interval containing the points a and b, $a < b$, such that $y(a) = y(b) = 0$, then

$$4 < (b-a)\int_a^b \big|q(s)\big|\,\mathrm{d}s. \tag{5.4.2}$$

Since from the appearance of the above inequality, various proofs, generalizations, extensions and improvements have appeared in the literature. In this section we are concerned with inequalities related to Lyapunov's inequality established by Hartman [145], Patula [361], Kwong [176] and Harris [143] for second-order differential equations.

In [145, p. 345] Hartman has given the following theorem.

THEOREM 5.4.1. *Let $q(t)$ be real-valued and continuous for $a \leqslant t \leqslant b$. Let $m(t) \geqslant 0$ be a continuous function for $a \leqslant t \leqslant b$ and*

$$\gamma_m = \inf \frac{m(t)}{(t-a)(b-t)} \quad \text{for } a < t < b. \tag{5.4.3}$$

If a real-valued nontrivial solution $y(t)$ of (5.4.1) *has two zeros, then*

$$\int_a^b m(t) q^+(t)\,\mathrm{d}t > \gamma_m (b-a), \tag{5.4.4}$$

where $q^+(t) = \max\{q(t), 0\}$, in particular,

$$\int_a^b (t-a)(b-t) q^+(t)\,\mathrm{d}t > b-a. \tag{5.4.5}$$

PROOF. Assume that (5.4.1) has a nontrivial solution with two zeros on $[a, b]$. Since $q^+(t) \geqslant q(t)$, the equation

$$y'' + q^+(t) y = 0 \tag{5.4.6}$$

is a Sturm majorant for (5.4.1) and hence has a nontrivial solution $y(t)$ with two zeros $t = \alpha, \beta$ on $[a, b]$ (see [145, p. 334]). Since $y'' = -q^+ y$, it follows that (see [145, p. 328])

$$(\beta - \alpha) y(t) = (\beta - t) \int_\alpha^t (s-\alpha) q^+(s) y(s)\,\mathrm{d}s + (t-\alpha) \int_t^\beta (\beta - s) q^+(s) y(s)\,\mathrm{d}s.$$

Suppose that α, β are successive zeros of y and that $y(t) > 0$ for $\alpha < t < \beta$. Choose $t = t_0$ so that $y(t_0) = \max y(t)$ on (α, β). The right-hand side is increased if $y(s)$ is replaced by $y(t_0)$. Thus dividing by $y(t_0) > 0$ gives

$$(\beta - \alpha) < (\beta - t) \int_\alpha^t (s-\alpha) q^+(s)\,\mathrm{d}s + (t-\alpha) \int_t^\beta (\beta - s) q^+(s)\,\mathrm{d}s,$$

where $t = t_0$. Since $\beta - t \leqslant \beta - s$ for $t \geqslant s$ and $t - \alpha \leqslant s - \alpha$ for $s \geqslant t$,

$$\beta - \alpha < \int_\alpha^\beta (\beta - s)(s-\alpha) q^+(s)\,\mathrm{d}s. \tag{5.4.7}$$

Finally, note that $(t-a)(b-t)/(b-a) \geqslant (t-\alpha)(\beta-t)/(\beta-\alpha)$ for $a \leqslant \alpha \leqslant t \leqslant \beta \leqslant b$; in fact, differentiation with respect to β and α shows that $(t-\alpha)(\beta-t)/(\beta-\alpha)$ increases with β if $t \geqslant \alpha$ and decreases with α if $t \leqslant \beta$. Hence (5.4.5) follows from the last inequality (5.4.7). The relation (5.4.4) is a consequence of (5.4.3) and (5.4.5). The proof is complete. □

Since $(t-a)(b-t) \leqslant (b-a)^2/4$, the choice $m(t) = 1$ in Theorem 5.4.1 gives the following corollary.

COROLLARY 5.4.1 (Lyapunov [201]). *Let $q(t)$ be real-valued and continuous on $a \leqslant t \leqslant b$. A necessary condition for* (5.4.1) *to have a nontrivial solution $y(t)$ possessing two zeros is that*

$$\int_a^b q^+(t)\,\mathrm{d}t > \frac{4}{b-a}.$$

One of the nice purposes of (5.4.2) is that one may obtain a lower bound for the distance between two consecutive zeros of a solution of (5.4.1) by means of the integral measurement of q.

In [361] Patula (see also [62]) has given the following useful variant of Lyapunov's inequality.

THEOREM 5.4.2. *Let $y(t)$ be a solution of* (5.4.1), *where $y(a) = y(b) = 0$, and $y(t) \neq 0$, $t \in (a,b)$. Let c be a point in (a,b) where $|y(t)|$ is maximized. Then*

(i) $$\int_a^c q^+(t)\,\mathrm{d}t > \frac{1}{c-a},$$

(ii) $$\int_c^b q^+(t)\,\mathrm{d}t > \frac{1}{b-c},$$

(iii) $$\int_a^b q^+(t)\,\mathrm{d}t > \frac{b-a}{(b-c)(c-a)}.$$

PROOF. Writing $q(t) = q^+(t) - q^-(t)$, $q^-(t) = -\min\{q(t), 0\}$ and integrating (5.4.1) yields

$$y'(t) - y'(c) = \int_c^t q^-(s)y(s)\,\mathrm{d}s - \int_c^t q^+(s)y(s)\,\mathrm{d}s.$$

Note that $y'(c) = 0$. Another integration gives

$$y(t) - y(c) = \int_c^t (t-s)q^-(s)y(s)\,\mathrm{d}s - \int_c^t (t-s)q^+(s)y(s)\,\mathrm{d}s. \tag{5.4.8}$$

Let $t = b$, so that $y(b) = 0$. Equation (5.4.8) implies that

$$y(b) - y(c) = \int_c^b (b-s)q^-(s)y(s)\,\mathrm{d}s - \int_c^b (b-s)q^+(s)y(s)\,\mathrm{d}s$$

or

$$y(c) + \int_c^b (b-s)q^-(s)y(s)\,\mathrm{d}s = \int_c^b (b-s)q^+(s)y(s)\,\mathrm{d}s.$$

We may assume without loss of generality that $y(t) \geqslant 0$, $t \in [a,b]$. Thus we have

$$y(c) \leqslant \int_c^b (b-s)q^+(s)y(s)\,\mathrm{d}s < (b-c)\int_c^b q^+(s)y(s)\,\mathrm{d}s.$$

Since $y(s) \leqslant y(c)$ if $s \in [a,b]$, it implies

$$1 < (b-c)\int_c^b q^+(s)\,\mathrm{d}s,$$

which in turn implies

$$\int_c^b q^+(s)\,\mathrm{d}s > \frac{1}{b-c}.$$

This result proves part (ii). Part (i) follows in a similar fashion except that in equation (5.4.8) one now replaces t by a. The sum of (i) and (ii) yields part (iii) and the proof is complete. □

One way to view Theorem 5.4.2 is that it imposes some restrictions on the location of the point c and thus the maximum of $|y(t)|$ in $[a,b]$. That is, $\int_a^b q^+(t)\,\mathrm{d}t$ is a finite number. But

$$\lim_{c\to a^+} \frac{b-a}{(b-c)(c-a)} = \lim_{c\to b^-} \frac{b-a}{(b-c)(c-a)} = \infty.$$

Thus c cannot be too close to a or b. Also it is interesting to note that $(b-a)/((b-c)(c-a)) \geqslant 4/(b-a)$. This result means that under the hypotheses of Theorem 5.4.2, Corollary 5.4.1 follows.

As a consequence of Theorem 5.4.2, in [361] Patula has given the following theorem.

THEOREM 5.4.3. *Suppose $q^+(t) \in L^p[0,\infty)$, $1 \leqslant p < \infty$. If (5.4.1) is oscillatory and if $y(t)$ is any solution, then the distance between consecutive zeros of $y(t)$ must become infinite.*

PROOF. Suppose not. Then there exists a solution $y(t)$ with its sequence of zeros $\{t_n\}$, which has a subsequence $\{t_{n_k}\}$ such that $|t_{n_{k+1}} - t_{n_k}| \leqslant M < \infty$ for all k. Let s_{n_k} be a point in $(t_{n_k}, t_{n_{k+1}})$ where $|y(t)|$ is maximized. Then $|s_{n_k} - t_{n_k}| < M$ for all k. Since $q^+(t) \in L^p(0,\infty)$, $1 \leqslant p < \infty$, choose k so large that

$$\left(\int_{t_{n_k}}^{\infty} q^+(t)^p\,\mathrm{d}t\right)^{1/p} \leqslant M^{-1-1/r}, \qquad \frac{1}{p} + \frac{1}{r} = 1.$$

From Theorem 5.4.2, part (i), we have

$$\int_{t_{n_k}}^{s_{n_k}} q^+(t)\,\mathrm{d}t > \frac{1}{s_{n_k} - t_{n_k}}.$$

Thus

$$\begin{aligned}
1 &< (s_{n_k} - t_{n_k}) \int_{t_{n_k}}^{s_{n_k}} q^+(t)\,\mathrm{d}t \\
&< (s_{n_k} - t_{n_k}) \left(\int_{t_{n_k}}^{s_{n_k}} q^+(t)^p\,\mathrm{d}t\right)^{1/p} (s_{n_k} - t_{n_k})^{1/r} \\
&< (s_{n_k} - t_{n_k})^{1+1/r} \left(\int_{t_{n_k}}^{\infty} q^+(t)^p\,\mathrm{d}t\right)^{1/p} \\
&< M^{1+1/r} M^{-1-1/r} \quad \Longrightarrow \quad 1 < 1,
\end{aligned}$$

a contradiction. The proof is complete. □

The classical result of Lyapunov is usually formulated in connection with disconjugacy. Hence a violation of inequality (5.4.2) implies that (5.4.1) is disconjugate in $[a,b]$. In [176] Kwong strengthened Lyapunov's inequality by introducing the idea of disfocality. Below, by "a solution" we always mean "a nontrivial one". It is well known that between any two zeros of a solution y of (5.4.1) there is a zero of y'. We may thus decompose the interval (a,b) between zeros of y into the union of the intervals (a,ξ) and $[\xi,b)$, where $y'(\xi) = 0$. It is possible now to construct inequalities similar to (5.4.2) on the intervals (a,ξ) and $[\xi,b)$ separately. Following Kwong [176], (5.4.1) is right disfocal on the interval $[a,b]$ if the solution of (5.4.1) with $y'(a)=0$ has no zeros in $[a,b]$. Left disfocality is defined in a similar way. Equation (5.4.1) is disconjugate in an interval $[a,b]$ if and only if there exists a point $c \in [a,b]$ such that (5.4.1) is right disfocal in $[c,b]$ and left disfocal in $[a,c]$. Thus Lyapunov's result follows from the following stronger result. If (5.4.1) is not disfocal in an interval $[a,c]$, then

$$\int_a^c q^+(t)\,\mathrm{d}t > \frac{1}{c-a}. \tag{5.4.9}$$

This approach has been employed by Kwong in [176] to extend Lyapunov's inequality.

In [176] Kwong has given the following necessary inequality for disfocality.

THEOREM 5.4.4. *If* (5.4.1) *has a solution such that* $y'(0)=y(c)=0$, $0<c$, *then*

$$\int_0^c Q^+(t)\,\mathrm{d}t = \int_0^c (c-t)q^+(t)\,\mathrm{d}t > 1, \tag{5.4.10}$$

where $Q^+(t)=\int_0^t q^+(s)\,\mathrm{d}s$.

PROOF. The idea that the two integrals in (5.4.10) are equal is an elementary fact of double integration. □

Let us make two reductions. We may first assume that y has no zeros in $[0,c)$. Suppose that the theorem has been proved for this case. In the case that y has zeros in $[0,c)$, let $\bar{c}$ be the smallest zero. Then we have $\int_0^{\bar{c}} Q^+(t)\,\mathrm{d}t > 1$ from which (5.4.10) follows. Next we may assume that $q \geqslant 0$, so that $q^+=q$. In the contrary case, we consider the equation

$$z''(t)+q^+(t)z(t)=0, \tag{5.4.11}$$

and one of its solutions z such that $z'(0)=0$. It follows from a form of the Sturmain comparison theorem (notice that the potential q^+ of the new equation (5.4.11) dominates that of (5.4.1)) that z has a zero $\bar{c}$ in $(0,c)$. The result for positive potentials then gives, for equation (5.4.11), $\int_0^{\bar{c}} Q^+(t)\,\mathrm{d}t > 1$ from which (5.4.10) follows.

The following corollaries of Theorem 5.4.4 can be used in the study of disconjugacy criterion, which may be considered as the further extensions of Hartman's improvement of Lyapunov's result [145, p. 346].

COROLLARY 5.4.2. *If, for all $t \in [a, b]$, the following inequality holds*

$$\int_a^t \frac{(s-a)q^+(s)\,\mathrm{d}s}{t-a} + \int_t^b \frac{(b-s)q^+(s)\,\mathrm{d}s}{b-t} \leqslant \frac{1}{t-a} + \frac{1}{b-t},$$

then (5.4.1) *is disconjugate in* (a, b).

COROLLARY 5.4.3. *If, for some point $c \in [a, b]$,*

$$\int_a^c (t-a)q^+(t)\,\mathrm{d}t \leqslant 1 \quad \textit{and} \quad \int_c^b (b-t)q^+(t)\,\mathrm{d}t \leqslant 1,$$

then (5.4.1) *is disconjugate in* $[a, b]$.

In [143] Harris has given further extensions of Kwong's results in [176]. In [143] Theorem 5.4.4 is stated as follows.

THEOREM A. *If y is a solution of* (5.4.1) *with $y'(0) = 0$ and $y(c) = 0$, then*

$$\int_0^c \int_0^t q^+(r)\,\mathrm{d}r\,\mathrm{d}t > 1. \tag{5.4.12}$$

This result may be paraphrased to state that if the inequality of (5.4.12) *is violated then* (5.4.1) *is right disfocal on* $[0, c)$.

In [143] Harris has given the following keener result which also uses both positive and negative parts of $q(t)$.

THEOREM 5.4.5. *Let $\gamma(\cdot)$ denote a function with the properties*

(i) $\gamma(0) = 0$,
(ii) $\gamma(\,)$ *is differentiable on* $[0, c]$.

Set

$$Q(t) = q(t) - \gamma(t) + \gamma(t)^2 \quad \textit{and}$$

$$A(c) = \sup_{0 \leqslant x \leqslant c} \left| \int_0^x \exp\left\{2\int_t^x \gamma(s)\,\mathrm{d}s\right\} Q(t)\,\mathrm{d}t \right|,$$

$$B(c) = \sup_{0 \leqslant x \leqslant c} \int_0^x \exp\left\{2\int_t^x \gamma(s)\,\mathrm{d}s\right\} \mathrm{d}t.$$

If $4A(c)B(c) < 1$, *then* (5.4.1) *is right disfocal on* $[0, c)$.

COROLLARY 5.4.4. *If* $4c \sup_{0 \leqslant x \leqslant c} |\int_0^x q(t)\,\mathrm{d}t| < 1$, *then* (5.4.1) *is right disfocal on* $[0, c)$.

PROOF. We set $\gamma(t) = 0$ for $t \in [0, c)$ in Theorem 4.4.5. □

COROLLARY 5.4.5. *If*

$$B(c) = \sup_{0 \leqslant x \leqslant c} \int_0^c \exp\left\{2 \int_t^x \int_0^s q(r)\,\mathrm{d}r\,\mathrm{d}s\right\} \mathrm{d}t$$

and

$$A(c) = \sup_{0 \leqslant x \leqslant c} \int_0^x \exp\left\{2 \int_t^x \int_0^s q(r)\,\mathrm{d}r\,\mathrm{d}s\right\} \left(\int_0^t q(s)\,\mathrm{d}s\right)^2 \mathrm{d}t,$$

then (5.4.1) *is right disfocal on* $[0, c)$ *if* $4A(c)B(c) < 1$.

PROOF. We set $\gamma(t) = \int_0^t q(s)\,\mathrm{d}s$. □

COROLLARY 5.4.6. *If*

$$4c \exp\left\{\int_0^c \left(\int_0^s q(r)\,\mathrm{d}r\right)^+ \mathrm{d}s\right\} \int_0^c \left(\int_0^t q(s)\,\mathrm{d}s\right)^2 \mathrm{d}t < 1,$$

then (5.4.1) *is right disfocal on* $[0, c)$.

PROOF. The proof follows from Corollary 5.4.5. □

In [143] Harris has given an iterated form of Theorem A by means of a trivial observation.

Let y denote a solution of (5.4.1) with $y'(0) = 0$ and $y(c) = 0$. We may suppose without loss of generality that c is the least positive zero of y and $y(t) > 0$ for $t \in [0, c)$. It is also sufficient by the Sturmain comparison theorem to consider only the case $q(t) = q^+(t)$.

We integrate (5.4.1) between 0 and t to obtain

$$-y'(t) = \int_0^t q^+(s) y(s)\,\mathrm{d}s. \tag{5.4.13}$$

An integration over $[0, c]$ then yields

$$\begin{aligned} y(0) &= \int_0^c \int_0^t q^+(s) y(s)\, \mathrm{d}s\, \mathrm{d}t \\ &\leqslant y(0) \int_0^c \int_0^t q^+(s)\, \mathrm{d}s\, \mathrm{d}t. \end{aligned} \tag{5.4.14}$$

This result leads to Kwong's proof of Theorem A.

Suppose now that we integrate (5.4.13) over the interval from s to c and obtain

$$y(s) = \int_s^c \int_0^\tau q^+(r) y(r)\, \mathrm{d}r\, \mathrm{d}\tau.$$

Substitution into (5.4.13) now gives

$$y'(t) = \int_0^t q^+(s) \int_s^c \int_0^\tau q^+(r) y(r)\, \mathrm{d}r\, \mathrm{d}\tau\, \mathrm{d}s,$$

and an integration over $[0, c]$ yields

$$\begin{aligned} y(0) &= \int_0^c \int_0^t q^+(s) \int_s^c \int_0^\tau q^+(r) y(r)\, \mathrm{d}r\, \mathrm{d}\tau\, \mathrm{d}s\, \mathrm{d}t \\ &\leqslant y(0) \int_0^c \int_0^t q^+(s) \int_s^c \int_0^\tau q^+(r)\, \mathrm{d}r\, \mathrm{d}\tau\, \mathrm{d}s\, \mathrm{d}t. \end{aligned}$$

We thus deduce that if $y'(0) = 0$ and $y(c) = 0$ then

$$1 \leqslant \int_0^c \int_0^t q^+(s) \int_s^c \int_0^\tau q^+(r)\, \mathrm{d}r\, \mathrm{d}\tau\, \mathrm{d}s\, \mathrm{d}t. \tag{5.4.15}$$

In order to compare (5.4.15) with Theorem A, we let

$$\Phi(s) = \int_s^c \int_0^\tau q^+(r)\, \mathrm{d}r\, \mathrm{d}\tau.$$

Inequality (5.4.15) represents an improvement over Theorem A if $\Phi(s) < 1$. We write

$$\begin{aligned} \Phi(s) &= \int_s^c \left\{ \int_0^s q^+(r)\, \mathrm{d}r + \int_s^\tau q^+(r)\, \mathrm{d}r \right\} \mathrm{d}\tau \\ &= (c - s) \int_0^s q^+(r)\, \mathrm{d}r + \int_s^c (c - r) q^+(r)\, \mathrm{d}r \\ &= \int_0^c \Psi(s, r) q^+(r)\, \mathrm{d}r, \end{aligned} \tag{5.4.16}$$

where

$$\Psi(s,r) = \begin{cases} c-s & \text{if } 0 \leqslant r \leqslant s, \\ c-r & \text{if } s < r \leqslant c. \end{cases}$$

We note that $0 \leqslant \Psi(s,r) \leqslant c-r$, and using this upper bound in (5.4.16) we have

$$\Phi(s) \leqslant \int_0^c (c-r)q^+(r)\,\mathrm{d}r = \int_0^c \int_0^r q^+(s)\,\mathrm{d}s\,\mathrm{d}r. \tag{5.4.17}$$

This result is inconclusive since, by Theorem A, the right-hand side of (5.4.17) is greater than 1. On the other hand, if we use the upper bound, $\Psi(s,r) \leqslant c-s$, in (5.4.16) we deduce that

$$\Phi(s) \leqslant (c-s) \int_0^c q^+(r)\,\mathrm{d}r,$$

which may be less than 1.

This process may be iterated and leads to the result that if $y'(0) = 0$ and $y(c) = 0$ then for any integer n,

$$\int_0^c \int_0^{t_0} q^+(t_1) \int_{t_1}^c \int_0^{t_2} q^+(t_3) \cdots \int_{t_{2n+1}}^c \int_0^{t_{2n+2}} q^+(t_{2n+3})\,\mathrm{d}t_{2n+3} \cdots \mathrm{d}t_0 \geqslant 1.$$

PROOF OF THEOREM 5.4.5. Let $y(\cdot)$ denote a solution of (5.4.1) with $y'(0) = 0$ and $\gamma(\cdot)$ a differentiable function to be chosen later subject to

$$\gamma(0) = 0. \tag{5.4.18}$$

We follow the approach of Harris [142] and use γ to derive a regularizing transformation of (5.4.1). We write

$$r(x) = -\left(\frac{y'}{y} - \gamma\right), \tag{5.4.19}$$

so that by (5.4.18),

$$r(0) = 0, \tag{5.4.20}$$

and after substitution in (5.4.1),

$$r' = Q + 2\gamma r + r^2, \tag{5.4.21}$$

where $Q = q - \gamma' + \gamma^2$. We rearrange (5.4.21) as

$$r' - 2\gamma r = Q + r^2,$$

and integration yields

$$r(x)=\int_0^x \exp\left\{2\int_t^x \gamma(s)\,\mathrm{d}s\right\}Q(t)\,\mathrm{d}t+\int_0^x \exp\left\{2\int_t^x \gamma(s)\,\mathrm{d}s\right\}r^2(t)\,\mathrm{d}t. \tag{5.4.22}$$

Let

$$A(X)=\sup_{0\leqslant x\leqslant X}\left|\int_0^x \exp\left\{2\int_t^x \gamma(s)\,\mathrm{d}s\right\}Q(t)\,\mathrm{d}t\right|,$$

$$B(X)=\sup_{0\leqslant x\leqslant X}\int_0^x \exp\left\{2\int_t^x \gamma(s)\,\mathrm{d}s\right\}\mathrm{d}t,$$

$$R(X)=\sup_{0\leqslant x\leqslant X}\left|r(x)\right|.$$

It is clear from (5.4.22) that

$$\left|r(x)\right|\leqslant A(X)+B(X)R(X)^2 \quad \text{for } x\in[0,X]$$

and thus

$$R(X)\leqslant A(X)+B(X)R(X)^2. \tag{5.4.23}$$

□

LEMMA 5.4.1. *If X is such that $4A(X)B(X)<1$, then*

$$R(X)<2A(X) \quad \textit{for } x\in[0,X].$$

PROOF. We know that $R(0)=0$ so if the result were false there would be a least value of x, x_0, say, for which $R(x_0)=2A(x_0)$; thus from (5.4.23),

$$\begin{aligned}2A(x_0)&\leqslant A(x_0)+B(x_0)R(x_0)^2\\&=A(x_0)\big(1+4A(x_0)B(x_0)\big),\end{aligned}$$

which gives a contradiction. □

In particular, Lemma 5.4.1 shows that $4A(c)B(c)<1$ then

$$\left|\frac{y'(x)}{y(x)}-\gamma(x)\right|\leqslant 2A(c), \quad x\in[0,c].$$

Thus, if $\gamma(\cdot)$ is bounded for $x\in[0,c]$, then y has no zeros in $[0,c]$.

5.5 Extensions of Lyapunov's Inequality

In this section we deal with inequalities similar to Lyapunov's inequality established by Harris and Kong [144] and Brown and Hinton [46]. Consider the linear second-order differential equation

$$y'' + q(t)y = 0, \tag{5.5.1}$$

where q is a real-valued function belonging to L^1_{loc}. In [144] Harris and Kong extended the Lyapunov inequality given in Corollary 5.4.1 in such a way as to use the negative part of q to obtain keener bound.

The following lemma given in [144] is needed in further discussion.

LEMMA 5.5.1. *If y is a solution of* (5.5.1) *satisfying $y'(d) = 0$, $y(b) = 0$, and $y(t) > 0$ and $y'(t) \leqslant 0$ for $t \in (d, b)$, then*

$$\sup_{d \leqslant t \leqslant b} \int_d^t q(s)\, \mathrm{d}s > 0.$$

PROOF. Suppose the contrary. Then $\int_d^t q(s)\, \mathrm{d}s \leqslant 0$ for $t \in [d, b]$. Let $Q(t) = \int_d^t q(s)\, \mathrm{d}s$, and define the Riccati variable

$$r(t) = -\frac{y'(t)}{y(t)}. \tag{5.5.2}$$

We thus have $r(d) = 0$, $\lim_{t \to b^-} r(t) = \infty$ and $r(t) \geqslant 0$ for $t \in (d, b)$. It follows from (5.5.1) that

$$r'(t) = q(t) + r^2(t), \tag{5.5.3}$$

whence

$$r(t) = Q(t) + \int_d^t r^2(s)\, \mathrm{d}s.$$

In a similar way, if z is the nontrivial solution of the equation $z'' = 0$ with $z'(d) = 0$ and $R(t) = -z'(t)/z(t)$, then

$$R(t) = \int_d^t R(s)^2\, \mathrm{d}s,$$

so that $R(t) = 0$ for all $t \in [d, \infty)$. As a simple consequence of the general theory of integral inequalities we see that $r(t) \leqslant R(t) = 0$ for $t \in [d, b)$, thus contradicting the fact that $\lim_{t \to b^-} r(t) = \infty$. The proof is complete. □

The main results established in [144] are given in the following theorems.

THEOREM 5.5.1. *Let y denote a nontrivial solution of* (5.5.1) *satisfying* $y'(d) = 0$, $y(b) = 0$, *and* $y(t) \neq 0$ *for* $t \in [d, b)$. *Then*

$$(b-d) \sup_{d \leqslant t \leqslant b} \left| \int_d^t q(s)\,\mathrm{d}s \right| > 1. \tag{5.5.4}$$

Moreover, if there are no extreme values of y in (d, b), *then*

$$(b-d) \sup_{d \leqslant t \leqslant b} \int_d^t q(s)\,\mathrm{d}s > 1. \tag{5.5.5}$$

PROOF. We assume, without loss of generality, that $y(t) > 0$ for $t \in [d, b)$. With r defined by (5.5.2), we set

$$w(t) = \int_d^t r^2(s)\,\mathrm{d}s \quad \text{for } t \in [d, b). \tag{5.5.6}$$

Thus $r(d) = w(d) = 0$ and from (5.5.3) $\lim_{t \to b^-} r(t) = \lim_{t \to b^-} w(t) = \infty$ because

$$r(t) = \int_d^t q(s)\,\mathrm{d}s + w(t) \quad \text{for } t \in [d, b). \tag{5.5.7}$$

Set $Q^* = \sup_{d \leqslant t \leqslant b} |\int_d^t q(s)\,\mathrm{d}s|$ and observe that

$$|r(t)| \leqslant Q^* + w(t)$$

so that

$$w'(t) = r^2(t) \leqslant \left(Q^* + w(t)\right)^2,$$

that is,

$$\frac{w'(t)}{(Q^* + w(t))^2} \leqslant 1. \tag{5.5.8}$$

Integrating (5.5.8) over $[d, b]$ we obtain

$$-\frac{1}{Q^* + w(t)}\bigg|_d^b \leqslant b - d,$$

which implies that $1/Q^* \leqslant b - d$ or $(b-d)Q^* \geqslant 1$. We remark that equality cannot hold, for otherwise $|Q(t)| = |\int_d^t q(s)\,\mathrm{d}s| = Q^*$ almost everywhere on $[d, b)$, which contradicts the fact that Q is continuous and $Q(d) = 0$.

If d is the largest extreme point of y in $[d,b)$, then $y'(t) \leqslant 0$ and thus $r(t) \geqslant 0$ for $t \in [d,b)$. Set $Q_* = \sup_{d \leqslant t \leqslant b} \int_d^t q(s)\,\mathrm{d}s$. By Lemma 5.5.1, $Q_* > 0$ and from (5.5.7),

$$0 \leqslant r(t) \leqslant Q_* + w(t).$$

The proof of the second part of theorem now follows in a way similar to that of the first. □

THEOREM 5.5.2. *Let y denote a nontrivial solution of* (5.5.1) *satisfying $y(a) = 0$, $y'(c) = 0$, and $y(t) \neq 0$ for $t \in (a,c]$. Then*

$$(c-a) \sup_{a \leqslant t \leqslant c} \left| \int_t^c q(s)\,\mathrm{d}s \right| > 1. \tag{5.5.9}$$

Moreover, if there are no extreme values of y in (a,c), then

$$(c-a) \sup_{a \leqslant t \leqslant c} \int_t^c q(s)\,\mathrm{d}s > 1. \tag{5.5.10}$$

The proof is similar to the proof of Theorem 5.5.1 and is omitted.

COROLLARY 5.5.1. *If*

$$(b-d) \sup_{d \leqslant t \leqslant b} \left| \int_d^t q(s)\,\mathrm{d}s \right| \leqslant 1,$$

then (5.5.1) *is right disfocal on $[d,b)$.*

If

$$(c-a) \sup_{a \leqslant t \leqslant c} \left| \int_t^c q(s)\,\mathrm{d}s \right| \leqslant 1,$$

then (5.5.1) *is left disfocal on $(a,c]$.*

THEOREM 5.5.3. *Let a and b denote two consecutive zeros of a nontrivial solution y of* (5.5.1). *Then there exist two disjoint subintervals of $[a,b]$, I_1 and I_2 satisfying*

$$(b-a) \int_{I_1 \cup I_2} q(s)\,\mathrm{d}s > 4 \tag{5.5.11}$$

and

$$\int_{[a,b]-(I_1 \cup I_2)} q(s)\,\mathrm{d}s \leqslant 0. \tag{5.5.12}$$

PROOF. Let c and d denote the least and greatest extreme points of y on $[a,b]$, respectively. If there is only one zero of y' in (a,b), then c and d coincide. Then $y'(d)=0$, $y(b)=0$, and $y'(t)\neq 0$ for $t\in[d,b]$. By Theorem 5.5.1, inequality (5.5.5) holds. Thus there exists $b_1\in(d,b]$ such that

$$\int_d^{b_1} q(s)\,\mathrm{d}s > \frac{1}{b-d} \quad \text{and} \quad \int_d^{b_1} q(s)\,\mathrm{d}s \geqslant \int_d^b q(s)\,\mathrm{d}s.$$

Similarly, we can choose $a_1\in[a,c)$ such that

$$\int_{a_1}^{c} q(s)\,\mathrm{d}s > \frac{1}{c-a} \quad \text{and} \quad \int_{a_1}^{c} q(s)\,\mathrm{d}s \geqslant \int_a^c q(s)\,\mathrm{d}s.$$

Let $I_1=[d,b_1]$ and $I_2=[a_1,c]$, then

$$\begin{aligned}(b-a)\int_{I_1\cup I_2} q(s)\,\mathrm{d}s &\geqslant \big[(b-d)+(c-a)\big]\left(\int_d^{b_1} q(s)\,\mathrm{d}s+\int_{a_1}^{c} q(s)\,\mathrm{d}s\right)\\ &> \big[(b-d)+(c-a)\big]\left(\frac{1}{b-d}+\frac{1}{c-a}\right)\\ &= 2+\frac{c-a}{b-d}+\frac{b-d}{c-a}\geqslant 4\end{aligned}$$

and (5.5.11) is verified. It is also easy to see that $\int_{b_1}^{b} q(s)\,\mathrm{d}s\leqslant 0$ and $\int_a^{a_1} q(s)\times\mathrm{d}s\leqslant 0$. To verify (5.5.12) it is sufficient to show that $\int_c^d q(s)\,\mathrm{d}s\leqslant 0$. In fact, since $y'(c)=y'(d)=0$, we have $r(c)=r(d)=0$. From (5.5.3),

$$0=r(d)-r(c)=\int_c^d q(s)\,\mathrm{d}s+\int_c^d r^2(s)\,\mathrm{d}s.$$

This result means that $\int_c^d q(s)\,\mathrm{d}s\leqslant 0$ and hence that (5.5.12) holds. □

COROLLARY 5.5.2. *Suppose that, for every two disjoint subintervals, I_1 and I_2, of $[\alpha,\beta]$, we have*

$$(\beta-\alpha)\int_{I_1\cup I_2} q(s)\,\mathrm{d}s\leqslant 4. \tag{5.5.13}$$

Then (5.5.1) *is disconjugate on* $[\alpha,\beta]$.

PROOF. Suppose the contrary, then there exists a nontrivial solution y of (5.5.1) with $y(a)=y(b)=0$ for $\alpha\leqslant a<b\leqslant\beta$. Without loss of generality we assume

that $y(t) \neq 0$ for $t \in (a, b)$. By Theorem 5.5.3, there exist two disjoint intervals, I_1 and I_2, of $[a, b] \subset [\alpha, \beta]$ with

$$(b-a)\int_{I_1 \cup I_2} q(s)\,\mathrm{d}s > 4.$$

Hence, $(\beta - \alpha)\int_{I_1 \cup I_2} q(s)\,\mathrm{d}s > 4$, which gives a contradiction. □

COROLLARY 5.5.3. *Suppose that a nontrivial solution of* (5.5.1) *has N zeros in $[a, b]$ for $N \geqslant 2$. There exist $2N$ disjoint subintervals of $[a, b]$, I_{ij} for $i = 1, \ldots, N$, $j = 1, 2$, such that*

$$N < \frac{1}{2}\left[(b-a)\int_I q(s)\,\mathrm{d}s\right]^{1/2} + 1, \tag{5.5.14}$$

and

$$\int_{[a,b]-I} q(s)\,\mathrm{d}s \leqslant 0, \tag{5.5.15}$$

where $I = \bigcup_{i=1}^{N} \bigcup_{j=1}^{2} I_{ij}$.

PROOF. Let t_i, $i = 1, \ldots, N$, be the zeros of y in $[a, b]$. By Theorem 5.5.3, for $i = 1, \ldots, N-1$, there are two disjoint subintervals of $[t_i, t_{i+1}]$, I_{i1} and I_{i2}, with

$$\int_{I_{i1} \cup I_{i2}} q(s)\,\mathrm{d}s > \frac{4}{t_{i+1} - t_i} \tag{5.5.16}$$

and

$$\int_{[t_i, t_{i+1}] - (I_{i1} \cup I_{i2})} q(s)\,\mathrm{d}s \leqslant 0. \tag{5.5.17}$$

We sum (5.5.16) for i from 1 to $N-1$ and see that

$$\int_I q(s)\,\mathrm{d}s > 4\sum_{i=1}^{N-1} \frac{1}{t_{i+1} - t_i},$$

and by the inequality for harmonic mean

$$\int_I q(s)\,\mathrm{d}s > \frac{4(N-1)^2}{t_N - t_1} \geqslant \frac{4(N-1)^2}{b-a},$$

whence

$$(N-1)^2 < \frac{b-a}{4}\int_I q(s)\,\mathrm{d}s.$$

This result implies (5.5.14). From (5.5.17) it is easy to deduce (5.5.15). □

REMARK 5.5.1. Corollary 5.5.3 provides an extension of the result [145, Corollary 5.2, p. 347] in that the negative part of q to achieve a sharper bound is used.

In [46] Brown and Hinton studied the two problems concerning the equation

$$y'' + q(x)y = 0, \quad a \leqslant x \leqslant b, \tag{5.5.1'}$$

where q is real and $q \in L(a, b)$:

(i) obtain lower bounds for the spacing of zeros of a solution, and

(ii) obtain lower bounds for the spacing $\beta - \alpha$ for a solution y of (5.5.1′) satisfying $y(\alpha) = y'(\beta) = 0$ or $y'(\alpha) = y(\beta) = 0$.

In [46] results which relate to problems (i) and (ii) are given by using the following versions of the Opial-type inequalities.

LEMMA 5.5.2. *If f is absolutely continuous on $[a, b]$ with $f(a) = 0$ and $s \in L^2(a, b)$, then*

$$\int_a^b s(x)\left|f(x)\right|\left|f'(x)\right| \mathrm{d}x \leqslant k \int_a^b \left|f'(x)\right|^2 \mathrm{d}x, \tag{5.5.18}$$

where

$$k = \frac{1}{\sqrt{2}}\left(\int_a^b s(t)^2 (t - a)\, \mathrm{d}t\right)^{1/2}, \tag{5.5.19}$$

with equality if and only if $f \equiv 0$ (or f is linear and s is constant).

REMARK 5.5.2. Inequality (5.5.18) is a special case of an inequality obtained by Beesack and Das (see [4]). If we replace $f(a) = 0$ in Lemma 5.5.2 by $f(b) = 0$, then (5.5.18) holds where k in (5.5.19) is given by

$$k = \frac{1}{\sqrt{2}}\left(\int_a^b s(t)^2 (b - t)\, \mathrm{d}t\right)^{1/2}. \tag{5.5.20}$$

The following version of the Opial inequality is also used in [46].

LEMMA 5.5.3. *If f is absolutely continuous on $[a, b]$ with $f(a) = 0$ or $f(b) = 0$ and $1 \leqslant p \leqslant 2$, then*

$$\int_a^b \left|f(x)\right|^p \left|f'(x)\right|^p \mathrm{d}x \leqslant K(p)(b - a)\left(\int_a^b \left|f'(x)\right|^2 \mathrm{d}x\right)^p, \tag{5.5.21}$$

where

$$K(p)=\begin{cases}\frac{1}{2}, & p=1,\\ \frac{4}{\pi^2}, & p=2,\\ \frac{2-p}{2p}\left(\frac{1}{p}\right)^{2p-2}I^{-p}, & 1<p<2,\end{cases} \tag{5.5.22}$$

with

$$I=\int_0^1\left\{1+\frac{2(p-1)}{2-p}t\right\}^{-2}\{1+(p-1)t\}^{1/p-1}\,\mathrm{d}t.$$

For $p=1$, equality holds in (5.5.21) *only for f linear.*

REMARK 5.5.3. Lemma 5.5.3 has immediate application to the case where $f(a)=f(b)=0$. Choose $c=(a+b)/2$ and apply (5.5.21) to $[a,c]$ and $[c,b]$ then add to obtain that

$$\begin{aligned}&\int_a^b |f(x)|^p|f'(x)|^p\,\mathrm{d}x\\ &\quad\leqslant K(p)\left(\frac{b-a}{2}\right)\left\{\left(\int_a^c |f'(x)|^2\,\mathrm{d}x\right)^p+\left(\int_c^b |f'(x)|^2\,\mathrm{d}x\right)^p\right\}\\ &\quad\leqslant K(p)\left(\frac{b-a}{2}\right)\left\{\left(\int_a^b |f'(x)|^2\,\mathrm{d}x\right)\right\}^p.\end{aligned} \tag{5.5.23}$$

For $p=1$, (5.5.23) is strict unless f is linear in each of the subintervals $[a,c]$ and $[c,b]$.

The main results established in [46] are given in the following theorems.

THEOREM 5.5.4. *Suppose y is a nontrivial solution of* (5.5.1′) *which satisfies $y(a)=y'(b)=0$. Then*

$$1<2\int_a^b Q(x)^2(x-a)\,\mathrm{d}x, \tag{5.5.24}$$

where $Q(x)=\int_x^b q(t)\,\mathrm{d}t$. If $y'(a)=y(b)=0$, then

$$1<2\int_a^b Q(x)^2(b-x)\,\mathrm{d}x, \tag{5.5.25}$$

where $Q(x)=\int_a^x q(t)\,\mathrm{d}t$.

PROOF. We first establish (5.5.24). Multiplying (5.5.1′) by y and integrating by parts gives

$$\begin{aligned}
\int_a^b y'(x)^2\,dx &= \int_a^b q(x)y(x)^2\,dx \\
&= \int_a^b -Q'(x)y(x)^2\,dx \\
&= \int_a^b 2Q(x)y(x)y'(x)\,dx \\
&\leqslant 2\int_a^b |Q(x)||y(x)||y'(x)|\,dx \\
&\leqslant \frac{2}{\sqrt{2}}\left(\int_a^b Q(x)^2(x-a)\,dx\right)^{1/2}\int_a^b y'(x)^2\,dx, \qquad (5.5.26)
\end{aligned}$$

by (5.5.18) and (5.5.19) of Lemma 5.5.2. The inequality is strict since y linear implies $y \equiv 0$ as $y(a) = y'(b) = 0$. By canceling $\int_a^b y'(x)^2\,dx$ and squaring, we obtain (5.5.24). The proof of (5.5.25) is similar using integration by parts and (5.5.18) and (5.5.20) instead of (5.5.19). □

REMARK 5.5.4. By using the maximum of $|Q|$ on $[a,b]$ in (5.5.24) and (5.5.25), integrating and then taking a square root, we see that

$$1 < (b-a)\max_{a\leqslant x\leqslant b}\left|\int_x^b q(t)\,dt\right| \tag{5.5.27}$$

when $y(a) = y'(b) = 0$, and

$$1 < (b-a)\max_{a\leqslant x\leqslant b}\left|\int_a^x q(t)\,dt\right| \tag{5.5.28}$$

when $y'(a) = y(b) = 0$, which are the inequalities obtained by Harris and Kong [144].

The following result similar to Theorem 5.5.4 given in [46] may be obtained by application of Lemma 5.5.3.

THEOREM 5.5.5. *Suppose y is a nontrivial solution of* (5.5.1′) *which satisfies $y(a) = y'(b) = 0$, $1 \leqslant p \leqslant 2$, and p' is the conjugate index of p, that is,*

$\frac{1}{p}+\frac{1}{p'}=1$. *Then*

$$1 \leqslant 2K(p)^{1/p}(b-a)^{1/p}\left(\int_a^b |Q(x)|^{p'}\,\mathrm{d}x\right)^{1/p'}, \tag{5.5.29}$$

where $Q(x)=\int_x^b q(t)\,\mathrm{d}t$; *if* $y'(a)=y(b)=0$ *then* (5.5.29) *is true with* $Q(x)=\int_a^x q(t)\,\mathrm{d}t$. *In either case* $K(p)$ *is given by* (5.5.22). *For* $p=1$ *the inequality is strict. For* $p=1$ *the* p' *norm of* Q *in* (5.5.29) *becomes* $\max|Q(x)|$, $a\leqslant x\leqslant b$.

PROOF. In the case $y(a)=y'(b)=0$ from the proof of Theorem 5.5.4, we have that

$$\int_a^b y'(x)^2\,\mathrm{d}x \leqslant 2\int_a^b |Q(x)||y(x)||y'(x)|\,\mathrm{d}x. \tag{5.5.30}$$

By application of Hölder's inequality and Lemma 5.5.2 to (5.5.30), we get that

$$\begin{aligned}\int_a^b y'(x)^2\,\mathrm{d}x &\leqslant 2\left(\int_a^b |Q(x)|^{p'}\,\mathrm{d}x\right)^{1/p'}\left(\int_a^b |y(x)y'(x)|^{p}\,\mathrm{d}x\right)^{1/p}\\ &\leqslant 2K(p)^{1/p}(b-a)^{1/p}\left(\int_a^b |Q(x)|^{p'}\,\mathrm{d}x\right)^{1/p'}\int_a^b y'(x)^2\,\mathrm{d}x,\end{aligned}$$

with strict inequality for $p=1$. Canceling $\int_a^b y'(x)^2\,\mathrm{d}x$ yields (5.5.29). A similar argument yields (5.5.29) with $Q(x)=\int_a^x q(t)\,\mathrm{d}t$ when $y'(a)=y(b)=0$. □

REMARK 5.5.5. Note that, for $p=1$, (5.5.29) in the $y(a)=y'(b)=0$ case is the same as (5.5.27) and in the $y'(a)=y(b)=0$ case the same as (5.5.28). Theorems 5.5.4 and 5.5.5 yield sufficient conditions for disfocality of (5.5.1′), that is, sufficient conditions so that there does not exist a nontrivial solution y of (5.5.1′) satisfying either $y(a)=y'(b)=0$ or $y'(a)=y(b)=0$.

As an application of (5.5.23) in [46] the following Lyapunov-type inequality is given.

THEOREM 5.5.6. *Suppose* y *is a nontrivial solution of* (5.5.1′) *which satisfies* $y(a)=y(b)=0$, $1\leqslant p\leqslant 2$, *and* $Q'(x)=q(x)$ *on* $[a,b]$. *Then*

$$1 \leqslant 2K(p)^{1/p}\left(\frac{b-a}{2}\right)^{1/p}\left(\int_a^b |Q(x)|^{p'}\,\mathrm{d}x\right)^{1/p'}, \tag{5.5.31}$$

with $K(p)$ *given by* (5.5.22). *For* $p=1$ *the inequality is strict.*

PROOF. As in the proof of Theorem 5.5.4, multiplying (5.5.1′) by y and integration by parts yields that

$$\int_a^b y'(x)^2\,\mathrm{d}x = \int_a^b q(x)y(x)^2\,\mathrm{d}x = -2\int_a^b Q(x)y(x)y'(x)\,\mathrm{d}x. \tag{5.5.32}$$

By application of Hölder's inequality and (5.5.23) to (5.5.32), we get that

$$\begin{aligned}
&\int_a^b y'(x)^2\,\mathrm{d}x \\
&\quad\leqslant 2\left(\int_a^b |Q(x)|^{p'}\,\mathrm{d}x\right)^{1/p'}\left(\int_a^b |y(x)y'(x)|^{p}\,\mathrm{d}x\right)^{1/p} \\
&\quad\leqslant 2K(p)^{1/p}\left(\frac{b-a}{2}\right)^{1/p}\left(\int_a^b |Q(x)|^{p'}\,\mathrm{d}x\right)^{1/p'}\int_a^b y'(x)^2\,\mathrm{d}x,
\end{aligned} \tag{5.5.33}$$

from which (5.5.31) follows. For $p=1$ the inequality is strict since a solution of (5.5.1′) cannot be linear on each of the intervals $[a, \frac{a+b}{2}]$, $[\frac{a+b}{2}, b]$ as this implies a discontinuity of y'. □

5.6 Lyapunov-Type Inequalities I

In 1970, Eliason [100] established a Lyapunov-type inequality for a second-order possibly singular nonlinear differential equation of the form

$$\big(r(t)y'\big)' + p(t)yf(y) = 0, \tag{5.6.1}$$

which is more general than (5.4.1). The conditions assumed in [100] on r, p and f are as follows:

(i) on an interval $[a,b]$ under consideration, r' and p are real and continuous, and $r>0$;

(ii) for $y\neq 0$, $f(y)$ is real, even, positive and continuous; and

(iii) on each interval of the form $(0, M]$ for $M>0$, there exists a $\nu>0$ such that $y^{\nu+1}f(y)$ is strictly increasing in y and has zero limit at 0.

By a solution of (5.6.1) we mean a real continuous function $y(t)$ which satisfies (5.6.1) when $y(t)\neq 0$.

The following lemma given in [100] is needed in order to establish a Lyapunov-type inequality for equation (5.6.1).

LEMMA 5.6.1. *Let $r(t)$, $y(t)$ and $y'(t)$ be differentiable on an interval (a,b) with $r(t) > 0$ and $y(t) > 0$. Suppose that $\lim_{t\to a^+,b^-} y(t) = 0$ and $\lim_{t\to a^+,b^-} (r(t) \times y'(t))' y^\nu(t) = 0$ for some $\nu > 0$. Then $\lim_{t\to a^+,b^-} r(t)y'(t)y^\nu(t) = 0$.*

PROOF. We shall establish the limit at a only. Let $g(t) = r(t)y'(t)$ and assume the conclusion is false. This result being the case since $\lim_{t\to a^+} y(t) = 0$, we may assume there is an $\varepsilon_0 > 0$ and a sequence $t_n \to a^+$ such that

$$\left|g(t_n)y^\nu(t_n)\right| \geqslant \varepsilon_0. \tag{5.6.2}$$

Now, we may assume that, for some $\delta_1 > 0$,

$$\left|g'(u)y^\nu(u)\right| < \frac{\varepsilon_0}{2(b-a)} \tag{5.6.3}$$

for all $u \in (a, a+\delta_1)$.

Choose n_0 such that $n \geqslant n_0$ implies $t_n \in (a, a+\delta_1)$. For these n, we will consider the two possibilities of $g(t_n) > 0$ and $g(t_n) < 0$.

In the first case where $g(t_n) > 0$, we have $y'(t_n) > 0$ and so let $s_n = a + \delta_1$ if $y'(u) > 0$ on $(t_n, a+\delta_1)$ or let s_n be the least zero of $y'(u)$ on $(t_n, a+\delta_1)$, otherwise. Clearly it follows that

$$0 \leqslant g(s_n) \leqslant \left|g(a+\delta_1)\right| \quad \text{for each } n \geqslant n_0. \tag{5.6.4}$$

Also, by the mean value theorem, we have that

$$g(t_n) = g(s_n) + \left[g'(\xi_n)\right](t_n - s_n) \tag{5.6.5}$$

for some $\xi_n \in (t_n, s_n)$. Thus since $y^\nu(t_n) \leqslant y^\nu(\xi_n)$, we have from (5.6.3)–(5.6.5) that

$$\begin{aligned} \left|g(t_n)\right|\left|y^\nu(t_n)\right| &\leqslant \left|g(s_n)\right|\left|y^\nu(t_n)\right| + \left|g'(\xi_n)\right||t_n - s_n|\left|y^\nu(t_n)\right| \\ &\leqslant \left|g(a+\delta_1)\right|\left|y^\nu(t_n)\right| + \left|g'(\xi_n)\right|\left|y^\nu(\xi_n)\right|(b-a) \\ &< \left|g(a+\delta_1)\right|\left|y^\nu(t_n)\right| + \frac{\varepsilon_0}{2}. \end{aligned} \tag{5.6.6}$$

The second possibility is to consider those $n \geqslant n_0$ where $g(t_n) < 0$. Clearly, since $y(t) > 0$ on (a,b) and since $\lim_{t\to a^+} y(t) = 0$, there exists a $v_n \in (a, t_n)$ such that $y'(v_n) = 0$ and $y'(u) < 0$ on (v_n, t_n). Again, by the mean value theorem, we have

$$g(t_n) = \left[g'(\theta_n)\right](t_n - v_n)$$

for some $\theta_n \in (v_n, t_n)$. Thus since $y^\nu(\theta_n) > y^\nu(t_n)$ we have

$$\begin{aligned}\left|g(t_n)y^\nu(t_n)\right| &\leqslant \left|g'(\theta_n)\right|(b-a)\left|y^\nu(\theta_n)\right| \\ &< \frac{\varepsilon_0}{2} + \left|g(a+\delta_1)\right|\left|y^\nu(t_n)\right|. \end{aligned} \tag{5.6.7}$$

Now, since $\lim_{t\to a^+} y(t) = 0$, we can choose an $n_1 \geqslant n_0$ such that $n \geqslant n_1$ implies that $|g(a+\delta_1)||y^\nu(t_n)| < \varepsilon_0/2$. This result together with (5.6.5) and (5.6.6) leads to a contradiction of (5.6.2). Thus the conclusion of the lemma is true. □

The following theorem and corollary are established in [100].

THEOREM 5.6.1. *Let $y(t)$ be a solution of* (5.6.1) *with consecutive zeros at $a < b$. Assume* (i) *and* (ii) *are satisfied. Let $M = \sup\{|y(t)|\colon\ t \in (a,b)\}$, and suppose $\nu \geqslant 1$ satisfies* (iii) *on* $(0, M]$. *Then*

$$16\nu(\nu+1)^{-2} < f(M)\int_a^b r^{-1}\,\mathrm{d}t \int_a^b p^+\,\mathrm{d}t. \tag{5.6.8}$$

PROOF. Assume without loss of generality that $0 < y(t) \leqslant M$ on (a,b) and let $t_0 \in (a,b)$ be such that $y(t_0) = M$.

With $\lambda = (\nu+1)/2$ and $\nu \geqslant 1$ it follows that $y^{\lambda-1}(t)$ is continuous on (a,b). Also $y'(t)$ is continuous on (a,b) so that we may consider improper integrals in the following computations.

First we have

$$\lambda^{-1}M^\lambda = \lambda^{-1}y^\lambda(t_0) = \int_{a^+}^{t_0} y^{\lambda-1}y'\,\mathrm{d}t \leqslant \int_{a^+}^{t_0} y^{\lambda-1}|y'|\,\mathrm{d}t.$$

This equation together with a similar argument on $[t_0, b]$ yields

$$2\lambda^{-1}M^\lambda \leqslant \int_{a^+}^{b^-} y^{\lambda-1}|y'|\,\mathrm{d}t.$$

The following equalities and inequalities provide the main part of the argument establishing the theorem. The fourth equality is due to Lemma 5.6.1. We compute

$$\begin{aligned}16M^{\nu+1}(\nu+1)^{-2} &= \left(2M^\lambda\lambda^{-1}\right)^2 \\ &\leqslant \left(\int_{a^+}^{b^-} y^{\lambda-1}|y'|\,\mathrm{d}t\right)^2\end{aligned}$$

$$
\begin{aligned}
&= \left(\int_{a+}^{b-} r^{1/2} y^{\lambda-1} |y'| r^{-1/2} \, \mathrm{d}t \right)^2 \\
&\leqslant \int_{a+}^{b-} r y^{2(\lambda-1)} y'^2 \, \mathrm{d}t \int_{a+}^{b-} r^{-1} \, \mathrm{d}t \\
&= \left[\nu^{-1} r y^{\nu} y' \right]_{a+}^{b-} \int_{a+}^{b-} r^{-1} \, \mathrm{d}t \\
&\quad + \int_{a+}^{b-} -\nu^{-1} (r y')' y^{\nu} \, \mathrm{d}t \int_{a+}^{b-} r^{-1} \, \mathrm{d}t \\
&= \int_{a+}^{b-} \nu^{-1} p y^{\nu+1} f(y) \, \mathrm{d}t \int_{a}^{b} r^{-1} \, \mathrm{d}t \\
&< \nu^{-1} M^{\nu+1} f(M) \int_a^b p^+ \, \mathrm{d}t \int_a^b r^{-1} \, \mathrm{d}t.
\end{aligned}
$$

The last strict inequality can be established by using the continuously increasing property of $y^{\nu+1} f(y)$ due to (iii), the continuity of y and y' on (a, b) and the fact that p must be positive on some interval where $y' > 0$.

When $0 < \nu < 1$ holds in (iii) the above theorem can also be established; however, since $16\nu/(\nu+1)^2 < 4$ for $\nu \in (0, 1)$ the following corollary yields a better result. This result can be established by noting the fact that if $\nu_1 > 0$ satisfies (iii) on $(0, M]$ then any $\nu_2 > \nu_1$ also satisfies (iii). □

COROLLARY 5.6.1. *Let $y(t)$ be as in Theorem* 5.6.1 *except assume here that* $0 < \nu < 1$, *then*

$$
4 < f(M) \int_a^b r^{-1} \, \mathrm{d}t \int_a^b p^+ \, \mathrm{d}t. \tag{5.6.9}
$$

In [100, p. 465] it is noted that (5.6.9) is sharp and also shown that $16\nu(\nu+1)^{-2}$ in (5.6.8) cannot be replaced by a constant greater than 4.

In 1974, Eliason [102] established Lyapunov inequalities and bounds on solutions of the nonlinear second-order differential equations of the forms:

$$
\big(r(t) y'(t) \big)' + p(t) f\big(y(t) \big) = 0 \tag{5.6.10}
$$

and

$$
y''(t) + m(t) y'(t) + n(t) f\big(y(t) \big) = 0, \tag{5.6.11}
$$

under the conditions

(H_0) The real-valued functions r, r' and p are continuous on a nontrivial interval J of reals, and $r(t) > 0$ for $t \in J$;

(H_1) $f : \mathbb{R} \to \mathbb{R}$ is continuously differentiable and odd with $f'(y) > 0$ for all real y;

(H_2) The real-valued functions m and n are continuous on a nontrivial interval J of reals.

Multiplying (5.6.11) by

$$r(t) \equiv \exp\left[\int_\alpha^t m(s)\,\mathrm{d}s\right] \quad \text{for } t \in J, \tag{5.6.12}$$

where $\alpha \in J$ is fixed, we obtain (5.6.10) and the relation

$$p(t) = r(t)n(t). \tag{5.6.13}$$

In the special case when $f(y) \equiv y$, Fink and Mary [119] (see also [205]) established that if $a < b$ in J are consecutive zeros of a nontrivial solution of (5.6.11), then

$$(b-a)\int_a^b n^+ - 4\exp\left[-\left(\frac{1}{2}\right)\int_a^b |m|\right] > 0. \tag{5.6.14}$$

Below, the bounds are expressed in terms of a maximum value of the solution and integral functionals involving the coefficients.

For reals $d < e$ we let

$$\begin{aligned} R(d,e;p) &= \sup_{d\leqslant x\leqslant e}\int_x^e p, & L(d,e;p) &= \sup_{d\leqslant x\leqslant e}\int_d^x p, \\ S(d,e;p) &= \sup_{d\leqslant u\leqslant v\leqslant e}\int_u^v p, & I(d,e;p) &= \inf_{d\leqslant u\leqslant v\leqslant e}\int_u^v p. \end{aligned} \tag{5.6.15}$$

Clearly, we have

$$-\int_d^e p^- \leqslant F(d,e;p) \leqslant \int_d^e p^+ \tag{5.6.16}$$

holding for F denoting R, L, S or I. Also for fixed e, R and S decrease monotonically as d increases. Other obvious monotonicity properties of L, S and I will be used without explicitly stating them here. By studying relationships (5.6.16) more closely one may also see how the inequalities become strict in certain cases when p is not of constant sign on $[d,e]$.

Two inequalities improving (5.6.14) are

$$(b-a)\int_a^b n^+ - 4\exp\left\{\left(\frac{1}{2}\right)\left[I(a,b;m) - S(a,b;m)\right]\right\} > 0; \tag{5.6.17}$$

and, when $m \equiv 0$ and the solution y is positive on (a, b) and, for some $c \in (a, b)$, satisfies

$$(c-t)y'(t) \geqslant 0 \quad \text{for } t \in [a,b], \tag{5.6.18}$$

$$(b-a)S(a,b;n) > 4. \tag{5.6.19}$$

By (5.6.16), the improvement of (5.6.17) follows from

$$I(a,b;m) - S(a,b;m) \geqslant \int_a^b -\left(m^- + m^+\right) = -\int_a^b |m|. \tag{5.6.20}$$

Strict inequality holds here, for example, when $a = 0$, $b = 4\pi$ and $m(t) = \sin kt$ where k is a positive integer. In fact, we here have the rather interesting phenomena that $-\int_a^b |m|$ remains constants while $I(a,b;m) - S(a,b;m) \to 0$ as $k \to \infty$.

Bounds on solutions and related inequalities. We first consider a solution y of (5.6.10) where $y'(c) = 0$ for some $c \in J$. By integrating twice and applying an integration by parts, for $x \in J$, we have

$$y(c) - y(x) = \int_x^c \left[r(x)\right]^{-1}\left\{\left(\int_t^c p(\tau)\,\mathrm{d}\tau\right) f\left(y(t)\right) + \int_t^c \left(\int_s^c p(\tau)\,\mathrm{d}\tau\right) f'\left(y(s)\right) y'(s)\,\mathrm{d}s\right\}\mathrm{d}t. \tag{5.6.21}$$

By the oddness of f, if $y(c) \neq 0$, we may assume $y(c) > 0$; and throughout, between consecutive zeros we will assume a solution is positive. Thus if $x < c$ and if y is positive and monotone increasing on $(x, c]$ we may conclude from (5.6.21) that

$$\begin{aligned} y(c) - y(x) &\leqslant \int_x^c \left[r(t)\right]^{-1} R(t,c;p)\left[f\left(y(t)\right) + f\left(y(c)\right) - f\left(y(t)\right)\right]\mathrm{d}t \\ &= f\left(y(c)\right)\int_x^c \left[r(t)\right]^{-1} R(t,c;p)\,\mathrm{d}t \\ &\leqslant f\left(y(c)\right) R(x,c;p)\int_x^c \left(\frac{1}{r}\right). \end{aligned} \tag{5.6.22}$$

Furthermore, by (H_1) and (5.6.21), if $y(x) < y(c)$, then y' and p must both be positive on some subinterval of $[x, c]$. As a result it may be argued that the inequalities in (5.6.22) are strict in this case.

By a similar argument, if $x > c$ and if y is a positive and monotone decreasing on $[c, x]$, then

$$\begin{aligned} y(c) - y(x) &\leqslant f\big(y(c)\big) \int_c^x \big[r(t)\big]^{-1} L(c, t; p)\, \mathrm{d}t \\ &\leqslant f\big(y(c)\big) L(c, x; p) \int_c^x \left(\frac{1}{r}\right), \end{aligned} \tag{5.6.23}$$

where the same conclusions on strictness apply here if $y(x) < y(c)$.

The inequalities in (5.6.22) and (5.6.23) clearly yield lower bounds on the solution y. They will next be used to place implicit lower bounds on the distance from c to the first possible zero of y lying to the left or right of c.

Suppose, then, that $a < b$ in J are two consecutive zeros of a solution y and suppose $c \in (a, b)$ satisfies (5.6.18), where, as is understood, y is positive on (a, b). With $f_1(y) = f(y)/y$ for $y \neq 0$, (5.6.22) and (5.6.23), respectively, yield

$$\begin{aligned} 1 &< f_1\big(y(c)\big) \int_a^c \big[r(t)\big]^{-1} R(t, c; p)\, \mathrm{d}t \\ &< f_1\big(y(c)\big) R(a, c; p) \int_a^c \left(\frac{1}{r}\right) \end{aligned} \tag{5.6.24}$$

and

$$\begin{aligned} 1 &< f_1\big(y(c)\big) \int_c^b \big[r(t)\big]^{-1} L(c, t; p)\, \mathrm{d}t \\ &< f_1\big(y(c)\big) L(c, b; p) \int_c^b \left(\frac{1}{r}\right). \end{aligned} \tag{5.6.25}$$

The inequalities provided by the extremes of (5.6.24) and (5.6.25) improve those of Mary [205, Theorem 7] when

$$R(a, c; p) < \int_a^c p^+ \quad \text{or} \quad L(c, b; p) < \int_c^b p^+, \tag{5.6.26}$$

respectively, and of course, (5.6.18) hold.

We now consider a problem of "distance between zeros". By using different variables of integration and then multiplying, from (5.6.24) and (5.6.25), we obtain the Lyapunov inequalities

$$\begin{aligned}1 &< f_1^2(y(c)) \int_a^c \int_c^b [r(u)r(v)]^{-1} R(u,c;p)L(c,v;p)\,dv\,du \\ &\leqslant f_1^2(y(c))4^{-1} \int_a^c \int_c^b [r(u)r(v)]^{-1} [S(u,v;p)]^2\,dv\,du \\ &< f_1^2(y(c))4^{-2}[S(a,b;p)]^2 \left(\int_a^b \left(\frac{1}{r}\right)\right)^2. \end{aligned} \tag{5.6.27}$$

The second inequality above follows from $\alpha\beta \leqslant 4^{-1}(\alpha+\beta)^2$ and

$$0 \leqslant R(u,c;p) + L(c,v;p) \leqslant S(u,v;p).$$

The third inequality follows from monotonicity properties of S and

$$\int_a^c \left(\frac{1}{r}\right) \int_c^b \left(\frac{1}{r}\right) \leqslant 4^{-1}\left(\int_a^b \left(\frac{1}{r}\right)\right)^2.$$

Inequality (5.6.19) is now a special case of (5.6.27) by simply taking square roots in (5.6.27) where, of course, $f_1(v) \equiv 1$.

In order to obtain (5.6.17), we consider $a < b$ to be two consecutive zeros of a solution y of (5.6.11) where y is positive on (a,b). Then, for some $a < c_1 \leqslant c_2 < b$, we have $y'(c_1) = y'(c_2) = 0$ and y is monotone on $(a, c_1]$ and on $[c_2, b)$.

Using (5.6.12) and (5.6.13) the first inequality of (5.6.24) yields

$$\begin{aligned}1 &< f_1(y(c_1)) \int_a^{c_1} \exp\left[-\int_a^t m(w)\,dw\right] \\ &\qquad \times \max_{t\leqslant s\leqslant c_1} \int_s^{c_1} \exp\left[\int_\alpha^u m(w)\,dw\right] n(u)\,du\,dt \\ &= f_1(y(c_1)) \int_a^{c_1} \max_{t\leqslant s\leqslant c_1} \int_s^{c_1} \exp\left[\int_t^u m(w)\,dw\right] n(u)\,du\,dt \\ &\leqslant f_1(y(c_1)) \int_a^{c_1} \exp[L(t,c_1;m)] \int_t^{c_1} n^+(u)\,du\,dt. \end{aligned} \tag{5.6.28}$$

In the linear case, the inequality provided by the extremes of (5.6.28) improves inequality given in [119].

By (5.6.25), we also obtain

$$1 < f_1\big(y(c_2)\big) \int_{c_2}^{b} \exp\big[R(c_2, t; m)\big] \int_{c_2}^{t} n^+(u)\,\mathrm{d}u\,\mathrm{d}t. \tag{5.6.29}$$

Thus with

$$Q = \max\big\{f_1\big(y(c_1)\big), f_1\big(y(c_2)\big)\big\}, \tag{5.6.30}$$

by (5.6.28) and (5.6.29), using different variables of integration and multiplying we have

$$\begin{aligned}1 &< Q^2 \int_a^{c_1}\int_{c_2}^{b} \big\{\exp\big[L(u, c_1; m) + R(c_2, v; m)\big]\big\} \\ &\qquad \times \left(\int_u^{c_1}\int_{c_2}^{v} n^+(x)n^+(z)\,\mathrm{d}z\,\mathrm{d}x\right)\mathrm{d}v\,\mathrm{d}u \\ &\leqslant 4^{-1}Q^2 \int_a^{c_1}\int_{c_2}^{b}\left\{\exp\left[\int_u^v m - I(u, v; m)\right]\right\}\left(\int_u^v n^+\right)^2 \mathrm{d}v\,\mathrm{d}u \\ &< 4^{-2}Q^2\big\{\exp\big[S(a, b; m) - I(a, b; m)\big]\big\}\left(\int_a^b n^+\right)^2 (b-a)^2. \end{aligned} \tag{5.6.31}$$

The inequalities follow from the definitions and properties of L, R, S and I, along with modifications of the argument used to establish (5.6.27).

In the linear case where $Q = 1$, by taking square roots of (5.6.31), we obtain (5.6.17).

We now summarize the above results.

THEOREM 5.6.2. *Let y be a solution of* (5.6.10) *satisfying $y'(c) = 0$ and $y(c) > 0$ for some $c \in J$. Then, for $x < c$ ($x > c$), as long as y is positive and monotone increasing on $(x, c]$ (monotone decreasing on $[c, x)$), the inequalities in* (5.6.22) ((5.6.23)) *provide lower bounds on $y(x)$ which are expressed in terms of $y(c)$ and integral functionals as defined by* (5.6.15) *involving the coefficients r and p of* (5.6.10). *They are strict if $y(x) < y(c)$.*

As a result, inequalities (5.6.24) ((5.6.25)), provide implicit lower bounds on the distance from c to the first possible zero a (b) of y lying to the left (right) of c. They improve previous results when (5.6.26) and (5.6.18) hold.

Inequalities (5.6.24) and (5.6.25), in turn, yield Lyapunov inequalities concerning the distance between consecutive zeros $a < b$ of a solution y of (5.6.10) or of (5.6.11), which is positive on (a, b). The first inequalities, provided by (5.6.27), relate to (5.6.10) and assume condition (5.6.18). The second ones provided by (5.6.31), relate to (5.6.11) and does not assume condition (5.6.18), and they improve previous results when inequality (5.6.31) is strict.

For various other results on Lyapunov-type inequalities for certain second-order functional differential equations and equations having delayed arguments, we refer to Eliason [103,104].

5.7 Lyapunov-Type Inequalities II

This section deals with some Lyapunov-type inequalities established by Pachpatte in [282,298,322,328]. In what follows, it is assumed that the solutions to the equations under consideration exist on $I \subset \mathbb{R}$ ($\mathbb{R}$ the set of reals) containing the points a, b $(a < b)$.

In [322] the following Lyapunov-type inequalities are established for the nonlinear second-order differential equations of the forms

$$\left(r(t)\left|y(t)\right|^{p}\left|y'(t)\right|^{p-2}y'(t)\right)' + q(t)\left|y(t)\right|^{2p-2}y(t) = 0, \tag{A}$$

$$\left(r(t)\left|y(t)\right|^{p-2}\left|y'(t)\right|^{p-2}y'(t)\right)' + q(t)\left|y(t)\right|^{2p-4}y(t) = 0, \tag{B}$$

where $t \in I$, $p \geqslant 2$ is a real constant, the function $r : I \to \mathbb{R}$ is C^1-smooth and $r > 0$, the function $q : I \to \mathbb{R}$ is continuous.

THEOREM 5.7.1. *Let $y(t)$ be a solution of* (A) *with $y(a) = y(b) = 0$ and $y(t) \neq 0$ for $t \in (a, b)$. Let $|y(t)|$ be maximized in a point $c \in (a, b)$. Then*

$$1 \leqslant \left(\int_a^b r^{-1/(p-1)}(s)\,\mathrm{d}s\right)^{p-1}\left(\int_a^b \left|q(s)\right|\mathrm{d}s\right), \tag{5.7.1}$$

$$1 \leqslant 2^p\left(\int_a^c r^{-1/(p-1)}(s)\,\mathrm{d}s\right)^{p-1}\left(\int_a^c \left|q(s)\right|\mathrm{d}s\right), \tag{5.7.2}$$

$$1 \leqslant 2^p\left(\int_c^b r^{-1/(p-1)}(s)\,\mathrm{d}s\right)^{p-1}\left(\int_c^b \left|q(s)\right|\mathrm{d}s\right). \tag{5.7.3}$$

PROOF. Let $M = \max|y(t)| = |y(c)|$, $c \in (a, b)$. By assumption, M is a positive constant. Since $y(a) = y(b) = 0$, we have

$$M^2 = \left|y(c)\right|^2 = 2\left|\int_a^c y(s)y'(s)\,\mathrm{d}s\right| \leqslant 2\int_a^c \left|y(s)\right|\left|y'(s)\right|\mathrm{d}s, \tag{5.7.4}$$

$$M^2 = \left|y(c)\right|^2 = 2\left|-\int_c^b y(s)y'(s)\,\mathrm{d}s\right| \leqslant 2\int_c^b \left|y(s)\right|\left|y'(s)\right|\mathrm{d}s, \tag{5.7.5}$$

implying

$$M^2 \leqslant \int_a^b |y(s)||y'(s)|\,ds = \int_a^b r^{-1/p}(s)r^{1/p}(s)|y(s)||y'(s)|\,ds. \tag{5.7.6}$$

By taking pth power on both sides of (5.7.6), applying Hölder's inequality with indices p, $p/(p-1)$, integrating by parts and using the fact that $y(t)$ is a solution of (A) such that $y(a) = y(b) = 0$, we have

$$\begin{aligned}
M^{2p} &\leqslant \left(\int_a^b r^{-1/(p-1)}(s)\,ds\right)^{p-1}\left(\int_a^b r(s)|y(s)|^p|y'(s)|^p\,ds\right)\\
&= \left(\int_a^b r^{-1/(p-1)}(s)\,ds\right)^{p-1}\left(\int_a^b \left(r(s)|y(s)|^p|y'(s)|^{p-2}y'(s)\right)y'(s)\,ds\right)\\
&= \left(\int_a^b r^{-1/(p-1)}(s)\,ds\right)^{p-1}\left(-\int_a^b \left(r(s)|y(s)|^p|y'(s)|^{p-2}y'(s)\right)'y(s)\,ds\right)\\
&= \left(\int_a^b r^{-1/(p-1)}(s)\,ds\right)^{p-1}\left(\int_a^b \left(q(s)|y(s)|^{2p-2}y(s)\right)y(s)\,ds\right)\\
&\leqslant \left(\int_a^b r^{-1/(p-1)}(s)\,ds\right)^{p-1}\left(\int_a^b \left(|q(s)||y(s)|^{2p}\right)ds\right)\\
&\leqslant \left(\int_a^b r^{-1/(p-1)}(s)\,ds\right)^{p-1}\left(M^{2p}\int_a^b |q(s)|\,ds\right).
\end{aligned} \tag{5.7.7}$$

Now, dividing both sides of (5.7.7) by M^{2p}, we get (5.7.1).

Inequalities in (5.7.2) and (5.7.3) follow in a similar fashion, except that now we take pth power on both sides of (5.7.4) and (5.7.5) and applying Hölder's inequality with indices p, $p/(p-1)$, integrating by parts and using the fact that $y(t)$ is a solution of (A) such that $y(a) = y(b) = 0$ and $y'(c) = 0$. The proof is complete. □

THEOREM 5.7.2. *Let $y(t)$ be a solution of* (B) *with $y(a) = y(b) = 0$ and $y(t) \neq 0$ for $t \in (a,b)$. Let $|y(t)|$ be maximized in a point $c \in (a,b)$. Then*

$$1 \leqslant \frac{1}{3}\left(\int_a^b r^{-1/(p-1)}(s)\,ds\right)^{p-1}\left(\int_a^b |q(s)|\,ds\right), \tag{5.7.8}$$

$$1 \leqslant \frac{1}{3}2^p\left(\int_a^c r^{-1/(p-1)}(s)\,ds\right)^{p-1}\left(\int_a^c |q(s)|\,ds\right), \tag{5.7.9}$$

$$1 \leqslant \frac{1}{3}2^p\left(\int_c^b r^{-1/(p-1)}(s)\,\mathrm{d}s\right)^{p-1}\left(\int_c^b |q(s)|\,\mathrm{d}s\right). \tag{5.7.10}$$

The proof can be completed by following the proof of Theorem 5.7.1 with suitable modifications.

In [328] Pachpatte has established Lyapunov-type inequalities for differential equations of the forms

$$\left(r(t)|y'|^{\alpha-1}y'\right)' + p(t)y' + q(t)y + f(t, y) = 0, \tag{C}$$

$$\left(r(t)|y'|^{\beta}|y'|^{\gamma-2}y'\right)' + p(t)y' + q(t)y + f(t, y) = 0, \tag{D}$$

where $t \in I$, $\alpha \geqslant 1$, $\beta \geqslant 0$, $\gamma \geqslant 2$ are real constants and $\gamma > \beta$, the functions $r, p, q : I \to \mathbb{R}$ are continuous, r and p are continuously differentiable and $r(t) > 0$, the function $f : I \times \mathbb{R} \to \mathbb{R}$ is continuous and satisfies the condition $|f(t, y)| \leqslant w(t, |y|)$, where $w : I \times \mathbb{R}_+ \to \mathbb{R}_+$ ($\mathbb{R}_+$ the set of nonnegative reals) is continuous and $w(t, u) \leqslant w(t, v)$ for $0 \leqslant u \leqslant v$.

THEOREM 5.7.3. *Let $y(t)$ be a solution of equation* (C) *with $y(a) = y(b) = 0$ and $y(t) \neq 0$ for $t \in (a, b)$. Let $|y(t)|$ be maximized at a point $c \in (a, b)$. Then*

$$\begin{aligned}1 \leqslant{}& \frac{1}{2^{\alpha+1}}\left(\int_a^b r^{-1/\alpha}(s)\,\mathrm{d}s\right)^{\alpha}\\ &\times\left(\frac{1}{M^{\alpha-1}}\int_a^b \left|q(s) - \frac{p'(s)}{2}\right|\mathrm{d}s + \frac{1}{M^{\alpha}}\int_a^b w(s, M)\,\mathrm{d}s\right),\end{aligned} \tag{5.7.11}$$

where $M = \max\{|y(t)|\colon a \leqslant t \leqslant b\}$.

PROOF. From the hypotheses, we have

$$M = |y(c)| = \left|\int_a^c y'(s)\,\mathrm{d}s\right| = \left|-\int_c^b y'(s)\,\mathrm{d}s\right|. \tag{5.7.12}$$

From (5.7.12) we observe that

$$2M \leqslant \int_a^b |y'(s)|\,\mathrm{d}s = \int_a^b r^{-1/(\alpha+1)}(s)r^{1/(\alpha+1)}(s)|y'(s)|\,\mathrm{d}s. \tag{5.7.13}$$

Now, raising both sides of (5.7.13) into $(\alpha + 1)$th power, using the Hölder inequality on the right-hand side of the resulting inequality with indices $(\alpha + 1)/\alpha$,

$\alpha + 1$, performing integration by parts and using the fact that $y(t)$ is a solution of equation (C) such that $y(a) = y(b) = 0$, we observe that

$$
\begin{aligned}
(2M)^{\alpha+1} &\leqslant \left(\int_a^b r^{-1/\alpha}(s)\,\mathrm{d}s\right)^{\alpha}\left(\int_a^b r(s)\left|y'(s)\right|^{\alpha+1}\mathrm{d}s\right) \\
&= \left(\int_a^b r^{-1/\alpha}(s)\,\mathrm{d}s\right)^{\alpha}\left(\int_a^b \left(r(s)\left|y'(s)\right|^{\alpha-1}y'(s)\right)y'(s)\,\mathrm{d}s\right) \\
&= \left(\int_a^b r^{-1/\alpha}(s)\,\mathrm{d}s\right)^{\alpha}\left(-\int_a^b \left(r(s)\left|y'(s)\right|^{\alpha-1}y'(s)\right)'y(s)\,\mathrm{d}s\right) \\
&= \left(\int_a^b r^{-1/\alpha}(s)\,\mathrm{d}s\right)^{\alpha} \\
&\quad \times \left(\int_a^b y(s)\left[p(s)y'(s) + q(s)y(s) + f\left(s, y(s)\right)\right]\mathrm{d}s\right) \\
&= \left(\int_a^b r^{-1/\alpha}(s)\,\mathrm{d}s\right)^{\alpha} \\
&\quad \times \left(\int_a^b \left(q(s) - \frac{p'(s)}{2}\right)y^2(s)\,\mathrm{d}s + \int_a^b y(s)f\left(s, y(s)\right)\mathrm{d}s\right) \\
&\leqslant \left(\int_a^b r^{-1/\alpha}(s)\,\mathrm{d}s\right)^{\alpha} \\
&\quad \times \left(\int_a^b \left|q(s) - \frac{p'(s)}{2}\right|\left|y(s)\right|^2\mathrm{d}s + \int_a^b \left|y(s)\right|\left|f\left(s, y(s)\right)\right|\mathrm{d}s\right) \\
&\leqslant \left(\int_a^b r^{-1/\alpha}(s)\,\mathrm{d}s\right)^{\alpha} \\
&\quad \times \left(\int_a^b M^2\left|q(s) - \frac{p'(s)}{2}\right|\mathrm{d}s + \int_a^b Mw(s, M)\,\mathrm{d}s\right). \qquad (5.7.14)
\end{aligned}
$$

Now, dividing both sides of (5.7.14) by $(2M)^{\alpha+1}$, we get the desired inequality in (5.7.11). The proof is complete. □

THEOREM 5.7.4. *Let $y(t)$ be a solution of equation* (D) *with* $y(a) = y(b) = 0$, *and* $y(t) \neq 0$ *for* $t \in (a,b)$. *Let* $|y(t)|$ *be maximized in a point* $c \in (a,b)$. *Then*

$$1 \leqslant \left(\int_a^b r^{-1/(\gamma-1)}(s)\,\mathrm{d}s\right)^{\gamma-1} \times \left(\frac{1}{M^{\beta+\gamma-2}}\int_a^b \left|q(s)-\frac{p'(s)}{2}\right|\mathrm{d}s + \frac{1}{M^{\beta+\gamma-1}}\int_a^b w(s,M)\,\mathrm{d}s\right), \tag{5.7.15}$$

where $M = \max|y(t)| = |y(c)|$, $c \in (a,b)$.

PROOF. From the hypotheses, we have

$$M^2 = 2\int_a^c y(s)y'(s)\,\mathrm{d}s = -2\int_a^c y(s)y'(s)\,\mathrm{d}s. \tag{5.7.16}$$

From (5.7.16) we observe that

$$\begin{aligned} M^2 &\leqslant \int_a^b |y(s)||y'(s)|\,\mathrm{d}s \\ &= \int_a^b \left(r^{-1/\gamma}(s)|y(s)|^{1-\beta/\gamma}\right)\left(r^{1/\gamma}(s)|y(s)|^{\beta/\gamma}|y'(s)|\right)\mathrm{d}s. \end{aligned} \tag{5.7.17}$$

The rest of the proof can be completed by taking the power γ to both sides of (5.7.17), using Hölder's inequality with indices $\gamma/(\gamma-1)$, γ, performing integration by parts, using the fact that $y(t)$ is a solution of equation (D) such that $y(a) = y(b) = 0$, and closely looking at the proof of Theorem 5.7.3. □

In [298] Pachpatte has derived Lyapunov-type inequalities for the differential equations of the forms

$$\left(r(t)h\left(y'(t)\right)\right)' + p(t)y(t)f\left(t, y(t)\right) = 0, \tag{E}$$

$$\left(r(t)h\left(y(t)\right)y'(t)\right)' + p(t)y(t)f\left(t, y(t)\right) = 0, \tag{F}$$

where the following conditions are assumed to hold:

(i) $r, p : I \to \mathbb{R}$ are continuous and r is positive and continuously differentiable on I;

(ii) $h \in C^1(\mathbb{R}, (0,\infty))$, $h(-x) = -h(x)$, $\operatorname{sgn} h(x) = \operatorname{sgn} x$, $x/h(x) \leqslant \beta$, where $\beta > 0$ is a constant and $\lim_{x\to 0} x/h(x)$ exists finitely;

(iii) $f : I \times \mathbb{R} \to \mathbb{R}$ is a continuous function such that $|f(t,y)| \leqslant w(t,|y|)$, where $w : I \times \mathbb{R}_+ \to \mathbb{R}_+$ is continuous and $w(t,u) \leqslant w(t,v)$ for $0 \leqslant u \leqslant v$.

THEOREM 5.7.5. *Assume that the hypotheses* (i)–(iii) *hold. Let* $y(t)$ *be a solution of* (E) *with* $y(a)=y(b)=0$, *and* $y(t)\neq 0$, $t\in(a,b)$. *Let* c *be a point in* (a,b) *at which* $|y(t)|$ *is maximized. Then*

$$4\leqslant \beta\left(\int_a^b \frac{1}{r(s)}\,\mathrm{d}s\right)\left(\int_a^b \left|p(s)\right|w(s,M)\,\mathrm{d}s\right), \tag{5.7.18}$$

$$1\leqslant \beta\left(\int_a^c \frac{1}{r(s)}\,\mathrm{d}s\right)\left(\int_a^c \left|p(s)\right|w(s,M)\,\mathrm{d}s\right), \tag{5.7.19}$$

$$1\leqslant \beta\left(\int_c^b \frac{1}{r(s)}\,\mathrm{d}s\right)\left(\int_c^b \left|p(s)\right|w(s,M)\,\mathrm{d}s\right), \tag{5.7.20}$$

where $M=\max|y(t)|=|y(c)|$, $c\in(a,b)$.

PROOF. By hypotheses, we have the equalities

$$M=\left|y(c)\right|=\left|\int_a^c y'(s)\,\mathrm{d}s\right|, \tag{5.7.21}$$

$$M=\left|y(c)\right|=\left|-\int_c^b y'(s)\,\mathrm{d}s\right|, \tag{5.7.22}$$

which imply

$$2M\leqslant \int_a^b \left|y'(s)\right|\mathrm{d}s. \tag{5.7.23}$$

Squaring both sides of (5.7.23) and using Schwarz inequality, the integration by parts and the fact that $y(t)$ is a solution of (E) with $y(a)=y(b)=0$, by hypotheses (i)–(iii), we have

$$\begin{aligned}
4M^2 &\leqslant \bigg(\int_a^b \left[r^{-1/2}(s)\left|h\left(y'(s)\right)\right|^{-1/2}\left|y'(s)\right|^{1/2}\right]\\
&\qquad \times\left[r^{1/2}(s)\left|h\left(y'(s)\right)\right|^{1/2}\left|y'(s)\right|^{1/2}\right]\mathrm{d}s\bigg)^2\\
&\leqslant \left(\int_a^b \frac{1}{r(s)}\frac{y'(s)}{h(y'(s))}\,\mathrm{d}s\right)\left(\int_a^b r(s)h\left(y'(s)\right)y'(s)\,\mathrm{d}s\right)\\
&\leqslant \beta\left(\int_a^b \frac{1}{r(s)}\,\mathrm{d}s\right)\left(\int_a^b r(s)h\left(y'(s)\right)y'(s)\,\mathrm{d}s\right)\\
&= \beta\left(\int_a^b \frac{1}{r(s)}\,\mathrm{d}s\right)\left(-\int_a^b \left(r(s)h\left(y'(s)\right)\right)'y(s)\,\mathrm{d}s\right)
\end{aligned}$$

$$= \beta\left(\int_a^b \frac{1}{r(s)}\,\mathrm{d}s\right)\left(\int_a^b p(s)y^2(s)f\big(s, y(s)\big)\,\mathrm{d}s\right)$$

$$\leqslant \beta\left(\int_a^b \frac{1}{r(s)}\,\mathrm{d}s\right)\left(M^2\int_a^b \big|p(s)\big|w(s, M)\,\mathrm{d}s\right). \tag{5.7.24}$$

Now, dividing both sides of (5.7.24) by M^2, we get (5.7.18).

Inequalities (5.7.19), (5.7.20) follow in similar fashion, but using, moreover, the condition $y'(c) = 0$. □

THEOREM 5.7.6. *Assume that the hypotheses* (i)–(iii) *hold. Let $y(t)$ be a solution of* (F), *with $y(a) = y(b) = 0$, and $y(t) \neq 0$, $t \in (a, b)$. Let c be a point in (a, b) at which $|y(t)|$ is maximized. Then*

$$2 \leqslant \beta\left(\int_a^b \frac{1}{r(s)}\,\mathrm{d}s\right)\left(\frac{1}{M}\int_a^b \big|p(s)\big|w(s, M)\,\mathrm{d}s\right), \tag{5.7.25}$$

$$\frac{1}{2} \leqslant \beta\left(\int_a^c \frac{1}{r(s)}\,\mathrm{d}s\right)\left(\frac{1}{M}\int_a^c \big|p(s)\big|w(s, M)\,\mathrm{d}s\right), \tag{5.7.26}$$

$$\frac{1}{2} \leqslant \beta\left(\int_c^b \frac{1}{r(s)}\,\mathrm{d}s\right)\left(\frac{1}{M}\int_c^b \big|p(s)\big|w(s, M)\,\mathrm{d}s\right), \tag{5.7.27}$$

where $M = \max|y(t)| = |y(c)|$, $c \in (a, b)$.

PROOF. By hypotheses, we have the equalities

$$M^2 = y^2(c) = 2\int_a^c y(s)y'(s)\,\mathrm{d}s, \tag{5.7.28}$$

$$M^2 = y^2(c) = -2\int_c^b y(s)y'(s)\,\mathrm{d}s, \tag{5.7.29}$$

which imply

$$M^2 \leqslant \int_a^b \big|y(s)\big|\big|y'(s)\big|\,\mathrm{d}s. \tag{5.7.30}$$

Squaring both sides of (5.7.30) and rewriting we have

$$M^4 \leqslant \left(\int_a^b \left[r^{-1/2}(s)\big|h\big(y(s)\big)\big|^{-1/2}\big|y(s)\big|^{1/2}\right]\right.$$
$$\left.\times\left[r^{1/2}(s)\big|h\big(y(s)\big)\big|^{1/2}\big|y(s)\big|^{1/2}\big|y'(s)\big|\right]\mathrm{d}s\right)^2.$$

The rest of the proof can be completed by closely looking at the proof of Theorem 5.7.5 with suitable modifications. □

The following result given in [282] deals with a Lyapunov-type inequality for the second-order linear finite difference equation

$$\Delta\big(r(n)\Delta y(n)\big) + p(n)y(n) = 0 \tag{G}$$

for $n \in I_\infty = \{a, a+1, a+2, \dots\}$, a is an integer, the operator Δ is defined by $\Delta y(n) = y(n+1) - y(n)$, $n \in I_\infty$, $y(n)$, $r(n)$, $c(n)$, $n \in I_\infty$, are real-valued functions and $r(n) > 0$. Let $I \subset I_\infty$ be defined by $I = \{a, a+1, a+2, \dots, a+m = b\}$, $m \geqslant 2$, we denote by I^0 the interior of I. Clearly, I^0 is nonempty.

THEOREM 5.7.7. *Let $y(n)$ be a solution of equation* (G) *such that $y(a) = y(b) = 0$, $y(n) \neq 0$ for $n \in I^0$. Let k be a point in I^0 where $|y(n)|$ is maximized. Then*

$$4 \leqslant \left(\sum_{n=a}^{b-1} \frac{1}{r(n)}\right)\left(\sum_{n=a}^{b-1} \big|p(n)\big|\right). \tag{5.7.31}$$

PROOF. Let $M = |y(k)|$, $k \in I^0$. It is obvious that

$$y(k) = \sum_{n=a}^{k-1} \Delta y(n), \tag{5.7.32}$$

$$y(k) = -\sum_{n-k}^{b-1} \Delta y(n). \tag{5.7.33}$$

From (5.7.32) and (5.7.33), we observe that

$$2M \leqslant \sum_{n=a}^{b-1} \big|\Delta y(n)\big|. \tag{5.7.34}$$

Now, squaring both sides of (5.7.34), using the Schwarz inequality, the following formula of summation by parts

$$\sum_{s=0}^{n-1} u(s)\Delta v(s) = \big(u(n)v(n) - u(0)v(0)\big) - \sum_{s=0}^{n-1} v(s+1)\Delta u(s), \tag{5.7.35}$$

and the fact that $y(n)$ is a solution of (G) with $y(a) = y(b) = 0$, we observe that

$$\begin{aligned}
4M^2 &\leqslant \left(\sum_{n=a}^{b-1} r^{-1/2}(n) r^{1/2}(n) \left|\Delta y(n)\right|\right)^2 \\
&\leqslant \left(\sum_{n=a}^{b-1} \frac{1}{r(n)}\right)\left(\sum_{n=a}^{b-1} \big(r(n)\Delta y(n)\big)\Delta y(n)\right) \\
&= \left(\sum_{n=a}^{b-1} \frac{1}{r(n)}\right)\left(-\sum_{n=a}^{b-1} y(n+1)\Delta\big(r(n)\Delta y(n)\big)\right) \\
&= \left(\sum_{n=a}^{b-1} \frac{1}{r(n)}\right)\left(\sum_{n=a}^{b-1} y(n+1) p(n) y(n)\right) \\
&\leqslant \left(\sum_{n=a}^{b-1} \frac{1}{r(n)}\right) M^2 \left(\sum_{n=a}^{b-1} \left|p(n)\right|\right).
\end{aligned} \tag{5.7.36}$$

Dividing both sides of (5.7.36) by M^2 we get the desired inequality in (5.7.32). The proof is complete. □

5.8 Lyapunov-Type Inequalities III

In this section we present Lyapunov-type inequalities for certain higher-order differential equations established by Hochstadt [150], Chen [57], Chen and Yeh [58] and Pachpatte [321]. We shall consider only those solutions of the equations considered here which exist on $I \subset \mathbb{R}$ ($\mathbb{R}$ the set of reals) containing the points a, b $(a < b)$ and are nontrivial.

We begin with the following Lyapunov-type inequality established by Hochstadt in [150] (see also [225]).

THEOREM 5.8.1. *Consider the differential equation*

$$y^{(n)} - p y^{(n-1)} - q y = 0, \quad n \geqslant 2, \tag{A_1}$$

where p and q are integrable on $[a, b]$. Suppose that a nontrivial solution of (A_1) *has at least n zeros on $[a, b]$. Then*

$$\left[(b-a)^{n-1} \int_a^b |q| \,\mathrm{d}t\right]^{1/n} + \frac{1}{n} \int_a^b |p| \,\mathrm{d}t \geqslant 2. \tag{5.8.1}$$

PROOF. In order to prove (5.8.1), we reduce (A_1) to a system by letting

$$x_i = y^{(i-1)}, \quad i = 1, 2, \ldots, n,$$

so that

$$\begin{aligned} x_i' &= x_{i+1}, \quad i = 1, 2, \ldots, n-1, \\ x_n' &= p x_n + q x_1. \end{aligned} \tag{5.8.2}$$

Since y vanishes n times on $[a, b]$, each x_i vanishes at least once on that interval. We can, therefore, split $[a, b]$ into two subintervals $[a, c]$ and $[c, b]$, where $a < c < b$, such that on each of them, each x_i vanishes at least once.

First, we shall consider the interval $[a, c]$, and let $\bar{x}_i$ denote the maximum of $|x_i|$ on that interval. Using (5.8.2) and the fact that each x_i vanishes at some point on $[a, c]$ we have

$$\bar{x}_i \leqslant \bar{x}_{i+1}(c-a), \quad i = 1, 2, \ldots, n-1, \tag{5.8.3}$$

$$|x_n| \leqslant \bar{x}_1 \int_a^c |q| \,\mathrm{d}t + \int_t^c |p||x_n| \,\mathrm{d}t, \tag{5.8.4}$$

where $x_n(c) = 0$. From (5.8.3) we see that

$$\bar{x}_1 \leqslant \bar{x}_n (c-a)^{n-1},$$

and combined with (5.8.4) we finally have

$$|x_n| \leqslant \bar{x}_n (c-a)^{n-1} \int_a^c |q| \,\mathrm{d}t + \int_t^c |p||x_n| \,\mathrm{d}t. \tag{5.8.5}$$

From (5.8.5), by means of Gronwall's inequality [145, p. 24], we find that

$$\bar{x}_n \leqslant \bar{x}_n (c-a)^{n-1} \int_a^c |q| \,\mathrm{d}t \exp\left(\int_a^c |p| \,\mathrm{d}t\right), \tag{5.8.6}$$

and finally,

$$\int_a^c |q| \,\mathrm{d}t \geqslant \frac{\exp(-\int_a^c |p| \,\mathrm{d}t)}{(c-a)^{n-1}}. \tag{5.8.7}$$

Similarly,

$$\int_c^b |q| \,\mathrm{d}t \geqslant \frac{\exp(-\int_c^b |p| \,\mathrm{d}t)}{(b-c)^{n-1}}. \tag{5.8.8}$$

Combine (5.8.7) and (5.8.8) and use the inequality

$$\frac{A^n}{a^{n-1}} + \frac{B^n}{b^{n-1}} \geqslant \frac{(A+B)^n}{(a+b)^{n-1}},$$

to obtain

$$(b-a)^{n-1}\int_a^b |q|\,dt \geqslant \left[\exp\left(-\frac{1}{n}\int_a^c |p|\,dt\right) + \exp\left(-\frac{1}{n}\int_c^b |p|\,dt\right)\right]^n. \tag{5.8.9}$$

In order to derive the required inequality (5.8.1), we use the fact that

$$\exp(-x) \geqslant 1 - x$$

in (5.8.9) and extract the nth root of both sides. Then

$$\begin{aligned}\left[(b-a)^{n-1}\int_a^b |q|\,dt\right]^{1/n} &\geqslant 2 - \frac{1}{n}\left(\int_a^c |p|\,dt + \int_c^b |p|\,dt\right)\\ &= 2 - \frac{1}{n}\int_a^b |p|\,dt,\end{aligned}$$

which is equivalent to (5.8.1). □

In [58] Chen and Yeh have given the Lyapunov-type inequality for the differential equation of the form

$$L_n x(t) + \sum_{i=1}^m p_i(t)x(t)f_i\big(x(t)\big) = q(t), \tag{A$_2$}$$

where the operators L_j are recursively defined by

$$L_0 x = x, \qquad L_j x = \frac{1}{r_j(t)}\frac{d}{dt}L_{j-1}x, \quad j = 1, 2, \ldots, n, r_n(t) = 1,$$

and

(i) $r_j(t) \in C(\mathbb{R}_+, \mathbb{R}_+ \setminus \{0\})$, $j = 1, 2, \ldots, n$;
(ii) $p_i(t), q(t) \in C(\mathbb{R}_+, \mathbb{R})$, $i = 1, 2, \ldots, m$, $p_i^+(t) \not\equiv 0$;
(iii) $f_i(y) \in C(\mathbb{R}, \mathbb{R})$, for $y > 0$, $f_i(y) = f_i(-y) > 0$, $i = 1, 2, \ldots, m$.

The main result established in [58] is given in the following theorem.

THEOREM 5.8.2. *Let $\alpha_1 > \alpha_2 > \cdots > \alpha_{n-1}$ be respectively the zeros of*

$$L_1x(t), \quad L_2x(t), \quad \ldots, \quad L_{n-1}x(t),$$

where $x(t)$ is a nontrivial solution of (A_2). *Suppose that $b < \alpha_{n-1}$ and $a > \alpha_1$ are zeros of $x(t)$. If*

$$\begin{aligned} M &= \max\left|x(t)\right| = \left|x(t_0)\right|, \quad t, t_0 \in (b, a), \\ K_i &= \max_{y\in[-M,M]} f_i(y), \quad i = 1, 2, \ldots, m. \end{aligned} \tag{5.8.10}$$

Then

$$1 < \int_b^{t_0} r_1(s_1) \int_{\alpha_1}^{s_1} r_2(s_2) \cdots \int_{\alpha_{n-1}}^{s_{n-1}} \left\{ \sum_{i=1}^m p_i^+(s) K_i + \frac{|q(s)|}{M} \right\} \mathrm{d}s\, \mathrm{d}s_{n-1} \cdots \mathrm{d}s_1, \tag{5.8.11}$$

$$1 < \int_{t_0}^{a} r_1(s_1) \int_{\alpha_1}^{s_1} r_2(s_2) \cdots \int_{\alpha_{n-1}}^{s_{n-1}} \left\{ \sum_{i=1}^m p_i^+(s) K_i + \frac{|q(s)|}{M} \right\} \mathrm{d}s\, \mathrm{d}s_{n-1} \cdots \mathrm{d}s_1, \tag{5.8.12}$$

$$2 < \int_b^{a} r_1(s_1) \int_{\alpha_1}^{s_1} r_2(s_2) \cdots \int_{\alpha_{n-1}}^{s_{n-1}} \left\{ \sum_{i=1}^m p_i^+(s) K_i + \frac{|q(s)|}{M} \right\} \mathrm{d}s\, \mathrm{d}s_{n-1} \cdots \mathrm{d}s_1. \tag{5.8.13}$$

PROOF. On repeated integration from equation (A_2), we get

$$\begin{aligned} \frac{x'(t)}{r_1(t)} &= L_1x(t) - L_1x(\alpha_1) \\ &= \int_{\alpha_1}^{t} r_2(s_2) \int_{\alpha_2}^{s_2} r_3(s_3) \cdots \\ &\quad \cdots \int_{\alpha_{n-1}}^{s_{n-1}} \left\{ \sum_{i=1}^m \left[p_i^-(s) - p_i^+(s)\right] x(s) f_i\big(x(s)\big) + q(s) \right\} \mathrm{d}s\, \mathrm{d}s_{n-1} \cdots \mathrm{d}s_1. \end{aligned}$$

Integrating it from t_0 to t we obtain

$$\begin{aligned} x(t) - x(t_0) &= \int_{t_0}^{t} r_1(s_1) \int_{\alpha_1}^{s_1} r_2(s_2) \cdots \\ &\cdots \int_{\alpha_{n-1}}^{s_{n-1}} \left\{ \sum_{i=1}^{m} \left[p_i^-(s) - p_i^+(s)\right] x(s) f_i\big(x(s)\big) + q(s) \right\} \mathrm{d}s \, \mathrm{d}s_{n-1} \cdots \mathrm{d}s_1. \end{aligned} \tag{5.8.14}$$

Let $t = a$ so that $x(a) = 0$. Hence equation (5.8.14) becomes

$$\begin{aligned} x(t_0) &+ \int_{t_0}^{a} r_1(s_1) \int_{\alpha_1}^{s_1} r_2(s_2) \cdots \\ &\cdots \int_{\alpha_{n-1}}^{s_{n-1}} \sum_{i=1}^{m} p_i^-(s) x(s) f_i\big(x(s)\big) \, \mathrm{d}s \, \mathrm{d}s_{n-1} \cdots \mathrm{d}s_1 \\ &= \int_{t_0}^{a} r_1(s_1) \int_{\alpha_1}^{s_1} r_2(s_2) \cdots \\ &\cdots \int_{\alpha_{n-1}}^{s_{n-1}} \left\{ \sum_{i=1}^{m} p_i^+(s) x(s) f_i\big(x(s)\big) - q(s) \right\} \mathrm{d}s \, \mathrm{d}s_{n-1} \cdots \mathrm{d}s_1. \end{aligned}$$

Without loss of generality, we may assume that $x(t) \geqslant 0$, $t \in [b, a]$. Thus, it follows from condition (iii) that

$$\begin{aligned} x(t_0) &\leqslant \int_{t_0}^{a} r_1(s_1) \int_{\alpha_1}^{s_1} r_2(s_2) \cdots \\ &\cdots \int_{\alpha_{n-1}}^{s_{n-1}} \left\{ \sum_{i=1}^{m} p_i^+(s) x(s) f_i\big(x(s)\big) - q(s) \right\} \mathrm{d}s \, \mathrm{d}s_{n-1} \cdots \mathrm{d}s_1 \end{aligned}$$

which by (5.8.10) implies

$$\begin{aligned} 1 &< \int_{t_0}^{a} r_1(s_1) \int_{\alpha_1}^{s_1} r_2(s_2) \cdots \\ &\cdots \int_{\alpha_{n-1}}^{s_{n-1}} \left\{ \sum_{i=1}^{m} p_i^+(s) K_i + \frac{|q(s)|}{M} \right\} \mathrm{d}s \, \mathrm{d}s_{n-1} \cdots \mathrm{d}s_1. \end{aligned}$$

This result proves (5.8.12). Similarly we can prove (5.8.11) except that in (5.8.14) we now replace t by b. The sum of (5.8.11) and (5.8.12) yields (5.8.13). □

In [57] Chen has given the Lyapunov-type inequality for the differential-difference equation of the form

$$\left(r(t)h\left(y'(t)\right)\right)^{n-1}+a(t)y(t)f\left(y\left(t-\sigma(t)\right)\right)=b(t), \tag{A$_3$}$$

where

(i) $a, b \in C(\mathbb{R}_+, \mathbb{R})$, $\mathbb{R}_+ = [0, \infty) \subset \mathbb{R}$ ($\mathbb{R}$ the set of reals);
(ii) $r \in C^{n-1}(\mathbb{R}_+, \mathbb{R})$ and $r > 0$;
(iii) $\sigma \in C(\mathbb{R}_+, (0, \infty))$ and $\sigma(t) \leqslant m$, where $m > 0$ is a constant;
(iv) $h \in C^1(\mathbb{R}, (0, \infty))$, $h(-x) = -h(x)$, $\operatorname{sgn} h(x) = \operatorname{sgn} x$, $x/h(x) \leqslant \beta$, where $\beta > 0$ is a constant and $\lim_{x\to 0} x/h(x)$ exists finitely;
(v) $f(x)$ is a continuous, even, real positive function on $\mathbb{R}$ and increasing on $\mathbb{R}_+$, with $f(0) = 0$.

The following Lyapunov-type inequality is established in [57].

THEOREM 5.8.3. *Assume that* $\alpha_1 > \alpha_2 > \cdots > \alpha_{n-3} > \alpha_{n-2}$ *are respectively zeros of*

$$\left(r(t)h\left(y'(t)\right)\right)', \quad \left(r(t)h\left(y'(t)\right)\right)'', \quad \ldots,$$
$$\left(r(t)h\left(y'(t)\right)\right)^{n-3}, \quad \left(r(t)h\left(y'(t)\right)\right)^{n-2},$$

where $y(t)$ *is a nontrivial solution of equation* (A$_3$). *Furthermore, suppose that* $t_1 < \alpha_{n-2}$ *and* $t_2 > \alpha_1$ *are zeros of* $y(t)$. *Let*

$$L = \sup\left\{y(t) \colon t \in (t_1 - m, t_2), t_1, t_2 > m\right\}$$

and

$$M = \sup\left\{\left|y(t)\right| \colon t \in [t_1, t_2]\right\}.$$

Then

$$4 \leqslant \beta \int_{t_1}^{t_2} \frac{\mathrm{d}t}{r(t)} \left\{ f(L) \int_{t_1}^{t_2} \frac{(t-t_1)^{n-2}}{(n-2)!} \left|a(t)\right| \mathrm{d}t + \frac{1}{M} \int_{t_1}^{t_2} \frac{(t-t_1)^{n-2}}{(n-2)!} \left|b(t)\right| \mathrm{d}t \right\}. \tag{5.8.15}$$

PROOF. Integration of (A_3) $n-2$ times gives

$$\begin{aligned}
&(-1)^n\big(r(t)h\big(y'(t)\big)\big)' \\
&\quad+\int_t^{\alpha_1}\int_{s_2}^{\alpha_2}\cdots\int_{s_{n-2}}^{\alpha_{n-2}} a(s)y(s)f\big(y\big(s-\sigma(s)\big)\big)\,\mathrm{d}s\,\mathrm{d}s_{n-2}\cdots\mathrm{d}s_2 \\
&=\int_t^{\alpha_1}\int_{s_2}^{\alpha_2}\cdots\int_{s_{n-2}}^{\alpha_{n-2}} b(s)\,\mathrm{d}s\,\mathrm{d}s_{n-2}\cdots\mathrm{d}s_2. \qquad (5.8.16)
\end{aligned}$$

Since $\alpha_1>\alpha_2>\cdots>\alpha_{n-3}>\alpha_{n-2}$, we obtain from (5.8.16),

$$\begin{aligned}
&\big|\big(r(t)h\big(y'(t)\big)\big)'\big| \\
&\quad\leqslant\int_t^{\alpha_1}\int_{s_2}^{\alpha_1}\cdots\int_{s_{n-2}}^{\alpha_1}\big|a(s)\big|\big|y(s)\big|\big|f\big(y\big(s-\sigma(s)\big)\big)\big|\,\mathrm{d}s\,\mathrm{d}s_{n-2}\cdots\mathrm{d}s_2 \\
&\qquad+\int_t^{\alpha_1}\int_{s_2}^{\alpha_1}\cdots\int_{s_{n-2}}^{\alpha_1}\big|b(s)\big|\,\mathrm{d}s\,\mathrm{d}s_{n-2}\cdots\mathrm{d}s_2,
\end{aligned}$$

which implies

$$\begin{aligned}
\big|\big(r(t)h\big(y'(t)\big)\big)'\big| &\leqslant \int_t^{\alpha_1}\frac{(s-t)^{n-3}}{(n-3)!}\big|a(s)\big|\big|y(s)\big|\big|f\big(y\big(s-\sigma(s)\big)\big)\big|\,\mathrm{d}s \\
&\quad+\int_t^{\alpha_1}\frac{(s-t)^{n-3}}{(n-3)!}\big|b(s)\big|\,\mathrm{d}s. \qquad (5.8.17)
\end{aligned}$$

Let $M=|y(t_0)|$, $t_0\in[t_1,t_2]$. Now,

$$M=\big|y(t_0)\big|=\left|\int_{t_1}^{t_0} y'(t)\,\mathrm{d}t\right|, \qquad (5.8.18)$$

$$M=\big|y(t_0)\big|=\left|-\int_{t_0}^{t_2} y'(t)\,\mathrm{d}t\right|, \qquad (5.8.19)$$

which implies

$$\begin{aligned}
2M &\leqslant \int_{t_1}^{t_2}\big|y'(t)\big|\,\mathrm{d}t \\
&=\int_{t_1}^{t_2}\frac{1}{(r(t))^{1/2}}\frac{|y'(t)|^{1/2}}{(h(y'(t)))^{1/2}}\big(r(t)\big)^{1/2}\big(h\big(y'(s)\big)\big)^{1/2}\big|y'(t)\big|^{1/2}\,\mathrm{d}t.
\end{aligned}$$

(5.8.20)

The rest of the proof can be completed by squaring both sides of (5.8.20), using the Schwarz inequality, integration by parts, the fact that $y(t)$ is a solution of (A_3) with $y(t_1) = y(t_2) = 0$, formula (5.8.17) and by closely looking at the proof of Theorem 5.7.5. □

In [321] Pachpatte has derived Lyapunov-type inequalities for the equations of the forms:

$$D^n\big[r(t)D^{n-1}\big[p(t)g\big(y'(t)\big)\big]\big] + y(t)f\big(t, y(t)\big) = Q(t), \tag{B$_1$}$$

$$D^n\big[r(t)D^{n-1}\big[p(t)h\big(y(t)\big)y'(t)\big]\big] + y(t)f\big(t, y(t)\big) = Q(t), \tag{B$_2$}$$

$$D^n\big[r(t)D^{n-1}\big[p(t)h\big(y(t)\big)g\big(y'(t)\big)\big]\big] + y(t)f\big(t, y(t)\big) = Q(t), \tag{B$_3$}$$

where $n \geqslant 2$ is an integer and $D^n = \frac{\mathrm{d}^n}{\mathrm{d}t^n}$. The conditions assumed on the functions involved in (B_1)–(B_3) are as follows.

(H_1) $r : I \to \mathbb{R}$ is C^n-smooth and $r > 0$; $p : I \to \mathbb{R}$ is C^{2n-1}-smooth and $p > 0$ and $Q : I \to \mathbb{R}$ is continuous;

(H_2) $g \in C^1(\mathbb{R}, (0, \infty))$, $g(-x) = -g(x)$, $\operatorname{sgn} g(x) = \operatorname{sgn} x$, $x/g(x) \leqslant \alpha$, $\alpha > 0$ is a constant and $\lim_{x\to 0} x/g(x)$ exists finitely;

(H_3) $h \in C^1(\mathbb{R}, (0, \infty))$, $h(-x) = -h(x)$, $\operatorname{sgn} h(x) = \operatorname{sgn} x$, $x/h(x) \leqslant \beta$, $\beta > 0$ is a constant and $\lim_{x\to 0} x/h(x)$ exists finitely;

(H_4) $f : I \times \mathbb{R} \to \mathbb{R}$ is a continuous function such that $|f(t, y)| \leqslant w(t, |y|)$, where $w : I \times \mathbb{R}_+ \to \mathbb{R}_+$ is continuous and $w(t, u) \leqslant w(t, v)$ for $0 \leqslant u \leqslant v$.

For simplification of details of presentation, we set

$$\begin{aligned} &E\big(t, m, z(s_{2n})\big) \\ &\quad = \int_t^{\alpha_2}\int_{s_3}^{\alpha_3}\cdots\int_{s_{n-1}}^{\alpha_{n-1}} \frac{1}{m(s_n)}\int_{s_n}^{\alpha_n}\int_{s_{n+1}}^{\alpha_{n+1}}\cdots \\ &\qquad \cdots\int_{s_{2n-1}}^{\alpha_{2n-1}} z(s_{2n})\,\mathrm{d}s_{2n}\,\mathrm{d}s_{2n-1}\cdots\mathrm{d}s_{n+1}\,\mathrm{d}s_n\,\mathrm{d}s_{n-1}\cdots\mathrm{d}s_3, \end{aligned} \tag{B$_4$}$$

where $n \geqslant 2$, $t \in (a, b)$, and $m(t) > 0$, $z(t) \geqslant 0$ are real-valued continuous functions defined on (a, b) and $\alpha_2, \alpha_3, \ldots, \alpha_{n-1}, \alpha_n, \alpha_{n+1}, \ldots, \alpha_{2n-1}$ are suitable points in (a, b). We denote by $\overline{E}(t, m, z(s_{2n}))$ the integral on the right-hand side of (B_4) with the upper limits $\alpha_2, \alpha_3, \ldots, \alpha_{n-1}, \alpha_n, \alpha_{n+1}, \ldots, \alpha_{2n-1}$ of the integrals all replaced by the greatest number from α_i, $i = 2, 3, \ldots, n-1, n, n+1, \ldots, 2n-1$.

The main results established in [321] are given in the following theorem.

THEOREM 5.8.4. (i) *Assume that the hypotheses* (H_1), (H_2) *and* (H_4) *hold. Let* $\alpha_2 > \alpha_3 > \cdots > \alpha_{n-1} > \alpha_n > \alpha_{n+1} > \cdots > \alpha_{2n-1}$ *be respectively zeros of* $D[p(t)g(y'(t))]$, $D^2[p(t)g(y'(t))], \ldots, D^{n-2}[p(t)g(y'(t))]$, $r(t)D^{n-1}[p(t)\times g(y'(t))]$, $D[r(t)D^{n-1}[p(t)g(y'(t))]], \ldots, D^{n-1}[r(t)D^{n-1}[p(t)g(y'(t))]]$, *where* $y(t)$ *is a nontrivial solution of* (B_1). *Suppose that* $a < \alpha_{2n-1}$ *and* $b > \alpha_2$ *are zeros of* $y(t)$. *Let* c *be a point in* (a, b) *where* $|y(t)|$ *is maximized. Then*

$$4 \leqslant \alpha\left(\int_a^b \frac{1}{p(s_2)}\,\mathrm{d}s_2\right) \times \left(\int_a^b \left[\overline{E}\big(s_2, r, w(s_{2n}, M)\big) + \frac{1}{M}\overline{E}\big(s_2, r, |Q(s_{2n})|\big)\right]\mathrm{d}s_2\right), \tag{5.8.21}$$

where $M = \max|y(t)| = |y(c)|$, $c \in (a, b)$.

(ii) *Assume that the hypotheses* (H_1), (H_3) *and* (H_4) *hold. Let* $\alpha_2 > \alpha_3 > \cdots > \alpha_{n-1} > \alpha_n > \alpha_{n+1} > \cdots > \alpha_{2n-1}$ *be respectively zeros of* $D[p(t)h(y(t))\times y'(t)]$, $D^2[p(t)h(y(t))y'(t)], \ldots, D^{n-2}[p(t)h(y(t))y'(t)]$, $r(t)D^{n-1}[p(t)\times h(y(t))y'(t)]$, $D[r(t)D^{n-1}[p(t)h(y(t))y'(t)]], \ldots, D^{n-1}[r(t)D^{n-1}[p(t)\times h(y(t))y'(t)]]$, *where* $y(t)$ *is a nontrivial solution of* (B_2). *Suppose that* $a < \alpha_{2n-1}$ *and* $b > \alpha_2$ *are zeros of* $y(t)$. *Let* c *be a point in* (a, b), *where* $|y(t)|$ *is maximized. Then*

$$2 \leqslant \beta\left(\int_a^b \frac{1}{p(s_2)}\,\mathrm{d}s_2\right) \times \left(\int_a^b \left[\frac{1}{M}\overline{E}\big(s_2, r, w(s_{2n}, M)\big) + \frac{1}{M^2}\overline{E}\big(s_2, r, |Q(s_{2n})|\big)\right]\mathrm{d}s_2\right), \tag{5.8.22}$$

where $M = \max|y(t)| = |y(c)|$, $c \in (a, b)$.

(iii) *Assume that the hypotheses* (H_1)–(H_4) *hold. Let* $\alpha_2 > \alpha_3 > \cdots > \alpha_{n-1} > \alpha_n > \alpha_{n+1} > \cdots > \alpha_{2n-1}$ *be respectively zeros of* $D[p(t)h(y(t))g(y'(t))]$, $D^2[p(t)h(y(t))g(y'(t))], \ldots, D^{n-2}[p(t)h(y(t))g(y'(t))]$, $r(t)D^{n-1}[p(t)\times h(y(t))g(y'(t))]$, $D[r(t)D^{n-1}[p(t)h(y(t))g(y'(t))]], \ldots, D^{n-1}[r(t)D^{n-1}[p(t)\times h(y(t))g(y'(t))]]$, *where* $y(t)$ *is a nontrivial solution of* (B_3). *Suppose that* $a < \alpha_{2n-1}$ *and* $b > \alpha_2$ *are zeros of* $y(t)$. *Let* c *be a point in* (a, b), *where* $|y(t)|$ *is maximized. Then*

$$2 \leqslant \alpha\beta\left(\int_a^b \frac{1}{p(s_2)}\,\mathrm{d}s_2\right) \times \left(\int_a^b \left[\frac{1}{M}\overline{E}\big(s_2, r, w(s_{2n}, M)\big) + \frac{1}{M^2}\overline{E}\big(s_2, r, |Q(s_{2n})|\big)\right]\mathrm{d}s_2\right), \tag{5.8.23}$$

where $M = \max|y(t)| = |y(c)|$, $c \in (a, b)$.

PROOF. (i) Integrating $2n - 2$ times equation (B_1), by hypotheses, we get

$$\begin{aligned}(-1)^{2n-2}\big[p(t)g\big(y'(t)\big)\big]' + E\big(t, r, y(s_{2n})f\big(s_{2n}, y(s_{2n})\big)\big) \\ = E\big(t, r, Q(s_{2n})\big).\end{aligned} \tag{5.8.24}$$

From the hypotheses, we have

$$M = |y(c)| = \left|\int_a^c y'(s_2)\,ds_2\right| = \left|-\int_a^b y'(s_2)\,ds_2\right|. \tag{5.8.25}$$

From (5.8.25) we observe that

$$\begin{aligned}2M &\leqslant \int_a^b |y'(s_2)|\,ds_2 \\ &= \int_a^b \Big(p^{-1/2}(s_2)|y'(s_2)|^{1/2}\big|g\big(y'(s_2)\big)\big|^{-1/2}\Big) \\ &\qquad \times \Big(p^{1/2}(s_2)|y'(s_2)|^{1/2}\big|g\big(y'(s_2)\big)\big|^{1/2}\Big)\,ds_2.\end{aligned} \tag{5.8.26}$$

By squaring both sides of (5.8.26), applying Schwarz inequality, integrating by parts, using the facts that $y(a) = y(b) = 0$, the solution $y(t)$ of (B_1) satisfies the equivalent integral equation (5.8.24) and hypotheses (H_1), (H_2) and (H_4), we have

$$\begin{aligned}4M^2 &\leqslant \left(\int_a^b \frac{1}{p(s_2)}\frac{y'(s_2)}{g(y'(s_2))}\,ds_2\right)\left(\int_a^b p(s_2)g\big(y'(s_2)\big)y'(s_2)\,ds_2\right) \\ &\leqslant \alpha\left(\int_a^b \frac{1}{p(s_2)}\,ds_2\right)\left(\int_a^b p(s_2)g\big(y'(s_2)\big)y'(s_2)\,ds_2\right) \\ &= \alpha\left(\int_a^b \frac{1}{p(s_2)}\,ds_2\right)\left(-\int_a^b \big[p(s_2)g\big(y'(s_2)\big)\big]'y(s_2)\,ds_2\right) \\ &\leqslant \alpha\left(\int_a^b \frac{1}{p(s_2)}\,ds_2\right)\left(\int_a^b |y(s_2)|\big|\big[p(s_2)g\big(y'(s_2)\big)\big]'\big|\,ds_2\right) \\ &\leqslant \alpha\left(\int_a^b \frac{1}{p(s_2)}\,ds_2\right)\left(\int_a^b |y(s_2)|\big[\overline{E}\big(s_2, r, |y(s_{2n})|\big|f\big(s_{2n}, y(s_{2n})\big)\big|\big)\right. \\ &\qquad\qquad \left. + \overline{E}\big(s_2, r, |Q(s_{2n})|\big)\big]\,ds_2\right)\end{aligned}$$

$$
\begin{aligned}
&\leqslant \alpha\left(\int_a^b \frac{1}{p(s_2)}\,\mathrm{d}s_2\right) \\
&\quad \times \left(\int_a^b M\left[\overline{E}\big(s_2, r, Mw(s_{2n}, M)\big) + \overline{E}\big(s_2, r, \left|Q(s_{2n})\right|\big)\right]\mathrm{d}s_2\right). \qquad (5.8.27)
\end{aligned}
$$

Now, dividing both sides of (5.8.27) by M^2, we get the required inequality in (5.8.21).

(ii) Integrating $2n-2$ times equation (B_2), by hypotheses, we get

$$
\begin{aligned}
&(-1)^{2n-2}\left[p(t)h\big(y(t)\big)y'(t)\right]' + E\big(t, r, y(s_{2n})f\big(s_{2n}, y(s_{2n})\big)\big) \\
&\quad = E\big(t, r, Q(s_{2n})\big). \qquad (5.8.28)
\end{aligned}
$$

From the hypotheses, we have

$$
M^2 = y^2(c) = 2\int_a^c y(s_2)y'(s_2)\,\mathrm{d}s_2 = -2\int_c^b y(s_2)y'(s_2)\,\mathrm{d}s_2. \qquad (5.8.29)
$$

From (5.8.29) we observe that

$$
\begin{aligned}
M^2 &\leqslant \int_a^b \left|y(s_2)\right|\left|y'(s_2)\right|\mathrm{d}s_2 \\
&= \int_a^b \left(p^{-1/2}(s_2)\left|y(s_2)\right|^{1/2}\left|h\big(y(s_2)\big)\right|^{-1/2}\right) \\
&\qquad \times \left(p^{1/2}(s_2)\left|h\big(y(s_2)\big)\right|^{1/2}\left|y(s_2)\right|^{1/2}\left|y'(s_2)\right|\right)\mathrm{d}s_2. \qquad (5.8.30)
\end{aligned}
$$

The rest of the proof follows by arguments similar to those in the proof of (i) given below inequality (5.8.26) with suitable changes.

(iii) Integrating $2n-2$ times equation (B_3), by hypotheses, we get

$$
\begin{aligned}
&(-1)^{2n-2}\left[p(t)h\big(y(t)\big)g\big(y'(t)\big)\right]' + E\big(t, r, y(s_{2n})f\big(s_{2n}, y(s_{2n})\big)\big) \\
&\quad = E\big(t, r, Q(s_{2n})\big). \qquad (5.8.31)
\end{aligned}
$$

As in the proof of (ii), we observe that

$$
\begin{aligned}
M^2 &\leqslant \int_a^b \left|y(s_2)\right|\left|y'(s_2)\right|\mathrm{d}s_2 \\
&= \int_a^b \left(p^{\;1/2}(s_2)\left|y(s_2)\right|^{1/2}\left|h\big(y(s_2)\big)\right|^{-1/2}\left|y'(s_2)\right|^{1/2}\left|g\big(y'(s_2)\big)\right|^{-1/2}\right) \\
&\qquad \times \left(p^{1/2}(s_2)\left|y(s_2)\right|^{1/2}\left|h\big(y(s_2)\big)\right|^{1/2}\left|y'(s_2)\right|^{1/2}\left|g\big(y'(s_2)\big)\right|^{1/2}\right)\mathrm{d}s_2.
\end{aligned}
$$

Inequality (5.8.23) follows in a fashion similar to that in the proofs of (i), (ii) with suitable modifications. The proof is complete. □

5.9 Miscellaneous Inequalities

5.9.1 Bobisud [34]

Consider the differential inequality

$$u'' + A(t)f(u) \leqslant 0 \tag{5.9.1}$$

and the differential equation

$$v'' + a(t)f(v) = 0, \tag{5.9.2}$$

on an interval $[a, \beta)$. Suppose that the following hypotheses hold.

(H_1) $f \in C^1(0,\infty) \cap C[0,\infty)$, $f(u) > 0$ and $f'(u) > 0$ for $u > 0$, $f(0) = 0$, f' nondecreasing for positive arguments;

(H_2) $a, A \in C[\alpha, \beta]$ with $A(t) \geqslant a(t) \geqslant 0$ on $[\alpha, \beta)$. Let u, v satisfy (5.9.1), (5.9.2), respectively, on $[\alpha, \beta)$ and be such that $u(\alpha) > v(\alpha) > 0$, $u > 0$ on $[\alpha, \beta)$, $v'(\alpha) \leqslant 0$, and

$$-\frac{u'(\alpha)}{f(u(\alpha))} > -\frac{v'(\alpha)}{f(v(\alpha))},$$

$$-\frac{u'(\alpha)}{f(v(\alpha))} \geqslant -\frac{v'(\alpha)}{f(v(\alpha))} + \int_\alpha^\beta a(s)\,\mathrm{d}s.$$

Then v does not vanish on $[\alpha, \beta)$.

5.9.2 Bobisud [34]

Consider the differential inequality (5.9.1) and the differential equation (5.9.2). Suppose that the following hypothesis holds

(H_3) $f \in C^1(0,\infty) \cap C[0,\infty)$, $f(u) > 0$ and $f'(u) > 0$ for $u > 0$, $f(0) = 0$, f' nonincreasing for positive arguments.

Let u, v satisfy (5.9.1), (5.9.2), respectively, on $[\alpha, \beta)$, where $A, a \in C[\alpha, \beta]$. Suppose further that $v(\alpha) \geqslant u(\alpha) > 0$, $u > 0$ on $[\alpha, \beta)$, and

$$-\frac{u'(\alpha)}{f(u(\alpha))} + \int_\alpha^t A(s)\,\mathrm{d}s > \left| -\frac{v'(\alpha)}{f(v(\alpha))} + \int_\alpha^t a(s)\,\mathrm{d}s \right|, \quad t \in [\alpha, \beta).$$

Then v does not vanish on $[\alpha, \beta)$ and

$$v(t) \geqslant u(t), \qquad -\frac{u'(t)}{f(u(t))} > \left|\frac{v'(t)}{f(v(t))}\right|, \qquad t \in [\alpha, \beta).$$

5.9.3 Wong [427]

Let f satisfy

(i) $f(x, y)$ is continuous for $x \geqslant 0$ and $y \geqslant 0$,
(ii) $f(x, y) > 0$ for each $x \geqslant 0$ and $y > 0$,
(iii) $f(x, y)$ is an increasing function of y for each $x \geqslant 0$.

Suppose u and v are respectively solutions of

$$u'' = f(x, u)u^{1+2\varepsilon}, \qquad u(0) = A, \ u'(0) = B,$$

and

$$v'' > f(x, v)v^{1+2\varepsilon}, \qquad v(0) = A, \ v'(0) = B,$$

for $0 \leqslant x < T$, where $\varepsilon > 0$ is a constant. Then $v(x) > u(x)$ for $0 < x < T$.

5.9.4 Wong [427]

Let f satisfy

(i) $f(x, y)$ is continuous for $x \geqslant 0$ and $y \geqslant 0$,
(ii) $f(x, y) > 0$ for each $x \geqslant 0$ and $y > 0$,
(iii) $f(x, y)$ is a decreasing function of y for each $x \geqslant 0$.

Let $L(x) = A + Bx$, where $A \geqslant 0$, $B \geqslant 0$ and $A^2 + B^2 > 0$.

Suppose u and v are respectively solutions of

$$u'' = f(x, u)u^{1+2\varepsilon}, \qquad u(0) = A, \ u'(0) = B,$$

and

$$v'' > f\big(x, L(x)\big)v^{1+2\varepsilon}, \qquad v(0) = A, \ v'(0) = B,$$

on $[0, T)$, where $\varepsilon > 0$ is a constant. Then $v(x) > u(x)$ on $(0, T)$.

5.9.5 Hartman [145]

Let $q(t)$ be real-valued and continuous for $0 \leqslant t \leqslant T$. Let $u(t) \not\equiv 0$ be a solution of

$$u'' + q(t)u = 0,$$

and N the number of its zeros on $0 < t \leqslant T$. Then

$$N < \frac{1}{2}\left(T \int_0^T q^+(t)\,\mathrm{d}t\right)^{1/2} + 1.$$

5.9.6 Fink and Mary [119]

Let a and b be successive zeros of a nontrivial solution to

$$y'' + gy' + fy = 0,$$

where f and g are integrable. Then

$$(b-a)\int_a^b f^+(x)\,\mathrm{d}x - 4\exp\left(-\frac{1}{2}\int_a^b \left|g(x)\right|\mathrm{d}x\right) > 0$$

and a fortiori

$$(b-a)\int_a^b f^+(x)\,\mathrm{d}x + 2\int_a^b \left|g(x)\right|\mathrm{d}x > 4.$$

5.9.7 Eliason [101]

Consider the nonlinear second-order differential equation of the form

$$y'' + p(x)y^{2n+1} = 0, \tag{5.9.3}$$

where n is a positive integer and p is a positive and continuous function on a compact interval of reals $[a, b]$ with $a < b$. Along with (5.9.3) consider Rayleigh quotient

$$J(y) = \frac{(\int_a^b y'^2\,\mathrm{d}x)^{n+1}}{(\int_a^b py^{2n+2}\,\mathrm{d}x)}, \tag{5.9.4}$$

where the domain of J is

$$D(J) = \left\{y \in D'[a, b]\colon\ y(a) = 0 \text{ and } y \neq 0 \text{ on } [a, b]\right\},$$

where $D'[a,b]$ is the set of all continuous real-valued functions having sectionally continuous derivatives on $[a,b]$. Let $\lambda_1(p)$ be the least positive value of $J(y)$ in (5.9.4) for $y \in D(J)$. Then

$$(b-a)^{n+1}\lambda_1(p)\int_a^b p\,\mathrm{d}x > 1.$$

Furthermore the inequality is sharp.

5.9.8 Wend [422]

Let $G(I)$ denote the class of all complex-valued, continuous, and nonzero functions $p(x)$ defined on $I\colon x_0 \leqslant x < \infty$ which has the further property that, for any three numbers a, b and c such that $x_0 \leqslant a < b < c < \infty$,

$$\left|\int_a^b \frac{1}{p(x)}\,\mathrm{d}x\right| < \left|\int_a^c \frac{1}{p(x)}\,\mathrm{d}x\right|,$$

$$\left|\int_b^c \frac{1}{p(x)}\,\mathrm{d}x\right| < \left|\int_a^c \frac{1}{p(x)}\,\mathrm{d}x\right|.$$

Suppose $p(x) \in G(I)$, and $a_1 < a_2 < \cdots < a_n$ are consecutive zeros of a solution of $(p(x)y')' + f(x)y = 0$, $a_1 \geqslant x_0$, where $f(x)$ is a complex-valued and continuous function on I. Then a_n must satisfy the inequalities

$$n-1 < \int_{x_0}^{a_n} |f(x)|\,\mathrm{d}x \int_{x_0}^{a_n} \frac{\mathrm{d}x}{|p(x)|},$$

$$n-1 < \int_{x_0}^{a_n} |f(x)|\left(\int_{x_0}^{x} \frac{\mathrm{d}t}{|p(t)|}\right)\mathrm{d}x,$$

$$n-1 < \int_{x_0}^{a_n} |f(x)|\left(\int_{x}^{a_n} \frac{\mathrm{d}t}{|p(t)|}\right)\mathrm{d}x.$$

5.9.9 Wend [422]

Suppose $f(x)$ is complex-valued and continuous on $I\colon x_0 \leqslant x < \infty$, $x_0 \geqslant 0$, and $\int_{x_0}^{\infty} |f(x)|\,\mathrm{d}x = N$. If $a_1 < a_2 < \cdots < a_n$ are n consecutive zeros of a solution of $y''' + f(x)y = 0$, $a_1 \geqslant x_0$, then

$$a_n > \sqrt{\left[(n-1) - \frac{(1+(-1)^n)}{2}\right]\Big/2N}, \quad n \geqslant 3.$$

5.9.10 Pachpatte [297]

Let $y(t)$ be a nontrivial solution of

$$\big(r(t)y'\big)' + p(t)y' + q(t)y + g(t)yf(t, y) = h(t)$$

with consecutive zeros at a and b, $a < b$. Assume that the following conditions hold.

(i) $r, p, q, g, h : I \to \mathbb{R}$ are continuous functions; r, p are continuously differentiable and $r(t) > 0$ (where $I \subset \mathbb{R}$),

(ii) The function $f : I \times \mathbb{R} \to \mathbb{R}$ is continuous and satisfies $|f(t, y)| \leqslant w(t, |y|)$, in which $w : I \times \mathbb{R}_+ \to \mathbb{R}_+$ is continuous and $w(t, u) \leqslant w(t, v)$ for $0 \leqslant u \leqslant v$.

Let $M = \sup\{|y(t)|\colon t \in (a, b)\}$. Then

$$4 \leqslant \left(\int_a^b \frac{1}{r(t)}\,\mathrm{d}t\right)\left(\int_a^b \left[\frac{1}{2}\left|p'(t)\right| + \left|q(t)\right| + \left|g(t)\right|w(t, M) + \frac{1}{M}\left|h(t)\right|\right]\mathrm{d}t\right).$$

5.9.11 Dahiya and Singh [75]

Let $y(t)$ be a solution of

$$\big(r(t)h\big(y'(t)\big)\big)' + p(t)y(t)f\big(y\big(t - \sigma(t)\big)\big) = 0$$

with consecutive zeros at $a < b$. Let $L = \sup\{y(t),\ t \in (a - m, b)\}$, $a, b > m$. Assume that the following conditions hold.

(i) $r \in C^1(\mathbb{R}_+, \mathbb{R})$, $r(t) > 0$; $p \in C(\mathbb{R}_+, \mathbb{R})$;

(ii) $\sigma \in C(\mathbb{R}_+, (0, \infty))$ and $\sigma(t) \leqslant m$, where $m > 0$ is a constant; $f(x)$ is a continuous, even, real positive function on $\mathbb{R}$ and increasing on $\mathbb{R}_+$ with $f(0) = 0$;

(iii) $h \in C^1(\mathbb{R}, (0, \infty))$, $h(-x) = -h(x)$, $\operatorname{sgn} h(x) = \operatorname{sgn} x$, $x/h(x) \leqslant \beta$, where $\beta > 0$ is a constant and $\lim_{x\to 0} x/h(x)$ exists finitely.

Suppose that $y(t)$ is not identically equal to zero on $[a, b]$, then

$$\frac{4}{\beta} \leqslant f(L)\left(\int_a^b \frac{1}{r(t)}\,\mathrm{d}t\right)\left(\int_a^b p^+(t)\,\mathrm{d}t\right),$$

where $p^+(t) = \max\{p(t), 0\}$.

5.9.12 Pachpatte [354]

Consider the differential equation

$$\big(r(t)\big|y'(t)\big|^{\alpha-1}y'(t)\big)' + q(t)\big|y(t)\big|^{\beta-1}y(t) = 0, \tag{5.9.5}$$

where $t \in I$, $\alpha \geqslant 1$, $\beta \geqslant 1$ are constants, the function $r : I \to \mathbb{R}$ is C^1-smooth, $r > 0$, and the function $q : I \to \mathbb{R}$ is continuous. Let $y(t)$ be a solution of (5.9.5) with $y(a) = y(b) = 0$ and $y(t) \neq 0$ for $t \in (a, b)$. Let $|y(t)|$ be maximized in a point $c \in (a, b)$. Then

$$1 \leqslant M^{\beta-\alpha}\left(\int_a^b r^{-1/\alpha}(s)\,\mathrm{d}s\right)^{\alpha}\left(\int_a^b \big|q(s)\big|\,\mathrm{d}s\right),$$

$$1 \leqslant 2^{\alpha+1}M^{\beta-\alpha}\left(\int_a^c r^{-1/\alpha}(s)\,\mathrm{d}s\right)^{\alpha}\left(\int_a^c \big|q(s)\big|\,\mathrm{d}s\right),$$

$$1 \leqslant 2^{\alpha+1}M^{\beta-\alpha}\left(\int_c^b r^{-1/\alpha}(s)\,\mathrm{d}s\right)^{\alpha}\left(\int_c^b \big|q(s)\big|\,\mathrm{d}s\right),$$

where $M = \max|y(t)| = |y(c)|$, $c \in (a, b)$.

5.9.13 Pachpatte [320]

Consider the differential equation

$$\big(r(t)\big|y'(t)\big|^{p-2}y'(t)\big)' + \big|y(t)\big|^{p-2}f\big(t, y(t)\big) = 0, \tag{5.9.6}$$

where $t \in I$ and $p \geqslant 2$ is a constant, and (i) $r : I \to \mathbb{R}$ is a continuously differentiable function and $r(t) > 0$ (where $I \subset \mathbb{R}$) (ii) $f : I \times \mathbb{R} \to \mathbb{R}$ is a continuous function such that $|f(t, y)| \leqslant w(t, |y|)$, where $w : I \times \mathbb{R}_+ \to \mathbb{R}_+$ is a continuous function and $w(t, u) \leqslant w(t, v)$ for $0 \leqslant u \leqslant v$. Let $y(t)$ be a solution of (5.9.6) with $y(a) = y(b) = 0$, and $y(t) \neq 0$, $t \in (a, b)$. Let c be a point in (a, b) where $|y(t)|$ is maximized. Then

$$1 \leqslant \left(\frac{1}{2}\right)^{p}\left(\int_a^b r^{-1/(p-1)}(s)\,\mathrm{d}s\right)^{p-1}\left(\int_a^b \frac{1}{M}w(s, M)\,\mathrm{d}s\right),$$

$$1 \leqslant \left(\int_a^c r^{-1/(p-1)}(s)\,\mathrm{d}s\right)^{p-1}\left(\int_a^c \frac{1}{M}w(s, M)\,\mathrm{d}s\right),$$

$$1 \leqslant \left(\int_c^b r^{-1/(p-1)}(s)\,\mathrm{d}s\right)^{p-1}\left(\int_c^b \frac{1}{M}w(s, M)\,\mathrm{d}s\right),$$

where $M = \max|y(t)| = |y(c)|$, $c \in (a, b)$.

5.9.14 Pachpatte [322]

Consider the following differential equation

$$\big(f(t,y)y'\big)' + g(t,y) = 0, \tag{5.9.7}$$

under the conditions:

(i) $f(t,y) \in C^1(I \times \mathbb{R}, (0,\infty))$, is an odd function with respect to y, $\operatorname{sgn} f(t,y) = \operatorname{sgn} y$, $y/f(t,y) \leqslant r(t)$, where $I \subset \mathbb{R}$, $r \in C(I,(0,\infty))$ and $\lim_{y\to 0} y/f(t,y)$ exists finitely;

(ii) $g \in C(I \times \mathbb{R}, \mathbb{R})$ and $|g(t,y)| \leqslant w(t,|y|)$, where $w \in C(I \times \mathbb{R}_+, \mathbb{R}_+)$ such that $w(t,u) \leqslant w(t,v)$ for $0 \leqslant u \leqslant v$.

Let $y(t)$ be a solution of (5.9.7) with $y(a) = y(b) = 0$, and $y(t) \neq 0$ for $t \in (a,b)$. Let c be a point in (a,b), where $|y(t)|$ is maximized. Then

$$1 \leqslant \frac{1}{2}\left(\int_a^b r(s)\,\mathrm{d}s\right)\left(\frac{1}{M^2}\int_a^b w(s,M)\,\mathrm{d}s\right),$$

where $M = \max|y(t)| = |y(c)|$, $c \in (a,b)$.

5.9.15 Pachpatte [331]

Consider the finite difference equation

$$\Delta\big(r(n)\big|\Delta y(n)\big|^{\alpha-1}\Delta y(n)\big) + c(n)\big|y(n)\big|^{\beta-1}y(n) = 0, \tag{5.9.8}$$

where $n \in I_\infty = \{a, a+1, a+2, \dots\}$, a is an integer, $\alpha \geqslant 1$, $\beta \geqslant 1$ are constants, $\Delta y(n) = y(n+1) - y(n)$ for $n \in I_\infty$, $r(n)$, $c(n)$, $n \subset I_\infty$ are real-valued functions and $r(n) > 0$. Define a subset I of I_∞ by $I = \{a, a+1, a+2, \dots, a+m = b\}$, $m \geqslant 2$, denote by I^0 the nonempty interior of I. Let $y(n)$, $n \in I$, be a solution of equation (5.9.8) such that $y(a) = y(b) = 0$, $y(n) \neq 0$ for $n \in I^0$. If k is a point in I^0 where $|y(n)|$ is maximized, then

$$1 \leqslant M^{\beta-\alpha}\left(\sum_{s=a}^{b-1} r^{-1/\alpha}(s)\right)^{\alpha}\left(\sum_{s=a}^{b-1}\big|c(s)\big|\right),$$

where $M = \max|y(n)| = |y(k)|$, $k \in I^0$.

5.9.16 Levin [188]

Let $q(t)$ be a real-valued continuous function on $[a,b]$ such that relative to the differential equation

$$(-1)^n y^{(2n)}(t) - q(t)y(t) = 0,$$

there exists on $[a,b]$ a pair of conjugate points, then

$$\int_a^b q^+(t)\,dt > 4^{2n-1}(2n-1)\frac{[(n-1)!]^2}{(b-a)^{2n-1}},$$

where $q^+(t) = \max\{q(t), 0\}$.

5.9.17 Chen [57]

Consider the differential equation

$$y^{(n)}(t) + a(t)y(t) = 0, \tag{5.9.9}$$

where $a \in C(\mathbb{R}_+, \mathbb{R})$. Assume that $\alpha_2 > \alpha_3 > \cdots > \alpha_{n-1}$ are zeros of $y''(t)$, $y'''(t), \ldots, y^{(n-1)}(t)$, respectively, where $y(t)$ is a nontrivial solution of equation (5.9.9). Let $t_1 < \alpha_{n-1}$ and $t_2 > \alpha_2$ be zeros of $y(t)$. Then

$$\frac{4}{t_2 - t_1} \leqslant \int_{t_1}^{t_2} \frac{(t-t_1)^{n-2}}{(n-2)!}\left|a(t)\right| dt.$$

5.9.18 Wend [422]

Suppose $f(x)$ is continuous and complex-valued function on I: $x_0 \leqslant x < \infty$, $x_0 \geqslant 0$. If $a_1 < a_2 < \cdots < a_n$ are n consecutive zeros of a solution of equation

$$y^{(k)} + f(x)y = 0,$$

$a_1 \geqslant x_0 \geqslant 0$, $n = kq + r \geqslant k$, then

$$1 < \int_{a_j}^{a_{j+k-1}} \left|g(x,s)\right|\left|f(x)\right| dx, \quad j = 1, 2, \ldots, n-k+1,$$

where $g(x,s)$ is Green's function for the system

$$y^{(k)} = 0, \qquad y(a_j) = y(a_{j+1}) = \cdots = y(a_{j+k-1}) = 0.$$

5.9.19 Pachpatte [319]

Consider the following differential equation

$$\left(\left|y'(t)\right|^{p-2} y'(t)\right)^{n-1} + q(t)\left|y(t)\right|^{p-2} y(t) = 0, \quad p \geqslant 2,\ n \geqslant 3, \tag{5.9.10}$$

where q is real-valued and continuous function on $I = [0, \infty)$. Set

$$E\big(t, h(s)\big) = \int_t^{\alpha_1} \int_{s_2}^{\alpha_2} \cdots \int_{s_{n-2}}^{\alpha_{n-2}} h(s)\,\mathrm{d}s\,\mathrm{d}s_{n-2} \cdots \mathrm{d}s_3\,\mathrm{d}s_2, \tag{5.9.11}$$

where $h(t)$ is a real-valued nonnegative continuous function defined on I and $\alpha_1, \alpha_2, \dots, \alpha_{n-2}$ are suitable points in I. Denote by $\overline{E}(t, h(s))$ the integral on the right-hand side of (5.9.11) when the upper limits $\alpha_1, \alpha_2, \dots, \alpha_{n-2}$ of integrals are all replaced by the greatest number from α_i, $i = 1, 2, \dots, n-2$. Let $\alpha_1 > \alpha_2 > \cdots > \alpha_{n-2}$ be, respectively, zeros of $(|y'(t)|^{p-2} y'(t))'$, $(|y'(t)|^{p-2} y'(t))''$, $\dots$, $(|y'(t)|^{p-2} y'(t))^{(n-2)}$, where $y(t)$ is a nontrivial solution of (5.9.10), let $a < \alpha_{n-2}$ and $b > \alpha_1$ be zeros of $y(t)$, and $|y(t)|$ is maximized in $c \in (a, b)$. Then

$$1 \leqslant 2^{-p}(b-a)^{p-1} \int_a^b \overline{E}\big(s_1, |q(s)|\big)\,\mathrm{d}s_1,$$

$$1 \leqslant (c-a)^{p-1} \int_a^c \overline{E}\big(s_1, |q(s)|\big)\,\mathrm{d}s_1,$$

$$1 \leqslant (b-c)^{p-1} \int_c^b \overline{E}\big(s_1, |q(s)|\big)\,\mathrm{d}s_1.$$

5.9.20 Pachpatte [339]

Consider the differential equation

$$\begin{aligned}&\Big(r_{n-1}(t)\Big(r_{n-2}(t)\Big(\cdots\Big(r_2(t)\Big(r_1(t)\left|y'(t)\right|^{\alpha-1} y'(t)\Big)'\Big)'\cdots\Big)'\Big)'\Big)' \\ &\quad + q(t)\left|y(t)\right|^{\beta-1} y(t) = 0,\end{aligned} \tag{5.9.12}$$

where $n \geqslant 2$, $t \in I = [0, \infty)$, $\alpha \geqslant 1$, $\beta \geqslant 1$ are constants, the functions $r_i : I \to \mathbb{R}$, $i = 1, 2, \dots, n-1$, are sufficiently smooth and $r_i(t) > 0$ and the function $q : I \to \mathbb{R}$ is continuous. Set

$$\begin{aligned}&E\big[t, \bar{r}, h(s)\big] \\ &\quad = E\big[t, r_2, r_3, r_4, \dots, r_{n-1}, h(s)\big]\end{aligned}$$

$$= \frac{1}{r_2(t)} \int_t^{\alpha_1} \frac{1}{r_3(s_2)} \int_{s_2}^{\alpha_2} \frac{1}{r_4(s_3)} \cdots$$

$$\cdots \int_{s_{n-3}}^{\alpha_{n-3}} \frac{1}{r_{n-1}(s_{n-2})} \int_{s_{n-2}}^{\alpha_{n-2}} h(s)\,\mathrm{d}s\,\mathrm{d}s_{n-2} \cdots \mathrm{d}s_3\,\mathrm{d}s_2, \quad (5.9.13)$$

where $t \in I$, $r_i(t)$ are as defined above and $h(t)$ is a real continuous function defined on I. Denote by $\overline{E}[t, \bar{r}, h(s)]$ the integral on the right-hand side of (5.9.13), when the upper limits α_i, all replaced by the greatest of α_i. Let $\alpha_1 > \alpha_2 > \cdots > \alpha_{n-2}$ be respectively zeros of

$$\left(r_1(t)|y'(t)|^{\alpha-1}y'(t)\right)', \quad \left(r_2(t)\left(r_1(t)|y'(t)|^{\alpha-1}y'(t)\right)'\right)', \ldots,$$
$$\left(r_{n-2}(t)\left(\cdots\left(r_2(t)\left(r_1(t)|y'(t)|^{\alpha-1}y'(t)\right)'\right)'\cdots\right)'\right)',$$

where $y(t)$ is a nontrivial solution of (5.9.12), let $a < \alpha_{n-2}$ and $b > \alpha_1$ be zeros of $y(t)$ and $|y(t)|$ is maximized in $c \in (a, b)$. Then

$$1 \leqslant M^{\beta-\alpha}\left(\int_a^b r_1^{-1/\alpha}(s_1)\,\mathrm{d}s_1\right)^{\alpha}\left(\int_a^b \overline{E}\left[s_1, \bar{r}, |q(s)|\right]\mathrm{d}s_1\right),$$

$$1 \leqslant 2^{\alpha+1}M^{\beta-\alpha}\left(\int_a^c r_1^{-1/\alpha}(s_1)\,\mathrm{d}s_1\right)^{\alpha}\left(\int_a^c \overline{E}\left[s_1, \bar{r}, |q(s)|\right]\mathrm{d}s_1\right),$$

$$1 \leqslant 2^{\alpha+1}M^{\beta-\alpha}\left(\int_c^b r_1^{-1/\alpha}(s_1)\,\mathrm{d}s_1\right)^{\alpha}\left(\int_c^b \overline{E}\left[s_1, \bar{r}, |q(s)|\right]\mathrm{d}s_1\right),$$

where $M = \max|y(t)| = |y(c)|$, $c \in (a, b)$.

5.10 Notes

The results given in Theorems 5.2.1 and 5.2.2 are the further extensions of well-known Sturm's theorem, established in 1960 by Levin [187]. Lemmas 5.2.1 and 5.2.2 and Theorems 5.2.3 and 5.2.4 are taken from Kreith [171]. Theorems 5.2.5 and 5.2.6 are due to Ladas [177]. Theorems 5.3.1–5.3.4 are the further generalizations of Levin's comparison theorems and are taken from Lalli and Jahagirdar [180,181]. Theorems 5.3.5–5.3.8 are due to Pachpatte [327] which deals with Levin-type comparison theorems related to certain second-order differential equations.

Theorem 5.4.1 and Corollary 5.4.1 are taken from Hartman [145]. Theorems 5.4.2 and 5.4.3 are taken from Patula [361], see also Cohen [62] for similar results. Theorem 5.4.4 and Corollaries 5.4.2 and 5.4.3 are due to Kwong [176]. Theorem 5.4.5 and Corollaries 5.4.4–5.4.6 are taken from Harris [143]. Lemma 5.5.1,

Theorems 5.5.1–5.5.3 and Corollaries 5.5.1–5.5.3 are taken from Harris and Kong [144], and Lemmas 5.5.2 and 5.5.3, Theorems 5.5.4 and 5.5.5 and Corollaries 5.5.4 and 5.5.5 are taken from Brown and Hinton [46]. Lemma 5.6.1, Theorem 5.6.1, Corollary 5.6.1 and Theorem 5.6.2 are due to Eliason [100,102].

Theorems 5.7.1 and 5.7.2 are taken from Pachpatte [322]. Theorems 5.7.3 and 5.7.4 are due to Pachpatte [328] and Theorems 5.7.5 and 5.7.6 are taken from Pachpatte [298] while Theorem 5.7.7 is taken from Pachpatte [282]. Theorem 5.8.1 is due to Hochstadt [150], Theorem 5.8.2 is due to Chen and Yeh [58], Theorem 5.8.3 is due to Chen [57] and Theorem 5.8.4 is taken from Pachpatte [321]. Section 5.9 contains some miscellaneous inequalities of Levin and Lyapunov type.

References

[1] Abramovich, S., B. Mond and J.E. Pečarić, Sharpening Jensen's inequality and a majorization theorem, J. Math. Anal. Appl. 214 (1997), 721–728.

[2] Adams, R.A., Some integral inequalities with applications to the imbedding of Sobolev spaces defined over irregular domains, Trans. Amer. Math. Soc. 178 (1973), 401–429.

[3] Adams, R.A., Sobolev Spaces, Academic Press, New York, 1985.

[4] Agarwal, R.P. and P.Y.H. Pang, Opial Inequalities with Applications in Differential and Difference Equations, Kluwer Academic Publishers, Dordrecht, 1995.

[5] Agarwal, R.P. and P.Y.H. Pang, Remarks on the generalization of Opial's inequality, J. Math. Anal. Appl. 190 (1995), 559–577.

[6] Agarwal, R.P. and Q. Sheng, Sharp integral inequalities in n independent variables, Nonlinear Anal. 26 (1996), 179–210.

[7] Agarwal, R.P., J. Pečarić and I. Brnetić, Improved integral inequalities in n independent variables, Comput. Math. Appl. 33 (1997), 27–38.

[8] Allegretto, W., Nonoscillation theory of elliptic equations of order $2n$, Pacific J. Math. 64 (1976), 1–16.

[9] Alzer, H., On a problem of Littlewood, J. Math. Anal. Appl. 199 (1996), 403–408.

[10] Alzer, H., Some integral and discrete inequalities in several independent variables, Math. Comput. Modelling 25 (1997), 97–104.

[11] Alzer, H., On Opial-type inequality involving higher-order derivatives of two functions, Appl. Math. Lett. 10 (1997), 123–128.

[12] Anastassiou, G.A., Opial-type inequalities for linear differential operators, Math. Inequal. Appl. 1 (1998), 193–200.

[13] Andersen, K.F., Weighted inequalities for the Stieltjes transformation and Hilbert double series, Proc. Roy. Soc. Edinburgh Sect. A 86 (1980), 75–84.

[14] Andersen, K.F. and H.P. Heinig, Weighted norm inequalities for certain integral operators, SIAM J. Math. Anal. 14 (1983), 834–844.

[15] Barnes, D.C., Some complements of Hölder's inequality, J. Math. Anal. Appl. 26 (1969), 82–87.

[16] Beckenbach, E.F. and R. Bellman, Inequalities, Springer-Verlag, Berlin–New York, 1970.

[17] Beesack, P.R., Hardy's inequality and its extensions, Pacific J. Math. 11 (1961), 39–61.

[18] Beesack, P.R., Integral inequalities involving a function and its derivative, Amer. Math. Monthly 78 (1971), 705–741.

[19] Beesack, P.R., On inequalities complementary to Jensen's, Canad. J. Math. 35 (1983), 324–338.

[20] Beesack, P.R. and J.E. Pečarić, On Jessen's inequality for convex functions, J. Math. Anal. Appl. 110 (1985), 536–552.

[21] Bennett, D.M., An extension of Rellich's inequality, Proc. Amer. Math. Soc. 106 (1989), 987–993.

[22] Bennett, G., An inequality suggested by Littlewood, Proc. Amer. Math. Soc. 100 (1987), 474–476.

[23] Bennett, G., Some elementary inequalities, Quart. J. Math. Oxford 38 (1987), 401–425.

[24] Bennett, G., Some elementary inequalities, Quart. J. Math. Qxford 39 (1988), 385–400.

[25] Benson, D.C., Inequalities involving integrals of functions and their derivatives, J. Math. Anal. Appl. 17 (1967), 292–308.

[26] Bergh, J., A converse inequality of Hölder type, Math. Z. 215 (1994), 205–208.

[27] Bernis, F., Compactness of the support for some nonlinear elliptic problems of arbitrary order in dimension N, Comm. Partial Differential Equations 9 (1984), 271–312.

[28] Berwald, L., Verallgemeinerung eines Mittelwertsatzes von J. Favard für positive konkave Funktionen, Acta Math. 79 (1947), 17–37.

[29] Block, H.D., Discrete analogues of certain integral inequalities, Proc. Amer. Math. Soc. 8 (1957), 852–859.

[30] Bloom, S., First and second order Opial inequalities, Studia Math. 126 (1997), 27–50.

[31] Boas, R.P. Jr., The Jensen–Steffensen inequality, Univ. Beograd Publ. Elektrotehn. Fak. Ser. Mat. Fiz. Nos 302–319 (1970), 1–8.

[32] Boas, R.P. Jr., Some integral inequalities related to Hardy's inequality, J. Anal. Math. 23 (1970), 53–63.

[33] Boas, R.P. Jr. and C.O. Imoru, Elementary convolution inequalities, SIAM J. Math. Anal. 6 (1975), 457–471.

[34] Bobisud, L.E., Comparison and oscillation theorems for nonlinear second-order differential equations and inequalities, J. Math. Anal. Appl. 37 (1970), 5–14.

[35] Bobisud, L.E., Steady-state turblent flow with reaction, Rocky Mountain J. Math. 21 (1991), 993–1007.

[36] Bobisud, L.E., Existence of solutions of some nonlinear diffusion problems, J. Math. Anal. Appl. 168 (1992), 413–424.

[37] Borell, C., Inverse inequalities for concave or generalized concave functions, Research Report No. 44, Dept. of Math., Uppsala Univ., 1972.

[38] Borell, C., Inverse Hölder inequalities in one and several dimensions, J. Math. Anal. Appl. 41 (1973), 300–312.

[39] Borell, C., Integral inequalities for generalized concave or convex functions, J. Math. Anal. Appl. 43 (1973), 419–440.

[40] Bradley, J.S., Hardy inequalities with mixed norms, Canad. Math. Bull. 21 (1978), 405–408.

[41] Brenner, J.L. and H. Alzer, Integral inequalities for concave functions with applications to special functions, Proc. Roy. Soc. Edinburgh Sect. A 118 (1991), 173–192.

[42] Breuer, S. and D. Gottlieb, Hille–Wintner type oscillation criteria for linear ordinary differential equations of second order, Ann. Polon. Math. 30 (1975), 257–262.

[43] Brezis, H. and E.H. Lieb, Sobolev inequalities with reminder terms, J. Funct. Anal. 62 (1985), 73–86.

[44] Brnetić, I. and J. Pečarić, Some new Opial type inequalities, Math. Inequal. Appl. 1 (1998), 385–390.

[45] Brown, B.M. and W.D. Evans, On an extension of Copson's inequality for finite series, Proc. Roy. Soc. Edinburgh Sect. A 121 (1992), 169–183.

[46] Brown, R.C. and D.W Hinton, Opial's inequality and oscillation of 2-nd order equations, Proc. Amer. Math. Soc. 125 (1992), 1123–1129.

[47] Bullen, P.S., Error estimates for some elementary quadrature rules, Univ. Beograd Publ. Elektrotehn. Fak. Ser. Mat. Fiz. Nos 602–633 (1978), 97–103.

[48] Bullen, P.S., D.S. Mitrinović and P.M Vaasić, Means and Their Inequalities, Reidel, Dordrecht–Boston–Lancaster–Tokyo, 1988.

[49] Buşe, C., S.S. Dragomir and D. Barbu, The convergence of some sequences connected to Hadamard's inequality, Demonstratio Math. 29 (1996), 53–59.

[50] Calderon, A.P. and A. Zygmund, Local properties of solutions of elliptic partial differential equations, Studia Math. 20 (1961), 171–225.

[51] Calvert, J., Some generalizations of Opial's inequality, Proc. Amer. Math. Soc. 18 (1967), 72–75.

[52] Chan, L.Y., Some extensions of Hardy's inequality, Canad. Math. Bull. 22 (1979), 165–169.

[53] Chang, S.Y.A., J.M. Wilson and T.H. Wolff, Some weighted norm inequalities concerning the Schrödinger operators, Comment. Math. Helv. 60 (1985), 217–246.

[54] Chanillo, S. and R.L. Wheeden, Inequalities for Peano maximal functions and Marcinkiewicz integrals, Duke Math. J. 50 (1983), 573–603.

[55] Chanillo, S. and R.L. Wheeden, L^p estimates for fractional integrals and Sobolev inequalities with applications to Schrödinger operators, Comm. Partial Differential Equations 10 (1985), 1077–1116.

[56] Chanillo, S. and R.L. Wheeden, Weighted Poincaré and Sobolev inequalities and estimates for weighted Peano maximal functions, Amer. J. Math. 107 (1995), 1191–1226.

[57] Chen, L.S., A Lyapunov inequality and forced oscillations in general nonlinear nth order differential–difference equations, Glasgow Math. J. 18 (1977), 161–166.

[58] Chen, L.S. and C.C. Yeh, Note on distance between zeros of the nth order differential equations, Atti. Accad. Naz. Lincei 61 (1976), 217–221.

[59] Cheung, W.S., On Opial-type inequalities in two variables, Aequationes Math. 38 (1989), 236–244.

[60] Cheung, W.S., Some new Opial-type inequalities, Mathematica 37 (1990), 136–142.

[61] Cheung, W.S., Some generalized Opial-type inequalities, J. Math. Anal. Appl. 162 (1991), 317–321.

[62] Chon, J.H.E., Consecutive zeros of solutions of ordinary second order differential equations, J. London Math. Soc. 5 (1972), 465–468.

[63] Ciesielski, Z., A note on some inequalities of Jensen's type, Ann. Polon. Math. 4 (1958), 269–274.

[64] Clark, S. and D. Hinton, Lyapunov inequality for linear Hamiltonian systems, Math. Inequal. Appl. 1 (1998), 201–209.

[65] Cochran, J.A. and C.S. Lee, Inequalities related to Hardy's and Heinig's, Math. Proc. Cambridge Philos. Soc. 96 (1984), 1–7.

[66] Coddington, E.A., An Introduction to Ordinary Differential Equations, Prentice-Hall, Englewood Cliffs, NJ, 1961.

[67] Coddington, E.A. and N. Levinson, Theory of Ordinary Differential Equations, McGraw-Hill, New York, 1955.

[68] Coifman, R.R. and C.L. Fefferman, Weighted norm inequalities for maximal functions and singular integrals, Studia Math. 51 (1974), 241–250.

[69] Copson, E.T., Note on series of positive terms, J. London Math. Soc. 2 (1927), 9–12 and 3 (1928), 49–51.

[70] Copson, E.T., Some integral inequalities, Proc. Roy. Soc. Edinburgh Sect. A 75 (1975/1976), 157–164.

[71] Copson, E.T., Two series inequalities, Proc. Roy. Soc. Edinburgh Sect. A 83 (1979), 109–114.

[72] Courant, R. and D. Hilbert, Methods of Mathematical Physics, Vol. I, Wiley, New York, 1953.

[73] Crooke, P.S., On two inequalities of Sobolev type, Appl. Anal. 3 (1974), 345–358.

[74] Curtain, R.F., Foundations of functional analysis theory, Control Theory and Topics in Functional Analysis, Vol. I, International Atomic Energy Agency, Vienna, 1976, pp. 101–176.

[75] Dahiya, R.S. and B. Singh, A Lyapunov inequality and nonoscillation theorem for a second order nonlinear differential–difference equation, J. Math. Phys. Sci. 7 (1973), 163–170.

[76] Das, K.M., Comparison and monotonicity theorems for second order nonlinear differential equations, Acta Math. Hungar. 15 (1964), 449–456.

[77] Das, K.M., An inequality similar to Opial's inequality, Proc. Amer. Math. Soc. 22 (1969), 258–261.

[78] Davies, G.S. and G. Peterson, On an inequality of Hardy's (II), Quart. J. Math. Oxford 15 (1964), 165–169.

[79] Delbosco, D., Sur une inégalite de la norme, Univ. Beograd Publ. Elektrotehn. Fak. Ser. Mat. Fiz. Nos 678–715 (1980), 206–208.

[80] Del Pino, M. and R. Manásevich, Oscillation and non-oscillation for $(|u'|^{p-2}u')' + a(t)|u|^{p-2}u = 0$, $p > 1$, Houston J. Math. 14 (1988), 173–177.

[81] Del Pino, M., M. Elgueta and R. Manásevich, A homotopic deformation along p of Leray–Schauder degree result and existence for $(|u'|^{p-2}u')' + f(t,u) = 0$, $u(0) = u(T) = 0$, $p > 1$, J. Differential Equations 80 (1989), 1–13.

[82] Dragomir, S.S., Two refinements of Hadamard's inequalities, Coll. Sci. Pap. Fac. Sci. Kragujevac 11 (1990), 23–26.

[83] Dragomir, S.S., Some refinements of Ky Fan's inequality, J. Math. Anal. Appl. 163 (1992), 317–321.

[84] Dragomir, S.S., Two mappings in connection to Hadamard's inequalities, J. Math. Anal. Appl. 167 (1992), 49–56.

[85] Dragomir, S.S., Some refinements of Jensen's inequality, J. Math. Anal. Appl. 168 (1992), 518–522.

[86] Dragomir, S.S., Some integral inequalities for differentiable convex functions, Contrib. Sec. Math. Tech. Sci. 13 (1992), 13–17.

[87] Dragomir, S.S., Refinements of Hadamard's inequality for isotonic linear functionals, Tamkang J. Math. 24 (1993), 101–106.

[88] Dragomir, S.S. and N.M. Ionescu, On some inequalities for convex-dominated functions, Anal. Numer. Theor. Approx. 19 (1990), 21–27.

[89] Dragomir, S.S. and N.M. Ionescu, Some remarks on convex functions, Anal. Numer. Theor. Approx. 21 (1992), 31–36.

[90] Dragomir, S.S. and D.M. Milošević, A sequence of mappings connected with Jensen's inequality and applications, Mat. Vesnik 44 (1992), 113–121.

[91] Dragomir, S.S. and J. Sándor, Some inequalities for uniformly-convex functions, Mathematica 34 (1992), 133–138.

[92] Dragomir, S.S., Y.J. Cho and S.S. Kim, Inequalities of Hadamard's type for Lipschitzian mappings and their applicaitions, J. Math. Anal. Appl. 245 (2000), 489–501.

[93] Dragomir, S.S., J.E. Pečarić and L. Persson, Some inequalities of Hadamard type, Research Report No. 13, Dept. Appl. Math., Lulea Univ. of Technology, 1994.

[94] Dragomir, S.S., J.E. Pečarić and J. Sándor, A note on the Jensen–Hadamard inequalities, Anal. Numer. Theor. Approx. 19 (1990), 29–34.

[95] Dubinskii, J.A., Some integral inequalities and the solvability of degenerate quasilinear elliptic systems of differential equations, Math. Sb. 64 (1964), 458–480.

[96] Ebihara, Y. and T. Nambu, A remark on the initial value problems for some quasilinear parabolic equations, Math. Rep. Kyushu Univ. 11 (1977), 47–51.

[97] Edelson, A.L. and J.D. Schuur, Nonoscillatory solutions of $(rx^{(n)})^{(n)} \pm f(t,x)x = 0$, Pacific J. Math. 109 (1983), 313–325.

[98] Edmunds, D.E. and B. Opic, Weighted Poincaré and Friedrichs inequalities, J. London Math. Soc. 47 (1993), 79–96.

[99] Edmunds, D.E., B. Opic and J. Rákosník, Poincaré and Friedrichs inequalities in abstract Sobolev spaces, II, Math. Proc. Cambridge Philos. Soc. 115 (1994), 159–173.

[100] Eliason, S.B., A Lyapunov inequality for a certain second order nonlinear differential equation, J. London Math. Soc. 2 (1970), 461–466.

[101] Eliason, S.B., Comparison theorems for second order nonlinear differential equations, Quart. Appl. Math. 35 (1971), 391–402.

[102] Eliason, S.B., Lyapunov inequalities and bounds on solutions of certain second order equations, Canad. Math. Bull. 17 (1974), 499–504.

[103] Eliason, S.B., Lyapunov type inequalities for certain second order functional differential equations, SIAM J. Appl. Math. 27 (1974), 180–199.

[104] Eliason, S.B., Distance between zeros of certain differential equations having delayed arguments, Ann. Mat. Pura Appl. CVI (1975), 273–291.

[105] Elliott, E.B., A simple exposition of some recently proved facts as to convergency, J. London Math. Soc. 1 (1926), 93–96.

[106] Evans, W.D. and W.N. Everitt, A return to the Hardy–Littlewood integral inequality, Proc. Roy. Soc. London Ser. A 380 (1982), 447–486.

[107] Evans, W.D. and W.N. Everitt, On an inequality of Hardy–Littlewood type, I, Proc. Roy. Soc. Edinburgh Sect. A 101 (1985), 131–140.

[108] Fabes, E.B., C.E. Kenig and R.P. Serapioni, The local regularity of solutions of degenerate elliptic equations, Comm. Partial Differential Equations 7 (1982), 77–116.

[109] Fan, K., O. Taussky and J. Todd, Discrete analogues of inequalities of Wirtinger, Monatsh. Math. 59 (1955), 73–90.

[110] Farwig, R. and D. Zwick, Some divided difference inequalities for n-convex functions, J. Math. Anal. Appl. 108 (1985), 430–437.

[111] Favard, J., Sur les valeures moyennes, Bull. Sci. Math. 57 (1933), 54–64.

[112] Fefferman, C.L. and E.M. Stein, H^p spaces of several variables, Acta Math. 129 (1972), 137–193.

[113] Feinberg, J.M., Some Wirtinger-like inequalities, SIAM J. Math. Anal. 10 (1979), 1258–1271.

[114] Fejér, L., Über die Fourierreihen, II, Math.-Naturwiss. Anz. Ungarn Akad. Wiss. 24 (1906), 369–390.

[115] Fink, A.M., On the zeros of $y'' + py = 0$ with linear, convex and concave p, J. Math. Pures Appl. 46 (1967), 1–10.

[116] Fink, A.M., Discrete inequalities of generalized Wirtinger type, Aequationes Math. 11 (1974), 31–39.

[117] Fink, A.M., On Opial's inequality for $f^{(n)}$, Proc. Amer. Math. Soc. 115 (1992), 177–181.

[118] Fink, A.M., Two inequalities, Univ. Beograd Publ. Elektrotehn. Fak. Ser. Mat. Fiz. 6 (1995), 48–49.

[119] Fink, A.M. and D.F. St. Mary, On an inequality of Nehari, Proc. Amer. Math. Soc. 21 (1969), 640–642.

[120] Friedman, A., Partial Differential Equations of Parabolic Type, Prentice-Hall, Englewood Cliffs, NJ, 1964.

[121] Friedman, A., Partial Differential Equations, Holt, Reinhart and Winston, New York, 1969.

[122] Fuchs, L., A new proof of an inequality of Hardy–Littlewood–Pólya, Math. Tidsskr. 13 (1947), 53–54.

[123] Gabler, S., Folgenkonvexe Funktionen, Manuscripta Math. 29 (1979), 29–47.

[124] Galbraith, A.S., On the zeros of solutions of ordinary differential equations of the second order, Proc. Amer. Math. Soc. 17 (1966), 333–337.

[125] Garabedian, P.R., Partial Differential Equations, Wiley, New York, 1964.

[126] Gatto, A.E. and R.L. Wheeden, Sobolev inequalities for products of powers, Trans. Amer. Math. Soc. 314 (1989), 727–743.

[127] Ghidaglia, J.M., M. Marion and R. Temam, Generalization of the Sobolev–Lieb–Thirring inequalities and applications to the dimension of attractors, Differential Integral Equations 1 (1988), 1–21.

[128] Gill, P.M., C.E.M. Pearce and J.E. Pečarić, Hadamard's inequalities for r-convex functions, J. Math. Anal. Appl. 215 (1997), 461–470.

[129] Godunova, E.K. and V.I. Levin, On an inequality of Maroni (in Russian), Mat. Zametki 2 (1967), 221–224.

[130] Godunova, E.K. and V.I. Levin, Neravenstva dlja funkcii shirokogo klassa soderzhashchego vypuklye, monotonnye i nekotorye drugie vidy funkcii, Vychisl. Mat. i Mat. Fiz., Mežvuzov. Sb. Nauč. Trudov, MGPL, Moscow, 1985, pp. 138–142.

[131] Graef, J.R., T. Kushano, H. Onose and P.W. Spikes, On the asymptotic behavior of oscillatory solutions of functional differential equations, Funkcial. Ekvac. 26 (1983), 11–16.

[132] Grimmer, R.C. and P. Waltman, A comparison theorem for a class of nonlinear differential inequalities, Monatsh. Math. 72 (1968), 133–136.

[133] Grüss, G., Über das Maximum des absoluten Betrages von $\frac{1}{b-a}\int_a^b f(x)g(x)\,dx - \frac{1}{(b-a)^2}\int_a^b f(x)\,dx\int_a^b g(x)\,dx$, Math. Z. 39 (1935), 215–226.

[134] Hadamard, J., Etude sur les propriétés des fonctions entières et en particulier d'une fonction considérée par Riemann, J. Math. Pures Appl. 9 (1893), 171–215.

[135] Hammer, P.C., The mid-point method of numerical integration, Math. Mag. 31 (1957/1958), 193–195.

[136] Hardy, G.H., Note on a theorem of Hilbert, Math. Z. 6 (1920), 314–317.

[137] Hardy, G.H., Notes on some points in the integral calculus, Messenger of Math. 57 (1928), 12–16.

[138] Hardy, G.H., Divergent Series, Cambridge Univ. Press, 1948.

[139] Hardy, G.H. and J.E. Littlewood, Elementary theorems concerning power series with positive coefficients and moment constants of positive functions, J. Reine Angew. Math. 157 (1927), 141–158.

[140] Hardy, G.H. and J.E. Littlewood, Some integral inequalities connected with the calculus of variations, Quart. J. Math. Oxford 3 (1932), 241–252.

[141] Hardy, G.H., J.E. Littlewood and G. Pólya, Inequalities, Cambridge Univ. Press, 1934.

[142] Harris, B.J., On the oscillation of solutions of linear differential equations, Mathematica 31 (1984), 214–226.

[143] Harris, B.J., On an inequality of Lyapunov for disfocality, J. Math. Anal. Appl. 146 (1990), 495–500.

[144] Harris, B.J. and Q. Kong, On the oscillation of differential equations with an oscillatory coefficient, Trans. Amer. Math. Soc. 347 (1995), 1831–1839.

[145] Hartman, P., Ordinary Differential Equations, Wiley, New York, 1964.

[146] Heing, H.P., Some extensions of Hardy's inequality, SIAM J. Math. Anal. 6 (1975), 698–713.

[147] Heing, H.P. and L. Maligranda, Chebyshev inequality in function spaces, Real Anal. Exchange 17 (1991/1992), 211–247.

[148] Heing, H.P. and V.D. Stepanov, Weighted Hardy inequalities for increasing functions, Canad. J. Math. 45 (1993), 104–116.

[149] Hellwing, G., Differential Operators of Mathematical Physics, Addison-Wesley, Boston, MA, 1967.

[150] Hochstadt, H., On an inequality of Lyapunov, Proc. Amer. Math. Soc. 22 (1969), 282–284.

[151] Hölder, O., Über einen Mittelwerthssatz, Nachr. Ges. Wiss. Göttingen, 1889, pp. 38–47.

[152] Horgan, C.O., Integral bounds for solutions of nonlinear reaction–diffusion equations, J. Appl. Math. Phys. (ZAMP) 28 (1977), 197–204.

[153] Horgan, C.O., Plane entry flows and energy estimates for the Navier–Stokes equations, Arch. Ration. Mech. Anal. 68 (1978), 359–381.

[154] Horgan, C.O., Eigenvalue estimates and the trace theorem, J. Math. Anal. Appl. 69 (1979), 231–242.

[155] Horgan, C.O. and P.R. Nachlinger, On the domain of attraction for steady states in heat conduction, Internat. J. Engrg. Sci. 14 (1976), 143–148.

[156] Horgan, C.O. and W.E. Olmstead, Exponential decay estimates for a class of nonlinear Dirichlet problems, Arch. Ration. Mech. Anal. 71 (1979), 221–235.

[157] Horgan, C.O. and L.T. Wheeler, Spatial decay estimates for the Navier–Stokes equations with applications to the problem of entry flow, SIAM J. Appl. Math. 35 (1978), 97–116.

[158] Hua, L.K., On an inequality of Opial, Scientia Sinica 14 (1965), 789–790.

[159] Hurri, R., Poincaré domains in R^n, Ann. Acad. Sci. Fenn. Ser. A I Math. Dissertationes 71 (1988), 1–41.

[160] Hurri, R., The weighted Poincaré inequalities, Math. Scand. 67 (1990), 145–160.

[161] Imoru, C.O., On some integral inequalities related to Hardy's, Canad. Math. Bull. 20 (1977), 307–312.

[162] Izumi, M. and S. Izumi, On some inequalities for Fourier series, J. Anal. Math. 21 (1968), 277–291.

[163] Izumi, M., S. Izumi and G.M. Petersen, On Hardy's inequality and its generalization, Tôhoku Math. J. 21 (1969), 601–613.

[164] Jensen, J.L.W.V., Om konvexe Funktioner og Uligheder Mellem Middelværdier, Nyt. Tidsskr. Math. 16B (1905), 49–69.

[165] Jensen, J.L.W.V., Sur les fonctions convexes et les inégalités entre les valeurs moyennes, Acta Math. 30 (1906), 175–193.

[166] Jessen, B., Bemærkninger om konvekse Funktioner og Uligheder imellem Middelærdier, I, Mat. Tidsskr. B (1931), 17–28.

[167] Johnson, P.D. Jr. and R.N. Mohapatra, Inequalities involving lower triangular matrices, Proc. London Math. Soc. 41 (1980), 83–137.

[168] Kadlec, J. and A. Kufner, Characterization of functions with zero traces by integrals with weight functions, II, Časopis Pěst. Mat. 92 (1967), 16–28.

[169] Karlin, S. and Z. Ziegker, Some inequalities for generalized concave functions, J. Approx. Theory 13 (1975), 276–293.

[170] Knopp, K., Über die maximalen Abstände und Verhältnisse verschiedener Mittelwerte, Math. Z. 39 (1935), 768–776.

[171] Kreith, K., Comparison theorems for nonselfadjoint differential equations based on integral inequalities, Proc. Amer. Math. Soc. 34 (1972), 105–109.

[172] Kreith, K., Oscillation Theory, Lecture Notes in Math., Vol. 324, Springer-Verlag, Berlin–New York, 1973.

[173] Kufner, A. and B. Opic, How to define reasonably weighted Sobolev spaces, Comment. Math. Univ. Carolin. 25 (1984), 537–554.

[174] Kufner, A., O. John and S. Fučik, Function Spaces, Noordhoff Internat. Publishers, Layden, 1977.

[175] Kulenovic, M.R. and M.K. Grammatikopoules, On the asymptotic behavior of second order differential inequalities with alternating coefficients, Math. Nachr. 98 (1980), 317–327.

[176] Kwong, M.K., On Lyapunov's inequality for disfocality, J. Math. Anal. Appl. 83 (1981), 486–494.

[177] Ladas, G., On oscillation and boundedness of solutions of nonlinear differential equations, Bull. Soc. Math. Grece 10 (1969), 48–54.

[178] Ladyzhenskaya, O.A., The Mathematical Theory of Viscous Incompressible Flow, Gordon and Breach, New York, 1969.

[179] Ladyzhenskaya, O.A. and N.N. Uraltseva, Linear and Quasilinear Elliptic Equations, Academic Press, New York, 1969.

[180] Lalli, B.S. and R.P. Jahagirdar, On Levin's comparison theorem, Bull. Soc. Math. Grece 13 (1972), 129–132.

[181] Lalli, B.S. and R.P. Jahagirdar, Comparison theorems of Levin type, J. Math. Anal. Appl. 49 (1975), 705–709.

[182] Landu, E., A note on a theorem concerning series of positive terms, J. London Math. Soc. 1 (1926), 38–39.

[183] Lee, C.M., On a discrete analogue of inequalities of Opial and Yang, Canad. Math. Bull. 11 (1968), 73–77.

[184] Lee, C.S., On some generalizations of inequalities of Opial, Yang and Shum, Canad. Math. Bull. 23 (1980), 71–80.

[185] Leighton, W., On the zeros of solutions of a second order linear differential equations, J. Math. Pures Appl. 44 (1965), 297–310.

[186] Leindler, L., Generalization of inequalities of Hardy and Littlewood, Acta Sci. Math. 31 (1970), 279–285.

[187] Levin, A.J., A comparison principle for second-order differential equations, Soviet Math. Dokl. 1 (1960), 1313–1316.

[188] Levin, A.J., Distribution of the zeros of solutions of a linear differential equation, Soviet Math. Dokl. 5 (1964), 818–821.

[189] Levinson, N., Generalization of an inequality of Ky Fan, J. Math. Anal. Appl. 8 (1964), 133–134.

[190] Levinson, N., Generalizations of an inequality of Hardy, Duke Math. J. 31 (1964), 389–394.

[191] Lewis, R.T., Singular elliptic operators of second order with purely discrete spectra, Trans. Amer. Math. Soc. 271 (1982), 653–666.

[192] Liberman, G.M., Interior gradient bounds for non-uniformly parabolic equations, Indiana Univ. Math. J. 32 (1983), 579–601.

[193] Lieb, E. and W. Thirring, Inequalities for the moments of the eigenvalues of the Schrödinger Hamiltonian and their relation to Sobolev inequalities, Studies in Mathematical Physics (E. Lieb, S. Stmon and A. Wightman, Eds), Princeton Univ. Press, Princeton, NJ, 1976, pp. 269–303.

[194] Lin, T.P., The power mean and the logarithmic mean, Amer. Math. Monthly 81 (1979), 879–883.

[195] Littlewood, J.E., Some new inequalities and unsolved problems, Inequalities (O. Shisha, Ed.), Academic Press, New York, 1967, pp. 151–162.

[196] Love, E.R., Generalizations of Hardy's and Copson's inequalities, J. London Math. Soc. 30 (1984), 431–440.

[197] Love, E.R., Inequalities like Opial's inequality, Rocznik Naukowo-Dydaktyczny, Z. 97, Prace Mathematiczne, Vol. XI, 1985, Wydaw. Nauk. WSP, Kraków, pp. 109–118.

[198] Love, E.R., Generalizations of Hardy's integral inequality, Proc. Roy. Soc. Edinburgh Sect. A 100 (1985), 237–262.

[199] Love, E.R., Inequalities related to those of Hardy and of Cochran and Lee, Math. Proc. Cambridge Philos. Soc. 99 (1986), 395–408.

[200] Lupaş, A., A generalization of Hadamard's inequality for convex functions, Univ. Beograd Publ. Elektrotehn. Fak. Ser. Mat. Fiz. Nos 544–576 (1976), 115–121.

[201] Lyapunov, A.M., Probléme general de la stabilité du mouvement, Princeton Univ. Press, Princeton, NJ, 1947.

[202] Maligranda, L., Concavity and convexity of rearrangement, Comment. Math. Prace Mat. 32 (1992), 85–90.

[203] Maligranda, L., J.E. Pečarić and L.E. Persson, On some inequalities of the Grüss–Barnes and Borell type, J. Math. Anal. Appl. 187 (1994), 306–323.

[204] Maligranda, L., J.E. Pečarić and L.E. Persson, Weighted Favard and Berwald inequalities, J. Math. Anal. Appl. 190 (1995), 248–262.

[205] Mary, D.F.St., Some oscillation and comparison theorems for $(r(t)y')' + p(t)y = 0$, J. Differential Equations 5 (1969), 314–323.

[206] Maźja, V.G., Sobolev Spaces, Springer-Verlag, Berlin–New York, 1985.

[207] McShane, E.J., Jensen's inequality, Bull. Amer. Math. Soc. 43 (1937), 521–527.

[208] Michlin, S.G., Variational Methods in Mathematical Physics (in Russian), Moscow, 1957; English transl.: Oxford, 1965.

[209] Mikhailov, V.P., Partial Differential Equations, Mir, Moscow, 1978.

[210] Mingze, G., On Hilbert's inequality and its applications, J. Math. Anal. Appl. 212 (1997), 316–323.

[211] Mitrinović, D.S., Analytic Inequalities, Springer-Verlag, Berlin–New York, 1970.

[212] Mitrinović, D.S. and I.B. Lacković, Hermite and convexity, Aequationes Math. 28 (1985), 225–232.

[213] Mitrinović, D.S. and J.E. Pečarić, On inequalities of Hilbert and Widder, Proc. Edinburgh Math. Soc. 34 (1991), 411–414.

[214] Mitrinović, D.S. and P.M. Vasić, The centroid method in inequalities, Univ. Beograd Publ. Elektrotehn. Fak. Ser. Mat. Fiz. Nos 498–541 (1975), 3–16.

[215] Mitrinović, D.S., J.E. Pečarić and A.M. Fink, Inequalities Involving Functions and Their Integrals and Derivatives, Kluwer Academic Publishers, Dordrecht, 1991.

[216] Mitrinović, D.S., J.E. Pečarić and A.M. Fink, Classical and New Inequalities in Analysis, Kluwer Academic Publishers, Dordrecht, 1992.

[217] Mohapatra, R.N. and D.C. Russell, Integral inequalities related to Hardy's inequality, Aequationes Math. 26 (1985), 199–207.

[218] Mond, B. and J.E. Pečarić, Remarks on Jensen's inequality for operator convex functions, Ann. Univ. Mariae Curie-Skłodwska Sect. A XLVII (1993), 96–103.

[219] Mond, B. and J.E. Pečarić, Inequalities for semi-convex matrix functions, J. Math. Anal. Appl. 185 (1994), 367–377.

[220] Mond, B. and J.E. Pečarić, A companion to Fink's inequality, Octogon Math. Mag. 5 (1997), 17–18.

[221] Muckenhoupt, B., Hardy's inequality with weights, Studia Math. XLIV (1972), 31–38.

[222] Muckenhoupt, B., Weighted norm inequalities for the Hardy maximal function, Trans. Amer. Math. Soc. 165 (1972), 207–226.

[223] Mulholland, H.P., The generalization of certain inequality theorems involving powers, Proc. London Math. Soc. 33 (1932), 481–516.

[224] Myjak, J., Boundary value problems for nonlinear differential and difference equations of the second order, Zeszyty Nauk. Uniw. Jagiello. Prace Mat. 15 (1971), 113–123.

[225] Nehari, Z., On an inequality of Lyapunov, Studies in Mathematical Analysis and Related Topics, Stanford Univ. Press, Stanford, CA, 1962, pp. 256–261.

[226] Nemeth, J., Generalizations of the Hardy–Littlewood inequality, Acta Sci. Math. (Szeged) 32 (1971), 295–299.

[227] Neuman, E., Inequalities involving multivariate convex functions, II, Proc. Amer. Math. Soc. 109 (1990), 965–974.

[228] Neuman, E. and J.E. Pečarić, Inequalities involving multivariate convex functions, J. Math. Anal. Appl. 137 (1989), 541–549.

[229] Nirenberg, L., On elliptic partial differential equations, Ann. Scuola Norm. Sup. Pisa 13 (1959), 116–162.

[230] Olech, C., A simple proof of a certain result of Z. Opial, Ann. Polon. Math. 8 (1960), 61–63.

[231] Opial, Z., Sur une inégalité, Ann. Polon. Math. 8 (1960), 29–32.

[232] Opic, B., Necessary and sufficient conditions for imbaddings in weighted Sobolev spaces, Časopis Pést Mat. 114 (1989), 165–175.

[233] Pachpatte, B.G., On Opial type inequalities in two independent variables, Proc. Roy. Soc. Edinburgh Sect. A 100 (1985), 263–270.

[234] Pachpatte, B.G., Discrete analogues of some inequalities ascribed to Wirtinger, Utilitas Math. 28 (1985), 137–143.

[235] Pachpatte, B.G., On certain discrete inequalities in two independent variables, Soochow J. Math. 11 (1985), 91–95.

[236] Pachpatte, B.G., On Poincaré type integral inequalities, J. Math. Anal. Appl. 114 (1986), 111–115.

[237] Pachpatte, B.G., On two inequalities of the Serrin type, J. Math. Anal. Appl. 116 (1986), 193–199.

[238] Pachpatte, B.G., On a class of new integral inequalities, J. Math. Anal. Appl. 117 (1986), 318–325.

[239] Pachpatte, B.G., On Opial-type integral inequalities, J. Math. Anal. Appl. 120 (1986), 547–556.

[240] Pachpatte, B.G., On certain integral inequalities related to Opial's inequality, Period. Math. Hungar. 17 (1986), 119–125.

[241] Pachpatte, B.G., On Wirtinger–Opial type integral inequalities, Tamkang J. Math. 17 (1986), 1–6.

[242] Pachpatte, B.G., On some new multidimensional discrete inequalities, Tamkang J. Math. 17 (1986), 21–29.

[243] Pachpatte, B.G., A note on an inequality ascribed to Wirtinger, Tamkang J. Math. 17 (1986), 69–73.

[244] Pachpatte, B.G., A note on a new integral inequality, Soochow J. Math. 12 (1986), 63–65.

[245] Pachpatte, B.G., On some new generalizations of Opial's inequality, Demonstratio Math. 19 (1986), 281–291.

[246] Pachpatte, B.G., On Sobolev type integral inequalities, Proc. Roy. Soc. Edinburgh Sect. A 103 (1986), 1–14.

[247] Pachpatte, B.G., A note on Poincaré's integral inequality, An. Stiinţ. Univ. "Al.I. Cuza" Iaşi 32 (1986), 35–36.

[248] Pachpatte, B.G., On some new integral inequalities in several independent variables, Chinese J. Math. 14 (1986), 69–79.

[249] Pachpatte, B.G., A note on two multidimensional integral inequalities, Utilitas Math. 30 (1986), 123–129.

[250] Pachpatte, B.G., On certain multidimensional discrete inequalities, Chinese J. Math. 14 (1986), 185–195.

[251] Pachpatte, B.G., On multidimensional integral inequalities involving three functions, Soochow J. Math. 12 (1986), 67–78.

[252] Pachpatte, B.G., On some extensions of Levinson's generalizations of Hardy's inequality, Soochow J. Math. 13 (1987), 203–210.

[253] Pachpatte, B.G., On Hardy type integral inequalities, Tamkang J. Math. 18 (1987), 27–41.

[254] Pachpatte, B.G., On a new class of Hardy type inequalities, Proc. Roy. Soc. Edinburgh Sect. A 105 (1987), 265–274.

[255] Pachpatte, B.G., On some variants of Hardy's inequality, J. Math. Anal. Appl. 124 (1987), 495–501.

[256] Pachpatte, B.G., On an integral inequality involving functions and their derivatives, Soochow J. Math. 13 (1987), 211–214.

[257] Pachpatte, B.G., On two inequalities similar to Opial's inequality in two independent variables, Period. Math. Hungar. 18 (1987), 137–141.

[258] Pachpatte, B.G., A note on Sobolev type inequalities in two independent variables, J. Math. Anal. Appl. 122 (1987), 114–120.

[259] Pachpatte, B.G., A note on multidimensional integral inequalities, J. Math. Anal. Appl. 122 (1987), 122–128.

[260] Pachpatte, B.G., On two independent variable Opial-type integral inequalities, J. Math. Anal. Appl. 125 (1987), 47–57.

[261] Pachpatte, B.G., On multidimensional Opial-type inequalities, J. Math. Anal. Appl. 126 (1987), 85–89.

[262] Pachpatte, B.G., A note on Opial and Wirtinger type discrete inequalities, J. Math. Anal. Appl. 127 (1987), 470–474.

[263] Pachpatte, B.G., On logarithmic Sobolev type inequalities, Chinese J. Math. 15 (1987), 69–79.

[264] Pachpatte, B.G., On two inequalities of the Sobolev type, Chinese J. Math. 15 (1987), 247–252.

[265] Pachpatte, B.G., A note on Poincaré and Sobolev type integral inequalities, Tamkang J. Math. 18 (1987), 1–7.

[266] Pachpatte, B.G., On Yang type integral inequalities, Tamkang J. Math. 18 (1987), 69–96.

[267] Pachpatte, B.G., On some new integral inequalities in two independent variables, J. Math. Anal. Appl. 129 (1988), 375–382.

[268] Pachpatte, B.G., On some integral inequalities similar to Hardy's inequality, J. Math. Anal. Appl. 129 (1988), 596–606.

[269] Pachpatte, B.G., On discrete inequalities of the Poincaré type, Period. Math. Hungar. 19 (1988), 227–233.

[270] Pachpatte, B.G., A note on discrete inequalities in several variables, Tamkang J. Math. 19 (1988), 1–6.

[271] Pachpatte, B.G., On certain two dimensional integral inequalities, Chinese J. Math. 17 (1989), 273–279.

[272] Pachpatte, B.G., On Wirtinger like discrete inequalities, Tamkang J. Math. 20 (1989), 211–219.

[273] Pachpatte, B.G., On a generalized Opial type inequality in two independent variables, An. Stiinţ. Univ. "Al.I. Cuza" Iaşi 35 (1989), 231–235.

[274] Pachpatte, B.G., On integral inequalities similar to Opial's inequality, Demonstratio Math. 22 (1989), 21–27.

[275] Pachpatte, B.G., On multidimensional Wirtinger type discrete inequalities, Tamkang J. Math. 20 (1989), 289–296.

[276] Pachpatte, B.G., On Sobolev–Lieb–Thirring type inequalities, Chinese J. Math. 18 (1990), 385–397.

[277] Pachpatte, B.G., A note on Copson's inequality involving series of positive terms, Tamkang J. Math. 21 (1990), 13–19.

[278] Pachpatte, B.G. and E.R. Love, On some new inequalities related to Hardy's integral inequality, J. Math. Anal. Appl. 149 (1990), 17–25.

[279] Pachpatte, B.G., A note on two dimensional integral inequality, Tamkang J. Math. 21 (1990), 211–213.

[280] Pachpatte, B.G., On Opial like discrete inequalities, An. Stiinţ. Univ. "Al.I. Cuza" Iaşi 36 (1990), 237–240.

[281] Pachpatte, B.G., A note on Lyapunov type inequalities, Indian J. Pure Appl. Math. 21 (1990), 45–49.

[282] Pachpatte, B.G., On Lyapunov type finite difference inequality, Tamkang J. Math. 21 (1990), 337–339.

[283] Pachpatte, B.G., A note on certain generalizations of Opial's inequality, An. Stiinţ. Univ. "Al.I. Cuza" Iaşi 36 (1990), 21–24.

[284] Pachpatte, B.G., Opial type inequality in several variables, Tamkang J. Math. 22 (1991), 7–11.

[285] Pachpatte, B.G., On Lieb and Thirring type discrete inequalities, Tamkang J. Math. 22 (1991), 145–151.

[286] Pachpatte, B.G., On some extensions of Rellich's inequality, Tamkang J. Math. 22 (1991), 259–265.

[287] Pachpatte, B.G., Opial type discrete inequalities in two variables, Tamkang J. Math. 22 (1991), 323–328.

[288] Pachpatte, B.G., A note on Rellich type inequalities, Libertas Math. 11 (1991), 105–115.

[289] Pachpatte, B.G., On Friedrichs and Trace type inequalities, The Chung Yuan J. 20 (1991), 1–5.

[290] Pachpatte, B.G., On some variants of Sobolev's inequality, Soochow J. Math. 17 (1991), 121–129.

[291] Pachpatte, B.G., Asymptotic behavior of oscillatory solutions of a second order functional differential equation, Proc. Nat. Acad. Sci. India Sect. A 61 (1991), 501–506.

[292] Pachpatte, B.G., A many variable generalization of Hardy's inequality concerning a series of terms, Tamkang J. Math. 23 (1992), 349–354.

[293] Pachpatte, B.G., On multivariate Hardy type inequalities, An. Stiinţ. Univ. "Al.I. Cuza" Iaşi 38 (1992), 355–361.

[294] Pachpatte, B.G., On an inequality of Opial type in two variables, Indian J. Pure Appl. Math. 23 (1992), 657–661.

[295] Pachpatte, B.G., A note on certain inequalities related to Hardy's inequality, Indian J. Pure Appl. Math. 23 (1992), 773–776.

[296] Pachpatte, B.G., On inequalities of the Opial type, Demonstratio Math. 25 (1992), 35–45.

[297] Pachpatte, B.G., On an inequality of Lyapunov for a certain second order differential equation, The Chung Yuan J. 21 (1992), 5–7.

[298] Pachpatte, B.G., On the zeros of solutions of certain differential equations, Demonstratio Math. 25 (1992), 825–833.

[299] Pachpatte, B.G., On certain two dimensional Hardy type integral inequalities, J. Ramanujan Math. Soc. 7 (1992), 93–102.

[300] Pachpatte, B.G., A note on certain new series inequality, Ann. Univ. Timişoara Ser. Ştiinţ. Mat. 31 (1993), 95–100.

[301] Pachpatte, B.G., A note on certain new integrodifferential inequalities, Stud. Univ. Babeş-Bolyai Math. 38 (1993), 25–29.

[302] Pachpatte, B.G., A note on generalized Opial type inequalities, Tamkang J. Math. 24 (1993), 229–235.

[303] Pachpatte, B.G., Some inequalities similar to Opial's inequality, Demonstratio Math. 26 (1993), 643–647.

[304] Pachpatte, B.G., On some inequalities of the Weyl type, An. Stiinţ. Univ. "Al.I. Cuza" Iaşi 40 (1994), 89–95.

[305] Pachpatte, B.G., Inequalities related to Hardy and Copson, An. Stiinţ. Univ. "Al.I. Cuza" Iaşi 40 (1994), 267–273.

[306] Pachpatte, B.G., A note on Hardy like integral inequalities, Tamkang J. Math. 25 (1994), 221–224.

[307] Pachpatte, B.G., On Hardy type discrete inequalities, Libertas Math. 14 (1994), 154–157.

[308] Pachpatte, B.G., On an inequality of Hardy–Littlewood type, J. Ramanujan Math. Soc. 9 (1994), 49–54.

[309] Pachpatte, B.G., A note on Weyl type discrete inequalities, J. Math. Anal. Appl. 188 (1994), 711–716.

[310] Pachpatte, B.G., On some generalizations of Hardy's integral inequality for functions of several variables, Demonstratio Math. 27 (1994), 43–51.

[311] Pachpatte, B.G., Discrete inequalities in three variables, Unpublished Manuscript, 1994.

[312] Pachpatte, B.G., A note on inequalities related to Opial's inequality, Indian J. Pure Appl. Math. 25 (1994), 1031–1037.

[313] Pachpatte, B.G., Inequalities involving functions of two variables, J. Sci. Marathwada Univ. 27 (1994), 53–54.

[314] Pachpatte, B.G., A Lyapunov type inequality for a certain second order differential equation, Proc. Nat. Acad. Sci. India Sect. A 64 (1994), 69–73.

[315] Pachpatte, B.G., On Hardy type integral inequalities for functions of two variables, Demonstratio Math. 28 (1995), 239–244.

[316] Pachpatte, B.G., Generalization of Opial's inequality, Octogon Math. Mag. 3 (1995), 4–6.

[317] Pachpatte, B.G., On Opial type inequalities involving higher order derivatives, J. Math. Anal. Appl. 190 (1995), 763–773.

[318] Pachpatte, B.G., A note on Opial type inequalities involving partial sums, Demonstratio Math. 28 (1995), 33–35.

[319] Pachpatte, B.G., A note on an inequality similar to Lyapunov's inequality, Demonstratio Math. 28 (1995), 915–920.

[320] Pachpatte, B.G., An inequality suggested by Lyapunov's inequality, Publ. Cent. Rech. Math. Pures Ser. I 26 (1995), 1–4.

[321] Pachpatte, B.G., On Lyapunov type inequalities for certain higher order differential equations, J. Math. Anal. Appl. 195 (1995), 527–536.

[322] Pachpatte, B.G., On some new inequalities related to the zeros of solutions of certain second order differential equations, (I) and (II), Bul. Inst. Politeh. Iaşi Mat. XLI (1995), 25–32 and 53–57.

[323] Pachpatte, B.G., Inequalities of Opial type that involve higher order derivatives, An. Stiinţ. Univ. "Al.I. Cuza" Iaşi 42 (1996), 367–371.

[324] Pachpatte, B.G., A note on Opial like integral inequalities, Publ. Cent. Rech. Math. Pures Ser. I 27 (1996), 13–17.

[325] Pachpatte, B.G., On an inequality of Lyapunov type, Octogon Math. Mag. 4 (1996), 29–31.

[326] Pachpatte, B.G., A note on some series inequalities, Tamkang J. Math. 27 (1996), 77–79.

[327] Pachpatte, B.G., On Levin type comparison theorems for certain second order differential equations, Acta Math. Scientia 17 (1997), 51–55.

[328] Pachpatte, B.G., Lyapunov type integral inequalities for certain differential equations, Georgian Math. J. 4 (1997), 139–148.

[329] Pachpatte, B.G., On an inequality related to the zeros of solutions of certain second order differential equations, Octogon Math. Mag. 5 (1997), 35–37.

[330] Pachpatte, B.G., A note on some inequalities analogues to Grüss inequality, Octogon Math. Mag. 5 (1997), 62–66.

[331] Pachpatte, B.G., Inequalities related to the zeros of solutions of certain second order difference equations, Stud. Univ. Babeş-Bolyai Math. XLII (1997), 91–96.

[332] Pachpatte, B.G., Inequalities for Differential and Integral Equations, Academic Press, New York, 1998.

[333] Pachpatte, B.G., On multivariate generalizations of Hardy's inequality, Libertas Math. 18 (1998), 23–47.

[334] Pachpatte, B.G., On some new inequalities similar to Hilbert's inequality, J. Math. Anal. Appl. 226 (1998), 166–179.

[335] Pachpatte, B.G., A note on Hilbert type inequality, Tamkang J. Math. 29 (1998), 293–298.

[336] Pachpatte, B.G., Discrete inequalities in three independent variables, Demonstratio Math. 21 (1998), 849–854.

[337] Pachpatte, B.G., A note on some extensions of Hardy's inequality, An. Stiinţ. Univ. "Al.I. Cuza" Iaşi 44 (1998), 95–100.

[338] Pachpatte, B.G., On some inequalities similar to Lyapunov's inequality, Bul. Inst. Politeh. Iaşi Math. XLIV (XLVIII) (1998), 21–30.

[339] Pachpatte, B.G., Integral inequalities associated to the zeros of solutions of certain higher order differential equations, Fasc. Math. 29 (1999), 87–97.

[340] Pachpatte, B.G., A note on an inequality similar to Opial's inequality, Tamkang J. Math. 30 (1999), 63–66.

[341] Pachpatte, B.G., A note on multivariate variants of Hardy's inequality, Fasc. Math. 30 (1999), 107–112.

[342] Pachpatte, B.G., Inequalities similar to the integral analogue of Hilbert's inequality, Tamkang J. Math. 30 (1999), 139–146.

[343] Pachpatte, B.G., On a new inequality analogues to Hilbert's inequality, Radovi Math. 9 (1999), 5–11.

[344] Pachpatte, B.G., On some generalizations of Hardy's integral inequality, J. Math. Anal. Appl. 234 (1999), 15–30.

[345] Pachpatte, B.G., On Sobolev–Visik–Dubinskii type inequalities, Tamkang J. Math. 30 (1999), 213–218.

[346] Pachpatte, B.G., On some generalized Opial type inequalities, Analytic and Geometric Inequalities and Their Applications (M. Rassias and H.M. Srivastava, Eds), Kluwer Academic Publishers, Dordrecht, 1999, pp. 301–322.

[347] Pachpatte, B.G., A note on two new discrete inequalities, Octogon Math. Mag. 7 (1999), 33–36.

[348] Pachpatte, B.G., A note on some new Opial type integral inequalities, Octogon Math. Mag. 7 (1999), 80–84.

[349] Pachpatte, B.G., On some integral inequalities involving convex functions, RGMIA Res. Rep. Coll. 3(3) (2000), 487–492.

[350] Pachpatte, B.G., A note on an inequality of Hilbert type, Octogon Math. Mag. 8 (2000), 23–26.

[351] Pachpatte, B.G., A note on Hadamard like inequalities, Octogon Math. Mag. 8 (2000), 37–42.

[352] Pachpatte, B.G., On an inequality similar to Hilbert's inequality, Bul. Inst. Politech. Iaşi Mat. XLVI (2000), 31–36.

[353] Pachpatte, B.G., Inequalities similar to certain extensions of Hilbert's inequality, J. Math. Anal. Appl. 243 (2000), 217–227.

[354] Pachpatte, B.G., Inequalities related to the zeros of solutions of certain second order differential equations, Facta Univ. Ser. Math. Inform. 16 (2001), 35–44.

[355] Pachpatte, B.G., On two independent variable Hardy type integral inequalities, Bul. Inst. Politech. Iaşi Mat. XLVIII (2002), 33–44.

[356] Pachpatte, B.G., On some inequalities for convex functions, RGMIA Res. Rep. Coll. 6(3) (2003), 1–9.

[357] Pachpatte, B.G., A note on variants of certain inequalities for convex functions, Tamkang J. Math. 34 (2003), 163–167.

[358] Pachpatte, B.G., Inequalities of Opial type in three independent variables, Tamkang J. Math. 35 (2004), 145–147.

[359] Pachpatte, B.G., Inequalities similar to Opial's inequality involving higher order derivatives, Tamkang J. Math., to appear.

[360] Pachpatte, B.G., On certain discrete inequalities involving functions of several variables, submitted.

[361] Patula, W.T., On the distance between zeros, Proc. Amer. Math. Soc. 52 (1975), 247–251.

[362] Payne, L.E., Uniqueness criteria for steady state solutions of the Navier–Stokes equations, Simpos. Internat. Appl. Anal. Fiz. Mat. (Cagliari-Sassari), Edizioni Cremonese, Rome, 1965, pp. 130–153.

[363] Pearcé, C.E.M and J.E. Pečarić, A continuous analogue and extension of Rado's formulae for convex and concave functions, Bull. Austral. Math. Soc. 53 (1996), 229–233.

[364] Pearcé, C.E.M and J.E. Pečarić, On some inequalities of Brenner and Alzer for concave functions, J. Math. Anal. Appl. 198 (1996), 282–288.

[365] Pearcé, C.E.M, J.E. Pečarić and V. Simie, Stolarsky means and Hadamard's inequality, J. Math. Anal. Appl. 220 (1998), 99–109.

[366] Pečarić, J.E., Inverse of Jensen–Steffensen's inequality, Glas. Math. 16 (1981), 229–233.

[367] Pečaric, J.E., A short proof of a variant of Jensen's inequality, J. Math. Anal. Appl. 87 (1982), 278–280.

[368] Pečarić, J.E., A simple proof of the Jensen–Steffensen inequality, Amer. Math. Monthly 91 (1984), 195–196.

[369] Pečarić, J.E., On some inequalities for functions with nondecreasing increments, J. Math. Anal. Appl. 98 (1984), 188–197.

[370] Pečarić, J.E., Notes on convex functions, General Inequalities, Internat. Ser. Numer. Math., Vol. 6, 1992, pp. 449–454.

[371] Pečarić, J.E., Remarks on two interpolations of Hadamard's inequalities, Contrib. Sec. Math. Tech. Sci. 13 (1992), 9–12.

[372] Pečarić, J.E., A multidimential generalization of weight convex functions, Comment. Math. 32 (1992), 123–131.

[373] Pečarić, J.E., On Jensen inequality for C–J-convex functions, Comment. Math. 33 (1993), 111–118.

[374] Pečarić, J.E., An Integral Inequality, Hadronic Press, Palm Harbor, FL, 1993, pp. 471–478.

[375] Pečarić, J.E., Remarks on an inequality of S. Sabler, J. Math. Anal. Appl. 184 (1994), 19–21.

[376] Pečarić, J.E. and P.R. Beesack, On Jessen's inequality for convex functions, II, J. Math. Anal. Appl. 118 (1986), 125–144.

[377] Pečarić, J.E. and L. Brnetić, Note on generalization of Godunova–Levin–Opial inequalities, Demonstratio Math. 30 (1997), 545–549.

[378] Pečarić, J.E. and L. Brnetić, Note on the generalization of the Godunova–Levin–Opial inequality in several independent variables, J. Math. Anal. Appl. 215 (1997), 274–282.

[379] Pečarić, J.E. and S.S. Dragomir, A refinement of Jensen inequality and applications, Stud. Univ. Babeş-Bolyai Math. 34 (1989), 15–19.

[380] Pečarić, J.E. and S.S. Dragomir, On some integral inequalities for convex functions, Bul. Inst. Politeh. Iaşi Math. 36 (1990), 19–23.

[381] Pečarić J.E. and S.S. Dragomir, A generalization of Hadamard's inequality for isotonic linear functionals, Radovi Math. 7 (1991), 103–107.

[382] Pečarić, J.E. and E.R. Love, Still more generalizations of Hardy's inequality, J. Austral. Math. Soc. Ser. A 58 (1995), 1–11.

[383] Pečarić, J.E., I. Perić and L.E. Persson, Integral inequalities for monotone functions, J. Math. Anal. Appl. 215 (1997), 235–251.

[384] Pečarić, J.E., F. Proschan and Y.L. Tong, Convex Functions, Partial Ordering, and Statistical Applications, Academic Press, San Diego, CA, 1992.

[385] Peterson, G.M., An inequality of Hardy's, Quart. J. Math. Oxford 13 (1962), 237–240.

[386] Pfeffer, A.M., On certain discrete inequalities and their continuous analogues, J. Res. Nat. Bur. Standards Ser. B 70 (1967), 221–231.

[387] Piepenbrink, J., Integral inequalities and theorems of Liouville type, J. Math. Anal. Appl. 2 (1969), 630–639.

[388] Pittenger, A.O., Inequalities between arithmetic and logarithmic means, Univ. Beograd Publ. Elektrotehn. Fak. Ser. Mat. Fiz. Nos 678–715 (1980), 15–18.

[389] Poincaré, H., Sur les équations de la physique mathématique, Rend. Circ. Mat. Palermo 8 (1884), 57–156.

[390] Popoviciu, T., Les fonctions convexes, Actualités Sci. Indust. No. 992, Paris, 1944.

[391] Popoviciu, T., La simplicité du rest dans certaines formules de quadrature, Mathematica (Cluj) 6 (1964), 157–184.

[392] Prenter, P.M., Splines and Variational Methods, Wiley, New York, 1975.

[393] Reid, W.T., A matrix Lyapunov inequality, J. Math. Anal. Appl. 32 (1970), 424–434.

[394] Reid, W.T., A generalized Lyapunov inequality, J. Differential Equations 13 (1973), 182–196.

[395] Reid, W.T., Interpolation between a trace formula and Lyapunov type inequalities, J. Differential Equations 23 (1977), 448–459.

[396] Rellich, F., Perturbation Theory of Eigenvalue Problems, Gordon and Breach, New York, 1969.

[397] Roberts, A.W. and D.E. Varberg, Convex Functions, Academic Press, New York and London, 1973.

[398] Rosen, G., Minimum value for c in the Sobolev inequality, $\|\phi^3\| \leqslant c\|\nabla\phi\|^3$, SIAM J. Appl. Math. 21 (1971), 30–32.

[399] Rozanova, G.I., Integral inequalities with derivatives and with arbitrary convex functions (in Russian), Moskow Gos. Red. Inst. Vcen Zap. 460 (1972), 58–65.

[400] Salem, Sh., On Opial type inequalities in three independent variables, Kyungpook Math. J. 36 (1996), 63–72.

[401] Sándor, J., Some integral inequalities, Elem. Math. 43 (1988), 177–180.

[402] Sándor, J., On the Jensen–Hadamard inequality, Stud. Univ. Babeş-Bolyai Math. 36 (1991), 9–15.

[403] Schmincke, U.W., Essential self-adjointness of a Schrödinger operator with strongly singular potential, Math. Z. 124 (1972), 47–50.

[404] Schwarz, H.A., Über ein die Flachen kleinsten Flacheninhalts betreffendes Problem der Variationsrechnung, Acta Soc. Sci. Fenn. 15 (1885), 315–362.

[405] Serrin, J., The initial value problem for the Navier–Stokes equations, Nonlinear Problems (R.E. Langer, Ed.), Univ. Wisconsin Press, Madison, WI, 1963, pp. 69–78.

[406] Showalter, R.E., Hilbert Space Methods for Partial Differential Equations, Pitman, New York, 1977.

[407] Shum, D.T., On integral inequalities related to Hardy's, Canad. Math. Bull. 14 (1971), 225–230.

[408] Singh, B., Forced oscillations in general ordinary differential equations, Tamkang J. Math. 6 (1975), 5–11.

[409] Sinnamon, G.J., Weighted Hardy and Opial-type inequalities, J. Math. Anal. Appl. 160 (1991), 434–445.

[410] Sobolev, S.L., Some Applications of Functional Analysis in Mathematical Physics, Izdat. Leningrad. Gos. Univ., Leningrad, 1950; English transl.: Trans. Math. Monographs, Vol. 7, Amer. Math. Soc., Providence, RI, 1963.

[411] Steffensen, J.F., On certain inequalities and method of approximation, J. Inst. Actuaries 51 (1919), 274–297.

[412] Stolz, O., Grundzüge der Differential und Integralrechnung, Vol. 1, Leipzig, 1893, pp. 35–36.

[413] Stredulinsky, W., Weighted Inequalities and Degenerate Elliptic Partial Differential Equations, Lecture Notes in Math., Vol. 1074, Springer-Verlag, New York, 1984.

[414] Sturm, C., Sur les équations differentielles linearies du second ordre, J. Math. Pures Appl. 1 (1836), 106–186.

[415] Swanson, C.A., Nonoscillation criteria for elliptic equations, Canad. Math. Bull. 12 (1969), 275–280.

[416] Swanson, C.A., Comparison and Oscillation Theory of Linear Differential Equations, Academic Press, New York, 1968.

[417] Świstochowski, Z., On the inequality of the type $A_1^{-1} + A_2^{-1} \geqslant 4(A_1 + A_2)^{-1}$, Ann. Univ. Mariae Curie-Skłodowska Sect. A (1985), 159–163.

[418] Talenti, G., Best constants in Sobolev's inequality, Ann. Mat. Pura Appl. 110 (1976), 353–372.

[419] Traple, J., On a boundary value problem for systems of ordinary differential equations of second order, Zeszyty Nauk. Univ. Jagiello. Prace Mat. 15 (1971), 159–168.

[420] Vasić, P.M. and I.B. Lacković, Some complements to the paper "On an inequality for convex functions", Univ. Beograd Publ. Elektrotehn. Fak. Ser. Mat. Fiz. Nos 544–576 (1976), 59–62.

[421] Visik, M.I., On the solvability of the first boundary value problem for non-linear elliptic systems of differential equations, Dokl. Akad. Nauk SSSR 134 (1960), 749–752.

[422] Wend, D.V.V., On the zeros of solutions of some linear complex differential equations, Pacific J. Math. 10 (1960), 713–722.

[423] Weyl, H., The Theory of Groups and Quantum Mechanics, English transl., Dover, New York, 1931.

[424] Widder, D.V., An inequality related to one of Hilbert's, J. London Math. Soc. 4 (1929), 194–198.

[425] Willett, D., The existence–uniqueness theorem for an n-th order linear ordinary differential equation, Amer. Math. Monthly 75 (1968), 174–178.

[426] Wong, J.S.W., A discrete analogue of Opial's inequality, Canad. Math. Bull. 10 (1967), 115–118.

[427] Wong, P.K., Bounds for solutions to a class of nonlinear second-order differential equations, J. Differential Equations 7 (1970), 139–146.

[428] Yang, G.S., On a certain result of Z. Opial, Proc. Japan Acad. 42 (1966), 78–83.

[429] Yang, G.S., Inequality of Opial-type in two variables, Tamkang J. Math. 13 (1982), 255–259.

[430] Yang, G.S., A note on some integrodifferential inequalities, Soochow J. Math. 9 (1983), 231–236.

[431] Yang, G.S., A note on an inequality similar to Opial's inequality, Tamkang J. Math. 18 (1987), 101–104.

[432] Yang, E.H., Opial-type integral inequalities involving several arbitrary increasing convex functions, J. Math. Anal. Appl. 195 (1995), 495–505.

[433] Yeh, C.C., Levin's comparison theorems for second order nonlinear differential equations and inequalities, Math. Japon. 36 (1991), 703–710.

[434] Zygmund, A., Trigometric Series, Vol. I, Cambridge Univ. Press, 1959.

Index

Printed in the United Kingdom
by Lightning Source UK Ltd.
134279UK00002B/1-24/A

9 780444 517951